Handbook of Bayesian Variable Selection

Chapman & Hall/CRC
Handbooks of Modern Statistical Methods

Series Editor
Garrett Fitzmaurice, *Department of Biostatistics, Harvard School of Public Health, Boston, MA, U.S.A.*

The objective of the series is to provide high-quality volumes covering the state-of-the-art in the theory and applications of statistical methodology. The books in the series are thoroughly edited and present comprehensive, coherent, and unified summaries of specific methodological topics from statistics. The chapters are written by the leading researchers in the field and present a good balance of theory and application through a synthesis of the key methodological developments and examples and case studies using real data.

Published Titles
Handbook of Discrete-Valued Time Series
Richard A. Davis, Scott H. Holan, Robert Lund, and Nalini Ravishanker
Handbook of Big Data
Peter Bühlmann, Petros Drineas, Michael Kane, and Mark van der Laan
Handbook of Spatial Epidemiology
Andrew B. Lawson, Sudipto Banerjee, Robert P. Haining, and María Dolores Ugarte
Handbook of Neuroimaging Data Analysis
Hernando Ombao, Martin Lindquist, Wesley Thompson, and John Aston
Handbook of Statistical Methods and Analyses in Sports
Jim Albert, Mark E. Glickman, Tim B. Swartz, and Ruud H. Koning
Handbook of Methods for Designing, Monitoring, and Analyzing Dose-Finding Trials
John O'Quigley, Alexia Iasonos, and Björn Bornkamp
Handbook of Quantile Regression
Roger Koenker, Victor Chernozhukov, Xuming He, and Limin Peng
Handbook of Statistical Methods for Case-Control Studies
Ørnulf Borgan, Norman Breslow, Nilanjan Chatterjee, Mitchell H. Gail, Alastair Scott, and Chris J. Wild
Handbook of Environmental and Ecological Statistics
Alan E. Gelfand, Montserrat Fuentes, Jennifer A. Hoeting, and Richard L. Smith
Handbook of Approximate Bayesian Computation
Scott A. Sisson, Yanan Fan, and Mark Beaumont
Handbook of Graphical Models
Marloes Maathuis, Mathias Drton, Steffen Lauritzen, and Martin Wainwright
Handbook of Mixture Analysis
Sylvia Frühwirth-Schnatter, Gilles Celeux, and Christian P. Robert
Handbook of Infectious Disease Data Analysis
Leonhard Held, Niel Hens, Philip O'Neill, and Jacco Walllinga
Handbook of Meta-Analysis
Christopher H. Schmid, Theo Stijnen, and Ian White
Handbook of Forensic Statistics
David L. Banks, Karen Kafadar, David H. Kaye, and Maria Tackett
Handbook of Statistical Methods for Randomized Controlled Trials
KyungMann Kim, Frank Bretz, Ying Kuen K. Cheung, and Lisa Hampson
Handbook of Measurement Error Models
Grace Yi, Aurore Delaigle, and Paul Gustafson
Handbook of Multiple Comparisons
Xinping Cui, Thorsten Dickhaus, Ying Ding, and Jason C. Hsu
Handbook of Bayesian Variable Selection
Mahlet G. Tadesse and Marina Vannucci

For more information about this series, please visit: https://www.crcpress.com/Chapman--HallCRC-Handbooks-of-Modern-Statistical-Methods/book-series/CHHANMODSTA

Handbook of Bayesian Variable Selection

Edited by
Mahlet G. Tadesse
Marina Vannucci

CRC Press is an imprint of the
Taylor & Francis Group, an **informa** business
A CHAPMAN & HALL BOOK

First edition published 2022
by CRC Press
6000 Broken Sound Parkway NW, Suite 300, Boca Raton, FL 33487-2742

and by CRC Press
2 Park Square, Milton Park, Abingdon, Oxon, OX14 4RN

Library of Congress Cataloging-in-Publication Data

Names: Tadesse, Mahlet G., editor. | Vannucci, Marina, editor.
Title: Handbook of Bayesian variable selection / [edited by] Mahlet G.
Tadesse, Marina Vannucci.
Description: First edition. | Boca Raton : CRC Press, 2022. | Includes
bibliographical references and index. |
Identifiers: LCCN 2021031721 (print) | LCCN 2021031722 (ebook) | ISBN
9780367543761 (hardback) | ISBN 9780367543785 (paperback) | ISBN 9781003089018 (ebook)
Subjects: LCSH: Bayesian statistical decision theory. | Variables
(Mathematics) | Regression analysis.
Classification: LCC QA279.5 .H363 2022 (print) | LCC QA279.5 (ebook) |
DDC 519.5/42--dc23/eng/20211004
LC record available at https://lccn.loc.gov/2021031721
LC ebook record available at https://lccn.loc.gov/2021031722

ISBN: 9780367543761 (hbk)
ISBN: 9780367543785 (pbk)
ISBN: 9781003089018 (ebk)

DOI: 10.1201/9781003089018

Publisher's note: This book has been prepared from camera-ready copy provided by the authors.

To our families

Contents

Preface

Variable selection is the process of identifying a subset of relevant predictors to include in a model. This topic has been the focus of much research over the past 30 years with the ubiquity of high-dimensional data in various fields of sciences, engineering and economics. Methods for variable selection can be categorized into those based on hypothesis testing and those that perform penalized parameter estimation. In the Bayesian framework, the former use Bayes factors and posterior model probabilities, while the latter specify shrinkage priors that induce sparsity. The literature on Bayesian variable selection has experienced substantial development since the early 1990s. However, aside from a couple of early monographs, no attempt has been made to unify the literature into a comprehensive textbook. Our goal with this handbook is to fill this gap by providing an overview of Bayesian methods for variable selection, while covering recent developments and related topics, such as variable selection in decision trees and edge selection in graphical models. The handbook is comprised of 19 carefully edited chapters divided into four parts: spike-and-slab priors; continuous shrinkage priors; extensions to various modeling; other approaches to Bayesian variable selection. The various chapters are devoted to theoretical, methodological and computational aspects with a focus on practical implications and some real data examples. R code for reproducing some of the examples are provided in the online supplement to the handbook.

Part I is devoted to theoretical and computational aspects of spike-and-slab priors for regression models. These priors use latent binary indicators to induce mixtures of two distributions on the regression coefficients of the model, one distribution peaked around zero (spike) to identify the zero coefficients, and the other a flat distribution (slab) to capture the non-zero elements. Two constructions have been developed in parallel in the literature. The discrete spike-and-slab formulation uses a mixture of a point mass at zero and a flat prior, while the continuous spike-and-slab prior uses a mixture of two continuous distributions. Stochastic search Markov chian Monte Carlo (MCMC) algorithms are used to explore the large model space and variable selection is achieved through marginal posterior inclusion probabilities. Part I consists of five chapters, one introducing discrete spike-and-slab priors for linear settings with a focus on prior constructions and computational implementation, two others devoted to theoretical aspects of discrete and continuous spike-and-slab priors respectively, one chapter discussing the continuous spike-and-slab LASSO, and a final one reviewing adaptive MCMC methods for efficient posterior sampling.

Part II reviews theoretical and computational aspects of continuous shrinkage priors, a class of unimodal distributions that promote shrinkage of small regression coefficients towards zero, similarly to frequentist penalized regression methods that accomplish regularization by maximizing the log-likelihood function subject to a penalty. Indeed, the least absolute shrinkage and selection operator (lasso) estimate can be interpreted as a Bayesian posterior mode estimate using independent Laplace priors for the regression coefficients. Bayesian penalization methods, however, do not shrink regression coefficients to be exactly zero, as frequentist penalization methods do. Instead, the variable selection is carried out using credible intervals for the regression coefficients or by defining a selection criterion on the posterior samples. Many continuous shrinkage priors can be parametrized as a scale mixture of normal distributions, which facilitates the MCMC implementation. Part II is comprised of three chapters, one devoted to theoretical guarantees of global-local shrinkage priors with a focus on the horseshoe prior, one covering computational aspects and MCMC methods, and a third one reviewing an approach to derive sparse posterior summaries for global-local shrinkage priors.

Part III contains six chapters addressing extensions of shrinkage priors to various modeling frameworks. A first chapter reviews Bayesian model averaging approaches in causal inference. Another one is devoted to extensions of spike-and-slab priors to multivariate regression models. The next three chapters cover approaches to spatial regression models, structured additive distributional regression, and state-space and time-varying parameter models. The final chapter discusses extensions of continuous spike-and-slab priors to the problem of edge estimation in single and multiple Gaussian graphical models.

Part IV is comprised of five chapters covering other approaches to Bayesian variable selection. One chapter is devoted to variable selection approaches that use Bayes factors by considering distinct subsets of covariates as hypotheses to be tested. Another chapter discusses sparsity and power trade-offs in Bayesian model selection. Two other chapters cover variable selection approaches for Bayesian additive regression trees and decision tree ensembles. The final chapter reviews a stochastic partitioning method for variable selection in mixture of regression models.

The idea of an edited book on Bayesian variable selection originated from a conversation with Rob Calver of Chapman & Hall/CRC at a Joint Statistical Meetings (JSM) conference a few years ago. After some thoughts and extensive discussions, we came up with a lineup of topics and a list of about 20 possible chapters. We invited specific authors for each of the chapters and, luckily, almost everybody accepted our invitation. As a result, we were able to secure contributions from outstanding scholars in the field. By the time we completed the invitation process and distributed guidelines to the authors, the COVID-19 pandemic had exploded all around the world and the handbook became our "pandemic project". We received the first version of all chapters by the end of summer 2020, we oversaw the review process and provided extensive comments to the authors. We received a second version of all chapters in Spring 2021. Although we never met in person during the entire time of the project, we worked together for countless hours, during virtual Zoom meetings. A particular effort on our part was made in trying to unify notation and format of the chapters as much as possible across the handbook. We are extremely grateful to all the contributors for sticking with their commitment and for being extremely responsive during the two rounds of review of their chapters. The handbook was completed in record time by May 2021.

We thank those who have contributed to the handbook by writing the chapters and taking part in the review process. Students and postdocs participating in the Bayesian research group at Rice University in the Fall 2020 semester provided many useful comments on several of the chapters. We also thank Rob Calver and Vaishali Singh at Chapman & Hall/CRC for their support throughout the project.

Finally, we are forever grateful for the love and support of our extended families, which span across several generations and across three continents.

Mahlet G. Tadesse, Washington D.C.

Marina Vannucci, Houston, TX

June 2021

Biography

Mahlet G. Tadesse is Professor and Chair in the Department of Mathematics and Statistics at Georgetown University, USA. Her research over the past two decades has focused on Bayesian modeling for high-dimensional data with an emphasis on variable selection methods and mixture models. She also works on various interdisciplinary projects in genomics and public health. She is a recipient of the Myrto Lefkopoulou Distinguished Lectureship award, an elected member of the International Statistical Institute and an elected fellow of the American Statistical Association.

Marina Vannucci is Noah Harding Professor of Statistics at Rice University, USA. Her research over the past 25 years has focused on the development of methodologies for Bayesian variable selection in linear settings, mixture models and graphical models, and on related computational algorithms. She also has a solid history of scientific collaborations and is particularly interested in applications of Bayesian inference to genomics and neuroscience. She has received an NSF CAREER award and the Mitchell prize by ISBA for her research, and the Zellner Medal by ISBA for exceptional service over an extended period of time with long-lasting impact. She is an elected Member of ISI and RSS and an elected fellow of ASA, IMS, AAAS and ISBA.

List of Contributors

Antonelli, Joseph
University of Florida
USA

Bai, Ray
University of South Carolina
USA

Bhattacharya, Anirban
Texas A&M University
USA

Bondell, Howard D.
University of Melbourne
Australia

Bottolo, Leonardo
University of Cambridge
UK

Carvalho, Carlos M.
University of Texas at Austin
USA

Dominici, Francesca
Harvard T.H. Chan School of Public Health
USA

Du, Junliang
Florida State University
USA

Frühwirth-Schnatter, Sylvia
Vienna University of Economics and Business
Austria

Garcia-Donato, Gonzalo
Universidad de Castilla-La Mancha
Spain

George, Edward I.
University of Pennsylvania
USA

Griffin, Jim E.
University College London
UK

Hahn, Richard P.
Arizona State University
USA

Johndrow, James
University of Pennsylvania
USA

Knaus, Peter
Vienna University of Economics and Business
Austria

Kneib, Thomas
Georg-August-Universität Göttingen
Germany

Linero, Antonio R.
University of Texas at Austin
USA

McCulloch, Robert E.
Arizona State University
USA

Monni, Stefano
American University of Beirut
Lebanon

Narisetty, Naveen N.
University of Illinois at Urbana-Champaign
USA

Pati, Debdeep
Texas A&M University
USA

Peterson, Christine B.
Univerisity of Texas MD Anderson Cancer Center
USA

Reich, Brian J.
North Carolina State University
USA

Richardson, Sylvia
University of Cambridge
UK

Ročková, Veronika
University of Chicago
USA

Rossell, David
Pompeu Fabra University
Spain

Rubio, Francisco Javier
University College London
UK

Ruffieux, Hélène
University of Cambridge
UK

Staicu, Ana-Maria
North Carolina State University
USA

Steel, Mark F.J.
University of Warwick
UK

Stingo, Francesco C.
University of Florence
Italy

Tadesse, Mahlet G.
Georgetown University
USA

van der Pas, Stéphanie
Amsterdam UMC
The Netherlands

Vannucci, Marina
Rice University
USA

Wagner, Helga
Johannes Kepler Universität Linz
Austria

Wiemann, Paul
Georg-August-Universität Göttingen
Germany

Yu, Weichang
University of Melbourne
Australia

Zhang, Yan Dora
University of Hong Kong
HK SAR China

Zhou, Shuang
Arizona State University
USA

List of Symbols

Regression & Variable Selection Notation

n	Sample size/number of observations
p	Number of candidate covariates
q	Number of response variables for multivariate outcome
y_i	Univariate response for i-th observation
$\boldsymbol{y} = (y_1, \ldots, y_n)^T$	Vector of univariate response for n samples
$\boldsymbol{y}_i = (y_{i1}, \ldots, y_{iq})^T$	q-vector of multivariate response for i-th observation
$\boldsymbol{Y} = (\boldsymbol{y}_1, \ldots, \boldsymbol{y}_n)^T$	$n \times q$ matrix of multivariate responses
$\boldsymbol{x}_i = (x_{i1}, \ldots, x_{ip})^T$	p-vector of covariates for i-th observation
$\boldsymbol{X} = (\boldsymbol{x}_1, \ldots, \boldsymbol{x}_n)^T$	$n \times p$ matrix of covariates
$\boldsymbol{\beta} = (\beta_1, \ldots, \beta_p)^T$	p-vector of regression coefficients
α or $\boldsymbol{\alpha}$	Intercept parameter or regression coefficients for covariates forced into model
$\sigma^2, \boldsymbol{\Sigma}$	Residual variance in univariate/multivariate linear regression
γ_j	Latent binary selection indicator for j-th covariate
$\boldsymbol{\gamma} = (\gamma_1, \ldots, \gamma_p)^T$	p-vector of binary variable selection indicators
$p_\gamma = \sum_{j=1}^p \gamma_j$	Number of selected covariates
$\boldsymbol{X}_\gamma$	$n \times p_\gamma$ submatrix of $\boldsymbol{X}$ corresponding to variables chosen by $\boldsymbol{\gamma}$
$\boldsymbol{\beta}_\gamma$	p_γ-vector of regression coefficients for variables selected by $\boldsymbol{\gamma}$
δ_0	Dirac measure at 0 for discrete spike-and-slab prior
$\boldsymbol{\lambda}, \tau$	Local and global shrinkage hyperparameters

Mathematical Functions & Matrix Notation

$\log(x)$	Natural logarithm function
$\exp(x)$	Exponential function
$B(\alpha, \beta)$	Beta function
$\Gamma(\alpha)$	Gamma function
$\mathbb{I}_A(x)$	Indicator function of a set A
$\mathbb{I}\{x = a\}$	Indicator function for an event
$\boldsymbol{I}_n$	Identity matrix of dimension $n \times n$
$\mathbf{1}_n$	n-vector of ones
$\text{Diag}(x_1, \ldots, x_d)$	Diagonal matrix with elements $x_1, \ldots, x_d$
$\boldsymbol{A}^T$	Transpose of matrix $\boldsymbol{A}$
$\boldsymbol{A}^{-1}$	Inverse of matrix $\boldsymbol{A}$
$\lvert\boldsymbol{A}\rvert$	Determinant of matrix $\boldsymbol{A}$
tr(A)	Trace of matrix $\boldsymbol{A}$
$E[Y]$	Expectation of random variable Y
$Var(Y)$	Variance of random variable Y
$Cov(Y)$	Variance-covariance matrix of random variable Y
$\bar{y}$	Sample mean
s_y^2	Sample variance
$\boldsymbol{S}_y$	Sample covariance matrix

Probability Functions & Distributions

$f(\boldsymbol{y}\|\boldsymbol{\theta})$	Probability density/mass function of $\boldsymbol{y}$
$p(\boldsymbol{\theta})$	Prior density of $\boldsymbol{\theta}$
$p(\boldsymbol{\theta}\|\boldsymbol{y})$	Posterior density of $\boldsymbol{\theta}$ given $\boldsymbol{y}$
$\mathcal{L}$	Likelihood function
ℓ	Log-likelihood function
$\phi(\cdot)$	Pdf of standard normal distribution
$\Phi(\cdot)$	Cdf of standard normal distribution
$\mathcal{B}er(\pi)$	Bernoulli distribution
$\mathcal{B}eta(a,b)$	Beta distribution
$\mathcal{B}in(n,\pi)$	Binomial distribution
$\mathcal{C}auchy(a,b)$	Cauchy distribution
$\mathcal{E}xp(\lambda)$	Exponential distribution
$\mathcal{G}(a,b)$	Gamma distribution
χ^2_ν	Chi-square distribution
$\mathcal{IG}(a,b)$	Inverse gamma distribution
$1/\chi^2_\nu$	Inverse chi-square distribution
$\mathcal{L}ap(\mu,\sigma)$	Laplace distribution
$\mathcal{N}(\mu,\sigma^2)$	Normal distribution
$\mathcal{N}_n(\boldsymbol{\mu},\boldsymbol{\Sigma})$	n-variate normal distribution
$\mathcal{P}oi(\lambda)$	Poisson distribution
$t_\nu(\mu,\sigma^2)$	Student-t distribution
$\mathcal{U}(a,b)$	Uniform distribution
$\mathcal{W}(\nu,\boldsymbol{S})$	Wishart distribution
$\mathcal{W}^{-1}(\nu,\boldsymbol{S})$	Inverse Wishart distribution

Part I

Spike-and-Slab Priors

1

Discrete Spike-and-Slab Priors: Models and Computational Aspects

Marina Vannucci

Rice University, Houston, TX (USA)

CONTENTS

A large body of research has been devoted to variable selection in recent years. Bayesian methods have been successful in applications, particularly in settings where the amount of measured variables can be much greater than the number of observations. This chapter reviews mixture priors that employ a point mass distribution at zero for variable selection in regression settings. The popular stochastic search Markov chain Monte Carlo (MCMC) algorithm with add-delete-swap moves is described. Posterior inference and prediction via Bayesian model averaging are briefly discussed. Regression models for non-Gaussian data, including binary, multinomial, survival and compositional count data, are also addressed. Prior constructions that take into account specific structures in the covariates are reviewed. These have been particularly successful in applications as they allow the integration of different sources of external information into the analysis. A discussion of variational algorithms for scalable inference concludes the chapter. Throughout, some emphasis is given to the author's contributions.

DOI: 10.1201/9781003089018-1

1.1 Introduction

Variable selection, also known as feature selection, has been an important topic in the statistical literature for the past several decades, with numerous papers published in both theory and practice. Finding a subset of features that best explain an outcome of interest is often an important aspect of the data analysis, as it allows for simpler interpretation, avoids overfitting and multicollinearity and provides insights into the mechanisms generating the data. Variable selection is especially important when the number of potential predictors is substantially larger than the sample size.

In linear regression settings, modern approaches to variable selection include criteria-based methods, such as AIC/BIC [36], penalized likelihood methods which shrink to zero coefficients of unimportant covariates [41], and Bayesian approaches that use shrinkage priors to induce sparsity, such as mixtures of two distributions (spike-and-slab priors) [6, 17, 18, 23, 37] and unimodal continuous shrinkage priors [10, 43, 46]. With spike-and-slab priors, a latent binary vector is introduced to index the possible subsets of predictors and used to induce mixture priors of two components on the regression coefficients, one peaked at zero (spike) and the other one a diffuse distribution (slab). Posterior inference is carried out via stochastic search MCMC techniques to identify the high-probability models, and variable selection is performed based on the posterior model probabilities. This chapter is devoted in particular to *discrete* spike-and-slab constructions, which employ a point mass distribution at zero as the spike component.

Bayesian methods for variable selection have several appealing features. They allow rich modeling via MCMC stochastic search strategies and incorporate optimal model averaging prediction; they extend naturally to multivariate responses and many linear and non-linear settings; they can handle the "small n-large p" setting, i.e., situations where the number of covariates is larger than the sample size; they allow the use of priors that incorporate past and collateral information into the model.

This chapter is organized as follows. Section 1.2 briefly reviews discrete spike-and-slab priors for variable selection in linear regression, including the popular stochastic search MCMC algorithm with add-delete-swap moves, for posterior inference, and a brief discussion of Bayesian model averaging, for prediction purposes. Section 1.3 addresses regression models for non-Gaussian data, including binary, multinomial and survival outcomes. It also covers model settings for compositional count data. Section 1.4 reviews prior constructions that take into account specific structures in the covariates, together with examples of modern biomedical studies in genomics and neuroimaging that have motivated those constructions. Section 1.5 discusses variational inference strategies for scalable inference. Final remarks are given in Section 1.6.

1.2 Spike-and-Slab Priors for Linear Regression Models

In the classical multiple linear regression model, a continuous response, y_i, is modeled via a linear combination of p covariates, $\mathbf{x}_i = (x_1, \ldots, x_p) \in \mathbb{R}^p$, as

$$y_i = \alpha + \mathbf{x}_i^T \boldsymbol{\beta} + \epsilon_i, \quad i = 1, \ldots, n, \tag{1.1}$$

with $\epsilon_i \sim \mathcal{N}(0, \sigma^2)$, $\boldsymbol{\beta} = (\beta_1, \ldots, \beta_p)^T$ the vector of regression coefficients and α the baseline or intercept. The variable selection problem arises when it is believed that not all p covariates are important in explaining changes of the response and identification of the important

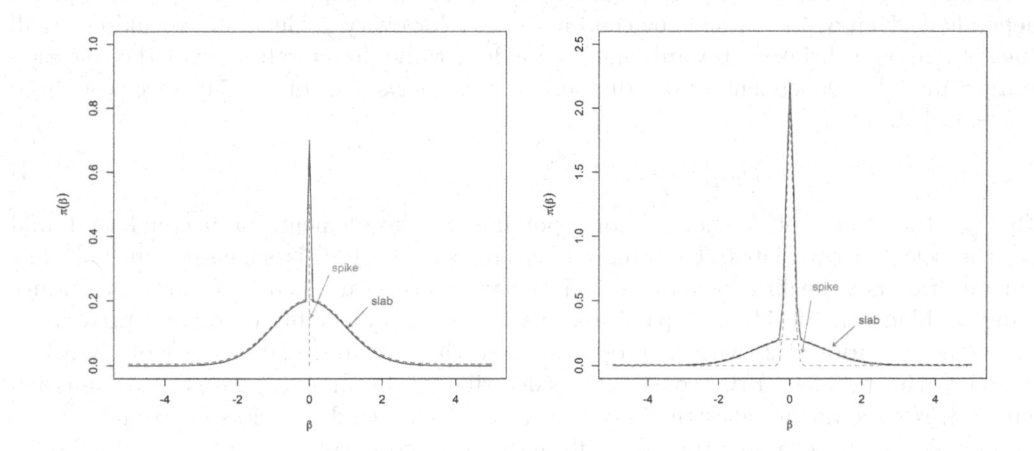

FIGURE 1.1
Spike-and-slab mixture priors for Bayesian variable selection. *Left:* The discrete construction (solid line) is a mixture of a point mass at zero (spike; dashed line) and a diffuse prior (slab; dotted line). *Right:* The continuous construction (solid line) is a mixture of two normal distributions, one peaked around zero (dashed line) and the other with a large variance (dotted line).

predictors is one of the goals of the analysis. Clearly, setting to zero some of the regression coefficients in (1.1) is equivalent to excluding the corresponding subset of predictors from the model. In the Bayesian paradigm this can be achieved by imposing sparsity-inducing mixture priors, known as *spike-and-slab* priors, on the β_j coefficients [17, 18, 23, 37]. This formulation introduces a latent vector $\boldsymbol{\gamma} = (\gamma_1, \ldots, \gamma_p)$ of binary indicators

$$\gamma_j = \begin{cases} 1 & \text{if variable } j \text{ is included in model,} \\ 0 & \text{otherwise.} \end{cases}$$

Two prior constructions have been developed in parallel in the statistical literature. This chapter focuses on the *discrete* construction, which employs a mixture prior distribution on β_j with a point mass at zero, see Figure 1.1, as

$$\beta_j|\sigma^2, \gamma_j \sim (1-\gamma_j)\delta_0(\beta_j) + \gamma_j \mathcal{N}(0, h_j\sigma^2), \tag{1.2}$$

for $j = 1, \ldots, p$, with $\delta_0(\cdot)$ the Dirac function at $\beta_j = 0$ and the h_j's a set of hyperparameters. Here, $\gamma_j = 0$ excludes the j-th variable from the model since the prior on the corresponding coefficient β_j is a point mass distribution at 0, while $\gamma_j = 1$ includes the predictor into the model, leading to a normal prior on β_j. Mixture priors of type (1.2) for the linear regression setting were originally proposed by [37] and made popular by [18]. The prior formulation is completed with an independent conjugate inverse-gamma priors on σ^2 and a Gaussian prior on the intercept α,

$$\alpha|\sigma^2 \sim \mathcal{N}(\alpha_0, h_0\sigma^2), \quad \sigma^2 \sim IG(\nu/2, \lambda/2), \tag{1.3}$$

with α_0, h_0, ν and λ hyperparameters to be chosen. Setting $\alpha_0 = 0$ and taking $h_0 \to \infty$ results in a vague prior on the intercept, so that mean centering the predictors sets the posterior mean for α at $\bar{\mathbf{y}}$.

Common choices for the hyperparameters h_j's in (1.2) assume that the β_j's are *a priori* independent given $\boldsymbol{\gamma}$, for example by choosing $h_j = c$ for every j. Generally speaking, small values of c induce shrinkage towards smaller models, while larger values favor the selection of larger models. Dependent priors that use the Zellner's g-prior of [74] have also been considered [24, 60],

$$\boldsymbol{\beta}_{(\gamma)}|\sigma^2 \sim \mathcal{N}(0, c(\mathbf{X}'_{(\gamma)}\mathbf{X}_{(\gamma)})^{-1}\sigma^2), \tag{1.4}$$

with $\boldsymbol{\beta}_{(\gamma)}$ the subset of coefficients corresponding to the elements of $\boldsymbol{\gamma}$ equal to 1 and $\mathbf{X}_{(\gamma)}$ the selected covariates. The range of values $c \in [10, 100]$ is suggested in [60]. The Zellner's prior is appealing because of its intuitive interpretation. It can, however, induce mixing problems in the MCMC, particularly with subsets of highly correlated predictors. [6] investigated prior (1.2) with h_j proportional to the j-th diagonal element of $(\mathbf{X}'\mathbf{X})^{-1}$, to alleviate this problem. Prior constructions described so far are conjugate to the Gaussian likelihood. Non-conjugate constructions, that assume independent priors on $\boldsymbol{\beta}$ and σ^2, are also possible, see for example [18]. We will revisit these constructions in Section 1.2.1.

The discrete construction (1.2) differs from the *continuous* spike-and-slab prior, which instead employs a mixture of two continuous components, typically two Gaussian distributions, one concentrated around zero and the second one more spread out over plausible large values [17, 23], see Figure 1.1. Unlike with the continuous construction, discrete priors of type (1.2) effectively exclude non-selected variables from the calculation of the likelihood. While optimality properties of continuous spike-and-slab priors have been studied fairly extensively [24, 28, 53], theoretical guarantees for the discrete construction in the linear regression setting (1.1) have become available only recently, due to the seminal work of [6], and include optimality results for the Zellner g-prior construction (1.4), see [73].

Prior construction (1.2) requires the choice of a prior distribution on $\boldsymbol{\gamma}$. The simplest and most common choice adopted in the literature is a product of independent Bernoulli's with common parameter ω as

$$p(\boldsymbol{\gamma}|\omega) = \prod_{j=1}^{p} \omega^{\gamma_j}(1-\omega)^{1-\gamma_j}, \tag{1.5}$$

that leads to $p\omega$ being the number of variables expected *a priori* to be included in the model. Uncertainty on ω can be modeled by imposing a Beta hyperprior, $\omega \sim Beta(a, b)$, with a, b to be chosen. If inference on ω is not of interest, it can be integrated out to simplify the MCMC implementation. A weakly-informative prior can be obtained by setting $a = b = 1$, resulting in the prior expected mean value to be $m = a/(a+b) = .5$. An attractive feature of the Beta-Binomial prior construction is that it imposes an *a priori* multiplicity penalty, as argued in [57]. The intuition behind this is that the marginal prior on $\boldsymbol{\gamma}$ contains a non-linear penalty which is a function of p and therefore, as p grows, with the number of true variables remaining fixed, the posterior distribution of ω concentrates near 0. A limitation of the Beta-Binomial construction is that it assumes that the inclusion indicators are stochastically independent. Alternative priors, that exploit complex dependence structures between covariates, as induced by underlying biological processes and/or networks, have been motivated by specific applications to data from biomedical studies. Some of these prior constructions will be described in Section 1.4.

Conjugate discrete spike-and-slab prior constructions have been extended by [6, 7] to multivariate linear regression models with q response outcomes. Their construction selects variables as relevant to either all or none of the q responses. [31] proposed multivariate constructions based on partition models that allow each covariate to be relevant for subsets and/or individual response variables. Other flexible multivariate prior formulations, that allow to select covariates for individual responses, were proposed by [52, 63]. See Section 1.3 for an example of such construction in a multivariate count data model setting.

1.2.1 Stochastic Search MCMC

Let us consider the linear setting (1.1) with the discrete spike-and-slab prior construction described in the previous section. The choice of conjugate priors makes it possible to integrate out the model parameters and obtain the relative posterior distribution of $\boldsymbol{\gamma}$ as

$$p(\boldsymbol{\gamma}|\mathbf{y}, \mathbf{X}) \propto f(\mathbf{y}|\boldsymbol{\gamma}, \mathbf{X})p(\boldsymbol{\gamma}). \tag{1.6}$$

This distribution allows to identify the "best" models as those with highest posterior probabilities. When a large number of predictors make the full exploration of the model space unfeasible, MCMC methods can be used as stochastic searches to explore the posterior distribution and identify models with high posterior probability. Marginalization (1.6), jointly with a QR deletion–addition algorithm for fast updating in the calculation of the marginal likelihood, leads to efficient MCMC schemes for posterior inference, see [18] for the univariate regression setting and [7] for multivariate regression. A commonly used algorithm is a Metropolis-Hastings scheme readapted from the MC^3 algorithm proposed by [34] in the context of model selection for discrete graphical models. It consists of *add-delete-swap* moves that allow the exploration of the posterior space by visiting a sequence of models where, at each step, the new model differs from the previously visited one by the inclusion and/or exclusion of one or two variables. More specifically, given a randomly chosen starting value, $\boldsymbol{\gamma}^0$, at a generic iteration the new model is generated from the previous one by randomly choosing one of the following transition moves:

1. (Adding or deleting) Randomly pick one of the p indices in $\boldsymbol{\gamma}^{old}$ and change its value. This results in either including a new variable in the model or in deleting a variable currently included.

2. (Swapping) Draw independently and at random a 0 and a 1 in $\boldsymbol{\gamma}^{old}$ and switch their values. This results in both the inclusion of a new variable in the model and the deletion of a currently included one.

By indicating with $\boldsymbol{\gamma}^{new}$ the candidate model, the acceptance probability is calculated as

$$\min\left[\frac{p(\boldsymbol{\gamma}^{new}|\mathbf{y}, \mathbf{X})}{p(\boldsymbol{\gamma}^{old}|\mathbf{y}, \mathbf{X})}, 1\right]. \tag{1.7}$$

Therefore, if the new candidate model has a higher probability than the current one, the chain moves to the new configuration. If not, then the move is still possible, but now only with a certain probability. Note that the acceptance probability (1.7) depends on an "exact" ratio, since the constants of proportionality from (1.6) cancel out. This allows the search to quickly move towards better models. The stochastic search results in a list of visited models, $\boldsymbol{\gamma}^{(0)}, \ldots, \boldsymbol{\gamma}^{(T)}$, and their corresponding relative posterior probabilities. Variable selection can then be achieved either by looking at $\boldsymbol{\gamma}$ vectors with largest joint posterior probabilities among the visited models or, marginally, by calculating frequencies of inclusion for each γ_j and then choosing those γ_j's with frequencies exceeding a given cut-off value. A common choice is a cut-off value of 0.5, which results in the median probability model [1]. Methods based on expected false discovery rates can also be employed, as suggested in [39].

Gibbs sampling schemes are also possible, see for example [18]. However, these schemes typically sample all variable indicators γ_j's at each iteration, unlike Metropolis schemes that allow a more efficient exploration of the space of only the relevant variables. This is particularly important in situations of sparsity of the true model. Improved stochastic MCMC schemes have been proposed, to achieve a faster exploration of the posterior space. See, for example, the *shotgun* algorithm of [22] and the evolutionary Monte Carlo schemes, combined with a parallel tempering step that prevents the chain from getting stuck in local modes, proposed by [5]. A Correlation-Based Stochastic Search method, the hybrid-CBS algorithm, which comprises add-delete-swap moves specifically designed to accommodate correlations among the covariates, was proposed by [29]. Adaptive schemes that specifically aim at improving the mixing of the MCMC chain have been investigated by [21, 30].

When non-conjugate priors are used, the marginalization of the model parameters is no longer possible and those parameters need to be sampled as part of the MCMC algorithm. Initial attempts employed the reversible jump algorithm of [20], to handle the varying dimensionality of the parameter vector, see for example [18]. Later, [19] showed that the reversible jump can be formulated in terms of a mixture of singular distributions, implying that the algorithm is the same as an MCMC algorithm that jointly samples parameters and binary indicators. This is key to designing efficient MCMC algorithms for variable selection in non-conjugate settings, particularly for the case of non-Gaussian data and, more generally, complex models for which conjugate prior formulations may not be available. For example, this idea was used by [56] to design add-delete-swap algorithms that jointly update parameters and selection indicators in a variable selection approach that incorporates Gaussian processes within a generalized linear model framework. We will see an example of a joint sampler for $(\boldsymbol{\beta}, \boldsymbol{\gamma})$ in Section 1.3, within a model setting for multivariate count data.

1.2.2 Prediction via Bayesian Model Averaging

Prediction is an important aspect of the inference in linear regression settings. Given the list of models visited by the stochastic search, $\boldsymbol{\gamma}^{(0)}, \ldots, \boldsymbol{\gamma}^{(T)}$, prediction of a future observation y^f can be done based on the selected models, either via least squares on single models, or by *Bayesian model averaging* (BMA) [50], which accounts for the uncertainty in the selection process by averaging over a set of *a posteriori* likely models. For example, for model (1.1) with prior (1.2) and posterior (1.6), BMA calculates the expected value of the predictive distribution $p(y^f|\mathbf{y}, \mathbf{X}^f)$, averaging over a set of configurations of $\boldsymbol{\gamma}$ with weights given by the posterior probabilities of these configurations as

$$\hat{y}^f = \sum_{\boldsymbol{\gamma}} \left(\hat{\alpha} + \mathbf{X}^f_{(\boldsymbol{\gamma})} \hat{\boldsymbol{\beta}}_{(\boldsymbol{\gamma})} \right) p(\boldsymbol{\gamma}|\mathbf{y}, \mathbf{X}), \tag{1.8}$$

with $\mathbf{X}^f_{(\boldsymbol{\gamma})}$ the covariates corresponding to the elements of $\boldsymbol{\gamma}$ equal to 1, $\hat{\alpha} = \bar{\mathbf{y}}$ and $\hat{\boldsymbol{\beta}}_{\boldsymbol{\gamma}} = (\mathbf{X}'_{(\boldsymbol{\gamma})}\mathbf{X}_{(\boldsymbol{\gamma})} + \mathbf{H}^{-1}_{(\boldsymbol{\gamma})})^{-1}\mathbf{X}'_{(\boldsymbol{\gamma})}\mathbf{y}$, with $\mathbf{H}$ a diagonal matrix with diagonal elements the hyperparameters h_j's of the slab component in (1.2). Typically, only the best k configurations among those visited by the MCMC, according to their posterior probabilities, are used in the summation.

1.3 Spike-and-Slab Priors for Non-Gaussian Data

Spike-and-slab mixture priors for variable selection have been extended beyond Gaussian data to other model settings that express a response variable as a linear combination of the predictors. A unified treatment of the class of generalized linear models (GLM) of [35] presents some challenges. Conditional densities in the general GLM framework cannot be obtained directly and the resulting mixture posterior may be difficult to sample from using standard MCMC methods due to multimodality. Some attempts were done by [49], who proposed approximate Bayes factors, and by [40], who developed a method to jointly select variables and the link function.

Several contributions exist on extending spike-and-slab priors to specific models in the GLM class, in particular models for binary and multinomial outcomes and parametric accelerated failure time (AFT) models for survival outcomes. For example, probit models with multinomial outcomes were considered by [59] and AFT models by [58]. In these settings, data augmentation approaches allow to express the model in a linear framework, with latent responses $\mathbf{z}$, and conjugate priors are used to integrate the regression coefficients out, obtaining the marginal likelihood $f(\mathbf{y}|\boldsymbol{\gamma}, \mathbf{X}, \mathbf{z})$, and facilitating the implementation of MCMC schemes that update $\boldsymbol{\gamma}$ conditional upon $\mathbf{z}$. For other settings, where marginalization of the regression coefficients is not possible, joint updates of coefficients and selection indicators can be performed and, whenever possible, coupled with data augmentation schemes for more efficient samplers. Examples include logistic and negative binomial regression models, for which the Pólya-Gamma (PG) data augmentation schemes developed by [45, 47] can be used to implement Gibbs samplers with PG updates on the latent variables followed by Gaussian updates on the regression coefficients. See [69] for recent work that combines these augmentation schemes with the add-delete-swap scheme of [56], as part of a variable selection approach to non-homogeneous hidden Markov models. Also, adaptive MCMC schemes for variable selection in logistic and AFT regression models were investigated in [68].

1.3.1 Compositional Count Data

[67] considered a Dirichlet-multinomial (DM) regression framework for compositional count data and demonstrated how to embed spike-and-slab priors for variable selection. Compositional count data $\mathbf{y}_i = (y_{i1}, \ldots, y_{iq})$ sum up to a fixed amount. A suitable distribution for this data is the multinomial

$$\mathbf{y}_i \sim \text{Multinomial}(\dot{y}_i|\mathbf{p}_i), \tag{1.9}$$

with $\dot{y}_i = \sum_{k=1}^{q} y_{ik}$, and $\mathbf{p}_i$ defined on the q-dimensional simplex

$$S^{q-1} = \{(p_{i1}, \ldots, p_{iq}) : p_{ik} \geq 0, \forall k, \sum_{k=1}^{q} p_{ik} = 1\}.$$

A Dirichlet conjugate prior can be imposed on the probability parameter vector, $\mathbf{p}_i \sim$ Dirichlet($\boldsymbol{\phi}_i$), with q-dimensional vector $\boldsymbol{\phi_i} = (\phi_{ik} > 0, \forall k)$, and then $\mathbf{p}_i$ can be integrated out to obtain the DM model $\mathbf{y}_i \sim \text{DM}(\boldsymbol{\phi}_i)$. The DM model allows more flexibility than the multinomial when encountering overdispersion, as it induces an increase in the variance by a factor of $(\dot{y}_i + \dot{\phi}_i)/(1 + \dot{\phi}_i)$.

Covariate effects can be incorporated into the DM model via a log-linear link on the concentration parameters $\boldsymbol{\phi}_i$'s, by setting $\lambda_{ik} = \log(\phi_{ik})$ and assuming

$$\lambda_{ik} = \alpha_k + \sum_{j=1}^{p} \beta_{jk}\, x_{ij}, \tag{1.10}$$

where $\boldsymbol{\beta_k} = (\beta_{1k}, \ldots, \beta_{pk})^T$ represents the covariates' potential relations with the kth compositional outcome, and α_k is a outcome-specific intercept term. Exponentiating (1.10) ensures positive hyperparmeters for the Dirichlet distribution. For variable selection purposes, the number of potential models to choose from is 2^{pq}, and grows quickly even for a small number of covariates. [67] introduced a set of q latent p-dimensional vectors $\boldsymbol{\gamma_k} = (\gamma_{1k}, \ldots, \gamma_{pk})$ of inclusion indicators. Thus, $\gamma_{jk} = 1$ indicates that the j-th covariate is associated with the k-th compositional outcome, and 0 otherwise. A discrete spike-and-slab prior on β_{jk} is then written as

$$\beta_{jk}|\gamma_{jk} \sim (1 - \gamma_{jk})\delta_0(\beta_{jk}) + \gamma_{jk}\mathcal{N}(0, r_k^2), \tag{1.11}$$

where the hyperparameters r_k^2 can be set large to impose a diffuse prior for the regression coefficients included in the model. This multivariate spike-and-slab prior, that allows to identify covariates associated with individual responses, is similar to constructions used by [52, 63] in linear regression settings for Gaussian data with multiple responses.

Posterior inference is carried out via stochastic search MCMC algorithms. Here, the regression coefficients β_{jk}'s cannot be integrated out and, therefore, need to be jointly updated with the inclusion indicators, following [56]. [67] employed this strategy within a Gibbs scheme that scans through the γ_{jk}'s and uses adaptive sampling on the β_{jk}'s. [28] incorporated the joint update within an add-delete Metropolis-Hastings within Gibbs sampling scheme that updates each α_k and a randomly selected $(\gamma_{jk}, \beta_{jk})$ at every iteration. The joint update works as follows:

- *Between-Model Step* – Randomly select a γ_{jk}.
  - Add: If the covariate is currently excluded ($\gamma_{jk} = 0$), change it to $\gamma'_{jk} = 1$. Then sample a new $\beta'_{jk} \sim \mathcal{N}(\beta_{jk}, c)$ with c fixed to a chosen value. Accept proposal with probability
  $$\min\left\{\frac{f(\mathbf{y}|\boldsymbol{\alpha}, \boldsymbol{\beta}', \boldsymbol{\gamma}', \mathbf{X})p(\beta'_{jk}|\gamma'_{jk})p(\gamma'_{jk})}{f(\mathbf{y}|\boldsymbol{\alpha}, \boldsymbol{\beta}, \boldsymbol{\gamma}, \mathbf{X})p(\gamma_{jk})}, 1\right\}.$$
  - Delete: If the covariate is currently included ($\gamma_{jk} = 1$), change it to $\gamma'_{jk} = 0$ and set $\beta'_{jk} = 0$. Accept proposal with probability
  $$\min\left\{\frac{f(\mathbf{y}|\boldsymbol{\alpha}, \boldsymbol{\beta}', \boldsymbol{\gamma}', \mathbf{X})p(\gamma'_{jk})}{f(\mathbf{y}|\boldsymbol{\alpha}, \boldsymbol{\beta}, \boldsymbol{\gamma}, \mathbf{X})p(\beta_{jk}|\gamma_{jk})p(\gamma_{jk})}, 1\right\}.$$
- *Within-Model Step* – Propose a $\beta'_{jk} \sim \mathcal{N}(\beta_{jk}, c)$ for each covariate currently selected in the model ($\gamma_{jk} = 1$). Accept each proposal with probability
  $$\min\left\{\frac{f(\mathbf{y}|\boldsymbol{\alpha}, \boldsymbol{\beta}', \boldsymbol{\gamma}, \mathbf{X})p(\beta'_{jk}|\gamma_{jk})}{f(\mathbf{y}|\boldsymbol{\alpha}, \boldsymbol{\beta}, \boldsymbol{\gamma}, \mathbf{X})p(\beta_{jk}|\gamma_{jk})}, 1\right\}.$$
  This within-model step is not required for ergodicity but allows to perform a refinement of the parameter space within the existing model, for faster convergence.

As customary with spike-and-slab priors, variable selection is performed based on the marginal posterior probabilities of inclusion (PPIs).

Recently, there has been a renewed interest in the biomedical community on statistical models for compositional count data, in particular due to the availability of high-throughput sequencing technologies that have enabled researchers to characterize the composition of the microbiome by quantifying its richness, diversity and abundances. Human microbiome research aims to understand how microbiome communities interact with their host, respond to their environment, and influence disease. DM regression models allow to appropriately handle the compositional structure of the data and accommodate the overdispersion induced by sample heterogeneity and varying proportions among samples. While model formulation (1.9) assumes that counts are negatively correlated, extensions exist that allow more general correlation structures between counts, such as the Dirichlet-tree multinomial model, that deconstructs the model into the product of multinomial distributions for each of the subtrees in a tree [71]. The `R` package `MicroBVS`, accompanying [28], comprises of a suite of regression models for compositional data, including DM and Dirichlet-tree multinomial regression models. It also implements the joint model of [27] that includes a phenotypical outcome to investigate how the microbiome may affect the relation between covariates and phenotypic responses. MCMC algorithms are written in C++, to increase performance time,

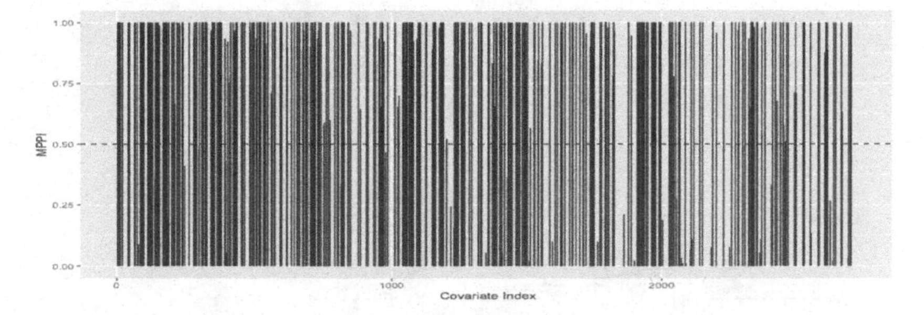

FIGURE 1.2
Analysis of data on dietary intake and microbiome via a regression model for compositional count data with variable selection [28]. Plot of PPIs of associations between microbial taxa and dietary factors, with each of the $p = 97$ dietary factors having a unique inclusion indicator for each of the $q = 28$ taxa.

and accessed through `R` wrapper functions. The package includes a vignette with worked out examples using simulated data and access to open-source data used in the papers. As an example, let us consider the analysis of a benchmark data set collected to study the relation between dietary intake and the human gut microbiome [72]. Briefly, the data used consist of counts on $q = 28$ microbial taxa obtained from 16S rRNA sequencing and a set of $p = 97$ dietary intake covariates derived from diet information collected using a food frequency questionnaire on $n = 98$ subjects. In this analysis, the DM regression model was fit to the data assuming a Beta-Binomial prior with $a = 1$ and $b = 999$ on the inclusion indicators γ_{jk} of prior (1.11). At convergence of the MCMC, about 398 of the roughly 2700 terms would be selected with a threshold of 0.5 on the PPIs, see Figure 1.2. Heatmaps of the selected positive and negative associations are shown in Figure 1.2. For this application, knowledge of the identified relations between microbial composition and nutrients may help researchers design tailored interventions to help maintain a healthy microbiome community [72].

1.4 Structured Spike-and-Slab Priors for Biomedical Studies

Spike-and-slab variable selection priors have found important applications in biomedical studies. In high-throughput genomic, for example, linear models are routinely used to relate large sets of biomarkers to disease-related outcomes, and variable selection methods are employed to identify the significant predictors. In neuroimaging, as another example, functional magnetic resonance imaging (fMRI) is used to measure blood flow changes across the whole brain and linear models are employed to detect (i.e, select) brain regions that activate in response to external stimuli. For these applications, extensions of spike-and-slab prior constructions have been motivated by specific characteristics of the data. For example, [63] put forward a graphical model formulation of a multivariate regression model where target genes (the outcomes) are regulated by microRNAs (the covariates), which are small RNA sequences located upstream of the genes. In the proposed model formulation, spike-and-slab priors allow to identify gene-microRNA interactions, therefore inferring a biological network. In place of the independent Bernoulli priors of type (1.5), the authors assumed a

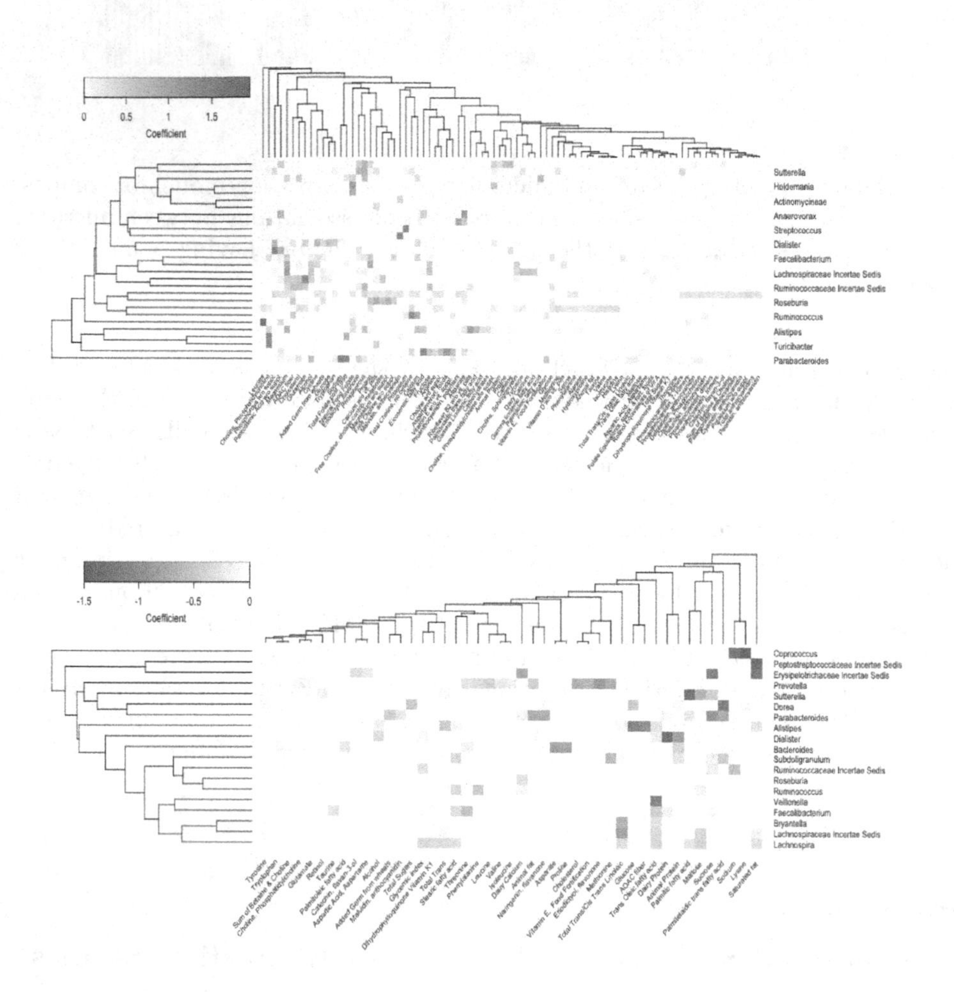

FIGURE 1.3
Analysis of data on dietary intake and microbiome via a regression model for compositional count data with variable selection [28]. Heatmaps of selected positive (upper plot) and negative (lower plot) associations.

logistic prior construction on γ of the form

$$P(\gamma_{jk} = 1|\tau, \eta) = \frac{\exp(\eta + \tau s_{jk})}{1 + \exp(\eta + \tau s_{jk})}, \tag{1.12}$$

that incorporates a set of available scores s_{jk} of possible association between gene-microRNA pairs, as obtained from external sequence/structure information. The prior assumes that the γ_{jk}'s are stochastically independent given τ and reduces to $p(\gamma_{jk} = 1) = \exp(\eta)/(1+\exp(\eta))$ when all $s_{jk} = 0$. Probit constructions that incorporate external information can also be used and have been investigated in other integrative settings [11, 48]. These constructions, while accounting for external information, still assume independence between inclusion indicators. However, in many practical applications, researchers may be interested in incorporating prior information on the dependence structure between covariates, as captured by an underlying biological process and/or a correlation network. Below, prior constructions on γ that account for such information are briefly described.

1.4.1 Network Priors

Network priors account for known relations among covariates in the form of a graph. For example, in genomics, when covariates are chosen as gene expression levels, a network of gene–gene interactions may be known based on biological information on known pathways (i.e., groups of genes). Here, individual genes are represented by nodes in the network and relations between them by edges. This network structure can be captured via a Markov random field (MRF) prior [3], also known as Ising prior, on the binary indicator vector γ as

$$P(\gamma|d, e) \propto \exp(d\mathbf{1}'\gamma + e\gamma' G\gamma), \tag{1.13}$$

with $\mathbf{G}$ a $p \times p$ adjacency matrix that represents the relations between covariates, that is, with elements $g_{jj'} = 1$ if variables j and j' have a direct link in the network, and $g_{jj'} = 0$ otherwise. Hyperparameters $d \in R$ and $e > 0$ control the global probability of inclusion and the influence of neighbors' inclusion on a covariate's inclusion, respectively. According to parametrization (1.13), a covariate's inclusion probability will increase if neighboring covariates in the known network are also included in the model. The prior simplifies to the independent Bernoulli($\exp(d)/(1 + \exp(d))$) for $e = 0$. MRF priors of type (1.13) have been employed in linear models for genomic applications to aid the identification of predictive genes by [32]. Also, [62] considered a linear model that predicts a phenotype based on predictors synthesizing the activity of genes belonging to same pathways. The prior model encodes information on gene–gene networks via a MRF prior, as retrieved from available databases, and inference results in the identification of both relevant pathways and subsets of genes. Among more recent applications, [31] considered a linear model with multivariate responses to identify the joint effect of pollutants on DNA methylation outcomes via structured spike-and-slab priors that leverage the dependence across markers. In all these papers, authors show how small increments of the parameter e in (1.13) can drastically increase the number of selected covariates and provide guidelines on how to select suitable values and/or prior distributions for this parameter.

In situations where the network structure $\mathbf{G}$ in prior (1.13) is unknown, it can be inferred from the data using priors and learning algorithms for undirected graphical models. [44] used this strategy to obtain a Bayesian modeling approach for linear regression settings that simultaneously performs variable selection while learning the dependence relations between covariates. In this setting, the matrix of covariates $\mathbf{X}$ is treated as random and the joint distribution of $(\mathbf{y}, \mathbf{X})$ is factorized as

$$f(\mathbf{y}, \mathbf{X}) = f(\mathbf{y}|\mathbf{X})f(\mathbf{X}), \tag{1.14}$$

with $f(\mathbf{y}|\mathbf{X})$ defining the linear regression model. Assuming Gaussianity of the $\mathbf{x}_i$'s, we have $\mathbf{x}_i \sim \mathcal{N}_p(\mathbf{0}, \boldsymbol{\Omega})$, where $\boldsymbol{\Omega} = \boldsymbol{\Sigma}^{-1}$ is a $p \times p$ precision matrix. Thus, the presence of edge $g_{jj'} = 1$ in graph $\mathbf{G}$ corresponds to $\omega_{jj'} \neq 0$ in $\boldsymbol{\Omega}$. [70] proposed a prior for $\boldsymbol{\Omega}$ that assumes continuous spike-and-slab distributions on the off-diagonal elements and exponential distributions for the diagonal components as

$$p(\boldsymbol{\Omega}|\mathbf{G}, v_0, v_1, \theta) \propto \prod_{j<j'} \mathcal{N}(\omega_{jj'}|0, v_{jj'}^2) \prod_{j} \mathcal{E}xp\,(\omega_{jj}|\theta/2)\, I_{\{\boldsymbol{\Omega}\in M^+\}}, \tag{1.15}$$

with $v_{jj'}^2 = v_1$ if $g_{jj'} = 1$, and $v_{jj'}^2 = v_0$ if $g_{jj'} = 0$, with $v_0 << v_1$, and where $\mathcal{E}xp(\cdot|\theta/2)$ represents an exponential distribution with mean $2/\theta$. The term $I_{\{\boldsymbol{\Omega}\in M^+\}}$ restricts the prior to the space of symmetric-positive definite matrices. The model is completed with a prior for $\mathbf{G}$, for example as a simple product of independent Bernoulli's on the $g_{jj'}$'s elements, with a common parameter π to represent the prior probability of inclusion for an individual edge. A specification of π that reflects prior beliefs of sparsity is recommended by [70]. Also, setting $\theta = 1$ implies a relatively vague prior for ω_{jj} when the data are standardized prior to analysis. For posterior inference, [44] incorporated two additional steps, two update $\boldsymbol{\Omega}$ and $\mathbf{G}$, following [70], within a stochastic search MCMC scheme for linear settings, to obtain simultaneous variable selection and estimation of a graph between covariates. Also, [28] extended these methods to the regression models for compositional count data discussed in Section 1.3.1. Both prior options, (1.13) with $\mathbf{G}$ known and (1.13) with prior (1.15) on $\boldsymbol{\Omega}|\mathbf{G}$, are available in the `R` package `MicroBVS`.

1.4.2 Spiked Nonparametric Priors

Other extensions of spike-and-slab priors include constructions that employ nonparametric priors [14, 25, 55]. One construction uses a mixture of a point mass at zero and a nonparametric slab, typically a Dirichlet process (DP) prior [15] with a continuous distribution as its centering distribution. Such construction clusters parameters together when information in the data provides evidence of a common effect. This, in turn, allows to borrow information across covariates, resulting in improved selection and estimation [14]. In [8], this construction is referred to as an "outer" spike-and-slab nonparametric prior, as opposed to the "inner" prior of [25], which is a DP prior where the base measure is modeled as a mixture of a point mass at zero and a diffuse measure. The inner prior formulation does not share information across covariates, but rather clusters vectors of regression coefficients across observations. Recent applications of outer discrete nonparametric constructions include covariate dependent random partition models [2] and dynamic extensions for spatio-temporal dynamic models with random effects [12].

Let us illustrate the outer construction via an application to functional magnetic resonance imaging (fMRI) data, another area of successful applications of models that employ spike-and-slab priors. In a typical task-based fMRI experiment, the whole brain is scanned at multiple time points while the subject is presented with a series of stimuli. Each scan is arranged as a 3D array of volume elements (or "voxels"), and the experiment returns time series data acquired at each voxel. Let $\mathbf{y}_{i\nu} = (y_{i\nu 1}, \ldots, y_{i\nu T})^T$ be the vector of the time series data at voxel ν, with $\nu = 1, \ldots, V$, for subject i. Common modeling approaches for the analysis of task-based fMRI data rely on the general linear model formulation originally proposed by [16]

$$\mathbf{y}_{i\nu} = \mathbf{X}_{i\nu}\boldsymbol{\beta}_{i\nu} + \varepsilon_{i\nu},\ \varepsilon_{i\nu} \sim \mathcal{N}_T(0, \boldsymbol{\Sigma}_{i\nu}), \tag{1.16}$$

where $\mathbf{X}_{i\nu}$ is a known $T \times K$ design matrix (for K stimuli) modeled as the convolution of the stimulus patterns with a hemodynamic response function that accounts for the delay of the response with respect to the stimulus onset. The task of the inference is to detect those voxels that activate in response to the stimuli, which is equivalent to inferring the non-zero regression coefficients in (1.16). Spatial correlation among brain voxels can be accounted for in the prior construction. Examples include spike-and-slab priors that incorporate MRF priors on the selection indicators, to account for neighboring correlation among voxels, and nonparametric slabs that capture spatial correlation among possibly distant voxels [61, 76]. Let us consider the simpler case $K = 1$, i.e., one stimulus. A spiked nonparametric prior on $\beta_{i\nu}$ can be written as

$$\beta_{i\nu}|\gamma_{i\nu}, G_0 \sim (1 - \gamma_{i\nu})\delta_0(\beta_{i\nu}) + \gamma_{i\nu} G_0, \tag{1.17}$$

where G_0 denotes a Dirichlet process prior with $\mathcal{N}(0, \tau)$ as the centering distribution. With multiple subjects, a hierarchical Dirichlet Process (HDP) prior can be specified as the nonparametric slab, inducing clustering among voxels within a subject on one level of the hierarchy and between subjects on the second level, as

$$\begin{aligned} \beta_{i\nu}|\gamma_{i\nu}, G_i &\sim (1 - \gamma_{i\nu})\delta_0(\beta_{i\nu}) + \gamma_{i\nu} G_i \\ G_i|\eta_1, G_0 &\sim DP(\eta_1, G_0) \\ G_0|\eta_2, P_0 &\sim DP(\eta_2, P_0) \\ P_0 &= \mathcal{N}(0, \tau), \end{aligned} \tag{1.18}$$

with τ fixed, η_1, η_2 the mass parameters and P_0 the base measure. With this prior formulation, the subject-specific distribution G_i varies around a population-based distribution G_0, which is centered around a known parametric model P_0. The mass parameters η_1 and η_2 control the variability of the distribution of the coefficients at the subject and population level, respectively. This construction enables the model to borrow information from subjects exhibiting similar activation patterns while also capturing spatial correlation among distant voxels.

For model (1.16) with prior (1.18), [76] implemented an MCMC algorithm that combines add-delete-swap moves with sampling algorithms for HDP models that use auxiliary parameters for cluster allocation [64]. To ensure scalability, the authors also investigated an alternative approach that uses variational inference combined with an importance sampling procedure [9]. For inference, spatial maps of the activated brain regions for each subject can be produced by thresholding the PPIs of the $\gamma_{i\nu}$'s and corresponding posterior β-maps can be obtained based on the estimated regression coefficients. As an additional feature, the use of the nonparametric HDP prior construction can be exploited to obtain a clustering of the subjects for possible discovery of differential activations. The methods have been implemented in the user-friendly Matlab GUI `NPBayes-fMRI` [26], see Figures 1.4 and 1.5 for some of the available features.

1.5 Scalable Bayesian Variable Selection

Despite the flexibility offered by spike-and-slab priors, and the availability of clever data augmentation schemes, computational algorithms for posterior inference in regression models remain a challenge, particularly for model settings with a large number of predictors. Below we review alternative strategies to sampling algorithms given by variational inference methods.

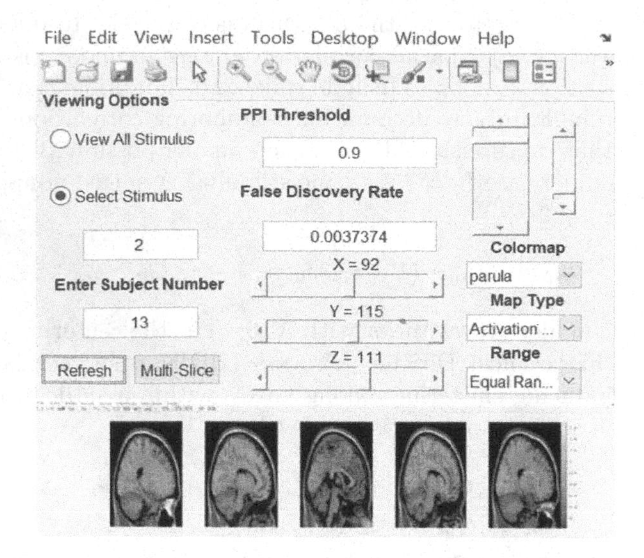

FIGURE 1.4
Matlab GUI NPBayes-fMRI: User friendly software that implements a nonparametric Bayesian spatio-temporal general linear model for task-based multi-subject fMRI data. Subject-level visualization interface and corresponding activation β-maps, for subject 13, stimulus 2 and PPI threshold of .9 (adapted from Kook *et al.* [26]).

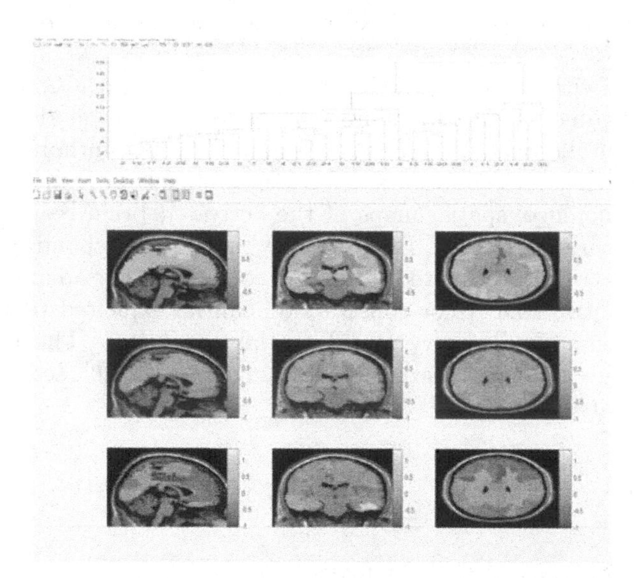

FIGURE 1.5
Matlab GUI NPBayes-fMRI: User friendly software that implements a nonparametric Bayesian spatio-temporal general linear model for task-based multi-subject fMRI data. Dendrogram and cluster-level β-maps obtained with three clusters (adapted from Kook *et al.* [26]).

1.5.1 Variational Inference

Variational Bayes approaches turn inference into an optimization problem, making posterior inference scalable and computationally faster than sampling-based MCMC methods [4]. Typically, variational approaches provide good estimates of mean parameters; however, they tend to underestimate posterior variances and the correlation structure of the data. This shortcoming can be an acceptable trade-off in variable selection problems. For example, [76] performed a comparative study of an MCMC and a variational algorithm for a same linear model and show on simulated data that the variational scheme reduces the computational cost without compromising accuracy in both the detection and the estimation of the non-zero coefficients.

Variational inference (VI) works by specifying a family of approximate distributions $\mathcal{Q}$, which are densities over model parameters and latent variables that depend on free parameters $\boldsymbol{\xi}$, and then using gradient descent to find the values of $\boldsymbol{\xi}$ that minimize the Kullback-Leibler (KL) divergence between the approximate distribution and the true posterior. Let us indicate with W the set of model parameters and latent variables. As discussed in [4], minimizing the KL divergence is equivalent to maximizing the Evidence Lower BOund (ELBO), which is defined as

$$\text{ELBO} = E_{\boldsymbol{\xi}}[\log p(\mathbf{y}, W)] - E_{\boldsymbol{\xi}}[\log q(W)], \tag{1.19}$$

with $p(\mathbf{y}, W)$ the joint distribution of the observed data and the latent variables and parameters, and $q(W)$ the variational distribution of the variables in W. Clearly, the complexity of the approximating class $q(W)$ determines the complexity of the optimization procedure.

The most common approach to obtain an approximating distribution within a VI scheme is mean field approximation, which assumes that the approximating distribution factorizes over some partition of the parameters. This approach is widely used with spike-and-slab priors [9, 66, 76]. In particular, the variational distribution for $(\boldsymbol{\beta}, \boldsymbol{\gamma})$ is assumed to factorize as

$$q\left(\boldsymbol{\beta}, \boldsymbol{\gamma} \mid \boldsymbol{\xi}\right) = \prod_{j=1}^{p} q\left(\beta_j, \gamma_j; \xi_j\right), \tag{1.20}$$

with

$$q\left(\beta_j, \gamma_j; \xi_j\right) = \begin{cases} \psi_j \mathcal{N}(\beta_j \mid \mu_j, s_j^2) & \text{if } \gamma_j = 1 \\ (1 - \psi_j)\, \delta_0\left(\beta_j\right) & \text{otherwise,} \end{cases} \tag{1.21}$$

and free parameters $\xi_j = \left(\psi_j, \mu_j, s_j^2\right)$. A coordinate ascent algorithm can then be implemented to maximize the ELBO by setting the partial derivatives equal to zero. After initializing the free parameters, the algorithm updates each component of ξ_j given all the others, iteratively, until convergence of the ELBO is met. The ELBO is further maximized by finding optimal values for the remainder of the model parameters. VI schemes can be combined with importance sampling procedures, to integrate over some of the model parameters, and/or data augmentation schemes, to implement efficient closed-form VI updates that exploit the conditional conjugacy of latent parameters [9, 41, 75]. At convergence, PPIs are approximated via variational distribution values and thresholded to select covariates. Corresponding regression coefficients are estimated as the variational distribution values at convergence. Variational approaches are only suitable for point estimation and do not allow to assess uncertainty about the estimates. Additionally, in situations with correlated covariates, performances can be sensitive to initialization and can result in poor estimation [51].

Recently, hybrid schemes that combine VI steps on $(\boldsymbol{\beta}, \boldsymbol{\gamma})$ with expectation-maximization (EM) estimation steps on latent variables and other model parameters have also been investigated [42]. As noted by [4], the first term of the ELBO is the object to optimize in EM. One could therefore consider EM approaches as a special case of variational inference, where the variational distributions are point masses.

1.6 Conclusion

Bayesian approaches offer a coherent and flexible framework for variable selection. This chapter has reviewed discrete spike-and-slab priors for linear settings, with a focus on prior constructions and computational aspects. Theoretical properties of these priors will be discussed in the next chapter. This will be followed by treatments of the continuous spike-and-slab priors, which employ mixtures of two unimodal distributions and require careful choices of the variance parameters that separate important variables from noisy ones.

Spike-and-slab priors are sometimes referred to as two-group priors, in contrast to the one-group unimodal continuous shrinkage priors, which will be covered in the second part of the handbook. An advantage of spike-and-slab priors over continuous shrinkage priors is that, in addition to the sparse estimation of the regression coefficients, they produce PPIs for each covariate. Another advantage is that the flexibility of the constructions allows to incorporate structural information among the covariates via the prior choice on the latent indicators, for example as the network priors described in this chapter. The disadvantages are obviously in the computations, particularly in high dimensions, as stochastic search MCMC algorithms need to explore a large posterior space of possible models. Some solutions are offered by optimization procedures, such as the EMVS of [54], for continuous spike-and-slab priors, and by the variational algorithms reviewed in this chapter. These methods, however, can only produce point estimates.

Spike-and-slab priors have been extended to a wide variety of modeling frameworks, such as multivariate regression models, state-space and time-varying coefficient models, as well as to edge selection in graphical models. These topics will be discussed in the third part of the handbook.

Software

The `R` package `MicroBVS`, written by Matthew Koslovsky, is available at `https://github.com/mkoslovsky/MicroBVS`. The user-friendly Matlab GUI `NPBayes-fMRI`, written by Eric Kook, is available at `https://github.com/marinavannucci/NPBayes\_fMRI`.

Acknowledgements

Many thanks to Matthew Koslovsky for producing Figures 1.2 and 1.3.

Bibliography

[1] M.M. Barbieri and J.O. Berger. Optimal predictive model selection. *The Annals of Statistics*, 32:870–897, 2004.

[2] W. Barcella, M. De Iorio, G. Baio, and J. Malone-Lee. Variable selection in covariate dependent random partition models: An application to urinary tract infection. *Statistics in Medicine*, 35:1373–89, 2016.

[3] J. Besag. Spatial interaction and the statistical analysis of lattice systems. *Journal of the Royal Statistical Society, Series B*, 36:192–236, 1974.

[4] D.M. Blei, A. Kucukelbir, and J.D. McAuliffe. Variational inference: A review for statisticians. *Journal of the American Statistical Association*, 112(518):859–877, 2017.

[5] L. Bottolo and S. Richardson. Evolutionary stochastic search. *Bayesian Analysis*, 5(3):583–618, 2010.

[6] P.J. Brown, M. Vannucci, and T. Fearn. Multivariate Bayesian variable selection and prediction. *Journal of the Royal Statistical Society, Series B*, 60(3):627–641, 1998.

[7] P.J. Brown, M. Vannucci, and T. Fearn. Bayes model averaging with selection of regressors. *Journal of the Royal Statistical Society, Series B*, 64:519–536, 2002.

[8] A. Canale, A. Lijoi, B. Nipoti, and I. Prünster. On the Pitman-Yor process with spike and slab base measure. *Biometrika*, 104(3):681–697, 2017.

[9] P. Carbonetto and M. Stephens. Scalable variational inference for Bayesian variable selection in regression, and its accuracy in genetic association studies. *Bayesian Analysis*, 7(1):73–108, 2012.

[10] C.M. Carvalho, N.G. Polson, and J.G. Scott. The horseshoe estimator for sparse signals. *Biometrika*, 97:465–480, 2010.

[11] A. Cassese, A. Guindani, and M. Vannucci. A Bayesian integrative model for genetical genomics with spatially informed variable selection. *Cancer Informatics*, 13(S2):29–37, 2014.

[12] A. Cassese, W. Zhu, M. Guindani, and M. Vannucci. A Bayesian nonparametric spiked process prior for dynamic model selection. *Bayesian Analysis*, 14(2):553–572, 2019.

[13] I. Castillo, J. Schmidt-Hieber, and A. Van der Vaart. Bayesian linear regression with sparse priors. *The Annals of Statistics*, 43(5):1986–2018, 2015.

[14] D.B. Dunson, A.H. Herring, and S.M. Engel. Bayesian selection and clustering of polymorphisms in functionally-related genes. *Journal of the American Statistical Association*, 103:534–546, 2008.

[15] T.S. Ferguson. A Bayesian analysis of some nonparametric problems. *The Annals of Statistics*, pages 209–230, 1973.

[16] K.J. Friston, P. Jezzard, and R. Turner. Analysis of functional MRI time-series. *Human Brain Mapping*, 1(2):153–171, 1994.

[17] E.I. George and R.E. McCulloch. Variable selection via Gibbs sampling. *Journal of the American Statistical Association*, 85:398–409, 1993.

[18] E.I George and R.E. McCulloch. Approaches for Bayesian variable selection. *Statistica Sinica*, pages 339–373, 1997.

[19] R. Gottardo and A.E. Raftery. Markov chain Monte Carlo with mixtures of mutually singular distributions. *Journal of Computational and Graphical Statistics*, 17:949–975, 2008.

[20] P.J. Green. Reversible jump Markov chain Monte Carlo computations and Bayesian model determination. *Biometrika*, 82:711–732, 1995.

[21] J.E. Griffin, K. Latuszynski, and M.F.J. Steel. In search of lost (mixing) time: Adaptive Markov chain Monte Carlo schemes for Bayesian variable selection with very large p. *Biometrika*, to appear, 2020.

[22] C. Hans, A. Dobra, and M. West. Shotgun stochastic search for "large p" regression. *Journal of the American Statistical Association*, 102 (478):507–516, 2007.

[23] H. Ishwaran and J.S. Rao. Spike and slab variable selection : frequentist and Bayesian strategies. *The Annals of Statistics*, 33:730–773, 2005.

[24] H. Ishwaran and J.S. Rao. Consistency of spike and slab regression. *Statistics and Probability Letters*, 81:1920–1928, 2011.

[25] S. Kim, D.B. Dahl, and M. Vannucci. Spiked Dirichlet process prior for Bayesian multiple hypothesis testing in random effects models. *Bayesian Analysis*, 4(4):707–732, 2009.

[26] J.H. Kook, M. Guindani, L. Zhang, and M. Vannucci. NPBayes-fMRI: Nonparametric Bayesian general linear models for single- and multi-subject fMRI data. *Statistics in Biosciences*, 11(1):3–21, 2019.

[27] M.D. Koslovsky, K.L. Hoffman, C.R. Daniel, and M. Vannucci. A Bayesian model of microbiome data for simultaneous identification of covariate associations and prediction of phenotypic outcomes. *Annals of Applied Statistics*, 14(3):1471–1492, 2020.

[28] M.D. Koslovsky and M. Vannucci. MicroBVS: Dirichlet-tree multinomial regression models with Bayesian variable selection – an R package. *BMC Bioinformatics*, 21:301:DOI 10.1186/s12859–020–03640–0, 2020.

[29] D.W. Kwon, M.T. Landi, M. Vannucci, H.J. Issaq, D. Prieto, and R.M. Pfeiffer. An efficient stochastic search for Bayesian variable selection with high-dimensional correlated predictors. *Computational Statistics and Data Analysis*, 55(10):2807–2818, 2011.

[30] D.S. Lamnisos, J.E. Griffin, and M.F.J. Steel. Adaptive Monte Carlo for Bayesian variable selection in regression models. *Journal of Computational and Graphical Statistics*, 22:729–748, 2013.

[31] K.H. Lee, M.G. Tadesse, A.A. Baccarelli, J. Schwartz, and B.A. Coull. Multivariate Bayesian variable selection exploiting dependence structure among outcomes: Application to air pollution effects on DNA methylation. *Biometrics*, 73:232–241, 2017.

[32] F. Li and N. Zhang. Bayesian variable selection in structured high-dimensional covariate space with application in genomics. *Journal of the American Statistical Association*, 105:1978–2002, 2010.

[33] F. Liang, R. Paulo, G. Molina, M.A. Clyde, and J.O. Berger. Mixture of g-priors for Bayes variable selection. *Journal of the American Statistical Association*, 103:410–423, 2008.

[34] D. Madigan and J. York. Bayesian graphical models for discrete data. *International Statistical Review*, 63:215–232, 1995.

[35] P. McCullagh and J.A. Nelder. *Generalized Linear Models, second edition.* Chapman & Hall, London, 1989.

[36] A. Miller. *Subset Selection in Regression.* Chapman & Hall/CRC, Bocan Raton: Florida, 2002.

[37] T.J. Mitchell and J.J. Beauchamp. Bayesian variable selection in linear regression. *Journal of the American Statistical Association*, 83:1023–1036, 1988.

[38] N.N. Narisetty and X. He. Bayesian variable selection with shrinking and diffusing priors. *The Annals of Statistics*, 42:789–817, 2014.

[39] M.A. Newton, A. Noueiry, D. Sarkar, and P. Ahlquist. Detecting differential gene expression with a semiparametric hierarchical mixture model. *Biostatistics*, 5(2):155–176, 2004.

[40] I. Ntzoufras, P. Dellaportas, and J.J. Forster. Bayesian variable and link determination for generalised linear models. *Journal of Statistical Planning and Inference*, 111:165–180, 2003.

[41] J.T. Ormerod, C. You, and S. Müller. A variational Bayes approach to variable selection. *Electronic Journal of Statistics*, 11(2):3549–3594, 2017.

[42] N. Osborne, C.B. Peterson, and M. Vannucci. Latent network estimation and variable selection for compositional data via variational EM. *Journal of Computational and Graphical Statistics*, in press, 2021.

[43] T. Park and G. Casella. The Bayesian lasso. *Journal of the American Statistical Association*, 103(482):681–686, 2008.

[44] C.B. Peterson, F.C. Stingo, and M. Vannucci. Joint Bayesian variable and graph selection for regression models with network-structured predictors. *Statistics in Medicine*, 35(7):1017–1031, 2016.

[45] J.W. Pillow and J. Scott. Fully Bayesian inference for neural models with negative-binomial spiking. In F. Pereira, C. J. C. Burges, L. Bottou, and K. Q. Weinberger, editors, *Advances in Neural Information Processing Systems 25*, pages 1898–1906. 2012.

[46] N.G. Polson and J.G. Scott. Shrink globally, act locally: sparse Bayesian regularization and prediction. *Bayesian Statistics*, 9:501–538, 2010.

[47] N.G. Polson, J.G. Scott, and J. Windle. Bayesian inference for logistic models using Pólya–Gamma latent variables. *Journal of the American Statistical Association*, 108(504):1339–1349, 2013.

[48] M.A. Quintana and D.V. Conti. Integrative variable selection via Bayesian model uncertainty. *Statistics in Medicine*, 32:4938–4953, 2013.

[49] A.E. Raftery. Approximate Bayes factors and accounting for model uncertainty in generalized linear models. *Biometrika*, 83:251–266, 1996.

[50] A.E. Raftery, D. Madigan, and J.A. Hoeting. Bayesian model averaging for linear regression models. *Journal of the American Statistical Association*, 92(437):179–191, 1997.

[51] K. Ray and B. Szabó. Variational Bayes for high-dimensional linear regression with sparse priors. *Journal of the American Statistical Association*, 0(0):1–12, 2021.

[52] S. Richardson, L. Bottolo, and J.S. Rosenthal. Bayesian models for sparse regression analysis of high dimensional data. In *Bayesian Statistics 9*, pages 539–569, 2010.

[53] V. Rockova. Bayesian estimation of sparse signals with a continuous spike-and-slab prior. *The Annals of Statistics*, 46(1):401–437, 2018.

[54] V. Rockova and E.I. George. EMVS: The EM approach to Bayesian variable selection. *Journal of the American Statistical Association*, 109:828–846, 2014.

[55] T. Savitsky and M. Vannucci. Spiked Dirichlet process priors for Gaussian process models. *Journal of Probability and Statistics*, article ID 201489, 2010.

[56] T. Savitsky, M. Vannucci, and N. Sha. Variable selection for nonparametric Gaussian process priors: Models and computational strategies. *Statistical Science*, 26(1):130–149, 2011.

[57] J.G. Scott and J.O. Berger. Bayes and empirical-Bayes multiplicity adjustment in the variable selection problem. *The Annals of Statistics*, 38(5):2587–2619, 2008.

[58] N. Sha, M.G. Tadesse, and M. Vannucci. Bayesian variable selection for the analysis of microarray data with censored outcome. *Bioinformatics*, 22(18):2262–2268, 2006.

[59] N. Sha, M. Vannucci, M.G. Tadesse, P.J. Brown, I. Dragoni, N. Davies, T.C. Roberts, A. Contestabile, N. Salmon, C. Buckley, and F. Falciani. Bayesian variable selection in multinomial probit models to identify molecular signatures of disease stage. *Biometrics*, 60(3):812–819, 2004.

[60] M. Smith and R. Kohn. Nonparametric regression using Bayesian variable selection. *Journal of Econometrics*, 75:317–343, 1996.

[61] M. Smith, B. Putz, B.D. Auer, and L.D. Fahrmeir. Assessing brain activity through spatial Bayesian variable selection. *Neuroimage*, 20:802–815, 2003.

[62] F.C. Stingo, Y.A. Chen, M.G. Tadesse, and M. Vannucci. Incorporating biological information in Bayesian models for the selection of pathways and genes. *Annals of Applied Statistics*, 5(3):1978–2002, 2011.

[63] F.C. Stingo, Y.A. Chen, M. Vannucci, M. Barrier, and P.E. Mirkes. A Bayesian graphical modeling approach to microRNA regulatory network inference. *Annals of Applied Statistics*, 4(4):2024–2048, 2010.

[64] Y.W. Teh, M.I. Jordan, M.J. Beal, and D.M. Blei. Hierarchical Dirichlet processes. *Journal of the American Statistical Association*, 101(476), 2006.

[65] R. Tibshirani. Regression shrinkage and selection via the lasso. *Journal of the Royal Statistical Society Series B*, 58:267–288, 1996.

[66] M.K. Titsias and M. Lázaro-Gredilla. Spike and slab variational inference for multi-task and multiple kernel learning. In *Advances in Neural Information Processing Systems*, pages 2339–2347, 2011.

[67] W.D. Wadsworth, R. Argiento, M. Guindani, J. Galloway-Pena, S.A. Shelburne, and M. Vannucci. An integrative Bayesian Dirichlet-multinomial regression model for the analysis of taxonomic abundances in microbiome data. *BMC Bioinformatics*, 18(1):94, 2017.

[68] K.Y.Y. Wan and J.E. Griffin. An adaptive MCMC method for Bayesian variable selection in logistic and accelerated failure time regression models. *Statistics and Computing*, 31(6), 2021.

[69] E.T. Wang, S. Chiang, Z. Haneef, V.R. Rao, R. Moss, and M. Vannucci. Bayesian nonhomogeneous hidden Markov model with variable selection for investigating drivers of seizure risk cycling. *Annals of Applied Statistics revised*, 2021.

[70] H. Wang. Scaling it up: Stochastic search structure learning in graphical models. *Bayesian Analysis*, 10(2):351–377, 2015.

[71] T. Wang and H. Zhao. A Dirichlet-tree multinomial regression model for associating dietary nutrients with gut microorganisms. *Biometrics*, 73(3):792–801, 2017.

[72] G.D. Wu, J. Chen, C. Hoffmann, K. Bittinger, Y.-Y. Chen, S.A. Keilbaugh, M. Bewtra, D. Knights, W.A. Walters, and R. Knight. Linking long-term dietary patterns with gut microbial enterotypes. *Science*, 334(6052):105–108, 2011.

[73] Y. Yang, M.J. Wainwright, and M.I. Jordan. On the computational complexity of high-dimensional Bayesian variable selection. *The Annals of Statistics*, 44(6):2497–2532, 2016.

[74] A. Zellner. On assessing prior distributions and Bayesian regression analysis with g-prior distributions. In P.K. Goel and A. Zellner, editors, *Bayesian Inference and Decision Techniques: Essays in Honor of Bruno de Finetti*, pages 233–243. North-Holland/Elsevier, 1986.

[75] C.-X. Zhang, S. Xu, and J.-S. Zhang. A novel variational Bayesian method for variable selection in logistic regression models. *Computational Statistics & Data Analysis*, 133:1–19, 2019.

[76] L. Zhang, M. Guindani, F. Versace, J.M. Engelmann, and M. Vannucci. A spatio-temporal nonparametric Bayesian model of multi-subject fMRI data. *Annals of Applied Statistics*, 10(2):638–666, 2016.

2

Recent Theoretical Advances with the Discrete Spike-and-Slab Priors

Shuang Zhou
Arizona State University (USA)

Debdeep Pati
Texas A&M University (USA)

CONTENTS

The key to inference in settings where the dimensionality of the parameter space is larger than the number of samples is to assume a parsimonious structure underlying the data generating process. While Bayesian hierarchical models offer a unified and coherent framework for structured modeling and inference, the properties of the posterior distribution and the

DOI: 10.1201/9781003089018-2

exact role of the hierarchy have begun to be understood in the last few years. In this chapter, we review some of the recent theoretical works aiming to provide theoretical guarantees of discrete spike-and-slab priors. As one moves away from simple parametric models, understanding properties of a posterior distribution poses a stiff technical challenge and the prior choice (e.g., spike-and-slab) assumes a more fundamental role. Besides, full Bayesian computation using these hierarchical models often poses a significant computation bottleneck, necessitating the development of approximate Bayes methods such as variational inference.

2.1 Introduction

The advent of sophisticated data acquisition techniques in many fields has triggered the development of innovative statistical methods identify relevant predictors associated with a response, among a large set of available predictors, often in situations with small sample sizes. This "large p, small n" paradigm is arguably the most researched topic in statistics in the last decade. The seemingly impossible task of inferring large number of parameters compels us to investigate and exploit lower dimensional structures underlying the data generating process. The most popular way of enforcing "structure" is to assume the model parameter space is *sparse*. Accordingly, the overwhelming emphasis in the current statistical literature on high-dimensional problems has been on producing a sparse point estimate based on penalization or thresholding [7, 21, 27]. There is a rich theoretical literature justifying the optimality properties of penalization approaches [27], with fast algorithms [12] and compelling applied results leading to their routine use.

In the Bayesian context, ℓ_1 and ℓ_2 regularization methods are equivalent to placing zero-mean double-exponential and Gaussian priors respectively on the parameter vector. The solutions of the corresponding optimization problems are precisely the mode of the Bayesian posterior distribution. In this context, sparsity favoring mixture spike-and-slab priors, with separate control on the signal and noise coefficients, have been proposed [6, 10, 11, 19], with huge success in practical applications. In this chapter, we consider discrete spike-and-slab constructions, for which optimality properties have been understood fairly recently [10, 11].

In the case of fixed dimensional parametric models, the Bernstein–von Mises (BvM) [32] phenomenon asserts that under mild conditions on the likelihood and the prior, the posterior distribution asymptotically assumes a Gaussian shape, centered at the maximum likelihood or some other efficient estimator, and with covariance the inverse Fisher information matrix. Thus, posterior credible sets asymptotically coincide with frequentist confidence sets, providing justification to Bayesian uncertainty characterization from a frequentist perspective. The BvM phenomenon implies that the prior information is asymptotically washed away by the data and thus practitioners starting with different priors would eventually obtain the same inference. However, the situation is markedly different when the parameter space is large (high-dimensional or infinite dimensional), *with the prior choice assuming a fundamentally more important role.* As a simple illustration, consider sparsity favoring priors on a vector of regression coefficients which a priori allow each variable to be included with probability π. The seemingly reasonable choice of $\pi = 1/2$ leads to an exponentially small prior probability of $(1/2)^p$ being assigned to the null model, making it impossible to override the prior informativeness with the information in the data when p is large [31].

It is thus necessary to construct a prior distribution in high-dimensional settings that respects the underlying sparsity assumption. However, since the underlying sparsity structure is not entirely known, it is important to construct a prior distribution that leads to adaptive inference, meaning the inference is *optimal* regardless of the true sparsity structure. A general recipe to achieve this is through a hierarchical specification, where parameters in a lower level of the hierarchy are assigned hyper-priors, and integrating over the hyper-prior distribution induces dependence and structure in a probabilistic fashion. Posterior distributions arising from such careful hierarchical specifications often possess frequentist optimality properties. For example, the posterior is said to concentrate at a minimax optimal rate if the posterior mass assigned to a minimax optimal neighborhood of the true parameter converges to one as the sample size increases. Bayesian hierarchical models commonly lead to adaptive minimax optimal procedures, where the procedure automatically adapts to unknown sparsity/smoothness of parameters via posterior model averaging over the unknown sparsity/smoothness hyper-parameters [10, 11, 48].

This chapter reviews recent developments in the construction of prior distributions that lead to optimal posterior concentration, sparse support recovery and BvM phenomenon in Gaussian sequence models (§2.2) and in linear regression (§2.3). We also review theoretical extensions to generalized linear models (§2.4) and to variational inference in linear models (§2.5).

2.2 Optimal Recovery in Gaussian Sequence Models

We first introduce some notations which will be used throughout the rest of this chapter. Given sequences a_n, b_n, we denote $a_n = O(b_n)$ if there exists a global constant C such that $a_n \leq Cb_n$ and $a_n = o(b_n)$ if $a_n/b_n \to 0$ as $n \to \infty$. We also use "$\lesssim$" and "$\gtrsim$" to denote "less than" and "greater than" up to a finite multiple, respectively. Given two constant $a, b \in \mathbb{R}$, we write "$a \wedge b$" and "$a \vee b$" as the smaller value and the larger value between a and b, respectively.

For a vector $\mathbf{x} \in \mathbb{R}^r$, we define the ℓ_q-norm of $\mathbf{x}$ as $\|\mathbf{x}\|_q = (\sum_{i=1}^r |x_i|^q)^{1/q}$ for any $1 \leq q < \infty$. When $q = 2$, it coincides with the Euclidean norm and the subscript may be omitted. We also define the ℓ_∞-norm as $\|\mathbf{x}\|_\infty = \max_{1\leq i \leq r} |x_i|$.

For a subset $S \subset \{1, \ldots, n\}$, let $|S|$ denote the cardinality of S and define $\boldsymbol{\theta}_S = (\theta_j : j \in S)$ for a vector $\boldsymbol{\theta} \in \mathbb{R}^n$, and also denote $\boldsymbol{\theta}_{S^c} = \{\theta_j, j \in S^c\}$ where $S^c = \{1, \ldots, n\} \backslash S$. We write $\text{supp}(\boldsymbol{\theta})$ as the *support* of θ. For convenience, we may use the notation $\boldsymbol{\theta} = (\boldsymbol{\theta}_S, \boldsymbol{\theta}_S^c)$ that allows the elements to be in a correct order for any subset S.

2.2.1 Minimax Rate in *Nearly Black* Gaussian Mean Models

Reconstruction of a *sparse* mean vector from noisy observations under the assumption of an unknown level of *sparsity* has received considerable attention in the literature. To describe the set-up, let $\boldsymbol{y} = (y_1, \ldots, y_n) \in \mathbb{R}^n$ be such that

$$y_i = \theta_i + \epsilon_i, \quad \text{i} = 1, \ldots, n \tag{2.1}$$

where $\epsilon_i \sim \mathcal{N}(0, \sigma^2)$ and $\boldsymbol{\theta} = (\theta_1, \ldots, \theta_n)$ is an *unknown* vector of means, which is assumed to be possibly *sparse*. *Sparsity* is defined using the class of *nearly black* vectors, defined as

$$l_0[p_n] = \{\boldsymbol{\theta} \in \mathbb{R}^n : \#(1 \leq i \leq n : \theta_i \neq 0) \leq p_n\}$$

where p_n is assumed to be $o(n)$, as $n \to \infty$. The extent of sparsity, measured by the constant p_n, is assumed unknown.

Our goal is to recover the true parameter vector $\boldsymbol{\theta}_0 \in l_0[p_n]$ from the observation vector $\boldsymbol{y}$. Of particular interest is the *minimax* risk bound or estimate error rate relative to ℓ_q metrics [13, 18]. It is well-known that the minimax risk for this class of problems as $n, p_n \to \infty$ with $p_n = o(n)$ is given by

$$\inf_{\hat{\boldsymbol{\theta}}} \sup_{\boldsymbol{\theta} \in \mathrm{l}_0[p_n]} \mathrm{P}_{n,\boldsymbol{\theta}} \|\hat{\boldsymbol{\theta}} - \boldsymbol{\theta}\|^2 = 2p_n \log(\tfrac{n}{p_n})\{1 + \mathrm{o}(1)\}.$$

Here the infimum is taken over all estimators $\hat{\boldsymbol{\theta}} = \hat{\boldsymbol{\theta}}(\boldsymbol{y})$, and $\mathrm{P}_{n,\boldsymbol{\theta}}$ denotes taking the expectation under the assumption that $\boldsymbol{y}$ is $\mathcal{N}_n(\boldsymbol{\theta}, \boldsymbol{I})$-distributed. More generally, one may achieve reconstruction relative to the ℓ_q metric for $0 < q \leq 2$, defined (without qth root) by

$$d_q(\boldsymbol{\theta}, \boldsymbol{\theta}') = \sum_{i=1}^{n} |\theta_i - \theta_i'|^q. \tag{2.2}$$

According to [10], for $q < 2$ this "metric" is more sensitive to small variations in the coordinates than the square Euclidean metric, which is d_2. From [18] the minimax rate over $\ell_0[p_n]$ for d_q is known to be of the order

$$r_{n,q}^* = p_n\{\log(n/p_n)\}^{q/2}. \tag{2.3}$$

Popular classical methods on minimax reconstruction in Gaussian sequence models include [5, 15, 25].

Since we shall be interested in Bayesian inference on $\boldsymbol{\theta}$, we shall construct a prior π_n and investigate the behavior of the posterior as a function of the sample size n. To this end, the posterior $\Pi_n(\cdot \mid \boldsymbol{y})$ is said to contract at a rate ϵ_n for sequence of positive numbers ϵ_n, if

$$\Pi_n(\mathcal{B}(\boldsymbol{\theta}_0; M_n\epsilon_n) \mid \boldsymbol{y}) \to 1$$

in probability for any sequence $M_n \to \infty$, as $n \to \infty$, where $\mathcal{B}(\boldsymbol{\theta}_0; \epsilon_n) = \{\boldsymbol{\theta} : d_q(\boldsymbol{\theta}, \boldsymbol{\theta}_0) < \epsilon_n\}$. We shall call a Bayes procedure to be optimal if the posterior contraction rate matches with the minimax rate.

2.2.2 Optimal Bayesian Recovery in ℓ_q-norm

Methods within the Bayesian framework deal with the sparsity by choosing appropriate prior distributions. In the sparse setting, priors are usually constructed via a hierarchical scheme that enforces model selection and parameter updating, separately. A popular choice is the discrete spike-and-slab prior $(1-\gamma)\,\delta_0(\cdot) + \gamma\, g(\cdot)$, which takes a point-mass mixture with a Dirac measure (the "spike") at zero used to model the negligible coordinates and a diffuse/ heavy-tailed density to model the non-negligible coordinates (the "slab"). The mixing proportion parameter may be endowed with a hyper prior to update according to the unknown level of the sparsity. Here, we review the main results of [10] that provided sufficient conditions of the point-mass mixture type of priors for optimal posterior contraction rate and variable selection consistency of the resulting posterior distributions.

For a full Bayesian inference in model (2.1), it is of particular interest to know which priors yield a posterior distribution that concentrates most of its mass on balls around $\boldsymbol{\theta}_0$ of d_q radius of order $r_{n,q}^*$ in (2.3), or close relatives as $p_n(\log n)^r$ that loose (only) a logarithmic factor. Two key features of good priors are: i) the priors $\pi_n(p_n)$ on the dimension p_n should express the sparsity of the mean vector while still giving sufficient weight to the true level of sparsity. The first part will be achieved by allowing an *exponential decrease* of $\pi_n(p_n)$ with the dimension, and the second part requires a necessary lower bound on the decay rate of $\pi_n(p_n)$. ii): the priors on the non-zero coordinates should have heavy tails that are at least not lighter than Laplace. On the contrary, if the priors have lighter tails such as the Gaussian, the resulting posterior attains a significantly lower contraction rates if true parameter vectors θ_0 are not close to the origin.

A hierarchical scheme is considered to construct the prior Π_n on $\mathbb{R}^n$:

(P1) A *dimension* p is chosen according to a prior probability measure π_n on the set $\{0, 1, 2, \ldots, n\}$.

(P2) Given p, a subset $S \subset \{1, \ldots, n\}$ of size $|S| = p$ is chosen uniformly at random from the $\binom{n}{p}$ subsets of size p.

(P3) Given (p, S), a vector $\boldsymbol{\theta}_S = (\theta_i : i \in S)$ is chosen from a probability distribution with Lebesgue density g_S on $\mathbb{R}^p$ (if $p \geq 1$) and this is extended to $\boldsymbol{\theta} \in \mathbb{R}^n$ by setting the remaining coordinates $\boldsymbol{\theta}_{S^c}$ equal to 0.

In (P1), the prior on dimension $\pi_n(p)$ is assumed to be positive for any integer p. For simplicity, g_S in (P3) uniformly denotes the density for every set of a given dimension $|S|$.

Given the prior Π_n and assuming $\boldsymbol{y}$ is generated from the Gaussian error model $\mathcal{N}_n(\boldsymbol{\theta}, I_n)$, Bayes' rule yields the *posterior distribution* denoted by $\Pi_n(\cdot \mid \boldsymbol{y})$. The probability $\Pi_n(B \mid \boldsymbol{y})$ of a Borel set $B \subset \mathbb{R}^n$ under the posterior distribution can be written as

$$\frac{\sum_{p=0}^{n} \pi_n(p)/\binom{n}{p} \sum_{|S|=p} \int_{(\boldsymbol{\theta}_S, \mathbf{0}) \in B} \Pi_{i \in S} \phi(y_i - \theta_i) \Pi_{i \notin S} \phi(y_i) g_S(\boldsymbol{\theta}_S) d\boldsymbol{\theta}_S}{\sum_{p=0}^{n} \pi_n(p)/\binom{n}{p} \sum_{|S|=p} \int \Pi_{i \in S} \phi(y_i - \theta_i) \Pi_{i \notin S} \phi(y_i) g_S(\boldsymbol{\theta}_S) d\boldsymbol{\theta}_S}, \tag{2.4}$$

where $(\boldsymbol{\theta}_S, \mathbf{0})$ is a shorthand notation that denotes a n-dimensional vector $\boldsymbol{\theta}$ whose element $\theta_i = 0$ for all $i \in S^c$, for any $S \subset \{1, \ldots, n\}$. The properties of the posterior distribution are studied under the frequentist assumption that the vector $\boldsymbol{y} = (y_1, \ldots, y_n)$ is generated from a true multivariate normal distribution with mean vector $\boldsymbol{\theta}_0$ and covariance matrix the identity matrix.

Posterior contraction results consist of both the dimensionality recovery and reconstruction of the mean vector $\boldsymbol{\theta}$. For theoretical investigation, let us first consider the case that the non-zero coordinates are independent under (P3) (the densities g_S are of product form over the coordinates in S), as it requires simpler conditions on the prior. We now introduce the assumption on the decay rate of π_n on the dimension.

Assumption (Exponential decrease): In the context of $\ell_0[p_n]$-classes, we say that the prior π_n on dimension has *exponential decrease* if, for some constant $C > 0$ and $D < 1$,

$$\pi_n(p) \leq D\pi_n(p-1), \qquad p > Cp_n. \tag{2.5}$$

If the condition is also satisfied with $C = 0$, then the prior on dimension is considered to have *strict* exponential decrease.

Theorem 2.2.1. (Dimension). *If π_n has exponential decrease* (5.3) *and g_S is a product of $|S|$ copies of a univariate density g, with mean zero and finite second moment, then there exists $M > 0$ such that, as p_n, $n \to \infty$,*

$$\sup_{\boldsymbol{\theta}_0 \in \ell_0[p_n]} P_{n,\boldsymbol{\theta}_0} \Pi_n(\boldsymbol{\theta} : |S_{\boldsymbol{\theta}}| > M\, p_n \mid \boldsymbol{y}) \to 0.$$

The theorem shows an *exponential decrease* is sufficient for the posterior distribution to concentrate towards subspaces of dimension at most a multiple of p_n. However the posterior will not necessarily concentrate on a fixed Mp_n-dimensional subspace. In order for a good reconstruction of the full mean vector $\boldsymbol{\theta}$, it also requires a lower bound on the prior mass given to the space of "correct" dimension. The following assumption ensures that *exponential decrease* is not too harsh to overshoot the space of "correct" dimension.

Assumption (Lower bound): For some $c > 0$, we assume

$$\pi_n(p_n) \gtrsim \exp\big(- c p_n \log(n/p_n)\big). \tag{2.6}$$

The lower bound assumption prevents the decay rate of π_n to become faster than an exponential decrease. In addition, good recovery requires also appropriate prior densities g_S on the non-zero coordinates. The tail behavior of g_S is especially important as light-tailed g_S in (P3) may lead to an unwanted shrinkage effect, even in the average recovery of the parameter as $n \to \infty$, yielding suboptimal behavior for true parameters $\boldsymbol{\theta}$ that are far from the origin. This shrinkage effect can be prevented by choosing priors g_S with sufficiently heavy tails.

Let us first consider the case that the non-zero coordinates are independent under (P3). To incorporate the tail condition, assume that the densities take the form of e^h for any function $h : \mathbb{R} \to \mathbb{R}$ satisfying

$$|h(x) - h(y)| \lesssim 1 + |x - y| \qquad \forall x, y \in \mathbb{R}. \tag{2.7}$$

Obviously uniformly Lipshitz functions satisfy this condition, which include Laplace and Student densities. It also covers other smooth densities e^h which have a function h that is bounded in a neighborhood of the origin and uniformly Lipschitz outside the neighborhood, such as densities with polynomial tails, or densities of the form $c_\alpha e^{-|x|^\alpha}$ for some $\alpha \in (0, 1]$.

Theorem 2.2.2. (Recovery). *If π_n has exponential decrease* (5.3) *and g_S is a product of $|S|$ univariate densities of the form e^h with mean zero and finite second moment and h satisfying* (2.7)*, then for any $q \in (0, 2]$, for r_n satisfying*

$$r_n^2 \geq \{p_n \log(n/p_n)\} \vee \log \frac{1}{\pi_n(p_n)} \tag{2.8}$$

and sufficiently large M, as $p_n, n \to \infty$ such that $p_n/n \to 0$,

$$\sup_{\boldsymbol{\theta}_0 \in \ell_0[p_n]} P_{n,\boldsymbol{\theta}_0} \Pi_n(\boldsymbol{\theta} : d_q(\boldsymbol{\theta}, \boldsymbol{\theta}_0) > M r_n^q \, p_n^{1-q/2} \mid \boldsymbol{y}) \to 0,$$

where $P_{n,\boldsymbol{\theta}_0}$ denotes taking the expectation with respect to the true data-generating distribution $\mathcal{N}_n(\boldsymbol{\theta}_0, \boldsymbol{I}_n)$.

The theorem asserts that the *full* posterior distribution contracts at the obtained rate uniformly over $\ell_0[p_n]$. For $q = 2$, the rate reduces to r_n^2 and one may consider the rate $(r_n^2)^{q/2} \, p_n^{1-q/2}$ as an adjusted one between the norms d_2 and d_q. Given the form of the lower bound in (2.8), the full posterior distribution attains the minimax rate if the second term of the lower bound satisfies $\log(1/\pi_n(p_n)) \lesssim p_n \log(n/p_n)$ as in assumption (2.6). If the decrease is faster than (2.6), then the rate of contraction may be slower than the minimax rate.

We conclude this subsection by drawing a connection of the sparse vector estimation problem in a sequence model (2.1) with the nonparametric regression literature using wavelet shrinkage and wavelet thresholding estimators. Expanding a suitably smooth function on a wavelet basis, the problem of nonparametric regression can be recast into a sparse vector estimation where many of these wavelet coefficients are assumed to be zero to reflect the underlying sparsity of the function. [14, 17] introduced non-linear wavelet estimators in nonparametric regression through thresholding which typically amounts to term-by-term assessment of estimates of coefficients in the empirical wavelet expansion of the unknown function. If an estimate of a coefficient is sufficiently large in absolute value, then the corresponding term in the empirical wavelet expansion is retained (or shrunk toward to zero by an amount equal to the threshold); otherwise it is omitted. The shrinkage can be tuned to attain minimax rates over a wide range of function classes [16].

2.2.3 Optimal Contraction Rate for Other Variants of Priors

Let us consider the case θ_is are dependent. It is easy to see that the marginal condition (2.7) will not apply to g_S, thus a joint condition is instead imposed on the logarithm of g_S.
Assumption: For every $S' \subset S \subset \{1,\ldots,n\}$ and a universal constant c_1, assume densities g_S in (P3) satisfy

$$\begin{aligned} \log g_S(\boldsymbol{\theta}) - \log g_S(\boldsymbol{\theta}') &\le c_1|S| + \frac{1}{64}\|\boldsymbol{\theta} - \boldsymbol{\theta}'\|^2 \qquad \forall \boldsymbol{\theta}, \boldsymbol{\theta}' \in \mathbb{R}^S, \\ \log g_S(\boldsymbol{\theta}) - \log g_{S'}(\pi_{S'}\boldsymbol{\theta}) &\le c_1|S| + \frac{1}{64}\|\pi_{S-S'}\boldsymbol{\theta}\|^2 \qquad \forall \boldsymbol{\theta} \in \mathbb{R}^S. \end{aligned} \tag{2.9}$$

where the projection $\pi_S : \mathbb{R}^n \to \mathbb{R}^S$ is defined by $\pi_S\boldsymbol{\theta} = \boldsymbol{\theta}_S$ and the constant 64 is chosen for technical purposes. For partition $S = S_1 \cup S_2$, we denote by $\boldsymbol{\theta} = (\boldsymbol{\theta}_1, \boldsymbol{\theta}_2)$ the corresponding partition of $\boldsymbol{\theta} \in \mathbb{R}^S$ and by $g_{S_1,S_2}(\boldsymbol{\theta}_1, \boldsymbol{\theta}_2) = g_S(\boldsymbol{\theta})$ the corresponding density. It is desirable that the density $g_{S_1,S_2}(\boldsymbol{\theta}_1, \boldsymbol{\theta}_2)$ within a given subspace S can factorize according to an arbitrary partition $S = S_1 \cup S_2$ without imposing a large scaling factor, allowing the posterior to separate the non-zero coordinates from zero coordinates. The following assumption describes the factorization property of g_S.

Assumption: We assume that there exist $C, m_1 > 0$ and, for any S_2, probability densities γ_{S_2} on $\mathbb{R}^{S_2}$, such that for any $\boldsymbol{\theta}_2 \in \mathbb{R}^{S_2}$ and $S_1 \subset S_2^c$,

$$\sup_{\boldsymbol{\theta}_1 \in \mathbb{R}^{S_1}} \frac{g_{S_1,S_2}(\boldsymbol{\theta}_1, \boldsymbol{\theta}_2)}{g_{S_1}(\boldsymbol{\theta}_1)} \le C m_1^{|S_1|+|S_2|} \gamma_{S_2}(\boldsymbol{\theta}_2). \tag{2.10}$$

Theorem 2.2.3. (Recovery). *Suppose π_n has strict exponential decrease, that is, satisfies* (5.3) *with $C = 0$ and some $D > 0$. The assertion of Theorems 2.2.1 and 2.2.2 are also true if the densities g_S are not product densities, but general densities with finite second moments that satisfy* (2.9) *and* (2.10) *with $Dm_1 < 1$, and m_1 the constant in* (2.10).

At the end of this section, we briefly discuss another particular form of π_n that shares similar features of priors defined in (5.3). For positive constants a, b, let,

$$\pi_n(p) \propto e^{-a\,p\,\log(bn/p)}, \tag{2.11}$$

for $p \in \{1,\ldots,n\}$. (2.11) is inversely proportional to the number of models of size p given the fact that $e^{p\log(n/p)} \le \binom{n}{p} \le e^{p\log(ne/p)}$, a quantity that could be viewed as the *model complexity* for a given dimension p. Thus this prior is desirable to downsize the model complexity. This prior is usually considered a "complexity prior".

Theorem 2.2.4. (Complexity prior). *If the densities g_S have finite second moments, satisfy* (2.9) *for some constant c_1, and the prior π_n satisfies $\pi_n(p) \lesssim e^{-a\,p\,\log(bn/p)}$ for any p, and for some $a \ge 1$ and $b \ge e^{7+2c_1}$, then, for r_n satisfying, for any $1 \le p_n \le n$ and any $r \ge 1$,*

$$\sup_{\boldsymbol{\theta}_0 \in \ell_0[p_n]} P_{n,\boldsymbol{\theta}_0}\Pi_n(\boldsymbol{\theta} : \|\boldsymbol{\theta} - \boldsymbol{\theta}_0\| > 45r_n + 10r \mid \boldsymbol{y}) \lesssim e^{-r^2/10}, \tag{2.12}$$

This prior is widely applied to more general sparse statistical learning problems, such as linear regression model or generalized linear models. One can refer to later sections for a detailed review.

2.2.4 Slow Contraction Rate for Light-tailed Priors

Condition (2.7) or (2.9) on the priors g_S for non-zero coefficients ensures that the posterior does not shrink to the center of the prior too much, preventing the prior from assigning little mass over the large true mean vector. It necessitates the tail of density g not to be too light. Theorem 2.2.4 takes the product priors and shows that choosing marginal densities proportional to $y \mapsto e^{-|y|^\alpha}$ for some $\alpha > 1$ leads to a slow contraction rate for large true vectors $\boldsymbol{\theta}_0$. The theorem is formulated in an asymptotic setting with a sequence of true vectors $\{\boldsymbol{\theta}_0^n\}$. In particular, let us assume the parameter vector $\boldsymbol{\theta}_0^n$ tends to infinity faster than the optimal rate, that is,

$$\|\boldsymbol{\theta}_0^n\|^2 \gg p_n \log(n/p_n), \tag{2.13}$$

where p_n denotes the number of non-zero coordinates of $\boldsymbol{\theta}_0^n$. The following theorem shows the posterior places no mass asymptotically on balls of radius a multiple of $\|\boldsymbol{\theta}_0^n\|$ around the true parameter. If considering a small ball of radius less than optimal rate centered at the truth, the Bayesian estimator will not be better than a zero estimator which would achieve the optimal rate automatically.

Theorem 2.2.5. *Assume that the densities g_S are products of S univariate densities proportional to $y \mapsto e^{-|y|^\alpha}$ and the prior π_n satisfies* (5.3) *for some $c > 0$: (i) If $\alpha \geq 2$ and $\|\boldsymbol{\theta}_0^n\|^2 / (p_n \log(n/p_n)) \to \infty$, then for sufficiently small $\eta > 0$, as $n \to \infty$,*

$$P_{n,\boldsymbol{\theta}_0^n} \Pi_n(\theta : \|\boldsymbol{\theta} - \boldsymbol{\theta}_0^n\| \leq \eta \|\boldsymbol{\theta}_0^n\| \mid \boldsymbol{y}) \to 0 \tag{2.14}$$

(ii) If $1 \leq \alpha \leq 2$ and $(\rho_{0,\alpha}^n)^2/(p_n \log(n/p_n)) \to \infty$, then for sufficiently small $\eta > 0$, as $n \to \infty$,

$$P_{n,\boldsymbol{\theta}_0^n} \Pi_n(\boldsymbol{\theta} : \|\boldsymbol{\theta} - \boldsymbol{\theta}_0^n\| \leq \eta \rho_{0,\alpha}^n \mid \boldsymbol{y}) \to 0 \tag{2.15}$$

Theorem 2.2.5 presents the problematic behavior of the posterior distribution with Gaussian priors (or densities with lighter tails) for signals with large magnitudes $\|\boldsymbol{\theta} - \boldsymbol{\theta}_0^n\|$.

2.3 Sparse Linear Regression Model

The linear regression problem under the "large p small n" regime is arguably the most researched topic in the last few decades. The frequentist methods are mostly penalization-orientated, imposing the model sparsity via a penalty term on the parameter space added to the log-likelihood. Various penalties are studied in the literature such as [20, 39, 54, 56], to name a few. The level of sparsity is usually employed through the regularization parameter, with larger values of the regularization parameter promoting stronger model sparsity. In the Bayesian framework, the sparsity is typically induced by a prior defined within the hierarchical framework (P1)–(P3) described in §2.2, of which the spike-and-slab priors are a special case.

Here, we review the seminal work of [11] on the theoretical property of the discrete spike-and-slab prior in terms of optimal support recovery, posterior contraction rate and consistent variable selection. In addition, we also review the recent work of [52] who proposed a different type of spike-and-slab priors where the slab is a dependent prior induced by a Zellner's g-prior.

We begin by introducing some notations which will be used for the rest of the chapter. For probability measures P, Q that are absolutely continuous with the Lebesgue measure, we define the total variance distance between P and Q as $\|P-Q\|_{TV} = \sup_{A\in\mathbb{R}} |P(A)-Q(A)|$. For their Radon–Nikodym densities p, q, we denote the Kullback–Leibler divergence between p and q by $\mathcal{D}(p\|q) = \int p\log(p/q)$. For two arbitrary densities g, h, we write $g \otimes h$ as a Dirac product of g and h. We denote the inner product associated with Euclidean space by $\langle\cdot,\cdot\rangle$. We use $\circ$ to denote the composition of functions. For any arbitrary set A, we denote the indicator function restricted to A by $\mathbb{I}_A(\cdot)$. For any matrix $\mathbf{B}$, we denote by $\mathbf{B}^{\mathrm{T}}$ its transpose. If $\mathbf{B}$ is a square matrix, we denote by $\lambda_{\min}(\mathbf{B}), \lambda_{\max}(\mathbf{B})$ its smallest and largest eigenvalues, respectively. We use $\propto$ to stand for "proportional to", and use "$\gg$" to denote "much more than" and "$\ll$" to denote "much less than".

2.3.1 Prior Construction and Assumptions

We consider estimation of a parameter $\boldsymbol{\beta} \in \mathbb{R}^{p_n}$ in the linear regression model

$$\boldsymbol{y} = \boldsymbol{X}\boldsymbol{\beta} + \boldsymbol{\varepsilon}, \tag{2.16}$$

where $\boldsymbol{X}$ is a $n \times p_n$ design matrix and $\boldsymbol{\varepsilon}$ is an n-variate standard normal vector. We focus on a *sparse* setting in high dimension, where $n \ll p_n$ and the parameter $\boldsymbol{\beta}$ is assumed to be sparse. We are interested in the Bayesian inference under the "frequentist" assumption that the data $\boldsymbol{y}$ is distributed according to a given (sparse) parameter $\boldsymbol{\beta}^*$. The expectation under the previous distribution is denoted by $\mathbb{E}_{\boldsymbol{\beta}^*}$.

Following the hierarchical scheme introduced in §2.2, a prior Π_n on $(S, \boldsymbol{\beta})$ can be constructed via steps (P1)-(P3). First, a dimension $s \in \{1, \ldots, p_n\}$ is drawn according to a prior π_n in (P1). One may then draw a set $S \subset \{1, \ldots, p_n\}$ of cardinality $|S| = s$ randomly. In (P3), we adopt the same notation g_S to denote the prior density on the non-zero coefficients denoted by $\boldsymbol{\beta}_S = \{\beta_i : i \in S\}$, and assign a Dirac measure δ_0 at zero on $\boldsymbol{\beta}_{S^c} = \{\beta_i : i \in S^c\}$. This procedure is summarized as

$$(S, \boldsymbol{\beta}) \mapsto \pi_p(|S|)\frac{1}{\binom{p_n}{|S|}} g_S(\boldsymbol{\beta}_S)\delta_0(\boldsymbol{\beta}_{S^c}). \tag{2.17}$$

For simplicity, we confine our attention to the case where g_S is a product of densities over the coordinates in S, for g a fixed continuous density on $\mathbb{R}$. For theoretical purpose, our main interest is on the properties of π_p and g that deliver good recovery results and variable selection consistency of the posterior distribution.

For a vector $\boldsymbol{\beta} \in \mathbb{R}^p$ and a set $S \subset \{1, 2, \ldots, p_n\}$ of indices, let $\boldsymbol{\beta}_S$ denote the vector $(\beta_i)_{i\in S} \in \mathbb{R}^S$, and $|S|$ the cardinality of S. The support of the parameter $\boldsymbol{\beta}$ is the set $S_{\boldsymbol{\beta}} = \{i : \beta_i \neq 0\}$. Similarly, let S^* denote the support of the true parameter $\boldsymbol{\beta}^*$, with cardinality $s^* := |S^*|$. We write $s = |S|$ if there is no ambiguity to which set S is referred to. For design matrix $\boldsymbol{X}$, we denote by $\boldsymbol{X}_{\cdot,i}$ the ith column of $\boldsymbol{X}$, and

$$\|\boldsymbol{X}\| = \max_{i=1,\ldots,p} \|\boldsymbol{X}_{\cdot,i}\|_2 = \max_{i=1,\ldots,p} (\boldsymbol{X}^{\mathrm{T}}\boldsymbol{X})_{i,i}^{1/2}. \tag{2.18}$$

Given the prior Π_n defined in (2.17), Baye's formula gives the following expression of the posterior distribution $\Pi_n[\cdot \mid \boldsymbol{y}]$ under model (2.16). For any Borel set B of $\mathbb{R}^{p_n}$,

$$\Pi_n[B \mid \boldsymbol{y}] = \int_B e^{-\|\boldsymbol{y}-\boldsymbol{X}\boldsymbol{\beta}\|_2^2/2} d\Pi_n(\boldsymbol{\beta}) \Big/ \int e^{-\|\boldsymbol{y}-\boldsymbol{X}\boldsymbol{\beta}\|_2^2/2} d\Pi_n(\boldsymbol{\beta}). \tag{2.19}$$

For theoretical purposes, it is reasonable to expect that π_p and g share similar properties that lead to optimal recovery in Gaussian mean models. One common choice of g_S is considering a product of $|S|$ Laplace densities $\beta \to 2^{-1}\lambda \exp(-\lambda|\beta|)$. Usually, the (inverse) scale parameter λ changes with p_n within the following range,

$$\frac{\|\boldsymbol{X}\|}{p_n} \leq \lambda \leq 2\bar{\lambda}, \qquad \bar{\lambda} = 2\|\boldsymbol{X}\|\sqrt{\log p_n}, \tag{2.20}$$

where $\|\boldsymbol{X}\|$ is defined in (2.18). The quantity $\bar{\lambda}$ coincides with the usual value of the regularization parameter λ of the LASSO [39]. Larger value of $\bar{\lambda}$ imposes more sparsity such that the LASSO estimator tends to shrink more coordinates β_i to zero. However, in Bayesian setup, sparsity is induced by model selection through the prior π_p. As the Laplace prior densities only model the non-zero coefficients, smaller values of λ are favored in Bayesian setting. One may refer to [11] for examples of small values or fixed values of λ under different Bayesian regression settings.

The following assumption provides a "sandwiched" bound on the decay rate of the prior π_p.

Assumption (Prior dimension): There are constants $A_1, A_2, A_3, A_4 > 0$ with

$$A_1 p_n^{-A_3}\pi_p(s-1) \leq \pi_p(s) \leq A_2 p_n^{-A_4}\pi_p(s-1), \qquad s = 1, \ldots, p_n. \tag{2.21}$$

The assumption entails that the decay rate of π_p is exponential for any $s \in \{1, \ldots, p_n\}$.

Example (Complexity prior): Assumption (2.21) is met by the priors of the following form, for constants $a, c > 0$,

$$\pi_p(s) \approx c^{-s} p_n^{-as}, \qquad s = 0, 1, \ldots, p_n. \tag{2.22}$$

These priors are also considered "complexity priors". We remark that a similar prior is defined (2.11) in §2.2 in the case $p_n = o(n)$.

Example (Discrete spike-and-slab prior): Consider a spike-and-slab prior, where the coordinates $\beta_1, \ldots, \beta_p$ are i.i.d. draws from a mixture $(1-\gamma)\,\delta_0 + \gamma\, g$, of a Dirac measure δ_0 at zero and a Laplace distribution g. It is easy to see such prior satisfies the form of (2.21) with π_p chosen to be a binomial distribution with parameter p_n and r. Assigning a $\mathcal{B}eta\,(1, p_n^a)$ hyper-prior with some $a > 1$, the overall prior satisfies the condition (2.21).

2.3.2 Compatibility Conditions on the Design Matrix

When fitting model (2.16) in high-dimensional setting ($p_n > n$), the estimation procedure becomes more complicated as the parameter $\boldsymbol{\beta}$ is non-identifiable due to non-invertibility of design matrix $\boldsymbol{X}$. There is a rich literature (e.g., [7, 47]) studying the sufficient conditions relative to the invertibility of the design matrix and/or its interplay with the *sparsity* of parameter $\boldsymbol{\beta}$. We shall introduce few "sparse invertibility" conditions that are commonly used in sparse regressions under the non-Bayesian setting [7], some of which are adapted to the Bayesian setup.

Definition (Compatibility): The compatibility number of model $S \subset \{1, \ldots, p\}$ is given by

$$\phi(S) := \inf\left\{\frac{\|\boldsymbol{X\beta}\|_2|S|^{1/2}}{\|\boldsymbol{X}\|\|\boldsymbol{\beta}_S\|_1} : \|\boldsymbol{\beta}_{S^c}\|_1 \leq c\|\boldsymbol{\beta}_S\|_1, \boldsymbol{\beta}_S \neq 0\right\}. \tag{2.23}$$

The constant c is greater than 1, commonly taken to be 7. The compatible number compares the ℓ_2-norm of the predictive vector $\boldsymbol{X\beta}$ to the ℓ_1-norm of the parameter $\boldsymbol{\beta}_S$. Comparison under inconsistent norms may cause the factor $|S|^{1/2}$ in the numerator through the Cauchy–Schwarz inequality, as $\|\boldsymbol{\beta}_S\|_1 \leq |S|^{1/2}\|\boldsymbol{\beta}_S\|_2$.

We also remark that the compatible number is defined for each model S and it involves a full parameter $\boldsymbol{\beta}$. Besides, the condition $\|\boldsymbol{\beta}_{S^c}\|_1 \leq c\|\boldsymbol{\beta}_S\|_1$ within the infimum is a common condition referred to as "*restricted nullspace property*" [49]. The next two definitions concern sparse vectors only, and the defined compatibility numbers are uniform in vectors up to a given dimension.

Definition (Uniform compatibility in sparse vectors): The compatibility number in vectors of dimension s is defined as

$$\bar{\phi}(s) := \inf\left\{\frac{\|\boldsymbol{X\beta}\|_2|S_{\boldsymbol{\beta}}|^{1/2}}{\|\boldsymbol{X}\|\|\boldsymbol{\beta}_S\|_1} : 0 \neq |S_{\boldsymbol{\beta}}| \leq s\right\}. \tag{2.24}$$

Definition (Smallest scaled sparse singular value): The smallest scaled singular value of dimension s is defined as

$$\tilde{\phi}(s) := \inf\left\{\frac{\|\boldsymbol{X\beta}\|_2}{\|\boldsymbol{X}\|\|\boldsymbol{\beta}_S\|_2} : 0 \neq |S_{\boldsymbol{\beta}}| \leq s\right\}. \tag{2.25}$$

For recovery we shall impose that these numbers for $s = s^*$ (or a multiple of) are bounded away from zero. By the Cauchy–Schwarz inequality, $\|\boldsymbol{\beta}\|_1 \leq |S_{\boldsymbol{\beta}}|^{1/2}\|\boldsymbol{\beta}\|_2$, one can obtain $\tilde{\phi}(s) \leq \bar{\phi}(s)$ for any $s > 0$. The stronger assumptions on the design matrix relative to $\tilde{\phi}(s)$ will be used for recovery with respect to the ℓ_2-norm, whereas the numbers $\bar{\phi}(s)$ suffice for ℓ_1-reconstruction.

We now introduce a stronger invertibility condition referred to as the "mutual coherence" of the regression matrix, which is the maximum correlation between its columns.

Definition (Mutual coherence). The mutual coherence number is

$$\mathrm{mc}(\boldsymbol{X}) := \max_{1\leq i\neq j\leq p} \frac{|\langle \boldsymbol{X}_{.,i}, \boldsymbol{X}_{.,j}\rangle|}{\|\boldsymbol{X}_{.,i}\|_2\|\boldsymbol{X}_{.,j}\|_2}. \tag{2.26}$$

A model S is said to satisfy the "*mutual coherence condition*" if there exists some arbitrary constant $K > 0$ such that $|S| \leq 1/(K\mathrm{mc}(\boldsymbol{X}))$. A deeper inspection on the relationship of the introduced indices (refer to [11, 36]) shows that the *mutual coherence condition* is stronger than conditions in terms of compatibility number or "restricted eigenvalue conditions". As shown in the following theorem on *recovery*, the *mutual coherence condition* is useful for reconstruction with respect to stronger norms such as ℓ_∞-norm, and it can be also used to bound other indices. Among other commonly considered conditions such as "restricted eigenvalue condition" and "irrepresentability", the compatibility number has been shown to be the weakest condition for prediction and reconstruction by the LASSO for the ℓ_1- and ℓ_2-norms, (refer to sections 6.13 and 7.5 in [7] for extensive discussions on the relationship of different conditions).

Next, we will discuss asymptotic properties of the *full* posterior distribution resulting from the prior (2.17). Given an *exponential decrease* assumption on π_p and a heavy-tail assumption on densities of non-zero coordinates, the posterior distribution enjoys (near-) optimal contraction rate for both prediction error and reconstruction of the parameter relative to different metrics. Consistent variable selection is achieved under two different prior schemes. Moreover, we shall see that the posterior distribution behaves asymptotically as a mixture of Bernstein-von Mises type approximations to submodels.

2.3.3 Posterior Contraction Rate

The first theorem gives an upper bound on the dimensionality of the support of posterior distribution.

Theorem 2.3.1. (Support recovery). *If λ satisfies* (2.20), *and π_p satisfies* (2.21), *then with $s^* = |S_{\boldsymbol{\beta}^*}|$ and for any $M > 2$,*

$$\sup_{\boldsymbol{\beta}^*} \mathbb{E}_{\boldsymbol{\beta}^*} \left(\boldsymbol{\beta} : |S_{\boldsymbol{\beta}}| > s^* + \frac{M}{A_4} \left(1 + \frac{16}{\phi(S^*)^2} \frac{\lambda}{\bar{\lambda}} \right) s^* \,\middle|\, \boldsymbol{y} \right) \to 0. \tag{2.27}$$

Theorem 2.3.1 asserts that the posterior distribution is asymptotically supported on subspaces of dimensions that are a multiple of the true dimension s^*. A stronger compatibility number imposes a smaller lower bound on the dimensionality of the posterior. In the case $\lambda \ll \bar{\lambda}$, the dominating term of lower bound on the dimensionality is $1 + M/A_4$, where A_4 is the constant defined in (2.21) that dictates the decreasing rate of the upper bound for π_p, and $M > 2$ is any arbitrary constant. Choosing sufficiently large A_4 such that $1 + M/A_4$ close to 1 (which induces a faster decay rate of π_p on the model size), the theorem leads to the conclusion that the posterior distribution concentrates on the set of models whose dimensionality does not exceed the true dimension s^*.

The next theorem provides rates of contraction of the posterior distribution both regarding prediction error $\|\boldsymbol{X}\boldsymbol{\beta} - \boldsymbol{X}\boldsymbol{\beta}^*\|_2$ and regarding the parameter $\boldsymbol{\beta}$ relative to the ℓ_1- and ℓ_2- and ℓ_∞-distances. First define

$$\bar{\psi}(S) = \bar{\phi}\left(\left(2 + \frac{3}{A_4} + \frac{33}{\phi(S)^2} \frac{\lambda}{\bar{\lambda}} \right) |S| \right),$$

$$\tilde{\psi}(S) = \tilde{\phi}\left(\left(2 + \frac{3}{A_4} + \frac{33}{\phi(S)^2} \frac{\lambda}{\bar{\lambda}} \right) |S| \right).$$

Quantities $\bar{\psi}(S^*)$ and $\tilde{\psi}(S^*)$ are compatibility numbers for the sparse vectors of the (relative) effective dimension (one may consider it as the dimensionality of a union of posterior support and the true model), that receive posterior probability mass converging to 1, as $n, p_n \to \infty$, according to Theorem 2.3.1. As shown in the following theorem, both $\bar{\phi}(S^*), \bar{\psi}(S^*)$, (resp. $\tilde{\phi}(S^*), \tilde{\psi}(S^*)$) play an important role in determining the posterior contraction rate for recovery.

Theorem 2.3.2. (Recovery). *If λ satisfies* (2.20), *and π_p satisfies* (2.21), *then for sufficiently large M, with $S^* = S_{\boldsymbol{\beta}^*}$,*

$$\sup_{\boldsymbol{\beta}^*} \mathbb{E}_{\boldsymbol{\beta}^*}\Pi \left(\boldsymbol{\beta} \,:\, \|\boldsymbol{X}(\boldsymbol{\beta} - \boldsymbol{\beta}^*)\|_2 > \frac{M}{\bar{\psi}(S^*)} \frac{\sqrt{|S^*| \log p_n}}{\phi(S^*)} \,\middle|\, \boldsymbol{y} \right) \to 0,$$

$$\sup_{\boldsymbol{\beta}^*} \mathbb{E}_{\boldsymbol{\beta}^*}\Pi \left(\boldsymbol{\beta} \,:\, \|\boldsymbol{\beta} - \boldsymbol{\beta}^*\|_1 > \frac{M}{\bar{\psi}(S^*)^2} \frac{|S^*| \sqrt{\log p_n}}{\|X\| \phi(S^*)^2} \,\middle|\, \boldsymbol{y} \right) \to 0,$$

$$\sup_{\boldsymbol{\beta}^*} \mathbb{E}_{\boldsymbol{\beta}^*}\Pi \left(\boldsymbol{\beta} \,:\, \|\boldsymbol{\beta} - \boldsymbol{\beta}^*\|_2 > \frac{M}{\tilde{\psi}(S^*)^2} \frac{\sqrt{|S^*| \log p_n}}{\|\boldsymbol{X}\| \phi(S^*)} \,\middle|\, \boldsymbol{y} \right) \to 0.$$

Furthermore, for every $c_0 > 0$, and $d_0 < c_0^2(1 + 2/A_4)^{-1}/8$ and s_n with $\lambda s_n \sqrt{\log p_n}/\|\boldsymbol{X}\| \to 0$, for sufficiently large M,

$$\sup_{\substack{\boldsymbol{\beta}^* : \phi(S^*) \geq c_0, \tilde{\psi}(S^*) \geq c_0 \\ |S^*| \leq s_n, |S^*| \leq d_0 \mathrm{mc}(\boldsymbol{X})^{-1}}} \mathbb{E}_{\boldsymbol{\beta}^*}\Pi \left(\boldsymbol{\beta} : \|\boldsymbol{\beta} - \boldsymbol{\beta}^*\|_\infty > M \frac{\sqrt{\log p_n}}{\|\boldsymbol{X}\|} \,\middle|\, \boldsymbol{y} \right) \to 0.$$

These asymptotic results hold uniformly for all true parameter $\boldsymbol{\beta}^*$. We remark that the first assertion is derived based on Bayesian oracle results (see Theorem 3 in [11]), and the second and third assertions are easily derived from the first assertion by applying the definitions of $\bar{\psi}(S^*), \tilde{\psi}(S^*)$. The fourth assertion shows that ℓ_∞-reconstruction requires the strongest "*mutual coherence condition*" on the true model. In addition, if the true model is compatible relative to different norms then ℓ_∞-norm contraction rate is free of compatibility numbers.

2.3.4 Variable Selection Consistency

The optimal recovery results show that a Bayesian procedure initially lets the prior to induce a distribution on a set of models, and allows the posterior to update the prior masses given to these models through the prior and the likelihood. Eventually the posterior distribution concentrates on the true parameter $\boldsymbol{\beta}^*$. However it does not prevent the posterior from including inactive coefficients of small magnitude, and thus will not recover the true model S^* even though the reconstruction might be optimal. The following theorem gives a closer inspection on the asymptotic support of the posterior distribution.

Theorem 2.3.3. (Selection: no supersets). *If* λ *satisfies* (2.20) *and* π_p *satisfies* (2.21) *with* $A_4 > 1$, *then for every* $c_0 > 0$ *and any* $s_n \leq p_n^a$ *with* $s_n\lambda\sqrt{\log p_n}/\|\boldsymbol{X}\| \to 0$ *and* $a < A_4 - 1$,

$$\sup_{\substack{\boldsymbol{\beta}^*:\phi(S^*)\geq c_0\\ |S^*|\leq s_n, \tilde{\psi}(S^*)\geq c_0}} \mathbb{E}_{\boldsymbol{\beta}^*}\Pi(\boldsymbol{\beta}: S_{\boldsymbol{\beta}} \supset S_{\boldsymbol{\beta}^*}, S_{\boldsymbol{\beta}} \neq S_{\boldsymbol{\beta}^*} \mid \boldsymbol{y}) \to 0.$$

Theorem 2.3.3 shows the posterior distribution concentrates on the subspace which contains no strict superset of the true model S^*, regardless of the magnitudes of the non-zero coordinates in $\boldsymbol{\beta}^*$. However the posterior support may only include a subset of the true model or include other inactive coordinates that causes possible *false discovery*. It is hard to recover the true model without putting any condition on the magnitude of true non-zero coordinates.

The next theorem shows the posterior distribution will detect all coordinates of $\boldsymbol{\beta}^*$ of magnitude above some threshold, given by the optimal rate for ℓ_1- and ℓ_2-reconstructions. If a true coordinate is of a magnitude less than the contraction rate, then a "zero-estimator" can achieve the optimal recovery, and overall the full posterior distribution may be supported only on a subset of the true model while still maintaining the optimal recovery. We also remark that the role of such thresholds may be similar to "*beta-min condition*", which is a sufficient condition for consistent variable selection of the LASSO estimator.

Theorem 2.3.4. (Selection). *If* λ *satisfies* (2.20), *and* π_p *satisfies* (2.21), *then for sufficiently large* M,

$$\inf_{\boldsymbol{\beta}^*} \mathbb{E}_{\boldsymbol{\beta}^*}\Pi\left(\boldsymbol{\beta}: S_{\boldsymbol{\beta}} \supset \left\{i: |\boldsymbol{\beta}_i^*| \geq \frac{M}{\bar{\psi}(S^*)^2}\frac{|S^*|\sqrt{\log p_n}}{\|\boldsymbol{X}\|\phi(S^*)^2}\right\} \,\middle|\, \boldsymbol{y}\right) \to 1.$$

$$\inf_{\boldsymbol{\beta}^*} \mathbb{E}_{\boldsymbol{\beta}^*}\Pi\left(\boldsymbol{\beta}: S_{\boldsymbol{\beta}} \supset \left\{i: |\boldsymbol{\beta}_i^*| \geq \frac{M}{\tilde{\psi}(S^*)^2}\frac{\sqrt{|S^*|\log p_n}}{\|\boldsymbol{X}\|\phi(S^*)}\right\} \,\middle|\, \boldsymbol{y}\right) \to 1.$$

Furthermore, for every $c_0 > 0$, *and* $d_0 \leq c_0^2(1 + 2/A_4)^{-1}/8$, *and any* s_n *with* $\lambda s_n\sqrt{\log p_n}/\|\boldsymbol{X}\| \to 0$,

$$\inf_{\substack{\boldsymbol{\beta}^*:\phi(S^*)\geq c_0, \tilde{\psi}(S^*)\geq c_0\\ |S^*|\leq s_n, |S^*|\leq d_0\mathrm{mc}(\boldsymbol{X})^{-1}}} \mathbb{E}_{\boldsymbol{\beta}^*}\Pi\left(\boldsymbol{\beta}: S_{\boldsymbol{\beta}} \supset \left\{i: |\boldsymbol{\beta}_i^0| \geq \frac{M\sqrt{\log p_n}}{\|\boldsymbol{X}\|}\right\} \,\middle|\, \boldsymbol{y}\right) \to 1.$$

We remark that the thresholds appearing in the first two assertions become smaller if the compatibility number become larger. If the true model satisfies "*mutual coherence condition*", the third assertion shows the corresponding threshold may be free of the compatibility number. It is not hard to see that Theorem 2.3.3 and Theorem 2.3.4 jointly imply that under the assumptions of the theorems the posterior distribution consistently selects the *correct* model if all non-zero coordinates of $\boldsymbol{\beta}^*$ are bounded away from 0 by the thresholds given in Theorem 2.3.4. For M as in the preceding theorem, define

$$\widetilde{B} = \left\{ \boldsymbol{\beta} : \min_{i \in S_{\boldsymbol{\beta}}} |\boldsymbol{\beta}_i| \geq \frac{M}{\widetilde{\psi}(S_{\boldsymbol{\beta}})^2} \frac{\sqrt{|S_{\boldsymbol{\beta}} \log p_n|}}{\|\boldsymbol{X}\| \phi(S_{\boldsymbol{\beta}})} \right\}. \tag{2.28}$$

We can define the set $\bar{B}$ in a similar manner by replacing $\widetilde{\psi}$ in the preceding display with $\bar{\psi}$. Now we give the strong variable selection result based on the preceding Theorems.

Corollary 2.3.5. (Consistent model selection). *If* λ *satisfies* (2.20), *and* π_p *satisfies* (2.21) *with* $A_4 > 1$, *and* $s_n \leq p_n^{\alpha}$ *such that* $a < A_4 - 1$ *and* $s_n \lambda \sqrt{\log p_n}/\|\boldsymbol{X}\| \to 0$, *then, for every* $c_0 > 0$,

$$\inf_{\substack{\boldsymbol{\beta}^* \in \widetilde{B} : \phi(S^*) \geq c_0 \\ |S^*| \leq s_n, \widetilde{\psi}(S^*) \geq c_0}} \mathbb{E}_{\boldsymbol{\beta}^*}(\beta : S_{\boldsymbol{\beta}} = S_{\boldsymbol{\beta}^*} \mid \boldsymbol{y}) \to 1. \tag{2.29}$$

The same is true with $\widetilde{B}$ *and* $\widetilde{\phi}$ *replaced by* $\bar{B}$ *and* $\bar{\phi}$.

Under the assumption that the true parameter $\boldsymbol{\beta}^* \in \widetilde{B}$, or say, $\boldsymbol{\beta}^*$ satisfies the "*beta-min*" condition, one can achieve consistent posterior model selection, that means the model with the largest posterior mass is model selection consistent in the frequentist sense. This result implies a stronger posterior contraction for the non-significant coordinates such that their posterior is bounded by the threshold defined in $\widetilde{B}$.

2.3.5 Variable Selection with Discrete Spike and Zellner's g-Priors

The key features of the discussed discrete spike-and-slab prior are fast decay rate on the model size and heavy-tailed prior on the non-zero coordinates. In this section, we briefly review another type of spike and slab prior with a g-prior slab based on a recent work [52].

Let us consider the regression model (2.16) and assume $p_n > n$. As before, the data is assumed to have a true distribution $\mathcal{N}(\boldsymbol{X}\boldsymbol{\beta}^*, I_n)$, with $\boldsymbol{\beta}^*$ denoting the vector of true parameters. Here we use $\boldsymbol{\gamma} = \{\gamma_j, j = 1, \ldots, p_n\} \in \{0,1\}^{p_n}$ to denote a binary indicator vector, and use the notation $\boldsymbol{\beta}_\gamma = \{\beta_j : \gamma_j = 1\}$ to denote the vector of non-zero coordinates of $\boldsymbol{\beta}$ selected by $\boldsymbol{\gamma}$. The model is introduced via a hierarchical structure

$$\begin{aligned} \boldsymbol{y} &= \boldsymbol{X}\boldsymbol{\beta} + \boldsymbol{\varepsilon}, \quad \boldsymbol{\varepsilon} \sim \mathcal{N}(\mathbf{0}, \phi^{-1} I_n), \quad \pi(\phi) \propto 1/\phi, \\ \boldsymbol{\beta}_\gamma \mid \gamma &\sim \mathcal{N}(\mathbf{0}, g\phi^{-1}(\boldsymbol{X}_\gamma^{\mathrm{T}} \boldsymbol{X}_\gamma)^{-1}), \quad \pi(\gamma) \propto p_n^{-\kappa|\gamma|} \mathbb{I}\{|\gamma| \leq s^0\}. \end{aligned} \tag{2.30}$$

We denote by $|\boldsymbol{\gamma}| = \sum_{j=1}^{p_n} \gamma_j$ the number of non-zero entries of $\boldsymbol{\gamma}$, and denote by $\boldsymbol{X}_\gamma$ the γ-indexed design matrix. One can consider (2.30) as a $\boldsymbol{\gamma}$-indexed linear model with a precision parameter ϕ endowed with an improper prior. Model (2.30) is equivalent to having a dependent multivariate normal prior for the non-zero coordinates conditional on picking the subset, appearing as a special case of the dependent prior in §2.2. More specifically, the dependence is induced through a Zellner's g-prior on $\boldsymbol{\beta}_\gamma$, where $g > 0$ controls the degree of dispersion. The sparsity favoring prior on $\boldsymbol{\gamma}$ obeys the condition (2.21). The $p_n^{-\kappa}$ decay rate for the marginal probability of including each covariate imposes a vanishing prior probability on the models of diverging sizes. The parameter s^0 is a pre-specified upper bound on the maximum number of important covariates.

In order for a strong variable selection consistency to hold, the "*beta-min*" condition on the true parameter $\boldsymbol{\beta}^*$ is required. Denote the lower bound on the minimum magnitude of $\boldsymbol{\beta}^*$ by some constant $C_\beta > 0$ depending on (n, p_n). We then define the set

$$S = S(C_\beta) := \{j : |\beta_j^*| \geq C_\beta\}.$$

Note that the element of $\boldsymbol{\beta}_{S^c}^*$ are not necessarily zero but bounded below by C_β in absolute value and

$$\left\| n^{-1/2} \boldsymbol{X} \boldsymbol{\beta}^* \right\|_2^2 \leq g \frac{\log p}{n}, \qquad \left\| n^{-1/2} \boldsymbol{X}_{S^c} \boldsymbol{\beta}_{S^c}^0 \right\|_2^2 \leq \tilde{L}_0 \frac{\log p}{n}, \tag{2.31}$$

for some $\tilde{L} > 0$.

The design matrix is assumed to be normalized so that $\|\boldsymbol{X}_j\|_2^2 = n$ for all $j = 1, \ldots, p_n$, which is commonly done in regression models. With $\boldsymbol{z} \sim \mathcal{N}(\boldsymbol{0}, I_n)$, there exist constants $\nu \in (0, 1]$ and $L < \infty$ such that $L\nu > 4$ and

Assumption 1 (Lower restricted eigenvalue):

$$\min_{|S| \leq s} \lambda_{\min} \left(\frac{1}{n} \boldsymbol{X}_S^{\mathrm{T}} \boldsymbol{X}_S \right) \geq \nu, \qquad \text{and} \tag{2.32}$$

Assumption 2 (Sparse projection condition):

$$\mathbb{E}_Z \left[\max_{|S| \leq s} \max_{k \in \{1, \ldots, p\} \setminus S} \frac{1}{\sqrt{n}} |\langle (I - \Phi_S) \boldsymbol{X}_k, \boldsymbol{z} \rangle| \right] \leq \frac{1}{2} \sqrt{L\nu \log p}, \tag{2.33}$$

Assumption 3 (Sparsity control): Setting $s^0 := p$, assume the true sparsity s^* is bounded as $\max\{1, s^*\} \leq (n/\log p - 8\tilde{L})/32$.

Assumption 4 (Choice of hyper-parameters): The noise hyper-parameter g and the sparsity penalty hyper-parameter $\kappa > 2$ are chosen such that

$$g \asymp p^{2\alpha} \quad \text{for some} \quad \alpha \geq 1/2 \quad \text{and} \quad \kappa + \alpha \geq 4(L + \tilde{L}) + 2. \tag{2.34}$$

We remark that the above assumptions are quite commonly assumed for establishing Bayesian variable selection consistency. We conclude this section with the following result.

Theorem 2.3.6. (Variable selection consistency). *Suppose that Assumptions 1, 2 with* $s = 2(\kappa + \alpha + \tilde{L} + 1) \max\{1, s^*\}$, *and Assumptions 3 and 4 hold. If the threshold* C_β *satisfies*

$$C_\beta^2 \geq 128\nu^{-2}(L + \tilde{L} + \kappa + \alpha) \frac{\log p}{n}, \tag{2.35}$$

then we have $\Pi(S^* \mid \boldsymbol{y}) \geq 1 - c_1 p^{-1}$ *with probability at least* $1 - c_2 p^{-c_3}$. *The probability is with respect to the true data-generating model.*

2.3.6 Bernstein-von Mises Theorem for the Posterior Distribution

In this section, we investigate the nature of the asymptotic shape of the posterior. Under the spike-and-slab prior, we shall see that the posterior distribution can be approximated by a random mixture of normal distributions. Under consistent variable selection of the true model, this mixture collapses to a single normal distribution. The result is derived under what is referred to as *small lambda regime*,

$$\frac{\lambda}{\|\boldsymbol{X}\|} |S_{\boldsymbol{\beta}^*}| \sqrt{\log p_n} \to 0. \tag{2.36}$$

For a Laplace density g, a smaller value of λ imposes non-informative prior on the non-zero coordinates of the parameter. In this setting, the prior (2.17) imposes flat-tailed priors on the non-zero coordinates and highly concentrated priors on the rest of coordinates, allowing the data to update only the distribution of the non-zero coordinates via a Bayesian procedure.

For a given model $S \subset \{1, \ldots, p_n\}$, let $\boldsymbol{X}_S$ denote the $n \times |S|$-submatrix of the regression matrix $\boldsymbol{X}$ consisting of the columns $\boldsymbol{X}_{\cdot,i}$ with $i \in S$, and let $\hat{\boldsymbol{\beta}}_{(S)}$ be a least square estimator in the restricted model $\boldsymbol{y} = \boldsymbol{X}_S \boldsymbol{\beta}_S + \boldsymbol{\varepsilon}$, that is,

$$\hat{\boldsymbol{\beta}}_{(S)} \in \operatorname{argmin}_{\boldsymbol{\beta}_S \in \mathbb{R}^S} \|\boldsymbol{y} - \boldsymbol{X}_S \boldsymbol{\beta}_S\|_2^2.$$

If the restricted model is correctly specified, then $\hat{\boldsymbol{\beta}}_{(S)}$ would have a $\mathcal{N}(\boldsymbol{\beta}_S^*, (\boldsymbol{X}_S^{\mathrm{T}} \boldsymbol{X}_S)^{-1})$ distribution. By Bernstein–von Mises theorem, the posterior distribution would have a $\mathcal{N}(\hat{\boldsymbol{\beta}}_{(S)}, (\boldsymbol{X}_S^{\mathrm{T}} \boldsymbol{X}_S)^{-1})$ distribution asymptotically. In the present setting, due to Theorem 2.3.1 and Theorem 2.3.2, the posterior distribution of $\boldsymbol{\beta}$ can be approximated by a random mixture

$$\Pi^{\infty}(\cdot \mid \boldsymbol{y}) = \sum_{S \in \mathcal{S}^*} \hat{\omega}_S \mathcal{N}(\hat{\boldsymbol{\beta}}_{(S)}, (\boldsymbol{X}_S^{\mathrm{T}} \boldsymbol{X}_S)^{-1}) \otimes \delta_{S^c}, \tag{2.37}$$

where δ_{S^c} denotes the Dirac measure at 0 for $\boldsymbol{\beta}_{S^c}^*$, the weights $(\hat{\omega}_S)_S$ satisfy

$$\hat{\omega}_S \propto \frac{\pi_p(s)}{\binom{p_n}{s}} \left(\frac{\lambda}{2}\right)^s (2\pi)^{s/2} |\boldsymbol{X}_S^{\mathrm{T}} \boldsymbol{X}_S|^{-1/2} e^{(1/2)\|\boldsymbol{X}_S \hat{\boldsymbol{\beta}}_{(S)}\|_2^2} \mathbb{I}_{S \in \mathcal{S}^*} \tag{2.38}$$

and, for a sufficiently large M,

$$\mathcal{S}^* = \left\{ S \ : \ |S| \leq \left(2 + \frac{4}{A_4}\right) |S_{\boldsymbol{\beta}^*}|, \ \|\boldsymbol{\beta}_{S^c}^*\|_1 \leq M |S_{\boldsymbol{\beta}^*}| \sqrt{\log p_n} / \|\boldsymbol{X}\| \right\}. \tag{2.39}$$

The collection of models $\mathcal{S}^*$ can be considered as a "neighborhood" of the true model, allowing a mild extension of model size from the true model and allowing true zero parameters to be away from 0 by a small margin. The weights $(\hat{\omega}_S)_S$ can be considered as a data-dependent probability weight that is proportional to a product of the prior mass or model S and the predictive likelihood of model S.

Theorem 2.3.7. (Bernstein-von Mises). *If* λ *satisfies* (2.20), *and* π_p *satisfies* (2.21), *then for every* $c_0 > 0$ *and any* s_n *with* $s_n \lambda \sqrt{\log p_n} / \|\boldsymbol{X}\| \to 0$,

$$\sup_{\substack{\boldsymbol{\beta}^* : \phi(S^*) \geq c_0 \\ |S^*| \leq s_n, \bar{\psi}(S^*) \geq c_0}} \mathbb{E}_{\boldsymbol{\beta}^*} \|\Pi(\cdot \mid \boldsymbol{y}) - \Pi^{\infty}(\cdot \mid \boldsymbol{y})\|_{TV} \to 0.$$

The random mixture of normal approximation is a consequence of the fact that the posterior distribution concentrates its mass on a "neighborhood" of the support of true parameters, which contains subspaces of dimension no larger than the optimal dimension that is close to the true model, rather than a fixed s^*-dimensional subspace. In addition, relying on strong variable selection consistency, the limiting distribution of posterior distribution will reduce to a point mass mixture with a single normal distribution.

Corollary 2.3.8. (Limit under strong model selection). *Under the combined assumptions of Corollary 2.3.5 and Theorem 2.3.7,*

$$\sup_{\substack{\boldsymbol{\beta}^* \in \tilde{B} : \phi(S^*) \geq c_0 \\ |S^*| \leq s_n, \bar{\psi}(S^*) \geq c_0}} \mathbb{E}_{\boldsymbol{\beta}^*} \|\Pi(\cdot \mid \boldsymbol{y}) - \mathcal{N}(\hat{\boldsymbol{\beta}}_{S^*}, (\boldsymbol{X}_{S^*}^{\mathrm{T}} \boldsymbol{X}_{S^*})^{-1}) \otimes \delta_{S^{*c}}\|_{TV} \to 0.$$

A direct application of distributional approximation is to provide frequentist justification for Bayesian credible intervals (CI). Let $\mathrm{CL}_i(\alpha)$ denote the posterior quantile credible interval of the ith covariate. Then Corollary 2.3.8 implies

$$\mathbb{P}_{\beta^*}(\beta_i^* \in \mathrm{CL}_i(\alpha)) \to 1-\alpha, \quad \text{if } i \in S^*,$$

$$\mathbb{P}_{\beta^*}(0 \in \mathrm{CL}_i(\alpha)) \to 1, \quad \text{if } i \in S^{*c},$$

for any $\alpha \in (0,1)$. Note that the frequentist coverage of Bayesian credible interval holds uniformly for all non-zero β_i^* given $\boldsymbol{\beta}^* \in \widetilde{B}$.

Finally, we briefly discuss the "*large lambda regime*", assuming $\lambda = 4\|\boldsymbol{X}\|\sqrt{\log p_n}$. Under the strict variable-selection condition, the limiting shape of the posterior under a "*large lambda regime*" will also be a point-mass mixture with a normal distribution with the LASSO estimator as the center. We briefly state the result and refer to [11] for more details.

Theorem 2.3.9. (Bernstein-von Mises, large lambda). *Let* $\lambda = 4\|\boldsymbol{X}\|\sqrt{\log p_n}$. *If* $|S^*| = o(p_n^{1-\delta} \wedge \mathrm{mc}(\boldsymbol{X})^{-1})$ *and* $\min_{i\in S^*}|\beta_j^*| \geq A\sqrt{s^*\log p}/\|\boldsymbol{X}\|$, *for some sufficiently large A, then*

$$\mathbb{E}_{\beta^*}\|\Pi(\cdot \mid \boldsymbol{y}) - \mathcal{N}(\hat{\boldsymbol{\beta}}_{S^*}^{\mathrm{LASSO}}, (\boldsymbol{X}_{S^*}^{\mathrm{T}}\boldsymbol{X}_{S^*})^{-1}) \otimes \delta_{S^{*c}}\|_{TV} \to 0,$$

where $\hat{\boldsymbol{\beta}}_{S^*}^{\mathrm{LASSO}}$ *denotes the LASSO estimator of the restricted model.*

2.4 Extension to Generalized Linear Models

The class of generalized linear models (GLM) [37] is a flexible generalization of ordinary linear regression that allows for response variables to accommodate error distributions which are non-additive and non-Gaussian. The GLM generalizes linear regression by allowing the linear model to be related to the response variable via a link function. Although primarily restricted to a lower dimensional setting, Bayesian approaches for GLM have been very popular since the 90s with the wide adoption of Markov chain Monte Carlo (MCMC) methods [12].

As discussed in §2.2 and §2.3, sparsity favoring mixture priors with separate control on the signal and noise coefficients have been proposed primarily for linear model. Although in principle such methods can be used for GLMs, accompanying theoretical justification on optimal estimation in the high-dimensional case is primarily available in the context of linear models. Analogous results for generalized linear models in the high-dimensional case are comparatively sparse, with the exception of [31]. However, special cases from the GLM family, including high-dimensional logistic regression using a pseudo likelihood [2] and using shrinkage priors [62], are available. Very recently, [30] considered posterior contraction in GLMs using complexity priors on the model space as described in §2.4. Their results make use of the same identifiability and compatibility assumptions as in [11] to deliver optimal posterior contraction rates, albeit with a growth restriction on the true coefficient vector.

Here, we shall elaborate on the recent work by [26], who developed a framework to study posterior contraction in high-dimensional clipped GLMs using complexity priors that involve a Laplace prior on the non-zero coefficients. The clipped GLM class deviates slightly from the standard GLM construction in that we allow the effect of linear term $\boldsymbol{x}^{\mathrm{T}}\boldsymbol{\beta}$ in the argument of the log-partition function to "clip" away from the singularities of the function. The clipped GLM directly subsumes high-dimensional linear, polynomial and logistic regression, while also incorporating variants of Poisson, negative Binomial (and similar) regressions, which are identical from a practical standpoint to the standard Poisson/negative binomial regressions. One primary advantage of using a clipping function is that it obviates the need for any growth assumption on the true coefficient vector. The assumptions for obtaining adaptive rate-optimal posterior contraction are specifically designed for the clipped GLMs which can be viewed as appropriate generalization of the identifiability and compatibility assumptions of [11] in the linear model case. Finally, the prior dependence on the true parameter can be completely eliminated making our results rate-adaptive.

2.4.1 Construction of the GLM Family

For both univariate and multivariate observations, one of the most widely used and well-structured family of models is the exponential family. We discuss this briefly with the example of univariate observations and real valued parameter. The exponential family takes the form

$$f(y \mid \theta) = h(y) \exp\left[\theta T(y) - A(\theta)\right], \; y \in \mathcal{Y} \subset \mathbb{R}, \tag{2.40}$$

where $\theta \in \Theta \subset \mathbb{R}$ is the parameter of interest, $h(\cdot) : \mathcal{Y} \to \mathbb{R}$ is called the base measure, $A(\cdot) : \Theta \to \mathbb{R}$ is the convex log-partition function and $T(\cdot) : \mathcal{Y} \to \mathbb{R}$ is called the sufficient statistic for estimating the parameter θ. This form is known as the canonical form of an exponential family. Many standard distributions, like the Bernoulli and Gaussian with known variance, Poisson, negative Binomial, among many others, follow model (2.40). $A'(\cdot)$ and $A''(\cdot)$ are thus known as the mean and variance functions respectively, and $A'(\cdot)$ can be assumed to strictly increasing on its domain. An interesting property of exponential families is that it affords a neat expression of Kullback–Leibler (KL) divergence in terms of the Bregman divergence of log-partition function $A(\cdot)$.

A GLM assumes that the response variable comes from an exponential family member, and models a function of the mean through a linear function of covariates, i.e., as $\boldsymbol{x}^{\mathrm{T}}\boldsymbol{\beta}$. The said function, denoted by $g(\cdot) : \text{range}[A'(\cdot)] \to \mathbb{R}$, is termed the link function. With n observations and p_n covariates, $\boldsymbol{X}$ makes up the design matrix, whose i-th row is denoted by $\boldsymbol{x}_i^{\mathrm{T}}$. Thus, for every $i = 1, \dots n$, GLM prescribes the transition θ to $\boldsymbol{\beta}$ as

$$g^{-1}\left(\boldsymbol{x}_i^{\mathrm{T}}\boldsymbol{\beta}\right) = A'(\theta), \text{ equivalently}\, \theta = \left(g \circ A'\right)^{-1}\left(\boldsymbol{x}_i^{\mathrm{T}}\boldsymbol{\beta}\right). \tag{2.41}$$

As we shall see in §2.4.2, (2.41) motivates modeling the original parameter θ using $A''(\cdot)$, and not through $A'(\cdot)$, leading to the definition of clipping function $\eta(\cdot)$ and clipped GLM family.

2.4.2 Clipped GLM and Connections to Regression Settings

We start with the canonical rank-one exponential family of distributions, where the canonical parameter $\boldsymbol{\theta}$ is expressed through a function of covariates. However, in contrast to GLM, we choose to represent

$$\theta = \eta\left(\boldsymbol{x}^{\mathrm{T}}\boldsymbol{\beta}\right),$$

where $\eta(\cdot)$, termed as the clipping function, depends only on $A''(\cdot)$. In cGLM, we consider log-partition functions $A(\cdot)$ that satisfy: i) that $A''(\cdot)$ exists everywhere in the domain of $A(\cdot)$; and ii) $\mathcal{I}_A(b) := \{t \in \mathbb{R} : 0 \leq A''(t) \leq b\}$ is an interval on the real line for any $b \in (0, \infty]$.

All the standard examples of exponential families satisfy these simple properties. We now turn to the clipping functions we use in cGLM, which play an intermediary role, sitting between $A(\cdot)$ and the i-th linear term $\boldsymbol{x}_i^{\mathrm{T}}\boldsymbol{\beta}$. We motivate the choice of clipping functions with some examples. Since $\boldsymbol{\beta} \in \mathbb{R}^{p_n}$, the linear term $\boldsymbol{x}_i^{\mathrm{T}}\boldsymbol{\beta} \in \mathbb{R}$, whereas the log-partition function A can have strict interval subsets of the real line as their support. These types of log-partition functions have a single *pole* (r_0 such that $\lim_{x\to r_0} A(x) = \infty$) on the real line. For instance, for Negative Binomial, $A(t) = -q\log(1-\exp(t)), t<0$ with q denoting known number of failures, so that $r_0 = 0$. For Exponential, $A(t) = -\log(-t), t<0$ so that $r_0 = 0$. For an extensive set of examples one can refer to [26].

The clipping function's first role is to ensure that $\eta_i \equiv \eta\left(\boldsymbol{x}_i^{\mathrm{T}}\boldsymbol{\beta}\right)$, which acts as an argument to $A(\cdot)$ to have the same range as the domain of $A(\cdot)$. Its second role, which turns out to be the central point of our hyper-parameter assumption, is to control the growth of $A''(\cdot)$, specifically to allow a local quadratic majorizability of $A(\cdot)$. Bernoulli and Gaussian (with known variance) have universally bounded $A''(\cdot)$. However, for Poisson, which has $A(t) = \exp(t), t \in \mathbb{R}$, or for distributions that have a pole in their log-partition function, $\eta(\cdot)$ should be assumed to be clipping the linear term $\boldsymbol{x}_i^{\mathrm{T}}\boldsymbol{\beta}$ away from $+\infty$ and r_0 respectively, or $\pm\infty$ and poles of log-partition functions of GLMs. Here we illustrate one choice of the clipping function for Poisson: $\eta(t) = \mathcal{C}_0 - \log(1+\exp(-t+\mathcal{C}_0))$, where $\mathcal{C}_0$ is a large positive absolute constant (see Figure 2.1, where $\mathcal{C}_0 = 10$).

The clipping function $\eta(\cdot)$ can be defined as injective and Lipschitz, as all our examples show. These two properties play an important role in identifiability of the model, as is discussed in the next section. Secondly, the practitioner can choose the constant $\mathcal{C}_0$ beforehand, and their choice is independent of the observed data or the true parameter. We now summarize the defining properties of clipping functions $\eta(\cdot)$ used in cGLM.

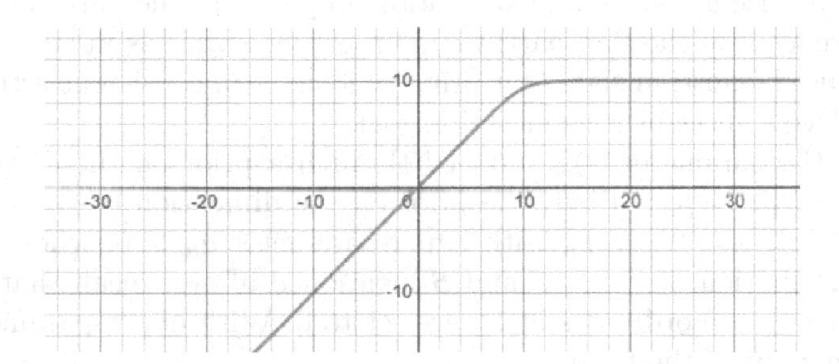

FIGURE 2.1
Graph of $y = 10 - \log(1+\exp(-x+10))$.

Clipping function condition: There exists constant $\mathcal{M}_0(A) > 0$ depending on $A(\cdot)$, so that $\eta(\cdot)$ satisfies

$$\eta(\cdot): \mathbb{R} \to \mathcal{I}_A\left(\frac{\mathcal{M}_0^2(A)}{2}\right), \text{ Lipschitz, injective.} \tag{2.42}$$

We now describe our data-generating model. For $i = 1,\dots n$, $y_i \in \mathcal{V} \subset \mathbb{R}$ are independent data points with $\boldsymbol{x}_i \in \mathbb{R}^{p_n}$ as the covariate, $\boldsymbol{\beta} \in \mathbb{R}^{p_n}$ as the parameter of interest and $\boldsymbol{\beta}^*$ denoting the true parameter value. Let $\eta_i \equiv \eta\left(\boldsymbol{x}_i^{\mathrm{T}}\boldsymbol{\beta}\right), \eta_i^* \equiv \eta\left(\boldsymbol{x}_i^{\mathrm{T}}\boldsymbol{\beta}^*\right)$ and let $\boldsymbol{X}$ denote the covariate matrix or design matrix, with the vector $\boldsymbol{x}_i^{\mathrm{T}}$ representing the i-th row of $\boldsymbol{X}$. The sufficient statistic is $T_i \equiv T(y_i)$, the base measure by $h(y_i)$ and the density for the i-th data point is denoted by $f(y_i \mid \eta_i)$. The i-th log-partition function is denoted by $A(\eta_i)$. We denote by S^* the true model, i.e., on S^{*c}, the co-ordinates of $\boldsymbol{\beta}^*$ are 0. Also, we shall denote by $\operatorname{supp}(\boldsymbol{\beta})$ the number of non-zero entries in $\boldsymbol{\beta}$, and by $\boldsymbol{\beta}_S$ the same vector as $\boldsymbol{\beta}$ with the co-ordinates in S^c set to zero. $L_n(\boldsymbol{\eta}, \boldsymbol{\eta}^*)$ stands for the log-likelihood ratio, which is expressed in terms of its two parts; $Z_n(\boldsymbol{\eta}, \boldsymbol{\eta}^*)$ is the centered stochastic term and while $\mathcal{D}_n(\boldsymbol{\eta}^*||\boldsymbol{\eta})$ denotes the KL divergence, both based on $y^{(n)}$. Thus we have the following:

$$\begin{aligned}
&f\left(y_i \mid \eta_i\right)=h\left(y_i\right) \exp \left(T_i \eta_i-A\left(\eta_i\right)\right), \ i=1, \ldots n, \\
&\mathcal{D}_i(\eta_i^*||\eta_i) := A\left(\eta_i\right)-A\left(\eta_i^*\right)-\left(\eta_i-\eta_i^*\right) A^{\prime}\left(\eta_i^*\right), \ \mathcal{D}_n(\boldsymbol{\eta}^*||\boldsymbol{\eta}) := \sum_{i=1}^n \mathcal{D}_i(\eta_i^*||\eta_i), \\
&Z_i(\eta_i, \eta_i^*) := \left(T_i-\mathbb{E} T_i\right)\left(\eta_i-\eta_i^*\right), \ Z_n(\boldsymbol{\eta}, \boldsymbol{\eta}^*) := \sum_{i=1}^n Z_i(\eta_i, \eta_i^*), \\
&L_n(\boldsymbol{\eta}, \boldsymbol{\eta}^*) := Z_n(\boldsymbol{\eta}, \boldsymbol{\eta}^*)-\mathcal{D}_n(\boldsymbol{\eta}^*||\boldsymbol{\eta}).
\end{aligned} \tag{2.43}$$

GLM and cGLM are interchangeable from the standpoint of practical implementation. Recall from (2.4.2) that we model the canonical parameter θ of the exponential family underlying cGLM as $\theta=\eta\left(\boldsymbol{x}_i^{\mathrm{T}} \boldsymbol{\beta}\right)$. For logistic regression, the native parameter here is the probability of success $p \in (0,1)$, while the canonical parameter is $\theta=\log(p/(1-p)) \in \mathbb{R}$. Thus, choosing Bernoulli for the exponential family and then, similar to linear regression, taking $\eta(t)=t, \ t \in \mathbb{R}$ as the clipping function, gives us the standard logistic regression setup. Here, $\mathcal{V}=\{0,1\}$. For other regression settings refer to [26].

2.4.3 Construction of Sparsity Favoring Prior

The sparsity favoring prior on the high-dimensional $\boldsymbol{\beta}$ is motivated by [10, 11] and follows the construction of discrete spike-and-slab prior proposed in the early references [10, 11]. The crucial difference is in the slab part; we use a Laplace prior as in [10, 11] instead of the more commonly used Gaussian slab. Our primary focus is in consistent estimation of $\boldsymbol{\beta}$ and a Spike-and-Laplace suffices in achieving this goal.

The prior on the parameter $\boldsymbol{\beta}$ is induced through a prior on the duo $(S, \boldsymbol{\beta})$, where S denotes a subset of $\{1, \ldots p_n\}$. First, the prior on the dimension $0 \leq s \leq p_n$ is chosen to be $\omega_n(s)=C_n p_n^{-a_n s}, \ s=0, \ldots, p_n$ with hyper-parameter $a_n>0$, where C_n is chosen to normalize the distribution. For any $\boldsymbol{\beta}$ and S mentioned above, recall that $\boldsymbol{\beta}_S$ denotes the same vector $\boldsymbol{\beta}$, but the co-ordinates in S^c are set to 0. With hyper-parameter $\lambda_n>0$, the full prior is taken to be of the form

$$\begin{aligned}
\Pi_n(S, \boldsymbol{\beta}) &:= \omega_n(|S|) . \binom{p_n}{|S|}^{-1} . \left(\frac{\lambda_n}{2}\right)^{|S|} . \exp(-\lambda_n\|\boldsymbol{\beta}_S\|_1) . \ \delta_0\left(\boldsymbol{\beta}_{S^c}\right) \\
&= C_n . \binom{p_n}{|S|}^{-1} . \left(\frac{\lambda_n}{2 p_n^{a_n}}\right)^{|S|} . \exp(-\lambda_n\|\boldsymbol{\beta}_S\|_1) . \ \delta_0\left(\boldsymbol{\beta}_{S^c}\right),
\end{aligned} \tag{2.44}$$

where $\|.\|_1$ denotes ℓ_1-norm of Euclidean vectors, $|S|$ denotes cardinality of the set S and δ_0 denotes the degenerate distribution. The prior on the main parameter of interest, $\boldsymbol{\beta}$, is given by

$$\Pi_n(\boldsymbol{\beta}) := \sum_{S \subset\{1, \ldots n\}} \Pi_n(S, \boldsymbol{\beta}),$$

and the posterior probability of a general $B \subset \mathbb{R}^{p_n}$ is

$$\Pi_n(B \mid \boldsymbol{y}) := \frac{\int_B \exp[L_n(\boldsymbol{\eta}, \boldsymbol{\eta}^*)] \Pi_n(\boldsymbol{\beta}) d\boldsymbol{\beta}}{\int \exp[L_n(\boldsymbol{\eta}, \boldsymbol{\eta}^*)] \Pi_n(\boldsymbol{\beta}) d\boldsymbol{\beta}}.$$

Our choice for model weights $\omega_n(\cdot)$ is known as a complexity prior in [11] (also see the example in §2.3). We thus induce sparsity in the posterior through our prior choice. We point out that other ways of specifying and generalizing the prior [10, 11] have the same effect on the posterior as our prior. We place independent Laplace signals for the non-zero coordinates. One can find dependent priors such as those discussed in §2.2.3 or in §2.3.5.

2.4.4 Assumptions on Data Generating Distribution and Prior

Our assumptions on the likelihood stem from that on the KL divergence term, while assumptions about the prior come from assumptions on the hyper-parameters λ_n and a_n. These assumptions also dictate the possible values of true $\boldsymbol{\beta}^*$, uniformly over which we shall state our results. We first present identifiability and compatibility (IC) conditions, and connect them to uniformly adaptive statements about the posterior. Next, we address the choice of hyper-parameters that avoid any dependence of the prior on the true $\boldsymbol{\beta}^*$. We start by describing some order conditions, which shall help us define the rest of the assumptions.

(A1) Order assumptions on sample size and parameter dimension. Since we work with a high-dimensional problem, a natural condition is $p_n > n$ where $n, p_n \to \infty$. Now define a deterministic sequence of positive reals $\{b_n\}$, such that

$$b_n = o\left(\frac{n}{\log p_n}\right). \tag{2.45}$$

We shall focus on those true β^* vectors that satisfy $1 \leq s^* \leq b_n$. This gives us, among other things, the important relations: $s^* \log p_n \to \infty$ and $(s^* \log p_n)/n \to 0$ as $n \to \infty$. It is also important that $s^* \nrightarrow 0$, which forces us to have $\log p_n = o(n)$. This shows that $b_n = O(1)$ is a valid choice, satisfying (2.45). We work with n large enough so that $b_n \log p_n < n$ for all our calculations. Also, note that $p_n > n$ implies $3b_n < p_n$ for large enough n.

(A2) Identifiability and compatibility (IC) assumptions. The ability of the log-likelihood term $L_n(\boldsymbol{\eta}, \boldsymbol{\eta}^*)$ to create a separation between the true value of $\boldsymbol{\beta}^*$ from any other $\boldsymbol{\beta}$ is a fundamental criterion in posterior contraction analysis, and is termed as the identifiability criterion. Again, the natural measure of discrepancy in cGLM model is the Kullback–Leibler divergence $\mathcal{D}_n(\boldsymbol{\eta}^* \| \boldsymbol{\eta})$, and since we work with Laplace signals in our prior, it is a natural demand to connect $\mathcal{D}_n(\boldsymbol{\eta}^* \| \boldsymbol{\eta})$ with the ℓ_1 distance, making them compatible. The requirements of compatibility and identifiability are simultaneously met by enforcing a lower bound on the KL divergence in terms of ℓ_1 distance between the β's i.e., $\|\boldsymbol{\beta}_2 - \boldsymbol{\beta}_1\|_1$ for $\boldsymbol{\beta}_1, \boldsymbol{\beta}_2 \in \mathbb{R}^{p_n}$. We express this through the IC (Model) and IC (Dimension) assumptions, essentially requiring existence of a model S and a dimension s, where $S \subset \{1, \ldots p_n\}$ and $s = 3b_n, \ldots p_n$ and they satisfy a certain lower bound property through the KL term. These assumptions not only generalize the compatibility assumptions made in [11] (also in §2.3), but also link them to identifiability of the truth.

IC (Model) Assumption: There exists at least one non-null model $S \subset \{1, \ldots p_n\}$ and the corresponding quantity $\phi_1(A, \boldsymbol{X}, S) > 0$, such that for any $\boldsymbol{\beta}_1, \boldsymbol{\beta}_2 \in \mathbb{R}^{p_n}$, we have

$$\begin{aligned} \boldsymbol{\beta}_{1S} &\neq \boldsymbol{\beta}_{2S}, \|\boldsymbol{\beta}_{2S^c} - \boldsymbol{\beta}_{1S^c}\|_1 < 7\|\boldsymbol{\beta}_{2S} - \boldsymbol{\beta}_{1S}\|_1 \\ \Rightarrow \quad \mathcal{D}_n(\boldsymbol{\eta}_1 \| \boldsymbol{\eta}_2) &\geq \frac{n\phi_1^2(A, \boldsymbol{X}, S)}{|S|} \|\boldsymbol{\beta}_{2S} - \boldsymbol{\beta}_{1S}\|_1^2. \end{aligned}$$

The quantity $\phi_1(A, \boldsymbol{X}, S)$ denotes the compatibility number of model S. For an original definition one may refer to (2.23) in §2.3. The subscript 1 of ϕ_1 emphasizes we are working with constraints in the ℓ_1 distance, as seen above. Intuitively, a general $\boldsymbol{\beta}$, that is close to the truth $\boldsymbol{\beta}^*$ in ℓ_1 norm, will tend to have smaller absolute values in the true noise co-ordinates

S^{*c}, and hence such a $\boldsymbol{\beta}$ will tend to satisfy $\|\boldsymbol{\beta}_{S^{*c}} - \boldsymbol{\beta}^*_{S^{*c}}\|_1 < 7\|\boldsymbol{\beta}_{S^*} - \boldsymbol{\beta}^*_{S^*}\|_1$ or equivalently $\|\boldsymbol{\beta}_{S^{*c}}\|_1 < 7\|\boldsymbol{\beta}_{S^*} - \boldsymbol{\beta}^*\|_1$. It is precisely in this scenario that we shall need the IC (Model) assumption, i.e., $\phi_1(A, X, S^*) > 0$ so that the KL term creates a separation of the true and non-true $\boldsymbol{\beta}$'s that are close in ℓ_1 distance. The IC (Model) assumption will be crucially used in our proof of Theorem 2.4.1.

Now consider the following subset of the parameter space:

$$\mathcal{B}_{1,n} := \{\boldsymbol{\beta} \in \mathbb{R}^{p_n} : \phi_1 (A, \boldsymbol{X}, \mathrm{supp}\,(\boldsymbol{\beta})) > 0\}.$$

Based on the previous discussion, we would need the true $\boldsymbol{\beta}^* \in \mathcal{B}_{1,n}$, and due to IC (Model) assumption, $\mathcal{B}_{1,n}$ is non-null. Also, given any A and $\boldsymbol{X}$, the quantity $\phi_1(A, \boldsymbol{X}, S)$ can only take finitely many values as S varies over subsets of $\{1, \ldots p_n\}$, all of those values being positive for $S = S^*$. This gives us the quantity, for any non-null $\mathcal{B} \subset \mathcal{B}_{1,n}$,

$$\phi_{\mathcal{B}}(A, \boldsymbol{X}) := \inf \left\{ \phi_1(A, \boldsymbol{X}, S^*) : \beta^* \in \mathcal{B} \right\} > 0. \tag{2.46}$$

This quantity, with a special choice of $\mathcal{B}$ as laid out in the ensuing discussion, plays an important role in Theorem 2.4.2. We now turn our attention to the IC (Dimension) assumption.

IC (Dimension) Assumption: There exists at least one $s \in \{3b_n, \ldots p_n\}$, and a corresponding quantity $\phi_0(A, \boldsymbol{X}, s) > 0$, such that for any $\boldsymbol{\beta}_1, \boldsymbol{\beta}_2 \in \mathbb{R}^{p_n}$, we have

$$\begin{aligned} \boldsymbol{\beta}_1 &\neq \boldsymbol{\beta}_2, |\,\mathrm{supp}\,(\boldsymbol{\beta}_2 - \boldsymbol{\beta}_1)| \leq s \quad \text{implies} \\ \mathcal{D}_n(\boldsymbol{\eta}_1 \| \boldsymbol{\eta}_2) &\geq \frac{n\phi_0^2(A, \boldsymbol{X}, s)}{|\,\mathrm{supp}\,(\boldsymbol{\beta}_2 - \boldsymbol{\beta}_1)|} \cdot \|\boldsymbol{\beta}_2 - \boldsymbol{\beta}_1\|_1^2. \end{aligned}$$

The subscript 0 of ϕ_0 emphasizes we are working with constraints in the ℓ_0 distance. Similar to IC (Model), the intuition behind IC (Dimension) is to guarantee that whenever a general $\boldsymbol{\beta}$ matches on most of the co-ordinates with true $\boldsymbol{\beta}^*$, i.e., their ℓ_0 distance is small, the KL term should be able to separate them.

The IC (Dimension) assumption, coupled with the IC (Model) assumption, form one of the central conditions in the proof of our posterior contraction statement, and we shall call it the IC (Joint) condition. First, consider the set

$$\mathcal{B}_{0,n} := \{\overline{\phi}_0\,(A, \boldsymbol{X}, 3.|\,\mathrm{supp}\,(\boldsymbol{\beta})|) > 0\},$$

where, for any $s \in \{1, \ldots p_n\}$,

$$\overline{\phi}_0(A, \boldsymbol{X}, s) := \inf \left\{ \frac{\sqrt{s\mathcal{D}_n(\boldsymbol{\eta}^* \| \boldsymbol{\eta})}}{\|\boldsymbol{\beta} - \boldsymbol{\beta}^*\|_1} : |\,\mathrm{supp}\,(\boldsymbol{\beta} - \boldsymbol{\beta}^*)| \leq s,\ \boldsymbol{\beta} \neq \boldsymbol{\beta}^* \right\}. \tag{2.47}$$

Now observe that $\overline{\phi}_0(A, \boldsymbol{X}, s)$ is decreasing in s, by definition, for any fixed A and $\boldsymbol{X}$. Now, due to by IC (Dimension), we have $\overline{\phi}_0(A, \boldsymbol{X}, 3b_n) > 0$, which shows $\overline{\phi}_0(A, \boldsymbol{X}, 3.|\,\mathrm{supp}\,(\beta)|) > 0$ whenever $|\,\mathrm{supp}\,(\boldsymbol{\beta})| \leq b_n$. We thus have

$$\mathcal{B}_{0,n} \supset \{\boldsymbol{\beta} \in \mathbb{R}^{p_n} : 0 < |\,\mathrm{supp}\,(\boldsymbol{\beta})| \leq b_n\} =: \mathcal{B}_{2,n}, \tag{2.48}$$

which is a desirable relation based on the discussion at the start of this section. We are now ready to state

IC (Joint) Assumption:

$$\mathcal{B}_n := \mathcal{B}_{1,n} \cap \mathcal{B}_{2,n} \text{ is non-empty.}$$

A direct and vital consequence of this assumption is $\phi_{\mathcal{B}_n}(A, X) > 0$, as seen from (2.46) by choosing $\mathcal{B} = \mathcal{B}_n$. As we shall see, the statements of our results in Theorem 2.4.1 and Theorem 2.4.2 are uniformly adaptive over $\boldsymbol{\beta}^* \in \mathcal{B}_n$, i.e.,

$$\sup_{\boldsymbol{\beta}^* \in \mathcal{B}_n} \mathbb{P}\left(\|\boldsymbol{\beta} - \boldsymbol{\beta}^*\|_1 > \varepsilon_{n,1} \mid \boldsymbol{y}\right) \to 0 \quad \text{as} \quad n \to \infty.$$

We now discuss the pivotal role of clipping function $\eta(\cdot)$ in the IC assumptions. We require the geometries of the likelihood and the prior to match up in terms of the parameter $\boldsymbol{\beta}$; $\mathcal{D}_n(\boldsymbol{\eta}_1 \| \boldsymbol{\eta}_2)$ captures the discrepancy among $\boldsymbol{\beta}$'s in the likelihood, while the ℓ_1 gap does the same for the Laplace signals in the prior. To have a posterior contraction statement in ℓ_1 distance, it is necessary for the $\mathcal{D}_n(\boldsymbol{\eta}_1 \| \boldsymbol{\eta}_2)$ to grow with $\|\boldsymbol{\beta}_2 - \boldsymbol{\beta}_1\|_1$, at least in sparsity restricted sense, and that is what the IC (Model) and IC (Dimension) assumptions reflect. Clipping function $\eta(\cdot)$, being an intermediary of $A(\cdot)$ and linear term $\boldsymbol{x}_i^{\mathrm{T}}\boldsymbol{\beta}$, must also reflect this growth, and hence has to be necessarily injective. The Lipschitz nature of $\eta(\cdot)$ allows us to translate gaps between $\boldsymbol{\eta}$'s to gaps between $\boldsymbol{\beta}$'s.

(A3) Hyper-parameter selection. Since we aim to avoid prior dependence on the truth, choosing the hyper-parameter λ_n, a_n should only take into account the sample size n, parameter dimension p_n, covariate matrix $\boldsymbol{X}$ and log-partition function $A(\cdot)$. Our assumptions must allow us to forgo use of any prior knowledge of the truth $\boldsymbol{\beta}^*$ while hyper-parameter selection. Choice of λ_n is significantly inter-twined with the log-partition function $A(\cdot)$ as well as the clipping function $\eta(\cdot)$. As in [11], λ_n needs to scale with some function of the design matrix $\boldsymbol{X}$, and since the covariate information from $\boldsymbol{X}$ is fed into the log-partition function through $\eta(\cdot)$, the choice of λ_n depends on $A(\cdot), \eta(\cdot)$ and $\boldsymbol{X}$. Based on this, consider the bound

$$\sup_{\boldsymbol{\beta}^* \in \mathcal{B}_{2,n}} \max_{1 \le i \le n} \sup \left\{ A''(\gamma) : |\gamma - \eta_i^*| \le \sqrt{\frac{s^* \log p_n}{n}} \right\} \le \mathcal{M}_0^2(A),$$

which essentially gives us local control over $A''(\eta_i)\ \forall\, i = 1, \ldots n$, uniformly over $\boldsymbol{\beta}^* \in \mathbb{R}^{p_n}$. The proof of this statement basically uses two main points. Firstly, since $s^* \le b_n$ for $\boldsymbol{\beta}^* \in \mathcal{B}_n$, we have $(s^* \log p_n)/n \to 0$ by (2.45), which allows us to have shrinking neighborhoods around every η_i^*. Secondly, based on the behavior of $A''(\cdot)$, the clipping function $\eta(\cdot)$ restricts the set of arguments passed to $A(\cdot)$, thus controlling the growth of $A''(\cdot)$.

Now, define the quantities

$$\begin{aligned} &\mathcal{M}_1(A) := \left(1 \wedge \mathcal{M}_0^{-1}(A)\right)^{-1}, \|\boldsymbol{X}\|_{(\infty,\infty)} := \max\{\boldsymbol{X}_{i,j} : i = 1, \ldots n, j = 1 \ldots . p_n\}, \\ &\mathcal{M}(A, \boldsymbol{X}) := \|\boldsymbol{X}\|_{(\infty,\infty)} \mathcal{M}_1(A). \end{aligned} \tag{2.49}$$

We can now state our assumption on the hyper-parameter λ_n:

Assumption $\mathcal{L}_0$:

$$\frac{\mathcal{M}(A, \boldsymbol{X})}{p_n} \le \lambda_n \le \mathcal{M}(A, \boldsymbol{X})\sqrt{\log p_n}.$$

This bound, which we utilize in all our Theorems, generalizes the hyper-parameter bounds mentioned in [11] (also see §2.3), as well as avoids prior dependence on the truth. Existence of $\mathcal{M}_0(A) > 0$, through which $\mathcal{M}(A, \boldsymbol{X})$ is defined in (2.49), is guaranteed by (2.42), and it acts as a pre-fixed constant quantity that the practitioner can choose based solely on $A(\cdot)$, and then choose clipping function $\eta(\cdot)$. This, in turn, shows that the choice of hyper-parameter λ_n depends solely on the three quantities $(A, \boldsymbol{X}, b_n)$. This makes our hyper-parameter choice of λ_n free of the truth.

We turn our attention to hyper-parameter a_n, which controls how fast the model weights $\omega_n(\cdot)$ decay. First, define

$$\mathcal{E}_1 := 8\left(1 + \frac{49\mathcal{M}^2(A, \boldsymbol{X})}{8\phi^2_{\mathcal{B}_n}(A, \boldsymbol{X})}\right), \tag{2.50}$$

which is an adaptive choice, as well as free of any knowledge of the truth, owing to (2.46) and IC (Joint) assumption. For mild demands, like in Theorem 2.4.1, $a_n > 1$ suffices. On the contrary, for the weak model selection result, we need to choose a_n that supports very strong down-weighting of larger models, namely $a_n \geq 1 + 2b_n\mathcal{E}_1$. One can note from (2.45) why this choice of a_n heavily penalizes larger models. Lastly, the choice $a_n \geq 1 + \mathcal{E}_1$, which is much milder than our previous choice, is sufficient for the posterior contraction result in Theorem 2.4.2. It is crucial to note that just like λ_n, our choice of hyper-parameter a_n avoids any knowledge of true $\boldsymbol{\beta}^*$.

2.4.5 Adaptive Rate-Optimal Posterior Contraction Rate in ℓ_1-norm

The contraction result has three main ingredients i) obtaining lower bound of the marginal likelihood ii) ensuring that the posterior does not spread its mass too far away from the true sparsity s^* and iii) exploit IC (Model) and IC (Dimension) to obtain separation of the likelihood from the truth. We elaborate on the later two points here. It is expected that the posterior would reflect this prior property, which amounts to the posterior having vanishingly low probability of exceeding a certain dimension. Theorem 2.4.1 does exactly that, showing that the posterior should be at least as sparse as the true $\boldsymbol{\beta}^*$, up to multiplicative constants. Sparsity is quantified using $|\operatorname{supp}(\boldsymbol{\beta})|$ and is compared with s^*, the true level of sparsity in $\boldsymbol{\beta}^*$.

Theorem 2.4.1. (Posterior dimension and weak model selection.) *Let $a_n > 1$ and λ_n satisfy assumption $\mathcal{L}_0$. Let $n, p_n \to \infty$ and $p_n > n$. Based on (2.45), consider large enough n so that $b_n \log p_n < n$. Let assumptions IC (Model) and IC (Joint) hold, and consider the non-null set $\mathcal{B}_n$. Then, with quantity $\phi_1(A, \boldsymbol{X}, S)$ given by IC (Model), and $\mathcal{M}(A, \boldsymbol{X})$ as in (2.49), we have for all sufficiently large n,*

$$\sup_{\boldsymbol{\beta}^* \in \mathcal{B}_n} \mathbb{E}\left[\Pi_n\left(|\operatorname{supp}(\boldsymbol{\beta})| > s^*\left[1 + \frac{8}{a_n - 1}\left(1 + \frac{49\mathcal{M}^2(A, \boldsymbol{X})}{8\phi_1^2(A, \boldsymbol{X}, S^*)}\right)\right]\middle| \boldsymbol{y}\right)\right]$$
$$\to 0 \quad \text{as} \quad n \to \infty.$$

The statement of the theorem is presented in an asymptotic fashion, but is true for every n large enough, satisfying the order assumptions. For simplicity, let us define the quantity $\mathcal{E}_1^* := 8\left(1 + 49\mathcal{M}^2(A, \boldsymbol{X})/8\phi_1^2(A, \boldsymbol{X}, S^*)\right)$ so that Theorem 2.4.1 is a statement about the posterior probability of the set $\{|\operatorname{supp}(\boldsymbol{\beta})| > s^*(1 + \mathcal{E}_1^*/(a_n - 1))\}$. It is important to note that we have used $\boldsymbol{\beta}^* \in \mathcal{B}_n$ implies $\phi_1(A, \boldsymbol{X}, S^*) > 0$. Owing to IC (Joint), (2.46) and the choice $\mathcal{B} = \mathcal{B}_n$, we can have from Theorem 2.4.1,

$$\sup_{\boldsymbol{\beta}^* \in \mathcal{B}_n} \mathbb{E}\left[\Pi_n\left(|\operatorname{supp}(\boldsymbol{\beta})| > s^*\left[1 + \frac{8}{a_n - 1}\left(1 + \frac{49\mathcal{M}^2(A, \boldsymbol{X})}{8\phi_{\mathcal{B}_n}^2(A, X)}\right)\right]\middle| \boldsymbol{y}\right)\right]$$
$$\to 0 \quad \text{as} \quad n \to \infty.$$

By the definition of $\mathcal{E}_1$ in (2.50) and its analogy with $\mathcal{E}_1^*$, we now work with the posterior probability of $\{|\operatorname{supp}(\boldsymbol{\beta})| > s^*(1 + \mathcal{E}_1/(a_n - 1))\}$. This allows to us to choose the hyper-parameter a_n as $a_n \geq 1 + 2b_n\mathcal{E}_1$, which is a truth-free choice.

With $\mathcal{E}_1$ as in (2.50), if hyper-parameter a_n in the prior satisfies $a_n \geq 1 + 2b_n\mathcal{E}_1$ in addition to the hypotheses of Theorem 2.4.1, we have

$$\sup_{\boldsymbol{\beta}^* \in \mathcal{B}_n} \mathbb{E}\left[\Pi_n\left(\operatorname{supp}(\boldsymbol{\beta}) \supsetneqq S^* \middle| \boldsymbol{y}\right)\right] \to 0 \quad \text{as} \quad n \to \infty. \tag{2.51}$$

(2.51) is a straightforward consequence of Theorem 2.4.1, the fact that $s^* \leq b_n$ for $\boldsymbol{\beta}^* \in \mathcal{B}_n$, and the observation that $\{\operatorname{supp}(\boldsymbol{\beta}) \supsetneqq S^*\} \subset \{|\operatorname{supp}(\boldsymbol{\beta})| > s^* + 1/2\}$.

We now turn our attention to the central result of this chapter, which is a truth adaptive statement about ℓ_1-contraction of the posterior distribution. Essentially, it gives the radius of the smallest possible ℓ_1 ball around true $\boldsymbol{\beta}^*$, whose posterior probability vanishes with large n. Define the quantity

$$\mathcal{E}_2 := 6 + \frac{12\mathcal{M}^2(A, \boldsymbol{X})}{\overline{\phi}_0^{\,2}(A, \boldsymbol{X}, 3b_n)}, \tag{2.52}$$

which can be observed to be truth-free. By describing the aforementioned radius in terms of $a_n, \mathcal{E}_2, p_n$ and n, we have the following.

Theorem 2.4.2. (Adaptive posterior contraction in ℓ_1 metric.) *Let hyper-parameter a_n satisfy $a_n \geq 1 + \mathcal{E}_1$ for $\mathcal{E}_1$ as in (2.52), and hyper-parameter λ_n satisfy assumption $\mathcal{L}_0$. Let $n, p_n \to \infty$ and $p_n > n$. Based on (2.45), consider large enough n so that $b_n \log p_n < n$. Let assumptions IC (Model), IC (Dimension) and IC (Joint) hold, and consider the non-null set $\mathcal{B}_n$. Then, with quantity $\mathcal{E}_2$ given by (2.52) and $\mathcal{M}(A, \boldsymbol{X})$ as in (2.49), we have for all sufficiently large n,*

$$\sup_{\boldsymbol{\beta}^* \in \mathcal{B}_n} \mathbb{E}\left[\Pi_n\left(\|\boldsymbol{\beta} - \boldsymbol{\beta}^*\|_1 > \frac{2s^*(1 + a_n + \mathcal{E}_2)}{\mathcal{M}(A, \boldsymbol{X})}\sqrt{\frac{\log p_n}{n}}\middle| \boldsymbol{y}\right)\right] \to 0. \tag{2.53}$$

It is important to note that the contraction rate linearly increases with a_n and as long as a_n is chosen to be a constant larger than $1 + \mathcal{E}_1$, the rate is unaffected. However, if one chooses a stronger penalty on the model space to achieve weak model selection consistency, the rate of contraction in ℓ_1 norm becomes slower unless the upper bound b_n on the number of true non-zero coefficients is assumed to be a constant.

2.5 Optimality Results for Variational Inference in Linear Regression Models

Variational inference [33, 50] is a widely-used tool for approximating complicated probability densities, especially those arising as posterior distributions from complex hierarchical Bayesian models such as those employing spike-and-slab priors. Variational inference turns the sampling/inference problem into an optimization problem, where a closest member, relative to the Kullback–Leibler divergence, in a family of approximate densities is picked out as a proxy to the target density. It has been empirically observed in many applications that variational inference operates orders of magnitude faster than MCMC for achieving the same approximation accuracy. Moreover, compared to MCMC, variational inference tends to be easier to scale to big data due to its inherent optimization nature, and can take advantage of modern optimization techniques such as stochastic optimization [34, 35] and distributed optimization [1]. While MCMC is known to produce (almost) exact samples from the target density for ergodic chains [43], statistical guarantees of variational inference have begun to be known only recently [40, 53, 55]. In the following, we start with a brief review of variational inference and then specialize to the context of linear models with discrete spike-and-slab priors.

Let P_θ denote a prior distribution on $\theta \in \Theta$ with density function p_θ. In a Bayesian framework, all inference is based on the augmented posterior density $p(\theta \,|\, \boldsymbol{y})$ given by

$$p(\theta \,|\, \boldsymbol{y}) \propto f(\boldsymbol{y} \,|\, \theta)\, p_\theta(\theta). \tag{2.54}$$

Let Γ denote a pre-specified family of density functions over Θ that can be either parameterized by some "variational parameters", or required to satisfy some structural constraints. The goal of variational inference is to approximate this conditional density $p(\theta \,|\, \boldsymbol{y})$ by finding the closest member of this family in KL divergence to the conditional density $p(\theta \,|\, \boldsymbol{y})$ of interest, that is, computing the minimizer

$$\begin{aligned} \widehat{q}_\theta &:= \arg\min\nolimits_{q_\theta \in \Gamma} D\big[\, q_\theta(\cdot) \,\big|\big|\, p(\cdot \,|\, \boldsymbol{y}) \big] \\ &= \arg\min\nolimits_{q_\theta \in \Gamma} \Big\{ - \underbrace{\int_\Theta q_\theta(\theta) \, \log \frac{f(\boldsymbol{y} \,|\, \theta)\, p_\theta(\theta)}{q_\theta(\theta)} \, d\theta}_{L(q_\theta)} \Big\} \end{aligned} \tag{2.55}$$

where the last step follows by using Bayes' rule and the fact that the marginal density $f(\boldsymbol{y})$ does not depend on θ and q_θ. The function $L(q_\theta)$ inside the argmin-operator above (without the negative sign) is called the evidence lower bound (ELBO, [6]) since it provides a lower bound to the log evidence $\log f(\boldsymbol{y})$,

$$\log f(\boldsymbol{y}) = L(q_\theta) + D\big[\, q_\theta(\cdot) \,\big|\big|\, p(\cdot \,|\, \boldsymbol{y}) \big] \geq L(q_\theta), \tag{2.56}$$

where the equality holds if and only if $q_\theta = p(\cdot \,|\, \boldsymbol{y})$. The decomposition (2.56) provides an alternative interpretation of variational inference to the original derivation from Jensen's inequality [33]; minimizing the KL divergence over the variational family Γ is equivalent to maximizing the ELBO over Γ.

For technical simplicity, we focus on the fractional posterior framework [3], where we conduct inference using the fractional posterior distribution

$$P_\alpha(\theta \in B \,|\, \boldsymbol{y}) = \frac{\int_B \big[f(\boldsymbol{y} \,|\, \theta)\big]^\alpha \, p_\theta(\theta)\, d\theta}{\int_\Theta \big[f(\boldsymbol{y} \,|\, \theta)\big]^\alpha \, p_\theta(\theta)\, d\theta},$$

for any measurable subset of Θ, which is obtained by combining the α-fractional likelihood function $\big[f(\boldsymbol{y}\,|\,\theta)\big]^{\alpha}$ with the prior p_θ using Bayes' rule.

Under the fractional posterior framework, we investigate statistical properties of the following variational approximation to the α-fractional posterior distribution,

$$\begin{aligned}\widehat{q}_\theta &:= \arg\min_{q_\theta\in\Gamma} D\big[q_\theta \,||\, p_\alpha(\cdot\,|\,\boldsymbol{y})\big]\\ &= \arg\min_{q_\theta\in\Gamma}\Big\{-\int_\Theta q_\theta(\theta)\,\log\frac{\big[f(\boldsymbol{y}\,|\,\theta)\big]^{\alpha}\,p_\theta(\theta)}{q_\theta(\theta)}\,d\theta\Big\}\\ &= \arg\min_{q_\theta\in\Gamma}\Big\{-\alpha\int_\Theta q_\theta(\theta)\,\log\frac{f(\boldsymbol{y}\,|\,\theta)}{f(\boldsymbol{y}\,|\,\theta^*)}\,d\theta + D(q_\theta\,||\,p_\theta)\Big\},\end{aligned} \tag{2.57}$$

where recall that Γ is the variational family of distributions. Here, we adopt the frequentist perspective by assuming that there is a true data generating model $P_{\theta^*}^{(n)}$ that generates the data $\boldsymbol{y}$, and θ^* will be referred to as the true parameter, or simply truth. We added $f(\boldsymbol{y}\,|\,\theta^*)$ in the denominator in (2.57) to illustrate how the variational approximation $\widehat{q}_\theta$ works–on the one hand it tries to maximize the likelihood function so that the first term in the last line of (2.57) becomes small; on the other hand the regularization term $\mathcal{D}(q_\theta\,||\,p_\theta)$ prevents it from over-fitting the data.

We continue to consider the Bayesian linear model (2.16) in the high-dimensional regime where $p_n \gg n$. Let $s^* \ll n$ denote the sparsity level, i.e., the number of non-zero coefficients, of the true regression parameter $\boldsymbol{\beta}^*$. Following [6], we introduce a latent indicator variable $\gamma_j = \mathbb{I}\{\beta_j \neq 0\}$ for each β_j to indicate whether the jth covariate X_j is included in the model, and call $\boldsymbol{\gamma} = (\gamma_1,\ldots,\gamma_{p_n}) \in \{0,1\}^{p_n}$ the latent indicator vector. We use the notation $\boldsymbol{\beta}_\gamma$ to denote the vector of non-zero components of $\boldsymbol{\beta}$ selected by $\boldsymbol{\gamma}$, that is $\boldsymbol{\beta}_\gamma = (\beta_j : \gamma_j = 1)$. Consider the following sparsity inducing hierarchical prior $p_{\boldsymbol{\beta},\gamma}$ over $(\boldsymbol{\beta},\,\boldsymbol{\gamma})$:

$$\begin{aligned}&\gamma_j \overset{iid}{\sim} \frac{1}{p_n}\,\delta_1 + \Big(1-\frac{1}{p_n}\Big)\,\delta_0, \quad j = 1,\ldots,p_n,\\ &\boldsymbol{\beta}_\gamma\,|\,\boldsymbol{\gamma} \sim p_{\beta\,|\,\gamma}, \quad \text{and} \quad \sigma \sim p_\sigma,\end{aligned} \tag{2.58}$$

where $p_{\beta\,|\,1} = \mathcal{N}(\beta; 0, \tau^2)$ with the hyper-parameter τ^2 and $p_{\beta\,|\,0} = \delta_0(\beta)$. The prior probability of $\{\gamma_j = 1\}$ is chosen as ${p_n}^{-1}$ so that on an average only $O(1)$ covariates are included in the model. Let $\boldsymbol{\gamma}^*$ denote the indicator vector associated with the truth $\boldsymbol{\beta}^*$.

By viewing the latent variable indicator vector $\boldsymbol{\gamma}$ as a parameter, we apply the block mean-field approximation [8] by using the family

$$q(\boldsymbol{\beta},\,\sigma,\,\gamma) = q_\sigma(\sigma)\prod_{j=1}^{p_n} q_{\gamma_j,\beta_j}(\gamma_j,\beta_j) \tag{2.59}$$

to approximate the joint α-fractional posterior distribution of $\theta = (\boldsymbol{\beta},\,\sigma,\,\gamma)$ with $\widehat{q}_\theta(\theta) = \widehat{q}_\sigma(\sigma)\prod_{j=1}^{p_n}\widehat{q}_{\gamma_j,\beta_j}(\gamma_j,\beta_j)$. Although we have a high-dimensional latent variable vector γ, the latent variable is associated with the parameter $\boldsymbol{\beta}$, and not with the observation $\boldsymbol{y}$. It turns out that the spike-and-slab prior with Gaussian slab is particularly convenient for computation—it is "conjugate" in that the resulting variational approximation falls into the same spike-and-slab family [8]. The following result appears as Corollary 4.1 in [53].

Theorem 2.5.1. *Suppose $p_{\boldsymbol{\beta}\,|\,\gamma^*}$ is continuous and thick at $\boldsymbol{\beta}^*_{\gamma^*}$, and p_σ is continuous and thick at σ^*. If $s\log p_n/n \to 0$ as $n\to\infty$, then it holds with probability tending to one as $n\to\infty$ that*

$$\Big\{\int h^2\big[p(\cdot\,|\,\theta)\,||\,p(\cdot\,|\,\theta^*)\big]\,\widehat{q}_\theta(\theta)\,d\theta\Big\}^{1/2} \lesssim \sqrt{\frac{s}{n\,\min\{\alpha,1-\alpha\}}\log(p_n\,n)},$$

where $\widehat{q}_\theta(\theta)$ is defined in (2.57).

Theorem 2.5.1 implies a convergence rate $\sqrt{n^{-1}\, s \log(p_n n)}$ of the variational-Bayes estimator $\widehat{\beta}_{\mathrm{VB},\alpha}$ under the restricted eigenvalue condition [4], which is the minimax rate up to log terms for high-dimensional sparse linear regression. It is important to note that Theorem 2.5.1 holds under very mild conditions on the prior and does not rely on having closed-form updates of any particular algorithm.

Very recently, [41] showed that under compatibility conditions on the design matrix described in §2.3 and using spike and Laplace prior with a complexity penalty on the model space, the variational estimate of the coefficients converges to the sparse truth at the optimal rate and gives optimal prediction of the response vector. Considering the hierarchical prior with

$$g_S = \prod_{i \in S} \mathrm{Laplace}(\theta_i; \lambda).$$

and a stronger mean-field approximation than the one considered in (2.59)

$$q(\boldsymbol{\beta}, \gamma) = \prod_{j=1}^{p_n} [\gamma_i \mathcal{N}(\mu_i, \tau_i^2) + (1 - \gamma_i)\, \delta_0].$$

[41] showed that an analogous result to Theorem 2.3.2 holds with the posterior replaced by the variational optimizer and with the same assumptions on the data generation mechanism.

2.6 Discussion

This chapter revisits some of the recent theoretical developments with discrete spike-and-slab priors in Gaussian sequence model, linear regression and generalized linear models. A particular emphasis is given to studying posterior contraction rates, support recovery (variable selection in regression) and Bernstein von-Mises results. A recent interesting theoretical direction is to understand how posterior distributions for spike-and-slab priors work for inference in terms of confidence sets [9]. In particular, it is of importance to understand whether the Bayesian credible sets are indeed proper frequentist confidence sets and have an optimal diameter.

The chapter also reviews some optimality results of variational inference. A general extension of variational inference to generalized linear models is lacking in the literature and the techniques are available on a case-by-case basis. In this regard, the idea of convex duality, which has been used in the context of logistic regression [29] with optimality results available only recently [24], is very useful. Extension to the high-dimensional case is found in a recent preprint [42]. We anticipate more research in this area in the near future.

Acknowledgments

Dr. Pati acknowledges support from NSF DMS (1854731, 1916371) and NSF CCF 1934904 (HDR-TRIPODS).

Bibliography

[1] A. Ahmed, M. Aly, J. Gonzalez, S. Narayanamurthy, and A. Smola. Scalable inference in latent variable models. In *International conference on Web search and data mining (WSDM)*, volume 51, pages 1257–1264, 2012.

[2] Y. Atchadé. On the contraction properties of some high-dimensional quasi-posterior distributions. *The Annals of Statistics*, 45(5):2248–2273, 2017.

[3] A. Bhattacharya, D. Pati, and Y. Yang. Bayesian fractional posteriors. *The Annals of Statistics*, 47(1):39–66, 2019.

[4] P. J. Bickel, Y. Ritov, and A. B. Tsybakov. Simultaneous analysis of Lasso and Dantzig selector. *The Annals of Statistics*, 37(4):1705–1732, 2009.

[5] L. Birgé and P. Massart. Gaussian model selection. *Journal of the European Mathematical Society*, 3(3):203–268, 2001.

[6] D. M. Blei, A. Kucukelbir, and J. D. McAuliffe. Variational inference: A review for statisticians. *Journal of the American Statistical Association*, 112(518):859–877, 2017.

[7] P. Bühlmann and S. Van De Geer. *Statistics for high-dimensional data: methods, theory and applications.* Springer Science & Business Media, 2011.

[8] P. Carbonetto and M. Stephens. Scalable variational inference for Bayesian variable selection in regression, and its accuracy in genetic association studies. *Bayesian Analysis*, 7(1):73–108, 2012.

[9] I. Castillo and B. Szabó. Spike and slab empirical Bayes sparse credible sets. *Bernoulli*, 26(1):127–158, 2020.

[10] I. Castillo and A. van der Vaart. Needles and straw in a haystack: Posterior concentration for possibly sparse sequences. *The Annals of Statistics*, 40(4):2069–2101, 2012.

[11] I. Castillo, J. Schmidt-Hieber, and A. Van der Vaart. Bayesian linear regression with sparse priors. *The Annals of Statistics*, 43(5):1986–2018, 2015.

[12] D. K. Dey, S. K. Ghosh, and B. K. Mallick. *Generalized linear models: A Bayesian perspective.* CRC Press, 2000.

[13] D. Donoho and I. M. Johstone. Minimax risk over ℓ_p-balls for ℓ_q-error. *Probability Theory Related Fields*, 99:277–303, 21994.

[14] D. L. Donoho. De-noising by soft-thresholding. *IEEE transactions on information theory*, 41(3):613–627, 1995.

[15] D. L. Donoho and I. Johnstone. *Minimax risk over ℓ_p-balls.* Department of Statistics, University of California, 1989.

[16] D. L. Donoho and I. M. Johnstone. Minimax estimation via wavelet shrinkage. *The Annals of Statistics*, 26(3):879–921, 1998.

[17] D. L. Donoho and J. M. Johnstone. Ideal spatial adaptation by wavelet shrinkage. *Biometrika*, 81(3):425–455, 1994.

[18] D. L. Donoho, I. M. Johnstone, J. C. Hoch, and A. S. Stern. Maximum entropy and the nearly black object. *Journal of the Royal Statistical Society: Series B (Statistical Methodology)*, 54(1):41–81, 1992. ISSN 00359246.

[19] B. Efron, T. Hastie, I. Johnstone, and R. Tibshirani. Least angle regression. *The Annals of Statistics*, 32(2):407–499, 2004.

[20] J. Fan and R. Li. Variable selection via nonconcave penalized likelihood and its Oracle properties. *Journal of the American Statistical Association*, 96(456):1348–1360, 2001.

[21] J. Friedman, T. Hastie, and R. Tibshirani. *The elements of statistical learning*, volume 1. Springer series in statistics New York, 2001.

[22] E. I. George and R. E. McCulloch. Variable selection via Gibbs sampling. *Journal of the American Statistical Association*, 88(423):881–889, 1993.

[23] E. I. George and R. E. McCulloch. Approaches for Bayesian variable selection. *Statistica sinica*, pages 339–373, 1997.

[24] I. Ghosh, A. Bhattacharya, and D. Pati. Statistical optimality and stability of tangent transform algorithms in logit models. *arXiv preprint arXiv:2010.13039*, 2020.

[25] G. K. Golubev. Reconstruction of sparse vectors in white Gaussian noise. *Problems of Information Transmission*, 38(1):65–79, 2002.

[26] B. S. Guha and D. Pati. Adaptive posterior convergence in sparse high dimensional clipped generalized linear models. *arXiv preprint arXiv:2103.08092*, 2021.

[27] T. Hastie, R. Tibshirani, and M. Wainwright. *Statistical learning with sparsity: the Lasso and generalizations*. Chapman and Hall/CRC, 2015.

[28] H. Ishwaran and J. S. Rao. Spike and slab variable selection: Frequentist and Bayesian strategies. *The Annals of Statistics*, 33(2):730–773, 2005.

[29] T. S. Jaakkola and M. I. Jordan. Bayesian parameter estimation via variational methods. *Statistics and Computing*, 10(1):25–37, 2000.

[30] S. Jeong and S. Ghosal. Posterior contraction in sparse generalized linear models. *Biometrika*, 108 (2):367–379, 2020.

[31] W. Jiang. Bayesian variable selection for high dimensional generalized linear models: convergence rates of the fitted densities. *The Annals of Statistics*, 35(4):1487–1511, 2007.

[32] I. M. Johnstone. High dimensional Bernstein-von Mises: simple examples. *Institute of Mathematical Statistics Collections*, 6:87, 2010.

[33] M. I. Jordan, Z. Ghahramani, T. S. Jaakkola, and L. K. Saul. An introduction to variational methods for graphical models. *Machine Learning*, 37(2):183–233, 1999.

[34] D. Kingma and J. Ba. Adam: A method for stochastic optimization. *arXiv preprint arXiv:1412.6980*, 2014.

[35] H. J. Kushner and G. G. Yin. Stochastic approximation algorithms and applications. 1997.

[36] K. Lounici. Sup-norm convergence rate and sign concentration property of Lasso and Dantzig estimators. *Electronic Journal of Statistics*, 2:90–102, 2008.

[37] P. McCullagh. *Generalized linear models*. Routledge, 2018.

[38] T. J. Mitchell and J. J. Beauchamp. Bayesian variable selection in linear regression. *Journal of the American Statistical Association*, 83(404):1023–1032, 1988.

[39] N. N. Narisetty and X. He. Bayesian variable selection with shrinking and diffusing priors. *The Annals of Statistics*, 42(2):789–817, 04 2014. doi: 10.1214/14-AOS1207.

[40] D. Pati, A. Bhattacharya, and Y. Yang. On statistical optimality of variational Bayes. In *International Conference on Artificial Intelligence and Statistics*, pages 1579–1588. PMLR, 2018.

[41] K. Ray and B. Szabó. Variational Bayes for high-dimensional linear regression with sparse priors. *Journal of the American Statistical Association*, pages 1–12, 2021.

[42] K. Ray, B. Szabó, and G. Clara. Spike and slab variational Bayes for high dimensional logistic regression. *arXiv preprint arXiv:2010.11665*, 2020.

[43] C. P. Robert. *Monte carlo methods*. Wiley Online Library, 2004.

[44] J. G. Scott and J. O. Berger. Bayes and empirical-Bayes multiplicity adjustment in the variable-selection problem. *The Annals of Statistics*, 38(5):2587–2619, 2010.

[45] Z. Shang and M. K. Clayton. Consistency of bayesian linear model selection with a growing number of parameters. *Journal of Statistical Planning and Inference*, 141(11): 3463–3474, 2011.

[46] R. Tibshirani. Regression shrinkage and selection via the Lasso. *Journal of the Royal Statistical Society: Series B (Statistical Methodology)*, 58(1):267–288, 1996.

[47] S. A. Van De Geer and P. Bühlmann. On the conditions used to prove Oracle results for the Lasso. *Electronic Journal of Statistics*, 3:1360–1392, 2009.

[48] A. W. van der Vaart and J. H. van Zanten. Adaptive Bayesian estimation using a Gaussian random field with inverse gamma bandwidth. *The Annals of Statistics*, 37(5B):2655–2675, 2009.

[49] M. J. Wainwright. *High-dimensional statistics: A non-asymptotic viewpoint*, volume 48. Cambridge University Press, 2019.

[50] M. J. Wainwright and M. I. Jordan. Graphical models, exponential families, and variational inference. *Foundations and Trends® in Machine Learning*, 1(1–2):1–305, 2008.

[51] R. Wei and S. Ghosal. Contraction properties of shrinkage priors in logistic regression. *Journal of Statistical Planning and Inference*, 207:215–229, 2020.

[52] Y. Yang, M. J. Wainwright, M. I. Jordan, et al. On the computational complexity of high-dimensional bayesian variable selection. *Annals of Statistics*, 44(6):2497–2532, 2016.

[53] Y. Yang, D. Pati, and A. Bhattacharya. α-variational inference with statistical guarantees. *The Annals of Statistics*, 48(2):886–905, 2020.

[54] C.-H. Zhang. Nearly unbiased variable selection under minimax concave penalty. *The Annals of Statistics*, 38(2):894–942, 2010.

[55] F. Zhang and C. Gao. Convergence rates of variational posterior distributions. *The Annals of Statistics*, 48(4):2180–2207, 2020.

[56] H. Zou and T. Hastie. Regularization and variable selection via the elastic net. *Journal of the Royal Statistical Society: Series B (Statistical Methodology)*, 67(2):301–320, 2005.

3

Theoretical and Computational Aspects of Continuous Spike-and-Slab Priors

Naveen N. Narisetty

University of Illinois at Urbana-Champaign (USA)

CONTENTS

Continuous spike-and-slab priors are commonly used for high-dimensional variable selection and shrinkage in the Bayesian framework. Gaussian and Laplace spike-and-slab priors are two of the most popular among such priors which have been widely used in a variety of settings. This chapter will provide an overview of some of the recent theoretical advances for the posteriors based on these priors in terms of both estimation accuracy and model selection consistency. Moreover, theoretical insights on the variable selection consistency using a general class of priors that include the Gaussian and Laplace ones will be provided. These theoretical results provide specific conditions on the spike-and-slab prior distributions with a general base density for achieving variable selection consistency. In particular, the requirements on the spike-and-slab prior distributions are characterized by their relative magnitudes at the origin and at the tails. Recent advances on scalable computational approaches for posterior computation will be discussed. This chapter will conclude with a discussion on some applications of continuous spike-and-slab priors beyond the linear regression models.

DOI: 10.1201/9781003089018-3

3.1 Introduction

Variable selection is a fundamental problems in statistics and is particularly important when the number of variables is very large. There has been a rich body of work on penalization approaches for variable selection following the celebrated LASSO [37]. Other penalization methods for high-dimensional variable selection include smoothly clipped absolute deviation (SCAD) [8], adaptive LASSO [42], minimum concave penalty (MCP) [41], and many variations of these methods. [9, 38] provide reviews of penalization methods for variable selection.

There have been several innovative methods for variable selection using the Bayesian approach as well. Several prior choices and their corresponding theoretical properties for model selection have been investigated in the Bayesian literature [3, 5, 6, 12, 13, 15, 17, 21, 23, 25, 29]. This chapter deals with a special class of prior distributions called continuous spike-and-slab priors commonly used for variable selection. [12] introduced Gaussian spike-and-slab priors with a computational motivation of utilizing Gibbs sampling for posterior computation. The theoretical and computational properties corresponding to such priors are extensively studied in the literature [15, 16, 23, 29] while other choices of continuous spike-and-slab priors such as the spike-and-slab LASSO priors have been recently [28] developed.

In spite of the apparent differences between Bayesian and frequentist approaches, prior distributions in the Bayesian framework have direct correspondence with penalty functions in frequentist penalization methods. This naturally calls upon the question of prior choice and how different priors influence the theoretical and computational aspects of the posterior distribution. In this chapter, we will discuss some recent developments in the literature corresponding to continuous spike-and-slab priors. In addition, we will also provide some novel results that give insights on how different choices of continuous spike-and-slab priors impact modeling and theoretical properties for high-dimensional variable selection.

3.2 Variable Selection in Linear Models

Let us consider the linear regression model

$$\mathbf{y}_{n\times 1} = \mathbf{X}_{n\times p}\boldsymbol{\beta}_{p\times 1} + \varepsilon_{n\times 1}.$$

The most natural likelihood used in the Bayesian framework for linear regression comes from specifying a Gaussian distribution on the errors. That is,

$$\mathbf{y} \mid (\mathbf{X}, \boldsymbol{\beta}) \sim \mathcal{N}(\mathbf{X}\boldsymbol{\beta}, \sigma^2\mathbf{I}).$$

When the main objective is to select the variables corresponding to the non-zero components of $\boldsymbol{\beta}$, this problem can be formulated as a selection problem by defining binary indicator variables γ_j to indicate whether β_j is non-zero. The binary vector $\boldsymbol{\gamma} = (\gamma_1, \cdots, \gamma_p)$ would then correspond to a subset of variables that identifies the non-zero components of $\boldsymbol{\beta}$. In the Bayesian framework, a prior on the binary vector $\boldsymbol{\gamma}$ followed by a prior on the regression vector $\boldsymbol{\beta}$ conditional on $\boldsymbol{\gamma}$ makes it possible to perform both the tasks of estimation of $\boldsymbol{\beta}$ using the posterior distribution of $\boldsymbol{\beta}$ given the data and the selection of relevant variables by considering the posterior distribution of the binary vector $\boldsymbol{\gamma}$ given data.

For performing the first task of estimation for $\boldsymbol{\beta}$, the posterior distribution $p(\boldsymbol{\beta} \mid \mathbf{y})$ can be used in multiple ways such as using the posterior mean, posterior mode, etc. Also, there are different ways the posterior distribution $p(\boldsymbol{\gamma} \mid \mathbf{y})$ can be used for variable selection.

One approach is to use the maximum a posteriori (MAP) model for variable selection. This approach maximizes the posterior distribution of $\boldsymbol{\gamma}$ given the data, that is, it finds the variables corresponding to

$$\arg\max_{\boldsymbol{k}} P[\boldsymbol{\gamma} = \boldsymbol{k} \mid \mathbf{y}],$$

where $\boldsymbol{k}$ is a binary vector with ones corresponding to the active covariates and zeroes corresponding to inactive covariates, and $P[\boldsymbol{\gamma} = \boldsymbol{k} \mid \mathbf{y}]$ is the posterior probability of the binary vector $\boldsymbol{k}$. While it may be computationally challenging to evaluate and optimize this posterior distribution in high dimensions, a common computational strategy is to use iterative methods [14, 40] that aim to obtain a model having posterior probability close to the MAP model. Another approach called median probability thresholding thresholds the marginal posterior probabilities at the value of 0.5. That is, the set of variables having marginal posterior probabilities larger than 0.5 $\{j : P[\gamma_j = 1 \mid \mathbf{y}] > 0.5\}$ are selected. [2] studied this approach extensively along with corresponding theoretical properties. Instead of the threshold of 0.5 in the median probability model, a data adaptive threshold could be better suited for variable selection in some circumstances, which can be chosen by using some criterion function such as the Bayesian Information Criterion (BIC). This strategy was used by [23, 24].

For achieving good performance for either tasks, it is to be expected that the choices of the priors for both $\boldsymbol{\beta}$ and $\boldsymbol{\gamma}$ are crucial. One of the most common choices for the prior on $\boldsymbol{\gamma}$ is given by placing independent Bernoulli priors on the components γ_j's. That is,

$$P[\gamma_j = 1] = 1 - P[\gamma_j = 0] = \theta,$$

where θ is a hyperparameter that is either pre-specified or a further hyperprior on it is placed. For instance, [23, 33, 40] treat θ as a hyperparameter and provide conditions on θ for achieving appropriate variable selection properties. [33] discuss a fully Bayesian approach by placing a Beta prior on θ as well as a different approach using empirical Bayes that uses the data to estimate the hyperparameter θ.

Another common prior specification is to place a joint prior on the binary vector $\boldsymbol{\gamma}$ as follows

$$p(\boldsymbol{\gamma}) \propto \binom{p}{p_{\gamma}}^{-1} g(p_{\gamma}),$$

where p_γ is the number of non-zero components of $\boldsymbol{\gamma}$, also referred to as the size of the model. [5] used the following form for $g(\cdot)$:

$$g(p_\gamma) \propto c^{-p_\gamma} p^{-a p_\gamma}, \quad c, a > 0, \tag{3.1}$$

which induces an exponentially decreasing prior mass as a function of the model size and induces sparsity.

For the prior on $\boldsymbol{\beta}$, there are primarily two classes of prior distributions classified based on the prior on the null hypothesis that the corresponding variable is not relevant: (i) discrete (spike) priors that place a point-mass spike prior which is a degenerate distribution having all the mass at 0, and (ii) continuous (spike) priors which take the view that the magnitude of the regression coefficient under the null hypothesis is "small" and practically not relevant but it need not be exactly zero. [22] pioneered spike-and-slab prior specification. In the next section, we will provide an elaborate discussion on the alternative approach of using continuous spike priors as this is the main focus of the current chapter.

3.3 Continuous Spike-and-Slab Priors

Continuous spike-and-slab priors take the following form

$$\beta_j \mid \gamma_j = 0 \sim p_0(\cdot), \quad \beta_j \mid (\gamma_j = 1) \sim p_1(\cdot), \tag{3.2}$$

where both p_0 and p_1 are continuous distributions. The prior p_0 still focuses the majority of its probability mass around zero and the interpretation of this is that under the null hypothesis the regression coefficient is negligible but can still be non-zero. The prior p_1 is generally a flatter distribution that assigns more probability mass to larger values of the parameters as this prior corresponds to the signals.

A foremost example of the continuous spike-and-slab prior framework was proposed by [12]:

$$\begin{gathered}
\mathbf{y} \mid (\boldsymbol{X}, \boldsymbol{\beta}, \sigma^2) \sim \mathcal{N}(\boldsymbol{X}\boldsymbol{\beta}, \sigma^2 \boldsymbol{I}), \\
\beta_j \mid (\sigma^2, \gamma_j = 0) \sim \mathcal{N}(0, \sigma^2\tau_0^2), \quad \beta_j \mid (\sigma^2, \gamma_j = 1) \sim \mathcal{N}(0, \sigma^2\tau_1^2), \\
P(\gamma_j = 1) = 1 - P(\gamma_j = 0) = \theta, \\
\sigma^2 \sim \mathcal{IG}(\alpha_1, \alpha_2),
\end{gathered} \tag{3.3}$$

where $0 < \tau_0^2 < \tau_1^2 < \infty$, are the multiplicative factors for the variances of the spike-and-slab priors, respectively that may be tuned, and $\mathcal{IG}(\alpha_1, \alpha_2)$ is the inverse-Gamma distribution with shape parameter α_1 and scale parameter α_2. The intuition behind this set-up is that the covariates with zero or very small coefficients will be identified with zero values for γ_j, and the active covariates will have $\gamma_j = 1$. The posterior probabilities of the latent variables γ_j are used to identify the active covariates. A discussion of the requirements on these prior parameters is deferred to Section 3.3.1 where the motivation for such prior choices is also provided.

3.3.1 Shrinking and Diffusing Priors

[23] endorsed the use of sample size dependent prior parameters for the hierarchical model (3.3) for achieving model selection consistency properties. To motivate such choices for the prior parameters, let us consider the case where the design matrix $\boldsymbol{X}$ is orthogonal, i.e., $\boldsymbol{X}'\boldsymbol{X} = n\boldsymbol{I}$. For this simple case, [23] argued that thresholding the posterior probabilities of $P[\gamma_j = 1 \mid \mathbf{y}]$ is equivalent to a thresholding of the least squares estimators $\hat{\beta}_j$, that is,

$$P[\gamma_j = 1 \mid \mathbf{y}] > 0.5 \quad \Leftrightarrow \quad |\hat{\beta}_j| > \sqrt{\frac{2\,(a_1 \log(1-\theta) - a_0 \log\theta)}{(a_0^{-2} - a_1^{-2})}}, \tag{3.4}$$

where a_0, a_1 depend on τ_0, τ_1 as $a_k = \sqrt{\sigma^2/n + \tau_k^2}$, for $k = 0, 1$. When all the prior hyperparameters are fixed and do not depend on n, the threshold for $|\hat{\beta}_j|$ will also be a fixed quantity and leads to inconsistent variable selection as the threshold should instead decrease to zero with the sample size to attain consistency.

To resolve this inconsistency issue, the spike prior's variance was first set to go to zero so that $\tau_0^2 = o(1/n)$, and the other parameters τ_1^2 and θ were allowed to depend on the sample size n and the covariate dimension p to yield a threshold that leads to strong selection consistency. More specifically, the prior choices studied by [23] are as follows:

$$n\tau_0^2 \to 0; \quad n\tau_1^2 \approx (n + p^{(2+\epsilon)}); \quad \theta \approx \frac{1}{p}, \tag{3.5}$$

where the notation $n\tau_0^2 \to 0$ indicates that with sample size $n \to \infty$, $n\tau_0^2$ decreases to zero, the notation $\approx$ indicates same order and $\epsilon > 0$ is any fixed positive number. With such choices, the threshold on $|\hat{\beta}_j|$ provided by (3.4) will have sufficient magnitude, which is in the order of $\sqrt{\frac{\log p}{n}}$ for the orthogonal case, to ensure strong selection consistency. The theoretical results of [23] cover the general design setting while allowing the covariate dimension p to be nearly exponentially large as a function of the sample size n.

One important point to note is that while the prior choices for the three hyperparameters considered by Equation (3.5) are sufficient to ensure strong selection consistency, they are not necessary conditions. In fact, there are multiple combinations for the choices of these parameters that would ensure consistency as the same threshold on the magnitude of $|\hat{\beta}_j|$ can be reached with different choices. Moreover, the choices considered in Equation (3.5) allow the spike prior variance τ_0 to be arbitrarily close to zero which corresponds to point mass spike prior in the limit. This reinforces the point that the use of continuous spike priors has its major motivation from a computational standpoint as it allows the use of a simple Gibbs sampler for posterior sampling as originally proposed by [12]. For the theoretical discussions in this chapter, we will also include the models with a point mass spike prior as a limiting case of the continuous spike-and-slab prior of the model given by Equation (3.3). We will also consider more general choices for the priors going beyond the specific choice of the Gaussian spike-and-slab priors considered so far.

3.3.2 Spike-and-Slab LASSO

Another popular choice for the continuous spike-and-slab prior is the spike-and-slab LASSO prior proposed by [30]. The spike-and-slab LASSO prior places a two-component mixture of Laplace priors on the regression parameters as follows:

$$\beta_j \mid (\gamma_j = 0) \sim \text{LP}(\lambda_0), \quad \beta_j \mid (\gamma_j = 1) \sim \text{LP}(\lambda_1), \tag{3.6}$$

where $\text{LP}(\lambda)$ is the Laplace distribution with pdf given by $\psi(\beta|\lambda) = \frac{\lambda}{2}\exp\{-\lambda|\beta|\}$, and $\lambda_0 \gg \lambda_1$. Figure 3.1 provides examples of spike-and-slab Gaussian and Laplace priors. The main difference between the Gaussian and Laplace versions is the heavier tails of the Laplace priors in comparison to the Gaussian priors.

In contrast to the previous literature that mainly focused on using the posterior distribution of the $\boldsymbol{\gamma}$'s for variable selection, [29] considered using the maximum a posterior (MAP) estimator corresponding to the posterior $\boldsymbol{\beta} \mid (\mathbf{y}, \boldsymbol{X})$. The authors proposed a novel EM algorithm for obtaining this MAP estimator and studied the corresponding theoretical properties. While this is fundamentally a different approach as the focus is on a summary from the posterior of $\boldsymbol{\beta}$ as opposed to that of $\boldsymbol{\gamma}$, the prior framework is quite useful even if we would like to use the posterior of $\boldsymbol{\gamma}$ for model selection. For the discussion in this chapter, even when we consider the spike-and-slab LASSO priors, we focus on model selection using the posterior of $\boldsymbol{\gamma}$.

3.4 Theoretical Properties

As we discussed so far, different choices for the form of the priors p_0 and p_1 can be considered such as the Gaussian priors [12, 13, 15, 20] and the Laplace prior [10, 29]. Gaussian spike-and-slab priors are a common choice not only in model selection but also in many applications including graphical models, inverse covariance matrix estimation, etc. On the other

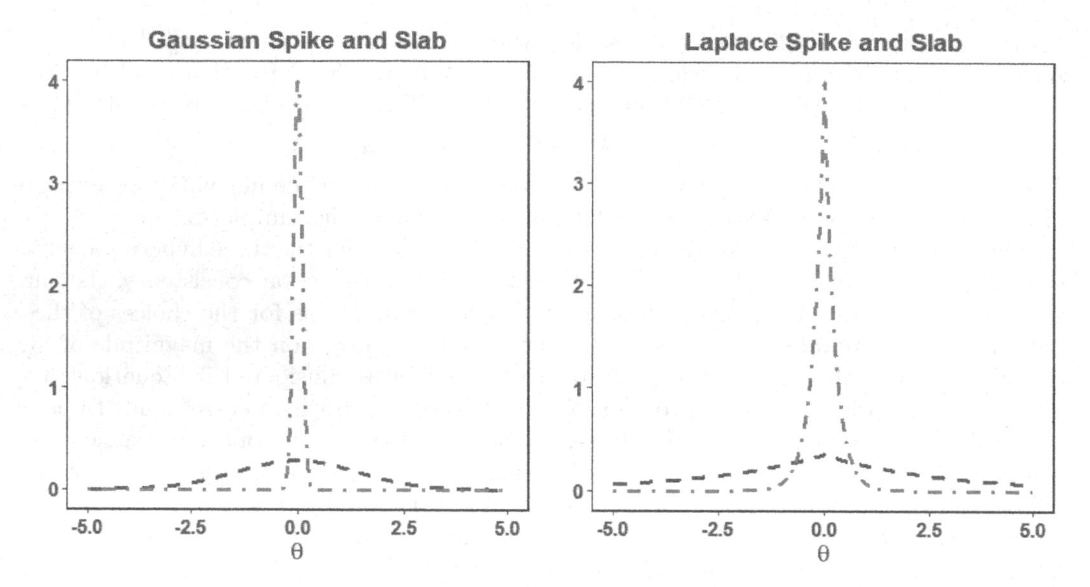

FIGURE 3.1
Examples of continuous spike-and-slab priors: spike-and-slab Gaussian priors, and spike-and-slab LASSO priors. In each example, the distribution with a sharp peak around zero corresponds to the spike prior and assigns majority of its probability mass around zero. The slab prior on the other hand distributes majority of its mass over parameter values away from zero.

hand, there is not a significant computational reason to restrict these priors to be Gaussian as many other priors such as Laplace and t priors can be easily incorporated into the posterior sampling algorithms such as Gibbs samplers. Therefore, there is a great amount of importance on understanding the theoretical and computational properties associated with these prior choices. In this chapter, we will investigate some preliminary theoretical directions in understanding the general version of continuous spike-and-slab priors in terms of their model selection consistency.

3.4.1 Variable Selection Consistency

For the linear regression model with Gaussian spike-and-slab priors, [15, 16] investigated certain theoretical properties of the corresponding posterior mean and of a model selection procedure based on thresholding the posterior mean. While their results provide insights about the shrinkage properties of the spike-and-slab priors, these results were applicable only for the low dimensional case where the dimension p is not larger than the sample size n. Moreover, model selection is more commonly performed using posterior probabilities of the binary indicators γ's. For non-local priors, a thorough investigation of the posterior distribution of the γ's was conducted by [17] when the dimension of the covariate vector is smaller than the sample size. An insightful observation they made is that the posterior corresponding to local priors, which can be heuristically defined as priors that place non-vanishing probability mass around a small neighborhood of zero, give rise to posterior inconsistency for model selection. While Gaussian spike-and-slab priors may appear to be local priors if their prior parameters are fixed, they retain the desirable properties of non-local priors when their prior parameters are allowed to depend on the sample size. Motivated by this, [23] thoroughly investigated the model selection properties of Gaussian spike-and-slab priors in the high-dimensional setting when the number of covariates can be nearly exponentially larger than the sample size. In particular, using conditions similar to 3.5 on the prior parameters, these results assure that model selection consistency can be obtained for a general non-orthogonal design matrix.

3.4.2 Novel Insights

There are several natural and important questions that often arise in the context of variable selection using the Bayesian spike-and-slab prior framework. For instance, it is useful to know which kind of priors are better suited in terms of variable selection performance from a theoretical standpoint. Specifically, it is helpful to know if there is anything special with the commonly used Gaussian or Laplace spike-and-slab priors or can variable selection consistency be achieved using other choices of priors.

We will consider a relatively simple but novel theoretical investigation in this chapter to provide at least partial answers to some of these questions. We will start by considering a general prior setting within the continuous spike-and-slab prior framework that includes both the Gaussian and the Laplace priors. That is, we will consider priors that are indexed by a scale parameter with scale parameters being τ_0 for the spike and τ_1 for the slab. More specifically, our model will be

$$\beta_j \mid (\gamma_j = 0) \sim p_0 \equiv f_{\tau_0}, \quad \beta_j \mid (\gamma_j = 1) \sim p_1 \equiv f_{\tau_1}, \tag{3.7}$$

where the priors p_0 and p_1 have densities f_{τ_0} and f_{τ_1}, respectively. More specifically, we consider spike-and-slab priors taking the form

$$p_0(x) = \frac{1}{\tau_0} f(x/\tau_0); \;\; p_1(x) = \frac{1}{\tau_1} f(x/\tau_1), \tag{3.8}$$

for some given base density f. Such priors are commonly used in the literature. For instance, when the base density f_0 is Gaussian this gives us the approaches taken by [12, 15, 23]. If the base density is Laplace, then we obtain the spike-and-slab LASSO prior specification of [30]. In general, we assume f is continuous, symmetric, and unimodal density. Naturally, we also require $\tau_1 > \tau_0$.

Our goal is to investigate the requirement on the prior parameters for a general form distribution f which need not be either normal or Laplace to achieve model selection consistency. Incorporating heavy tailed prior distributions in our theoretical study is of particular interest as it is not well-studied how the tails of the prior distributions impact model selection consistency properties. We consider the case where the number of covariates $p_n \leq n$, and assume that the design matrix $\boldsymbol{X}$ is orthogonal for the theoretical investigation. We also assume σ^2 to be known and equal to identity. Although this is a simplistic set-up, this simple case provides motivation for the necessity of sample size dependent prior parameters as well as an insight into the mechanism of model selection using these priors. At this moment, we do not impose any assumptions on the prior distributions.

For the theoretical variable selection consistency, we will consider two versions of consistency—weak selection consistency (WSC) and strong selection consistency (SSC). The first one WSC requires the posterior probabilities to converge uniformly to the right values, that is,

$$\min_{j=1,\cdots,p} P[\gamma_j = t_j \mid \mathbf{y}] \xrightarrow{\text{P}} 1, \tag{3.9}$$

where $t_j = 0/1$ indicates whether the jth covariate is inactive or not, respectively. The second version SSC requires the posterior probability on the true model to converge to one, that is,

$$P[\boldsymbol{\gamma} = \boldsymbol{t} \mid \mathbf{y}] \xrightarrow{\text{P}} 1, \tag{3.10}$$

where $\boldsymbol{t} = (t_1, \ldots, t_p)$ indicates the set of true variables.

The following additional notations will be used for the theoretical discussions. For sequences a_n and b_n, $a_n \sim b_n$ means $\frac{a_n}{b_n} \to c$ for some $c > 0$, $b_n \succeq a_n$ (or $a_n \preceq b_n$) means $b_n = O(a_n)$, and $b_n \succ a_n$ (or $a_n \prec b_n$) means $b_n = o(a_n)$.

We now specify the following conditions used to study selection consistency of the spike-and-slab priors.

Condition 1 (True model). $\mathbf{y} \mid \boldsymbol{X} \sim N(\boldsymbol{X}_t \boldsymbol{\beta}_t, \boldsymbol{I})$ *where the size of the true model* $p_t = o(\frac{n}{\log p})$.

Condition 2 (Identifiability). $\min_{i \in t} |\beta_i| \geq m_n := \sqrt{\frac{K\sigma^2 \log p}{n}}$ *for large enough* $K > 0$.

We first define the following quantities based on the prior distributions p_0 and p_1:

$$C_n := \sup_{|a| \leq r_n; |b| \leq r_n} \frac{\theta \; p_1(a)}{(1-\theta) \; p_0(b)}; \quad T_n := \inf_{m_n \leq |a|; m_n \leq |b|; |a-b| \leq \epsilon_n} \frac{\theta \; p_1(a)}{(1-\theta) \; p_0(b)},$$

where ϵ_n is such that $\frac{1}{\sqrt{n}} \leq \epsilon_n \to 0$.

Remark 3.4.1. *C_n and T_n represent how far apart the priors p_0 and p_1 are at the center (i.e., around zero) and at the tails, respectively. Intuitively, the quantity C_n measures the difference in peakedness of the spike prior in comparison to the slab prior. The faster C_n goes to zero, the more peaked p_0 is compared to p_1. The quantity T_n measures how flat the tails of p_1 are in comparison with that of the tails of prior p_0. The larger T_n is the faster the tails decay.*

Condition 3 (Prior parameters). *We shall assume the following two conditions on the priors:*
(a)

$$C_n \to 0; \quad T_n \to \infty.$$

(b)

$$p \; C_n \to 0; \quad \frac{T_n}{p_t} \to \infty.$$

The conditions 3(a) and 3 (b) present a generalization for the shrinking and diffusing nature of the prior parameters as discussed in [23]. In particular, they provide a specific requirement on how the continuous spike-and-slab priors should behave in terms of their relative magnitudes at the origin and at the tails for achieving variable selection consistency. In particular, the condition on C_n converging to zero requires that the relative magnitude of the slab prior compared to the spike prior should converge to zero near the origin whereas the relative magnitude should converge to infinity at the tails. The requirements on the rates for these convergences are stricter in Condition 3 (b) as it would provide strong selection consistency whereas 3 (a) is sufficient for weak selection consistency as implied by the following theorem.

Theorem 3.4.2. *Under Conditions 1–3 (a), we have weak selection consistency. That is,* $\min_{j=1,\cdots,p} P[\gamma_j = t_j \mid \mathbf{y}] \xrightarrow{\text{P}} 1$. *In addition, if 3(b) holds we have strong selection consistency of the posterior distribution, i.e.,* $P[\boldsymbol{\gamma} = \mathbf{t} \mid \mathbf{y}] \xrightarrow{\text{P}} 1$.

The above theorem provides very general conditions under which the posterior from the spike-and-slab prior specification has model selection consistency. For specifying the conditions explicitly in terms of the base density f, we first define quantities analogous to C_n and T_n. Define

$$L_n := \frac{\theta \tau_0 f(0)}{(1-\theta)\tau_1 f\left(\frac{r_n}{\tau_0}\right)}, \quad U_n := \frac{\theta \tau_0 f\left(\frac{m_n}{\tau_1}\right)}{(1-\theta)\tau_1 \left(f\left(\frac{m_n}{2\tau_0}\right) + \tau_0 \exp\left\{-\frac{nm_n^2}{2}\right\}\right)},$$

where $r_n := \sqrt{\frac{(2+\epsilon)\sigma^2 \log p}{n}}$, $\epsilon > 0$.

Condition 4 (Prior parameters with a base density). *The condition has two parts:*
(a)

$$\tau_0 \preceq r_n \to 0; \quad L_n \to 0; U_n \to \infty,$$

(b)

$$\tau_0 \preceq r_n \to 0; \quad pL_n \to 0; \frac{U_n}{p_t} \to \infty,$$

where the notation $a \preceq b$ means that $a/b \to 0$.

Theorem 3.4.3. *Under Conditions 1-2, 4 (a), we have weak selection consistency. That is,* $\min_{j=1,\cdots,p} P[\gamma_j = t_j \mid \mathbf{y}] \xrightarrow{\text{P}} 1$. *In addition, if 4(b) holds we have strong selection consistency of the posterior distribution, i.e.,* $P[\gamma = \mathbf{t} \mid \mathbf{y}] \xrightarrow{\text{P}} 1$.

We would like to make the following observations based on the results so far. The first observation is that weak selection consistency holds easily for many base densities. Based on Theorem 2, WSC holds for any base density f which is unimodal and decreasing on the positive part if the hyperparameters are such that

$$m_n f(m_n/2\tau_0) \leq \frac{\theta\tau_0}{(1-\theta)}, \quad \tau_1 \succeq m_n$$

and τ_1 is large enough to let U_n go to infinity. Such a τ_0 and θ can be easily obtained for essentially all commonly used base densities implying that weak selection consistency can be achieved easily as long as the hyperparameters are appropriately chosen. However, it is more difficult for the SSC to be satisfied due to the stronger conditions required. For some example base densities, we shall investigate if the SSC is possible to hold for some choices of τ_0 and τ_1.

Another observation is that the requirement on minimum signal strength is different for different base densities to achieve strong selection consistency. Condition 4 (b) assumed for strong selection consistency implicitly requires that $pp_t \frac{f\left(\frac{m_n}{2\tau_0}\right) + \exp\left\{-\frac{nm_n^2}{2}\right\}}{f\left(\frac{r_n}{\tau_0}\right)} \to 0$. This places a requirement on the minimum signal m_n that depends on the base density function $f(\cdot)$ for strong selection consistency. As we will see in the examples, different base densities imply different requirements on the minimum signal strength.

Finally, we can also observe that the conditions placed on τ_1 depend on the base density f and heavy tailed base densities require a weaker condition on how large τ_1 should be. For requiring a slow rate for τ_1, we would like to have $f\left(\frac{r_n}{\tau_0}\right)$ large because the numerator in L_n involves the product of τ_1 and $f\left(\frac{r_n}{\tau_0}\right)$. In particular, this implies that the rate of tail decay of f at r_n/τ_0 governs the magnitude of τ_1 in an inverse relationship. The larger the tail, the weaker the requirement on the magnitude of τ_1 is.

3.4.3 Examples

We shall now consider specific examples for the prior distribution f and consider how τ_1 changes accordingly. We will consider two specific choices for the spike parameter τ_0 namely, $\tau_0 = \frac{1}{\sqrt{n}}$ and $\tau_0 = r_n$ and then consider the corresponding requirement on the other hyperparameters for these two choices.

Gaussian Spike-and-Slab Priors

If f is standard Gaussian, then $f(x) \propto \exp\{\frac{-x^2}{2}\}$ and the implications of the conditions are as follows. For $\tau_0 = \frac{1}{\sqrt{n}}$, $L_n \sim \frac{\theta}{(1-\theta)\sqrt{n}\phi(\sqrt{2c\log p})} \sim \frac{\theta p^c}{(1-\theta)\sqrt{n}}$ and $U_n \geq \frac{\theta p^K}{(1-\theta)\sqrt{n}}$, where c is a positive constant, K depends on the minimal signal strength as:

$$K = \frac{nm_n^2}{2\log p},$$

or equivalently

$$m_n = \sqrt{\frac{2K\log p}{n}}.$$

Therefore, both the conditions for WSC and SSC are satisfied if K is a large enough fixed constant. Therefore, $n\tau_1^2$ has to be at least in the order of $\theta^2 p^{2c}$ which increases polynomially in p. Note that the prior rates proposed for Gaussian priors in [23] satisfy this.

For $\tau_0 = r_n$, $L_n \sim \frac{\theta r_n}{(1-\theta)}$ and $U_n \geq \frac{\theta\exp\{K^2/2\}}{(1-\theta)}$. Therefore, WSC holds true for any K. However, for SSC to hold true, we would need that

$$K \geq \sqrt{2\log(pr_n)},$$

which is a stronger restriction on minimum signal strength than fixed K, the optimal rate.

Spike-and-Slab LASSO Priors

If f is Laplace (double exponential), then the base density is $f(x) \propto \exp\{-|x|\}$ and will imply the following requirements. For $\tau_0 = \frac{1}{\sqrt{n}}$, $L_n \sim \frac{\theta}{(1-\theta)\sqrt{n}f(\sqrt{2c\log p})} \sim \frac{\theta\exp\{\sqrt{2c\log p}\}}{(1-\theta)\sqrt{n}}$ and $U_n \geq \frac{\theta\exp\{\sqrt{2K\log p}\}}{(1-\theta)\sqrt{n}}$. Therefore, both the conditions for WSC hold true. For SSC to hold true, we would need $K > c\sqrt{\log p}$, for a large enough constant c.

For $\tau_0 = r_n$, $L_n \sim \frac{\theta r_n}{(1-\theta)}$ and $U_n \geq \frac{\theta\exp\{\sqrt{2K\log p}\}}{(1-\theta)}$. Therefore, WSC holds true for any K. However, for SSC to hold true, we would need

$$K \geq \sqrt{2\log(pr_n)}.$$

This is a stronger restriction on minimum signal strength than fixed K, the optimal rate. The strength of this requirement depends on the magnitude of r_n which is usually in the order of $\sqrt{\log p/n}$.

Spike-and-slab Student's t Priors

Suppose f is a t_ν distribution, that is, $f(x) \propto (1 + x^2/\nu)^{-(\nu+1)/2}$. For $\tau_0 = \frac{1}{\sqrt{n}}$, we have $f^2(\sqrt{2c\log p}) \sim (\log p)^{-(\nu+1)/2}$. Therefore, $n\tau_1^2$ has to be at least in the order of $\theta^2(\log p)^{(\nu+1)/2}$ which increases poly-log in p. In particular, for $\nu = 1$, we have the Cauchy prior and $n\tau_1^2$ can be as small as $\log p$ in order. In comparison to the Gaussian base distribution that requires polynomial order in p, this is a much weaker requirement. On the other hand, $U_n \geq \frac{\theta(K \log p)^{(\nu+1)/2}}{(1-\theta)\sqrt{n}}$. So for the SSC condition to hold, we would need K to be at least as large as p, which is prohibitive as a condition on the minimum signal condition.

For $\tau_0 = r_n$, we will need

$$K \geq (pr_n)^{2/(1+\nu)},$$

for strong selection consistency. This can again be restrictive in terms of the requirement on minimum signal strength.

Based on the above discussion, it is clearly the case that the heavier the tail of f is, the weaker the requirement on τ_1 is. However, the disadvantage of the heavier tailed base priors is that we will need a stronger requirement on the minimum signal strength for achieving strong selection consistency. To summarize the key theoretical observations and to answer the theoretical questions of interest, we can make the following conclusions based on the requirements on the prior hyperparameters.

Variable selection consistency results in terms of the weak selection consistency can be easily achieved with a general spike-and-slab prior. With appropriate conditions on the prior hyperparameters, weak selection consistency can hold true irrespective of the form of the base density. On the other hand, for achieving strong selection consistency, the prior hyperparameters and the form of the base density play a crucial role. Heavier tailed priors require weaker restrictions on the prior hyperparameters compared to light tailed distributions for achieving strong selection consistency. However, this comes at a cost of requiring stronger signal strength for heavier tailed priors.

The theoretical results provide a guidance on the ranges to be used for the prior hyperparameter values in order to achieve weak or strong selection consistency for each base density. While we only considered theoretical aspects, the choice for the form of the prior and its hyperparameters will also depend on other aspects such as computational convenience and scalability. It will be a fruitful future exercise to study the theoretical properties of continuous spike-and-slab priors and their computational properties simultaneously to help practitioners decide optimal choices of priors.

We would like to note here that in the context of point-mass spike priors, [5] found heavier tailed slab priors to be more suitable for obtaining their posterior concentration results as heavier tailed priors require milder constraints on the signal strength. Although we find heavier tailed priors to impose weaker restrictions on the prior hyperparameters, they require stronger conditions on the signal strength in our analysis. The results of [5] cannot be directly compared with ours as our focus is on strong model selection consistency and more importantly, the continuous spike prior plays an important role in the theoretical analysis. More specifically, we currently consider the same tailed priors for both the spike and the slab priors. However, this is not necessary and more flexibility can be obtained if we consider lighter tailed priors for the spike prior and heavier tailed priors for the slab prior. This is quite a natural way to place the spike-and-slab priors as the spike prior needs to be more concentrated around zero and the slab priors needs to be more flat. To the best of our knowledge, this possibility of light tailed spike and heavy tailed slab prior has not been explored in the literature and needs to be studied in the future.

3.5 Computations

Efficient and scalable computation is an important aspect of statistical procedures especially for large datasets. Gibbs sampling algorithms are quite commonly used for computation of Bayesian posterior distributions in the context of continuous spike-and-slab priors. For Bayesian variable selection, [12, 13] pioneered Gibbs sampling based computational approaches for Gaussian spike-and-slab priors. With the model (3.3), the standard Gibbs sampling algorithm they considered would take the following form:

The full conditional distribution of $\boldsymbol{\beta}$ is given by $\boldsymbol{\beta} \mid (\boldsymbol{\gamma}, \mathbf{y}, \boldsymbol{X}) \sim N(\boldsymbol{V}\boldsymbol{X}^{\top}\mathbf{y}, \sigma^2\boldsymbol{V})$, where $\boldsymbol{V} = (\boldsymbol{X}^{\top}\boldsymbol{X} + \boldsymbol{D}_{\gamma})^{-1}$, and $D_{\gamma} = \text{Diag}(\gamma\tau_1^{-2} + (1-\gamma)\tau_0^{-2})$.

The full conditional distributions for γ_j are independent across different j and take the following form:

$$P(\gamma_j = 1 \mid (\boldsymbol{\beta}, \sigma^2, \mathbf{y}, \boldsymbol{X})) = \frac{\theta\phi(\beta_j, 0, \sigma^2\tau_1^2)}{\theta\phi(\beta_j, 0, \sigma^2\tau_1^2) + (1-\theta)\phi(\beta_j, 0, \sigma^2\tau_0^2)}.$$

Sampling from these distributions is straightforward but the main challenge is that sampling from a p dimensional multivariate normal distribution for $\boldsymbol{\beta}$ is computationally intensive especially when p is large. Sampling the $\boldsymbol{\beta}$ directly as above would impose a p^3 order computational complexity which is not desirable. With the motivation of easing this computational burden, [24] proposed a scalable Gibbs sampling algorithm called the Skinny Gibbs which we will briefly outline now.

3.5.1 Skinny Gibbs for Scalable Posterior Sampling

Skinny Gibbs algorithm modifies the standard Gibbs sampler for reducing the computational burden. The Skinny Gibbs algorithm splits the $\boldsymbol{\beta}$ into two parts in each Gibbs iteration, corresponding to the "active" (with the current $\gamma_j = 1$) and "inactive" (with the current $\gamma_j = 0$) parts of the binary vector $\boldsymbol{\gamma}$. Due to sparsity in $\boldsymbol{\beta}$, the active part typically has a low dimension, and can be sampled from a multivariate normal distribution. The inactive part has a high dimension, and is sampled from a normal distribution with independent marginals. To adjust for the loss of dependency induced by this modification, an adjustment step in the sampling of $\boldsymbol{\gamma}$ is added which will make sure that the resultant posterior distribution corresponding to Skinny Gibbs also has the same desired variable selection consistency properties. While the Skinny Gibbs algorithm was proposed in the context of Gaussian spike-and-slab priors, it can be generalized to other non-Gaussian spike-and-slab priors as we shall describe in this section. We will first describe the algorithm for Gaussian spike-and-slab priors and then provide a generalization.

More specifically, the Skinny Gibbs sampler proceeds as follows, after an initialization. **Step 1 (for sampling $\boldsymbol{\beta}$).** Define the index sets A and I as the active (corresponding to $\gamma_j = 1$) and the inactive (corresponding to $\gamma_j = 0$) sets and decompose $\boldsymbol{\beta} = (\boldsymbol{\beta}_A, \boldsymbol{\beta}_I)$ so that $\boldsymbol{\beta}_A$ and $\boldsymbol{\beta}_I$ contain the components of $\boldsymbol{\beta}$ corresponding to $\gamma_j = 1$ and $\gamma_j = 0$, respectively. Similarly rearrange the design matrix $\boldsymbol{X} = [\boldsymbol{X}_A, \boldsymbol{X}_I]$. Then, the vector $\boldsymbol{\beta}$ is sampled as:

$$\boldsymbol{\beta}_A \mid (\mathbf{y}, \boldsymbol{\gamma}) \sim N(\boldsymbol{m}_A, \sigma^2 V_A^{-1}), \qquad \boldsymbol{\beta}_I \mid (\mathbf{y}, \boldsymbol{\gamma}) \sim N(0, \sigma^2 \boldsymbol{V}_I^{-1}), \tag{3.11}$$

where $\boldsymbol{V}_A = (\boldsymbol{X}_A'\boldsymbol{X}_A + \tau_1^{-2}\boldsymbol{I})$, $\boldsymbol{m}_A = \boldsymbol{V}_A^{-1}\boldsymbol{X}_A'Y$, and $\boldsymbol{V}_I = \text{Diag}(\boldsymbol{X}_I'\boldsymbol{X}_I) + \tau_0^{-2}\boldsymbol{I} = (n + \tau_0^{-2})\boldsymbol{I}$.

Step 2 (for sampling γ). Generate γ_j $(j = 1, \cdots, p)$ conditioned on the remaining components of γ using the following conditional odds:

$$\begin{aligned}&\frac{P[\gamma_j = 1 \mid \gamma_{-j}, \boldsymbol{\beta}, \mathbf{y}]}{P[\gamma_j = 0 \mid \gamma_{-j}, \boldsymbol{\beta}, \mathbf{y}]}\\ &= \frac{\theta\phi(\beta_j, 0, \tau_1^2)}{(1-\theta)\phi(\beta_j, 0, \tau_0^2)} \times \exp\left\{\beta_j \boldsymbol{X}_j'(\mathbf{y} - \boldsymbol{X}_{C_j}\boldsymbol{\beta}_{C_j})\right\},\end{aligned} \tag{3.12}$$

where γ_{-j} is the γ vector without the jth component, and C_j is the index set corresponding to the active components of γ_{-j}, i.e., $C_j = \{k : k \neq j, \gamma_k = 1\}$.

The intuition behind the Skinny Gibbs sampler is that in Step 1, the update of $\boldsymbol{\beta}$ is modified such that the coefficients corresponding to $\gamma_j = 1$ and those corresponding to $\gamma_j = 0$ (denoted by β_I) are sampled independently, and the components of β_I are updated independently so that large matrix computations are avoided. That is, Skinny Gibbs modifies the precision matrix V_γ as

$$\boldsymbol{V}_\gamma = \begin{pmatrix} \boldsymbol{X}_A'\boldsymbol{X}_A + \tau_1^{-2}\boldsymbol{I} & \boldsymbol{X}_A'\boldsymbol{X}_I \\ \boldsymbol{X}_I'\boldsymbol{X}_A & \boldsymbol{X}_I'\boldsymbol{X}_I + \tau_0^{-2}\boldsymbol{I} \end{pmatrix}$$

$$\Downarrow$$

$$\begin{pmatrix} \boldsymbol{X}_A'\boldsymbol{X}_A + \tau_1^{-2}\boldsymbol{I} & 0 \\ 0 & (\mathrm{Diag}(\boldsymbol{X}_I'\boldsymbol{X}_I) + \tau_0^{-2})\boldsymbol{I} \end{pmatrix}.$$

The precision matrix is heavily modified in Step 1 which can alter the original Gibbs sampler. To compensate for the loss of dependence structure that the modification induces, step 2 of the Skinny Gibbs algorithm is used to adjust for this dependence structure by introducing dependence into the sampling of the γ_j's. In spite of this modification, the theoretical results of [24] guarantee that the Skinny Gibbs algorithm has strong model selection consistency property.

This technique developed for Skinny Gibbs can be generalized to many modeling settings where the likelihood or priors involved can be written as mixtures of normal distributions. For instance, [24] studied a Skinny Gibbs algorithm applied to logistic regression which can be further generalized to any likelihood that can be represented as mixtures of normal distributions. Another generalization is to prior settings where the prior distributions can be written as scale mixtures of normal distributions.

It is worth noting here that, [3] developed an alternative approach to scale up the Gibbs sampling algorithms which involves large multivariate normal distributions. Their approach is based on exploring properties of matrices to sample from the original high-dimensional normal distribution. This algorithm has linear order complexity in terms of p, it has a quadratic complexity in terms of the sample size n while Skinny Gibbs has a linear order complexity in n. Their strategy is quite useful in practice especially for the large p small n case and can also be utilized within the Skinny Gibbs sampler for sampling the active components of $\boldsymbol{\beta}$.

We shall now provide a simple demonstration of the Skinny Gibbs algorithm on a sample data example. We consider the data by [18] from an experiment to study the genetics of two inbred mouse populations (B6 and BTBR). The data include gene expression levels of $n = 60$ mice. We consider the physiological phenotype called SCD1 measured by quantitative real-time PCR. The gene expression data and the phenotypic data are available at GEO (http://www.ncbi.nlm.nih.gov/geo; accession number GSE3330). Following the same

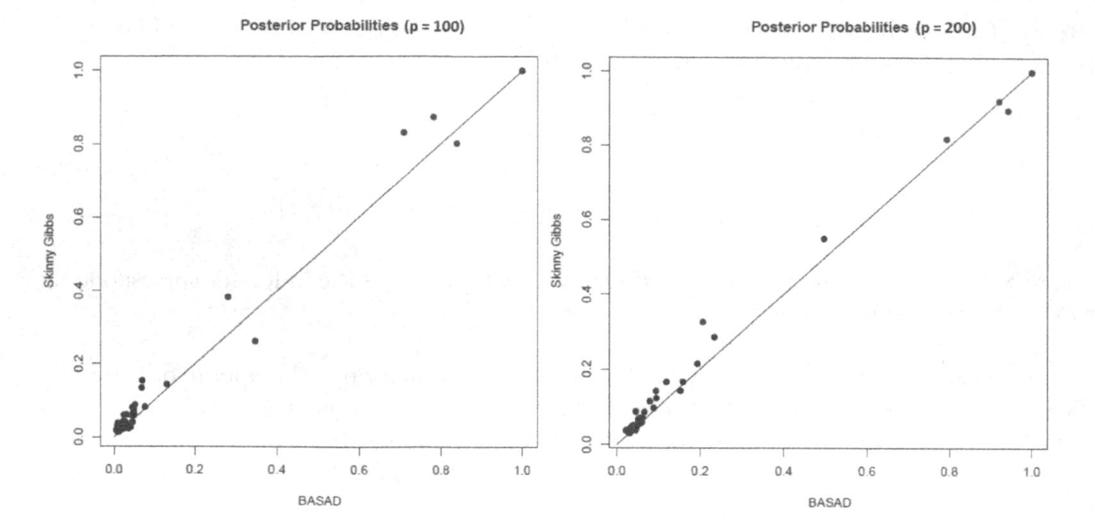

FIGURE 3.2
Comparison of the marginal posterior probabilities for the gene expression data example. Marginal posterior probabilities for the standard Gibbs algorithm described in [23], denoted as BASAD, are plotted on X-axis and the probabilities for the Skinny Gibbs algorithm [24] are plotted on Y-axis. The left panel corresponds to the case with $n = 60, p = 100$ and the right panel corresponds to the case with $n = 60, p = 200$.

screening procedure described in [23], we first screen the genes to include $p = 100$ and $p = 200$ covariates for our demonstration. The details regarding the choice of prior hyper-parameters are also as provided in [23] and can be obtained from the R code provided in the online supplementary material.

Figure 3.2 provides a comparison of the marginal posterior probabilities of the binary indicators γ_j obtained from the Gibbs sampling algorithm described in the beginning of Section 3.5 and the Skinny Gibbs algorithm described in Section 3.5.1. We can observe that the marginal posterior probabilities corresponding to the two algorithms are quite close to each other. This is in agreement with the theoretical properties of Skinny Gibbs assuring that Skinny Gibbs has desirable variable selection performance while being computationally scalable.

3.5.2 Skinny Gibbs for Non-Normal Spike-and-Slab Priors

The Skinny Gibbs technique can be generalized beyond the Gaussian spike-and-slab priors including the spike-and-slab LASSO priors. In fact, it can be generalized to any prior distribution that can be written as a scale mixture of normal distributions. As discussed in [39], this is a very large class and includes many commonly used distributions including the exponential and t distributions. We will now illustrate how Skinny Gibbs can be generalized.

Consider the model given by (3.7) where the spike-and-slab prior distributions are scale mixtures of normal distributions. That is,

$$f_\tau(u) = \int \phi\left(\frac{u}{\tau w}\right) g(w)dw,$$

where $\phi(\cdot)$ is the density of the standard normal distribution. Equivalently, if the prior distribution for a regression coefficient β_j follows the distribution $f_\tau(u)$, the prior distribution can be equivalently written as

$$\beta_j \mid W_j \sim N(0, \tau^2 W_j^2); \quad W_j \sim g(w),$$

for $j = 1, \cdots, p$.

With this prior, the corresponding Skinny Gibbs algorithm can be generalized by treating the W_j's corresponding to each parameter as latent variables which can be sampled within the Gibbs sampling algorithm.

Generalized Skinny Gibbs algorithm:

Step 1 (for sampling $\boldsymbol{\beta}$). Following the same notation as used for the Gaussian case, the conditional distribution for $\boldsymbol{\beta}$ is given by:

$$\boldsymbol{\beta}_A \mid (\boldsymbol{W}, \mathbf{y}, \boldsymbol{\gamma}) \sim N(\boldsymbol{m}_A, \sigma^2 \boldsymbol{V}_A^{-1}), \qquad \boldsymbol{\beta}_I \mid (\boldsymbol{W}, \mathbf{y}, \boldsymbol{\gamma}) \sim N(\mathbf{0}, \sigma^2 \boldsymbol{V}_I^{-1}), \tag{3.13}$$

where $V_A = (\boldsymbol{X}_A' \boldsymbol{X}_A + \tau_1^{-2} \boldsymbol{W}_A^{-2})$, $\boldsymbol{m}_A = \boldsymbol{V}_A^{-1} \boldsymbol{X}_A' Y$, and $\boldsymbol{V}_I = \text{Diag}\,(\boldsymbol{X}_I' \boldsymbol{X}_I) + \tau_0^{-2} \boldsymbol{W}_I^{-2}$, where the matrix $\boldsymbol{W} = Diag(W_1, \ldots, W_p)$ and the notation $\boldsymbol{W}_A$ and $\boldsymbol{W}_I$ denote the active and inactive part of the matrix $\boldsymbol{W}$, respectively. This sampling distribution is essentially a weighted version of the one corresponding to the Gaussian case (3.11) with the weights given by $\boldsymbol{W}$.

Step 2 (for sampling $\boldsymbol{\gamma}$). Generate γ_j $(j = 1, \ldots, p)$ conditioned on the remaining components of $\boldsymbol{\gamma}$ using the following conditional odds:

$$\begin{aligned} &\frac{P[\gamma_j = 1 \mid \boldsymbol{W}, \boldsymbol{\gamma}_{-j}, \boldsymbol{\beta}, \mathbf{y}]}{P[\gamma_j = 0 \mid \boldsymbol{W}, \boldsymbol{\gamma}_{-j}, \boldsymbol{\beta}, \mathbf{y}]} \\ &= \frac{\theta \phi(\beta_j, 0, \tau_{1,n}^2 W_j^2)}{(1-\theta)\phi(\beta_j, 0, \tau_{0,n}^2 W_j^2)} \times \exp\left\{\beta_j \boldsymbol{X}_j'(\mathbf{y} - \boldsymbol{X}_{C_j} \boldsymbol{\beta}_{C_j})\right\}, \end{aligned} \tag{3.14}$$

where we used the same notation as in Equation (3.12) for the Gaussian case.

Step 3 (for sampling W). The generalized version of Skinny Gibbs requires sampling of the weights W_j's. The density of the full conditional distribution of these weights is given by

$$f(W_j \mid (\boldsymbol{\beta}, \mathbf{y}, \boldsymbol{\gamma}) \propto g(W_j) \phi(\beta_j, 0, W_j^2 \tau_{\gamma_j}^2),$$

for $j = 1, \ldots, p$. Depending on the weight distribution g this full conditional may or may not be a standard distribution. For instance, for the spike-and-slab LASSO prior, the full conditional turns out to be an inverse Gamma distribution. Even if this is not a standard distribution, this can be easily sampled from using many standard techniques as this is a univariate distribution.

While we focused primarily on Gibbs sampling based strategies for posterior computation, an alternative method uses an EM algorithm based computational approach as proposed and generalized by [29, 30]. This approach can be used to obtain the maximum a posteriori estimator for the continuous parameter $\boldsymbol{\beta}$. Although this approach may not provide the posterior probabilities of the binary indicators γ_j, approximate posterior probabilities can be obtained by conditioning on the estimated value of the $\boldsymbol{\beta}$ parameter. That is, one can use $P[\gamma_j = 1 \mid \mathbf{y}, \hat{\boldsymbol{\beta}}]$ as an alternative to the posterior probabilities of the γ_j's. This approach is computationally advantageous and has been observed to perform well in practice [10, 30].

3.6 Generalizations

While the discussion in the chapter mainly focused on linear regression models, the applications of continuous spike-and-slab priors are well beyond the linear regression model. There is a lot of past and recent interest in using continuous spike-and-slab priors for more general modeling settings. [21] compared the statistical and computational performance of several choices of spike-and-slab priors using simulated data including both point mass spike priors and continuous spike priors. They observed that continuous spike priors tend to have better computational performance in comparison to the point mass spike priors.

There are several generalizations of the spike-and-slab priors to account for specific structures such as grouping of covariates, hierarchical structures, etc. In this section, we will provide references to some of the related work on this but we do not aim to provide a comprehensive review of all such papers. [7] proposed multivariate spike-and-slab LASSO prior for simultaneously performing variable selection and covariance selection. [32] used spike-and-slab t priors for function selection in structured additive regression models. [36] considered spike-and-slab LASSO priors for generalized linear models and used them for gene detection problems. [24] studied spike-and-slab Gaussian priors for logistic regression and devised a Skinny Gibbs algorithm for posterior computation. Their approach can accommodate other non-Gaussian likelihood settings as long as the likelihood can be written as a mixture of Gaussian distributions. [10] used spike-and-slab Laplace for Gaussian graphical models and studied their Bayesian regularization problems. [11] extended the spike-and-slab Laplace priors for joint modeling of multiple graphical models. For jointly estimating graphical structures of multiple graphs, [26] developed a Bayesian approach with priors on graph structures where spike-and-slab priors were placed on the prior parameters. [24] used spike-and-slab Gaussian priors for joint estimation of multiple graphical models.

[34] studied the use of continuous spike-and-slab priors for nowcasting economic time series data. [9] proposed time varying spike-and-slab prior processes to perform dynamic variable selection in time series regression and extended the application of spike-and-slab priors to time series data.

The aforementioned references focus on continuous spike-and-slab priors. Generalizations that use point mass spike priors have also been proposed. For instance, [40] used such spike-and-slab priors for accelerated failure time models in the context of survival data analysis. [27] develop a joint method to select covariates and graph structures present in the covariates for network valued covariates for regression problems. [1] used spatio-temporal spike-and-slab Gaussian process priors for modeling sparsity and structure in spatio-temporal data. [4] devised nonparametric spiked spatio-temporal process priors for performing dynamic variable selection.

3.7 Conclusion

Continuous spike-and-slab priors have been widely used for variable selection in the Bayesian context. This chapter provided a discussion of some of the existing theoretical results and computational strategies for performing Bayesian variable selection with continuous spike-and-slab priors. In addition, new theoretical results that demonstrate that model selection consistency properties can be achieved with a wide range of continuous spike-and-slab priors are provided. While the majority of the discussion in the chapter is limited to the linear regression setting, the continuous spike-and-slab prior framework has a lot of potential and scope for expanding its use to many other statistical modeling settings. We hope that future research will explore further theoretical and computational advances.

Appendix

Proof of Theorem 3.4.2:

Under this set-up, the joint posterior of $\boldsymbol{\beta}$ and $\boldsymbol{\gamma}$ can be written as:

$$\begin{aligned}
&P(\boldsymbol{\beta},\boldsymbol{\gamma} \mid \mathbf{y}) \\
&\propto \exp\left\{-\tfrac{1}{2\sigma^2}\|\mathbf{y}-\boldsymbol{X}\boldsymbol{\beta}\|_2^2\right\}\textstyle\prod_{i=1}^p \left((1-\theta)p_0(\beta_i)\right)^{1-\gamma_i}\left(\theta p_1(\beta_i)\right)^{\gamma_i} \\
&\propto \exp\left\{-\tfrac{1}{2\sigma^2}(\boldsymbol{\beta}'\boldsymbol{X}'\boldsymbol{X}\boldsymbol{\beta}-\boldsymbol{\beta}'\boldsymbol{X}'\mathbf{y})\right\}\textstyle\prod_{i=1}^p \left((1-\theta)p_0(\beta_i)\right)^{1-\gamma_i}\left(\theta p_1(\beta_i)\right)^{\gamma_i} \\
&\propto \exp\left\{-\tfrac{n}{2\sigma^2}\sum_{i=1}^p(\beta_i-\hat{\beta}_i)^2\right\}\textstyle\prod_{i=1}^p \left((1-\theta)p_0(\beta_i)\right)^{1-\gamma_i}\left(\theta p_1(\beta_i)\right)^{\gamma_i},
\end{aligned}$$

where $\hat{\beta}_i$ is the OLS estimator of β_i, i.e., $\hat{\beta}_i = \boldsymbol{X}_i'\mathbf{y}/n$.

Therefore the marginal posterior of γ_i is given by

$$P(\gamma_i \mid \mathbf{y}) \propto \int \exp\left\{-\frac{n}{2\sigma^2}(b-\hat{\beta}_i)^2\right\}\left((1-\theta_n)p_0(b)\right)^{1-\gamma_i}\left(\theta_n p_1(b)\right)^{\gamma_i} db,$$

and

$$P(\gamma_i = 0 \mid \sigma^2, \mathbf{y}) = \frac{(1-\theta)E_{\hat{\beta}_i}(p_0(B))}{(1-\theta)E_{\hat{\beta}_i}(p_0(B)) + \theta E_{\hat{\beta}_i}(p_1(B))}, \tag{3.15}$$

where $E_{\hat{\beta}_i}$ is the expectation under B following the normal distribution with mean $\hat{\beta}_i$ and variance σ^2/n. When p_0 and p_1 are Gaussian distributions, these expectations can be calculated explicitly. If p_0 and p_1 are not Gaussian, then it may not be possible to obtain explicit closed-form expressions for these probabilities. Suppose that p_1 and p_0 are Cauchy distributions with different scale parameters τ_1^2 and τ_0^2, respectively. As before, we typically require $\tau_0^2 = o(1/n)$ and $\tau_1^2 \succ \tau_0^2$. Consider the term

$$\begin{aligned}
E_{\hat{\beta}_i}(p_k(B)) &= \frac{1}{2p\sigma\tau_k}\int \exp\left\{-\frac{n}{2\sigma^2}(b-\hat{\beta}_i)^2\right\}p_k(b)db \\
&= \frac{1}{2p\sigma\tau_k}\int \exp\left\{-\frac{n}{2\sigma^2}(b-\hat{\beta}_i)^2\right\}\left(1+\frac{b^2}{\tau_k^2}\right)^{-1} db
\end{aligned}$$

Though this integral is hard to obtain a closed form expression for, we can bound this term to obtain the rates for τ_k^2. For achieving weak selection consistency, we would like to choose the prior parameters such that

$$P[\gamma_j = 0 \mid \mathbf{y}] \geq P[\gamma_j = 1 \mid \mathbf{y}], \text{ if } |\hat{\beta}_j| \leq r_n, \tag{3.16}$$

where r_n is the largest magnitude of spurious signal which can be upper bounded by $\sqrt{\frac{(2+\epsilon)\log p}{n}}$ with high probability. Therefore, we will denote

$$r_n := \sqrt{\frac{(2+\epsilon)\log p}{n}}.$$

In addition, we also want that

$$P(\gamma_i = 1 \mid \mathbf{y}) > P(\gamma_i = 0 \mid \mathbf{y}), \text{ if } |\hat{\beta}_i| > \sqrt{\frac{c \log p}{n}}, \tag{3.17}$$

for large enough c.

Another type of consistency we consider is the strong selection consistency which requires that $P[\boldsymbol{\gamma} = \boldsymbol{t} \mid \mathbf{y}] \xrightarrow{\text{P}} 1$. A sufficient condition to achieve this is to have $\sum_{j=1}^{p}(1 - P[\gamma_j = t_j \mid) \xrightarrow{\text{P}} 0$. This in turn holds true if $P[\gamma_j = t_j \mid >(1 - c'/p)$, for all j with high probability for c' small enough. To achieve such a consistency, τ_1 may need to be dependent on sample size n and number of variables p. This dependence of τ_1 on n and p for achieving could depend on the base density function f. We will investigate which choices for the hyperparameters will yield selection consistency results for a general base-density function.

Let us assume that $\hat{\beta}_i > 0$ without loss of generality. To see which choices of τ_1^2 would satisfy (3.16) and (3.17) for a given f, consider

$$\begin{aligned}
\frac{E_{\hat{\beta}_i}(p_1(B))}{E_{\hat{\beta}_i}(p_0(B))} &= \frac{\int \phi\left(\sqrt{n}(b - \hat{\beta}_i)\right) p_1(b) db}{\int \phi\left(\sqrt{n}(b - \hat{\beta}_i)\right) p_0(b) db} \\
&\leq \frac{\int \phi\left(\sqrt{n}(b - \hat{\beta}_i)\right) p_1(b) db}{\int_{|b - \hat{\beta}_i| \leq \frac{c_n}{\sqrt{n}}} \phi\left(\sqrt{n}(b - \hat{\beta}_i)\right) p_0(b) db} \\
&= \frac{\tau_0 \int \phi\left(\sqrt{n}(b - \hat{\beta}_i)\right) f(b/\tau_1) db}{\tau_1 \int \phi\left(\sqrt{n}(b - \hat{\beta}_i)\right) f(b/\tau_0) db} \\
&\leq \frac{\tau_0 \int \phi\left(\sqrt{n}(b - \hat{\beta}_i)\right) f(b/\tau_1) db}{\tau_1 \int_{|b - \hat{\beta}_i| \leq \frac{c_n}{\sqrt{n}}} \phi\left(\sqrt{n}(b - \hat{\beta}_i)\right) f(b/\tau_0) db} \\
&\preceq \frac{\tau_0 f(0)}{\tau_1 f\left(\frac{b_i}{\tau_0}\right)},
\end{aligned}$$

where $b_i = |\hat{\beta}_i| + \frac{c_n}{\sqrt{n}}$, and c_n is such that $c_n \to \infty, c_n = o(\sqrt{\log p})$. Since $\max_{i:\beta_i=0} b_i < r_n := \sqrt{\frac{(2+\epsilon)\sigma^2 \log p}{n}}$ with high probability, we have

$$\begin{aligned}
\max_{i:\beta_i=0} \frac{\theta E_{\hat{\beta}_i}(p_1(B))}{(1-\theta) E_{\hat{\beta}_i}(p_0(B))} &\preceq \frac{\theta p_1(0)}{(1-\theta) p_0(r_n)} \\
&\leq C_n \to 0,
\end{aligned}$$

where the last convergence follows due to Condition 3 (a).

$$\begin{aligned}
\max_{i:\beta_i=0} \frac{\theta E_{\hat{\beta}_i}(p_1(B))}{(1-\theta) E_{\hat{\beta}_i}(p_0(B))} &\preceq \frac{\theta \tau_0 f(0)}{(1-\theta)\tau_1 f\left(\frac{r_n}{\tau_0}\right)} \\
&\leq \frac{\theta \tau_0 f(0)}{(1-\theta)\tau_1 f\left(\frac{r_n}{\tau_0}\right)} \xrightarrow{\text{P}} 0,
\end{aligned} \tag{3.18}$$

where the last convergence follows due to Condition 3 (a).

For the active covariates,

$$\begin{aligned}
\frac{E_{\hat{\beta}_i}(p_1(B))}{E_{\hat{\beta}_i}(p_0(B))} &= \frac{\tau_0 \int \phi\left(\sqrt{n}(b-\hat{\beta}_i)\right) f(b/\tau_1) db}{\tau_1 \int \phi\left(\sqrt{n}(b-\hat{\beta}_i)\right) f(b/\tau_0) db} \\
&\geq \frac{\tau_0 f(|\hat{\beta}_i| + \frac{c_n}{\sqrt{n}}/\tau_1)}{\tau_1 (f(|\hat{\beta}_i| - \frac{c_n}{\sqrt{n}}/\tau_0) + w e^{-c_n^2/2})} \\
&\geq \frac{\tau_0 c}{\tau_1 (f(\beta_i/2\tau_0) + w e^{-c_n^2/2})}, \\
&\geq \frac{\tau_0 c}{\tau_1 (f(m_n/2\tau_0) + w e^{-c_n^2/2})},
\end{aligned}$$

with high probability. Therefore, from Condition 3 (a), we will have

$$\min_{i:\beta_i \neq 0} \frac{\theta E_{\hat{\beta}_i}(p_1(B))}{(1-\theta) E_{\hat{\beta}_i}(p_0(B))} \xrightarrow{\text{P}} \infty, \tag{3.19}$$

which proves the weak selection consistency.

Now, to show strong selection consistency, let us consider the posterior on the entire vector γ which is given by

$$\begin{aligned}
P(\gamma = \mathbf{t} \mid \mathbf{y}) &\propto \prod_{i=1}^{p} \int \exp\left\{-\frac{n}{2\sigma^2}(b-\hat{\beta}_i)^2\right\} \left((1-\theta)p_0(b)\right)^{1-t_i} \left(\theta p_1(b)\right)^{t_i} db \\
&= \prod_{i:t_i=1} \frac{\theta E_{\hat{\beta}_i}(p_1(B))}{(1-\theta)E_{\hat{\beta}_i}(p_0(B)) + \theta E_{\hat{\beta}_i}(p_1(B))} \\
&\quad \times \prod_{i:t_i=0} \frac{(1-\theta)E_{\hat{\beta}_i}(p_0(B))}{(1-\theta)E_{\hat{\beta}_i}(p_0(B)) + \theta E_{\hat{\beta}_i}(p_1(B))}
\end{aligned}$$

Using Equation (3.18) from the weak selection consistency proof, we have that

$$\begin{aligned}
\min_{i:t_i=0} \frac{(1-\theta)E_{\hat{\beta}_i}(p_0(B))}{(1-\theta)E_{\hat{\beta}_i}(p_0(B)) + \theta E_{\hat{\beta}_i}(p_1(B))} &\geq (1 - \max_{i:t_i=0} P(\gamma_i = 1 \mid \mathbf{y})) \\
&\geq \left(1 - \frac{\theta \tau_0 w}{(1-\theta)\tau_1 f\left(\frac{r_n}{\tau_0}\right)}\right).
\end{aligned}$$

As we have $\frac{p\theta\tau_0 w}{(1-\theta)\tau_1 f\left(\frac{r_n}{\tau_0}\right)} \to 0$ due to Condition 4, this would imply that

$$\prod_{i:t_i=0} \frac{(1-\theta)E_{\hat{\beta}_i}(p_0(B))}{(1-\theta)E_{\hat{\beta}_i}(p_0(B)) + \theta E_{\hat{\beta}_i}(p_1(B))} \geq \left(1 - \frac{p\theta\tau_0 w}{(1-\theta)\tau_1 f\left(\frac{r_n}{\tau_0}\right)}\right) \xrightarrow{\text{P}} 1.$$

For the other part with $t_i = 1$,

$$\min_{i:t_i=1} \frac{\theta E_{\hat{\beta}_i}(p_1(B))}{(1-\theta)E_{\hat{\beta}_i}(p_0(B)) + \theta E_{\hat{\beta}_i}(p_1(B))} \geq \min_{i:t_i=1} \frac{w_i}{1+w_i},$$

where $w_i = \frac{\theta E_{\hat{\beta}_i}(p_1(B))}{(1-\theta)E_{\hat{\beta}_i}(p_0(B))} \geq \frac{U_n}{\tau_1} \to \infty$ due to the condition on prior parameters and the results from the previous proof. Therefore, to show the product of p_t such numbers will go to one, it is sufficient to have that $\frac{p_t \tau_1}{U_n} \to 0$, which implies $P[\gamma = \mathbf{t}] \xrightarrow{\text{P}} 1$.

Proof of Theorem 3.4.3:
To prove this threorem, it is sufficient to prove that the conditions of the theorem imply the conditions required for Theorem 3.4.2. It can be shown easily that $C_n \leq L_n$. As τ_1 can be replaced with a slightly larger one, it is sufficient to have T_n defined here as

$$T_n := \inf_{m_n \leq |a|} \frac{\theta\, p_1(a)}{(1-\theta)\, p_0(a)},$$

because ϵ_n is small and the difference can be adjusted.

It can also be shown that $T_n \geq \frac{U_n}{\tau_1} \to \infty$ if we assume that the function $\frac{f(b/\tau_1)}{f(b/\tau_0)}$ is decreasing for $|b| \geq m_n$. Also, in U_n, we won't need to add the term $\exp\left\{-\frac{nm_n^2}{2}\right\}$ if we assume the Condition 5 on the tail of f_0 that $f_0(b) \geq \phi(\sqrt{n}b)$. In that case, $U_n := \frac{\theta\tau_0}{(1-\theta)f\left(\frac{m_n}{2\tau_0}\right)}$. Note that the $f(m_n/\tau_1)$ part in the numerator is not present because $\tau_1 \succeq m_n$. These arguments show that the conditions of Theorem 3.4.2 are satisfied.

Acknowledgement

The author gratefully acknowledges grant support from National Science Foundation grants NSF DMS-1811768 and NSF CAREER-1943500.

Bibliography

[1] M.R. Andersen, A. Vehtari, O. Winther, and L. K. Hansen. Bayesian inferene for spatio-temporal spike-and-slab priors. *The Journal of Machine Learning Research*, 18:1–58, 2017.

[2] M. M. Barbieri and J. O. Berger. Optimal predictive model selection. *Annals of Statistics*, 32:870–897, 2004.

[3] A. Bhattacharya, A. Chakraborty, and B. Mallick. Fast sampling with Gaussian scale mixture priors in high-dimensional regression. *Biometrika*, 103:985–991, 2016.

[4] A. Cassese, W. Zhu, M. Guindani, and M. Vannucci. A bayesian nonparametric spiked process prior for dynamic model selection. *Bayesian Analysis*, 14(2):553–572, 2019.

[5] I. Castillo, J. Schmidt-Hieber, and A. van der Vaart. Bayesian linear regression with sparse priors. *Annals of Statistics*, 43:1986–2018, 2015.

[6] I. Castillo and A. van der Vaart. Needles and straw in a haystack: Posterior concentration for possibly sparse sequences. *Annals of Statistics*, 40:2069–2101, 2012.

[7] S. K. Deshpande, V. Rockova, and E. I. George. Simultaneous variable and covariance selection with the multivariate spike-and-slab lasso. *Journal of Computational and Graphical Statistics*, 28(4):921–931, 2019.

[8] J. Fan and R. Li. Variable selection via nonconcave penalized likelihood and its oracle properties. *Journal of the American Statistical Association*, 96:1348–1360, 2001.

[9] J. Fan and Jinchi. Lv. A selective overview of variable selection in high dimensional feature space. *Statistica Sinica*, 20:101–148, 2010.

[10] L. Gan, N. N Narisetty, and F. Liang. Bayesian regularization for Graphical models with unequal shrinkage. *Journal of the American Statistical Association*, 114(527):1218–1231, 2019.

[11] L. Gan, X. Yang, N. N. Narisetty, and F. Liang. Bayesian joint estimation of multiple graphical models. In H. Wallach, H. Larochelle, A. Beygelzimer, F. d'Alché Buc, E. Fox, and R. Garnett, editors, *Advances in Neural Information Processing Systems 32*, pages 9802–9812. Curran Associates, Inc., 2019.

[12] E. I. George and R. E. McCulloch. Variable selection via Gibbs sampling. *Journal of the American Statistical Association*, 88:881–889, 1993.

[13] E. I. George and R. E. McCulloch. Approaches for Bayesian variable selection. *Statistica Sinica*, 7:339–373, 1997.

[14] C. Hans, A. Dobra, and M. West. Shotgun stochastic search for "large p" regression. *Journal of the American Statistical Association*, 102:507–516, 2007.

[15] H. Ishwaran and J. S. Rao. Spike and slab variable selection: Frequentist and Bayesian strategies. *Annals of Statistics*, 33:730–773, 2005.

[16] H. Ishwaran and J. S. Rao. Consistency of spike and slab regression. *Statistics and Probability Letters*, 81:1920–1928, 2011.

[17] V. E. Johnson and D. Rossell. Bayesian model selection in high-dimensional settings. *Journal of the American Statistical Association*, 107:649–660, 2012.

[18] H. Lan, M. Chen, J. B. Flowers, B. S. Yandell, D. S. Stapleton, C. M. Mata, E. T. Mui, M. T. Flowers, K. L. Schueler, K. F. Manly, R. W. Williams, K. Kendziorski, and A. D. Attie. Combined expression trait correlations and expression quantitative trait locus mapping. *PLoS Genetics*, 2:e6, 2006.

[19] Z. Li, T. Mccormick, and S. Clark. Bayesian joint spike-and-slab graphical lasso. In Kamalika Chaudhuri and Ruslan Salakhutdinov, editors, *Proceedings of Machine Learning Research*, volume 97, pages 3877–3885, Long Beach, California, USA, 09–15 Jun 2019. PMLR.

[20] F. Liang, Q. Song, and K. Yu. Bayesian subset modeling for high dimensional generalized linear models. *Journal of the American Statistical Association*, 108:589–606, 2013.

[21] G. Malsiner-Walli and H. Wagner. Comparing spike and slab priors for Bayesian variable selection. *Austrian Journal of Statistics*, 40(4):241–264, Feb. 2016.

[22] T. J. Mitchell and J. J. Beauchamp. Bayesian variable selection in linear regression. *Journal of the American Statistical Association*, 83:1023–1032, 1988.

[23] N. N. Narisetty and X. He. Bayesian variable selection with shrinking and diffusing priors. *Annals of Statistics*, 42:789–817, 2014.

[24] N. N. Narisetty, J. Shen, and X. He. Skinny Gibbs: A scalable and consistent Gibbs sampler for model selection. *Journal of the American Statistical Association*, 114:1205–1217, 2019.

[25] T. Park and G. Casella. The Bayesian lasso. *Journal of the American Statistical Association*, 103:681–686, 2008.

[26] C. Peterson, F. C. Stingo, and M. Vannucci. Bayesian inference of multiple Gaussian graphical models. *Journal of the American Statistical Association*, 110(509):159–174, 2015.

[27] C. B. Peterson, F. C. Stingo, and M. Vannucci. Joint Bayesian variable and graph selection for regression models with network-structured predictors. *Statistics in medicine*, 35(7):1017–1031, 2016.

[28] V. Ročková. Bayesian estimation of sparse signals with a continuous spike-and-slab prior. *Annals of Statistics*, 2018.

[29] V. Ročková and E. I. George. EMVS: The EM approach to Bayesian variable selection. *Journal of the American Statistical Association*, 109(506):828–846, 2014.

[30] V. Ročková and E. I. George. The spike-and-slab lasso. *Journal of the American Statistical Association*, 113:431–444, 2018.

[31] V. Ročková and K. McAlinn. Dynamic variable selection with spike-and-slab process priors. *Bayesian Analysis*, 2020. Advance publication.

[32] S. Scheipl, L. Fahrmeir, and T. Kneib. Spike-and-slab priors for function selection in structured additive regression models. *Journal of the American Statistical Association*, 107(500):1518–1532, 2012.

[33] G.S. Scott and J.O. Berger. Bayes and empirical-Bayes multiplicity adjustment in the variable-selection problem. *Annals of Statistics*, 38:2587–2619, 2010.

[34] S. L. Scott and H. R. J. Varian. Bayesian variable selection for nowcasting economic time series. In Avi Goldfarb, Shane M. Greenstein, and Catherine E. Tucker, editors, *Economic Analysis of the Digital Economy*, pages 119 – 135. University of Chicago Press, National Bureau of Economic Research, 2015.

[35] N. Sha, M. G. Tadesse, and M. Vannucci. Bayesian variable selection for the analysis of microarray data with censored outcomes. *Bioinformatics*, 22(18):2262–2268, 2006.

[36] Z. Tang, Y. Shen, X. Zhang, and N. Yi. The spike-and-slab lasso generalized linear models for prediction and associated genes detection. *Genetics*, 205(1):77–88, 2017.

[37] R. Tibshirani. Regression shrinkage and selection via the lasso. *Journal of the Royal Statistical Society, Series B*, 58:267–288, 1996.

[38] L. Wasserman and K. Roeder. High-dimensional variable selection. *Annals of Statistics*, 37:2178–2201, 2009.

[39] M. West. On scale mixtures of normal distributions. *Biometrika*, 74(3):646–648, 09 1987.

[40] Y. Yang, M. J. Wainwright, and M. I. Jordan. On the computational complexity of high-dimensional Bayesian variable selection. *Annals of Statistics*, 44:2497–2532, 2016.

[41] C. H. Zhang. Nearly unbiased variable selection under minimax concave penalty. *The Annals of Statistics*, 38:894–942, 2010.

[42] H. Zou. The adaptive lasso and its oracle properties. *Journal of the American Statistical Association*, 101:1418–1429, 2006.

4

Spike-and-Slab Meets LASSO: A Review of the Spike-and-Slab LASSO

Ray Bai

The University of South Carolina (USA)

Veronika Ročková

The University of Chicago (USA)

Edward I. George

The University of Pennsylvania (USA)

CONTENTS

High-dimensional data sets have become ubiquitous in the past few decades, often with many more covariates than observations. In the frequentist setting, penalized likelihood methods are the most popular approach for variable selection and estimation in high-dimensional data. In the Bayesian framework, spike-and-slab methods are commonly used as probabilistic constructs for high-dimensional modeling. Within the context of linear regression, [56] introduced the spike-and-slab LASSO (SSL), an approach based on a prior which provides

DOI: 10.1201/9781003089018-4

a continuum between the penalized likelihood LASSO and the Bayesian point-mass spike-and-slab formulations. Since its inception, the spike-and-slab LASSO has been extended to a variety of contexts, including generalized linear models, factor analysis, graphical models, and nonparametric regression. The goal of this chapter is to survey the landscape surrounding spike-and-slab LASSO methodology. First we elucidate the attractive properties and the computational tractability of SSL priors in high dimensions. We then review methodological developments of the SSL and outline several theoretical developments. We illustrate the methodology on both simulated and real datasets.

4.1 Introduction

High-dimensional data are now routinely analyzed. In these settings, one often wants to impose a low-dimensional structure such as sparsity. For example, in astronomy and other image processing contexts, there may be thousands of noisy observations of image pixels, but only a small number of these pixels are typically needed to recover the objects of interest [33, 34]. In genetic studies, scientists routinely observe tens of thousands of gene expression data points, but only a few genes may be significantly associated with a phenotype of interest. For example, [68] has confirmed that only seven genes have a non-negligible association with Type I diabetes. Among practitioners, the main objectives in these scenarios are typically: a) identification (or *variable selection*) of the non-negligible variables, and b) *estimation* of their effects.

A well-studied model for sparse recovery in the high-dimensional statistics literature is the normal linear regression model,

$$\boldsymbol{y} = \boldsymbol{X}\boldsymbol{\beta} + \boldsymbol{\varepsilon}, \qquad \boldsymbol{\varepsilon} \sim \mathcal{N}_n(\boldsymbol{0}, \sigma^2 \boldsymbol{I}_n), \tag{4.1}$$

where $\boldsymbol{y} \in \mathbb{R}^n$ is a vector of n responses, $\boldsymbol{X} = [\boldsymbol{x}_1, \dots, \boldsymbol{x}_p] \in \mathbb{R}^{n \times p}$ is a design matrix of p potential covariates, $\boldsymbol{\beta} = (\beta_1, \dots, \beta_p)^T$ is a p-dimensional vector of unknown regression coefficients, and $\boldsymbol{\varepsilon}$ is the noise vector. When $p > n$, we often assume that most of the elements in $\boldsymbol{\beta}$ are zero or negligible. Under this setup, there have been a large number of methods proposed for selecting and estimating the active coefficients in $\boldsymbol{\beta}$. In the frequentist framework, penalized likelihood approaches such as the least absolute shrinkage and selection operator (LASSO) [59] are typically used to achieve sparse recovery for $\boldsymbol{\beta}$. In the Bayesian framework, spike-and-slab priors are a popular approach for sparse modeling of $\boldsymbol{\beta}$.

The spike-and-slab LASSO (SSL), introduced by [56], forms a continuum between these penalized likelihood and spike-and-slab constructs. The spike-and-slab LASSO methodology has experienced rapid development in recent years, and its scope now extends well beyond the normal linear regression model (4.1). The purpose of this chapter is to offer a timely review of the SSL and its many variants. We first provide a basic review of frequentist and Bayesian approaches to high-dimensional variable selection and estimation under the model (4.1). We then review the spike-and-slab LASSO and provide an overview of its attractive properties and techniques to implement it within the context of normal linear regression. Next, we review the methodological developments of the SSL and some of its theoretical developments.

4.2 Variable Selection in High-Dimensions: Frequentist and Bayesian Strategies

We first review the frequentist penalized regression and the Bayesian spike-and-slab frameworks before showing how the spike-and-slab LASSO bridges the gap between them.

4.2.1 Penalized Likelihood Approaches

In the frequentist high-dimensional literature, there have been a variety of penalized likelihood approaches proposed to estimate $\boldsymbol{\beta}$ in (4.1). A variant of the penalized likelihood approach estimates $\boldsymbol{\beta}$ with

$$\widehat{\boldsymbol{\beta}} = \arg\max_{\boldsymbol{\beta}\in\mathbb{R}^p} -\frac{1}{2}\|\boldsymbol{y} - \boldsymbol{X}\boldsymbol{\beta}\|_2^2 + \text{pen}_\lambda(\boldsymbol{\beta}), \tag{4.2}$$

where $\text{pen}_\lambda(\boldsymbol{\beta})$ is a penalty function indexed by penalty parameter λ. Most of the literature has focused on penalty functions which are separable, i.e., $\text{pen}_\lambda(\boldsymbol{\beta}) = \sum_{j=1}^p \rho_\lambda(\beta_j)$. In particular, the popular least absolute shrinkage and selection operator (LASSO) penalty of [59] uses the function $\rho_\lambda(\beta_j) = -\lambda|\beta_j|$. Besides the LASSO and its many variants [7, 60, 71, 73, 74], other popular choices for $\rho_\lambda(\cdot)$ include non-concave penalty functions, such as the smoothly clipped absolute deviation (SCAD) penalty [19] and the minimax concave penalty (MCP) [70]. All of the aforementioned penalties threshold some coefficients to zero, thus enabling them to perform variable selection and estimation simultaneously. In addition, SCAD and MCP also mitigate the well-known estimation bias of the LASSO.

Any penalized likelihood estimator (4.2) also has a Bayesian interpretation in that it can be seen as a posterior mode under an independent product prior $p(\boldsymbol{\beta} \mid \lambda) = \prod_{j=1}^p p(\beta_j \mid \lambda)$, where $\text{pen}_\lambda(\boldsymbol{\beta}) = \log p(\boldsymbol{\beta} \mid \lambda) = \sum_{j=1}^p \log p(\beta_j \mid \lambda)$. In particular, the solution to the LASSO is equivalent to the posterior mode under a product of Laplace densities indexed by hyperparameter, λ:

$$p(\boldsymbol{\beta} \mid \lambda) = \prod_{j=1}^p \frac{\lambda}{2} e^{-\lambda|\beta_j|}. \tag{4.3}$$

This prior, known as the Bayesian LASSO, was first introduced by [48]. In [48], both fully Bayes and empirical Bayes procedures were developed to tune the hyperparameter λ in (4.3). The fully Bayes approach of placing a prior on λ, in particular, renders the Bayesian LASSO penalty *non*-separable. Thus, the fully Bayesian LASSO has the added advantage of being able to share information across different coordinates. Despite this benefit, [54] showed that the fully Bayesian LASSO cannot simultaneously adapt to sparsity *and* avoid the estimation bias issue of the original LASSO. In addition, [25] showed that the univariate Bayesian LASSO often undershrinks negligible coefficients, while overshrinking large coefficients. Finally, [16] also proved that the posterior under the Bayesian LASSO contracts at a suboptimal rate. In Sections 4.3–4.6, we will illustrate how the *spike-and-slab LASSO* mitigates these issues.

In addition to the spike-and-slab LASSO, other alternative priors have also been proposed to overcome the limitations of the Bayesian LASSO (4.3). These priors, known as global-local shrinkage (GL) priors, place greater mass around zero and have heavier tails than the Bayesian LASSO. Thus, GL priors shrink small coefficients more aggressively towards zero, while their heavy tails prevent overshrinkage of large coefficients. Some examples include the normal-gamma prior [26], the horseshoe prior [13], the generalized double Pareto prior [2], the Dirichlet-Laplace prior [10], and the normal-beta prime prior [4]. We refer the reader to [8] for a detailed review of GL priors.

4.2.2 Spike-and-Slab Priors

In the Bayesian framework, variable selection under the linear model (4.1) arises directly from probabilistic considerations and has frequently been carried out through placing spike-and-slab priors on the coefficients of interest. The spike-and-slab prior was first introduced by [42] and typically has the following form,

$$\begin{aligned} p(\boldsymbol{\beta} \mid \boldsymbol{\gamma}, \sigma^2) &= \prod_{j=1}^{p} \left[(1-\gamma_j)\delta_0(\beta_j) + \gamma_j p(\beta_j \mid \sigma^2) \right], \\ p(\boldsymbol{\gamma} \mid \theta) &= \prod_{j=1}^{p} \theta^{\gamma_j}(1-\theta)^{1-\gamma_j}, \qquad \theta \sim p(\theta), \\ \sigma^2 &\sim p(\sigma^2), \end{aligned} \tag{4.4}$$

where δ_0 is a point mass at zero used to model the negligible entries (the "spike"), $p(\beta_j \mid \sigma^2)$ is a diffuse and/or heavy-tailed density (rescaled by the variance σ^2) to model the non-negligible entries (the "slab"), $\boldsymbol{\gamma}$ is a binary vector that indexes the 2^p possible models, and $\theta \in (0,1)$ is a mixing proportion. The error variance σ^2 is typically endowed with a conjugate inverse gamma prior or an improper Jeffreys prior, $p(\sigma^2) \propto \sigma^{-2}$. With a well-chosen prior on θ, this prior (4.4) also automatically favors parsimonious models in high dimensions, thus avoiding the curse of dimensionality.

The point-mass spike-and-slab prior (4.4) is often considered "theoretically ideal," or a "gold standard" for sparse Bayesian problems [14, 49, 50]. In high dimensions, however, exploring the full posterior over the entire model space using point-mass spike-and-slab priors (4.4) can be computationally prohibitive, in large part because of the combinatorial complexity of updating the discrete indicators $\boldsymbol{\gamma}$. There has been some work to mitigate this issue by using either shotgun stochastic search (SSS) [12, 28] or variational inference (VI) [50] to quickly identify regions of high posterior probability.

As an alternative to the point-mass spike-and-slab prior, fully continuous spike-and-slab models have been developed. In these continuous variants, the point-mass δ_0 in (4.4) is replaced by a continuous density that is heavily concentrated about zero. The first such continuous relaxation was made by [23], who used a normal density with very small variance for the spike and a normal density with very large variance for the slab. Specifically, the prior for $\boldsymbol{\beta}$ in [23] is

$$p(\boldsymbol{\beta} \mid \boldsymbol{\gamma}, \sigma^2) = \prod_{j=1}^{p} \left[(1-\gamma_j)\mathcal{N}(0, \sigma^2\tau_0^2) + \gamma_j \mathcal{N}(0, \sigma^2\tau_1^2) \right], \tag{4.5}$$

where $0 < \tau_0^2 \ll \tau_1^2$. [23] developed a stochastic search variable selection (SSVS) procedure based on posterior sampling with Markov chain Monte Carlo (MCMC) and thresholding the posterior inclusion probabilities, $\Pr(\gamma_j = 1 \mid \boldsymbol{y}), j = 1, \ldots, p$. In practice, the "median thresholding" rule [6], i.e., $\Pr(\gamma_j = 1 \mid \boldsymbol{y}) > 0.5, j = 1, \ldots, p$, is often used to perform variable selection. [29, 45] further extended the model (4.5) by rescaling the variances τ_0^2 and τ_1^2 with sample size n in order to better control the amount of shrinkage for each individual coefficient.

To further reduce the computational intensiveness of SSVS, a deterministic optimization procedure called EM variable selection (EMVS) was developed by [53]. The EMVS procedure employs (4.5) as the prior for $\boldsymbol{\beta}$ and uses an EM algorithm to target the posterior mode for $(\boldsymbol{\beta}, \theta, \sigma)$. ([53] also consider continuous spike-and-slab models where the slab,

$\mathcal{N}(0, \sigma^2\tau_1^2)$, is replaced with a polynomial-tailed density, such as a Student's t or a Cauchy distribution, to prevent overshrinkage of the non-negligible entries in $\boldsymbol{\beta}$). Compared to SSS and SSVS, EMVS has been shown to more rapidly and reliably identify those sets of higher probability submodels which may be of most interest [53]. Recently, [35] proposed a general algorithmic framework for Bayesian variable selection with graph-structured sparsity which subsumes the EMVS algorithm as a special case. Like SSVS, these algorithms also require thresholding the posterior inclusion probabilities to perform variable selection. Letting $(\widehat{\boldsymbol{\beta}}, \widehat{\theta}, \widehat{\sigma})$ denote the posterior mode for $(\boldsymbol{\beta}, \theta, \sigma)$, [53] recommend using median thresholding, $\Pr(\gamma_j = 1 \mid \widehat{\boldsymbol{\beta}}, \widehat{\theta}, \widehat{\sigma}) > 0.5$, for selection.

4.3 The Spike-and-Slab LASSO

Having reviewed the penalized likelihood and spike-and-slab paradigms for sparse modeling in the normal linear regression model (4.1), we are now in a position to review the spike-and-slab LASSO (SSL) of [56]. The SSL forms a bridge between these two parallel developments, thereby combining the strengths of both approaches into a single procedure. Throughout this section and Section 4.4, we assume that $\boldsymbol{y}$ has been centered at zero to avoid the need for an intercept and that the design matrix $\boldsymbol{X}$ has been centered and standardized so that $\|\boldsymbol{x}_j\|_2^2 = n$ for all $1 \leq j \leq p$.

4.3.1 Prior Specification

The spike-and-slab LASSO prior is specified as

$$\begin{aligned} p(\boldsymbol{\beta} \mid \gamma) &= \prod_{j=1}^{p} \left[(1-\gamma_j)\psi(\beta_j \mid \lambda_0) + \gamma_j \psi(\beta_j \mid \lambda_1)\right], \\ p(\gamma \mid \theta) &= \prod_{j=1}^{p} \left[\theta^{\gamma_j}(1-\theta)^{1-\gamma_j}\right], \\ \theta &\sim \mathcal{B}eta(a, b), \end{aligned} \tag{4.6}$$

where $\psi(\beta \mid \lambda) = (\lambda/2)e^{-\lambda|\beta|}$ denotes the Laplace density with scale parameter λ. Figure 4.1 depicts the Laplace density for two different choices of scale parameter. We see that for large λ ($\lambda = 20$), the density is very peaked around zero, while for small λ ($\lambda = 1$), it is diffuse. Therefore, in our prior (4.6), we typically set $\lambda_0 \gg \lambda_1$, so that $\psi(\cdot \mid \lambda_0)$ is the "spike" and $\psi(\cdot \mid \lambda_1)$ is the "slab."

The original SSL model of [56] assumed known variance $\sigma^2 = 1$. [44] extended the SSL to the unknown variance case. As σ^2 is typically unknown, we consider the hierarchical formulation in [44] in this chapter and place an independent Jeffreys prior on σ^2,

$$p(\sigma^2) \propto \sigma^{-2}.$$

Note that unlike the mixture of normals (4.5), we do *not* scale the Laplace priors in $p(\boldsymbol{\beta} \mid \gamma)$ by σ^2. [44] showed that such scaling severely underestimates the variance σ^2 when $\boldsymbol{\beta}$ is sparse or when $p > n$, thus making the model prone to overfitting.

By choosing $\lambda_1 = \lambda_0$ in (4.6), we obtain the familiar LASSO ℓ_1 penalty. On the other hand, if $\lambda_0 \to \infty$, we obtain the "theoretically ideal" point-mass spike-and-slab (4.4) as a limiting case. Thus, a feature of the SSL prior is its ability to induce a non-concave continuum between the penalized likelihood and (point-mass) spike-and-slab constructs.

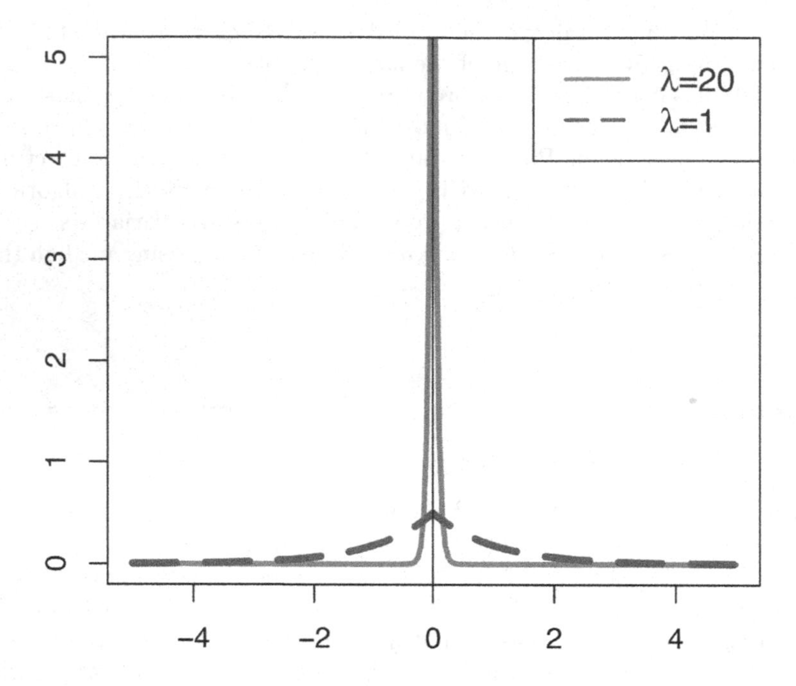

FIGURE 4.1
Plot of the central region for the Laplace density with two different choices of scale parameter.

Since it is a mixture of two Laplace distributions, the SSL prior (4.6) can be seen as a two-group refinement of the LASSO's ℓ_1 penalty on the coefficients. Thus, the posterior mode for $p(\boldsymbol{\beta} \mid \boldsymbol{y})$ under (4.6) is *exactly* sparse and can be used to perform simultaneous variable selection and parameter estimation. This automatic modal thresholding property offers an advantage over previous spike-and-slab formulations (4.4)–(4.5) which do not give exactly sparse estimates of the coefficients and which typically require *post hoc* thresholding of the posterior inclusion probabilities for selection.

It is well-known that the original LASSO [59] suffers from estimation bias, wherein coefficients with large magnitude are overshrunk. One may wonder what advantages the SSL (4.6) confers over penalized likelihood approaches such as the adaptive LASSO, SCAD, or MCP penalties [19, 70, 73] which are designed to mitigate the bias problem of the LASSO. In what follows, we discuss two major advantages of the SSL. First, we demonstrate that the SSL mixes two LASSO "bias" terms *adaptively* in such a way that either a very large amount of shrinkage is applied if $|\beta_j|$ is small or a very small amount of shrinkage is applied if $|\beta_j|$ is large. This is in contrast to the adaptive LASSO [73] and similar penalties which assign *fixed* coefficient-specific penalties and thus do not gear the coefficient-specific shrinkage towards these extremes. Second, the prior on θ in (4.6) ultimately renders the coordinates in $\boldsymbol{\beta}$ *dependent* in the marginal prior $p(\boldsymbol{\beta})$ and the SSL penalty *non*-separable. This provides the SSL with the additional ability to borrow information across coordinates and adapt to ensemble information about sparsity.

4.3.2 Selective Shrinkage and Self-Adaptivity to Sparsity

As noted in Section 4.2.1, any sparsity-inducing Bayesian prior can be recast in the penalized likelihood framework by treating the logarithm of the marginal prior $\log p(\boldsymbol{\beta})$ as a penalty function. The SSL penalty is defined as

$$\text{pen}(\boldsymbol{\beta}) = \log\left[\frac{p(\boldsymbol{\beta})}{p(\mathbf{0}_p)}\right], \tag{4.7}$$

where the penalty has been centered at $\mathbf{0}_p$, the p-dimensional zero vector, so that $\text{pen}(\mathbf{0}_p) = 0$ [56]. Using (4.7) and some algebra, the log posterior under the SSL prior (up to an additive constant) can be shown to be

$$L(\boldsymbol{\beta}, \sigma^2) = -\frac{1}{2\sigma^2}\|\boldsymbol{y} - \boldsymbol{X}\boldsymbol{\beta}\|_2^2 - (n+2)\log\sigma + \sum_{j=1}^{p} \text{pen}(\beta_j \mid \theta_j), \tag{4.8}$$

where for $j = 1, \ldots, p$,

$$\text{pen}(\beta_j \mid \theta_j) = -\lambda_1|\beta_j| + \log[p^\star_{\theta_j}(0)/p^\star_{\theta_j}(\beta_j)], \tag{4.9}$$

with

$$p^\star_{\theta_j}(\beta_j) = \frac{\theta_j\psi(\beta_j \mid \lambda_1)}{\theta_j\psi(\beta_j \mid \lambda_1) + (1-\theta_j)\psi(\beta_j \mid \lambda_0)}, \tag{4.10}$$

and

$$\theta_j = E[\theta \mid \boldsymbol{\beta}_{\setminus j}] = \int \theta p(\theta \mid \boldsymbol{\beta}_{\setminus j})d\theta. \tag{4.11}$$

When p is large, [56] noted that θ_j is very similar to $E[\theta \mid \boldsymbol{\beta}] = \int \theta p(\theta \mid \boldsymbol{\beta})d\theta$ for every $j = 1, \ldots, p$. Thus, for practical purposes, we replace the individual θ_j's in (4.8)–(4.11) with a single $\widehat{\theta} = E[\theta \mid \boldsymbol{\beta}]$ going forward.

The connection between the SSL and penalized likelihood methods is made clearer when considering the derivative of each singleton penalty $\text{pen}(\beta_j \mid \widehat{\theta})$ in (4.9). This derivative corresponds to an implicit bias term [56] and is given by

$$\frac{\partial\text{pen}(\beta_j \mid \widehat{\theta})}{\partial|\beta_j|} = -\lambda^\star_{\widehat{\theta}}(\beta_j), \tag{4.12}$$

where

$$\lambda^\star_{\widehat{\theta}}(\beta_j) = \lambda_1 p^\star_{\widehat{\theta}}(\beta_j) + \lambda_0[1 - p^\star_{\widehat{\theta}}(\beta_j)]. \tag{4.13}$$

The Karush-Kuhn-Tucker (KKT) conditions yield the following necessary condition for the global mode $\widehat{\boldsymbol{\beta}}$:

$$\widehat{\beta}_j = \frac{1}{n}\left[|z_j| - \sigma^2\lambda^\star_{\widehat{\theta}}(\widehat{\beta}_j)\right]_+ \text{sign}(z_j), \qquad j = 1, \ldots, p, \tag{4.14}$$

where $z_j = \boldsymbol{x}_j^T(\boldsymbol{y} - \sum_{k\neq j}\boldsymbol{x}_k\widehat{\beta}_k)$. Notice that the condition (4.14) resembles the soft-thresholding operator for the LASSO, except that it contains an adaptive penalty term $\lambda^\star_{\widehat{\theta}}$ for *each* coefficient. In particular, the quantity (4.13) is a weighted average of the two regularization parameters, λ_1 and λ_0, and the weight $p^\star_{\widehat{\theta}}(\beta_j)$. Thus, (4.13)-(4.14) show that the SSL penalty induces an *adaptive* regularization parameter which applies a different amount of shrinkage to each coefficient, unlike the original LASSO which applies the same shrinkage to every coefficient.

It is worth looking at the term $p^\star_{\widehat{\theta}}(\beta_j)$ more closely. In light of (4.10), this quantity can be viewed as a conditional probability that β_j was drawn from the slab distribution rather than the spike distribution, having seen the regression coefficient β_j. We have $p^\star_{\widehat{\theta}}(\beta_j) = \Pr(\gamma_j = 1 \mid \beta_j, \hat{\theta})$, where

$$p^\star_{\widehat{\theta}}(\beta_j) = \frac{1}{1 + \frac{(1-\hat{\theta})}{\hat{\theta}}\frac{\lambda_0}{\lambda_1}\exp\left[-|\beta_j|(\lambda_0 - \lambda_1)\right]} \tag{4.15}$$

is an *exponentially increasing* function in $|\beta_j|$. From (4.15), we see that the functional $p^\star_{\hat{\theta}}$ has a sudden increase from near-zero to near-one. Therefore, $p^\star_{\hat{\theta}}(\beta_j)$ gears $\lambda^\star_{\hat{\theta}}$ in (4.13) towards the extreme values λ_1 and λ_0, depending on the size of $|\beta_j|$. Assuming that λ_1 is sufficiently small (and hence, the slab $\psi(\beta_j \mid \lambda_1)$ is sufficiently diffuse), this allows the large coefficients to escape the overall shrinkage effect, in sharp contrast to the single Laplace distribution (4.3), where the bias issue remains even if a prior is placed on λ [54].

Apart from its selective shrinkage property, a second key benefit of the SSL model (4.6) is its *self-adaptivity* to the sparsity pattern of the data through the prior on the mixing proportion θ, $p(\theta) \sim \mathcal{B}eta(a, b)$. As mentioned previously, this prior ultimately renders the SSL penalty *non*-separable. Fully separable penalty functions, such as those described in Section 4.2.1, are limited by their inability to adapt to common features across model parameters because they treat these parameters independently. In contrast, treating θ (the expected proportion for non-negligible coefficients in $\boldsymbol{\beta}$) as random, allows for automatic adaptivity to different levels of sparsity. As shown in (4.10)–(4.11) (and replacing θ_j with $\hat{\theta}$ and $\boldsymbol{\beta}_{\setminus j}$ with $\boldsymbol{\beta}$), the mixing weight $p^\star_{\hat{\theta}}$ is obtained by averaging $p^\star_{\theta}(\cdot)$ over $p(\theta \mid \boldsymbol{\beta})$, i.e., $p^\star_{\hat{\theta}}(\beta) = \int_0^1 p^\star_{\theta}(\beta_j) p(\theta \mid \boldsymbol{\beta}) d\theta$. It is through this averaging that the SSL penalty (4.7) is given an opportunity to borrow information across coordinates and learn about the underlying level of sparsity in $\boldsymbol{\beta}$.

For the hyperparameters (a, b) in the beta prior on θ, [56] recommended the default choice of $a = 1, b = p$. By Lemma 4 of [56], this choice ensures that $E[\theta \mid \widehat{\boldsymbol{\beta}}] \sim \widehat{p}_\gamma / p$, where $\widehat{p}_\gamma$ is the number of non-zero coefficients in $\widehat{\boldsymbol{\beta}}$. Further, this choice of hyperparameters results in an automatic multiplicity adjustment [59] and ensures that θ is small (or that most of the coefficients belong to the spike) with high probability. Thus, the SSL also favors parsimonious models in high dimensions and avoids the curse of dimensionality.

4.3.3 The Spike-and-Slab LASSO in Action

Before delving into the implementation details of the SSL, we perform a small simulation study to illustrate the benefits of the adaptive shrinkage of SSL versus the non-adaptive shrinkage of the LASSO. We simulated data of $n = 50$ observations with $p = 12$ predictors generated as four independent blocks of highly correlated predictors. More precisely, n rows of our design matrix $\boldsymbol{X}$ were generated independently from a $\mathcal{N}_p(\mathbf{0}, \boldsymbol{\Sigma})$ distribution with block diagonal covariance matrix $\boldsymbol{\Sigma} = \text{bdiag}(\widetilde{\boldsymbol{\Sigma}}, \ldots, \widetilde{\boldsymbol{\Sigma}})$, where $\widetilde{\boldsymbol{\Sigma}} = \{\widetilde{\sigma}_{ij}\}_{i,j=1}^3, \widetilde{\sigma}_{ij} = 0.9$ if $i \neq j$ and $\widetilde{\sigma}_{ii} = 1$. The response was generated from $\boldsymbol{y} \sim \mathcal{N}_n(\boldsymbol{X}\boldsymbol{\beta}_0, \boldsymbol{I})$, with $\boldsymbol{\beta}_0 = (1.3, 0, 0, 1.3, 0, 0, 1.3, 0, 0, 1.3, 0, 0)'$. Note that only x_1, x_4, x_7, and x_{10} are non-null in this true model.

We fit both the SSL and the LASSO of [59] to this model. Figure 4.2 displays the coefficient paths for both SSL and LASSO as the spike parameter λ_0 in the SSL and the regularization parameter λ in the LASSO are increased. For the SSL, the spike parameter $\lambda_1 = 0.01$ is fixed throughout. Both the SSL and LASSO begin at $\lambda_0 = \lambda = 0$ with the same 12 (non-zero) ordinary least squares (OLS) estimates for $\boldsymbol{\beta}_0$. However, as λ_0 increases for the SSL, the eight smaller OLS estimates are gradually shrunk to zero by the SSL's spike. Meanwhile, the four large estimates are held steady by the SSL's slab, eventually stabilizing at values close to their OLS estimates. The SSL correctly selects the four non-zero coefficients in the true model, demonstrating its self-adaptivity to the true sparsity pattern of the data.

In contrast, Figure 4.2 also shows that as the LASSO's single penalty parameter λ increases, *all* twelve estimates are gradually shrunk to zero. This is because without a slab distribution to help hold the large values steady, the LASSO eventually shrinks all estimates to zero for a large enough λ. Additionally, due to the order in which the 12 estimates have

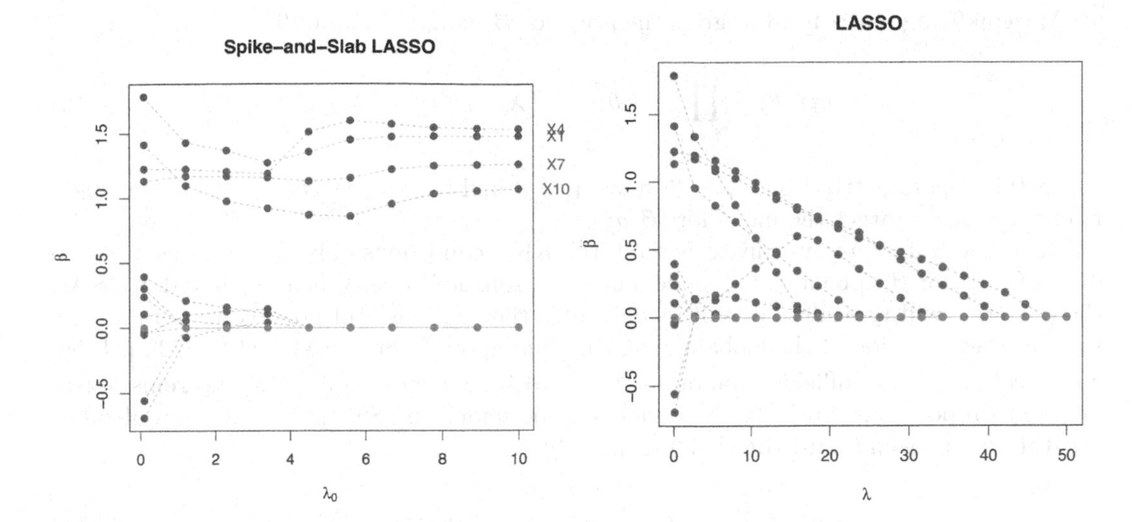

FIGURE 4.2
The coefficient paths of $\widehat{\beta}^{\text{SSL}}$ (left panel) and $\widehat{\beta}^{\text{LASSO}}$ (right panel) as λ_0 and λ respectively are increased. The connected points off the horizontal axis are the non-zero estimates. The points along the horizontal axis are the zero values where the negligible estimates disappear.

been thresholded to zero, no value of λ yields the correct subset selection $\{x_1, x_4, x_y, x_{10}\}$. In particular, if λ is chosen from cross-validation, the LASSO selects a subset of variables with four false positives.

Our small simulation study illustrates the advantage of the SSL over the LASSO. Specifically, because the LASSO applies the same amount of shrinkage to all regression coefficients, it may estimate a null model if its regularization parameter λ is too large. The SSL's two-group refinement of the LASSO penalty helps to mitigate this problem by facilitating selective shrinkage. In Section 4.6, we further illustrate the strong empirical performance of the SSL in high-dimensional settings when $p > n$.

4.4 Computational Details

We now turn our attention to implementation of the SSL model (4.6) under the normal linear regression model (4.1). The method described in Section 4.4.1 is implemented in the publicly available R package `SSLASSO` [57]. However, we also describe an alternative implementation approach in Section 4.4.3, which is amenable to situations outside of the Gaussian likelihood.

4.4.1 Coordinate-wise Optimization

As mentioned in Section 4.3 and shown in (4.14), the (global) posterior mode under the SSL prior (4.6) is exactly sparse, while avoiding the excessive bias issue for large coefficients. Therefore, we can obtain estimates for $\boldsymbol{\beta}$ by targeting the posterior mode.

Marginalizing out $\boldsymbol{\gamma}$ in (4.6) gives the prior for $\boldsymbol{\beta}$ (conditional on θ),

$$p(\boldsymbol{\beta} \mid \theta) = \prod_{j=1}^{p} \left[(1-\theta)\psi(\beta_j \mid \lambda_0) + \theta\psi(\beta_j \mid \lambda_1)\right]. \tag{4.16}$$

Using this reparametrization, [44, 56] developed a highly efficient coordinate ascent algorithm to quickly target the mode for $(\boldsymbol{\beta}, \sigma^2)$.

Since the SSL is a non-convex method, the KKT conditions only give a necessary condition (4.14) for $\widehat{\boldsymbol{\beta}}$ to be a global mode, but not a sufficient one. When $p > n$ and $\lambda_0 \gg \lambda_1$, the posterior will typically be multimodal. Nevertheless, it is still possible to obtain a refined characterization of the global mode. Building upon theory developed by [72], [44, 56] gave necessary *and* sufficient conditions for $\widehat{\boldsymbol{\beta}}$ to be a *global* mode. By Theorems 3-4 in [56] and Propositions 4-5 in [44], the global mode under the SSL prior (4.6) is a blend of soft-thresholding *and* hard-thresholding, namely

$$\widehat{\beta}_j = \frac{1}{n}\left[|z_j| - \sigma^2\lambda^{\star}_{\hat{\theta}}(\widehat{\beta}_j)\right]_+ \operatorname{sign}(z_j)\mathbb{I}(|z_j| > \Delta), \tag{4.17}$$

where $z_j = \boldsymbol{x}_j^T(\boldsymbol{y} - \sum_{k\neq j}\boldsymbol{x}_k\widehat{\beta}_k)$ and $\Delta \equiv \inf_{t>0}[nt/2 - \sigma^2 \text{pen}(t \mid \hat{\theta})/t]$. In [44], an approximation for Δ is given by

$$\Delta = \begin{cases} \sqrt{2n\sigma^2\log[1/p^{\star}_{\hat{\theta}}(0)]} + \sigma^2\lambda_1 & \text{if } g_{\hat{\theta}}(0) > 0, \\ \sigma^2\lambda^{\star}_{\hat{\theta}}(0) & \text{otherwise,} \end{cases}$$

where $g_{\theta}(x) = [\lambda^{\star}_{\theta}(x) - \lambda_1]^2 + (2n/\sigma^2)\log p^{\star}_{\theta}(x)$. The generalized thresholding operator (4.17) allows us to eliminate many suboptimal local modes from consideration through the threshold Δ. This refined characterization also facilitates a highly efficient coordinate ascent algorithm [39] to find the global mode, which we now detail.

After initializing $(\Delta^{(0)}, \boldsymbol{\beta}^{(0)}, \theta^{(0)}, \sigma^{2(0)})$, the coordinate ascent algorithm iteratively updates these parameters until convergence. The update for the threshold Δ at the tth iteration is

$$\Delta^{(t)} = \begin{cases} \sqrt{2n\sigma^{2(t-1)}\log[1/p^{\star}_{\hat{\theta}^{(t-1)}}(0)]} + \sigma^{2(t-1)}\lambda_1 & \text{if } g_{\theta^{(t-1)}}(0) > 0, \\ \sigma^{2(t-1)}\lambda^{\star}_{\theta^{(t-1)}}(0) & \text{otherwise.} \end{cases}$$

Next, $\boldsymbol{\beta}$ is updated as

$$\beta_j^{(t)} \leftarrow \frac{1}{n}\left(|z_j| - \lambda^{\star}_{\hat{\theta}^{(t-1)}}(\widehat{\beta}_j^{(t-1)})\right)_+ \operatorname{sign}(z_j)\mathbb{I}(|z_j| > \Delta^{(t)}).$$

Using the approximation for $E[\theta \mid \widehat{\boldsymbol{\beta}}]$ in Lemma 4 of [56], the update for $\hat{\theta}$ is

$$\hat{\theta}^{(t)} \leftarrow \frac{a + \widehat{p}_{\gamma}^{(t)}}{a + b + p},$$

where $\widehat{p}_{\gamma}^{(t)}$ is the number of non-zero entries in $\boldsymbol{\beta}^{(t)}$. Finally, the update for σ^2 is

$$\sigma^{2(t)} \leftarrow \frac{\|\boldsymbol{y} - \boldsymbol{X}\boldsymbol{\beta}^{(t)}\|_2^2}{n+2}.$$

4.4.2 Dynamic Posterior Exploration

The performance of the SSL model depends on good choices for the hyperparameters (λ_0, λ_1) in (4.6). To this end, [56] recommend a "dynamic posterior exploration" strategy in which the slab hyperparameter λ_1 is held fixed at a small value and the spike hyperparameter λ_0 is gradually increased along a ladder of increasing values, $\{\lambda_0^1, \ldots, \lambda_0^L\}$. The algorithm is not very sensitive to the specific choice of λ_1, provided that the slab is sufficiently diffuse. For each λ_0^s in the ladder for the spike parameters, we reinitialize $(\Delta^{(0)}, \boldsymbol{\beta}^{(0)}, \theta^{(0)}, \sigma^{2(0)})$ using the MAP estimates for these parameters from the previous spike parameter λ_0^{s-1} as a "warm start."

This sequential reinitialization strategy allows the SSL to more easily find the global mode. In particular, when $(\lambda_1 - \lambda_0)^2 < 4$ and σ^2 is fixed, the objective (4.8) is convex. The intuition here is to use the solution to the convex problem as a "warm" start for the non-convex problem (when $\lambda_0 \gg \lambda_1$). As we increase λ_0, the posterior becomes "spikier," with the spikes absorbing more and more of the negligible parameters. Meanwhile, keeping λ_1 fixed at a small value allows the larger coefficients to escape the pull of the spike. For large enough λ_0, the algorithm will eventually stabilize so that further increases in λ_0 do not change the solution. In Section 4.6, we illustrate this with plots of the SSL solution paths.

Additionally, as noted by [44], some care must also be taken when updating σ^2. When $p > n$ and $\lambda_0 \approx \lambda_1$, the model can become saturated, causing the residual variance to go to zero. To avoid this suboptimal mode at $\sigma^2 = 0$, [44] recommend fixing σ^2 until the λ_0 value in the ladder at which the algorithm starts to converge in less than 100 iterations. Then, $\boldsymbol{\beta}$ and σ^2 are simultaneously updated for the next largest λ_0 in the sequence. The complete algorithm for coordinate-wise optimization with dynamic posterior exploration is given in Section 4 of the supplementary material in [44]. This algorithm is implemented in the R package SSLASSO.

4.4.3 EM Implementation of the Spike-and-Slab LASSO

The coordinate ascent algorithm of Section 4.4.1 specifically appeals to the theoretical framework of [72] to search for the global SSL mode $\widehat{\boldsymbol{\beta}}$. An alternative approach, also proposed by [56], is to use an EM algorithm in the vein of EMVS [53]. Again treating the latent variables $\boldsymbol{\gamma}$ in (4.6) are treated as "missing" data, this EM implementation of the SSL proceeds as follows.

In the E-step at the tth iteration, we compute $E[\tau_j \mid \boldsymbol{y}, \boldsymbol{\beta}^{(t-1)}, \theta^{(t-1)}, \sigma^{2(t-1)}] = p^{\star}_{\theta^{(t-1)}}(\beta_j^{(t-1)})$, where $p^{\star}_{\theta}$ is as in (4.10). The M-step then iterates through the following updates:

$$\boldsymbol{\beta}^{(t)} \leftarrow \arg\max_{\boldsymbol{\beta} \in \mathbb{R}^p} \left\{ -\frac{1}{2}\|\boldsymbol{y} - \boldsymbol{X}\boldsymbol{\beta}\|_2^2 - \sum_{j=1}^{p} \sigma^{2(t-1)} \lambda^{\star}_{\theta^{(t-1)}}(\beta_j^{(t-1)})|\beta_j| \right\},$$

$$\theta^{(t)} \leftarrow \frac{\sum_{j=1}^{p} p^{\star}_{\theta^{(t-1)}}(\beta_j^{(t)}) + a - 1}{a + b + p - 2},$$

$$\sigma^{2(t)} \leftarrow \frac{\|\boldsymbol{y} - \boldsymbol{X}\boldsymbol{\beta}^{(t)}\|_2^2}{n + 2},$$

where $\lambda^{\star}_{\theta}(\beta) = \lambda_1 p^{\star}_{\theta}(\beta) + \lambda_0[1 - p^{\star}_{\theta}(\beta)]$. Note that the update for $\boldsymbol{\beta}^{(t)}$ is an adaptive LASSO regression with weights $\sigma^2 \lambda^{\star}_{\theta}$ and hence can be solved very efficiently using coordinate descent algorithms [20]. Like EMVS [53], the dynamic posterior exploration strategy detailed in Section 4.4.2 can be used to find a more optimal mode for $(\boldsymbol{\beta}, \sigma^2)$.

This EM approach can be straightforwardly adapted for other statistical models where the SSL prior (4.6) is used (such as the methods described in Section 4.7) but where the likelihood function differs and the theory of [72] is not applicable. Similar to the coordinate ascent algorithm described in Section 4.4.1, this EM algorithm may be sensitive to the initialization of $(\boldsymbol{\beta}^{(0)}, \theta^{(0)}, \sigma^{2(0)})$. The dynamic posterior exploration strategy described earlier can partly help to mitigate this issue, since the posterior starts out relatively flat when $\lambda_0 \approx \lambda_1$ but becomes "spikier" as λ_0 increases. By the time that the spikes have reappeared, the "warm start" solution from the previous λ_0 in the ladder should hopefully be in the basin of dominant mode. Other strategies such as running the algorithm for a wide choice of starting values or deterministic annealing can also aid in adding robustness against poor initializations [40, 53, 66].

4.5 Uncertainty Quantification

While the algorithms described in Section 4.4 can be used to rapidly target the modes of the SSL posterior, providing a measure of uncertainty for our estimates is a challenging task. In this section, we outline two possible strategies for the task of uncertainty quantification. The first is based on debiasing the posterior mode. The second involves posterior simulation.

4.5.1 Debiasing the Posterior Mode

One possible avenue for uncertainty quantification is to use debiasing [5, 30, 67, 71]. Let $\widehat{\boldsymbol{\Sigma}} = \boldsymbol{X}^T\boldsymbol{X}/n$ and let $\widehat{\boldsymbol{\Theta}}$ be an approximate inverse of $\widehat{\boldsymbol{\Sigma}}$. Note that when $p > n$, $\boldsymbol{X}$ is singular, so $\widehat{\boldsymbol{\Sigma}}^{-1}$ does not necessarily exist. However, we can still obtain a sparse estimate of the precision matrix $\widehat{\boldsymbol{\Theta}}$ for the rows of $\boldsymbol{X}$ by using techniques from the graphical models literature, e.g., the nodewise regression procedure in [41] or the graphical lasso [21]. We define the quantity $\widehat{\boldsymbol{\beta}}_d$ as

$$\widehat{\boldsymbol{\beta}}_d = \widehat{\boldsymbol{\beta}} + \widehat{\boldsymbol{\Theta}}\boldsymbol{X}^T(\boldsymbol{y} - \boldsymbol{X}\widehat{\boldsymbol{\beta}})/n. \tag{4.18}$$

where $\widehat{\boldsymbol{\beta}}$ is the MAP estimator of $\boldsymbol{\beta}$ under the SSL model. By [67], this quantity $\widehat{\boldsymbol{\beta}}_d$ has the following asymptotic distribution:

$$\sqrt{n}(\widehat{\boldsymbol{\beta}}_d - \boldsymbol{\beta}) \sim \mathcal{N}(\mathbf{0}, \sigma^2\widehat{\boldsymbol{\Theta}}\widehat{\boldsymbol{\Sigma}}\widehat{\boldsymbol{\Theta}}^T). \tag{4.19}$$

For inference, we replace the population variance σ^2 in (4.19) with the modal estimate $\widehat{\sigma}^2$ from the SSL model. Let $\widehat{\beta}_{dj}$ denote the jth coordinate of $\widehat{\boldsymbol{\beta}}_d$. We have from (4.19) that the $100(1-\alpha)\%$ asymptotic pointwise confidence intervals for $\beta_j, j = 1, \ldots, p$, are

$$[\widehat{\beta}_{dj} - c(\alpha, n, \widehat{\sigma}^2), \widehat{\beta}_{dj} + c(\alpha, n, \widehat{\sigma}^2)], \tag{4.20}$$

where $c(\alpha, n, \widehat{\sigma}^2) := \Phi^{-1}(1-\alpha/2)\sqrt{\widehat{\sigma}^2(\widehat{\boldsymbol{\Theta}}\widehat{\boldsymbol{\Sigma}}\widehat{\boldsymbol{\Theta}}^T)_{jj}/n}$ and $\Phi(\cdot)$ denotes the cumulative distribution function of $\mathcal{N}(0, 1)$.

Note that the posterior modal estimate $\widehat{\boldsymbol{\beta}}$ under the SSL prior already has much less bias than the LASSO estimator [59]. Therefore, the purpose of the debiasing procedure above is mainly to obtain an estimator with an asymptotically normal distribution from which we can construct asymptotic pointwise confidence intervals. While this procedure is asymptotically valid, [1] showed through numerical studies that constructing confidence intervals based on asymptotic arguments may provide coverage below the nominal level in finite samples, especially small samples. Therefore, it may be more ideal to use the actual SSL posterior $p(\boldsymbol{\beta} \mid \boldsymbol{y})$ for inference.

4.5.2 Posterior Sampling for the Spike-and-Slab LASSO

Fully Bayesian inference with the SSL can be carried out via posterior simulation. However, posterior sampling under spike-and-slab priors has continued to pose challenges. One immediate strategy for sampling from the SSL posterior is the SSVS algorithm of [23], described in Section 4.2.2. One can regard the Laplace distribution as a scale mixture of Gaussians with an exponential mixing distribution [48] and perform a variant of SSVS. Recently, several clever computational tricks have been suggested that avoid costly matrix inversions needed by SSVS by using linear solvers [9], low-rank approximations [32], or by disregarding correlations between active and inactive coefficients [46]. These techniques can be suitably adapted for fast posterior sampling of the SSL as well.

Intrigued by the speed of SSL mode detection, [47] explored the possibility of turning SSL into approximate posterior sampling by performing MAP optimization on many independently perturbed datasets. Building on Bayesian bootstrap ideas, they introduced a method for approximate sampling called Bayesian bootstrap spike-and-slab LASSO (BB-SSL) which scales linearly with both n and p. Beyond its scalability, they show that BB-SSL has strong theoretical support, matching the convergence rate of the original posterior in sparse normal-means and in high-dimensional regression.

4.6 Illustrations

In this section, we illustrate the SSL's potential for estimation, variable selection, and prediction on both simulated and real high-dimensional data sets.

4.6.1 Example on Synthetic Data

For our simulation study, we slightly modified the settings in [44]. We set $n = 100$ and $p = 1000$ in (4.1). The design matrix $\boldsymbol{X}$ was generated from a multivariate Gaussian distribution with mean $\mathbf{0}_p$ and a block-diagonal covariance matrix $\boldsymbol{\Sigma} = \text{bdiag}(\widetilde{\Sigma}, \dots, \widetilde{\Sigma})$, where $\widetilde{\Sigma} = \{\widetilde{\sigma}\}_{i,j=1}^{50}$, with $\widetilde{\sigma}_{ij} = 0.9$ if $i \neq j$ and $\widetilde{\sigma}_{ii} = 1$. The true vector of regression coefficients $\boldsymbol{\beta}_0$ was constructed by assigning regression coefficients $\{-3.5, -2.5, -1.5, 1.5, 2.5, 3.5\}$ to 6 entries located at the indices $\{1, 51, 101, 151, 201, 251\}$ and setting the remaining coefficients equal to zero. Hence, there were 20 independent blocks of 50 highly correlated predictors, where the first six blocks contained only one active predictor. We then generated the response $\boldsymbol{y}$ using (4.1), where the error variance was set as $\sigma^2 = 3$.

We compared the SSL with the LASSO [59], SCAD [19], and MCP [70]. The SSL method was applied using the R package `SSLASSO`. The competing methods were applied using the R package `ncvreg`. We repeated our experiment 500 times with new covariates and responses generated each time. For each experiment, we recorded the mean squared error (MSE) and mean prediction error (MPE), defined as

$$\text{MSE} = \frac{1}{p}\|\widehat{\boldsymbol{\beta}} - \boldsymbol{\beta}_0\|_2^2 \quad \text{and} \quad \text{MPE} = \frac{1}{n}\|\boldsymbol{X}(\widehat{\boldsymbol{\beta}} - \boldsymbol{\beta}_0)\|_2^2.$$

We also kept track of $\widehat{p}_\gamma$, or the size of the model selected by each of these methods. Finally,

	MSE	MPE	$\widehat{p}_\gamma$	FDR	FNR	MCC
SSL	**0.0067**	**0.701**	**6.05**	**0.0012**	0.187	**0.809**
	(0.0076)	(0.542)	(0.271)	(0.0010)	(0.160)	(0.162)
LASSO	0.011	1.14	33.38	0.028	**0.083**	0.387
	(0.0045)	(0.303)	(5.38)	(0.0055)	(0.109)	(0.062)
SCAD	0.011	0.985	12.74	0.0081	0.225	0.554
	(0.012)	(0.691)	(3.98)	(0.0043)	(0.187)	(0.178)
MCP	0.020	1.55	11.31	0.0077	0.395	0.447
	(0.016)	(0.849)	(2.70)	(0.0031)	(0.211)	(0.173)

TABLE 4.1
MPE, MPE, estimated model size, FDR, FNR, and MCC for SSL, LASSO, SCAD, and MCP. The results are averaged across 500 replications. In parentheses, we report the empirical standard errors.

we recorded the false discovery rate (FDR), the false negative rate (FNR), and the Matthews correlation coefficient (MCC) [38], defined respectively as

$$\text{FDR} = \frac{\text{FP}}{\text{TN}+\text{FP}}, \qquad \text{FNR} = \frac{\text{FN}}{\text{TP}+\text{FN}},$$
$$\text{MCC} = \frac{\text{TP}\times\text{TN} - \text{FP}\times\text{FN}}{\sqrt{(\text{TP}+\text{FP})(\text{TP}+\text{FN})(\text{TN}+\text{FP})(\text{TN}+\text{FN})}},$$

where TP, TN, FP, and FN denote the number of true positives, true negatives, false positives, and false negatives respectively. The MCC is a correlation coefficient between the predicted set of significant coefficients and the actual set of non-zero coefficients [38]. MCC has a range of -1 to 1, with -1 indicating completely incorrect selection (i.e., TP=TN=0) and 1 indicating completely correct variable selection (i.e., FP=FN=0). Models with MCC closer to 1 have higher selection accuracy. R code to reproduce these experiments is available in the online supplementary material.

Table 4.1 reports our results averaged across the 500 replications. We see that the SSL had the lowest average MSE and MPE, in addition to selecting (on average) the most parsimonious model. The LASSO (along with SCAD) had the second lowest MSE, but it tended to select far more variables than the other methods, leading to the highest FDR. In contrast, SSL had the lowest FDR and the highest MCC, indicating that the SSL had the best overall variable selection performance of all the methods. Our simulation study demonstrates that SSL achieves both parsimony *and* accuracy of estimation and selection.

Figure 4.3 illustrates the benefits of the dynamic posterior exploration approach outlined in Section 4.4.2. Specifically, Figure 4.3 plots the solution paths for the regression coefficients from one of our experiments as the spike hyperparameter λ_0 increases. We see that the SSL solution stabilizes fairly quickly (when λ_0 is less than 20), so that further increases in λ_0 do not change the solution. This demonstrates that dynamic posterior exploration offers a viable alternative to cross-validation. The R package SSLASSO provides the functionality to generate plots of these solution paths.

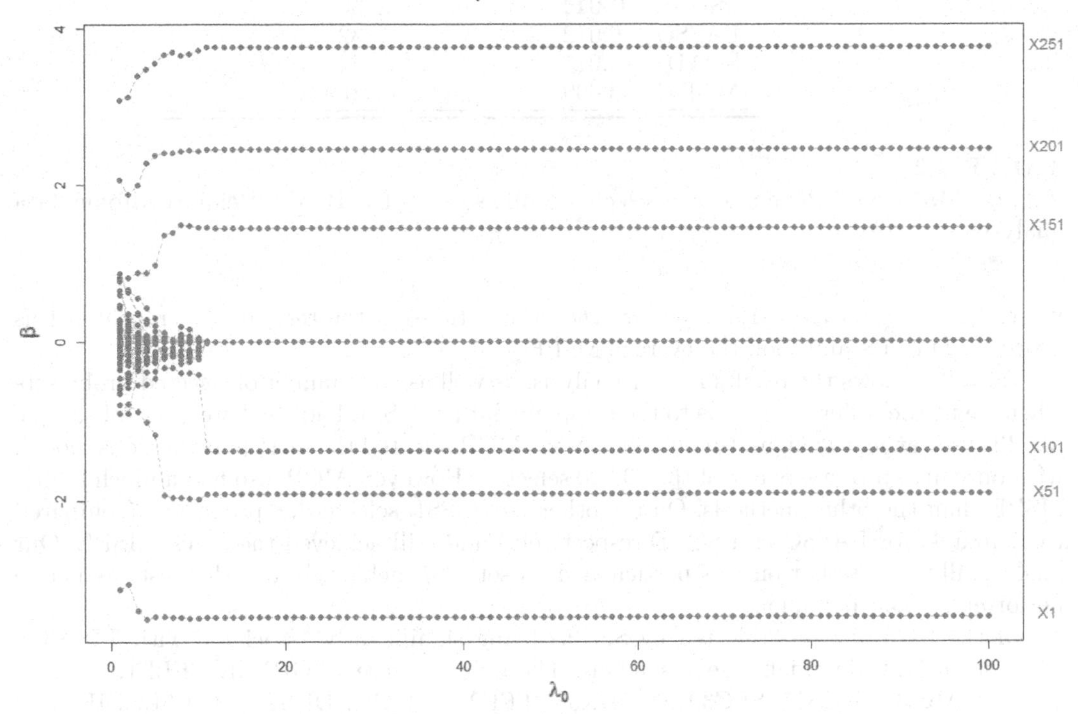

FIGURE 4.3
The solution paths for the SSL along the ladder of spike parameters λ_0. We see that the solution stabilizes after a certain point, so that further increases in λ_0 do not change the solution. The points along the horizontal axis are the zero values where the negligible estimates disappear.

4.6.2 Bardet-Beidl Syndrome Gene Expression Study

We now analyze a microarray data set consisting of gene expression measurements from the eye tissue of 120 laboratory rats[1]. The data was originally studied by [58] to investigate mammalian eye disease. In this data, the goal is to identify genes which are associated with the gene TRIM32. TRIM32 has previously been shown to cause Bardet-Biedl syndrome [17], a disease affecting multiple organs including the retina.

The original data consists of 31,099 probe sets. For our analysis, we included only the 10,000 probe sets with the largest variances in expression (on the log scale). This resulted in $n = 120$ and $p = 10,000$. We then fit the model (4.1) with an SSL penalty. We compared the SSL approach to LASSO, SCAD, and MCP.

To assess predictive accuracy, we randomly split the data set into 90 training observations and 30 test observations. We then fit the models on the training set and used the estimated $\widehat{\boldsymbol{\beta}}_{\text{train}}$ to compute the mean squared prediction error (MSPE) on the left-out test set,

$$\text{MSPE} = \frac{1}{30}\sum_{i=1}^{30}(y_{i,\text{test}} - \boldsymbol{x}_{i,\text{test}}^T\widehat{\boldsymbol{\beta}}_{\text{train}})^2,$$

[1]Data accessed from the Gene Expression Omnibus `www.ncbi.nlm.nih.gov/geo` (accession no. GSE5680).

	MSPE	Number of selected probe sets
SSL	**0.011**	28
LASSO	0.012	32
SCAD	0.015	44
MCP	3.699	9

TABLE 4.2
Average MSPE and the number of selected probe sets for the Bardet-Beidl Syndrome data analysis.

where $(\boldsymbol{x}_{i,\text{test}}, y_{i,\text{test}}), i = 1, \ldots, 30$, are the observations in the test set. We repeated this process 100 times and took the average MSPE.

Table 4.2 shows the results for our analysis, as well as the number of selected probe sets when we fit the different models to the complete data set. SSL had the lowest out-of-sample MSPE, indicating the highest predictive power. MCP selected the most parsimonious model, with only nine probe sets out of the 10,000 selected. However, MCP also had a much higher MSPE than the other methods. On the other hand, SSL selected 28 probe sets (compared to 32 and 44 for LASSO and SCAD respectively) and still achieved the lowest MSPE. Our analysis illustrates that on this particular data set, SSL achieved both the best predictive performance and parsimony.

Of the 28 probe sets selected by SSL as being significantly associated with TRIM32, 14 of them had identifiable gene symbols. These genes were SCGB1A1, CELF1, ASXL3, FGFR2, MOBP, TGM7, SLC39A6, DDX58, TFF2, CLOCK, DUS4L, HTR5B, BIK, and SLC16A6. In particular, according to `https://www.genecards.org` [2], SCGB1A1 is known to be an interacting protein for the TRIM32 gene. The other associations that we found may be useful for researchers in studying the genetic factors contributing to Bardet-Biedl syndrome.

4.7 Methodological Extensions

While we have focused on the normal linear regression model (4.1) in Sections 4.2–4.6, the spike-and-slab LASSO methodology has now been adopted in a variety of other statistical applications. In this section, we survey some of the extensions of the SSL to models beyond the normal linear regression framework.

Generalized linear models (GLMs). GLMs allow for a flexible generalization of the normal linear regression model (4.1) which can accommodate categorical and count data, in addition to continuous variables. Letting $\boldsymbol{x}_i \in \mathbb{R}^p$ denote a vector of covariates for the ith observation, GLMs assume that the mean of the response variable is related to the linear predictor via a link function,

$$E(y_i \mid \boldsymbol{x}_i) = h^{-1}(\boldsymbol{x}_i^T \boldsymbol{\beta}), \tag{4.21}$$

and that the data distribution is expressed as

$$f(\boldsymbol{y} \mid \boldsymbol{X}, \boldsymbol{\beta}, \varphi) = \prod_{i=1}^{n} f(y_i \mid \boldsymbol{x}_i, \boldsymbol{\beta}, \varphi), \tag{4.22}$$

[2]Accessed from `https://www.genecards.org/cgi-bin/carddisp.pl?gene=TRIM32` on October 11, 2020.

where φ is a dispersion parameter and the distribution $f(y_i \,|\, \boldsymbol{x}_i, \boldsymbol{\beta}, \varphi)$ can take various forms, including normal, binomial and Poisson distributions. Obviously, the normal linear model (4.1) is a special case of (4.21) with the identity link function $h(u) = u$.

[63] extended the SSL (4.3) to GLMs, including binary regression and Poisson regression, by placing the SSL prior (4.6) on the coefficients vector $\boldsymbol{\beta} \in \mathbb{R}^p$ in (4.21) and developing an EM algorithm to perform MAP estimation for $\boldsymbol{\beta}$. For inference with grouped variables in GLMs, [62] further employed group-specific sparsity parameters θ_g for each group of variables, instead of a single θ, as in (4.3).

Survival analysis. The SSL has also proven to be useful for predicting censored survival outcomes and detecting and estimating the effects of relevant covariates. Cox proportional hazards models are the most widely used method for studying the relationship between a censored survival response and an explanatory variable $\boldsymbol{x}_i \in \mathbb{R}^p$ [36]. This model assumes that the hazard function of survival time t takes the form,

$$h(t \,|\, \boldsymbol{x}_i) = h_0(t) \exp(\boldsymbol{x}_i^T \boldsymbol{\beta}). \tag{4.23}$$

[64] introduced the spike-and-slab LASSO Cox model which endows the coefficients $\boldsymbol{\beta}$ in (4.23) with the SSL prior (4.6). They developed an EM coordinate ascent algorithm to fit SSL Cox models. [61] further introduced the GssLASSO Cox model which incorporates grouping information by endowing each group of coefficients with a group-specific sparsity parameter θ_g instead of the single θ of (4.6).

Sparse factor analysis and biclustering. Factor models aim to explain the dependence structure among high-dimensional observations through a sparse decomposition of a $p \times p$ covariance matrix $\boldsymbol{\Omega}$ as $\boldsymbol{B}\boldsymbol{B}^T + \boldsymbol{\Sigma}$ where $\boldsymbol{B}$ is a $p \times K$ factor loadings matrix with $K \ll p$ and $\boldsymbol{\Sigma} = \text{diag}(\sigma_1^2, \ldots, \sigma_p^2)$. A generic latent factor model is

$$\boldsymbol{y}_i = \boldsymbol{B}\boldsymbol{\eta}_i + \boldsymbol{\varepsilon}_i, \qquad \boldsymbol{\varepsilon}_i \sim \mathcal{N}_p(\boldsymbol{0}, \boldsymbol{\Sigma}), \tag{4.24}$$

where $\boldsymbol{y}_i$ is a p-dimensional continuous response and $\boldsymbol{\eta}_i \sim \mathcal{N}_K(\boldsymbol{0}, \boldsymbol{I}_K)$ are unobserved latent factors. Many existing factor analysis approaches entail prespecification of the unknown factor cardinality K and *post hoc* rotations of the original solution to sparsity. For the factor model (4.24), [55] endowed the entries of the loading matrix $\boldsymbol{B}$ with independent SSL priors,

$$p(\beta_{jk} \,|\, \gamma_{jk}, \lambda_{0k}, \lambda_1) = (1 - \gamma_{jk})\psi(\beta_{jk} \,|\, \lambda_{0k}) + \gamma_{jk}\psi(\beta_{jk} \,|\, \lambda_1).$$

However, instead of endowing each of the indicators γ_{jk} with the usual beta-Bernoulli prior as in (4.6), [55] endowed these with the Indian buffet process (IBP) prior [27], which avoids the need to prespecify K. Further, [55] developed a parameter-expanded EM (PXL-EM) algorithm which employs *intermediate* orthogonal rotations rather than post hoc rotations. In addition to obtaining a sparse solution, the PXL-EM algorithm also converges much faster than the vanilla EM algorithm and offers robustness against poor initializations.

For the problem of biclustering, i.e., identifying clusters using only subsets of their associated features, [43] utilized the factor model (4.24) in which *both* the factors $\boldsymbol{\eta} = [\boldsymbol{\eta}_1^T, \ldots, \boldsymbol{\eta}_n^T] \in \mathbb{R}^{n \times K}$ and the loadings are sparse. To achieve a doubly sparse representation, [43] placed an SSL prior coupled with an IBP prior on the factors and an SSL prior coupled with a beta-Bernoulli prior on the loadings. An EM algorithm with a variational step was developed to implement spike-and-slab LASSO biclustering.

Graphical models. Suppose we are given data $\boldsymbol{Y} = (\boldsymbol{y}_1, \ldots, \boldsymbol{y}_n)^T$, where the $\boldsymbol{y}_i$'s are assumed to be iid p-variate random vectors distributed as $\mathcal{N}_p(\boldsymbol{0}, \boldsymbol{\Omega}^{-1})$ and $p > n$. In this setting, off-diagonal zero entries ω_{ij} encode conditional independence between variables i and j. To obtain a sparse estimate of $\boldsymbol{\Omega} = (\omega_{i,j})_{i,j}$, [22] introduced the following prior on

$\boldsymbol{\Omega}$:

$$p(\boldsymbol{\Omega}) = \prod_{i<j} [(1-\theta)\psi(\omega_{ij} \mid \lambda_0) + \theta\psi(\omega_{ij} \mid \lambda_1)] \prod_{i=1}^{p} [\tau e^{-\tau\omega_{ii}}] \mathbb{I}(\Omega \succ 0) \mathbb{I}(\|\boldsymbol{\Omega}\|_2 \leq B), \quad (4.25)$$

for some $\tau > 0, B > 0$. Here, $\boldsymbol{\Omega} \succ 0$ denotes that $\boldsymbol{\Omega}$ is positive-definite and $\|\boldsymbol{\Omega}\|_2$ denotes the spectral norm of $\boldsymbol{\Omega}$. The prior on $\boldsymbol{\Omega}$ (4.25) entails independent exponential priors on the diagonal entries and SSL priors on the off-diagonal entries. A similar prior formulation was considered in [18], except [18] did not constrain $\boldsymbol{\Omega}$ to lie in the space of $p \times p$ matrices with uniformly bounded spectral norm. [22] showed that constraining the parameter space for $\boldsymbol{\Omega}$ in such a way ensures that a) the corresponding optimization problem for the posterior mode is *strictly* convex, and b) the posterior mode is a symmetric positive definite matrix. [22] developed an EM algorithm to estimate the posterior mode of $p(\boldsymbol{\Omega} \mid \boldsymbol{Y})$.

The SSL prior was also extended to perform *joint* estimation of multiple related Gaussian graphical models by [37]. [37] leveraged similarities in the underlying sparse precision matrices and developed an EM algorithm to perform this joint estimation.

Seemingly unrelated regression models. In seemingly unrelated regression models, multiple correlated responses are regressed on multiple predictors. The multivariate linear regression is an important case. Letting $\boldsymbol{y}_i \in \mathbb{R}^q$ be the vector of q responses and $\boldsymbol{x}_i \in \mathbb{R}^p$ be the vector of p covariates, this model is

$$\boldsymbol{y}_i = \boldsymbol{x}_i^T \boldsymbol{B} + \boldsymbol{\varepsilon}_i, \qquad \boldsymbol{\varepsilon}_i \sim \mathcal{N}_q(\boldsymbol{0}, \boldsymbol{\Omega}^{-1}), \quad (4.26)$$

[18] introduced the *multivariate spike-and-slab LASSO* (mSSL) to perform joint selection and estimation from the $p \times q$ matrix of regressors $\boldsymbol{B}$ *and* the precision matrix $\boldsymbol{\Omega}$. To obtain a sparse estimate of $(\boldsymbol{B}, \boldsymbol{\Omega})$, [18] placed the SSL prior (4.6) on the individual entries $\beta_{jk}, 1 \leq j \leq p, 1 \leq k \leq q$ in $\boldsymbol{B}$ and a product prior similar to (4.25) on $\boldsymbol{\Omega}$ (except [18] did not constrain $\boldsymbol{\Omega}$ to have bounded spectral norm). An expectation/conditional maximization (ECM) algorithm was developed to perform this joint estimation.

Causal inference. In observational studies, we are often interested in estimating the causal effect of a treatment T on an outcome y, which requires proper adjustment of a set of potential confounders $\boldsymbol{x} \in \mathbb{R}^p$. When $p > n$, direct control for all potential confounders is infeasible and standard methods such as propensity scoring [51] often fail. In this case, it is crucial to impose a low-dimensional structure on the confounder space. Given data $(y_i, T_i, \boldsymbol{x}_i), i = 1, \ldots, n$, where T_i is the treatment effect, [1] estimated the (homogeneous) average treatment effect (ATE) $\Delta(t_1, t_2) = E(Y(t_1) - Y(t_2))$ by utilizing the model,

$$y_i \mid T_i, \boldsymbol{x}_i, \beta_0, \beta_t, \boldsymbol{\beta}, \sigma^2 \sim \mathcal{N}(0, \beta_0 + \beta_t T_i + \boldsymbol{x}_i^T \boldsymbol{\beta}, \sigma^2). \quad (4.27)$$

Under (4.27), the ATE is straightforwardly estimated as $\Delta(t_1, t_2) = (t_1 - t_2)\widehat{\beta}_t$. [1] endowed the coefficients of the confounders $\boldsymbol{\beta}$ with the SSL prior (4.6). In addition, [1] also weighted the sparsity parameter θ in (4.6) by raising θ to a power $w_j, j = 1, \ldots, p$, for each covariate, in order to better prioritize variables belonging to the slab (i.e., $\gamma_j = 1$) if they are also associated with the treatment. [1] further extended the model (4.27) to the more general case of heterogeneous treatment effects.

Regression with grouped variables. Group structure arises in many statistical applications. For example, in genomics, genes within the same pathway may form a group and act in tandem to regulate a biological system. For regression with grouped variables, we can model the response $\boldsymbol{y}$ as

$$\boldsymbol{y} = \sum_{g=1}^{G} \boldsymbol{X}_g \boldsymbol{\beta}_g + \boldsymbol{\varepsilon}, \qquad \boldsymbol{\varepsilon} \sim \mathcal{N}_n(\boldsymbol{0}, \sigma^2 \boldsymbol{I}_n), \quad (4.28)$$

where $\boldsymbol{\beta}_g \in \mathbb{R}^{m_g}$ is a coefficients *vector* of length m_g, and $\boldsymbol{X}_g$ is an $n \times m_g$ covariate matrix corresponding to group $g = 1, \ldots G$. Under model (4.28), it is often of practical interest to select non-negligible groups and estimate their effects. To this end, [5] introduced the *spike-and-slab group lasso* (SSGL). To regularize groups of coefficients, the SSGL replaces the univariate Laplace densities in the univariate SSL (4.6) with *multivariate* Laplace densities. The SSGL prior is

$$\begin{aligned} p(\boldsymbol{\beta} \mid \theta) = & \prod_{g=1}^{G} \left[(1-\theta)\boldsymbol{\Psi}(\boldsymbol{\beta}_g \mid \lambda_0) + \theta \boldsymbol{\Psi}(\boldsymbol{\beta}_g \mid \lambda_1) \right], \\ \theta \sim & \ \mathcal{B}eta(a, b), \end{aligned} \tag{4.29}$$

where $\boldsymbol{\Psi}(\boldsymbol{\beta}_g \mid \lambda) \propto \lambda^{m_g} e^{-\lambda \|\boldsymbol{\beta}_g\|_2}$ and $\lambda_0 \gg \lambda_1$. The SSGL (4.29) is a two-group refinement of an ℓ_2 penalty on groups of coefficients. Accordingly, the posterior mode under the SSGL thresholds entire groups of coefficients to zero, while simultaneously estimating the effects of non-zero groups and circumventing the estimation bias of the original group lasso [69]. [5] developed an efficient blockwise-coordinate ascent algorithm to implement the SSGL model.

Non-parameteric additive regression. The advent of the SSGL prior (4.29) paved the way for the spike-and-slab lasso methodology to be extended to nonparametric problems. [5] introduced the *nonparametric spike-and-slab lasso* (NPSSL) for sparse generalized additive models (GAMs). Under this model, the response surface is decomposed into the sum of univariate functions,

$$y_i = \sum_{j=1}^{p} f_j(x_{ij}) + \varepsilon_i, \qquad \varepsilon_i \overset{iid}{\sim} \mathcal{N}(0, \sigma^2). \tag{4.30}$$

In [5], each of the f_j's is approximated using a basis expansion, or a linear combination of basis functions $\mathcal{B}_j = \{g_{j1}, \ldots, g_{jd}\}$, i.e.,

$$f_j(x_{ij}) \approx \sum_{k=1}^{d} g_{jk}(x_{ij}) \beta_{jk}. \tag{4.31}$$

Under sparsity, most of the f_j's in (4.30) are assumed to be $f_j = 0$. This is equivalent to assuming that most of the weight vectors $\boldsymbol{\beta}_j = (\beta_{j1}, \ldots, \beta_{jd})^T$, $j = 1, \ldots, p$, in (4.31) are equal to $\mathbf{0}_d$. The NPSSL is implemented by endowing the basis coefficients $\boldsymbol{\beta} = (\boldsymbol{\beta}_1^T, \ldots, \boldsymbol{\beta}_p^T)^T$ with the SSGL prior (4.29) to simultaneously select and estimate non-zero functionals. In addition, [5] also extended the NPSSL to identify and estimate the effects of non-linear interaction terms $f_{rs}(X_{ir}, X_{is}), r \neq s$.

Functional regression. The spike-and-slab lasso methodology has also been extended to functional regression, where the response $y(t)$ is a function that *varies* over some continuum T (often time) A very popular model in this framework is the nonparametric varying coefficient model,

$$y_i(t) = \sum_{k=1}^{p} x_{ik}(t) \beta_k(t) + \varepsilon_i(t), \qquad t \in T, \tag{4.32}$$

where $y_i(t)$ and $x_{ik}(t)$ are time-varying responses and covariates respectively and $\varepsilon_i(t)$ is a zero-mean stochastic process which captures the within-subject temporal correlations for the ith subject. Under (4.32), the $\beta_k(t)$'s are smooth functions of time (possibly $\beta_k(t) = 0$ for all $t \in T$), and our primary interest is in estimation and variable selection from the $\beta_k(t)$'s, $k = 1, \ldots, p$.

[3] introduced the nonparametric varying coefficient spike-and-slab lasso (NVC-SSL) to simultaneously select and estimate the smooth functions $\beta_k(t), k = 1, \ldots, p$. Similarly as with the NPSSL, these functions are approximated using basis expansions of smoothing splines, and the basis coefficients are endowed with the SSGL prior (4.29). Unlike GAMs, however, the NVC-SSL model does *not* assume homoscedastic, independent error terms. Instead, the NVC-SSL model accounts for within-subject temporal correlations in its estimation procedure.

False discovery rate control with missing data. Sorted L-One Penalized Estimator (SLOPE) is an elaboration of the LASSO tailored to false discovery control by assigning more penalty to the larger coefficients [11]. SSL, on the other hand, penalizes large coefficients less and its false discovery rate is ultimately determined by a combination of the prior inclusion weight θ and penalties λ_1 and λ_0. [31] propose a hybrid procedure called adaptive Bayesian SLOPE, which effectively combines the SLOPE method (sorted l_1 regularization) together with the SSL method in the context of variable selection with missing covariate values. As with SSL, the coefficients are regarded as arising from a hierarchical model consisting of two groups: (1) the spike for the inactive and (2) the slab for the active. However, instead of assigning spike priors for each covariate, they propose a joint "SLOPE" spike prior which takes into account ordering of coefficient magnitudes in order to control for false discoveries.

4.8 Theoretical Properties

In addition to its computational tractability and its excellent finite-sample performance, the spike-and-slab LASSO has also been shown to provide strong theoretical guarantees. Although this chapter focuses mainly on methodology, we briefly outline a few of the major theoretical developments for spike-and-slab LASSO methods.

A common theme in Bayesian asymptotic theory is the study of the learning rate of posterior point estimates (such as the mean, median or mode) and/or of the full posterior. Working under the frequentist assumption of a "true" underlying model, the aim under the former is to study the *estimation* rate of point estimators under a given risk function, such as expected squared error loss. From a fully Bayes perspective, one may also be interested in the *posterior contraction rate*, or the speed at which the *entire* posterior contracts around the truth. In both cases, the frequentist minimax estimation rate is a useful benchmark, since the posterior cannot contract faster than this rate [24].

In a variety of contexts, including the Gaussian sequence model, sparse linear regression, and graphical models, the SSL global posterior mode has been shown to achieve the minimax estimation rate [22, 50, 54, 56]. From a fully Bayesian perspective, the *entire* posterior under SSL or SSL-type priors has *also* been shown to achieve (near) optimal posterior contraction rates in the contexts of the Gaussian sequence model, linear regression, regression with grouped variables, nonparametric additive regression, and functional regression [3, 5, 47, 50, 54, 56]. It is not necessarily the case that the posterior mode and the full posterior contract at the same rate [15, 16]. These theoretical results thus show that the SSL is optimal from *both* penalized likelihood *and* fully Bayesian perspectives.

4.9 Discussion

In this chapter, we have reviewed the spike-and-slab LASSO (4.6). The SSL forms a continuum between the penalized likelihood LASSO and the Bayesian point-mass spike-and-slab frameworks, borrowing strength from both constructs while addressing limitations of each. First, the SSL employs a *non*-separable penalty that self-adapts to ensemble information about sparsity and that performs selective shrinkage. Second, the SSL is amenable to fast maximum *a posteriori* finding algorithms which can be implemented in a highly efficient, scalable manner. Third, the posterior mode under the SSL prior can automatically perform both variable selection and estimation. Finally, uncertainty quantification for the SSL can be attained by either debiasing the posterior modal estimate or by utilizing efficient approaches to posterior sampling. Beyond linear regression, the spike-and-slab LASSO methodology is broadly applicable to a wide number of statistical problems, including generalized linear models, factor analysis, graphical models, and nonparametric regression.

Acknowledgments

This work was supported by funding from the University of South Carolina College of Arts & Sciences, the James S. Kemper Foundation Faculty Research Fund at the University of Chicago Booth School of Business, and NSF Grants DMS-1916245, DMS-1944740.

Bibliography

[1] J. Antonelli, G. Parmigiani, and F. Dominici. High-dimensional confounding adjustment using continuous spike and slab priors. *Bayesian Analysis*, 14(3):805–828, 09 2019.

[2] A. Armagan, D. B. Dunson, and J. Lee. Generalized double pareto shrinkage. *Statist. Sinica*, 23(1):119–143, 2013.

[3] R. Bai, M. R. Boland, and Y. Chen. Fast algorithms and theory for high-dimensional Bayesian varying coefficient models. *arXiv pre-print arXiv: 1907.06477*, 2020.

[4] R. Bai and M. Ghosh. On the beta prime prior for scale parameters in high-dimensional bayesian regression models. *Statistica Sinica (to appear)*, 2021.

[5] R. Bai, G. E. Moran, J. L. Antonelli, Y. Chen, and M. R. Boland. Spike-and-slab group lassos for grouped regression and sparse generalized additive models. *Journal of the American Statistical Association (to appear)*, 2020.

[6] M. M. Barbieri and J. O. Berger. Optimal predictive model selection. *The Annals of Statistics*, 32(3):870–897, 06 2004.

[7] A. Belloni, V. Chernozhukov, and L. Wang. Square-root lasso: pivotal recovery of sparse signals via conic programming. *Biometrika*, 98(4):791–806, 2011.

[8] A. Bhadra, J. Datta, N. G. Polson, and B. Willard. Lasso meets horseshoe: A survey. *Statist. Science*, 34(3):405–427, 2019.

[9] A. Bhattacharya, A. Chakraborty, and B. K. Mallick. Fast sampling with Gaussian scale mixture priors in high-dimensional regression. *Biometrika*, 103(4):985–991, 2016.

[10] A. Bhattacharya, D. Pati, N. S. Pillai, and D. B. Dunson. Dirichlet-laplace priors for optimal shrinkage. *Journal of the American Statistical Association*, 110(512):1479–1490, 2015. PMID: 27019543.

[11] M. Bogdan, E. van den Berg, C. Sabatti, W. Su, and E. J. Candés. SLOPE-adaptive variable selection via convex optimization. *The Annals of Applied Statistics*, 9(3):1103–1140, 2015.

[12] L. Bottolo and S. Richardson. Evolutionary stochastic search for Bayesian model exploration. *Bayesian Analysis*, 5(3):583–618, 09 2010.

[13] J. G. Scott C. M. Carvalho, N. G. Polson. The horseshoe estimator for sparse signals. *Biometrika*, 97(2):465–480, 2010.

[14] C. M. Carvalho, N. G. Polson, and J. G. Scott. Handling sparsity via the horseshoe. In David van Dyk and Max Welling, editors, *Proceedings of the Twelth International Conference on Artificial Intelligence and Statistics*, volume 5 of *Proceedings of Machine Learning Research*, pages 73–80, Hilton Clearwater Beach Resort, Clearwater Beach, Florida USA, 2009. PMLR.

[15] I. Castillo and R. Mismer. Empirical Bayes analysis of spike and slab posterior distributions. *Electronic Journal of Statistics*, 12(2):3953–4001, 2018.

[16] I. Castillo, J. Schmidt-Hieber, and A. van der Vaart. Bayesian linear regression with sparse priors. *The Annals of Statistics*, 43(5):1986–2018, 10 2015.

[17] A. P. Chiang, J. S. Beck, H.-J. Yen, M. K. Tayeh, T. E. Scheetz, R. E. Swiderski, D. Y. Nishimura, T. A. Braun, Kwang-Youn A. Kim, J. Huang, et al. Homozygosity mapping with SNP arrays identifies TRIM32, an E3 ubiquitin ligase, as a Bardet–Biedl syndrome gene (BBS11). *Proceedings of the National Academy of Sciences*, 103(16):6287–6292, 2006.

[18] S. K. Deshpande, V. Ročková, and E. I. George. Simultaneous variable and covariance selection with the multivariate spike-and-slab lasso. *Journal of Computational and Graphical Statistics*, 28(4):921–931, 2019.

[19] J. Fan and R. Li. Variable selection via nonconcave penalized likelihood and its oracle properties. *Journal of the American Statistical Association*, 96(456):1348–1360, 2001.

[20] J. Friedman, T. Hastie, and R. Tibshirani. Regularization paths for generalized linear models via coordinate descent. *Journal of Statistical Software*, 33(1):1–22, 2010.

[21] J. Friedman, T. Hastie and R. Tibshirani. Sparse inverse covariance estimation with the graphical lasso. *Biostatistics*, 9(3):432–441, 2007.

[22] L. Gan, N. N. Narisetty, and F. Liang. Bayesian regularization for graphical models with unequal shrinkage. *Journal of the American Statistical Association*, 114:1218–1231, 2018.

[23] E. I. George and R. E. McCulloch. Variable selection via Gibbs sampling. *Journal of the American Statistical Association*, 88(423):881–889, 09 1993.

[24] S. Ghosal, J. K. Ghosh, and A. W. van der Vaart. Convergence rates of posterior distributions. *The Annals of Statistics*, 28(2):500–531, 04 2000.

[25] P. Ghosh, X. Tang, M. Ghosh, and A. Chakrabarti. Asymptotic properties of bayes risk of a general class of shrinkage priors in multiple hypothesis testing under sparsity. *Bayesian Analysis*, 11(3):753–796, 2016.

[26] J. E. Griffin and P. J. Brown. Inference with normal-gamma prior distributions in regression problems. *Bayesian Analysis*, 5(1):171–188, 2010.

[27] T. L. Griffiths and Z. Ghahramani. The Indian buffet process: An introduction and review. *Journal fo Machine Learning Research*, 12:1185–1224, July 2011.

[28] C. Hans, A. Dobra, and M. West. Shotgun stochastic search for "large p" regression. *Journal of the American Statistical Association*, 102(478):507–516, 2007.

[29] H. Ishwaran and J. S. Rao. Spike and slab variable selection: Frequentist and bayesian strategies. *The Annals of Statistics*, 33(2):730–773, 04 2005.

[30] A. Javanmard and A. Montanari. Debiasing the lasso: Optimal sample size for gaussian designs. *The Annals of Statistics*, 46(6A):2593–2622, 2018.

[31] W. Jiang, M. Bogdan, J. Josse, B. Miasojedow, V. Rockova, and TraumaBase Group. Adaptive bayesian slope – high-dimensional model selection with missing values. *arXiv pre-print arXiv: 1907.06477*, 2019.

[32] J. Johndrow, P. Orenstein, and A. Bhattacharya. Scalable approximate MCMC algorithms for the horseshoe prior. *Journal of Machine Learning Research*, 21(73):1–61, 2020.

[33] I. M. Johnstone and B. W. Silverman. Needles and straw in haystacks: Empirical Bayes estimates of possibly sparse sequences. *The Annals of Statistics*, 32(4):1594–1649, 08 2004.

[34] I. M. Johnstone and B. W. Silverman. Empirical Bayes selection of wavelet thresholds. *The Annals of Statistics*, 33(4):1700–1752, 08 2005.

[35] Y. Kim and C. Gao. Bayesian model selection with graph structured sparsity. *arXiv preprint arXiv:1902.03316*, 2019.

[36] J. P. Klein and M. L. Moeschberger. *Survival Analysis Techniques for Censored and Truncated Data.* Second edition, 2003.

[37] Z. Li, T. Mccormick, and S. Clark. Bayesian joint spike-and-slab graphical lasso. In Kamalika Chaudhuri and Ruslan Salakhutdinov, editors, *Proceedings of the 36th International Conference on Machine Learning*, volume 97 of *Proceedings of Machine Learning Research*, pages 3877–3885, Long Beach, California, USA, 09–15 Jun 2019. PMLR.

[38] B.W. Matthews. Comparison of the predicted and observed secondary structure of t4 phage lysozyme. *Biochimica et Biophysica Acta (BBA) - Protein Structure*, 405(2):442–451, 1975.

[39] R. Mazumder, J. H. Friedman, and T. Hastie. Sparsenet: Coordinate descent with nonconvex penalties. *Journal of the American Statistical Association*, 106(495):1125–1138, 2011. PMID: 25580042.

[40] G. J. McLachlan and K. E. Basford. *Mixture models: Inference and applications to clustering.* Marcel Dekker, New York, 1988.

[41] N. Meinshausen and P. Bühlmann. High-dimensional graphs and variable selection with the lasso. *The Annals of Statistics*, 34(3):1436–1462, 2006.

[42] T.J. Mitchell and J.J. Beauchamp. Bayesian variable selection in linear regression. *Journal of the American Statistical Association*, 83(404):1023–1032, 1988.

[43] G. E. Moran, V. Ročková, and E. I. George. Spike-and-slab lasso biclustering. *The Annals of Applied Statistics (to appear)*, 2020.

[44] G. E. Moran, V. Ročková, and E. I. George. Variance prior forms for high-dimensional Bayesian variable selection. *Bayesian Analysis*, 14(4):1091–1119, 2019.

[45] N. N. Narisetty and X. He. Bayesian variable selection with shrinking and diffusing priors. *The Annals of Statistics*, 42(2):789–817, 04 2014.

[46] N. N. Narisetty, J. Shen, and X. He. Skinny Gibbs: A consistent and scalable Gibbs sampler for model selection. *Journal of the American Statistical Association*, 114(527):1205–1217, 2019.

[47] L. Nie and V. Ročková. Fast posterior sampling for the spike-and-slab LASSO. *arXiv pre-print arXiv: 2011.14279*, 2020.

[48] T. Park and G. Casella. The Bayesian lasso. *Journal of the American Statistical Association*, 103(482):681–686, 2008.

[49] N. G. Polson and L. Sun. Bayesian ℓ_0-regularized least squares. *Applied Stochastic Models in Business and Industry*, 35(3):717–731, 2019.

[50] K. Ray and B. Szabó. Variational Bayes for high-dimensional linear regression with sparse priors. *Journal of the American Statistical Association (to appear)*, 2020.

[51] P. R. Rosenbaum and D. B. Rubin. The central role of the propensity score in observational studies for causal effects. *Biometrika*, 70(1):41–55, 04 1983.

[52] V. Ročková. Bayesian estimation of sparse signals with a continuous spike-and-slab prior. *The Annals of Statistics*, 46(1):401–437, 2018.

[53] V. Ročková and E. I. George. EMVS: The EM approach to Bayesian variable selection. *Journal of the American Statistical Association*, 109(506):828–846, 2014.

[54] V. Ročková and E. I. George. Bayesian penalty mixing: The case of a non-separable penalty. In *Statistical Analysis for High-Dimensional Data - The Abel Symposium 2014*, pages 233–254. Springer, 2016.

[55] V. Ročková and E. I. George. Fast Bayesian factor analysis via automatic rotations to sparsity. *Journal of the American Statistical Association*, 111(516):1608–1622, 2016.

[56] V. Ročková and E. I. George. The spike-and-slab LASSO. *Journal of the American Statistical Association*, 113(521):431–444, 2018.

[57] V. Ročková and G. E. Moran. *SSLASSO: The Spike-and-Slab LASSO*, 2018. R package version 1.2-1.

[58] T. E. Scheetz, Kwang-Youn A. Kim, R. E. Swiderski, A. R. Philp, T. A. Braun, K. L. Knudtson, A. M. Dorrance, G. F. DiBona, J. Huang, T. L. Casavant, et al. Regulation of gene expression in the mammalian eye and its relevance to eye disease. *Proceedings of the National Academy of Sciences*, 103(39):14429–14434, 2006.

[59] J. G. Scott and J. O. Berger. Bayes and empirical-Bayes multiplicity adjustment in the variable-selection problem. *The Annals of Statistics*, 38(5):2587–2619, 2010.

[60] T. Sun and C.-H. Zhang. Scaled sparse linear regression. *Biometrika*, 99(4):879–898, 2012.

[61] Z. Tang, S. Lei, X. Zhang, Z. Yi, B. Guo, J. Y. Chen, Y. Shen, and N. Yi. Gsslasso Cox: a Bayesian hierarchical model for predicting survival and detecting associating genes by incorporating pathway information. *BMC Bioinformatics*, 20(1), 2019.

[62] Z. Tang, Y. Shen, Y. Li, X. Zhang, J. Wen, C. Qian, W. Zhuang, X. Shi, and N. Yi. Group spike-and-slab lasso generalized linear models for disease prediction and associated genes detection by incorporating pathway information. *Bioinformatics*, 34(6):901–910, 2018.

[63] Z. Tang, Y. Shen, X. Zhang, and N. Yi. The spike-and-slab lasso generalized linear models for prediction and associated genes detection. *Genetics*, 205(1):77–88, 2017.

[64] Z. Tang, Y. Shen, X. Zhang, and N. Yi. The spike-and-slab lasso Cox model for survival prediction and associated genes detection. *Bioinformatics*, 33(18):2799–2807, 05 2017.

[65] R. Tibshirani. Regression shrinkage and selection via the lasso. *Journal of the Royal Statistical Society: Series B (Statistical Methodology)*, 58:267–288, 1996.

[66] N. Ueda and R. Nakano. Deterministic annealing EM algorithm. *Neural Networks*, 11(2):271–282, 1998.

[67] S. van de Geer, P. Bühlmann, Y. Ritov, and R. Dezeure. On asymptotically optimal confidence regions and tests for high-dimensional models. *The Annals of Statistics*, 42(3):1166–1202, 2014.

[68] Wellcome Trust. Genome-wide association study of 14,000 cases of seven common diseases and 3000 shared controls. *Nature*, 447:661–678, 2007.

[69] M. Yuan and Y. Lin. Model selection and estimation in regression with grouped variables. *Journal of the Royal Statistical Society: Series B (Statistical Methodology)*, 68(1):49–67, 2006.

[70] C.-H. Zhang. Nearly unbiased variable selection under minimax concave penalty. *The Annals of Statistics*, 38(2):894–942, 04 2010.

[71] C.-H. Zhang and S. S. Zhang. Confidence intervals for low dimensional parameters in high dimensional linear models. *Journal of the Royal Statistical Society: Series B (Statistical Methodology)*, 76(1):217–242, 2014.

[72] C.-H. Zhang and T. Zhang. A general theory of concave regularization for high-dimensional sparse estimation problems. *Statistical Science*, 27(4):576–593, 2012.

[73] H. Zou. The adaptive lasso and its oracle properties. *Journal of the American Statistical Association*, 101(476):1418–1429, 2006.

[74] H. Zou and T. Hastie. Regularization and variable selection via the elastic net. *Journal of the Royal Statistical Society: Series B (Statistical Methodology)*, 67(2):301–320, 2005.

5

Adaptive Computational Methods for Bayesian Variable Selection

Jim E. Griffin

University College London (U.K.)

Mark F. J. Steel

University of Warwick (U.K.)

CONTENTS

Efficient computational methods to sample from the posterior distribution of models are key to the use of Bayesian variable selection in practice. Adaptive Monte Carlo techniques are a promising approach to build such algorithms. We review the use of these methods in generalized linear models with a particular focus on linear and logistic regression. We illustrate how these methods can be applied to simulated data and to two contrasting real-life examples. Firstly, we consider a high-dimensional example with an application to fine mapping in genomics with 10,995 observations and 5766 covariates (SNPs). Secondly, we consider an application to a complex model of environmental DNA which contains five logistic regressions and two sets of latent variables.

5.1 Introduction

Bayesian variable selection methods are attractive to screen variables for a relationship with a response variable, due to their theoretical properties and ability to account for model uncertainty. This approach treats the problem as high-dimensional regression with

DOI: 10.1201/9781003089018-5

many regressors and the prior distribution provides a natural way to favour sparse models where many regression coefficients are set equal to zero. In the Bayesian variable selection approach, each possible combination of inclusion or exclusion of a regressor is considered a model (exclusion of a variable is equivalent to setting a corresponding regression coefficient to zero). In high-dimensional settings, there will often be a lot of model uncertainty due to many competing models giving similar fits to the data. This model uncertainty can naturally be expressed through a posterior distribution. The use of posterior summaries such as posterior inclusion probabilities (the marginal posterior probabilities that each variable is included in the model) allow us to provide summaries of a variable's importance without making assumptions about the inclusion or exclusion of other variables in the model or about dependence between the regressors. Although theoretically appealing, posterior computation is particularly difficult with these methods since the space of possible models is a lattice whose dimension is given by the number of potential regressors. A common approach, which will be followed in this paper, uses Markov chain Monte Carlo (MCMC) methods to sample from the posterior distribution. These methods are commonly thought to mix poorly and this has been used as motivation for other methods such as the global–local mixture approach to Bayesian estimation of high-dimensional regression models, see [6] for a review. In this chapter, we will discuss and illustrate how adaptive Monte Carlo methods can be used to provide effective samplers from the posterior distribution in Bayesian variable selection problems.

We will initially focus on generalised linear models (GLMs) as a unifying framework for Gaussian and non-Gaussian regression models. Subsequently, we will concentrate on the specific examples of linear regression and logistic regression. Here, we assume that $\boldsymbol{y} = (y_1, \ldots, y_n)^T$ is a vector of observed dependent variables, and that $\boldsymbol{X}$ is the $(n \times p)$-dimensional matrix of independent variables where $x_{i1}, \ldots, x_{ip}$ are the corresponding values of p independent variables for the i-th observation. This is the design matrix for the full model with all p possible covariates. We are interested in variable selection and define $\boldsymbol{\gamma} = (\gamma_1, \ldots, \gamma_p)^T \in \Gamma = \{0, 1\}^p$ to be a vector of indicator variables with $\gamma_j = 1$ if the j-th variable is included in the model and $p_\gamma = \sum_{j=1}^p \gamma_j$. The GLM sampling model is then characterised by choosing a conditional distribution for $\boldsymbol{y} \mid \boldsymbol{X}$ with $\mathrm{E}(\boldsymbol{y} \mid \boldsymbol{X}) = \boldsymbol{\mu}_\gamma$ and $\mathrm{Var}(\boldsymbol{y} \mid \boldsymbol{X}) = \sigma^2 \boldsymbol{V}(\boldsymbol{\mu}_\gamma)$ (in some models, such as Bernoulli distributed data, σ^2 is assumed known) with

$$g(\boldsymbol{\mu}_\gamma) = \alpha \mathbf{1}_n + \boldsymbol{X}_\gamma \boldsymbol{\beta}_\gamma, \tag{5.1}$$

where $\boldsymbol{a}_n$ represents a n-dimensional column vector with all entries equal to a, and $\boldsymbol{X}_{\boldsymbol{\gamma}}$ is a $(n \times p_\gamma)$-dimensional data matrix formed using the included variables. Here $g(\cdot)$ is the link function. In the special case of the linear model, $g(\cdot)$ is the identity function and $\boldsymbol{V}(\cdot) = I_n$ so that we obtain the familar linear Gaussian model

$$\boldsymbol{y} = \alpha \mathbf{1}_n + \boldsymbol{X}_\gamma \boldsymbol{\beta}_\gamma + \boldsymbol{e}, \qquad \boldsymbol{e} \sim \mathcal{N}_n(\mathbf{0}_n, \sigma^2 I_n). \tag{5.2}$$

This is the so-called discrete spike-and-slab formulation [9, 15, 28]. An alternative approach is the continuous spike-and-slab [14, 24] which includes the variables for which $\gamma_j = 0$, denoted $\boldsymbol{X}_{-\gamma}$, in the linear prediction in (5.1) to give $g(\mu_\gamma) = \alpha \mathbf{1}_n + \boldsymbol{X}_\gamma \boldsymbol{\beta}_\gamma + \boldsymbol{X}_{-\gamma} \boldsymbol{\beta}_{-\gamma}$. This changes the interpretation of the inclusion variables to $\gamma_j = 1$ if the j-th variable is important whereas $\gamma_j = 0$ if that variable is unimportant. The prior for $\boldsymbol{\beta}_{-\gamma}$ is chosen to have a much smaller scale than the prior for $\boldsymbol{\beta}_\gamma$ which allows unimportant variables to have small (but non-zero) effects. MCMC algorithm using the discrete formulation work with $(p_\gamma \times p_\gamma)$-dimensional matrices and are usually much faster than ones using the continuous formulation (which involve $(p \times p)$-dimensional matrices) if p_γ is much smaller than p.

Throughout this chapter, we will assume the commonly used prior structure

$$p(\alpha, \sigma^2, \boldsymbol{\beta}_\gamma, \boldsymbol{\gamma}) \propto \sigma^{-2} p(\boldsymbol{\beta}_\gamma \mid \sigma^2, \boldsymbol{\gamma})\, p(\boldsymbol{\gamma}), \tag{5.3}$$

with $\boldsymbol{\beta}_\gamma \mid \sigma^2, \gamma \sim \mathcal{N}_{p_\gamma}(\mathbf{0}_{p_\gamma}, \sigma^2 \boldsymbol{V}_\gamma)$ and $p(\gamma) = h^{p_\gamma}(1-h)^{p-p_\gamma}$. The hyperparameter $0 < h < 1$ is the prior probability that a particular variable is included in the model and $\boldsymbol{V}_\gamma$ is often chosen as proportional to $(\boldsymbol{X}_\gamma^T \boldsymbol{X}_\gamma)^{-1}$, a g-prior, or to the identity matrix. The g-prior structure has been shown by [5] to satisfy many prior desiderata for the linear Gaussian model and was recommended as a natural extension for many cases of GLMs in [38]. The prior can be further extended with hyperpriors, for example, adopting $h \sim \mathcal{B}\text{eta}(a, b)$.

MCMC methods for linear models are greatly helped by the availability of an analytic form for the marginal likelihood $f(\boldsymbol{y} \mid \boldsymbol{X}_\gamma)$. This is usually not the case in other GLMs and has led to several different approaches. [34] introduced the idea of using Laplace-type approximations to the marginal likelihood in Bayesian variable selection. There are two further approaches which avoid approximation. Firstly, the MCMC sampler can be run on the joint distribution of models and regression coefficients. This is challenging since the dimension of the regression coefficients depends on the model. [31] discuss the use of reversible jump MCMC in GLMs allowing for both variable selection and estimation of the link function. Further developments in this direction can be seen in e.g., [13, 26, 32]. [17] shows that reversible jump MCMC can be formulated and generalised using a mixture of singular distributions. This idea is used by [39] to build an Add-Delete-Swap algorithm that jointly updates regression coefficients and inclusion variables. Secondly, latent variables $\boldsymbol{z}$ can be introduced so that $p(\boldsymbol{\beta}_\gamma, \boldsymbol{\gamma} | \boldsymbol{X}, \boldsymbol{y}) = \int p(\boldsymbol{\beta}_\gamma, \boldsymbol{\gamma}, \boldsymbol{z} | \boldsymbol{X}, \boldsymbol{y})\, d\boldsymbol{z}$ and the marginal likelihood $\int p(\boldsymbol{\beta}_\gamma, \boldsymbol{\gamma}, \boldsymbol{z} | \boldsymbol{X}, \boldsymbol{y})\, d\boldsymbol{\beta}_\gamma$ is available analytically. This approach was initially developed by [41] for probit regression models and extended to accelerate failure time models by [40].

5.1.1 Some Reasons to be Cheerful

MCMC for Bayesian variable selection has usually been considered a difficult computational task due to size of the space of all models (there are 2^p models if there are p possible variables) and perceived poor mixing of the widely-used Add-Delete-Swap (ADS) algorithm [9, 10]. These issues potentially become more worrying in large p problems. However, there are some reasons to think that MCMC for Bayesian variables selection in large p problems is feasible. Firstly, the use of sparsity prior distributions which assume that the size of models either does not grow or grows very slowly with p controls the number of models with non-negligible posterior probability. This avoids the MCMC algorithm having to traverse the whole of model space. The mixing of the ADS sampler is discussed in [44]. They show that the prior distribution on model space plays a key role in both variable selection consistency and rapid mixing of the ADS sampler in high-dimensional settings. Secondly, practitioners are often interested in low-dimensional summaries of the posterior on model space such as posterior inclusion probabilities or posterior predictive distributions. These are potentially much better approximated by an MCMC scheme than posterior probabilities of particular models (which will often have low posterior probabilities in large p problems and so need large numbers of MCMC samples to obtain reliable estimates). Thirdly, in large p, small n settings, where the sample size is relatively small, the posterior distribution is often quite flat with the ADS sampler enjoying relatively high acceptance probability. Since these are random walk type samplers which propose to only add and/or delete single variables at each iteration, high acceptance probabilities are associated with poor mixing and samplers which use more ambitious moves can lead to better mixing. This motivates the use of informed proposal distributions which can direct the sampler towards "good" (i.e., higher probability) models. For example, an informed proposal can be constructed by proposing models from a neighbourhood around the current model with suitable weights [45]. Alternatively, variables or combinations of variables with higher marginal posterior probabilities can be learnt during the MCMC run and the proposal adjusted to take this information into account. In this article, we review the use of adaptive Monte Carlo methods to implement this strategy.

5.1.2 Adaptive Monte Carlo Methods

MCMC methods construct a Markov chain whose stationary distribution is a chosen target distribution. In Bayesian statistics, this target distribution is the posterior distribution. This approach allows us to approximate characteristics of the posterior distribution using sample averages of functions of the values of the Markov chain. For example, posterior moments are approximated by sample moments and posterior quantiles are approximated by sample quantiles. This follows from the ergodicity of the chains and relies on properties of Markov chains. A wide range of methods (Metropolis-Hastings, Gibbs sampling, etc.) have been developed but it does not allow the arbitrary use of previous values of the Markov chain. This seems potentially wasteful since these previous values contain information about the posterior distribution and could be used to inform the behavior of the Markov chain. Adaptive Monte Carlo methods take MCMC methods and allow the chain to depend on all (or some) previous values of the chain. The approach allows us to develop algorithms with better performance than standard MCMC algorithms but there are some important design conditions needed to maintain ergodicity, see [2] for an introduction.

The development of adaptive Monte Carlo methods involves defining the adaptation of tuning parameters of the algorithm as the MCMC run progresses. They usually make use of theory about the behavior of MCMC algorithms and optimal choices of tuning parameters for particular types of target distributions. For example, Metropolis-Hastings random walk are often tuned to have an acceptance rate of 0.234. If the target distribution is a d-dimensional vector of independent normal random variables with the same variance, the acceptance rate of 0.234 is optimal as d tends to infinity [37]. Although the result holds for a very specific target distribution, it has also been shown to provide near optimal performance for a far wider class of target distributions. This result underlies the adaptive Metropolis-Hastings random walk [3]. The method uses a symmetric proposal centred on the current value of the chain and tunes the variance of the proposal as follows

$$s^{(t+1)} = s^{(t)} + t^{-\phi}(a_t - 0.234),$$

where $s^{(t)}$ is the variance of the proposal distribution used at iteration t, $1/2 < \phi \leq 1$ and a_t is the acceptance rate at the t-th iteration of the MCMC run. The update is motivated by the Robbins-Monro algorithm (see [36]) and will lead to the average acceptance rate to converge to 0.234. The approach has been shown to work well in a wide range of situations. This illustrates a general principle for the construction of adaptive proposals. An optimality result is found and used to adapt the proposal. The optimality result may only hold for a very specific example, but the procedure might still have good properties in more general settings.

5.2 Some Adaptive Approaches to Bayesian Variable Selection

Adaptive Monte Carlo methods are attractive for high-dimensional regression problems since we usually assume that only a small number of variables are relevant to predicting the response. Our posterior distribution provides information about which variables are likely (and unlikely) to be relevant and this information can be used to concentrate computational effort on important parts of the posterior distribution. The number of irrelevant variables will usually grow as the number of regressors increases. Methods which dedicate equal effort to all variables will then become less useful as the number of regressors increases. How to use this information effectively in constructing algorithms has been less clear and has lead to methods which have not been able to provide substantial computational improvements over simpler methods.

An early contribution in this direction was [30] who develop underlying theory for adaptive MCMC on finite state space and provide an adaptive Gibbs sampling scheme. In their scheme, the full conditional distribution $p(\gamma_j \mid \gamma_{-j}, \boldsymbol{X}, \boldsymbol{y})$ is approximated by a regression model using previous samples of γ. This allows faster evaluation than directly calculating the full conditional distribution. [35] consider multivariate regression problems which arise in quantitative trait loci modelling in genomics. They introduce the Hierarchical Evolutionary Stochastic Search (HESS) algorithm. This builds on the Evolutionary Stochastic Search (ESS) of [8], which uses powered versions of the posterior distribution in a parallel tempering algorithm. Moves between chains are proposed in a Metropolis-Hastings sampler using moves such as crossover and exchange, which are familiar from genetic algorithms in optimisation. The parameter updating for the regression for each response is controlled by an adaptive scheme where the parameters of the i-th response are chosen to be updated in proportion to the average number of variables included in the corresponding regression model. [27] describe an adaptive Metropolis-Hastings sampler where the standard Add-Delete-Swap sampler is extended by allowing more variables to be proposed to be changed. Their sampler generates $N \sim \mathcal{B}\text{in}(N_{max} - 1, \xi)$ where N_{max} is the maximum number of variables that can be changed in one move and then chooses at random one of Add, Delete or Swap moves. If Add is selected then $N+1$ new variables are proposed to be added to the model and, similarly, in the case of Delete $N+1$ currently included variables are proposed to be removed. If Swap is selected, $N+1$ new variables are proposed to be added and $N+1$ included variables are proposed to be deleted. They provide empirical evidence that tuning ξ to give an acceptance rate close to 0.234 provides near optimal effective sample size. They use this insight to adapt the tuning parameter ξ during the MCMC run using the Robbins-Monro updating scheme. [25] focus on the posterior distribution of β and approximate the marginal distribution of β_k by a mixture of a point-mass at zero (when a variable is excluded from the model) and a normal distribution (when a variable is included in the model). They assume independence between each β_k under the posterior distribution and adapt the parameters of the mixture distributions for each dimension by proposing an adaptive scheme which minimises the Kullback-Leibler divergence between the approximation and the marginal posterior distribution.

5.3 Two Adaptive Algorithms

Many of the adaptive Monte Carlo methods described in previous section concentrate on adapting the computational complexity of the samplers but do not adapt to the different marginal posterior probabilities of the variables. Alternatively, [19] consider a very general proposal distribution in a Metropolis-Hastings sampler where any combination of variables can be changed. The proposal changes the variables included in the model independently across variables (but depending on the current value of the inclusion vector γ). In particular, the value of γ'_j is proposed according to

- If the j-th variable is currently excluded from the model ($\gamma_j = 0$), the j-th variable is proposed to be added to the model with probability A_j.
- If the j-th variable is currently included in the model ($\gamma_j = 1$), the j-th variable is proposed to be removed from the model with probability D_j.

This implies that the proposal probability has the form

$$q_\eta(\gamma, \gamma') = \prod_{j=1}^{p} D_j^{\gamma_j(1-\gamma_j')}(1-D_j)^{\gamma_j\gamma_j'} A_j^{(1-\gamma_j)\gamma_j'}(1-A_j)^{(1-\gamma_j)(1-\gamma_j')},$$

where $\boldsymbol{\eta} = (A_1, \ldots, A_p, D_1, \ldots, D_p)$ are the tuning parameters of the proposal.

Choosing the tuning parameters $\boldsymbol{\eta}$ is a daunting task since its dimension is $2p$. [19] propose two methods for adaptively tuning their value during the run of the MCMC algorithm. Both methods exploit the following optimality result. Suppose that the posterior distribution of $\boldsymbol{\gamma}$ has the form

$$\pi(\boldsymbol{\gamma}) = \prod_{j=1}^{p} \pi_j^{\gamma_j}(1-\pi_j)^{1-\gamma_j},$$

which implies that the inclusion/exclusion of each variable is independent and that the posterior probability of including the j-th variables is π_j. Firstly, the choice

$$\frac{A_j}{D_j} = \frac{\pi_j}{1-\pi_j}, \qquad \text{for every } j = 1, \ldots, p, \tag{5.4}$$

implies that the Metropolis-Hastings acceptance probability is equal to one. The choices

$$A_j = \min\left\{1, \frac{\pi_j}{1-\pi_j}\right\}, \qquad D_j = \min\left\{1, \frac{1-\pi_j}{\pi_j}\right\} \tag{5.5}$$

lead to the largest possible value of $A_j + D_j$. [19] show that this is optimal in the sense that this provides the largest expected squared jumping distance and provides the smallest asymptotic variance of estimates of any function of $\boldsymbol{\gamma}$. The form of the proposal also makes intuitive sense. If π_j is close to one, the sampler rarely proposes to remove that variables from the model (D_j is close to 0) but will always propose to include the variables if it has been excluded (A_j is 1). Similarly, if π_j is close to 0, the sampler rarely proposes to add a variables to the model (A_j is close to 0) but will always propose to exclude the variables if it has been included (D_j is 1). If $\pi_j = 1/2$, A_j and D_j both equal 1 and this variable is proposed to be added to the model if excluded or removed from the model if included. In each case, this clearly represents the fastest possible mixing chain.

These results can be used to develop adaptive samplers for Bayesian variable selection. We will write $A_j^{(t)}$ and $D_j^{(t)}$ to represent the values of A_j and D_j used at the t-th iteration.

The first adaptive method that [19] introduce is the Exploratory Individual Adaptation (EIA) algorithm which uses the result in (5.4) to guide updates of A_j and D_j whilst also trying to maximise the expected squared jumping distance of the sampler. The updates are

$$\text{logit}_\epsilon A_j^{(t+1)} = \text{logit}_\epsilon A_j^{(t)} + t^{-\phi}\left(\gamma_j^{A^{(t)}} d_t(\tau_U) + \gamma_j^{D^{(t)}} d_t(\tau_L) - \gamma_j^{A^{(t)}}(1 - d_t(\tau_U))\right)$$

and

$$\text{logit}_\epsilon D_j^{(t+1)} = \text{logit}_\epsilon D_j^{(t)} + t^{-\phi}\left(\gamma_j^{D^{(t)}} d_i(\tau_U) + \gamma_j^{A^{(t)}} d_t(\tau_L) - \gamma_j^{D^{(t)}}(1 - d_t(\tau_U))\right),$$

where $d_t(\tau) = \text{I}(a_t \geq \tau)$, $1/2 < \phi \leq 1$ and $\text{logit}_\epsilon(x) = \log(x - \epsilon) - \log(1 - x - \epsilon)$ where $0 \leq \epsilon \leq 1/2$. Usually, ϵ is chosen to be small and this construction implies that $\epsilon \leq x \leq 1-\epsilon$. We also define

$$\gamma_j^{A^{(t)}} = \begin{cases} 1 & \text{if } \gamma_j' \neq \gamma_j^{(t)} \text{ and } \gamma_j^{(t)} = 0 \\ 0 & \text{otherwise} \end{cases}$$

and

$$\gamma_j^{D(t)} = \begin{cases} 1 & \text{if } \gamma_j' \neq \gamma_j^{(t)} \text{ and } \gamma_j^{(t)} = 1 \\ 0 & \text{otherwise} \end{cases},$$

which indicated whether the j-th variable has been proposed to be added or deleted from the model respectively. The EIA algorithm introduces two tuning parameters τ_U and τ_L, which determine the type of update that will be applied.

These updates of $A_j^{(t+1)}$ and $D_j^{(t+1)}$ move $A_j^{(t+1)}/D_j^{(t+1)}$ towards the optimal value in (5.4). An expansion step occurs if $a_t > \tau_U$ and both $A_j^{(t+1)}$ and $D_j^{(t+1)}$ are set larger than $A_j^{(t)}$ and $D_j^{(t)}$. Similarly, a shrinkage step occurs if $a_t < \tau_L$ and both $A_j^{(t+1)}$ and $D_j^{(t+1)}$ are set smaller than $A_j^{(t)}$ and $D_j^{(t)}$. Otherwise, a correction step occurs (when $\tau_L \leq a_t \leq \tau_U$) when the ratio are adjusted to be closer to A_j/D_j. See [19] for more details.

The second adaptive method proposed in [19] is the Adaptively Scaled Independence (ASI) sampler which uses a scaled version of the result in (5.5) to set $A_j^{(t+1)}$ and $D_j^{(t+1)}$. The method assumes that a Rao-Blackwellised estimate $\hat{\pi}_j^{(t)}$ of π_j has been calculated using the output of the first t iterations of the sampler. The proposal is

$$A_j^{(t+1)} = \zeta^{(t+1)} \min\left\{1, \frac{\kappa + (1-2\kappa)\hat{\pi}_j^{(t)}}{\kappa + (1-2\kappa)(1-\hat{\pi}_j^{(t)})}\right\}$$

and

$$D_j^{(t+1)} = \zeta^{(t+1)} \min\left\{1, \frac{\kappa + (1-2\kappa)(1-\hat{\pi}_j^{(t)})}{\kappa + (1-2\kappa)\hat{\pi}_j^{(t)}}\right\},$$

where $\zeta^{(t+1)}$ is an (adaptive) tuning parameter and $0 \leq \kappa \leq 1/2$ is usually chosen to be small (its inclusion avoids very small estimated probabilities). The adaptive parameter is tuned using

$$\text{logit}_\epsilon \zeta^{(t+1)} = \text{logit}_\epsilon \zeta^{(t)} + t^{-\phi}(a_t - 0.234),$$

where $1/2 < \phi \leq 1$ and a_t is the Metropolis-Hastings acceptance probability at the t-th iteration. We use the initialisation $\zeta^{(0)} = \min\{0.5, 1/(h\,p)\}$ which avoids very large initial model sizes being proposed.

The method relies on quickly finding a good estimate of the posterior inclusion probabilities π_j. This is achieved through the use of a Rao-Blackwellised estimate in place of the simpler sample average of the previous draws of γ_j and the use of multiple independent chains which use the same adapted tuning parameters in their proposals. This provides more representative draws from the posterior distribution to be used in the estimator.

In [19] both algorithms are found to outperform most available alternatives by some margin. The ASI algorithm is found to perform better than the EIA algorithm by between 2 and 100 times across a range of simulated data sets with 5000 variables. The difference is due to the number of iterations needed to tune the EIA algorithm (which uses $2p$ adaptive parameter) relative to the ASI algorithm (which uses only one adaptive parameter) and so these differences will tend to increase with p. Thus, we will mostly focus on the ASI algorithm in the sequel.

5.3.1 Linear Regression

The EIA and ASI methods work by adapting the proposal on model space (the space of possible values of γ). In linear regression models, the method can be used to directly sample from the posterior distribution $p(\gamma \mid \boldsymbol{y}, \boldsymbol{X})$ which is available analytically if a conjugate prior distribution is used or the prior structure in (5.3). The method also needs an analytic expression for $p(\gamma_j = 1 \mid \boldsymbol{y}, \boldsymbol{X}, \boldsymbol{\gamma}_{-j})$ which is available if $f(\boldsymbol{y} \mid \boldsymbol{\gamma}, \boldsymbol{X})$ is analytically available. Computationally efficient forms for $p(\gamma_j = 1 \mid \boldsymbol{y}, \boldsymbol{X}, \boldsymbol{\gamma}_{-j})$ are given by the supplementary material of [19] for independent priors and [16] for g-priors.

5.3.2 Non-Gaussian Models

Computational methods for Bayesian variable selection are often more complicated when a linear regression model is not appropriate since $f(\boldsymbol{y} \mid \boldsymbol{\gamma}, \boldsymbol{X})$ is then usually not available analytically. GLMs represent a particularly important class of such models and [42] consider the application of the ASI method in logistic and accelerated failure time regression models. There are three main approaches to the lack of analytic form for $f(\boldsymbol{y} \mid \boldsymbol{\gamma}, \boldsymbol{X})$. Firstly, we can introduce latent variables $\boldsymbol{z}$ which allow us to calculate $f(\boldsymbol{y} \mid \boldsymbol{\gamma}, \boldsymbol{X}, \boldsymbol{z})$ analytically. This allows the method for linear regression to be directly used but the update of $\boldsymbol{\gamma}$ is made conditional on $\boldsymbol{z}$ and the chain can mix slowly if $\boldsymbol{z}$ is highly correlated with $\boldsymbol{\gamma}$. Secondly, we can approximate $f(\boldsymbol{y} \mid \boldsymbol{\gamma}, \boldsymbol{X})$ using a Laplace approximation to this marginal likelihood. This method can introduce some bias into the output of the MCMC algorithm. The Laplace approximation can also be used to construct an unbiased estimate of $f(\boldsymbol{y} \mid \boldsymbol{\gamma}, \boldsymbol{X})$ which can be used in the pseudo-marginal MCMC approach developed by [1]. The computational efficiency of Laplace-based methods tends to improve relative to latent variable methods with sample size n since the Laplace approximation becomes more accurate whereas the latent variables methods become more computationally expensive (since the dimension of the latent variable is n). Thirdly, we can work directly with the likelihood $f(\boldsymbol{y} \mid \boldsymbol{\beta}_\gamma, \boldsymbol{X}_\gamma)$ and use a reversible jump MCMC sampler which jointly proposes γ and $\boldsymbol{\beta}_\gamma$ using the proposal $p(\boldsymbol{\beta}'_\gamma \mid \boldsymbol{\beta}_\gamma, \boldsymbol{\gamma}'.\boldsymbol{\gamma})p(\boldsymbol{\gamma}' \mid \boldsymbol{\gamma})$. The success of this method depends on finding a good proposal $p(\boldsymbol{\beta}'_\gamma \mid \boldsymbol{\beta}_\gamma, \boldsymbol{\gamma}'.\boldsymbol{\gamma})$ which is a challenging problem.

[42] discuss how these approaches can be used in logistic and accelerated failure time models. We will concentrate on logistic regression and consider how the three approaches can be applied to this model. The logistic regression model for the response $y_i \in \{0, 1\}$ assumes that for $i = 1, \ldots, n$

$$f(y_i = 1 \mid \boldsymbol{X}_\gamma) = \frac{\exp\{\mu_{\gamma,i}\}}{1 + \exp\{\mu_{\gamma,i}\}}, \qquad \mu_{\gamma,i} = \alpha + \boldsymbol{x}_{\gamma,i}\boldsymbol{\beta}_\gamma,$$

where $\boldsymbol{x}_{\gamma,i}$ is the i-th row of $\boldsymbol{X}_\gamma$.

There are several data augmentation approaches for logistic regression. The Pólya-gamma data augmentation method of [33] is well-established and expresses the model for y_i with an additional latent variable ω_i as

$$f(y_i = 1 \mid \omega_i, \boldsymbol{X}_\gamma) = 2^{-1/2} \exp\{\kappa_i \mu_{\gamma,i}\} \exp\{-\omega_i \mu_{\gamma,i}^2/2\},$$

where $\kappa_i = y_i - 1/2$ and $\omega_i \sim \text{PG}(1, 0)$ where $\text{PG}(a, b)$ represents the Pólya-gamma distribution (see [33] for a description of this distribution and efficient methods for simulation of Pólya-gamma distributed random variates). If we assume that $\boldsymbol{\beta}_\gamma \sim \mathcal{N}_{p_\gamma}(\mathbf{0}_{p_\gamma}, \boldsymbol{V}_\gamma)$, the introduction of the latent variables allows us to derive an analytic expression for $f(\boldsymbol{y} \mid \boldsymbol{\omega}, \boldsymbol{X}_\gamma)$ where $\boldsymbol{\omega} = (\omega_1, \omega_2, \ldots, \omega_n)$ which has the form

$$|\boldsymbol{V}_\gamma|^{-1/2} |\tilde{\boldsymbol{X}}_\gamma^T \tilde{\boldsymbol{X}}_\gamma + \boldsymbol{V}_\gamma^{-1}|^{-1/2} \exp\left\{-\frac{1}{2} \boldsymbol{K} \boldsymbol{X}_\gamma (\tilde{\boldsymbol{X}}_\gamma^T \tilde{\boldsymbol{X}}_\gamma + \boldsymbol{V}_\gamma^{-1})^{-1} \boldsymbol{X}_\gamma^T \boldsymbol{K}^T\right\},$$

where $\boldsymbol{K} = (\kappa_1, \ldots, \kappa_n)$, $\tilde{\boldsymbol{X}}_\gamma$ is a $(n \times p_\gamma)$-dimensional matrix with (i, j)-th element $X_{i,j}\sqrt{\omega_i}$. This also allows a Gibbs sampler to be constructed for $\boldsymbol{\gamma}$ and $\omega_1, \ldots, \omega_n$ in which $\boldsymbol{\gamma}$ is updated conditionally on $\omega_1, \ldots, \omega_n$ using the ASI sampler and $\omega_1, \ldots, \omega_n$ are updated by first sampling α and $\boldsymbol{\beta}_\gamma$ from their full conditional distribution and then sampling $\omega_i \mid \alpha, \boldsymbol{\beta}_\gamma$ from its full conditional for $i = 1.\ldots, n$. Full details of these steps are provided by [42].

The second approach uses the Laplace approximation to the posterior distribution $p(\alpha, \boldsymbol{\beta}_\gamma \mid \boldsymbol{X}_\gamma, \boldsymbol{y})$, which is

$$p_{Laplace}(\boldsymbol{\theta}_\gamma) = \mathcal{N}_{p_\gamma+1}(\hat{\boldsymbol{\theta}}_\gamma, \boldsymbol{\Sigma}_\gamma),$$

where $\hat{\boldsymbol{\theta}}_\gamma$ is the posterior mode of $\boldsymbol{\theta}_\gamma = (\alpha, \boldsymbol{\beta}_\gamma)$ and $\boldsymbol{\Sigma}_\gamma = (-\boldsymbol{H}_\gamma)^{-1}$, where $\boldsymbol{H}_\gamma$ is the Hessian of $\log f(\boldsymbol{y} \mid \boldsymbol{\theta}_\gamma) + \log p(\boldsymbol{\theta}_\gamma)$ evaluated at $\hat{\boldsymbol{\theta}}_\gamma$. The posterior mode can often be found quickly using an Iterated Reweighted Least Squares algorithm. Using this approximation, the marginal likelihood can be approximated by

$$f_{Laplace}(\boldsymbol{y} \mid \gamma) = |\boldsymbol{\Sigma}_\gamma|^{1/2} f(\boldsymbol{y} \mid \hat{\boldsymbol{\theta}}_\gamma)\, p(\hat{\boldsymbol{\theta}}_\gamma),$$

and used to define an approximate posterior distribution $p_{Laplace}(\gamma \mid \boldsymbol{y}) \propto p_{Laplace}(\boldsymbol{y} \mid \gamma)\, p(\gamma)$. Although approximate, it is widely used and is often similar to the actual posterior distribution $p(\gamma \mid \boldsymbol{y}) \propto f(\boldsymbol{y} \mid \gamma)\, p(\gamma)$. Alternatively, the Laplace approximation can be used to define an unbiased estimate of $f(\boldsymbol{y} \mid \gamma)$. The unbiased estimate is

$$\hat{f}(\boldsymbol{y} \mid \gamma) = \frac{1}{N} \sum_{i=1}^{N} \frac{f\left(\boldsymbol{y} \mid \alpha^{(i)}, \boldsymbol{\beta}_\gamma^{(i)}\right) p\left(\alpha^{(i)}, \boldsymbol{\beta}_\gamma^{(i)}\right)}{p_{Laplace}\left(\alpha^{(i)}, \boldsymbol{\beta}_\gamma^{(i)}\right)},$$

where $\left(\alpha^{(1)}, \boldsymbol{\beta}_\gamma^{(1)}\right), \ldots, \left(\alpha^{(N)}, \boldsymbol{\beta}_\gamma^{(N)}\right)$ are samples from $p_{Laplace}(\alpha, \boldsymbol{\beta}_\gamma)$. This unbiased estimate of the marginal likelihood can be used in a pseudo-marginal sampler (as in [1]) where $\hat{f}(\boldsymbol{y} \mid \gamma)$ is used in place of $f(\boldsymbol{y} \mid \gamma)$ in the usual Metropolis-Hastings acceptance probability. [42] also describe how this approach can be extended to a correlated pseudo-marginal approach (see [12]), which makes the mixing of the chain less sensitive to the Monte Carlo error of $\hat{f}(\boldsymbol{y} \mid \gamma)$.

The third approach uses reversible jump MCMC. The success of this approach depends on the form of the proposal of the regression coefficients conditional on the proposed model γ'. [26] looked at a wide-range of possible transition mechanisms and found the automatic generic sampler of [18] provides the most computationally efficient approach (in terms of effective sample size per unit of computational time). Interestingly, [42] show how the automatic generic sampler for logistic regression models can be expressed as a special case of the correlated pseudo-marginal approach. This suggests that correlated pseudo-marginal methods are a more attractive approach than reversible jump MCMC for exact Bayesian variable selection in logistic regression models.

5.4 Examples

5.4.1 Simulated Example: Linear Regression

To illustrate the ASI method compared to the ADS algorithm, we look at the relative performance of the two algorithms in an application to linear regression with simulated data. The data is generated using the approach of [44] which assumes the usual linear regression model,

$$\boldsymbol{Y} = \boldsymbol{X}\boldsymbol{\beta}^\star + \boldsymbol{e},$$

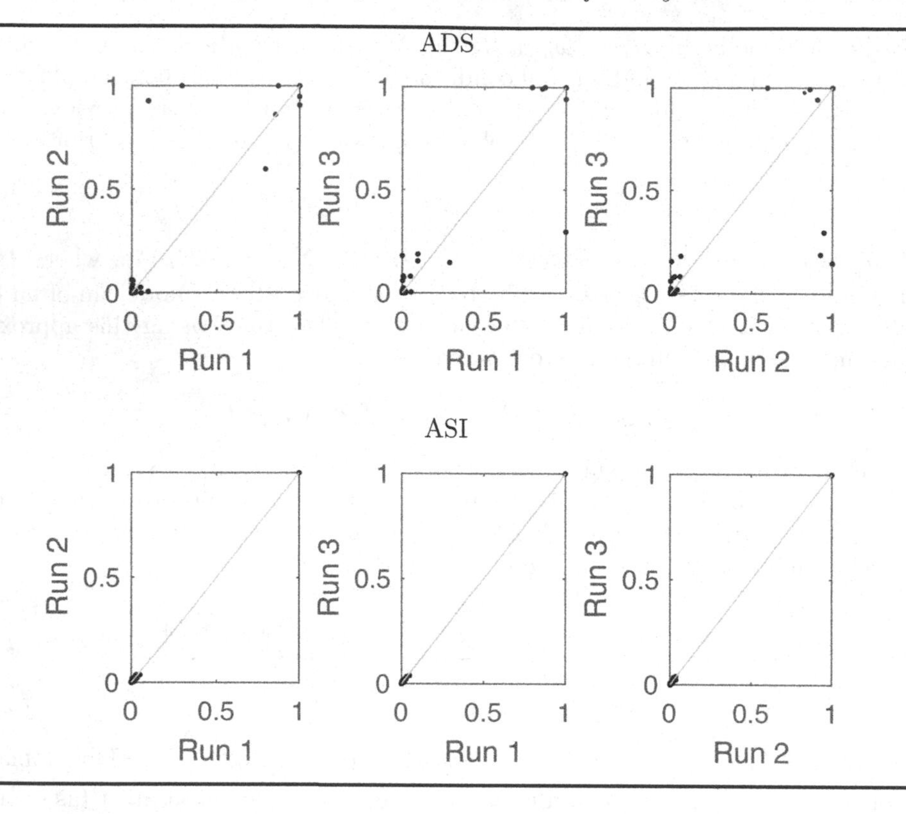

FIGURE 5.1
Simulated example: Scatter plots of the estimated marginal posterior inclusion probabilities from three different runs of the ASI algorithm and the ADS algorithm. Each run took the same time. The 45° line is also shown in each graph.

where $\boldsymbol{e} \sim \mathcal{N}_n(\mathbf{0}_n, I_n)$. Only the first ten regression coefficients are non-zero with

$$\boldsymbol{\beta}^\star = \text{SNR}\sqrt{\frac{\log p}{n}}(2, -3, 2, 2, -3, 3, -2, 3, -2, 3, 0, \ldots, 0)^T \in \mathbb{R}^p,$$

where SNR is the signal-to-noise ratio. The variables for the i-th observation are generated as $(x_{i1}, \ldots, x_{ip})^T \sim \mathcal{N}_p(\mathbf{0}_p, \mathbf{\Sigma})$ where $\Sigma_{jk} = 0.6^{|j-k|}$.

We compared the performance of the ADS and ASI algorithms by simulating a data set with 1000 observations and 50000 variables. Each algorithm was run three times with each run taking 15 minutes (using Matlab 2020a on an iMac with a 3.4 GHz Quad-Core i5 processor). The results of the three runs are shown in Figure 5.1. Clearly, the ASI algorithm leads to far more consistent estimates of the posterior inclusion probabilities across the three runs than the ADS algorithm. In fact, the ADS algorithm often doesn't converge. Performance was compared by first calculating a "gold standard" estimate of the PIPs using the ASI algorithm run for a total of 2,750,000 iterations and secondly calculating the mean squared error (MSE) of the estimated PIPs from each algorithm across three runs. Figure 5.2 shows the logarithm of the running MSE's. The j-th running MSE is the average of the MSE's for the j variables with the highest PIPs. In the ASI results, the running MSE for the first ten variables is zero since the PIPs from the three runs are all 1 which agrees with the gold standard. The results for the other PIPs show that MSE of ASI is on average

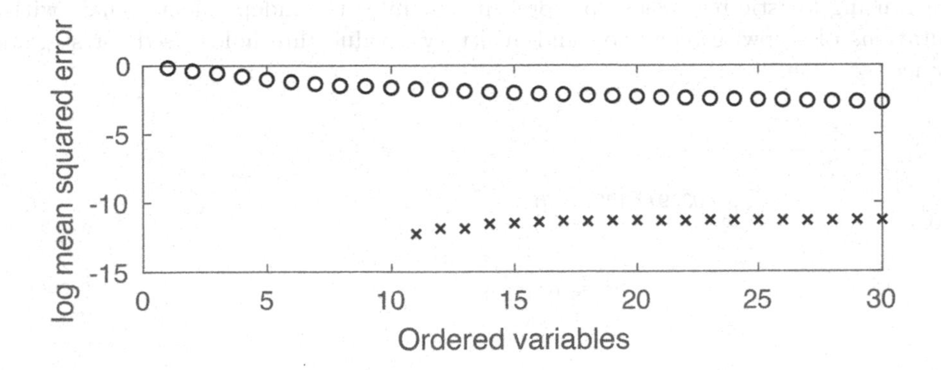

FIGURE 5.2
Simulated example: The logarithm of the running mean squared error for the thirty SNPs with the highest PIPs using the ADS algorithm (circles) and the ASI algorithm (crosses).

5000 times smaller than ADS on this example. This is due to the ASI algorithms ability to concentrate effort on more important variables.

5.4.2 Fine Mapping for Systemic Lupus Erythematosus

Fine mapping in genomics involves finding single-nucleotide polymorphisms (SNPs) in a small region of the genome which are associated with a particular trait. In binary traits (such as the presence or absence of a disease), the standard approach uses logistic regression with case/control as outcome and the SNPs as explanatory variables. The SNP data are coded as 0,1 or 2 copies of a reference allele (usually the allele observed to have the lowest frequency, the "minor allele"). Often, separate logistic regressions with individual SNPs are fitted generating a p-value for the inclusion of each SNP. These p-values for SNPs with similar genetic positions are usually highly correlated due to linkage disequilibrium and many methods exist for processing these p-values to determine a subset of SNPs associated with the disease, which are usually called causal SNPs. The association at a causal SNP and its surrounding correlated SNPs is referred to as a "signal". We expect there are many independently acting causal SNPs. The aim is to determine how many independent signals there are, and which SNPs are most likely to be the causal SNPs at each signal. An alternative approach is to use variable selection methods with all SNPs (e.g., stepwise regression [11], penalised regression [4, 23, 43] and Bayesian variable selection [21]).

We consider an example of finding SNPs which are associated with Systemic Lupus Erythematosus (SLE) using a case-control study. The data consist of genotypes taken from a genome-wide genetic case/control association study (4,036 cases and 6,959 controls) and has been previously analysed using stepwise regression in [29]. The cases are SLE patients and the controls are from a public repository with both groups of European ancestry. In this example we only look at genotypes in one genetic locus on Chromosome 1 around position 173 megabases called "TNFSF4", with 5771 SNPs in our data. The TNFSF4 locus is known to have two strong independent association signals from standard classical logistic regression using stepwise regression to determine the number of signals. Figure 5.4 shows the p-values (on a log scale) for the effect of each SNP in single SNP logistic regressions. There are two clear areas of the genome where the p-values tends to be much smaller (around 1.7325 and 1.7334 Mb) indicating that these may be areas of the genome with casual SNPs for SLE. Here, we applied Bayesian variable selection with the ASI adaptation scheme to these data using logistic regression to more directly infer the independent signals without the limitations of stepwise regression and arbitrary p-value thresholds used for statistical significance.

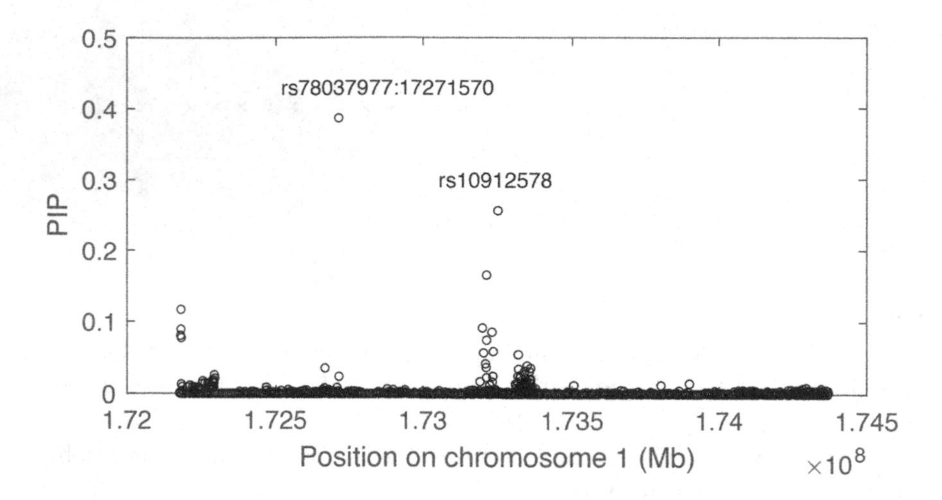

FIGURE 5.3
SLE example: Estimated marginal posterior inclusion probabilities (PIPs) of each SNP considered on chromosome 1 calculated using the ASI algorithm.

Bayesian variable selection was applied to this data with the following prior: $h \sim \mathcal{B}\text{eta}(1, 5766/5)$, which implies that the prior mean model size is 5, $\alpha \sim \mathcal{N}(0, 1)$ and $\boldsymbol{\beta}_\gamma \sim \mathcal{N}_{p_\gamma}(\mathbf{0}_{p_\gamma}, 1/4 I_{p_\gamma})$. The results of applying the Bayesian variable selection method are shown in Figure 5.3. There is clear evidence of three independent association signals, two of which contained the original two signals. The additional signal has a high-posterior inclusion probability for a single SNP (at 1.72715702 Mb). Its marginal p-value when tested as a single marker is $p \approx 0.0001$, which is not sufficiently small after accounting for multiple testing and suggests that the effect only becomes evident after adjusting for the other two signals. To validate this additional signal, the association between disease and this SNP was tested using an independent dataset (3,568 cases and 14,923 controls) and found to be strongly associated. The finding also have an interesting biological interpretation. The original two signals have been found to control the expression of TNFSF4, using data that correlated the SNP data with gene expression readings. The control of gene expression is local (as they are both close to the gene) and likely acts through the gene promotor. The new signal has a longer-range action affecting gene transcription though DNA looping which brings it closer to the gene promotor and affects binding of other transcription factors. This result was found by looking at Capture Hi-C data which demonstrated that the lead SNP at this new signal interacts with the promotor region of TNFSF4 in B cells. Our collaborators also found that the interaction between the lead SNP and the TNFSF4 promoter is strongest in Monocytes, Macrophages and B cells (naïve and total), which are the cell types expressing TNFSF4. In conclusion the approach validated two signals at TNFSF4 and added to the biological understanding by identifying a further weaker signal which was explained by distal action on gene expression.

To compare algorithmic performance, each algorithm was run three times with each run taking an hour (using the same set-up as the simulated example). The results of the three runs are shown in Figure 5.5. Clearly, the ASI algorithm leads to more consistent estimates of the posterior inclusion probabilities across the three runs than the ADS algorithm.

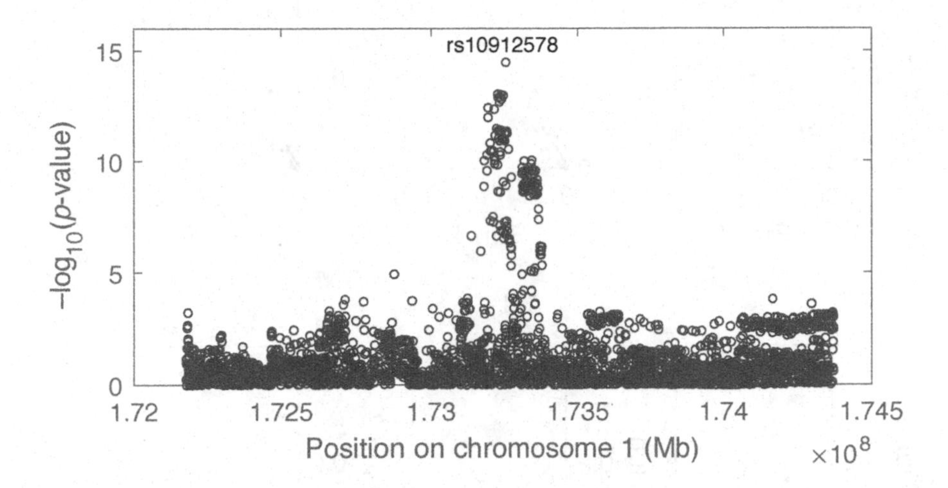

FIGURE 5.4
SLE example: p-values for the effect of each SNP considered on chromosome 1 derived from single SNP logistic regression fits.

We quantify the differences in the performance of the two algorithms by calculating running MSEs for the PIPs in the same way as the simulated example. Figure 5.6 shows the logarithm of the running MSEs for the thirty SNPs with the highest PIPs. The differences between the MSEs from the two algorithms tend to be larger for the most important SNPs with the average logarithmic difference for the top five SNPs being 3 ($e^3 = 20$). The difference decreases as PIPs becomes smaller but the thirtieth most important SNP has a difference of 2 ($e^2 = 7.4$). The MSE is a sum of the bias and the variance of the estimated PIPs. The variance of the estimated PIPs has the form $\text{PIP}(1 - \text{PIP})/n_{eff}$ where n_{eff} is the effective sample size and, in the absence of bias, the difference in logarithm of the MSEs should be constant across all SNPs. The larger MSE for larger PIPs therefore reflects the additional bias in the ADS estimates compared to the ASI estimates. This can be clearly seen in Figure 5.5 where the two most important SNPs have PIPs around 0.4 for all runs of the ASI algorithm but which vary much more widely with evidence of a downward bias for the ADS algorithm. This illustrates the typically faster convergence of the ASI algorithm compared to the ADS algorithm (since the Rao-Blackwellised estimates of the PIPs guide the sampler much more quickly towards higher posterior probability models) as well as better mixing.

5.4.3 Analysing Environmental DNA Data

Environmental DNA has emerged as an important tool for the assessment of aquatic biodiversity. The approach uses laboratory analysis to detect the presence of particular species in water samples. This approach is much cheaper than large-scale monitoring and has the potential to detect rare or cryptic species. However, these approaches are not error free and can suffer from both false positive and false negative errors which can occur at both water collection and laboratory analysis.

We consider the analysis of a single species using qPCR as the laboratory analysis technique. The qPCR runs are usually replicated and we denote the number of replicates

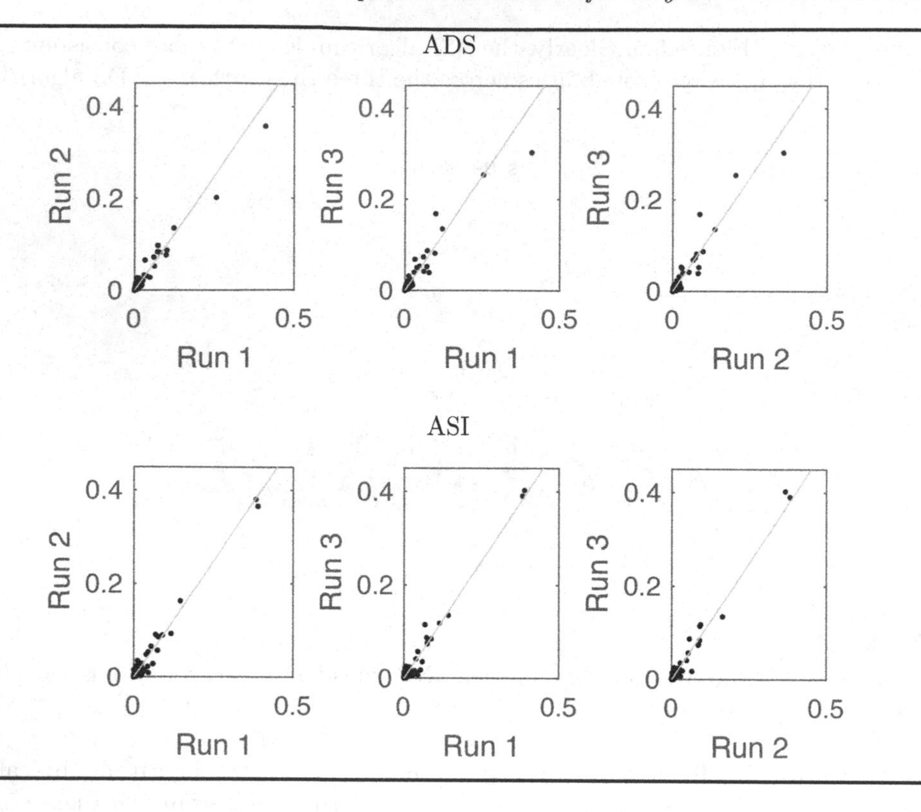

FIGURE 5.5
SLE example: Scatter plots of the estimated marginal posterior inclusion probabilities from three different runs of the ASI algorithm and the ADS algorithm. Each run took the same time. The 45° line is also shown in each graph.

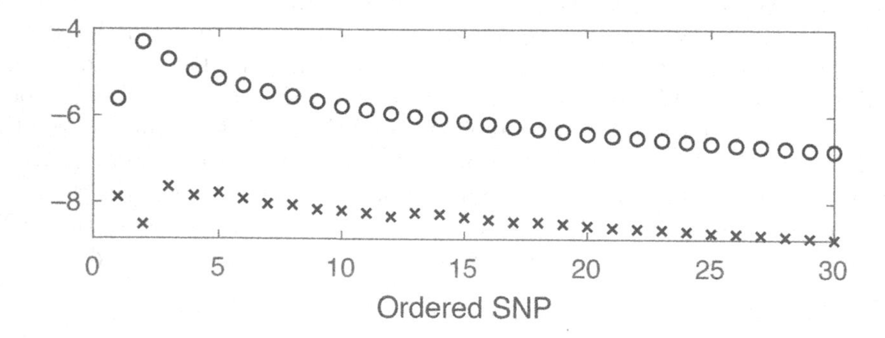

FIGURE 5.6
SLE example: The logarithm of the running mean squared error for the thirty SNPs with the highest PIPs using the ADS algorithm (circles) and the ASI algorithm (crosses).

by K. Water samples are collected at S sites on M occasions and y_{sm} is the number of positive qPCR replicates for the s-th site on the m-th occasion. A suitable model for these data [20, 22] has the following hierarchical structure, for $s = 1, \ldots, S$ and $m = 1, \ldots, M$,

$$y_{sm} \mid w_{sm} \sim \begin{cases} \mathcal{B}in(K, p_{11s}) & \text{if } w_{sm} = 1 \\ \mathcal{B}in(K, p_{10s}) & \text{if } w_{sm} = 0 \end{cases},$$

$$w_{sm} \sim \begin{cases} \mathcal{B}er(\theta_{11s}) & \text{if } z_s = 1 \\ \mathcal{B}er(\theta_{10s}) & \text{if } z_s = 0 \end{cases},$$

$$z_s \sim \mathcal{B}er(\psi_s),$$

and y_{sm}, w_{sm} and z_s are conditionally independent across sites and occasions. The parameter z_s is 1 if the species is present at the s-th site and w_{sm} is 1 if the DNA from the species is present in the sample collected on the m-th occasion at the s-th site. The parameters ψ_s, θ_{11s} and θ_{10s} control the probability of these events and p_{11s} is the probability of a positive qPCR run if DNA is present in a sample from the s-th location and, similarly, p_{10s} is the probability if DNA is absent. Covariates are observed for each site which can be included in the model by assuming that ψ_s, θ_{11s}, θ_{10s}, p_{11s} and p_{10s} are the logistic transform of separate linear predictors, e.g.

$$\text{logit}(\xi_s) = \mu^\xi + \sum_{j=1}^{p_\gamma^\xi} x_{\gamma,s,j}^\xi \beta_j^\xi, \tag{5.6}$$

where $\xi \in \{\psi, \theta_{11}, \theta_{10}, p_{11}, p_{10}\}$, γ^ξ is the vector of inclusion/exclusion variables for ξ, $x_{\gamma,s,j}^\xi$ is the element (s, j) of the design matrix of the included variables for ξ which is an $(S \times p_\gamma^\xi)$-dimensional matrix.

Inference in this model is challenging since there are a number of likelihood symmetries. These can be addressed by including additional data (see [22] for work in this direction). We follow [20] by using ϵ well ordered priors for the pairs (p_{11}, p_{10}) and $(\theta_{11s}, \theta_{10s})$. The prior will be explained in terms of a generic pair (p, q). This prior is defined in terms of three parameters a, b and ϵ where a and b control the prior expectations of p and q respectively and ϵ controls the probability that $p > q$. This allows us to express the prior belief that the probability of a false positive, q, is unlikely to be larger than the probability of a true positive, p, by choosing a large, b small and ϵ small. In regression problems, the form of the prior for a pair (p, q) is

$$\mu^p \sim \mathcal{N}(\mu_0^p, \Delta^p \phi_\alpha^p), \qquad \boldsymbol{\beta}^p \mid \gamma^p \sim \mathcal{N}_{p_\gamma^p}(\mathbf{0}_{p_\gamma^p}, \Delta^p \phi_\beta^p C^p),$$

$$\mu^q \sim \mathcal{N}(\mu_0^q, \Delta^q \phi_\alpha^q), \qquad \boldsymbol{\beta}^q \mid \gamma^q \sim \mathcal{N}_{p_\gamma^q}(\mathbf{0}_{p_\gamma^q}, \Delta^q \phi_\beta^q C^q),$$

where

$$\Delta^\xi = \frac{(\text{logit}(a) - \text{logit}(b))^2}{2(\Phi^{-1}(\epsilon))^2 \left(\phi_\alpha^\xi + \phi_\beta^\xi \sum_{i=1}^{p_\gamma^\xi} \sum_{i=1}^{p_\gamma^\xi} C_{ij}^\xi \Sigma_{ij}^\xi\right)},$$

$\boldsymbol{\Sigma}^\xi$ is the sample covariance matrix of the regressors included in the regression for ξ, ϕ_α^ξ and ϕ_β^ξ are hyperparameters controlling the relative variances of μ^ξ and $\boldsymbol{\beta}^\xi$, and $\xi \in \{p, q\}$.

The posterior distribution can be sampled using MCMC methods by introducing z_s and w_{sm} as latent variables in the sampler. Then inclusion/exclusion variables γ^ψ, $\gamma^{\theta_{11}}$, $\gamma^{\theta_{10}}$, $\gamma^{p_{11}}$ and $\gamma^{p_{10}}$ can be updated in a Gibbs sampler using separate ASI samplers for logistic regression models. More details about the computational scheme are given in [20].

We extend the analysis in [20] to include interactions in the linear predictors for the regressions for ψ, θ_{11}, θ_{10}, p_{11} and p_{10}. The data are samples collected on great crested newts and commissioned by Natural England [7]. Twelve qPCR replicates were performed per sample. Data were collected at 189 sites on one occasion each and presence was known at

15 of these sites (due to physical evidence of the animal during a visit to the site). Suppose that $k_s = 1$ if there is known presence of the species at site s and 0 otherwise, the model can be extended to include this information

$$k_s \mid z_s = 1 \sim \mathcal{B}er(\pi), \qquad p(k_s = 1 \mid z_s = 0) = 0.$$

We assume the weakly-informative prior $\pi \sim \mathcal{B}\text{eta}(0.5, 0.5)$. There were 19 environmental covariates collected for each site and these are given in the on-line supplementary material to [20]. Including all interactions gives us a total of $p = 190$ possible covariates. For each regression, we assume the prior $h \sim \mathcal{B}\text{eta}(1, \frac{186}{4})$ which implies a prior expectation that four regressors are included in the regression. We use the ϵ well ordered prior with $a = 0.8$, $b = 0.2$ and $\epsilon = 0.025$ for the pair $(\theta_{11}, \theta_{10})$ and with $a = 0.9$, $b = 0.1$ and $\epsilon = 0.001$ for the pair (p_{11}, p_{10}). This implies that we expected less false positive and false negatives in the laboratory test than the field experiments and is consistent with the prior belief of ecologists involved in the statistical analysis. The priors for the ψ regression are $\mu^{\psi} \sim \mathcal{N}(0, 4)$ and $\beta_j^{\psi} \mid \gamma_j^{\psi} = 1 \sim \mathcal{N}(0, 1/4)$.

(w_1, w_2, w_3)	ψ	θ_{11}	θ_{10}	p_{11}	p_{10}
(0.75, 0.75, 0.75)	5.7	3.4	1.3	0.4	0.3
(0.75, 0.75, 0.25)	6.6	2.8	1.4	0.9	0.8
(0.75, 0.25, 0.25)	6.7	2.1	1.5	0.8	0.8

TABLE 5.1
eDNA example: The relative effective sample sizes for three configurations of the adaptive sampler compared to Add-Delete-Swap.

This is a challenging example since the sampler runs on five logistic regression linked by latent variables and the inclusion of interactions induces strong correlation structure in the posterior distribution of the inclusion variables. For these reasons, rather than using ASI directly, we use a sampler where we propose the ASI proposal with probability w and from an Add-Delete-Swap proposal with probability $1 - w$. We consider different values of w for the different stages of the model: w_1 for ψ, w_2 for θ_{11} and θ_{10}, and w_3 for p_{11} and p_{10}. Table 5.1 shows the effective sample size relative to the Add-Delete-Swap, i.e., $\text{ESS}_{w_1,w_2,w_3}/\text{ESS}_{ADS}$ where ESS_{w_1,w_2,w_3} and ESS_{ADS} are the effective sample size with mixing proportions w_1, w_2 and w_3 and Add-Delete-Swap respectively. The results demonstrate that the mixing for ψ, θ_{11} and θ_{10} improves with all choices of w_1, w_2 and w_3 considered. The mixing is worse for both p_{11} and p_{10} with $w_1 = 0.75$, $w_2 = 0.75$, $w_3 = 0.75$ and similar to Add-Delete-Swap for the other two choices. The poor mixing of ASI for p_{11} and p_{10} is due to the strong correlations in the inclusion variables. The choice of $w_1 = 0.75$, $w_2 = 0.75$ and $w_3 = 0.25$ gives the best overall performance and so favours ASI proposals for ψ, θ_{11} and θ_{10} and ADS proposals for p_{11} and p_{10}.

The inclusion of interactions provide clearer results of the Bayesian variable selection than those given in [20]. There is little evidence about the variables which affect presence of the great crest newts (GCN), ψ, with the only regressors with posterior inclusion probabilities above 0.1 being the interactions of Shade and Permanance (0.14), Overhang and Permanence (0.12) and Macrophytes and Water Quality (0.11). This suggests that permanence and either shade or overhang may have an effect on the presence of GCN. There are no regressors with posterior inclusion probabilities above 0.1 for the regression of either θ_{11} and θ_{10} suggesting that the variables collected do not effect the collection of water samples. There are stronger results for the probability of true positive (p_{11}) and false positives (p_{10}).

For p_{11}, the interaction of pond density and rough grass has strong evidence of an effect (0.99), with the interactions of Shade and Area (0.25) and Shade and Width (0.23) also having some evidence of an effect (these suggest that a combination of shade and some measure of size of the pond has evidence of an effect). For p_{10}, the main effect of fish (0.45), and interactions of overhang and outflow (0.37), fish and water quality (0.27) and fish and scrub hedge (0.25). This suggests that the presence of fish and water quality contribute to the probability of false positives.

5.5 Discussion

Bayesian variable selection methods are attractive for high-dimensional regression problems where the large number of possible regressors often means that many combinations provide plausible explanations of the data. The posterior distribution over model space can be used to identify variables linked to a response without having to choose a particular model and provides a way to extract information in the presence of collinearity. In the fine mapping example, the SNPs with non-negligible posterior inclusion probabilities appear in several tight clusters and these clusters can be interpreted as potential sites for a causal SNP. Furthermore, the SNP rs78037977 is identified as potentially related to the response which was missed by methods such as stepwise regression which work through generating a sequence of models. The eDNA example illustrates the ability of the ASI proposal to mix well and to mix badly depending on the correlation structure of the posterior distribution in this relatively low-dimensional setting. A hybrid approach is able to provide similar (or much better) performance than the standard Add-Delete-Swap proposal.

Although the benefits of a Bayesian approach are well understood, the computation needed has been challenging. In this paper, we demonstrate the potential of adaptive Monte Carlo to improve the quality of computation in large p settings. We use information from the posterior distribution (in the form of posterior inclusion probabilities) to guide the sampler. This provides (often sizeable) improvements over the standard Add-Delete-Swap sampler in the two problems considered and has proven useful in a range of settings [19, 42]. The improvement tends to grow with the number of potential covariates. This is because the method avoids proposing models with variables which has very low marginal importance (and the number of these types of variables will grow as the number of potential covariates grows). However, the method does not take into account any correlation structure in the posterior distribution in model space and so can propose models which have relatively low likelihood values. One area for future work is to develop the ASI sampler to be more likely to propose models with relatively high likelihood values. This could be achieved by taking into account posterior correlations between the inclusions or exclusions of pairs of variables. A challenge is that there are p^2 correlations if there are p potential covariates and so all correlations cannot be stored if p is very large.

The success of the ASI approach is partly driven by the use of Rao-Blackwellised estimates of the posterior inclusion probability which are calculated during the run of the sampler. This allows the method to quickly distinguish between important and unimportant variables in the way that models are proposed. This only needs to be updated in the initial iterations of the sampler and can be quickly calculated for linear regression models. The approach can be extended to generalised linear models if latent variables can be introduced to convert the problem into a linear regression (e.g., using Pólya-gamma random variables for logistic regression models). This is a limitation since latent variable representations are not available for all generalised linear regression and generic approaches, such as those based on Laplace approximation, do not scale well with the number of potential covariates. The methods also have the potential to be extended to sampling over discrete spaces in models with similar structure, such as graphical models. The development of appropriate, generic approximations of the posterior inclusion probabilities is still an important problem and could allow the application of these adaptive Monte Carlo methods to a much wider range of models. The experience of the authors is that the performance of the algorithm is sensitive to the quality of the approximation and more work is needed to understand this effect. This could be used to either choose the number of samples used in the Rao-Blackwellised estimate or to use estimates which have lower quality but are computationally much cheaper.

Although the use of these algorithms for Bayesian variable selection are still at an early stage, they offer the potential to provide much better samples from the posterior distribution on model space. These can be used to provide better estimates of posterior inclusion probabilities or predictive distribution but also offer the potential to better understand the structure of the posterior distribution on model space, e.g., using the probability of including pairs of variables or to develop approximate credible sets of models. This would provide us with a clearer picture of posterior model uncertainty and better understanding how the relationships between variables interplays with their relationships with the response.

To conclude, we believe that adaptive Monte Carlo methods have the potential to provide computationally efficient posterior inference for variable selection problems and these will drive the development of methods for summarisation beyond maximum a posteriori and median model methods. This will provide users of these methods with a much richer understanding of the structure within the data.

Acknowledgements

The authors would like to acknowledge Professor Tim Vyse and Dr David Morris of King's College London for providing the data and biological meaning for the fine mapping and Natural England for the use of the data in the eDNA example.

Bibliography

[1] C. Andrieu and G. O. Roberts. The pseudo-marginal approach for efficient Monte Carlo computations. *The Annals of Statistics*, 37:697–725, 2009.

[2] C. Andrieu and J. Thoms. A tutorial on adaptive MCMC. *Statistics and Computing*, 18:343–373, 2008

[3] Y. F. Atchadé and J. S. Rosenthal. On Adaptive Markov Chain Monte Carlo Algorithms. *Bernoulli*, 11:815–828, 2005.

[4] K. L. Ayers and H. J. Cordell. SNP selection in genome-wide and candidate gene studies via penalized logistic regression. *Genetic Epidemiology*, 34:879–891, 2010.

[5] M.J. Bayarri, J.O. Berger, A. Forte, and G. García-Donato. Criteria for Bayesian model choice with application to variable selection. *The Annals of Statistics*, 40:1550–1577, 2012.

[6] A. Bhadra, J. Datta, N. G. Polson, and B. Willard. Lasso meets Horseshoe: A survey. *Statistical Science*, 34:405–427, 2019.

[7] D. Bormpoudakis, J. Foster, T. Gent, R. A. Griffiths, L. Russell, T. Starnes, J. Tzanopoulos, and J. Wilkinson. Developing models to estimate the occurrence in the English countryside of Great Crested Newts, a protected species under the Habitats Directive. Defra Project WC1108. Technical report, DICE, University of Kent, Canterbury, UK, 2016.

[8] L. Bottolo and S. Richardson. Evolutionary stochastic search for Bayesian model exploration. *Bayesian Analysis*, 5:583–618, 2010.

[9] P. J. Brown, M. Vannucci, and T. Fearn. Multivariate Bayesian variable selection and prediction. *Journal of the Royal Statistical Society, B*, 60:627–641, 1998.

[10] H. Chipman, E. I. George, and R. E. McCulloch. The practical implementation of Bayesian model selection. In P. Lahiri, editor, *Model Selection.* Hayward, 2001.

[11] H. J. Cordell and D. G. Clayton. A unified stepwise regression procedure for evaluating the relative effects of polymorphisms within a gene using case/control or family data: Application to HLA in Type 1 Diabetes. *American Journal of Human Genetics*, 70:124–141, 2002.

[12] G. Deligiannidis, A. Doucet, and M. K. Pitt. The correlated pseudomarginal method. *Journal of the Royal Statistical Society, Series B*, 80:839–870, 2018.

[13] J. J. Forster, R. C. Gill, and A. M. Overstall. Reversible jump methods for generalised linear models and generalised linear mixed models. *Statistics and Computing*, 22:107–120, 2012.

[14] E. I. George and R. E. McCulloch. Variable selection via Gibbs sampling. *Journal of the American Statistical Association*, 88:881–889, 1993.

[15] E. I. George and R. E. McCulloch. Approaches for Bayesian variable selection. *Statistica Sinica*, 7:339–373, 1997.

[16] J. Ghosh and M. A. Clyde. Rao-Blackwellisation for Bayesian variable selection and model averaging in linear and binary regression: A novel data augmentation approach. *Journal of the American Statistical Association*, 106:1041–1052, 2011.

[17] R. Gottardo and A. E. Raftery. Markov chain Monte Carlo with mixtures of mutually singular distributions. *Journal of Computational and Graphical Statistics*, 17:949–975, 2008.

[18] P. J. Green. Trans-Dimensional Markov chain Monte Carlo. In P. J. Green, N. L. Hjort, and S. Richardson, editors, *Highly Structured Stochastic Systems*, pages 179–198. Oxford University Press, Oxford, U. K., 2003.

[19] J. E. Griffin, K. Łatuszyński, and M. F. J. Steel. In search of lost mixing time: Adaptive Markov chain Monte Carlo schemes for Bayesian variable selection with very large p. *Biometrika*, 108:53–69, 2021.

[20] J. E. Griffin, E. Matechou, A. S. Buxton, D. Bormpoudakis, and R. A. Griffiths. Modelling environmental DNA data; Bayesian variable selection accounting for false positive and false negative errors. *Journal of the Royal Statistical Society, Series C*, 69:377–392, 2020.

[21] Y. Guan and M. Stephens. Bayesian variable selection regression for genome-wide association studies and other large-scale problems. *The Annals of Applied Statistics*, 5:1780–1815, 2011.

[22] G. Guillera-Arroita, J. J. Lahoz-Monfort, A. R. van Rooyen, A. R. Weeks, and R. Tingley. Dealing with false-positive and false-negative errors about species occurrence at multiple levels. *Methods in Ecology and Evolution*, 8:1081–1091, 2017.

[23] C. J. Hoggart, J. C. Whittaker, M. De Iorio, and D. J. Balding. Simultaneous analysis of all SNPs in genome-wide and re-sequencing association studies. *PLos Genetics*, 4:e1000130, 2008.

[24] H. Ishwaran and J. S. Rao. Spike and slab variable selection: Frequentist and Bayesian strategies. *Annals of Statistics*, 33:730–773, 2005.

[25] C. Ji and S. C. Schmidler. Adaptive Markov chain Monte Carlo for Bayesian variable selection. *Journal of Computational and Graphical Statistics*, 22:708–728, 2013.

[26] D. Lamnisos, J. E. Griffin, and M. F. J. Steel. Transdimensional sampling algorithms for Bayesian variable selection in classification problems with many more variables than observations. *Journal of Computational and Graphical Statistics*, 18(3):592–612, 2009.

[27] D. S. Lamnisos, J. E. Griffin, and M. F. J. Steel. Adaptive Monte Carlo for Bayesian variable selection in regression models. *Journal of Computational and Graphical Statistics*, 22:729–748, 2013.

[28] T. J. Mitchell and J. J. Beauchamp. Bayesian variable selection in linear regression. *Journal of the American Statistical Association*, 83:1023–1032, 1988.

[29] D. L. Morris, Y. Sheng, Y. Zhang, Wang Y.-F., Z. Zhu, P. Tombleson, L. Chen, D. S.-C. Graham, J. Bentham, R. Chen, X. Zuo, T. Wang, L. Wen, C. Yang, L. Liu, L. Yang, F. Li, X. Yin, S. Yang, L. Rönnblom, B. G. Fürnrohr, R. E. Voll, G. Schett,

N. Costedoat-Chalumeau, P. M. Gaffney, Y. L. Lau, X. Zhang, W. Yang, Y. Cui, and T. J. Vyse. Genome-wide association meta-analysis in Chinese and European individuals identifies ten new loci associated with systemic lupus erythematosus. *Nature Genetics*, 48:940–946, 2016.

[30] D. J. Nott and R. Kohn. Adaptive sampling for Bayesian variable selection. *Biometrika*, 92:747–763, 2005.

[31] I. Ntzoufras, P. Dellaportas, and J. J. Forster. Bayesian variable and link determination for generalised liner models. *Journal of Statistical Planning and Inference*, 111:165–180, 2003.

[32] M. Papathomas, P. Dellaportas, and V. G. S. Vasdekis. A novel reversible jump alogrithm for generalized linear models. *Biometrika*, 98:231–236, 2011.

[33] N. G. Polson, J. G. Scott, and J. Windle. Bayesian inference for logistic models using Pólya-Gamma latent variables. *Journal of the American Statistical Association*, 108:1339–1349, 2013.

[34] A. E. Raftery. Approximate Bayes factors and accounting for model uncertainty in generalised linear models. *Biometrika*, 83:251–266, 1996.

[35] S. Richardson, L. Bottolo, and J. S. Rosenthal. Bayesian models for sparse regression analysis of high dimensional data. *Bayesian Statistics*, 9:539–568, 2010.

[36] H. Robbins and S. Monro. A stochastic approximation method. *Annals of Mathemical Statistics*, 22:400–407, 1951.

[37] G. O. Roberts, A. Gelman, and W. R. Gilks. Weak convergence and optimal scaling of random walk Metropolis algorithms. *Annals of Applied Probability*, 7:110–120, 1997.

[38] D. Sabanés Bové and L. Held. Hyper-g priors for generalized linear models. *Bayesian Analysis*, 6:387–410, 2011.

[39] T. Savitsky, M. Vannucci, and N. Sha. Variable selection for nonparametric Gaussian process priors: Models and computational strategies. *Statistical Science*, 26:130–149, 2011.

[40] N. Sha, M. G. Tadesse, and M. Vannucci. Bayesian variable selection for the analysis of microarray data with censored outcome. *Bioinformatics*, 22:2262–2268, 2006.

[41] N. Sha, M. Vannucci, M. G. Tadesse, P. J. Brown, I. Dragoni, N. Davies, T. C. Roberts, A. Contestabile, M. Salmon, C. Buckley, and F. Falciani. Bayesian Variable Selection in Multinomial Probit Models to Identify Molecular Signatures of Disease Stage. *Biometrics*, 60:812–819, 2004.

[42] K. Y. Y. Wan and J. E. Griffin. An adaptive MCMC method for Bayesian variable selection in logistic and accelerated failure time regression models. *Statistics and Computing*, 31:6, 2021.

[43] T. T. Wu, Y. F. Chen, T. Hastie, E. Sobel, and K. Lange. Genome-wide association analysis by lasso penalized logistic regression. *Bioinformatics*, 25:714–721, 2009.

[44] Y. Yang, M. Wainwright, and M. I. Jordan. On the computational complexity of high-dimensional Bayesian variable selection. *Annals of Statistics*, 44:2497–2532, 2016.

[45] G. Zanella. Informed proposals for local MCMC in discrete spaces. *Journal of the American Statistical Association*, 115:852–865, 2020.

Part II

Continuous Shrinkage Priors

6

Theoretical Guarantees for the Horseshoe and Other Global-Local Shrinkage Priors

Stéphanie van der Pas

Amsterdam UMC (The Netherlands)

CONTENTS

Global-local shrinkage priors quickly found favor in the Bayesian community because of their excellent empirical behavior and the promise of fast implementations. Soon after, theoretical guarantees followed. Such priors allow for sharing of information about the number of signals through one global parameter, while local parameters allow for adjustments at the individual parameter level. Many global-local shrinkage priors are now known to lead to optimal posterior concentration rates, which justifies the use of their posterior means as estimators. While results on uncertainty quantification are still scarce, a select few global-local shrinkage priors have even been proven to lead to good coverage results, as well as to variable selection procedures with low false discovery rates. In this chapter, theoretical results and their conditions will be reviewed, as well as their implications for practice. The focus is on the normal means model and on the horseshoe prior, since its behavior is especially well understood, with some discussion of results for other global-local shrinkage priors.

DOI: 10.1201/9781003089018-6

6.1 Introduction

Assuming a sparse underlying structure makes many high-dimensional problems amenable to theoretical study, and is quite reasonable in a variety of applications. The truth is simplified to a degree acceptable to many if it is assumed that only a fraction of all genes is linked to a specific disease, or that only a fraction of pixels in an astronomical image is of interest. This is the type of sparsity assumption that will be made in this chapter, and the properties of a specific class of shrinkage priors, the global-local shrinkage priors, will be evaluated under such sparse circumstances.

6.1.1 Model and Notation

In this chapter the focus will be on the sparse *normal means model*, also known as the *sequence model*. We collect n observations and assume that they arise from independent but not identical normal distributions, as follows:

$$y_i = \mu_{0i} + \varepsilon_i, \quad i = 1, \ldots, n, \tag{6.1}$$

with $\varepsilon_i \sim \mathcal{N}(0, \sigma^2)$ i.i.d. and $\boldsymbol{\mu}_0 = (\mu_{01}, \mu_{02}, \ldots, \mu_{0n})$ an element of $\mathbb{R}^n$. Our goals will be to recover the true mean vector $\boldsymbol{\mu}_0$ and to describe the uncertainty in our estimate. The vector of observations $(y_1, y_2, \ldots, y_n)$ will alternatively be denoted by $\boldsymbol{y}^n$.

The sparse normal means model serves as an important test case for theoretical understanding of sparsity methods [2, 6, 9, 14, 16, 17, 29, 36, 50, 62–65]. It is a special case of sparse linear regression, with the design matrix equal to the $n \times n$ identity matrix. The true mean vector $\boldsymbol{\mu}_0$ is assumed to be sparse and there are many ways to make this assumption precise. In this chapter, the focus will be on vectors where the majority of the entries are exactly equal to zero and a few are non-zero. The non-zero means will be referred to as "signals" and are the needles we are looking for in the haystack mathematically represented by $\boldsymbol{\mu}_0$. We will assume that there are s_n non-zero means and will denote this assumption by writing $\boldsymbol{\mu}_0 \in \ell_0[s_n]$, where $\ell_0[s] = \{\boldsymbol{\mu} \in \mathbb{R}^n : \sum_{i=1}^n \mathbf{1}\{\mu_{0i} \neq 0\} = s\}$ is colloquially referred to as the class of "nearly black" vectors [36]. We do not know which of the means are non-zero, and realistically, we do not even know how many signals we are looking for. We will assume that the number of signals s_n grows to infinity as the number of observations increases towards infinity, and that this increase happens at a slower rate than the increase in observations, that is, $s_n \to \infty$ and $s_n/n \to 0$ as $n \to \infty$. The assumption $s_n \to \infty$ makes the problem more interesting within an asymptotic framework, since if we would assume instead that s_n is bounded by some constant as $n \to \infty$, then asymptotically the number of zero means would be so overwhelmingly large that the estimator that is simply equal to zero for every mean would perform very well.

6.1.2 Global-Local Shrinkage Priors and Spike-and-Slab Priors

Spike-and-slab type priors [41] have tremendous intuitive appeal, featuring a point mass at zero in anticipation of many means being exactly equal to zero, mixed with a "slab" distribution for the benefit of the non-zero means. Since spike-and-slab type priors are treated elsewhere in this book, we will at this time only recall their formulation so that we may draw comparisons later in this chapter. Each coordinate i is equipped with a prior of the form

$$(1 - \tau)\delta_0 + \tau g(\cdot), \tag{6.2}$$

where $\tau \in [0,1]$ represents the proportion of signals, δ_0 is the Dirac at zero and g the "slab". Optionally, although in practice usually, the parameter τ is endowed with a prior distribution as well. Provided one chooses a "slab" distribution with sufficiently heavy tails and a reasonable prior on τ, spike-and-slab priors are optimal in several senses [14–17]. One could wonder why we would study global-local shrinkage priors then, if the spike-and-slab priors work so well. There are several motivations to do so.

The idea that deployment of spike-and-slab priors would incur prohibitive computational expense, due to a need to search over a space of size 2^n, has often been mentioned as a motivation to study smooth shrinkage priors instead, but with clever algorithms there may be no strong computational reason to prefer one over the other within the context of the sparse normal means model, see e.g., [17, 22, 67, 70].

A second motivation for global-local shrinkage priors, as well as for other continuous shrinkage priors, is that they can be a better match for the expected type of sparsity, since the corresponding estimates can be close to zero but are with probability one not exactly equal to zero, unless one takes additional processing steps. If it is expected that many of the parameters will be negligible in a sense, but not exactly equal zero, then a prior with a lot of mass, but not a point mass, near zero has intuitive appeal.

For more motivation, one might be interested in global-local shrinkage priors out of scientific curiosity. Is it possible to achieve similar results as those obtained for spike-and-slab priors with a continuous alternative? The answer to that question is in many cases "yes", as we will see in this chapter.

Finally, one may simply observe that for one reason or another, some global-local shrinkage priors have already found favor with practitioners [3, 18, 24, 25, 30, 31, 37, 38, 42, 49, 66, 69, 71], and theoretical study will show whether their use is justified, and whether perhaps improvements in performance are possible. This motivation will be the guiding principle of this chapter; no effort will be made to convince anyone to use global-local shrinkage priors, but rather, accepting that someone is interested in using them, some relevant theoretical results are stated and practical implications discussed. The focus of this chapter will be on the horseshoe prior [13], since it has found popularity in practice [18, 24, 30, 31, 37, 38, 42, 49, 69] and is especially well understood from a theoretical point of view [19, 62–64]. Related results for other global-local shrinkage priors will be treated as well, although without a doubt there are some theoretical results which did not find their way into this chapter.

6.1.3 Performance Measures

The theoretical results considered in this chapter are of frequentist Bayesian nature [28], meaning that it is assumed that there exists some true sparse vector $\boldsymbol{\mu_0}$ underlying the data generating process, and the Bayesian procedures are evaluated on the extent to and precision with which they are able to find this true $\boldsymbol{\mu_0}$. The performance measures considered below may look decidedly Bayesian but in the end all lead to evaluation on traditional frequentist criteria: minimax rates, size and confidence level of confidence sets.

The main criteria on which global-local shrinkage priors are evaluated theoretically, are the ability of the corresponding posterior distribution (i) to recover the underlying sparse parameter and (ii) to quantify the uncertainty remaining on the exact value of the truth. The performance of a prior on those two criteria needs to be considered in combination with (iii) the assumptions on the true parameter $\boldsymbol{\mu_0}$ as well as on other quantities.

For (i) recovery, or estimation, some measure of center of the posterior distribution is typically used: the mean, median or mode. The quality of the estimate may be evaluated by finding or bounding the expected value of the difference between the truth and the estimate. That is, if we denote the estimator by $\widehat{\boldsymbol{\mu}}$, the object of study is $E_{\boldsymbol{\mu_0}}\|\widehat{\boldsymbol{\mu}} - \boldsymbol{\mu_0}\|$ for some norm $\|.\|$. In this chapter, we will mainly consider minimax rates, bounding $\sup_{\boldsymbol{\mu_0} \in D} E_{\boldsymbol{\mu_0}}\|\widehat{\boldsymbol{\mu}} - \boldsymbol{\mu_0}\|$

for some subset D of $\mathbb{R}^n$. The bound on the latter supremum is then compared to the known minimax rates for this problem, which with D a set of nearly black vectors and the norm equal to the squared ℓ_2 norm is $2s_n \log(n/s_n)$ [21]. More precisely: as $n, s_n \to \infty$:

$$\inf_{\widehat{\boldsymbol{\mu}} \in \mathbb{R}^n} \sup_{\boldsymbol{\mu} \in \ell_0[s]} \mathbb{E}_{\boldsymbol{\mu}} \|\boldsymbol{\mu} - \widehat{\boldsymbol{\mu}}\|_2^2 = 2s_n \log \frac{n}{s_n}(1 + o(1)).$$

In the normal means problem, global-local shrinkage priors admit a convenient representation of their posterior mean. Close study of this representation allows one to obtain satisfactory bounds on $\sup_{\boldsymbol{\mu}_0 \in D} E_{\boldsymbol{\mu}_0} \|\widehat{\boldsymbol{\mu}} - \boldsymbol{\mu}_0\|$, as we will see in Section 6.3.2.

Another approach towards evaluating the properties of an estimator of $\boldsymbol{\mu}_0$ is through establishing a posterior contraction rate. Let $\Pi_n(\cdot \mid \boldsymbol{y}^n)$ denote the posterior distribution based on n observations $y_1, \ldots, y_n$.

Definition (see e.g., [28]). A sequence η_n is a *posterior contraction rate* at the parameter $\boldsymbol{\mu}_0$ with respect to the semimetric d if $\Pi_n(\boldsymbol{\mu} : d(\boldsymbol{\mu}, \boldsymbol{\mu}_0) \geq M_n \eta_n \mid \boldsymbol{y}^n) \to 0$ in $P^n_{\boldsymbol{\mu}_0}$-probability, for every $M_n \to \infty$.

If a posterior contraction rate is proven, for example by verifying the conditions from [26, 27], then one may conclude that an estimator achieving the same rate can be derived from the posterior distribution [26, Theorem 2.5]. As a consequence, the rate of contraction of the posterior distribution can be no faster than the minimax estimation rate for the parameter under study.

For (ii) uncertainty quantification, credible sets are studied. A set $\widehat{C}_n = \widehat{C}_n(\boldsymbol{y}^n)$ is a subset of the parameter space, often taken such that the posterior mass on $\widehat{C}_n$ is at least a prescribed number. A first question one may ask is whether we may expect $\widehat{C}_n$ to contain the true $\boldsymbol{\mu}_0$. One way to approach this question is by verifying whether a credible set is *honest*, relative to a set $D \subset \mathbb{R}^n$, that is, by verifying whether for some prescribed confidence level $1 - \alpha$,

$$\liminf_{n \to \infty} \inf_{\boldsymbol{\mu}_0 \in D} P_{\boldsymbol{\mu}_0}(\boldsymbol{\mu}_0 \in \widehat{C}_n) \geq 1 - \alpha.$$

The above criterion is easily satisfied by setting $\widehat{C}_n = \mathbb{R}^n$, but such a choice would be about as helpful as the announcement by a delivery company that they will deliver your parcel anytime between now and the next 100 years. So as a second criterion, we evaluate the size of $\widehat{C}_n$. The posterior concentration rate already gives a hint as to the size of a credible set, but establishing posterior concentration is in itself not sufficient to declare a credible set honest and of useful size. In Section 6.4, a more detailed discussion on establishing the optimal size of a credible set is given.

An important aspect of any theoretical result is the (iii) conditions under which it is true. Ideally either very few conditions are necessary, or only conditions that seem reasonable in practice. In the sparse normal means problem, some results require a minimum size of the signals. That is, $|\mu_{0i}| \geq t$ needs to hold for any non-zero μ_{0i}. Intuitively it is easy to see why: the further the signals are separated from the noise, the easier it becomes to detect the signals. Conceptually this is comparable to an "effect size": the larger the effect size, the easier it is to find the effect. The motivation to make an assumption on the size of the entries of $\boldsymbol{\mu}_0$ can be because the proof technique demands it, or because the prior shrinks so strongly that only very large signals can survive the shrinkage. Other common conditions are that the variance σ^2 is known, or that the number of signals s_n is known. In the latter case the procedure is considered "non-adaptive" and is in most cases ruled out for deployment in practice, but studying this situation can be a useful starting point towards developing "adaptive" procedures, where good performance is maintained without access to the number of signals s_n. This is discussed further in the next section.

6.2 Global-Local Shrinkage Priors

Global-local shrinkage priors (e.g., [48]) are scale mixtures of normals, taking the form

$$\mu_i \mid \lambda_i^2, \tau^2, \sigma^2 \sim \mathcal{N}(0, \sigma^2\tau^2\lambda_i^2), \quad \lambda_i^2 \sim p(\lambda_i^2), \quad i = 1, \ldots, n, \tag{6.3}$$

where p is a prior, and specific choices of p lead to the large variety of global-local shrinkage priors in use today. A prominent example is the horseshoe [13], which arises by taking a half-Cauchy prior for p. More examples are given below.

The "global" in "global-local" refers to the parameter τ, which all coordinates μ_i have in common and drives the overall sparsity level. As will be made more precise in the theoretical results in Sections 6.3, 6.4 and 6.5, ideally τ is set equal to the proportion of true signals, that is s_n/n, or a value very close to it. It may thus be viewed as the analogue of the mixing proportion of the spike-and-slab prior, also denoted by τ in (6.2).

The "local" component consists of the λ_i^2's. Intuitively, these parameters can either counteract the shrinkage towards zero that is encouraged by small values of τ, or further enforce it. The former happens when λ_i^2 is realized as a relatively large value, the latter if λ_i^2 takes a value close to zero.

Optionally, priors on the global parameter τ and on the model variance σ^2 (see (6.1)) may be introduced. In most practical situations, this is a necessity, since it is rare that the number of signals (required for setting τ at a sensible value) or the model variance itself is known. In theoretical studies, for simplicity τ and especially σ^2 are frequently assumed known, and the latter typically set equal to one without loss of generality. When a prior on or an estimator for τ is studied and the resulting procedure is found to lead to good performance, the method is considered "adaptive", where the adaptivity is to the unknown sparsity level.

The shrinkage properties of global-local priors can be seen from several angles. One is directly from formulation (6.3). The prior is centered around zero, and unless the prior on $\boldsymbol{\lambda}^2 = (\lambda_1^2, \lambda_2^2, \ldots, \lambda_n^2)$ is very heavy-tailed, draws from the prior will tend to be close to zero. A second angle, and a more precise statement, appears when we consider the posterior distribution of each μ_i given not only the observation y_i but also the local variance component λ_i, which is

$$\mu_i \mid y_i, \lambda_i \sim \mathcal{N}\left(\frac{\tau^2\lambda_i^2}{1+\tau^2\lambda_i^2}y_i, \frac{\tau^2\lambda_i^2}{1+\tau^2\lambda_i^2}\sigma^2\right). \tag{6.4}$$

We introduce the transformation of λ_i into the "shrinkage factor" $f(\lambda_i) = \frac{1}{1+\tau^2\lambda_i^2}$ and adopt the setting where both σ^2 and τ^2 are assumed known. From (6.4), we find that we may express the posterior mean of μ_i in terms of the posterior mean of $f(\lambda_i)$, as follows

$$\mathbb{E}[\mu_i \mid y_i] = (1 - \mathbb{E}[f(\lambda_i) \mid y_i])y_i. \tag{6.5}$$

Since $f(\lambda_i)$, and thus its conditional expectation, is between zero and one, we observe from (6.5) that the posterior mean for each coordinate μ_i is between 0 and the observed value y_i. The dependence of the amount of shrinkage on λ_i is such that if λ_i is close to zero, then the posterior mean for that μ_i will also be close to zero, while as λ_i grows larger and larger, the posterior mean for the corresponding μ_i will become closer and closer to the observation y_i. Relationship (6.5) is illustrated in Figure 6.1 for the horseshoe prior, which is discussed further below.

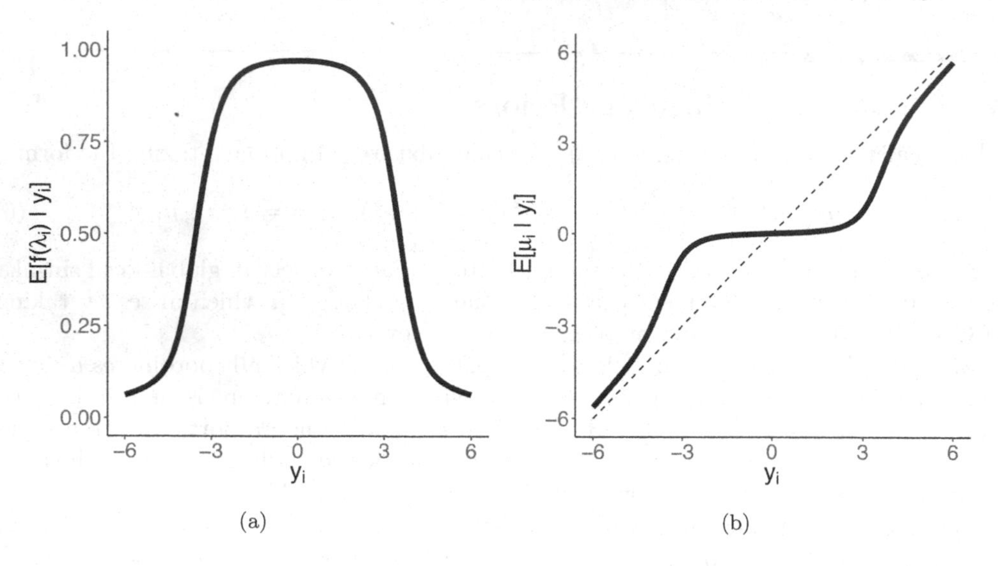

FIGURE 6.1
Illustration of relationship (6.5) for the horseshoe prior with $\tau = 0.05$ and $\sigma^2 = 1$. Figure 6.1(a) shows the shrinkage factor $\mathbb{E}[f(\lambda_i) \mid y_i]$ as a function of y_i; Figure 6.1(b) shows the posterior mean $\mathbb{E}[\mu_i \mid y_i]$ as a function of y_i (solid) and the identity (dashed). In Figure 6.1(a), values close to one represent near total shrinkage, corresponding to values close to zero in Figure 6.1(b). In the interval $[-2.5, 2.5]$ observations are shrunk almost to zero. Shrinkage rapidly decreases outside that interval.

In a sparse setting, a main requirement for a shrinkage prior is that it can separate the signals ($\mu_i \neq 0$) from the noise ($\mu_i = 0$). The prior on the local variance component λ_i determines how this balancing act is carried out, as can be seen from (6.5). This intuition becomes slightly more complicated but remains largely the same in the adaptive setting, where τ is unknown and its prior or estimator possibly interacts with the prior on the λ_i's.

A particularly popular and well-studied global-local shrinkage prior is the horseshoe prior [13]. Its prior on the λ_i's is designed precisely to achieve separation between signals and noise. This is perhaps not immediately apparent from the usual statement, where under the prior, $\lambda_i \sim C^+(0, 1)$ i.i.d. (i.e., the half-Cauchy prior on the positive reals, with density $p_\lambda(u) = \frac{2}{\pi(1+u^2)}$), but it comes into clear view when the prior on the shrinkage factor $f(\lambda_i) = \frac{1}{1+\tau^2\lambda_i^2}$ is inspected, for fixed τ. It is given by

$$p_{f(\lambda)}(u) = \frac{\tau}{\pi} \frac{1}{1 - (1 - \tau^2)u} u^{-1/2}(1 - u)^{-1/2},$$

for $u \in (0, 1)$. In the special case of $\tau = 1$, the prior density on f reduces to the density of the Beta(1/2, 1/2)-distribution, which resembles a horseshoe. This horseshoe-shape matches the idea of the balancing act that is to be performed through the λ_i's. As is illustrated in Figure 6.2, most of the prior mass of the horseshoe is placed on shrinkage factors either very close to zero or one, corresponding to a high prior probability of an observation originating from a signal or from noise respectively.

The case $\tau = 1$ is very special indeed, because theoretical study showed that τ should be more or less equal to the proportion of signals s_n/n (see Sections 6.3, 6.4 and 6.5 for more details). With that intuition, a value of $\tau = 1$ would correspond to a situation where all coordinates are relevant and there is no sparsity at all. By adjusting the value of τ to

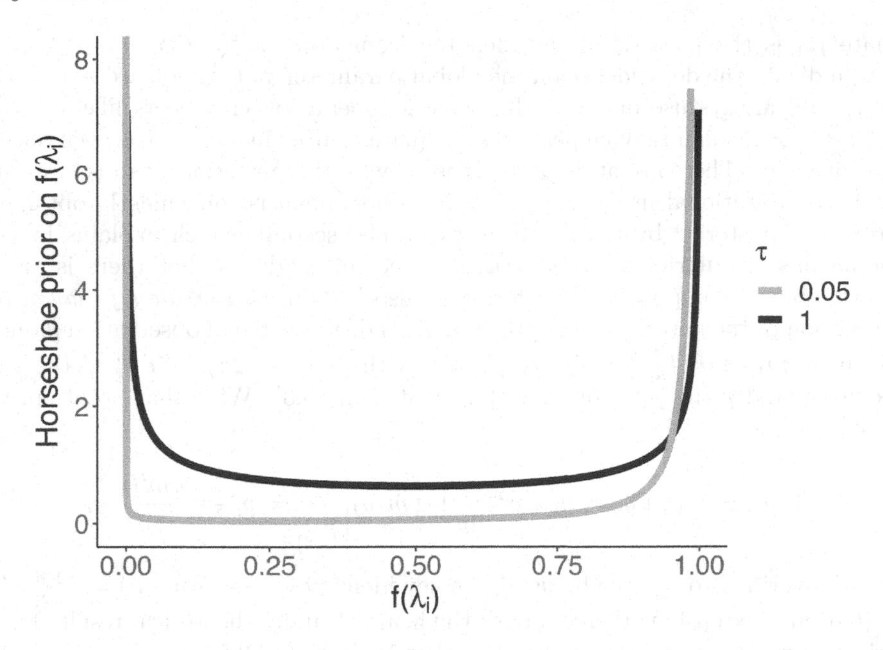

FIGURE 6.2
The horseshoe prior on the shrinkage factor $f(\lambda_i) = 1/(1 + \tau^2\lambda_i^2)$. With $\tau = 1$, it reduces to the Beta(1/2, 1/2)-prior (shown in black), placing high prior probability on shrinkage factors close to zero and one. When τ is decreased to 0.05 (shown in gray), prior mass on shrinkage factors close to one is increased.

values closer to zero, the prior mass on shrinkage factors close to one is increased (Figure 6.2), corresponding to an increased prior expectation of sparsity.

Different choices for the prior on λ_i give rise to many further examples of global-local shrinkage priors. This long list includes the normal-gamma prior [12], the normal-inverse-Gaussian prior [34], the normal-exponential-gamma [33], the Normal-Jeffreys [4, 23], Strawderman-Berger [7, 56], Laplace [45] and many others [1, 29, 48]. Some of these examples are discussed further below.

6.3 Recovery Guarantees

In this section as well as in Sections 6.4 and 6.5, a selection of Bayesian asymptotic theory for global-local shrinkage priors is presented in the setting of the sparse normal means problem, in which the observations are modeled as, recalling (6.1),

$$y_i = \mu_{0i} + \varepsilon_i, \quad i = 1, \ldots, n,$$

with the ε_i independently and identically normal distributed with mean zero and variance σ^2, and $\boldsymbol{\mu_0} = (\mu_{01}, \ldots, \mu_{0n})$ assumed to be sparse in the sense that s_n of the μ_{0i} are exactly equal to zero.

A first question one might like to have the answer to, is how close some aspect of the posterior comes to the true vector $\boldsymbol{\mu_0}$. The usual "aspect of the posterior" that is employed to estimate $\boldsymbol{\mu_0}$ is the posterior mean, denoted from now on by $\boldsymbol{T(y^n)} = \mathbb{E}[\boldsymbol{\mu} \mid \boldsymbol{y^n}]$ or $\boldsymbol{T_\tau(y^n)}$ to indicate the dependence on the global parameter τ. It is not necessarily the case that one should always use or study the posterior mean, other aspects like the posterior mode or posterior median may be perfectly adequate or may in some sense even outperform the posterior mean. There are at least two reasons why the posterior mean is often studied. The first is computational in nature, as the posterior mean is conveniently obtained from draws from the posterior by taking an average. The second, which explains in part the posterior mean's popularity as a theoretical object of study, is that there is an elegant formula available to compute it. This formula, also known as *Tweedie's formula*, requires the introduction of the notation $m(y_i)$, the marginal distribution of observation y_i under the prior, given by $m(y_i) = \int_\infty^\infty \phi\left(\frac{y_i-\mu_i}{\sigma}\right) p(\mu_i) d\mu_i$, with $\phi(u) = (2\pi)^{-1/2} e^{-u^2/2}$ the standard normal density and p the prior on μ_i as formulated in (6.3). With that notation, we may express [46]:

$$T(y_i) = \mathbb{E}[\mu_i \mid y_i] = y_i + \sigma^2 \frac{d}{dy} \log m(y)\bigg|_{y=y_i} = y_i + \sigma^2 \frac{m'(y_i)}{m(y_i)}, \tag{6.6}$$

which may be verified to be true by using the key identity $(\mu - y)\phi(y-\mu) = \sigma^2 \frac{d}{dy}\phi(y-\mu)$. Identity (6.6) and extensions thereof are at the heart of many theoretical results for global-local shrinkage priors, as will be explored further in Section 6.3.2.

The sparsity assumption which is commonly taken in theoretical study of global-local priors, and which we will focus on in this section as well, is the assumption that the true vector $\boldsymbol{\mu_0}$ is in the class of "nearly-black" vectors $\ell_0[s_n] = \{\boldsymbol{\mu} \in \mathbb{R}^n : \sum_{i=1}^n \mathbf{1}\{\mu_i \neq 0\} \leq s_n\}$ for some s_n, which may be known (non-adaptive case) or unknown (adaptive case). That is, at most s_n entries of $\boldsymbol{\mu_0}$ are non-zero ("signals") and all remaining entries are exactly equal to zero. This assumption is intuitively pleasant due to its clear separation of signals and noise, but perhaps at odds with the decision to use a continuous prior. As will become clear in Section 6.3.2, the possibility of splitting up the parameters into two groups, one of zero means and one of non-zero means, is convenient for bounding the mean squared error, but in principle the proof techniques can be adapted to allow other sparsity assumptions, such as those where assumptions are made on the sum $\sum_{i=1}^n |\mu_{0i}|^s$ for some $s > 0$. We will come back to this in Section 6.3.4.

6.3.1 Non-Adaptive Posterior Concentration Theorems

In the non-adaptive setting, we assume that the number of signals s_n, that is the number of non-zero means μ_{0i}, is available to us. We assume we know how many needles there are in the haystack, and just need to find their locations. The number of signals turns out to be very valuable information for setting the global parameter τ at a sensible level. The assumption of knowing s_n is quite a strong one to make, however. It is difficult to come up with a practical application where the number of non-zero coefficients is exactly known in advance. Nevertheless theory in this setting is very useful, as it reveals what value of τ to go after, which in turn provides guidance on which priors or estimators for τ to use in practice when the number of signals s_n is unknown. Moreover, the understanding of the behavior of the global-local shrinkage priors in this somewhat idealized setting is a firm base for theoretical results when no knowledge of s_n is assumed, as will be seen in Section 6.3.3.

We start by stating a result for the horseshoe prior, i.e., (6.3) with $\lambda_i \sim C^+(0,1)$. As it turns out, this result can be extended to many other global-local shrinkage priors as well. The more general result is stated and discussed after the more specific result for the horseshoe.

Theorem 6.3.1 (Theorem 3.3 of [64]). *Let Π_τ denote the posterior distribution of the horseshoe, and $\boldsymbol{T}_\tau$ the posterior mean. Set $\tau = (s_n/n)^\alpha$, for some $\alpha \geq 1$. Then, assuming $s_n = o(n)$,*

$$\sup_{\boldsymbol{\mu}_0 \in \ell_0[s_n]} E_{\boldsymbol{\mu}_0} \Pi_\tau \left(\boldsymbol{\mu} : \|\boldsymbol{\mu} - \boldsymbol{\mu}_0\|^2 > M_n s_n \log \frac{n}{s_n} \middle| \boldsymbol{y}^n \right) \to 0, \tag{6.7}$$

and

$$\sup_{\boldsymbol{\mu}_0 \in \ell_0[s_n]} E_{\boldsymbol{\mu}_0} \Pi_\tau \left(\boldsymbol{\mu} : \|\boldsymbol{\mu} - \boldsymbol{T}_\tau(\boldsymbol{y}^n)\|^2 > M_n s_n \log \frac{n}{s_n} \middle| \boldsymbol{y}^n \right) \to 0, \tag{6.8}$$

for every $M_n \to \infty$ as $n, s_n \to \infty$.

We derive at least three pieces of news from this theorem. Statements (6.7) and (6.8) provide us with upper bounds on the posterior contraction rates of the horseshoe around the true parameter vector $\boldsymbol{\mu}_0$ and around the posterior mean respectively. Since a posterior cannot contract faster than the minimax rate around the true mean vector ([26]), the bound on the rate of contraction in (6.7) is sharp. The rate in (6.8) is not necessarily sharp, as we'll come back to in Section 6.4. Thus, the first piece of news we conclude from Theorem 6.3.1 is that the horseshoe posterior mean achieves the minimax rate as an estimator of $\boldsymbol{\mu}_0$ (up to multiplicative constants).

The second piece of news is that it is important to make a good choice for the parameter τ. As long as we set τ equal to the proportion of signals s_n/n or some power thereof, we are guaranteed the aforementioned minimax rate. Close study of the proof of Theorem 6.3.1 reveals that we may also set $\tau = (s_n/n)\sqrt{\log(n/s_n)}$. In that case, the spread of the posterior is of the same order as the minimax risk. When we come to the adaptive results in Section 6.3.3, we will frequently refer to this slightly larger value and in the interest of uncertainty quantification, it is optimal to take this slightly larger value (see also, Section 6.4). However, the difference between s_n/n or $(s_n/n)\sqrt{\log(n/s_n)}$ is only a logarithmic factor, and the value $\tau = s_n/n$ has a pleasant interpretation as the proportion of signals.

The third piece of news is that optimal posterior contraction properties can be achieved with global-local shrinkage priors. Even though they do not contain a point mass at zero, apparently the large amount of mass near zero mimics that well enough, and the right balance between mass at zero and mass near the tails is achieved by the horseshoe.

The result for the horseshoe may lead one to wonder whether other global-local shrinkage priors enjoy similar proporties. A more general result on recovery by global-local shrinkage priors is available, asserting that the class of global-local shrinkage priors with which optimal posterior contraction rates can be achieved is much larger than just the horseshoe [65]. Posterior concentration for global-local shrinkage priors is guaranteed at nearly the minimax rate if (i) the prior's tails are not "too light", (ii) the prior puts "sufficient" mass close to zero, and (iii) the decay of the prior away from a neighborhood of zero is "fast". To make these conditions, described in [65], precise, we need the concept of *uniformly regular varying* functions. We say that a function g is uniformly regular varying if there exist constants $R, u_0 \geq 1$ such that

$$\frac{1}{R} \leq \frac{f(au)}{f(u)} \leq R, \quad \text{for all } a \in [1,2], \text{ and all } u \geq u_0.$$

Examples of uniformly regular varying functions include $f(u) = u^b$ or $f(u) = \log^b(u)$ with $b \in \mathbb{R}$ (e.g., take $R = 2^{|b|}$ and $u_0 = 2$). An example of a function that is not uniformly regular varying is $f(u) = e^u$. Uniformly regular varying functions are either everywhere positive or everywhere negative on $[u_0, \infty)$. The reciprocal of a uniformly regular varying function is again a uniformly regular varying function, as is the product of two uniformly regular varying functions.

The class of priors considered in [65] is of the form

$$\mu_i \mid \sigma_i^2 \sim \mathcal{N}(0, \sigma_i^2), \quad \sigma_i^2 \sim \widetilde{p}(\sigma_i^2), \quad i = 1, \ldots, n,$$

of which the global-local shrinkage priors (6.3) are an important special case. We now state the conditions under which recovery at the near-minimax rate is guaranteed. The first condition assures good recovery of the non-zero means.

Condition (i): Sufficiently heavy tails. The density of the prior on the squared local components λ_i^2 is uniformly regular varying and does not depend on n, and τ is set to $\tau = (s_n/n)^\alpha$ for some $\alpha \geq 0$.

A risk with any shrinkage prior is that too much focus is placed on the "shrinkage" part, translating to so much prior mass around zero that even very large observations cannot counteract this pull towards zero, and every mean is close to zero a posteriori. Condition (i) assures sufficient mass in the tails to adequately recover the non-zero means, in terms of ℓ_2 minimax rates. As is perhaps more apparent from the more general version of Condition (i) stated in [65], Condition (i) allows for at most exponentially fast decay of the tail of the prior on each μ_i, which is similar to the conditions found on the "slab" part of spike-and-slab priors in [17].

Conditions (ii) and (iii) are for the benefit of the zero means.

Condition (ii): Sufficient mass near zero. There exists a constant $c > 0$ such that $\int_0^1 \widetilde{p}(u)du \geq c$.

Condition (iii): Fast decay away from zero. Write $a_n = (s_n/n)\log(n/s_n)$ and $b_n = \sqrt{\log(n/s_n)}$. There exists a constant $d > 0$ such that

$$\int_{a_n}^{b_n^2} u\widetilde{p}(u)du + b_n^3 \int_{b_n^2}^{\infty} \frac{\widetilde{p}(u)}{\sqrt{u}}du + b_n \int_1^{b_n^2} \frac{\widetilde{p}(u)}{\sqrt{u}}du \leq d a_n.$$

In [65], two stronger conditions which imply both Condition (ii) and (iii) are stated. In specific examples, these stronger conditions are easier to verify.

Condition (ii) is intuitively expected, as to achieve the desired shrinkage behavior we will need a good amount of mass close to zero, and Condition (ii) specifies when we can be sure to have a "good amount". Condition (iii) offers a surprise, since it requires some control on the prior on the interval $[a_n, 1]$. That is surprising because when one thinks of the balancing act that needs to be accomplished for the two types of coefficients (zero and non-zero), ideally one would like to have total shrinkage for the zero means and no shrinkage at all for the non-zero means, translating to a lot of mass near zero and heavy tails respectively. So do we need a condition for the area in between zero and the tails? Apparently yes, since in [65] it is also proven (Theorem 2.2) that if the condition is relaxed so that we allow $\int_{a_n}^1 u\widetilde{p}(u)du \lesssim c_n$ for arbitrary $c_n \gg a_n$, then there exists a prior that meets all other conditions but fails to concentrate at the minimax rate. Thus, while the amount of mass close to zero and in the tails needs to be balanced out, the amount of mass in between should be minimal. Combining these conditions, we see that we need a large amount of mass close to zero, more so as the sparsity increases (i.e., when we have fewer signals), while maintaining sufficient mass in the tails for adequate recovery of the signals, with most of the mass in a shrinking interval close to zero. Under the conditions above, we obtain the following result, denoting by $\boldsymbol{T_\tau(y^n)}$ the vector that contains the posterior mean for each μ_i.

Theorem 6.3.2 (Theorem 2.1 of [65]). *Let Π_τ denote the posterior distribution resulting from a prior of the form (6.3) meeting Conditions (i)–(iii). Then, assuming $s_n = o(n)$,*

$$\sup_{\mu_0 \in \ell_0[s_n]} E_{\mu_0} \Pi_\tau \left(\mu : \|\mu - \mu_0\|^2 > M_n s_n \log \frac{n}{s_n} \middle| y^n \right) \to 0, \tag{6.9}$$

for every $M_n \to \infty$ as $n, s_n \to \infty$, and

$$\sup_{\mu_0 \in \ell_0[s_n]} E_{\mu_0} \|T_\tau(y^n) - \mu_0\|^2 \lesssim s_n \log \frac{n}{s_n}. \tag{6.10}$$

The first statement, (6.9) is a posterior contraction statement, showing that in the worst case, asymptotically and in expectation, all posterior mass will be contained in a ball of radius of the order $s_n \log(n/s_n)$. Up to multiplicative constants, this is the best we can hope for. By Theorem 2.5 of [26], we can conclude the existence of an estimator derived from the posterior that achieves the minimax rate. The second statement, (6.10) shows this directly: the posterior mean is a minimax optimal estimator for the true vector μ_0, again up to multiplicative constants.

A wide range of global-local shrinkage priors can be shown to meet the conditions for Theorem 6.3.2, thus assuring their optimal performance in a posterior contraction sense [65]. Among these are the class of priors from [29] as well as some priors outside of the class of global-local scale mixtures of normals, like the spike-and-slab lasso [50].

6.3.2 Proof Techniques

We describe the main proof ideas for the posterior contraction result just presented. For ease of exposition, we present the main proof ideas specialized to the horseshoe, so for Theorem 6.3.1. Full details may be found in [64]. The proof of the more general result, Theorem 6.3.2, is obtained by a similar argument.

At first glance, since the discussed theorems are in the field of Bayesian nonparametrics, one might expect the proof to rest on the "Ghosal Ghosh Van der Vaart" conditions, and to hence consist of a careful balancing out of entropy, prior mass and sieve conditions [26–28]. Rather a detailed study of the moments of the posterior distribution, in combination with Markov's inequality, suffices.

In Section 6.3, Tweedie's formula for the posterior mean was already highlighted in equation (6.6). Similarly, one may express the posterior variance as [46]:

$$Var(\mu_i \mid y_i) = \sigma^2 - \left(\sigma^2 \frac{m'(y_i)}{m(y_i)}\right)^2 + \sigma^4 \frac{m''(y_i)}{m(y_i)}. \tag{6.11}$$

In the specific case for the horseshoe, these expressions are given by:

$$T_\tau(y_i) = y_i \frac{\int_0^1 z^{1/2} \frac{1}{\tau^2 + (1-\tau^2)z} e^{\frac{y_i^2}{2\sigma^2} z} dz}{\int_0^1 z^{-1/2} \frac{1}{\tau^2 + (1-\tau^2)z} e^{\frac{y_i^2}{2\sigma^2} z} dz}; \tag{6.12}$$

$$Var(\mu_i \mid y_i) = \frac{\sigma^2}{y_i} T(y_i) - (T(y_i) - y_i)^2 + y_i^2 \frac{\int_0^1 (1-z)^2 z^{-1/2} \frac{1}{\tau^2 + (1-\tau^2)z} e^{\frac{y_i^2}{2\sigma^2} z} dz}{\int_0^1 z^{-1/2} \frac{1}{\tau^2 + (1-\tau^2)z} e^{\frac{y_i^2}{2\sigma^2} z} dz}.$$

The proof is organized around a split between the zero means ($\mu_{0i} = 0$) and non-zero means ($\mu_{0i} \neq 0$). This is a feasible strategy because the assumed type of sparsity allows this kind of division, and because the μ_i's are a priori and a posteriori independent, since information is only shared through the global parameter τ, which is assumed known in this non-adaptive setting.

A crucial quantity is the "threshold" $\zeta_\tau = \sqrt{2\sigma^2 \log(1/\tau)}$, as it will be seen that the behavior of the horseshoe's posterior mean is very different before and after this threshold.

Non-zero means: Suppose $\mu_{0i} \neq 0$. It will be argued that

$$\mathbb{E}_{\mu_{0i}}(T_\tau(y_i) - \mu_{0i})^2 \lesssim \sigma^2 + \zeta_\tau^2. \tag{6.13}$$

To this end, the left hand side is bounded by

$$\mathbb{E}_{\mu_{0i}}(T_\tau(y_i) - \mu_i)^2 \leq 2\sigma^2 + 2\left(\sup_y |T_\tau(y) - y|\right)^2.$$

A crucial ingredient is the following bound, based on Watson's lemma (see [40]):

$$|T_\tau(y) - y| \leq h(y, \tau),$$

where $h(y,\tau) \to \infty$ if $|y/\zeta_\tau| \to c$ as $\tau \to 0$, for any $c \leq 1$, and $h(y,\tau) \to 0$ otherwise. Hence, as $\tau \to 0$, we find: $\arg\max_y |T_\tau(y) - y| \lesssim \zeta_\tau$. Since $|T_\tau(y)| \leq |y|$, this implies $(\sup_y |T_\tau(y) - y|)^2 \lesssim \zeta_\tau^2$ and therefore also (6.13).

Zero means For $\mu_{0i} = 0$, we need to bound $\mathbb{E}_0 T_\tau(y_i)^2$. By two different arguments, one for observations below ζ_τ and one for observations larger than ζ_τ, we find

$$\mathbb{E}_0 T_\tau(y_i)^2 \lesssim t\sqrt{\log\frac{1}{\tau}},$$

provided $\tau < \exp(-\sigma^2/2)$. The arguments for $\mathbb{E}_0 T_\tau(y_i)^2$ if $|y_i| > \zeta_\tau$ are based on Mill's ratio and the identity $\frac{d}{dy}[-y\phi(y)] = y^2\phi(y) - \phi(y)$ (with ϕ the standard normal density).

The bound on $\mathbb{E}_0 T_\tau(y_i)^2$ if $|Y_i| \leq \zeta_\tau$ starts by evaluation of (6.12) after bounding $\exp(\frac{y^2}{2\sigma^2}z)$ below by 1 and above by $\exp(\frac{y^2}{2\sigma^2})$. This yields

$$T_\tau(y) \leq y e^{\frac{y^2}{2\sigma^2}} \frac{\tau}{1-\tau^2}\left(\frac{\sqrt{1-\tau^2}}{\arctan\left(\frac{\sqrt{1-\tau^2}}{\tau}\right)} - \tau\right).$$

Then, for $y_i \sim \mathcal{N}(0, \sigma^2)$:

$$\mathbb{E}_0\left[T_\tau(y_i)^2 \mathbf{1}_{\{|y_i| \leq \zeta_\tau\}}\right] \leq \int_{-\zeta_\tau}^{\zeta_\tau} y^2 e^{\frac{y^2}{\sigma^2}} f(\tau)^2 \frac{1}{\sqrt{2\pi\sigma^2}} e^{-\frac{y^2}{2\sigma^2}} dy = \frac{f(\tau)^2}{\sqrt{2\pi\sigma^2}} \int_{-\zeta_\tau}^{\zeta_\tau} y^2 e^{\frac{y^2}{2\sigma^2}} dy,$$

where $f(\tau) = \frac{\tau}{1-\tau^2}(\sqrt{1-\tau^2}/\arctan(\frac{\sqrt{1-\tau^2}}{\tau}) - \tau)$. As $\frac{d}{dy}\sigma^2 y e^{\frac{y^2}{2\sigma^2}} = \sigma^2 e^{\frac{y^2}{2\sigma^2}} + y^2 e^{\frac{y^2}{2\sigma^2}}$:

$$\int_{-\zeta_\tau}^{\zeta_\tau} y^2 e^{\frac{y^2}{2\sigma^2}} dy = \left[\sigma^2 y e^{\frac{y^2}{2\sigma^2}}\right]_{-\zeta_\tau}^{\zeta_\tau} - \int_{-\zeta_\tau}^{\zeta_\tau} \sigma^2 e^{\frac{y^2}{2\sigma^2}} dy \leq 2\sigma^2 \zeta_\tau e^{\frac{\zeta_\tau^2}{2\sigma^2}}.$$

Combining this bound with Shafer's inequality for the arctangent [53] leads to

$$\mathbb{E}_0\left[T_\tau(y_i)^2 \mathbf{1}_{\{|y_i| \leq \zeta_\tau\}}\right] \leq \sqrt{\frac{2}{\pi}}\sigma\frac{4}{9}\zeta_\tau \tau \lesssim \sqrt{\log\frac{1}{\tau}}\tau,$$

as desired.

Conclusion: Combining the bounds for the zero and non-zero means, we find

$$\sum_{i=1}^{n} \mathbb{E}_{\mu_{0i}}(T_\tau(y_i) - \mu_i)^2 \lesssim s_n\left(\sigma^2 + \log\frac{1}{\tau}\right) + (n - s_n)\tau\sqrt{\log\frac{1}{\tau}}.$$

Plugging in $\tau = (s_n/n)^\alpha$ and realizing that the minimax rate for this problem is $2\sigma^2 s_n \log(n/s_n)(1 + o(1))$ as $s_n/n \to 0$ [21], finishes the argument showing

$$\mathbb{E}_{\boldsymbol{\mu}_0}\|\boldsymbol{T}_\tau(\boldsymbol{y}^n) - \boldsymbol{\mu}\|^2 \asymp s_n \log\frac{n}{s_n}. \tag{6.14}$$

Applying Markov's inequality now leads to (6.8). To show (6.7), a similar study of the posterior variance is undertaken and combined with (6.14), again finished by Markov's inequality.

6.3.3 Adaptive Posterior Concentration Theorems

While the non-adaptive results from the previous section give valuable insights, we need to take a further step to understand the procedures as they are used in practice, when the number of signals s_n is not known. From now on, we will focus on the horseshoe [13], for which adaptive results are available [62, 63]. There are results both for empirical Bayes, when τ is estimated and then plugged into the prior, and for hierarchical Bayes, where τ is equipped with a prior. One insight that we learned from the non-adaptive results, is how to set τ. The interpretation of τ as the proportion of signals (perhaps up to a logarithmic factor) leads us towards selecting estimators or priors on $[0, 1]$.

From this point forward, we will assume that σ^2 is equal to one. For empirical Bayes with the horseshoe, an early result was available for the simple plug-in estimator

$$\widehat{\tau}_S = \max\left\{\frac{\sum_{i=1}^n \mathbf{1}\{|y_i| \geq \sqrt{c_1 \log n}\}}{c_2 n}, \frac{1}{n}\right\},$$

with $c_1 > 2, c_2 > 1$. This simple estimator uses the number of observations larger than the "universal threshold" of $\sqrt{2\log n}$ (see [36]) as an estimate of the number of signals. The lower bound of $1/n$ is introduced to avoid computational problems, and because with the intuition that τ should equal s_n/n, it corresponds to the assumption that there is at least one signal. In [64], it was shown to achieve a worst-case ℓ_2 risk of order $s_n \log n$, but no full concentration rate was shown yet.

A fully adaptive posterior concentration statement was shown later in [62], for empirical and hierarchical Bayes. For empirical Bayes, the maximum marginal likelihood estimator (MMLE) was studied. The MMLE is the maximum likelihood estimator one would use if the observations Y_i were distributed as the convolution of the prior on μ_i and the standard normal density. For the horseshoe, it is given by

$$\widehat{\tau}_M = \arg\max_{\tau \in [\frac{1}{n}, 1]} \prod_{i=1}^{n} \int_{-\infty}^{\infty} \phi(y_i - \mu) g_\tau(\mu) d\mu, \tag{6.15}$$

where $\phi(\cdot)$ is the standard normal density and g_τ is the marginal prior density for a μ_i, that is, $g_\tau(\mu) = \int_0^\infty \phi(\frac{\mu}{\lambda\tau})\frac{1}{\lambda\tau}\frac{2}{\pi(1+\lambda^2)}d\lambda$. Again, the estimate of τ is restricted to $[1/n, 1]$, for the intuitive and computational reasons described for the simple estimator $\widehat{\tau}_S$. Computation of the MMLE $\widehat{\tau}_M$ requires only one-dimensional optimization, as illustrated in Figure 6.3, and it is implemented in the R package `horseshoe` [61].

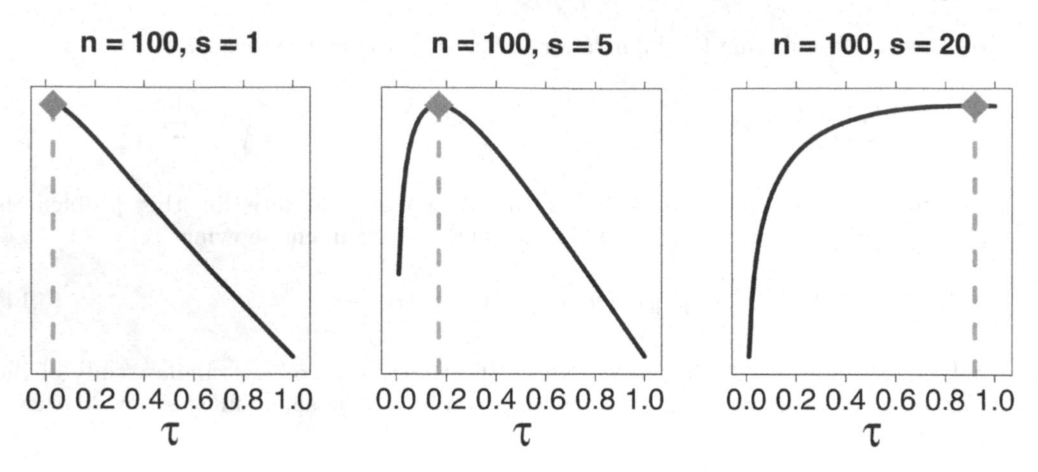

FIGURE 6.3
Illustration of the behavior of the MMLE. The logarithm of the quantity to be optimized in (6.15) is plotted as a function of τ, based on three simulated data sets consisting of 100 observations following the normal means model with, from left to right, 1, 5 or 20 means equal to 10 and the remaining means equal to zero. The red dot indicates the location of the maximum and the dotted line the corresponding value of the MMLE.

In the non-adaptive results, it was found that the optimal value for τ is $(s_n/n)\sqrt{\log(n/s_n)}$, in the sense that for this value, the posterior spread is of the same order as the minimax risk [64]. This value plays an important role in the adaptive results as well, and we will refer to it with $\tau_n(s_n)$, so $\tau_n(s_n) = (s_n/n)\sqrt{\log(n/s_n)}$.

In [62], it is shown that to achieve posterior contraction at the near minimax rate, a sufficient condition on a plug-in estimator $\widehat{\tau}_n$ of τ is

Condition 1. *There exists a constant $C > 0$ such that $\widehat{\tau}_n \in [1/n, C\tau_n(s_n)]$, with $P_{\boldsymbol{\mu}_0}$-probability tending to one, uniformly in $\boldsymbol{\mu}_0 \in \ell_0[s_n]$.*

So if the data is generated based on any nearly black $\boldsymbol{\mu}_0$, the plug-in estimator for τ should be at most a constant multiple of $\tau_n(s_n)$. Disregarding the log-factor, it means that the estimator should not overshoot the proportion of signals s_n/n by too much.

The contraction result is then stated as, with $\Pi_{\widehat{\tau}_n}$ denoting the empirical Bayes posterior:

Theorem 6.3.3 (Theorem 3.2 of [62]). *For any estimator $\widehat{\tau}_n$ of τ that satisfies Condition 1, the empirical Bayes posterior distribution contracts around the true parameter at the near-minimax rate: for any $M_n \to \infty$ and $s_n \to \infty$,*

$$\sup_{\boldsymbol{\mu}_0 \in \ell_0[s_n]} \mathbb{E}_{\boldsymbol{\mu}_0} \Pi_{\widehat{\tau}_n} \left(\boldsymbol{\mu} : \|\boldsymbol{\mu}_0 - \boldsymbol{\mu}\|_2 \geq M_n \sqrt{s_n \log n} \mid \boldsymbol{y}^n \right) \to 0.$$

In particular, this is true for $\widehat{\tau}_n$ equal to the MMLE.

The final statement of Theorem 6.3.3 follows because Condition 1 is verified for the MMLE (6.15). A benefit of using an empirical Bayes procedure is that it enables the computation of the posterior mean without the need for MCMC. One can simply plug in the empirical Bayes estimate of τ into expression (6.12), which may be computed by, for example, a quadrature routine.

Hierarchical Bayes solutions, where τ is equipped with a prior, can lead to near minimax contraction rates as well. Condition 2 below suffices and is similar in spirit to Condition 1 for empirical Bayes procedures.

Condition 2. *The prior density* p *on* τ *is supported inside* $[1/n, 1]$, *and*

$$\int_{t_n/2}^{t_n} p(\tau)\, d\tau \gtrsim e^{-cs_n}, \quad \text{for some } c \leq C_u/2,$$

where $t_n = C_u \pi^{3/2} \tau_n(s_n)$ *and* C_u *is a constant bounding a quantity related to the log-likelihood, see [62] for details.*

In essence, this condition is met if there is "sufficient" prior mass close to the optimal value of τ. An example of a prior on τ for which Condition 2 can be verified is the Cauchy prior truncated to $[1/n, 1]$, provided $s_n \geq C \log n$ for a sufficiently large C.

Theorem 6.3.4 (Theorem 3.7 of [62]). *If the prior on* τ *satisfies Condition 2, then the hierarchical Bayes posterior contracts to the true parameter at the near minimax rate: for any* $M_n \to \infty$ *and* $p_n \to \infty$,

$$\sup_{\boldsymbol{\mu}_0 \in \ell_0[s_n]} \mathbb{E}_{\boldsymbol{\mu}_0} \Pi(\boldsymbol{\mu} : \|\boldsymbol{\mu} - \boldsymbol{\mu}_0\|_2 \geq M_n \sqrt{s_n \log n} \mid \boldsymbol{y}^n) \to 0.$$

A weaker version of Condition 2 is also available, designed for the very sparse case where $s_n \lesssim \log n$, which leads to a statement similar to Theorem 6.3.4, with the somewhat worse rate $\sqrt{s_n} \log n$ instead of $\sqrt{s_n \log n}$.

Both the empirical and hierarchical Bayes options lead to similar conclusions, and the empirical Bayes posterior distribution with the MMLE plugged in for τ closely mimics the hierarchical Bayes posterior distribution. This phenomenon has been observed in other settings as well [52, 58].

6.3.4 Other Sparsity Assumptions

Throughout this chapter, results are stated under the assumption that most means are exactly equal to zero. Other types of sparsity assumptions occur in the literature (e.g., [17, 36]), for example that the mean vector is in an ℓ_p norm ball of small radius, i.e., $\boldsymbol{\mu}_0 \in \ell_p[r]$ for some $p \in (0, 2]$ and some small r, where $\ell_p[r] = \{\boldsymbol{\mu} \in \mathbb{R}^n : \sum_{i=1}^n |\mu_i|^p \leq r^p\}$. The cited results also cover minimax rates in ℓ_q norm, $q \in (0, 2]$, whereas the results in this chapter are for the ℓ_2 norm. Intuitively one would expect that global-local shrinkage priors, due to their continuous nature, would do especially well in such a setting, since the ℓ_p balls for $p > 0$ allow for a type of sparsity where the "irrelevant" means are not necessarily exactly zero. The proof technique outlined in Section 6.3.2, at first glance, very much relies on the ability to split the means into two groups: zero or non-zero. However on inspection it seems likely that similar bounds can be achieved by splitting the means into "small" and "large" groups, perhaps based on the "threshold" of $\sqrt{2 \log(1/\tau)}$, thus providing a path towards obtaining posterior concentration results for other sparsity assumptions.

6.3.5 Implications for Practice

Several implications can be gleaned from Theorems 6.3.1, 6.3.2, 6.3.3 and 6.3.4 as well as their proofs (see also Section 6.3.2).

Implication 1: The posterior mean is a good estimator of the underlying mean vector $\boldsymbol{\mu}_0$, from a minimax point of view. This is true not only for the horseshoe, but for a wide variety of global-local shrinkage priors. This follows directly from (6.10).

Implication 2: The posterior mean displays soft thresholding behavior, with the threshold around $\zeta_\tau := \sqrt{2 \log(1/\tau)}$. This is apparent from the proof for the horseshoe (see Section 6.3.2) and similar behavior was found through the proof of the more general case [65].

Observations that fall below this threshold are shrunk almost to zero, while larger observations are shrunk much less – the amount of shrinkage depends on the particular prior. This is illustrated for the horseshoe prior in Figure 6.4. In case of the horseshoe, observations well past the threshold are more or less untouched. This threshold at $\sqrt{2\log(1/\tau)}$ gives a hint towards the optimal value for τ, as it resembles the 'universal threshold' of $\sqrt{2\log n}$ [20].

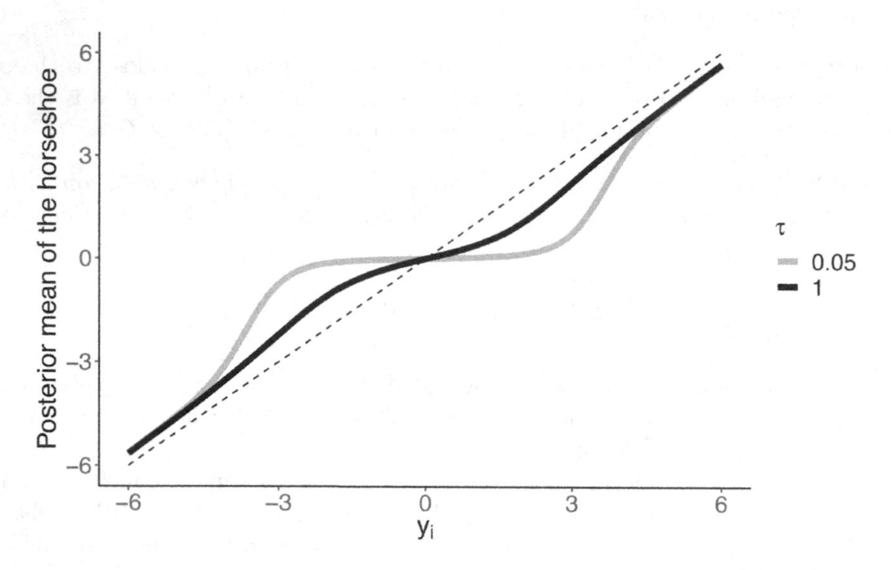

FIGURE 6.4
The relationship between the global parameter τ and the amount of shrinkage in the horseshoe posterior mean, with the identity (dashed line) for reference. The posterior mean is plotted as a function of the observation y_i. Decreasing τ towards zero leads to more shrinkage.

Implication 3: Optimally, τ is set to $(s_n/n)\sqrt{\log(n/s_n)}$ for the horseshoe, or, disregarding the log factor in favor of the simple interpretation as the proportion of signals, to s_n/n. For recovery, some power of s_n/n suffices as well, but we will see later in the discussion of uncertainty quantification that it is best to aim for a value close to $(s_n/n)\sqrt{\log(n/s_n)}$. This ideal value informs the priors on τ considered for a hierarchical Bayes formulation, leading to a preference for priors supported on $[0, 1]$. It also lends further justification to the interpretation of τ as a global shrinkage parameter, since it unifies the information available in the means through the quantity s_n.

Implication 4: With the horseshoe we can achieve (near) optimal posterior contraction rates and use the posterior mean as a (nearly) minimax optimal estimator even when we do not know how many signals there are. This adaptive property is very useful in practice. If a reasonable choice is made for the prior in a hierarchical Bayes setting (Condition 2), or the MMLE is used as a plug-in estimator in an empirical Bayes setting (verifying Condition 1), we still have good recovery guarantees despite us not knowing the number of signals s_n.

Implication 5: The tail behavior of a prior matters, but in terms of minimax recovery rates, a wide range is acceptable. The conditions in Section 6.3.1 allow for at most exponentially fast decay. The intuition behind this is that heavy tails prevent overshrinkage, so that true signals can still be identified. This is consistent with results for spike-and-slab priors, where the 'slab' was assumed to have at least exponential tails [17]. Based on Theorem 6.3.2, we do not have a preference for any of the priors meeting the conditions for the theorem.

In part, this may be because (6.10) indicates the rate up to multiplicative constants. It leaves open the possibility that some priors will have a better rate, once the multiplicative constant is accounted for. Moreover, for coverage, the situation is more delicate, as we will see in Section 6.4.

6.4 Uncertainty Quantification Guarantees

A first indication on the usefulness of credible sets for uncertainty quantification is given by the posterior contraction rate. It tells us how large we can expect the credible set to be. If the posterior contracts slowly, the credible sets can be too large to be useful. For example, based on the contraction rate alone, we can conclude that the credible sets of the Lasso [59] (when viewed as a Bayesian procedure, where each μ_i receives an i.i.d. Laplace prior) are not useful for uncertainty quantification as the full posterior contracts much slower than the mode (which is the well-known Lasso estimator) [15].

Posterior concentration rates give some indication about the size of credible sets, but do not by themselves imply that a credible set is honest. While a statement like (6.7) tells us that asymptotically, all posterior mass will be in a ball of radius of the order $s_n \log(n/s_n)$ around the true $\boldsymbol{\mu_0}$, it does not tell us where exactly the mass is located within that ball. It is, for example, entirely possible that the majority of the mass is contained in a smaller ball near the posterior mean $\boldsymbol{T_\tau}$ but at a distance from our target parameter $\boldsymbol{\mu_0}$. This is illustrated in Figure 6.5. Thus, additional arguments are required to capture the behavior of credible sets.

6.4.1 Credible Intervals

We first consider marginal credible intervals, that is, credible intervals for each individual μ_i. Such intervals can be constructed from the marginal posterior distributions. For fixed τ, the results are for intervals of the form

$$\widehat{C}_{ni}(L,\tau) = \{\mu_i : |\mu_i - T_\tau(y_i)| \leq c\hat{r}_i(\alpha,\tau)\}, \tag{6.16}$$

where $T_\tau(y_i) = \mathbb{E}[\mu_i \mid y_i, \tau]$ is the marginal posterior mean, c a positive constant, and $\hat{r}_i(\alpha,\tau)$ is determined so that, for a given $0 < \alpha \leq 1/2$, the following equality for the marginal posterior distribution holds

$$\Pi_\tau\big(\mu_i : |\mu_i - T_\tau(y_i)| \leq \hat{r}_i(\alpha,\tau) \mid y_i, \tau\big) = 1 - \alpha.$$

Again it is instructive to consider theoretical guarantees in case we have access to s_n and can set τ based on this information. While the result below is formulated for the horseshoe, it can be extended to the global-local shrinkage priors considered by [29], who consider mixtures of the form (6.3) with density $p(\lambda_i^2)$ given by

$$p(\lambda_i^2) = \frac{K}{\lambda_i^{2+2\alpha}} L(\lambda_i^2),$$

where $a \geq 1/2, K > 0$ and the function $L : (0,\infty) \mapsto (0,\infty)$ needs to be such that $\sup_{u>0} L(u) \leq c_1$ and $\inf_{u \geq u_0} L(u) \geq c_2$ for some $c_1, c_2, u_0 > 0$. This class contains not only the horseshoe prior, but also the inverse gamma prior, the half-t prior, the normal-exponential-gamma prior, the generalized double Pareto and the three parameter beta normal mixtures.

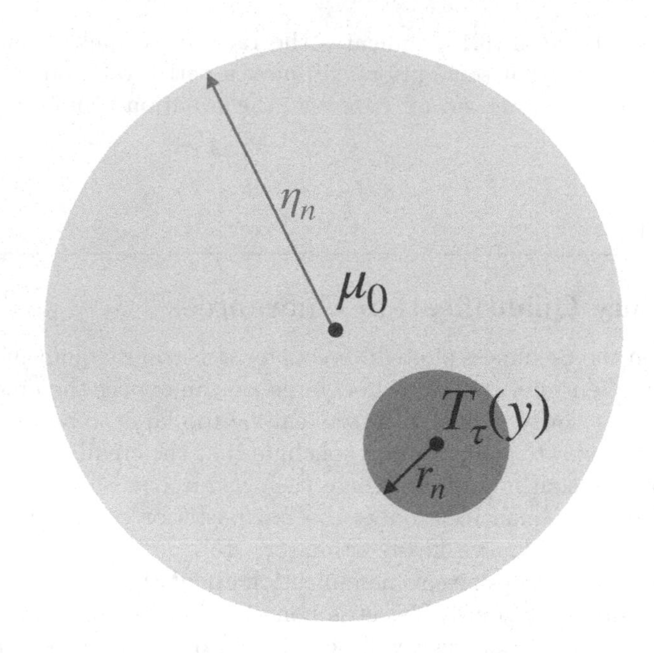

FIGURE 6.5
If all posterior mass is contained in a ball of radius η_n around $\boldsymbol{\mu}_0$, it is possible that a credible set containing, say, 95% of the posterior mass, is not honest. In the illustration, most of the posterior mass is in the ball of radius $r_n < \eta_n$ around the posterior mean $\boldsymbol{T_\tau(y)}$.

The coverage of the marginal credible intervals varies depending on the size of the true mean μ_{0i}. We can distinguish three regions. These regions combined do not cover $\mathbb{R}$ entirely, but their boundaries are very close. We define the regions for given $\tau \to 0$, positive constants c_S, c_M, c_L and numbers $c_\tau \uparrow \infty$ as $\tau \to 0$. Conditions on these numbers will be imposed shortly. First, we define the regions, which we can think of as "small", "medium" and "large" means.

$$\begin{aligned}
\mu_S &:= \{1 \le i \le n : |\mu_{0,i}| \le c_S\tau\},\\
\mu_M &:= \{1 \le i \le n : c_\tau\tau \le |\mu_{0,i}| \le c_M\zeta_\tau\},\\
\mu_L &:= \{1 \le i \le n : c_L\zeta_\tau \le |\mu_{0,i}|\}.
\end{aligned}$$

Here, $\zeta_\tau = \sqrt{2\log(1/\tau)}$ makes a reappearance as an important threshold. The takeaway from the more precise statement below is that the coverage of the credible intervals for zero or tiny (at most of order τ) means is as desired, and so is the coverage of the credible intervals for means of size roughly $\sqrt{2\log(n/s_n)}$ or larger. However, for "medium" non-zero means below the threshold, we expect the credible intervals not to cover them.

Let $|\cdot|$ denote the cardinality of a set, and let the $\widehat{C}_{ni}$ refer to the marginal credible intervals constructed from the horseshoe.

Theorem 6.4.1 (Theorem 1 of [63]). *Set $c_S > 0, c_M < 1, c_L > 1$ and take $c_\tau \uparrow \infty$ as $\tau \to 0$. Let c_n be a sequence $c_n \to c$ for some $0 \le c \le 1/2$ so that $\sqrt{\log(1/\tau)} \gg \sqrt{\log(1/c_n)}$ as $\tau \to 0$. Then:*

$$P_{\mu_0}\Big(\frac{1}{|\mu_S|}|\{i \in \mu_S : \mu_{0,i} \in \widehat{C}_{ni}(L_S,\tau)\}| \ge 1 - c_n\Big) \to 1, \tag{6.17}$$

$$P_{\mu_0}\big(\mu_{0,i} \notin \widehat{C}_{ni}(L,\tau)\big) \to 1, \quad \text{for any } L > 0 \text{ and } i \in \mu_M, \tag{6.18}$$

$$P_{\mu_0}\Big(\frac{1}{|\mu_L|}|\{i \in \mu_L : \mu_{0,i} \in \widehat{C}_{ni}(L_L,\tau)\}| \geq 1 - c_n\Big) \to 1, \tag{6.19}$$

where $L_S = (2.1/z_\alpha)\big[c_S + (2/c_n)\zeta_{c_n/2}\big]$ *and* $L_L = (1.1/z_\alpha)\zeta_{c_n/2}$.

In the adaptive setting, where s_n is unknown, a similar result is available (Theorem 2 in [63]). For empirical Bayes, it requires an estimator of τ which meets Condition 1. In particular, the MMLE will supply the desired result. For hierarchical Bayes, any prior meeting Condition 2 will do, if we are willing to assume $s_n \gtrsim \log n$.

Thus, the marginal credible intervals work well for very small means and means larger than the universal threshold, and not so well for intermediate values. The results for the marginal intervals match the design principle behind the horseshoe. It was designed with a pole at zero (for the zero means) and heavy tails (for the non-zero means), so with the sets μ_S and μ_L in mind. It is hard to tell the intermediate values in μ_M apart from the noise, because if we have n i.i.d. $\mathcal{N}(0,1)$ random variables, the expected value of the maximum of these n random variables is approximately the "universal threshold" $\sqrt{2\log n}$.

6.4.2 Credible Balls

In this context, a credible ball is a credible set for the entire vector $\boldsymbol{\mu_0} \in \mathbb{R}^n$. They are sets of the form

$$\widehat{C}_n(L,\tau) = \{\boldsymbol{\mu} : \|\boldsymbol{\mu} - \boldsymbol{T_\tau}(\boldsymbol{y^n})\|_2 \leq L\hat{r}(\alpha,\tau)\}, \tag{6.20}$$

with $\boldsymbol{T_\tau}$ the posterior mean, L a positive constant, and $\hat{r}(\alpha,\tau)$ is the radius, chosen such that for a given $\alpha \in (0,1)$

$$\Pi_\tau(\boldsymbol{\mu} : \|\boldsymbol{\mu} - \boldsymbol{T_\tau}(\boldsymbol{y^n})\|_2 \leq \hat{r}(\alpha,\tau) \mid \boldsymbol{y^n},\tau) = 1 - \alpha.$$

The constant L is introduced to ensure that the confidence level of the credible set matches the credible level, as in [57].

To achieve good coverage in the non-adaptive case, a mild requirement on τ suffices. The condition is: $n\tau/\zeta_\tau \to \infty$, which is met for example by setting $\tau = (s_n/n)\sqrt{\log(n/s_n)}$ as $s_n \to \infty$. Under this condition, it can be shown that the radius $\hat{r}(\alpha,\tau)$ is bounded below by $0.5\sqrt{n\tau\zeta_\tau}$ with $P_{\boldsymbol{\mu_0}}$-probability tending to one (Lemma 4.1 in [63]), which when combined with Markov's inequality and (6.14) leads to the following statement:

Theorem 6.4.2 (Theorem 4.1 in [63]). *If* $\tau \geq (s_n/n)\sqrt{\log(n/s_n)}$ *and* $\tau \to 0$ *and* $s_n \to \infty$ *with* $s_n = o(n)$, *then, there exists a large enough* $L > 0$ *such that*

$$\liminf_{n\to\infty} \inf_{\boldsymbol{\mu_0}\in\ell_0[s_n]} P_{\boldsymbol{\mu_0}}\big(\boldsymbol{\mu_0} \in \widehat{C}_n(L,\tau)\big) \geq 1 - \alpha.$$

Of course, in practice it is not very realistic that one knows how many signals there are. The theorem above is useful mainly because it gives us some idea for what to expect in the adaptive case.

There is some tension between honest uncertainty quantification and adaptation to sparsity [44]. In the adaptive case, there is a risk that using the data to estimate s_n will lead to over-shrinkage, because the estimator may not "see" the signals that are non-zero but below some detection threshold. For the horseshoe, it turns out to be sufficient that a fraction of the non-zero means is above a threshold given by $\sqrt{2\log(n/s_n)}$. To this end, the concept of self-similarity is adapted [11, 32, 43, 47, 51, 54, 57].

Condition 3 (self-similarity). *A vector $\boldsymbol{\mu_0} \in \ell_0[s]$ is called* self-similar *if*

$$\#(i : |\mu_{0,i}| \geq A\sqrt{2\log(n/s)}) \geq \frac{s}{C}. \tag{6.21}$$

for some $C \geq 1$ and $A > 1$.

This condition ensures that a sufficient amount of the non-zero means is large enough that τ can be estimated at the "correct" order of $(s_n/n)\sqrt{\log(n/s_n)}$. The condition might remind one of the "beta-min condition" for the adaptive Lasso, which imposes a lower bound on the non-zero means in the interest of consistent selection of the true signals [10, 60].

The self-similarity condition may be weakened to an "excessive bias" condition, which requires the sum of the squares of the smaller signals to be suitably dominated. We refer to [63] and [6] for more discussion on this weaker condition. In [63], it is established that adaptive credible sets based on either the MMLE or a hierarchical Bayes formulation with $\tau \sim p_n$ with p_n a probability density on $[1/n, n]$ bounded away from zero, attain the desired coverage and have a radius of at most order $\sqrt{s_n \log(n/s_n)}$ under such an excessive bias condition.

6.4.3 Implications for Practice

The presented results offer some good news, as well as room for improvement. The adaptive results for the marginal credible intervals provide a reliable way to quantify uncertainty for individual means, but there is a region of non-zero signals below the threshold of roughly $\sqrt{2\log(n/s_n)}$ that are too small for the horseshoe to pick up. Nevertheless, one can have high confidence in marginal intervals that indicate that a particular mean represents a signal.

In simulations [63], the best options for the horseshoe were identified as either empirical Bayes with the MMLE or hierarchical Bayes with a truncated Cauchy prior. Both are available in the R package `horseshoe` [61]. The two approaches yield very similar results, as was expected from the theoretical study of the two procedures, and a selection between the two may be based on considerations other than theoretical ones, for example based on computational considerations.

The practical value of credible balls is less evident in this setting.

6.5 Variable Selection Guarantees

In the context of the sparse normal means model (6.1), variable selection means finding the subset $\{i \in 1, \ldots, n : \mu_{0i} \neq 0\}$. Due to the continuous nature of the global-local shrinkage priors, none of the estimates of the means will be exactly equal to zero. Thus, there is the need to design a decision rule to perform variable selection, possibly incorporating more information from the posterior than just the estimate of the mean vector. Two options that build upon the results for recovery and coverage from Sections 6.3 and 6.4 are discussed below.

6.5.1 Thresholding on the Amount of Shrinkage

A procedure that is very simple and computationally efficient to carry out in practice was proposed by [13] and is based on relationship (6.5) between the posterior mean and the 'shrinkage factor' $f(\lambda_i) = (1 + \tau^2\lambda_i^2)^{-1}$. Writing $\widehat{f_i} = \mathbb{E}[f(\lambda_i) \mid y_i]$, we have $\mathbb{E}[\mu_i \mid y_i] = (1 - \widehat{f_i})y_i$.

The decision rule is then: consider μ_i a signal (non-zero) if $(1 - \widehat{f}_i) \geq 0.5$, and noise (zero) otherwise.

In a simulation study, [13] show that for the horseshoe, this procedure leads to good control of the number of false-positive classifications and note a "striking resemblance" between the weights $(1 - \widehat{f}_i)$ and the posterior probabilities from the spike-and-slab prior.

Again for the horseshoe, the type I and type II error probabilities and Bayes risk of this decision rule were studied [19] and it was found that, as long as τ can be set to $\tau = s_n/n$, the horseshoe attains close to the risk of the Bayes oracle. It has also been shown, in Theorem B.2 of the supplement to [63], that the decision rule has similar theoretical properties to those of marginal credible intervals described in Section 6.5.2.

6.5.2 Checking for Zero in Marginal Credible Intervals

A natural Bayesian option is to use the marginal credible intervals and follow a procedure reminiscent of frequentist confidence interval based hypothesis testing: if 0 is not in the marginal credible interval for μ_i, we declare it a signal, and otherwise we declare it noise. Performing model selection like this is easier than attaining good coverage of the marginal intervals. This is because the requirements for the signals are less strict, since to perform correct variable selection, the marginal intervals for the non-zero means only need to stay away from zero, but it is not necessary that they contain the true parameter value (although perhaps desired for other reasons). For the zero means, the coverage requirement and the variable selection property coincide.

For the horseshoe, this procedure was studied in several configurations [63]: adaptive with the MMLE, adaptive with hierarchical Bayes, and non-adaptive. In each setting, similar results were obtained: of the sets of true zeroes and "large" signals, only a vanishing fraction is classified incorrectly, with probability tending to one, while the fraction of "small" and "medium" signals that are incorrectly classified as noise tends to one. Here "small", "medium" and "large" refers to variations on the sets μ_S, μ_M, μ_L from Section 6.4.1. We denote these versions of the sets by $\widetilde{\mu}_S, \widetilde{\mu}_M, \widetilde{\mu}_L$. They are given by, abbreviating $\tau_n(s_n) = (s_n/n)\sqrt{\log(n/s_n)}$:

$$\widetilde{\mu}_S := \{1 \leq i \leq n : |\mu_{0,i}| \leq c_S/n\},$$
$$\widetilde{\mu}_M := \{1 \leq i \leq n : c_n \tau_n(s_n) \leq |\mu_{0,i}| \leq c_M \zeta_{\tau_n(s_n)}\},$$
$$\widetilde{\mu}_L := \{1 \leq i \leq n : c_L\sqrt{2\log n} \leq |\mu_{0,i}|\}.$$

In addition, we study the set of zeroes:

$$\mu_Z = \{1 \leq i \leq n : \mu_{0i} = 0\}$$

The precise statement for the adaptive case is, using some of the same notation as in Theorem 6.4.1:

Theorem 6.5.1 (Theorem 3 of [63]). *Suppose that $c_M < 1 < c_L$ and $c_n \uparrow \infty$ and let $L > 0$. For any sequence γ_n such that $\zeta^2_{\gamma_n} \ll \zeta^2_{\tau_n(s_n)}$, the following statements hold, with probability tending to one:*

(i) The number of selected parameters with $i \in \mu_Z$ divided by the total number $|\mu_Z|$ of zero parameters is at most γ_n.

(ii) The number of selected parameters in $i \in \widetilde{\mu}_L$ divided by $|\widetilde{\mu}_L|$ is at least $1 - \gamma_n$, i.e.,

$$P_{\mu_0}\Big(\frac{\#(\{i \in \widetilde{\mu}_L : 0 \notin \hat{C}_{ni}(L, \hat{\tau})\})}{\#(\widetilde{\mu}_L)} \geq 1 - \gamma_n\Big) \to 1,$$

and the same for the hierarchical Bayes intervals.

(iii) At most a fraction γ_n of the parameters within $i \in (\mu_Z^c \cap \widetilde{\mu}_S) \cup \widetilde{\mu}_M$ will be selected.

The implication for practice is that while the horseshoe misses "intermediate" signals, most of the discoveries made by the horseshoe will be true discoveries. As mentioned before, a similar result holds for the thresholding procedure of Section 6.5.1. In a simulation study, it was found that the thresholding procedure from Section 6.5.1 leads to more discoveries but also a higher False Discovery Rate compared to the credible interval based method. Both methods are implemented in the R package `horseshoe` [61].

6.6 Discussion

Theoretical guarantees for global-local shrinkage priors remains an active area of research. Many theoretical questions remain open. While some promising results have recently become available, e.g., [5, 55, 68], general theory for global-local shrinkage priors in models other than the normal means model would be welcome. There is some hope that it is possible to prove desirable behavior of the horseshoe in the linear regression setting, since such proofs have been given for the spike-and-slab priors [15] and at least within the normal means setting, the behavior of the horseshoe seems to mimic that of the spike-and-slab prior [17]. Another question of interest is allowing error distributions other than normal ones, and in particular allowing dependent error structures. In the meantime, it remains important to develop algorithms and software to deliver on the initial promise of computational convenience of global-local shrinkage priors (as in e.g., [8, 35, 39]), and to ensure that the theoretically good behavior finds its way to practice.

Bibliography

[1] D. F. Andrews and C. L. Mallows. Scale mixtures of normal distributions. *J. R. Stat. Soc. Ser. B Stat. Methodol.*, pages 99–102, 1974.

[2] A. Armagan, D. B. Dunson, and J. Lee. Generalized double Pareto shrinkage. *Statistica Sinica*, 23:119–143, 2013.

[3] M. Authier, C. Saraux, and C. Péron. Variable selection and accurate predictions in habitat modelling: a shrinkage approach. *Ecography*, 40(4):549–560, 2017.

[4] K. Bae and B. K. Mallick. Gene selection using a two-level hierarchical Bayesian model. *Bioinformatics*, 20(18):3423–3430, 2004.

[5] R. Bai and M. Ghosh. High-dimensional multivariate posterior consistency under global-local shrinkage priors. *Journal of Multivariate Analysis*, 167:157–170, 2018.

[6] E. Belitser and N. Nurushev. Needles and straw in a haystack: Robust confidence for possibly sparse sequences. *Bernoulli*, 26(1):191–225, 02 2020.

[7] J. Berger. A robust generalized Bayes estimator and confidence region for a multivariate normal mean. *The Annals of Statistics*, pages 716–761, 1980.

[8] A. Bhattacharya, A. Chakraborty, and B. K. Mallick. Fast sampling with Gaussian scale mixture priors in high-dimensional regression. *Biometrika*, pages 985–991, 2016.

[9] A. Bhattacharya, D. Pati, N. S. Pillai, and D. B. Dunson. Dirichlet–Laplace priors for optimal shrinkage. *Journal of the American Statistical Association*, 110(512):1479–1490, 2015.

[10] P. Bühlmann and S. van de Geer. *Statistics for High-Dimensional Data.* Springer-Verlag Berlin Heidelberg, 2011.

[11] A. Bull. Honest adaptive confidence bands and self-similar functions. *Electron. J. Statist.*, 6:1490–1516, 2012.

[12] F. Caron and A. Doucet. Sparse Bayesian nonparametric regression. In *Proceedings of the 25th International Conference on Machine Learning*, ICML '08, pages 88–95, New York, NY, USA, 2008. ACM.

[13] C. M. Carvalho, N. G. Polson, and J. G. Scott. The horseshoe estimator for sparse signals. *Biometrika*, 97(2):465–480, 2010.

[14] I. Castillo and R. Mismer. Empirical Bayes analysis of spike and slab posterior distributions. *Electron. J. Statist.*, 12(2):3953–4001, 2018.

[15] I. Castillo, J. Schmidt-Hieber, and A. van der Vaart. Bayesian linear regression with sparse priors. *Ann. Statist.*, 43(5):1986–2018, 10 2015.

[16] I. Castillo and B. Szabó. Spike and slab empirical Bayes sparse credible sets. *Bernoulli*, 26(1):127–158, 02 2020.

[17] I. Castillo and A. W. Van der Vaart. Needles and straw in a haystack: Posterior concentration for possibly sparse sequences. *Ann. Statist.*, 40(4):2069–2101, 2012.

[18] J. C. Chang, S. Vattikuti, and C. C. Chow. Probabilistically-autoencoded horseshoe-disentangled multidomain item-response theory models. *arXiv preprint arXiv:1912.02351*, 2019.

[19] J. Datta and J. K. Ghosh. Asymptotic properties of Bayes risk for the horseshoe prior. *Bayesian Analysis*, 8(1):111–132, 2013.

[20] D. L. Donoho and I. M. Johnstone. Ideal spatial adaptation by wavelet shrinkage. *Biometrika*, 81(3):425–455, 1994.

[21] D. L. Donoho, I. M. Johnstone, J. C. Hoch, and A. S. Stern. Maximum entropy and the nearly black object (with discussion). *Journal of the Royal Statistical Society. Series B (Methodological)*, 54(1):41–81, 1992.

[22] D. Dunson and T. Papamarkou. Discussions. *International Statistical Review*, 88(2): 321–324, 2020.

[23] M. Figueiredo. Adaptive sparseness for supervised learning. *IEEE transactions on pattern analysis and machine intelligence*, 25(9):1150–1159, 2003.

[24] L. Follett and C. Yu. Achieving parsimony in Bayesian vector autoregressions with the horseshoe prior. *Econometrics and Statistics*, 11:130–144, 2019.

[25] T. Ge, C.-Y. Chen, Y. Ni, Y.-C. A. Feng, and J.W. Smoller. Polygenic prediction via Bayesian regression and continuous shrinkage priors. *Nature Communications*, 10(1):1–10, 2019.

[26] S. Ghosal, J. K. Ghosh, and A. W. van der Vaart. Convergence rates of posterior distributions. *The Annals of Statistics*, 28(2):500–531, 2000.

[27] S. Ghosal and A. van der Vaart. Convergence rates of posterior distributions for noniid observations. *Annals of Statistics*, 35:192–223, 2007.

[28] S. Ghosal and A. van der Vaart. *Fundamentals of nonparametric Bayesian inference*, volume 44. Cambridge University Press, 2017.

[29] P. Ghosh and A. Chakrabarti. Asymptotic optimality of one-group shrinkage priors in sparse high-dimensional problems. *Bayesian Analysis*, 12(4):1133–1161, 2017.

[30] S. Ghosh, J. Yao, and F. Doshi-Velez. Model selection in Bayesian neural networks via horseshoe priors. *Journal of Machine Learning Research*, 20(182):1–46, 2019.

[31] M. Giacomazzo and Y. Kamarianakis. Bayesian estimation of subset threshold autoregressions: short-term forecasting of traffic occupancy. *Journal of Applied Statistics*, 47(13-15):2658–2689, 2020.

[32] E. Giné and R. Nickl. Confidence bands in density estimation. *Ann. Statist.*, 38(2):1122–1170, 2010.

[33] J. E. Griffin and P. J. Brown. Alternative prior distributions for variable selection with very many more variables than observations. *Technical Report, University of Warwick.*, 2005.

[34] J. E. Griffin and P. J. Brown. Inference with normal-gamma prior distributions in regression problems. *Bayesian Analysis*, 5(1):171–188, 2010.

[35] J. E. Johndrow, P. Orenstein, and A. Bhattacharya. Scalable approximate MCMC algorithms for the horseshoe prior. *Journal of Machine Learning Research*, 21(73):1–61, 2020.

[36] I. M. Johnstone and B. W. Silverman. Needles and straw in haystacks: Empirical Bayes estimates of possibly sparse sequences. *Ann. Statist.*, 32(4):1594–1649, 2004.

[37] D. Kohns and T. Szendrei. Horseshoe prior Bayesian quantile regression. *arXiv preprint arXiv:2006.07655*, 2020.

[38] G. B. Kpogbezan, M. A. van de Wiel, W. N. van Wieringen, and A. W. van der Vaart. Incorporating prior information and borrowing information in high-dimensional sparse regression using the horseshoe and variational Bayes. *arXiv preprint arXiv:1901.10217*, 2019.

[39] E. Makalic and D. F. Schmidt. A simple sampler for the horseshoe estimator. *IEEE Signal Processing Letters*, 23(1):179–182, 2015.

[40] P. D. Miller. *Applied Asymptotic Analysis*, volume 75 of *Graduate Studies in Mathematics*. The American Mathematical Society, 2006.

[41] T. J. Mitchell and J. J. Beauchamp. Bayesian variable selection in linear regression. *Journal of the American Statistical Association*, 83(404):1023–1032, 1988.

[42] M. Nalenz and M. Villani. Tree ensembles with rule structured horseshoe regularization. *The Annals of Applied Statistics*, 12(4):2379–2408, 2018.

[43] R. Nickl and B. T. Szabó. A sharp adaptive confidence ball for self-similar functions. *to appear in Stochastics Processes and their Applications*, 2014.

[44] R. Nickl and S. van de Geer. Confidence sets in sparse regression. *Ann. Statist.*, 41(6):2852–2876, 12 2013.

[45] R. Park and G. Casella. The Bayesian lasso. *J. Amer. Statist. Assoc.*, 103(482):681–686, 2008.

[46] L. R. Pericchi and A. F. M. Smith. Exact and approximate posterior moments for a normal location parameter. *Journal of the Royal Statistical Society. Series B (Methodological)*, 54(3):793–804, 1992.

[47] D. Picard and K. Tribouley. Adaptive confidence interval for pointwise curve estimation. *Ann. Statist.*, 28(1):298–335, 2000.

[48] N. G. Polson and J. G. Scott. Shrink globally, act locally: Sparse Bayesian regularization and prediction. In J.M. Bernardo, M.J. Bayarri, J.O. Berger, A.P. Dawid, D. Heckerman, A.F.M. Smith, and M. West, editors, *Bayesian Statistics 9*. Oxford University Press, 2010.

[49] R. Pong-Wong. Estimation of genomic breeding values using the horseshoe prior. In *BMC proceedings*, volume 8, page S6. Springer, 2014.

[50] V. Rocková. Bayesian estimation of sparse signals with a continuous spike-and-slab prior. *Ann. Statist.*, 46(1):401–437, 02 2018.

[51] J. Rousseau and B. Szabó. Asymptotic frequentist coverage properties of Bayesian credible sets for sieve priors. *Annals of Statistics*, 48(4):2155–2179, 2020.

[52] J. Rousseau and B. Szabó. Asymptotic behaviour of the empirical Bayes posteriors associated to maximum marginal likelihood estimator. *Ann. Statist.*, 45(2):833–865, 04 2017.

[53] R. E. Shafer. Elementary problems. Problem E 1867. *The American Mathematical Monthly*, 73(3):309, 1966.

[54] S. Sniekers and A. van der Vaart. Credible sets in the fixed design model with Brownian motion prior. *J. Statist. Plann. Inference*, 166:78–86, 2015.

[55] Q. Song and F. Liang. Nearly optimal Bayesian shrinkage for high dimensional regression. *arXiv preprint arXiv:1712.08964*, 2017.

[56] W. E. Strawderman. Proper Bayes minimax estimators of the multivariate normal mean. *The Annals of Mathematical Statistics*, 42(1):385–388, 1971.

[57] B. Szabó, A. W. van der Vaart, and J. H. van Zanten. Frequentist coverage of adaptive nonparametric Bayesian credible sets. *Ann. Statist.*, 43(4):1391–1428, 08 2015.

[58] B. T. Szabó, A. W. van der Vaart, and J. H. van Zanten. Empirical Bayes scaling of Gaussian priors in the white noise model. *Electron. J. Statist.*, 7:991–1018, 2013.

[59] R. Tibshirani. Regression shrinkage and selection via the Lasso. *J. R. Stat. Soc. Ser. B Stat. Methodol.*, 58(1):267–288, 1996.

[60] S. van de Geer, P. Bühlmann, and S. Zhou. The adaptive and the thresholded Lasso for potentially misspecified models (and a lower bound for the Lasso). *Electron. J. Statist.*, 5:688–749, 2011.

[61] S. van der Pas, J. Scott, A. Chakraborty, and A. Bhattacharya. *horseshoe: Implementation of the Horseshoe Prior*, 2016. R package version 0.2.0.

[62] S. van der Pas, B. Szabó, and A. van der Vaart. Adaptive posterior contraction rates for the horseshoe. *Electron. J. Statist.*, 11(2):3196–3225, 2017.

[63] S. van der Pas, B. Szabó, and A. van der Vaart. Uncertainty quantification for the horseshoe (with discussion). *Bayesian Anal.*, 12(4):1221–1274, 12 2017.

[64] S. L. van der Pas, B. J. K. Kleijn, and A. W. van der Vaart. The horseshoe estimator: Posterior concentration around nearly black vectors. *Electron. J. Statist.*, 8(2):2585–2618, 2014.

[65] S.L. van der Pas, J.-B. Salomond, and J. Schmidt-Hieber. Conditions for posterior contraction in the sparse normal means problem. *Electron. J. Statist.*, 10(1):976–1000, 2016.

[66] S. van Erp, D. L. Oberski, and J. Mulder. Shrinkage priors for Bayesian penalized regression. *J. Math. Psych.*, 89:31–50, 2019.

[67] T. van Erven and B. Szabó. Fast exact Bayesian inference for sparse signals in the normal sequence model. *Bayesian Anal.*, 2020. Advance publication.

[68] R. Wei and S. Ghosal. Contraction properties of shrinkage priors in logistic regression. *Journal of Statistical Planning and Inference*, 207:215–229, 2020.

[69] H.-C. Yang, Y. Xue, Y. Pan, Q. Liu, and G. Hu. Time fused coefficient SIR model with application to COVID-19 epidemic in the United States, Journal of Applied Statistics, DOI: 10.1080/02664763.2021.1936467, 2021

[70] Y. Yang, M. J. Wainwright, and M. I. Jordan. On the computational complexity of high-dimensional Bayesian variable selection. *Ann. Statist.*, 44(6):2497–2532, 12 2016.

[71] A. Yazdani and D. B. Dunson. A hybrid Bayesian approach for genome-wide association studies on related individuals. *Bioinformatics*, 31(24):3890–3896, 2015.

7

MCMC for Global-Local Shrinkage Priors in High-Dimensional Settings

Anirban Bhattacharya
Texas A&M University (USA)

James Johndrow
University of Pennsylvania (USA)

CONTENTS

In this chapter, we provide a review of MCMC computation with one-group or continuous shrinkage priors expressed as scale mixtures of Gaussians in the high-dimensional regression model. We primarily focus on blocked Gibbs samplers, which are popularly used in hierarchical Gaussian models. For sake of concreteness, we additionally focus on the popular horseshoe prior while pointing out generalizations whenever appropriate. Our discussions span various blocking strategies, computational complexities, numerical issues, and geometric convergence. We conclude with some topics for future research.

7.1 Introduction

Since the seminal work of [56] over 60 years ago, shrinkage estimation has been immensely successful in various statistical disciplines and continues to enjoy widespread attention. The radical discovery that propelled this area of research to prominence was the observation [56] that the maximum likelihood estimator for a multivariate Gaussian mean with isotropic covariance ceases to be admissible under the squared error loss when the dimension of the mean vector exceeds two. Specifically, consider the Gaussian sequence model

$$y_j = \mu_j + \epsilon_j, \quad \epsilon_j \sim \mathcal{N}(0, \sigma^2), \quad j = 1, \ldots, n, \tag{7.1}$$

where $\mathbf{y} = (y_1, \ldots, y_n)'$ represents the observed data vector and $\boldsymbol{\mu} = (\mu_1, \ldots, \mu_n)'$ is the parameter to be estimated. The inadmissibility result implies the existence of an estimator with a uniformly (over all possible values of the true parameter μ) smaller expected loss

DOI: 10.1201/9781003089018-7

(or risk) compared to the maximum likelihood estimator $\mathbf{y}$. The now famous James–Stein estimator [26] given by

$$\hat{\boldsymbol{\mu}}^{JS} = \left[1 - \frac{(n-2)\sigma^2}{\|\mathbf{y}\|^2}\right]\mathbf{y},$$

was later shown to uniformly dominate $\mathbf{y}$ in quadratic risk when $n > 2$, providing a constructive proof of the inadmissibility of the maximum likelihood estimator. The James–Stein estimator was found to be itself inadmissible, generating extensive research into admissibility of shrinkage estimators [11, 12, 57].

The key idea behind shrinkage is to reduce variance at the expense of incurring bias. For example, the James–Stein estimator shrinks the maximum likelihood estimator towards the origin (or some prior guess for the mean) in a *non-linear* fashion. This has a natural Bayesian interpretation as Bayesian procedures naturally induce shrinkage through the prior distribution. For the example in (7.1) above, assume an isotropic Gaussian prior on $\boldsymbol{\mu}$ given by $\boldsymbol{\mu} \mid \tau \sim \mathcal{N}(0, \tau^2\sigma^2 I_n)$, and parameterize $\tau^2 = (1-\omega)/\omega$ with $\omega \in (0,1)$. It is easy to verify that

$$E[\boldsymbol{\mu} \mid \omega, \mathbf{y}] = (1-\omega)\mathbf{y}$$

has the form of a linear shrinkage estimator with shrinkage factor ω. For any fixed value of ω, a simple calculation shows that risk of the linear shrinkage estimator $(1-\omega)\mathbf{y}$ fails to uniformly dominate the maximum likelihood estimator. Indeed, the form of the James–Stein estimator crucially points towards the importance of learning the shrinkage factor from the data. In a Bayesian paradigm, one may resort to an empirical Bayes approach or take a fully Bayes approach to endow the hyperparameter ω with a prior. Under the marginal likelihood $\mathbf{y} \mid \omega \sim \mathcal{N}(0, (\sigma^2/\omega)I_n)$ obtained by integrating over $\boldsymbol{\mu}$, one has $\omega\|\mathbf{y}\|^2/\sigma^2 \sim \chi^2_n$, and therefore $\widehat{\omega} = (n-2)\sigma^2/\|\mathbf{y}\|^2$ is an unbiased estimator of ω. This empirical Bayes estimate [16] exactly coincides with the shrinkage factor of the James–Stein estimator. For a fully Bayesian treatment, refer to [44], where risk bounds under a $\mathcal{B}eta(a,b)$ family of priors for ω are shown to closely match the risk of the James–Stein estimator for $a, b < 1$. The case $a = b = 1/2$ is often used as a default, and this corresponds to a half-Cauchy prior on the scale parameter τ.

7.2 Global-Local Shrinkage Priors

Motivated by modern applications such as in genomics, a number of sparsity favoring shrinkage priors have appeared in the literature over the past decade and a half. The ability to incorporate approximate sparsity is a salient feature of these second generation priors which differentiates them from isotropic Gaussian priors inducing Stein shrinkage. These continuous shrinkage priors are almost exclusively expressed as *global-local* scale mixtures of Gaussians [47]

$$\mu_j \overset{ind.}{\sim} \mathcal{N}(0, \lambda_j^2\tau^2), \quad \lambda_j \overset{iid}{\sim} \upsilon_L, \quad \tau \sim \upsilon_G,$$

where υ_L and υ_G are densities on $(0,\infty)$ absolutely continuous with respect to the Lebesgue measure. In the setup of (7.1), one now has

$$E[\mu_j \mid \mathbf{y}, \lambda_j, \tau] = (1-\omega_j)y_j, \quad \omega_j = \frac{1}{(1+\lambda_j^2\tau^2)},$$

implying the conditional posterior mean of μ_j is shrunk towards zero by a shrinkage factor ω_j specific to the jth coordinate which is controlled by the parameters λ_j and τ. Indeed, τ controls global shrinkage towards the origin while the local parameters λ_js allow coefficient specific deviations in the degree of shrinkage. For appropriate choices of υ_L and υ_G, this prior structure therefore possesses the flexibility to induce approximate sparsity in μ by shrinking a subset of its entries aggressively towards zero while retaining the larger ones intact. In this sense, the global-local shrinkage priors provide an approximation to the operating characteristics of discrete mixture priors [19, 31] that allow a subset of the parameters to be exactly zero, as well as spike-and-slab priors [6, 19, 49] which replace the point mass with a symmetric unimodal density. Due to this, these global-local priors are also referred to as one-group priors, mimicking the commonly used two-group prior moniker for discrete mixture priors or spike-and-slab priors.

Various choices of υ_L and υ_G have been proposed in the literature; examples include the relevance vector machine [58], normal/Jeffrey's prior [2], the Bayesian lasso [24, 39], the horseshoe [49], normal/gamma and normal/inverse-Gaussian priors [13, 15], generalized double Pareto priors [1]. More recent proposals such as the Dirichlet–Laplace prior of [6], the R2-D2 prior [64], and the horseshoe+ prior of [5] have introduced dependence between the local parameters and/or used a deeper hierarchy. Various favorable properties of a sub-class of these priors have been unveiled [6, 15, 40, 48, 49, 55, 60, 62] and it is now generally understood that for favorable performance in sparse settings, it is desirable to have heavier than exponential tails for υ_L and sufficient mass near zero for υ_G.

A major motivation for global-local shrinkage priors compared to point-mass mixtures is ostensibly lower computational complexity. However, the typical argument that the continuous shrinkage priors lead to block updating of parameters and avoid navigating the landscape of all possible sub-models only reveal part of the story. For one, not enforcing a hard sparsity constraint means all the regression coefficients and corresponding local scale parameters are retained through the entirety of the sampling, leading to expensive matrix computations unless care is exercised. Second, it is also common to observe slow mixing and high autocorrelations, with a major contributing factor being the global scale parameter τ [24, 45]. In recent years, there has been a concerted push towards understanding and alleviating these issues to develop improved MCMC algorithms as well as theoretically characterizing their convergence properties. Full posterior sampling is desirable to characterize uncertainty in estimation as well as model selection using post-processing schemes [6, 7, 29, 64]. In this chapter, we recount some of these key aspects in the development of Markov chain Monte Carlo algorithms to sample from the posterior distribution in high-dimensional regression models with global-local shrinkage priors. This continues to remain a largely active area with many open questions analyzing the convergence rates of the algorithms involved, and we point to some of these challenges in due place.

7.3 Posterior Sampling

We now turn to the Gaussian regression model with likelihood

$$f(\mathbf{y} \mid \mathbf{X}\boldsymbol{\beta}, \sigma^2) = (2\pi\sigma^2)^{-n/2} e^{-\frac{1}{2\sigma^2}(\mathbf{y}-\mathbf{X}\boldsymbol{\beta})'(\mathbf{y}-\mathbf{X}\boldsymbol{\beta})}, \tag{7.2}$$

where $\mathbf{X}$ is a $n \times p$ matrix of covariates, $\boldsymbol{\beta} \in \mathbb{R}^p$ is the vector of regression coefficients, and $\mathbf{y} \in \mathbb{R}^n$ is the response vector of length n. We begin with the general structure of a global-local prior [47] for $\boldsymbol{\beta}$,

$$\begin{aligned} \beta_j \mid \sigma^2, \boldsymbol{\lambda}, \tau &\overset{iid}{\sim} \mathcal{N}(0, \sigma^2\tau^2\lambda_j^2), \quad \lambda_j \overset{iid}{\sim} \upsilon_L, \quad j = 1, \ldots, p, \\ \tau &\sim \upsilon_G, \quad \sigma^2 \sim \text{InvGamma}(\omega/2, \omega/2). \end{aligned} \tag{7.3}$$

As before, υ_L and υ_G are densities on $(0, \infty)$ which designate the prior distributions on the local and global scale parameters, respectively. Including the residual variance σ^2 within the prior variance of $\boldsymbol{\beta}$ is standard practice as it lends adaptivity against varying signal-to-noise ratio. While we have parameterized the prior on $\boldsymbol{\beta}$ in terms of the local and global scale parameters in (7.3), we shall switch to a precision parameterization while discussing associated Gibbs samplers. To that end, define $\xi = \tau^{-2}$ and $\boldsymbol{\eta} = (\eta_1, \ldots, \eta_p)$ with $\eta_j = \lambda_j^{-2}$ for $j = 1, \ldots, p$ to be the global and local precision parameters respectively.

The primary interest here is in the joint posterior of $(\boldsymbol{\beta}, \sigma^2)$ which is analytically intractable in general. Among Markov chain Monte Carlo (MCMC) algorithms to sample from the joint posterior, a major focus has been on blocked Gibbs samplers which can exploit the conditional conjugacy. For example, [45] (henceforth, PSW) used a blocked Gibbs sampler to sample from the joint posterior of $(\boldsymbol{\beta}, \sigma^2, \boldsymbol{\eta}, \xi)$ under the horseshoe prior with steps given by

$$\begin{aligned} &\boldsymbol{\beta}, \sigma^2 \mid \xi, \boldsymbol{\eta} \\ &\xi \mid \boldsymbol{\eta}, \boldsymbol{\beta}, \sigma^2 \\ &\boldsymbol{\eta} \mid \xi, \sigma^2, \boldsymbol{\beta}. \end{aligned} \tag{7.4}$$

The full-conditional of $(\boldsymbol{\beta}, \sigma^2)$ has a multivariate normal-inverse gamma distribution, which can be sampled by first marginalizing over $\boldsymbol{\beta}$ to draw $\sigma^2 \mid \xi, \boldsymbol{\eta}$ from an inverse-gamma distribution, and then sampling $\boldsymbol{\beta} \mid \sigma^2, \xi, \boldsymbol{\eta}$ from its multivariate normal full-conditional distribution. The sampling steps for ξ and $\boldsymbol{\eta}$ are prior-specific – [45] recommended slice samplers for each of them under the horseshoe prior. Blocked Gibbs samplers of a similar flavor can also be found in [1, 6, 15, 24, 39, 49].

7.3.1 Sampling Structured High-Dimensional Gaussians

The full-conditional distribution of $\boldsymbol{\beta}$ in (7.4) is

$$\mathcal{N}_p\left((\mathbf{X}'\mathbf{X} + \boldsymbol{\Gamma}^{-1})^{-1}\mathbf{X}'\mathbf{y}, \sigma^2(\mathbf{X}'\mathbf{X} + \boldsymbol{\Gamma}^{-1})^{-1}\right), \tag{7.5}$$

where $\boldsymbol{\Gamma} = \text{diag}(\xi^{-1}\eta_j^{-1})$; in the variance parameterization $\boldsymbol{\Gamma} = \text{diag}(\tau^2\lambda_j^2)$. This assumes the familiar $\mathcal{N}(\mathbf{Q}^{-1}\mathbf{b}, \mathbf{Q}^{-1})$ parameterization of a multivariate Gaussian appearing routinely in Bayesian computation, with the precision matrix $\mathbf{Q} = (\mathbf{X}'\mathbf{X} + \boldsymbol{\Gamma}^{-1})/\sigma^2$, and $\mathbf{b} = \mathbf{X}'\mathbf{y}/\sigma^2$. A standard procedure to sample from a $\mathcal{N}_d(\mathbf{Q}^{-1}\mathbf{b}, \mathbf{Q}^{-1})$ due to [53] proceeds as

- Perform a Cholesky decomposition $\mathbf{Q} = \mathbf{L}\mathbf{L}'$, where $\mathbf{L}$ is lower triangular.
- Draw $\mathbf{z} \sim \mathcal{N}(0, I_d)$, solve $\mathbf{L}'\mathbf{u} = \mathbf{z}$ for $\mathbf{u}$.
- Solve $\mathbf{L}'\boldsymbol{\mu} = \mathbf{v}$ for $\boldsymbol{\mu}$, where $\mathbf{L}\mathbf{v} = \mathbf{b}$.
- Set $\boldsymbol{\beta} = \mathbf{u} + \boldsymbol{\mu}$.

Then, $\boldsymbol{\beta}$ produced as above has the desired $\mathcal{N}(\mathbf{Q}^{-1}\mathbf{b}, \mathbf{Q}^{-1})$ distribution. Observe that $\boldsymbol{\mu} = \mathbf{Q}^{-1}\mathbf{b}$ computes the mean, and $u \sim \mathcal{N}(0, \mathbf{Q}^{-1})$. A salient feature of this algorithm is that it avoids explicitly computing the inverse of $\mathbf{Q}$, and only requires a Cholesky factorization and a series of linear system solutions, both of which are more stable operations compared to a matrix inversion. This algorithm can be readily embedded inside any larger blocked Gibbs sampler to sample from the full-conditional of $\boldsymbol{\beta}$ in moderate dimensions. However, there is an unavoidable breakdown point as dimension increases due to the Cholesky factorization step being prohibitively expensive. This algorithm originated in applications involving Markov random fields where the precision matrix $\mathbf{Q}$ is typically sparse, and hence the Cholesky factor can be efficiently computed even if the ambient dimension is large. However, barring structured design matrices $\mathbf{X}$, the precision matrix $\mathbf{Q} = (\mathbf{X}'\mathbf{X} + \boldsymbol{\Gamma}^{-1})/\sigma^2$ is typically a dense matrix in the present scenario.

For a general $p \times p$ matrix, the Cholesky factorization has $O(p^3)$ computational complexity, which is the overall computational complexity of the above algorithm. In [6], an alternative algorithm was developed to sample from (7.5) whose steps are as follows:

- Sample $u \sim \mathcal{N}(0, \boldsymbol{\Gamma})$ and $\mathbf{f} \sim \mathcal{N}(0, I_n)$ independently.
- Set $\mathbf{v} = \mathbf{X}\mathbf{u} + \mathbf{f}$, and $\mathbf{v}^\star = \mathbf{M}^{-1}(\mathbf{y}/\sigma - \mathbf{v})$, where $\mathbf{M} = I_n + \mathbf{X}\boldsymbol{\Gamma}\mathbf{X}'$.
- Set $\boldsymbol{\beta} = \sigma(\mathbf{u} + \boldsymbol{\Gamma}X'\mathbf{v}^\star)$.

This algorithm has a data-augmentation flavor – to sample the complicated target (7.5) in $\mathbb{R}^p$, one begins by sampling $(\mathbf{u}, \mathbf{f})$ in $\mathbb{R}^{p+n}$ with a diagonal covariance matrix, and performs a set of linear transformations to arrive at $\boldsymbol{\beta}$. Indeed, we can expand out the steps of the algorithm and equivalently write

$$\boldsymbol{\beta} = \boldsymbol{\Gamma}\mathbf{X}'\mathbf{M}^{-1}\mathbf{y} + \sigma(\mathbf{u} - \boldsymbol{\Gamma}\mathbf{X}'\mathbf{M}^{-1}\mathbf{v}).$$

An application of the Woodbury matrix inversion formula shows that $\boldsymbol{\Gamma}\mathbf{X}'\mathbf{M}^{-1}\mathbf{y} = (\mathbf{X}'\mathbf{X} + \boldsymbol{\Gamma}^{-1})^{-1}\mathbf{X}'\mathbf{y}$, with the right hand side the mean of the Gaussian distribution in (7.5). The quantity $(\mathbf{u} - \boldsymbol{\Gamma}\mathbf{X}'\mathbf{M}^{-1}\mathbf{v})$ can be shown to have a $\mathcal{N}\big(0, (\mathbf{X}'\mathbf{X} + \boldsymbol{\Gamma}^{-1})^{-1}\big)$ distribution – that it has a mean-zero Gaussian distribution is immediately evident from the fact that $(\mathbf{u}, \mathbf{v})$ jointly have a zero-mean Gaussian distribution. Thus, one only needs to work out the expression for its covariance; the Woodbury inversion formula comes in handy again. Together, these observations imply that $\boldsymbol{\beta}$ obtained from the algorithm has the correct target distribution.

When $p > n$, the most expensive step turns out to be the matrix product $\mathbf{X}\boldsymbol{\Gamma}\mathbf{X}'$ required to form the matrix M. This matrix product takes $O(n^2 p)$ computational steps which dominates the $O(n^3)$ complexity of solving the $n \times n$ linear system to obtain $\mathbf{v}^\star$. Accordingly, the overall algorithm has complexity $O(n^2 p)$, which produces substantial speed-ups versus the $O(p^3)$ Cholesky-based implementation when $p \gg n$. Figure 7.1 shows a time-comparison between the two algorithms embedded inside the blocked Gibbs sampler of [45] by fixing $n = 100$ and varying p between 200 and 5000. Evidently, the larger p/n is, the bigger the gains from the algorithm of [6].

When p and n are both large but $p < n$, the sampling of $\boldsymbol{\beta}$ can be performed by solving a $p \times p$ system instead of an $n \times n$ system using a more recent algorithm due to [37]. The key observation underlying their algorithm is that if one samples $\mathbf{b} \sim \mathcal{N}\big(\mathbf{X}'\mathbf{y}, (\mathbf{X}'\mathbf{X} + \boldsymbol{\Gamma}^{-1})\big)$ and then solves the system $(\mathbf{X}'\mathbf{X} + \boldsymbol{\Gamma}^{-1})\boldsymbol{\beta} = \mathbf{b}$, then $\boldsymbol{\beta}$ has the desired distribution in (7.5). Accordingly, their algorithm assumes the form

- Independently sample $\mathbf{u} \sim \mathcal{N}(0, I_p)$ and $\mathbf{f} \sim \mathcal{N}(0, I_n)$, and set

$$\mathbf{b} = \mathbf{X}'\mathbf{y} + \mathbf{X}'\mathbf{f} + \boldsymbol{\Gamma}^{-1/2}\mathbf{u}.$$

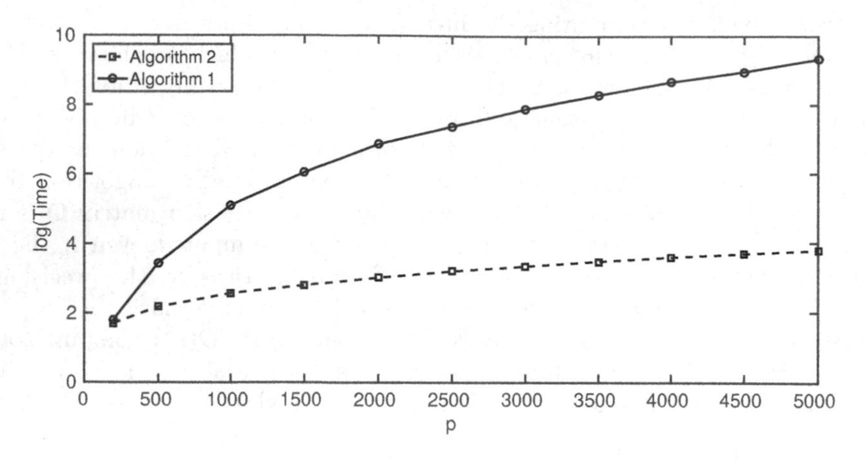

FIGURE 7.1
Algorithm 1 and 2, respectively denoting the stated algorithms from [53] and [6], were applied to sample from the conditional posterior of $\boldsymbol{\beta}$ in (7.5) within the blocked Gibbs sampling algorithm of [45] for the horseshoe prior on the regression coefficients in Gaussian linear regression. Logarithm of time (in seconds) to complete 6000 iterations of the Gibbs sampler is reported. The sample size n was fixed at 100 and the dimension p was varied from 500 to 5000, with 500 step size. To offer a comparison in the case $p \asymp n$, the case $p = 200$ is also included.

- Solve the $n \times n$ linear system

$$(\mathbf{X}'\mathbf{X} + \boldsymbol{\Gamma}^{-1})\boldsymbol{\beta} = \mathbf{b}.$$

Clearly, $\mathbf{b}$ obtained in the first step is Gaussian, with $E(\mathbf{b}) = \mathbf{X}'\mathbf{y}$ and $\text{cov}(\mathbf{b}) = \mathbf{X}'\mathbf{X} + \boldsymbol{\Gamma}^{-1}$. [37] further recommend solving the linear system using pre-conditioned conjugate gradient methods and analyze its convergence properties in the present setup.

The recent article of [22] takes an entirely different approach based on elliptical slice sampling [35]. The elliptical slice sampling is a general purpose algorithm to sample from distributions of the form

$$p(\boldsymbol{\beta}) \propto \mathcal{L}(\boldsymbol{\beta})\mathcal{N}(\boldsymbol{\beta}; 0, \boldsymbol{\Sigma}). \tag{7.6}$$

A canonical example, of course, is the case where $\mathcal{L}(\cdot)$ denotes a likelihood function and a zero-mean multivariate Gaussian prior is placed on the parameter $\boldsymbol{\beta}$. In this context, Metropolis–Hastings proposals

$$\widetilde{\boldsymbol{\beta}} = \rho\mathbf{z} + \sqrt{1-\rho^2}\,\boldsymbol{\beta}, \quad \mathbf{z} \sim \mathcal{N}(0, \boldsymbol{\Sigma}),$$

for $\rho \in [-1, 1]$ are known to possesss favorable empirical and theoretical properties. The parameter ρ is a step-size parameter here. Observe that unlike Metropolis random walk proposals with $\rho = 1$, the above AR(1) type proposal leaves the Gaussian prior invariant, i.e., if $\boldsymbol{\beta} \sim \mathcal{N}(0, \boldsymbol{\Sigma})$ and $\widetilde{\boldsymbol{\beta}} \mid \boldsymbol{\beta}$ is drawn as above, then marginally $\widetilde{\boldsymbol{\beta}} \sim \mathcal{N}(0, \boldsymbol{\Sigma})$. The elliptical slice sampler provides an automated and adaptive way to tune the step-size parameter. Specifically, a new location is generated on the random ellipse determined by the current state $\boldsymbol{\beta}$ and the auxiliary Gaussian draw $\mathbf{z}$ as

$$\widetilde{\boldsymbol{\beta}} = \mathbf{z}\sin\theta + \boldsymbol{\beta}\cos\theta$$

where the angle θ is uniformly generated from a $[\theta_{\min}, \theta_{\max}]$ interval which is shrunk exponentially fast until an acceptable state is reached.

In the present context, the full-conditional of $\boldsymbol{\beta}$ does not immediately admit a form as in (7.6) since the prior involved is non-Gaussian. The key idea of [22] is artificially injecting a Gaussian distribution to switch the roles of the prior and the likelihood. Specifically, with a general prior p on $\boldsymbol{\beta}$, they write

$$
\begin{aligned}
p(\boldsymbol{\beta} \mid -) &\propto \mathcal{N}(\mathbf{y}; X\boldsymbol{\beta}, \sigma^2)\, p(\boldsymbol{\beta}) \\
&\propto \mathcal{N}(\mathbf{y}; X\boldsymbol{\beta}, \sigma^2)\, \mathcal{N}(\boldsymbol{\beta}; 0, c^{-1}\sigma^2 I_p)\, \frac{p(\boldsymbol{\beta})}{\mathcal{N}(\boldsymbol{\beta}; 0, c^{-1}\sigma^2 I_p)} \\
&\propto \mathcal{N}\big(\boldsymbol{\beta}; \widehat{\boldsymbol{\beta}}_c, \sigma^2(\mathbf{X}'\mathbf{X} + cI_p)^{-1}\big)\, \frac{p(\boldsymbol{\beta})}{\mathcal{N}(\boldsymbol{\beta}; 0, c^{-1}\sigma^2 I_p)}.
\end{aligned}
$$

Here, $c > 0$ and $\widehat{\boldsymbol{\beta}}_c = (\mathbf{X}'\mathbf{X} + cI_p)^{-1}\mathbf{X}'\mathbf{y}$ is the ridge estimate. One can now set $L(\boldsymbol{\beta}) := p(\boldsymbol{\beta})/\mathcal{N}(\boldsymbol{\beta}; 0, c^{-1}\sigma^2 I_p)$ and proceed as usual. Observe that $\mathcal{N}\big(\boldsymbol{\beta}; \widehat{\boldsymbol{\beta}}_c, \sigma^2(\mathbf{X}'\mathbf{X} + cI_p)^{-1}\big)$ is a non-singular Gaussian distribution even if $\mathbf{X}'\mathbf{X}$ is a singular matrix (e.g., when $p > n$) since the ridge term makes it non-singular. This approach is very general as it requires minimal assumptions on p. In particular, one can integrate over the local parameters and work with the marginal prior on $\boldsymbol{\beta}$. Such an analytic marginalization can be carried out for most of the priors listed earlier, and circumvents the need to sample the local parameters.

7.3.2 Blocking can be Advantageous

The overall computational complexity of an MCMC algorithm is determined by the interplay between two main ingredients: (i) the per-iteration cost, i.e., the cost required to extend the Markov chain by one step, and (ii) the speed at which the Markov chain converges to its stationary distribution. In particular, even if the matrix computations required to extend the chain by one step are optimized, the design of the algorithm can still play a major role in dictating the overall efficiency. [45, Supplement] noted that the global precision parameter ξ tends to mix poorly for the horseshoe under (7.4). They considered both slice sampling and parameter expansion for sampling ξ, and found that parameter expansion only offered modest performance gains. A similar slow mixing of the global scale parameter has also been observed for the Bayesian lasso by [24].

Under the horseshoe prior, [27] suggested an update rule with more extensive blocking:

$$
\begin{aligned}
&\xi, \sigma^2, \boldsymbol{\beta} \mid \boldsymbol{\eta}, \\
&\boldsymbol{\eta} \mid \xi, \sigma^2, \boldsymbol{\beta}.
\end{aligned}
\tag{7.7}
$$

The joint sampling of $\xi, \sigma^2, \boldsymbol{\beta} \mid \boldsymbol{\eta}$ in the first step is carried out by sequentially sampling,

$$
[\xi \mid \boldsymbol{\eta}], \quad [\sigma^2 \mid \boldsymbol{\eta}, \xi], \quad [\boldsymbol{\beta} \mid \boldsymbol{\eta}, \xi, \sigma^2]. \tag{7.8}
$$

Analytic marginalizations are required to obtain $[\xi \mid \boldsymbol{\eta}]$ and $[\sigma^2 \mid \xi, \boldsymbol{\eta}]$ in closed form, which can be carried out in a tractable fashion in the Gaussian regression model. We shall abbreviate the above update rule by JOB subsequently. Code implementing the different blocking rules can be found in the online Supplementary Material of this edited book.

Figure 7.2 shows traceplots for $\log(\xi)$ from PSW and JOB obtained from a simulation with problem size $n = 500, p = 10,000$. Strikingly, the global parameter shows a diverging behavior under PSW, which disappears under the JOB algorithm. In addition to blocking, the algorithm of JOB carefully handles numerics with very small values of the local scales,

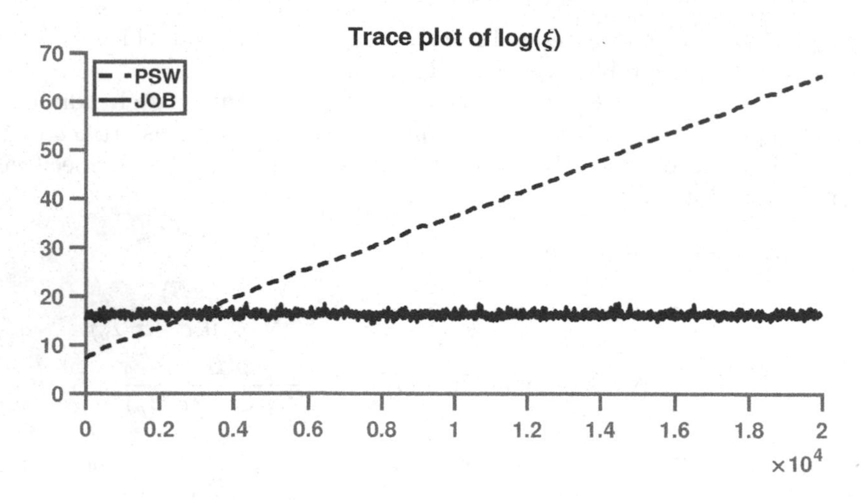

FIGURE 7.2
Trace plots for $\log(\xi)$ for PSW and JOB.

whereas previous implementations relied on various checks to control over- or under-flow of the local and global scales.

In addition to improved mixing, [27] exhibited that the statistical performance of their blocked sampler is also superior in terms of a number of different metrics, such as $\|\boldsymbol{\beta} - \boldsymbol{\beta}_{true}\|_2^2$ and $\|\mathbf{X}\boldsymbol{\beta} - \mathbf{X}\boldsymbol{\beta}_{true}\|_2^2$. In addition, a significant improvement is observed in the detection and recovery of "intermediate-sized" signals which lie near the boundary of detectable signals. The marginal posterior of such intermediate signals under the horseshoe is bimodal, with a mode at the origin and a second one away from zero. While this fact does not seem to have received a lot of attention – see, however, a brief comment at [15, pg. 114] – this is indeed desirable from an inferential perspective as it clearly represents the posterior uncertainty. Moreover, this is also a clear demonstration of how well the horseshoe (and other related shrinkage priors) approximate the posterior distribution under point mass priors with a non-zero mass at the origin. The top panel of Figure 7.3 shows trace plots from a path of length 20,000 for an entry of $\boldsymbol{\beta}$ whose true value of 0.5 is a quarter of the true residual standard deviation of $\sigma = 2$, while the bottom row shows a probability histogram of the marginal posterior. The left and right columns respectively correspond to PSW and JOB. It is apparent from the histogram that the old algorithm essentially fails to detect the second mode away from the origin, and places most of its mass near zero. Interestingly, an inspection of the trace plot reveals that after the first several iterations, the chain remains trapped near the origin and fails to escape it for the remainder of its length. The algorithm of JOB, on the other hand, consistently crosses the valley between the two modes.

While proposed for the horseshoe, the algorithm of JOB can be readily generalized to other shrinkage priors, at least in principle. The steps $[\sigma^2 \mid \boldsymbol{\eta}, \xi]$ and $[\boldsymbol{\beta} \mid \boldsymbol{\eta}, \xi, \sigma^2]$ remain exactly the same. The scalar parameter ξ was updated using a Gaussian random walk on $\log(\xi)$ with Metropolis correction in [27], and a similar strategy can be applied for other one-group shrinkage priors as well. Finally, the update of $\boldsymbol{\eta} \mid -$ can be generally written as

$$p(\boldsymbol{\eta} \mid \xi, \sigma^2, \boldsymbol{\beta}) \propto \prod_{j=1}^{p} p(\eta_j \mid \xi, \sigma^2, \beta_j) \propto \prod_{j=1}^{p} \eta_j^{1/2} e^{-m_j \eta_j} p_L(\eta_j),$$

where $m_j = \xi \beta_j^2 / 2$. For example, for the horseshoe, $p_L(\eta_j) \propto \eta_j^{-1/2}/(1+\eta_j)$ for $\eta_j > 0$, and the quantity inside the product reduces to $e^{-m_j \eta_j}/(1+\eta_j)$. PSW recommended sampling

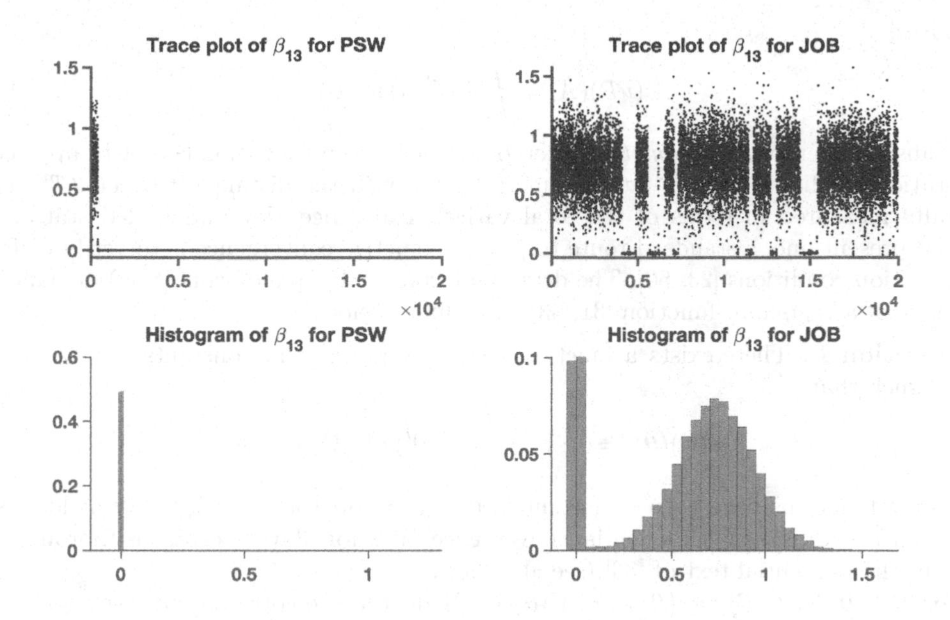

FIGURE 7.3
Trace plots and density estimates for one entry of $\boldsymbol{\beta}$.

from the density $e^{-mx}/(1+x)\,\mathbb{1}_{(0,\infty)}(x)$ using slice sampling while JOB used rejection sampling. While the local parameters are updated independently from a common univariate family of densities, one still needs to exercise care in cases such as the horseshoe, when the density family is a non-standard one and a vectorized implementation is not readily available in standard software. Since $m_j \propto \beta_j^2$, it takes a wide range of values across the variable index j, ranging from very small (for the null variables) to moderately large (for the non-null variables), and this can significantly impact the shape of the density proportional to $\eta_j^{1/2} e^{-m_j \eta_j} \pi_L(\eta_j)$. For example, for the horseshoe, the density $e^{-mx}/(1+x)\,\mathbb{1}_{(0,\infty)}(x)$ has effectively thick tails for all practical purpose for small m, while it closely resembles the exponential density with rate parameter m for large m. This is important for designing a rejection sampler – using an exponential proposal density, for example, leads to very poor acceptance rates in the horseshoe case when m_j is small. Importantly, one may easily need to sample 10^{10} times from this density family over the course of the entire sampler for large p, and even minor inefficiencies can massively add up. JOB used a careful piecewise approximation to the log-density function $-(mx + \log(1+x))$ and it was proved in [7] that the acceptance rate of this rejection sampler has a uniform lower bound as m ranges over the positive half-line. Numerically, this lower bound was estimated to be around 0.88.

7.3.3 Geometric Convergence

Let $\mathcal{P}$ denote the Markov transition kernel corresponding to a generic blocked Gibbs update rule for the state parameter $\boldsymbol{\theta} = (\boldsymbol{\beta}, \sigma^2, \boldsymbol{\eta}, \xi)$ living in the state space Θ. By design, $\mathcal{P}$ has stationary distribution $\nu^\star$, the posterior distribution of the state parameters. This means $\nu^\star \mathcal{P} = \nu^\star$, where we use standard convention (see, e.g., [23]) to denote the action of $\mathcal{P}$ on measures,

$$(\nu\mathcal{P})(A) = \int_{\Theta} \mathcal{P}(\boldsymbol{\theta}, A)\, \nu(d\boldsymbol{\theta}).$$

If a transition kernel $\mathcal{P}$ is *geometrically* ergodic, the m-step transition kernel $\mathcal{P}$' approaches the stationary distribution $\nu^\star$ at a geometric rate, with the distance between $\mathcal{P}^m$ and $\nu^\star$ typically measured in terms of the total variation distance. For unbounded state-spaces like the present one, a standard route to prove geometric convergence is to verify drift and minorization conditions [23, 50]. The drift condition (or Lyapunov condition) demands the existence of a *Lyapunov* function [31, 33, 50] defined below.

Assumption L. There exists a function $V : \Theta \to [0, \infty)$ and constants $0 < \gamma < 1$ and $K > 0$ such that

$$(\mathcal{P}V)(\boldsymbol{\theta}) := \int V(y)\mathcal{P}(\boldsymbol{\theta}, d\boldsymbol{\theta}') \leq \gamma V(\boldsymbol{\theta}) + K.$$

Lyapunov functions have been important in the study of stochastic stability at least since [31]; their role in proving exponential convergence rates for Markov chains is thoroughly set down in the influential text of [33] (see also [50]).

For $R > 0$, let $\mathcal{C}(R) := \{\boldsymbol{\theta} \in \Theta : V(\boldsymbol{\theta}) < R\}$ denote the corresponding sub-level set of the Lyapunov function. The minorization condition requires that

Assumption M. For every $R > 2K/(1-\gamma)$ there exists $\alpha \in (0, 1)$ (depending on R) such that

$$\sup_{\boldsymbol{\theta},\boldsymbol{\theta}' \in \mathcal{C}(R)} \|\delta_{\boldsymbol{\theta}}\mathcal{P} - \delta_{\boldsymbol{\theta}'}\mathcal{P}\|_{\text{TV}} \leq (1-\alpha), \tag{7.9}$$

where for probability distributions P and Q, $\|P - Q\|_{\text{TV}} := \sup_B |P(B) - Q(B)|$ denotes the total variation distance between P and Q.

Under Assumptions **L** and **M**, it can be shown that there exists $\bar{\alpha} \in (0, 1)$ and a positive function $C : \Theta \to \mathbb{R}_+$ such that for any $\boldsymbol{\theta} \in \Theta$,

$$\|\delta_{\boldsymbol{\theta}}\mathcal{P}^m - \nu^\star\|_{\text{TV}} \leq C(\boldsymbol{\theta})\, \bar{\alpha}^m, \tag{7.10}$$

implying the total variation distance between the m-step transition kernel $\delta_{\boldsymbol{\theta}}\mathcal{P}^m$ (with the chain initialized at $\boldsymbol{\theta}$) and the target $\nu^\star$ decreases geometrically in m. For a more precise statement, refer to [23, 50]. In addition to quantifying the speed of convergence, geometric ergodicity also implies a Markov chain central limit theorem [29], which is crucial for estimating uncertainty in MCMC-based approximations to posterior expectations.

While there is a substantial literature on establishing geometric ergodicity for Gibbs samplers in general, such results for global-local shrinkage priors have only began to appear more recently. [30] established the geometric ergodicity of the Bayes lasso Gibbs sampler of [39]; see also [47] for a related Gibbs sampler for the Bayes lasso where σ^2 is sampled after marginalizing over $\boldsymbol{\beta}$. [38] proved geometric ergodicity of the normal-gamma prior of [15] and the Dirichlet–Laplace prior of [6]. To the best of our knowledge, this was the first result for global-local shrinkage priors with heavier than exponential tails and an infinite spike at zero. [27] proved geometric ergodicity of their blocked Gibbs sampler for the horseshoe for $p \leq n$, and the same result for $p > n$ with an additional truncation of the prior density on the local precision parameters. Among other related results, see [51] for the Bayesian elastic net and [61] for the Bayesian fused lasso and group lasso.

[27] used a Lyapunov function given by

$$V(\boldsymbol{\eta}, \boldsymbol{\beta}, \sigma^2, \xi) = \frac{\|\mathbf{X}\boldsymbol{\beta}\|^2}{\sigma^2} + \xi^2 + \sum_{j=1}^{p} \left[\frac{\sigma^{2c}}{|\beta_j|^{2c}} + \frac{\eta_j^c |\beta_j|^c}{\sigma^c} + \eta_j^c \right]. \tag{7.11}$$

Interestingly, as a function of β_j^2/σ^2, V both grows at infinity and has a pole at zero, whereas most commonly encountered Lyapunov functions simply grow at infinity and are bounded on compact sets containing the origin. [38] previously had noted the importance of including negative fractional powers of β_j inside the Lyapunov function. This is almost exclusively a feature of the global-local shrinkage priors – because these priors have a pole at the origin, it becomes necessary to have all the β_j^2/σ^2 bounded away from both zero and infinity within the sub-level set to be able to establish the minorization condition.

7.4 Approximate MCMC

Approximations in Markov chain Monte Carlo methods are increasingly gaining popularity in Bayesian big data problems, where an exact but computationally expensive Markov transition kernel $\mathcal{P}$ is replaced with a more computationally amenable approximation/perturbation $\mathcal{P}_\epsilon$. There has been rapid algorithmic and theoretical development in this area in the recent past [4, 28, 32, 41, 52, 63]. Sub-sampling or mini-batching has been a prominent idea in such applications, where only a small subset of the data is used at each step to extend the Markov chain. Subsampling is best motivated when the number of samples is enormous (compared to the parameter dimension), so that a reasonable approximation to the exact transition kernel can be made using only a small portion of the data.

The situation, however, is markedly different when the parameter dimension p is comparable to or greater than the sample size n. Compared to the large n small p setup, the estimation problem is now statistically more challenging, computational considerations aside. In this scenario, using only a portion of the data typically causes large perturbations of the transition kernel and consequently results in a poor approximation to the posterior. Moreover, accurate approximations using minibatches typically require the construction of control variates [3, 4, 42], which can add significantly to the overall computational cost when the target is high-dimensional and its relevant directions are unknown *a priori*. Thus, fundamentally different approximation strategies are required in the high-dimensional setting, where the sparsity structure of the posterior must be exploited to arrive at meaningful approximations.

[36] developed a skinny Gibbs sampler for high-dimensional logistic regression with spike-and-slab priors where they partition the coefficient vector into active and inactive components, $\boldsymbol{\beta} = (\boldsymbol{\beta}_A; \boldsymbol{\beta}_I)$, depending on assignment to the slab or spike component, respectively. Then, the full-conditional $[\boldsymbol{\beta} \mid -]$ is approximated as $[\boldsymbol{\beta}_A \mid -] \otimes [\boldsymbol{\beta}_I \mid -]$, ignoring the dependence between the active and inactive coordinates. Since in a sparse setting one typically expects $|I| \gg |A|$, a further reduction is achieved by sampling the components of $\boldsymbol{\beta}_I$ independently of each other. Despite these multiple approximations, [36] showed that the resulting sampler correctly identifies the active set of variables asymptotically, under appropriate assumptions.

Under spike-and-slab priors, the latent component indicator naturally partitions the predictors into active and inactive group. However, for one-group shrinkage priors, such a decomposition is not naturally available. [27] proposed an approximation scheme (henceforth, JOB-approx) suited for one-group shrinkage priors, where costly matrix multiplications were replaced with a cheaper approximation exploiting the specific structure of one-group priors. The JOB-approx scheme was specifically designed for their blocked Gibbs sampler for the horseshoe, where the costliest step is the matrix multiplication $\mathbf{X}\boldsymbol{\Gamma}\mathbf{X}'$, which appears inside the updates of $\boldsymbol{\beta}$, σ^2, as well as the global parameter ξ. Recall that $\boldsymbol{\Gamma} = \text{diag}(\tau^2\lambda_j^2)$ is a diagonal matrix consisting of the product of the global and local variance parameters which gets updated at every step of the MCMC algorithm, and hence the matrix product $\mathbf{X\Gamma X'}$ needs to be computed at each step and cannot be pre-stored outside the MCMC loop. This matrix product has $O(n^2p)$ complexity which is substantial when (n, p) are both large.

To alleviate this computational burden at each step, [27] proposed approximating $\mathbf{X\Gamma X'}$ with $\mathbf{X\Gamma_\delta X'}$, where $\mathbf{\Gamma}_\delta$ is a *thresholded* version of $\mathbf{\Gamma}$,

$$\mathbf{\Gamma}_\delta = \text{diag}\Big((\xi\eta_j)^{-1}\mathbb{1}\big((\xi\eta_j)^{-1} > \delta\big)\Big) = \text{diag}\Big((\tau^2\lambda_j^2)\mathbb{1}(\tau^2\lambda_j^2 > \delta)\Big) \tag{7.12}$$

for a user-defined thresholding parameter $\delta > 0$. The motivation behind such a thresholding operation is as follows. The matrix product $\mathbf{X\Gamma X'}$ can be written as

$$\mathbf{X\Gamma X'} = \mathbf{X}_S\mathbf{\Gamma}_S\mathbf{X}_S' + \mathbf{X}_{S^c}\mathbf{\Gamma}_{S^c}\mathbf{X}_{S^c}',$$

where $S = \{j \in [p] : (\xi\eta_j)^{-1} > \delta\}$ and $\mathbf{X}_S$ denotes the $n \times |S|$ sub-matrix of $\mathbf{X}$ collecting the columns of $\mathbf{X}$ corresponding to the indices in S. Now, a global-local shrinkage prior with adaptive shrinkage properties like the horseshoe relies on aggressively shrinking the null coordinates of $\boldsymbol{\beta}$ towards the origin, while leaving the non-nulls unshrunk to various degrees depending on the signal strength. For a null β_j to be aggressively shrunk towards zero, the corresponding precisions $\xi\eta_j$s must be very large *a posteriori*, or equivalently, the inverse precision $(\xi\eta_j)^{-1} = \tau^2\lambda_j^2$ very small. Since $\mathbf{\Gamma}_{S^c}$ contains all these small inverse precision parameters, we can approximate $\mathbf{X\Gamma X'} \approx \mathbf{X}_S\mathbf{\Gamma}_S\mathbf{X}_S'$.

An important distinction between the JOB-approx algorithm and the skinny Gibbs of [36] is that the former preserves correlations between the variables in S and S^c in the approximation to the conditional covariance of the $\boldsymbol{\beta}$'s, whereas the latter zeros out these correlations. In fact, the conditional covariance matrix of the $\boldsymbol{\beta}$'s in the JOB-approx algorithm can be written as

$$\Sigma_\delta = \begin{bmatrix} (\mathbf{X}_S'\mathbf{X}_S + \mathbf{\Gamma}_S^{-1})^{-1} & -\mathbf{\Gamma}_S\mathbf{X}_S'\mathbf{M}_S^{-1}\mathbf{X}_{S^c}\mathbf{\Gamma}_{S^c} \\ -(\mathbf{\Gamma}_S\mathbf{X}_S'\mathbf{M}_S^{-1}\mathbf{X}_{S^c}\mathbf{\Gamma}_{S^c})' & \mathbf{\Gamma}_{S^c} \end{bmatrix}, \tag{7.13}$$

where $\mathbf{M}_S = (I_n + \mathbf{X}_S\mathbf{\Gamma}_S\mathbf{X}_S')$. Thus, a non-zero approximation is made to $\text{cov}(\boldsymbol{\beta}_S, \boldsymbol{\beta}_{S^c})$ in the JOB-approx algorithm, whereas in the skinny Gibbs algorithm of [36], these are set to zero.

Thresholding the small entries of $(\xi\eta_j)^{-1}$ has significant computational advantages. It was shown in [27] that carefully tracking down all instances of the matrix product $\mathbf{X\Gamma X'}$ entering the algorithm as well as appropriately modifying the ensuing matrix operations, the per-iteration cost of the JOB-approx algorithm is of order $(s_\delta^2 \vee p)n$, where

$$s_\delta = \sum_{j=1}^{p} \mathbb{1}\big((\xi\eta_j)^{-1} > \delta\big) = \sum_{j=1}^{p} \mathbb{1}(\tau^2\lambda_j^2 > \delta).$$

This is comparable to the per-iteration computational cost of coordinate descent algorithms for Lasso and Elastic Net [17, Sections 2.1, 2.2]. However, improvement in per-iteration complexity is only part of the story. In a problem of size $n = 2000$ and $p = 20,000$ where the true $\boldsymbol{\beta}$ consisted of a sparse sequence of signals of various sizes, the JOB-approx algorithm was approximately 50 times more efficient than the (exact) JOB algorithm, with the gain in efficiency measured in terms of *median effective samples per second.* For a univariate MCMC chain, the effective sample size per second refers to the effective sample size n_e divided by the runtime (in seconds), and provides a useful summary of the overall efficiency of the MCMC algorithm considered. For example, suppose algorithm A takes only 10 minutes to run 10,000 iterations while algorithm B takes an hour to run 10,000 iterations. From this, we can only conclude that algorithm A has a much better per-iteration complexity. It might well be possible that there are only 100 effectively independent samples from the path length of algorithm A and 6000 from algorithm B. Then, running both algorithms for an hour would result in about 600 effective samples from A and 6000 from B, implying that algorithm B is actually 10 times more efficient overall. [27] monitored the chains for the non-null (β_j, η_j)s, a randomly selected subset of the null (β_j, η_j)s, and (ξ, σ^2) and recorded the effective sample sizes per second for each of these chains. The median of these numbers came out to be 11.7 for the approximate algorithm compared to 0.22 for the old algorithm.

The speedup from this approximation is best when p is large relative to n and the truth is sparse or close to sparse, so that most entries of $\boldsymbol{\beta}$ are shrunk to near zero. Importantly, coefficients that are filtered out at iteration k need not be thresholded away at iteration $(k+1)$, and in practice the set of active variables considerably changed from one iteration to another. Note the thresholded coordinates are never actually set to zero or omitted, but rather sampled from a Gaussian that closely approximates the exact full conditional. Thus, the primary benefit of Bayesian methods for sparse regression is retained in the sense that estimates of uncertainty about the set of true signals are still valid. For example, the bimodal nature of the posterior for the intermediate-sized signal in Figure 7.3 was entirely retained by the approximate algorithm. Across a wide range of simulations, the approximate algorithm was found to have similar statistical performance as the exact algorithm both in terms of point estimation as well as coverage; see, for example, Figure 5 of [27]. [27] proceeded to apply the approximate algorithm to a GWAS dataset with hundreds of thousands of predictors and tens of thousands of samples.

Introducing approximations as above within an MCMC algorithm creates a perturbed version $\mathcal{P}_\epsilon$ of the original exact Markov transition kernel $\mathcal{P}$. Perturbation bounds for uniformly ergodic Markov operators date at least back to [34]. More recent work [28, 41, 52] has focused on the unbounded state space setting and the use of Lyapunov functions. The general goal here is to delineate conditions under which the approximate kernel $\mathcal{P}_\epsilon$ provides a good approximation to the exact kernel $\mathcal{P}$. It is often difficult to obtain exact relationships between e.g., $\|\mathcal{P}(\boldsymbol{\theta}, \cdot) - \mathcal{P}_\epsilon(\boldsymbol{\theta}, \cdot)\|_{\text{TV}}$ and the parameters such as δ in (7.12). However, [27] were able to establish that the approximation error $\|\mathcal{P}(\boldsymbol{\theta}, \cdot) - \mathcal{P}_\epsilon(\boldsymbol{\theta}, \cdot)\|_{\text{TV}}$ goes to zero at rate $\sqrt{\delta}$, which provides some practicable guidelines for choosing the tuning parameter δ.

7.5 Conclusion

The efficiency of an MCMC algorithm crucially relies on the per-iteration complexity as well as the mixing time. In high-dimensional settings, both of these aspects demand careful considerations. The infinite spike at the origin and the heavy tails of the continuous shrinkage priors are retained in the posterior distribution, rendering the target distribution not immediately amenable to sampling using off-the-shelf MCMC algorithms; see [9] for a detailed discussion. As we have illustrated in this chapter, carefully designed blocked Gibbs samplers can provide substantial improvements in mixing. Design of such blocked samplers in high-dimensional models, especially beyond the linear model, remains an active area.

Since the continuous shrinkage priors provide soft shrinkage, all the candidate variables remain active throughout the duration of the Markov chain. One therefore needs to exercise care with ensuing matrix manipulations to avoid the computations from getting prohibitively expensive. However, as we have demonstrated in this chapter, the per-iteration complexity can be high even after optimizing pieces of the algorithm, and the approximation schemes discussed here offer substantial gains. While computationally attractive, the more subtle theoretical properties of these approximation schemes pose exciting challenges. For example, characterizing the precise relationship between tuning parameters of various approximation schemes and the approximation error to the target measure of the exact Markov chain remains an open problem.

It is also worth noting that all the geometric convergence results discussed here are for fixed n and p, and that characterizing the dependence of the spectral gap $(1 - \bar{\alpha})$ on n and p largely remains an open problem. The typical drift and minorization argument is known to produce loose constants. For example, [47] noted that their bound on the geometric convergence rate tends to one at an exponential rate in n, p, which is far more pessimistic relative to empirical performance. [46] showed that published bounds on geometric convergence rates tending to one at an exponential rate with increasing dimension is a general phenomenon, underscoring the inherent difficulty of obtaining sharp bounds on geometric convergence rates of many kernels used in practice. Nevertheless, this continues to remain an active area of research in general, and hopefully there will be more progress in the near future on establishing favorable operating characteristics with increasing dimension for some of the sampling algorithms discussed here.

Bibliography

[1] A. Armagan, D. B. Dunson, and J. Lee. Generalized double Pareto shrinkage. *Statistica Sinica*, 23(1):119, 2013.

[2] K. Bae and B. K. Mallick. Gene selection using a two-level hierarchical Bayesian model. *Bioinformatics*, 20(18):3423–3430, 2004.

[3] J. Baker, P. Fearnhead, E.B. Fox, and C. Nemeth. Control variates for stochastic gradient MCMC. *Statistics and Computing 29*, 599–615, 2019.

[4] R. Bardenet, A. Doucet, and C. Holmes. On Markov chain Monte Carlo methods for tall data. *The Journal of Machine Learning Research*, 18(1):1515–1557, 2017.

[5] A. Bhadra, J. Datta, N.G. Polson, and B. Willard. The horseshoe+ estimator of ultra-sparse signals. *Bayesian Analysis*, 12(4):1105–1131, 2017.

[6] A. Bhattacharya, A. Chakraborty, and B.K. Mallick. Fast sampling with Gaussian scale mixture priors in high-dimensional regression. *Biometrika*, 103(4):985–991, 2016.

[7] A. Bhattacharya and J.E. Johndrow. Sampling local scale parameters in high dimensional regression models. *Handbook of Computational Statistics and Data Science (To appear), Thomas Lee Eds.*, 2021.

[8] A. Bhattacharya, D. Pati, N.S. Pillai, and D.B. Dunson. Dirichlet–Laplace priors for optimal shrinkage. *Journal of the American Statistical Association*, 110(512):1479–1490, 2015.

[9] N. Biswas, A. Bhattacharya, P.E. Jacob, and J.E. Johndrow. Coupled markov chain monte carlo for high-dimensional regression with half-t priors. *arXiv preprint arXiv:2012.04798*, 2020.

[10] H.D. Bondell and B.J. Reich. Consistent high-dimensional Bayesian variable selection via penalized credible regions. *Journal of the American Statistical Association*, 107(500):1610–1624, 2012.

[11] L.D. Brown. On the admissibility of invariant estimators of one or more location parameters. *The Annals of Mathematical Statistics*, 37(5):1087–1136, 1966.

[12] L.D. Brown. Admissible estimators, recurrent diffusions, and insoluble boundary value problems. *The Annals of Mathematical Statistics*, 42(3):855–903, 1971.

[13] F. Caron and A. Doucet. Sparse Bayesian nonparametric regression. In *Proceedings of the 25th international conference on Machine learning*, pages 88–95. ACM, 2008.

[14] C.M. Carvalho, N.G. Polson, and J.G. Scott. The horseshoe estimator for sparse signals. *Biometrika*, 97(2):465–480, 2010.

[15] J. Datta and J.K. Ghosh. Asymptotic properties of Bayes risk for the horseshoe prior. *Bayesian Analysis*, 8(1):111–132, 2013.

[16] B. Efron and C. Morris. Stein's estimation rule and its competitors? an empirical bayes approach. *Journal of the American Statistical Association*, 68(341):117–130, 1973.

[17] J. Friedman, T. Hastie, and R. Tibshirani. Regularization paths for generalized linear models via coordinate descent. *Journal of Statistical Software*, 33(1):1–22, 2010.

[18] E.I. George and R.E. McCulloch. Variable selection via gibbs sampling. *Journal of the American Statistical Association*, 88(423):881–889, 1993.

[19] E.I. George and R.E. McCulloch. Approaches for Bayesian variable selection. *Statistica sinica*, 7:339–373, 1997.

[20] J.E. Griffin and P.J. Brown. Inference with normal-gamma prior distributions in regression problems. *Bayesian Analysis*, 5(1):171–188, 03 2010.

[21] P. R. Hahn and C. M. Carvalho. Decoupling shrinkage and selection in Bayesian linear models: a posterior summary perspective. *Journal of the American Statistical Association*, 110(509):435–448, 2015.

[22] P.R. Hahn, J. He, and H.F. Lopes. Efficient sampling for Gaussian linear regression with arbitrary priors. *Journal of Computational and Graphical Statistics*, 28(1):142–154, 2019.

[23] M. Hairer and J.C. Mattingly. Yet another look at Harris' ergodic theorem for Markov chains. In *Seminar on Stochastic Analysis, Random Fields and Applications VI*, pages 109–117. Springer, 2011.

[24] C. Hans. Bayesian lasso regression. *Biometrika*, 96(4):835–845, 2009.

[25] H. Ishwaran and J.S. Rao. Spike and slab variable selection: frequentist and bayesian strategies. *Annals of statistics*, 33(2):730–773, 2005.

[26] W. James and C. Stein. Estimation with quadratic loss. In *Proceedings of the fourth Berkeley symposium on mathematical statistics and probability*, volume 1, pages 361–379, 1961.

[27] J. Johndrow, P. Orenstein, and A. Bhattacharya. Scalable Approximate MCMC Algorithms for the Horseshoe Prior. *Journal of Machine Learning Research (Under revision)*, 2019.

[28] J.E. Johndrow and J.C. Mattingly. Error bounds for approximations of Markov chains. *arXiv preprint arXiv:1711.05382*, 2017.

[29] G. L Jones. On the Markov chain central limit theorem. *Probability surveys*, 1:299–320, 2004.

[30] K. Khare and J.P. Hobert. Geometric ergodicity of the Bayesian lasso. *Electronic Journal of Statistics*, 7:2150–2163, 2013.

[31] R. Khasminskii. *Stochastic stability of differential equations*. Springer, 1980.

[32] A. Korattikara, Y. Chen, and M. Welling. Austerity in MCMC land: cutting the Metropolis-Hastings budget. In *International Conference on Machine Learning*, pages 181–189, 2014.

[33] S.P. Meyn and R.L. Tweedie. *Markov chains and stochastic stability*. Springer, 1993.

[34] A. Y. Mitrophanov. Sensitivity and convergence of uniformly ergodic Markov chains. *Journal of Applied Probability*, 42(4):1003–1014, 2005.

[35] I. Murray, R. Adams, and D. MacKay. Elliptical slice sampling. In *Proceedings of the thirteenth international conference on artificial intelligence and statistics*, pages 541–548, 2010.

[36] N.N. Narisetty, J. Shen, and X. He. Skinny gibbs: A consistent and scalable gibbs sampler for model selection. *Journal of the American Statistical Association*, pages 1–13, 2018.

[37] A. Nishimura and M.A. Suchard. Prior-preconditioned conjugate gradient for accelerated gibbs sampling in" large n & large p" sparse bayesian logistic regression models. *arXiv preprint arXiv:1810.12437*, 2018.

[38] S. Pal and K. Khare. Geometric ergodicity for Bayesian shrinkage models. *Electronic Journal of Statistics*, 8(1):604–645, 2014.

[39] T. Park and G. Casella. The Bayesian lasso. *Journal of the American Statistical Association*, 103(482):681–686, 2008.

[40] D. Pati, A. Bhattacharya, N.S. Pillai, and D. B. Dunson. Posterior contraction in sparse Bayesian factor models for massive covariance matrices. *The Annals of Statistics*, 42(3):1102–1130, 2014.

[41] N.S. Pillai and A. Smith. Ergodicity of approximate mcmc chains with applications to large data sets. *arXiv preprint arXiv:1405.0182*, 2014.

[42] M. Pollock, P. Fearnhead, A.M. Johansen, and G.O. Roberts. The scalable Langevin exact algorithm: Bayesian inference for big data. *arXiv preprint arXiv:1609.03436*, 2016.

[43] N.G. Polson and J.G. Scott. Shrink globally, act locally: Sparse Bayesian regularization and prediction. In J.M. Bernardo, M.J. Bayarri, and J.O. Berger, editors, *Bayesian Statistics*, volume 9, pages 501–538. Oxford, 2010.

[44] N.G. Polson and J.G. Scott. On the half-Cauchy prior for a global scale parameter. *Bayesian Analysis*, 7(4):887–902, 2012.

[45] N.G. Polson, J.G Scott, and J. Windle. The Bayesian bridge. *Journal of the Royal Statistical Society: Series B (Statistical Methodology)*, 76(4):713–733, 2014.

[46] B. Rajaratnam and D. Sparks. MCMC-based inference in the era of big data: A fundamental analysis of the convergence complexity of high-dimensional chains. *arXiv preprint arXiv:1508.00947*, 2015.

[47] B. Rajaratnam, D. Sparks, K. Khare, and L. Zhang. Scalable Bayesian shrinkage and uncertainty quantification in high-dimensional regression. *arXiv preprint arXiv:1703.09163*, 2017.

[48] P. Ray and A. Bhattacharya. Signal adaptive variable selector for the horseshoe prior. *arXiv preprint arXiv:1810.09004*, 2018.

[49] V. Ročková and E.I. George. The spike-and-slab lasso. *Journal of the American Statistical Association*, 113(521):431–444, 2018.

[50] J.S. Rosenthal. Minorization conditions and convergence rates for Markov chain Monte Carlo. *Journal of the American Statistical Association*, 90(430):558–566, 1995.

[51] V. Roy and S. Chakraborty. Selection of tuning parameters, solution paths and standard errors for bayesian lassos. *Bayesian Analysis*, 12(3):753–778, 2017.

[52] D. Rudolf and N. Schweizer. Perturbation theory for Markov chains via Wasserstein distance. *Bernoulli*, 24(4):2610–2639, 2018.

[53] H. Rue. Fast sampling of gaussian markov random fields. *Journal of the Royal Statistical Society. Series B, Statistical Methodology*, pages 325–338, 2001.

[54] J.G. Scott and J.O. Berger. Bayes and empirical-Bayes multiplicity adjustment in the variable-selection problem. *The Annals of Statistics*, 38(5):2587–2619, 2010.

[55] Q. Song and F. Liang. Nearly optimal bayesian shrinkage for high dimensional regression. *arXiv preprint arXiv:1712.08964*, 2017.

[56] C. Stein. Inadmissibility of the usual estimator for the mean of a multivariate normal distribution. In *Proceedings of the Third Berkeley symposium on mathematical statistics and probability*, volume 1(399), pages 197–206, 1956.

[57] C.M. Stein. Estimation of the mean of a multivariate normal distribution. *The Annals of Statistics*, pages 1135–1151, 1981.

[58] M. E. Tipping. Sparse Bayesian learning and the relevance vector machine. *The Journal of Machine Learning Research*, 1:211–244, 2001.

[59] S. van der Pas, B. Szabó, and A. van der Vaart. Adaptive posterior contraction rates for the horseshoe. *Electronic Journal of Statistics*, 11(2):3196–3225, 2017.

[60] S. van der Pas, B. Szabó, and A. van der Vaart. Uncertainty quantification for the horseshoe (with discussion). *Bayesian Analysis*, 12(4):1221–1274, 2017.

[61] D. Vats. Geometric ergodicity of gibbs samplers in bayesian penalized regression models. *Electronic Journal of Statistics*, 11(2):4033–4064, 2017.

[62] R. Wei and S. Ghosal. Contraction properties of shrinkage priors in logistic regression. *Journal of Statistical Planning and Inference*, 207:215–229, 2020.

[63] M. Welling and Y.W. Teh. Bayesian learning via stochastic gradient Langevin dynamics. In *Proceedings of the 28th International Conference on Machine Learning (ICML-11)*, pages 681–688, 2011.

[64] Y. Zhang, B.J. Reich, and H.D. Bondell. High dimensional linear regression via the R2-D2 shrinkage prior. *arXiv preprint arXiv:1609.00046*, 2016.

8

Variable Selection with Shrinkage Priors via Sparse Posterior Summaries

Yan Dora Zhang

The University of Hong Kong (Hong Kong SAR China)

Weichang Yu

The University of Melbourne (Australia)

Howard D. Bondell

The University of Melbourne (Australia)

CONTENTS

Shrinkage priors with the ability to perform both global and local shrinkage have become popular techniques for Bayesian regularization. These priors can typically be written as scale-mixtures of Gaussians, which enable the construction of posterior sampling schemes. While these priors are well-suited for estimation and parameter inference, they do not include variable selection as a by-product when posterior summaries such as means and covariances are extracted. [7] proposed an approach to post-process the posterior distribution of a parameter vector based on selecting the sparsest model among those contained within a given posterior credible region. By changing the coverage level of the credible region, they constructed a sequence of models to accommodate variable selection. In this chapter, we discuss the use of global-local shrinkage priors and combine them with variable selection using sparse posterior summaries in some common models.

DOI: 10.1201/9781003089018-8

8.1 Introduction

Consider a dataset $\{(\boldsymbol{x}_i, y_i)\}_{i=1}^n$ and the linear regression model

$$y_i = \boldsymbol{x}_i^T \boldsymbol{\beta} + \varepsilon_i, \tag{8.1}$$

where y_i is the response, $\boldsymbol{x}_i$ is the p-dimensional vector of covariates, $\boldsymbol{\beta} = (\beta_1, \cdots, \beta_p)^T$ is the vector of regression coefficients, and the error terms, ε_i, are assumed to be independent and identically distributed as $\mathcal{N}(0, \sigma^2)$. High-dimensional data with $p > n$ in this context is common in diverse areas of applications. However, the maximum likelihood estimate in this case does not uniquely exist. Therefore, a regularized estimation and/or variable selection procedure is needed. In the Bayesian framework, regularization may be achieved by assigning the regression coefficients with a shrinkage prior. Examples are discrete spike-and-slab priors which are constructed using the Dirac density function [4, 10, 11], continuous spike-and-slab priors using two slab functions [6, 19, 24], or continuous global-local shrinkage priors such as the generalized double Pareto shrinkage [2], Horseshoe [10, 49], normal-gamma [15], Horseshoe+ [5], normal beta-prime [3], Dirichlet-Laplace [6], and R2-D2 [35] priors.

The discrete spike-and-slab prior indexes each candidate variable with an indicator variable to denote its inclusion status. This is equivalent to assigning a mixture distribution with two components – a point mass (spike) at zero, and a continuous prior (slab) usually centered at zero. To avoid calculating posterior probabilities of each of the 2^p possible models, approximate solutions may be implemented such as stochastic search variable selection [6], shotgun stochastic search [17], variational Bayes [25], and expectation-maximization [30].

On the other hand, continuous shrinkage priors yield solutions that avoid a search over model space. However, they do not lead to a subset of *signal* variables as in the case of spike-and-slab priors. Additional posterior variable selection steps are required to induce sparsity in the posterior estimates, i.e., refined posterior summaries with some entries equal to 0. One straightforward approach is to choose a threshold on a posterior summary, such as the posterior t-statistic, which then identifies the selected signal variables [49]. Other approaches for performing variable selection through posterior summaries include penalized variable selection based on posterior credible regions [7], and decoupling shrinkage and selection [6].

In this chapter, we review variable selection via the use of posterior summaries and their applications to several models. Section 8.2 provides a detailed review of the penalized credible region selection method and an example of its application to a gene expression dataset. Section 8.3 describes other approaches based on posterior summaries in the literature. Section 8.4 provides details about the extension of the penalized credible region selection method to the logistic regression model. Section 8.5 discusses an application to the Gaussian graphical model. Section 8.6 examines the use of penalized credible regions for selecting confounding variables. Section 8.7 provides a recent application to a time-varying coefficient model. Section 8.8 summarizes the chapter and discusses possible future work.

8.2 Penalized Credible Region Selection

The general procedure of penalized credible region selection is introduced in this section, with specific cases followed in subsequent sections. A penalized credible region (PenCR) Bayesian variable selection method is proposed in [7]. The approach proceeds as follows.

First, the full model with all predictors is fitted with a continuous prior. Then, based on the posterior distribution a nested sequence of posterior credible regions is constructed by varying the level of the credible set. Within each credible region, the sparsest solution is then identified, yielding a sequence of estimated models.

Formally, let $||\boldsymbol{\beta}||_0$ denote the L_0 norm of the vector $\boldsymbol{\beta}$, i.e., the number of non-zero entries in $\boldsymbol{\beta}$. Let $\mathcal{C}_\alpha$ denote a $(1-\alpha)\times 100\%$ posterior credible region based on the user's choice of prior distribution. The proposed estimator for this fixed α is then given as

$$\widetilde{\boldsymbol{\beta}} = \arg\min_{\boldsymbol{\beta}} ||\boldsymbol{\beta}||_0 \quad \text{subject to} \quad \boldsymbol{\beta} \in \mathcal{C}_\alpha. \tag{8.2}$$

The credible region $\mathcal{C}_\alpha$ may be specified as any posterior region that has coverage $(1-\alpha)\times$ 100%, and is typically specified as the highest posterior density region, although such a specification is not necessary.

However, obtaining the solution to (8.2) involves a computationally exhaustive search over a possibly high-dimensional region. Moreover, the solution will be non-unique. To circumvent these obstacles, [7] replaced the L_0 norm with smoothed transition between L_0 and L_1 as in [22]. Consequently, the optimization problem given by (8.2) becomes

$$\widetilde{\boldsymbol{\beta}} = \arg\min_{\boldsymbol{\beta}} \sum_{j=1}^{p} \rho_a(|\beta_j|) \text{ subject to } \boldsymbol{\beta} \in \mathcal{C}_\alpha. \tag{8.3}$$

where

$$\rho_a(t) = \frac{(a+1)t}{a+t} = \left(\frac{t}{a+t}\right) + \left(\frac{a}{a+t}\right)t, \quad t \geq 0, \; a > 0.$$

Since $\lim_{a\to 0} \rho_a(t) = I[t \neq 0]$, we obtain our original optimization problem in (8.2) by letting $a \to 0$. Hence, we would like to choose $a = 0$ if it is computationally feasible. To make this choice of a feasible, we adopt the following linear approximation as suggested by [37]:

$$\rho_a(|\beta_j|) \approx \rho_a(|\widehat{\beta}_j|) + \rho_a'(\widehat{\beta}_j)(|\beta_j| - |\widehat{\beta}_j|), \tag{8.4}$$

where

$$\rho_a'(|\widehat{\beta}_j|) = \frac{a(a+1)}{(a+|\widehat{\beta}_j|)^2},$$

and, in this case, we set $\widehat{\beta}_j$ as the posterior mean.

By substituting this approximation into (8.4), omitting terms that are constant with respect to β, and setting $a = 0$, the optimization problem becomes

$$\widetilde{\boldsymbol{\beta}} = \arg\min_{\boldsymbol{\beta}} \sum_{j=1}^{p} |\widehat{\beta}_j|^{-2}|\beta_j| \text{ subject to } \boldsymbol{\beta} \in \mathcal{C}_\alpha. \tag{8.5}$$

Theoretical properties in terms of model selection consistency of the penalized credible region selection approach in both fixed-p and diverging-p regimes have been shown for the linear model in [7].

8.2.1 Gaussian Prior

For the linear regression model described in (8.1), the standard conjugate choice is to assign the coefficients with independent Gaussian priors such that

$$\boldsymbol{\beta} \mid \sigma^2, \tau \sim \mathcal{N}(0, (\sigma^2/\tau)\boldsymbol{I}_p), \tag{8.6}$$

where σ^2 is the error variance term as in (8.1), and τ is a fixed ratio of prior precision to error precision. The variance, σ^2, is often given a diffuse prior, such as an inverse-gamma, or other choice.

If τ is fixed, then regardless of the choice of prior for σ^2, the posterior distribution of $\boldsymbol{\beta}$ is elliptical, as it becomes a scale mixture of elliptical distributions, with density of the form

$$p(\boldsymbol{\beta} \mid \text{Data}) = H[(\boldsymbol{\beta} - \widehat{\boldsymbol{\beta}})^T \widehat{\boldsymbol{\Sigma}}^{-1} (\boldsymbol{\beta} - \widehat{\boldsymbol{\beta}})],$$

where $H(\cdot)$ is a monotone decreasing function, and $\widehat{\boldsymbol{\beta}}$ and $\widehat{\boldsymbol{\Sigma}}$ denote the posterior mean and covariance of $\boldsymbol{\beta}$ respectively.

Hence, for fixed τ, the highest density posterior credible regions have the elliptical form with

$$\mathcal{C}_\alpha = \{\boldsymbol{\beta} : (\boldsymbol{\beta} - \widehat{\boldsymbol{\beta}})^T \widehat{\boldsymbol{\Sigma}}^{-1} (\boldsymbol{\beta} - \widehat{\boldsymbol{\beta}}) \leq c_\alpha\}, \tag{8.7}$$

for some nonnegative c_α. Then, by replacing the L_0 penalization in (8.2) as discussed above in (8.4), the optimization problem is equivalent to

$$\widetilde{\boldsymbol{\beta}} = \arg\min_{\boldsymbol{\beta} \in \mathbb{R}^p} \left\{ (\boldsymbol{\beta} - \widehat{\boldsymbol{\beta}})^T \widehat{\boldsymbol{\Sigma}}^{-1} (\boldsymbol{\beta} - \widehat{\boldsymbol{\beta}}) + \lambda_\alpha \sum_{j=1}^{p} |\widehat{\beta}_j|^{-2} |\beta_j| \right\}, \tag{8.8}$$

where λ_α is a Lagrange multiplier having a one-to-one correspondence with α. Hence, the sequence of solutions to (8.8) can be directly obtained by plugging in the posterior mean and covariance of $\boldsymbol{\beta}$, followed by the use of any algorithm such as the Least Angle Regression (LARS) [12] that enables one to obtain the solution path for the L_1 constrained problem.

Here, the value of α is no longer needed as the original constraint optimization problem in (8.2) is re-expressed as an unconstrained optimization problem in (8.8). Given an extremely large value of λ_α, no variables will be selected into the model. As the value of λ_α decreases, the variables will be sequentially selected into the model. For example, consider the linear model with two predictors x_1 and x_2 as illustrated in Figure 8.1. As the value of α increases sequentially from 0.05 to 0.1 and further to 0.2, it corresponds to a decreasing sequence of λ_α. This leads to the credible region shrinking around the posterior mean, and thus not covering any sparse solutions. When $\alpha = 0.05$, the credible region covers the origin, resulting in a solution that is fully sparse, i.e., the null model. When $\alpha = 0.1$, the sparsest solution occurs at a point on the line $\beta_2 = 0$, and hence, the model contains x_1 only. When $\alpha = 0.2$, the credible interval is shrunk further around the posterior mean, and all points within the region have non-zero values for both x_1 and x_2. Hence, at this level, the model would be the full model containing both x_1 and $x2$.

We note that if a prior is placed on the value of τ, the posterior for $\boldsymbol{\beta}$ then becomes a mixture of elliptical distributions, where the mixing is over the mean as well. Hence, the posterior is no longer elliptical and the contours defined in equation (8.7) no longer align with the highest density regions. However, they still provide valid credible regions, albeit perhaps not obtaining the minimum volume. Thus, it is appropriate to either use other priors which we introduce in Section 8.2.2 to replace the normal prior, or apply PenCR to other models (see Sections 8.4–8.7).

8.2.2 Global-Local Shrinkage Priors

The original PenCR variable selection method proposed by [7] considered Gaussian priors without any strategic shrinkage effect. Here, it is possible for us to extend the method to other continuous shrinkage priors. We note that a rich class of continuous shrinkage priors may be expressed as a global-local scale mixture of normals as summarized in [47], i.e.,

$$\beta_j \mid \lambda_j, \tau \sim \mathcal{N}(0, \tau\lambda_j), \; \lambda_j \sim \pi(\lambda_j), \; (\tau, \sigma^2) \sim \pi(\tau, \sigma^2),$$

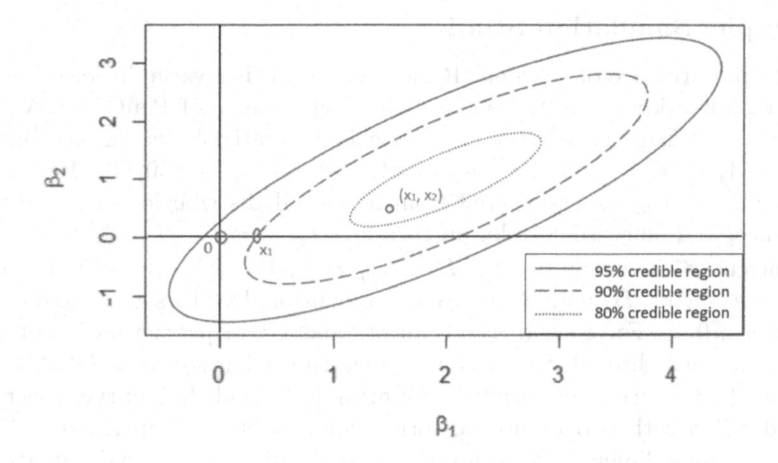

FIGURE 8.1
An example of the sequential nature of selection from PenCR. A sequence of three concentric credible regions, resulting in a sequence of models. The ellipses represent the boundaries of the respective credible regions, while the hollow dots mark the sparsest point within their respective region. In each case, the non-zero coordinate represents the variable that is selected in that model.

where τ represents the global shrinkage, and λ_j's are the local variance components. Some examples include the Normal-Gamma mixture [15], Horseshoe [10, 49], Horseshoe+ [5], generalized Beta [1], generalized double Pareto [2], Dirichlet-Laplace [6], normal-beta prime prior [3], and the R2-D2 prior [35].

Global-local shrinkage priors can also be incorporated into the PenCR variable selection framework. For the linear model as above, the Dirichlet-Laplace (DL) prior was coupled with the PenCR approach in [34], where they also derived concentration properties of the DL posterior. In particular, the DL prior proposes

$$\begin{aligned} \beta_j \mid \sigma, \lambda_j, \tau &\sim \mathcal{DE}(\sigma \lambda_j \tau), \\ (\lambda_1, \ldots, \lambda_p) &\sim \mathcal{D}ir(a, \ldots, a), \\ \tau &\sim \mathcal{G}(pa, 1/2), \end{aligned} \tag{8.9}$$

where $\mathcal{DE}(b)$ denotes a zero mean Laplace kernel with density $f(y) = (2b)^{-1} \exp\{-|y|/b\}$ for $y \in \mathbb{R}$, $\mathcal{D}ir(a, \ldots, a)$ is the Dirichlet distribution with concentration vector $(a, \ldots, a)$, and $\mathcal{G}(pa, 1/2)$ denotes a Gamma distribution with shape pa and rate $1/2$. Here, a small value of a leads to a majority of the posterior estimates of $(\lambda_1, \ldots, \lambda_p)$ to be near zero, with only a handful of entries with estimates substantially away from zero. The λ_j's are the local scales, allowing deviations in the degree of shrinkage. The parameter τ controls global shrinkage towards the origin and determines the tail behaviors of the marginal distribution of β_j's. We also assume a common $\mathcal{IG}(a_1, b_1)$ prior on the variance term σ^2, where a_1 and b_1 are the shape and rate parameters respectively.

In this framework, $\widehat{\beta}$ and $\widehat{\Sigma}$ in (8.8) are then the posterior mean and covariance matrix using the DL prior, respectively. The subsequent variable selection steps proceed as before using the elliptical contours for the credible region, although these are not the highest density contours, as discussed previously.

8.2.3 Example: Simulation Studies

To compare the performance of the PenCR method with other variable selection methods, [7] conducted a simulation study to compare the performance of PenCR, SSVS, LASSO, adaptive LASSO and Dantzig selector. The data is generated based on the linear model (8.1) with $\sigma^2 = 1$, $n = 60$, and the number of predictors $p \in \{50, 500, 1000\}$. The predictors $\boldsymbol{x}_i = (x_{i1}, \ldots, x_{ip})$ are generated from a normal distribution with mean $\mathbf{0}$, unit marginal variance, and pairwise correlation $\text{corr}(x_{ij_1}, x_{ij_2}) = \rho^{|j_1 - j_2|}$, where $\rho \in \{0.5, 0.9\}$. The true coefficient $\boldsymbol{\beta}^*$ is equals to $(\mathbf{0}_{10}^T, \boldsymbol{B}_1{}^T, \mathbf{0}_{20}^T, \boldsymbol{B}_2{}^T, \mathbf{0}_{p-40}^T)^T$, where $\mathbf{0}_k$ represents the k-dimensional zero vector, $\boldsymbol{B}_1$ and $\boldsymbol{B}_2$ are each 5-dimensional vectors generated componentwise from Uniform$(0, 1)$. For each method, an ordered set of predictors is obtained. The performance is assessed through the Receiver-Operating Characteristic (ROC) curve and Precision-Recall (PRC) curve plots. Plots of the mean ROC and PRC curves (over 200 runs) for PenCR and SSVS with two versions of priors for $p = 500$ are provided in Figure 8.2. In this moderately high dimensional setup, the PenCR outperforms SSVS greatly. Plots of the mean ROC and PRC curves (over 200 runs) for assessing the performance of PenCR and frequentist approaches for $p = 2000$ are provided in Figure 8.3. The PenCR method performs better than the LASSO and Dantzig method, especially in $\rho = 0.9$ case.

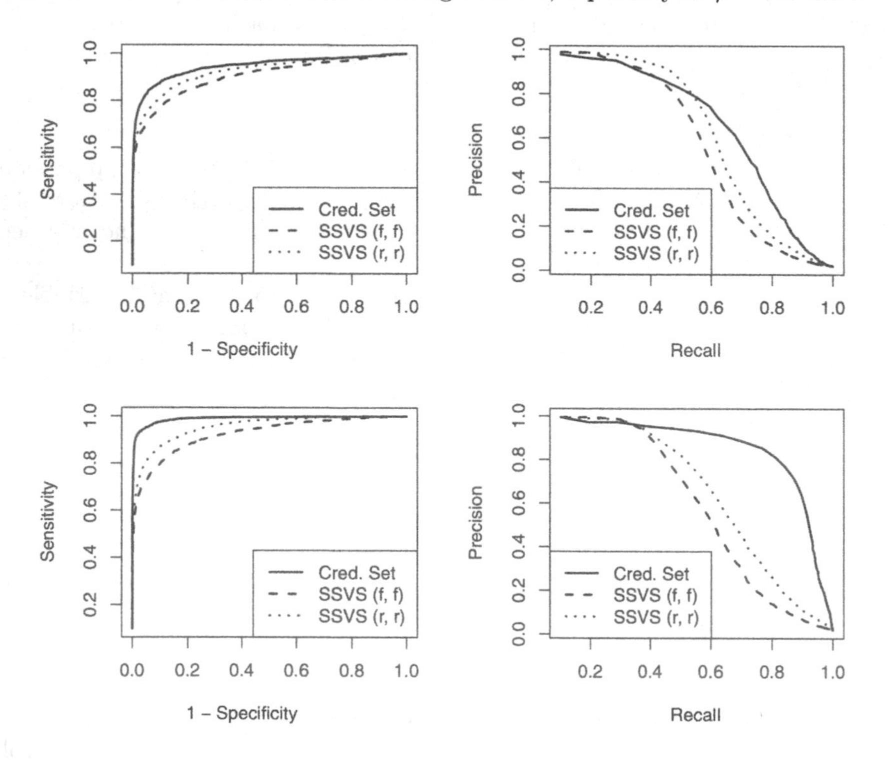

FIGURE 8.2
Mean ROC and PRC curves over the 200 datasets for $p = 500$ predictors, $n = 60$ observations. The first column is the ROC curve, and the second column is the PRC curve. The first row is for $\rho = 0.5$, while the second row is for $\rho = 0.9$. "Cred. Set" represents the PenCR method; "SSVS (f,f)" denotes SSVS with a fixed prior variance for the coefficients and fixed inclusion probability; "SSVS (r, r)" denotes SSVS with a prior with random variance for the coefficients and random prior inclusion probability. This figure corresponds to Figure 2 in [7].

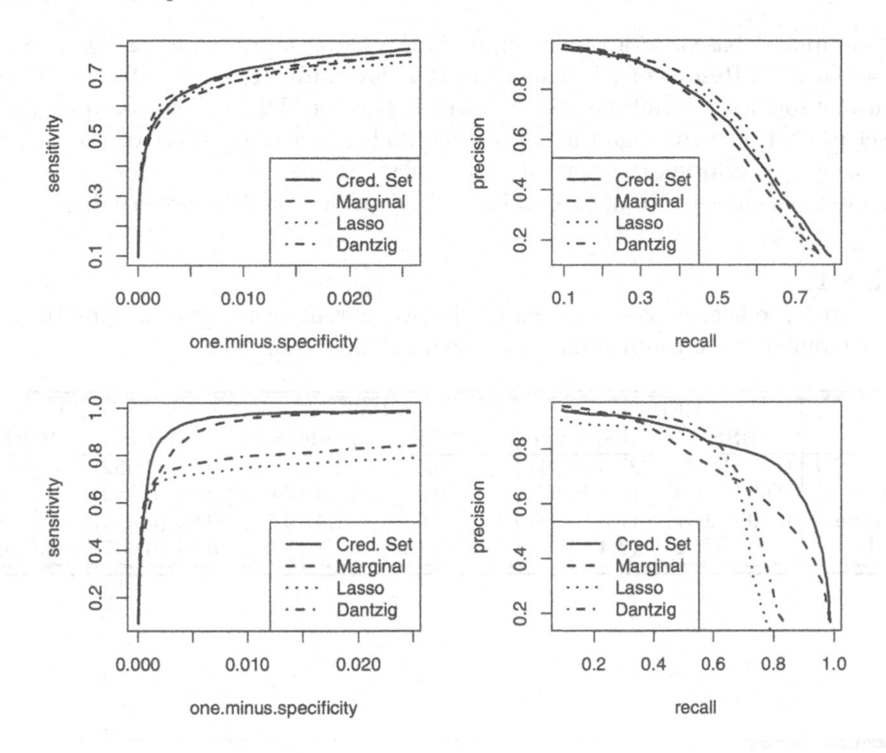

FIGURE 8.3
Mean ROC and PRC curves over the 200 datasets for $p = 2000$ predictors, $n = 60$ observations. The first column is the ROC curve, and the second column is the PRC curve. The first row is for $\rho = 0.5$, while the second row is for $\rho = 0.9$. "Cred. Set" represents the PenCR method; "Marginal" denotes the marginal PenCR method, i.e., constructing rectangular credible regions with setting bound marginally for each coefficient of posterior mean; "LASSO" denotes the LASSO method; "Dantzig" denotes the Dantzig selector [9]. This figure corresponds to Figure 3 in [7].

8.2.4 Example: Mouse Gene Expression Real-time PCR

In [34], they demonstrate the performance of the PenCR method with various shrinkage priors through a real application example. In a mouse gene expression experiment conducted by [21], there were 60 arrays to monitor the expression levels of 22,575 genes. There were 31 female and 29 male mice. Quantitative real-time *polymerase chain reaction* (PCR) was used to measure three physiological phenotypes – phosphoenopyruvate carboxykinase (PEPCK), glycerol-3-phosphate acyltransferase (GPAT), and stearoyl-CoA desaturase 1 (SCD1). The datasets are available for download from the Gene Expression Omnibus repository (`http://www.ncbi.nlm.nih.gov/geo|;accessionnumberGSE3330`).

The PenCR approach was applied for demonstration to select important genes affecting the three phenotypes. For comparison, three different prior choices are examined: (a) Gaussian, (b) Laplace, and (c) Dirichlet-Laplace. For an additional comparison, the frequentist LASSO approach is also implemented, which corresponds to the posterior mode for the Laplace. Each approach yields a sequence of models indexed by a tuning parameter. For each of the three PenCR approaches, the index parameter represents the level of the credible region, α. In each case, the tuning parameter is chosen via 5-Fold cross-validation (CV).

The performance is compared with three other prior choices over the three phenotype outcomes. For each outcome, the mean squared prediction error (MSPE) and selected model size are computed based on a random split of the dataset in a training set ($n = 50$) and testing set ($n = 5$). Results of this numerical study are summarized in Table 8.1. The PenCR selection method with shrinkage priors (normal, Laplace, DL) yields a smaller model size and lower prediction error than the LASSO method. In terms of choice of prior for PenCR, the DL prior outperforms the normal and Laplace priors in most cases. Based on these results, a default choice of a global-local shrinkage prior would be preferred.

TABLE 8.1
Mean squared prediction error and model size, with standard errors in parenthesis, based on 100 random splits of the real data as shown in Table 6 in [34].

	PEPCK		GPAT		SCD1	
	MSPE	Model Size	MSPE	Model Size	MSPE	Model Size
LASSO	0.54 (0.026)	25.8 (0.34)	1.43 (0.082)	24.4 (0.56)	0.55 (0.052)	26.1 (0.33)
Normal	0.66 (0.033)	16.8 (0.67)	1.30 (0.099)	16.3 (0.66)	0.71 (0.059)	10.8 (0.60)
Laplace	0.70 (0.037)	17.0 (0.78)	1.19 (0.086)	21.4 (0.56)	0.69 (0.054)	14.8 (0.82)
DL	0.49 (0.032)	18.4 (0.73)	1.37 (0.102)	13.1 (0.68)	0.54 (0.037)	14.0 (0.59)

8.3 Approaches Based on Other Posterior Summaries

The PenCR method introduced above is one way to conduct variable selection via a posterior summary. The Bayesian literature offers alternative approaches based on other posterior summary perspectives.

Decoupling shrinkage and selection from a posterior summary perspective

An alternative sparsification method known as decoupling shrinkage and selection (DSS), proposed in [6], takes the same approach of separating out the shrinkage by obtaining a posterior for the full model, and then processing that posterior to obtain a selected model. In this case, a direct tradeoff between prediction accuracy and parsimony is specified via a loss function. More specifically, the authors considered the minimization of a loss function – the sum of a parsimony-inducing penalty function and a squared prediction loss function:

$$\mathcal{L}(\boldsymbol{\beta}^s) = n^{-1}||\boldsymbol{X}\boldsymbol{\beta}^s - \boldsymbol{X}\bar{\boldsymbol{\beta}}||_2^2 + \lambda||\boldsymbol{\beta}^s||_0, \tag{8.10}$$

where $\bar{\boldsymbol{\beta}}$ is the posterior mean. To compute the minimizer with respect to $\boldsymbol{\beta}^s$, a simplistic approximation is to replace the L_0 norm with the L_1 norm, and then to apply LARS algorithm using $\bar{\boldsymbol{Y}} = \boldsymbol{X}\bar{\boldsymbol{\beta}}$ as the "data". To avoid the potential "double shrinkage", [6] also proposed a flexible norm (the adaptive LASSO penalty proposed by [36] and [22]) to replace the L_1 norm in (8.10) with another penalty, i.e.,

$$\mathcal{L}(\boldsymbol{\beta}^s) = n^{-1}||\boldsymbol{X}\boldsymbol{\beta}^s - \boldsymbol{X}\bar{\boldsymbol{\beta}}||_2^2 + \sum_{j=1}^{p} \frac{\lambda}{|\bar{\beta}_j|}|\beta_j^s|.$$

Again, the LARS algorithm can be applied to solve the objective function.

We note that this formulation mirrors the dual problem for PenCR as given in (8.8) since the squared prediction loss is given by the quadratic form defining the elliptical contours.

Consequently, it turns out that the resultant DSS optimization problem is equivalent to the PenCR objective in equation (8.8) if the prior on $\boldsymbol{\beta}$ is uninformative. In general, they may differ in the centering and covariance structure of the elliptical contours. As with the PenCR selection method, the DSS method is well-defined under any prior that leads to a proper posterior.

Sparsity using the Kullback-Leibler divergence

To obtain a sparse solution after the convergence of Markov Chain Monte Carlo (MCMC) algorithms for any continuous shrinkage priors, [26] proposed a method to achieve sparsification by minimizing a Kullback-Leibler (KL) divergence between an approximate normal distribution and the true posterior distribution which only depends on the posterior mean and covariance. It can also be shown that this method recovers the exact PenCR as defined in (8.2) at a specific penalty parameter value.

8.4 Model Selection for Logistic Regression

We now examine the use of PenCR for variable selection in other models aside from linear models. In these models, the posterior mean and covariance can be obtained through any applicable approach. Once those are obtained, then valid credible regions are formed via the elliptical contours as previously discussed. The extension of the PenCR variable selection method to the logistic regression model was proposed in [31], who suggested the use of the posterior ellipsoidal regions as the choice of valid credible regions. In this case, as mentioned in 1.2.1, the credible regions constructed by using the elliptical contours are not high density regions, however they remain valid exact credible regions. This follows from the fact that by varying the volume of the ellipsoid, any desired probability content can be achieved. The volume of the ellipsoid needed to contain a fixed probability content would, of course, be larger than if a high density region was constructed with the same probability content. However, the construction of a high density region and the subsequent search for a sparse vector within the region, in cases such as logistic regression, particularly in the high dimensional case, would not be computationally feasible.

Consider the logistic regression model

$$y_i \sim \text{Binomial}(n_i, \pi_i), \ \pi_i = \frac{\exp(\boldsymbol{x}_i^T \boldsymbol{\beta})}{1 + \exp(\boldsymbol{x}_i^T \boldsymbol{\beta})}, \ i = 1, \ldots, n. \tag{8.11}$$

This can then be coupled with any desired choice of prior. As an example, in [31], a Normal-Gamma prior was considered for $\boldsymbol{\beta}$ in (8.11) as follows:

$$\beta_j \mid \tau_j \sim \mathcal{N}(0, \tau_j), \ \tau_j \sim \mathcal{G}(\lambda, d), \ j = 1, \ldots, p, \tag{8.12}$$

where $\mathcal{G}(\lambda, d)$ represents the gamma distribution with shape λ and rate d. Details on the choice of hyperparameters have been extensively discussed in both [15] in general, and [31] for the specific case of logistic regression. The posterior mean and posterior covariance matrix generated based on such priors would then be plugged into (8.8) to perform variable selection. An MCMC approach can be used, and a fully-Gibbs implementation is provided via the use of a Polya–Gamma latent variable representation [27].

Example

A simulation example was considered in [31] to demonstrate comparisons between some variable selection approaches in logistic regression setup. The comparisons are based on the full sequence of models rather than just one selected model as in the previous case. This example is taken from [31] where the PenCR using two versions of the Normal-Gamma prior as well as the Laplace prior is compared with other variable selection methods: LASSO, forward selection, and screening. The two versions of the Normal-Gamma prior provide different apriori assumptions. One assumes a relatively sparse model, while the other assumes a dense model. Details of how the parameters in the Gamma distribution are chosen are given in [31]. For each method, the variable selection accuracy is measured by computing its ROC and PRC curves along its ordered solution path. We note that higher curves represent better selection performance.

The dimension of each simulated dataset is $n = 200$ and $p = 200$. The covariates are generated from $\boldsymbol{x}_i \sim \mathcal{N}(\mathbf{0}, \text{AR}(1; \rho))$, where the matrix AR refers to the autoregressive lag-1 correlation matrix with parameter ρ. Two decay rates are considered: $\rho = 0.8$ and $\rho = 0.5$. The non-zero coefficients occur equally spaced across the indices, in this case

$$\beta_j^* = \begin{cases} 1.8, & \text{if } j \in \{2, 10, 18, 26, 34, 42, 50, 58, 66\}; \\ 0, & \text{otherwise}. \end{cases}$$

The binary response are generated from

$$y_i \sim \text{Bernoulli}(\exp(\boldsymbol{x}_i^T \boldsymbol{\beta}^*)/\{1 + \exp(\boldsymbol{x}_i^T \boldsymbol{\beta}^*)\}).$$

There are 200 datasets generated and for each, the full solution path is obtained. The ROC and PRC curves are then computed. Figure 8.4 displays the averaged ROC and PRC curves over the 200 simulated datasets. The upper two plots show the results for $\rho = 0.8$ and lower for $\rho = 0.5$, with the left panel showing the PRC curve, and right panel showing the ROC. Based on Figure 8.4, we observe strong variable selection performance using the PenCR method relative to the other approaches.

8.5 Graphical Model Selection

The PenCR method of variable selection was extended to Bayesian graphical models in [20]. This extension involved a decoupling of the model fitting process from the covariance selection procedure.

Denote $\boldsymbol{X} = (\boldsymbol{x}_1, \ldots, \boldsymbol{x}_n)^T$ as the $n \times p$ dimensional data matrix. The goal of the graphical model analysis is to identify the conditional dependency structure. Using a Gaussian graphical model, conditional independence relationships can be read off from zero elements in the inverse covariance matrix. While zeros in the covariance matrix correspond to marginally independent variables, the zeros in the inverse covariance, or precision matrix, correspond to those that are conditionally independent given all of the other variables.

We specify the Bayesian Gaussian graphical model as follows:

$$\boldsymbol{x}_i \mid \boldsymbol{\Sigma} \sim \mathcal{N}(\mathbf{0}, \boldsymbol{\Sigma}), \quad \boldsymbol{\Sigma} \mid \boldsymbol{D}, \nu \sim \text{Inverse-Wishart}(\boldsymbol{D}, \nu), \tag{8.13}$$

where $\boldsymbol{D} = \text{diag}(d_1, \ldots, d_p)$, the scalar ν is the degrees of freedom, and $d_k \sim \mathcal{G}_k(\cdot)$ with $\mathcal{G}_k(\cdot)$, $k = 1, \ldots, p$, denoting mixing distributions that allow for various types of shrinkage across different scales.

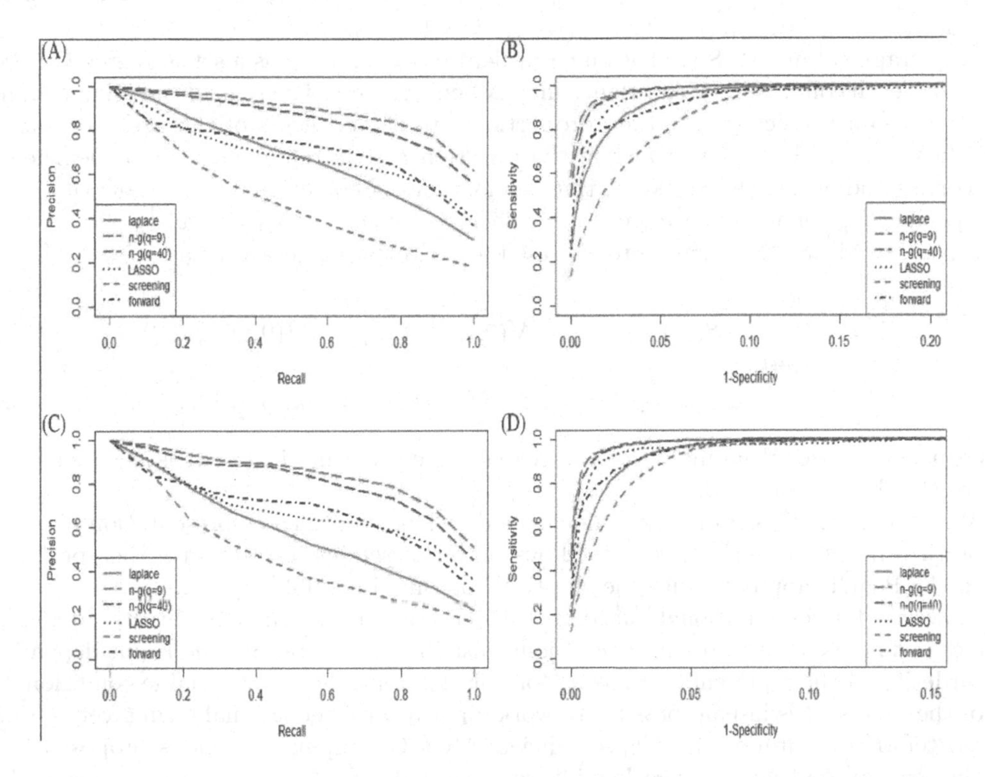

FIGURE 8.4
Mean PRC and ROC curves for 250 datasets with $n = 200$, $\rho = 0.8$, $p = 500$ for (A) & (B), and $n = 200$, $\rho = 0.5$, $p = 1000$ for (C) & (D). "laplace" denotes the model using a Laplace prior; "n-g ($q = 9$)" denotes the model using a Normal-Gamma prior with hyperparameter tuned to ensure that the number of important variables $q = 9$; "n-g ($q = 40$)" denotes the model using a Normal-Gamma prior with hyperparameter tuned to ensure that the number of important variables $q = 40$; "LASSO" denotes the general LASSO method; "screening" denotes the method considering each variable separately via its distribution in each of the binary groups; "forward" denotes the forward selection (Figure 5 in [31]).

For notational purposes, the covariance and precision matrices can be expressed as

$$\boldsymbol{\Sigma} \equiv \boldsymbol{\Sigma}_p = \begin{bmatrix} \boldsymbol{\Sigma}_{p-1,11} & \sigma_{p,21} \\ \boldsymbol{\sigma}_{p,12} & \sigma_{p,pp}, \end{bmatrix} \text{ with } \boldsymbol{\Sigma}_{k,11}^{-1} \equiv \boldsymbol{\Omega}_k = \begin{bmatrix} \boldsymbol{\Omega}_{11}^{k-1} & \omega_{k,21} \\ \boldsymbol{\omega}_{k,12} & \omega_{k,kk}, \end{bmatrix}$$

where $\boldsymbol{\Sigma}_{p-1,11}$ denotes the principal minor of dimension $p-1$ arising from the first $p-1$ rows and columns of $\boldsymbol{\Sigma}$, and $\boldsymbol{\Omega}_{11}^{k-1}$ denotes the principal minor of dimension $k-1$ for $\boldsymbol{\Omega}_k$. Let $\omega_{k,ij}$ denote the j-th element in the i-th row of $\boldsymbol{\Omega}_k$. The precision matrix can be decomposed as $\boldsymbol{\Sigma}^{-1} = \boldsymbol{T}'\boldsymbol{V}^{-1}\boldsymbol{T}$, with $\boldsymbol{V} = \text{diag}\{\omega_{p,pp}^{-1} \ldots, \omega_{1,11}^{-1}\}$ and $\boldsymbol{T}$ being a lower triangular matrix that has $t_{kj} = \omega_{k,kk}^{-1}\omega_{k,kj}$, $j < k$ and $t_{kk} = 1$, $k = 1, \ldots, p$.

The graphical model (8.13) has an equivalent representation as a set of regression problems with each column of the covariance matrix being regressed on each of the other columns. The regression coefficients are then proportional to the elements of the precision matrix. This formulation shows how a zero in the precision matrix represents a zero coefficient in the corresponding regression model, thus demonstrating conditional independence.

Specifically, denoting the regression coefficients as $\boldsymbol{\beta}_k = \{\beta_{kj} = -\omega_{p,kk}^{-1} \times \omega_{p,kj} : j = 1, \ldots, p,\ j \neq k\}$, $k = 1, \ldots, p$, we re-express the graphical model (8.13) as

$$x_{ik} = \sum_{j=1, j\neq k}^{p} x_{ij}\beta_{kj} + \epsilon_{ik},\ \epsilon_{ik} \sim \mathcal{N}(0, \omega_{p,kk}^{-1}),\ \beta_{kj} \sim \mathcal{N}(0, \omega_{p,kk}^{-1}/d_j),$$
$$\omega_{p,kk} \sim \mathcal{G}(b/2, d_k/2), \boldsymbol{D} \sim p(\boldsymbol{D}),\ j \neq k,\ j = 1, \ldots, p, \qquad (8.14)$$

where the conditional normality of the regression coefficients in (8.14) is ensured by the inverse-Wishart prior on $\boldsymbol{\Sigma}$.

We observe that (8.14) is now a linear model. This equivalent representation allows for theoretical results from [7] to be carried over into this setting if selection is then performed using the PenCR approach with the elliptical contoured credible regions.

However, for computational purposes, this representation cannot be directly applied, as it does not necessarily represent a valid joint distribution. Moreover, the result depends on the ordering of the representation, as its formulation represents one variable conditional on all of the others. It is instead possible to work directly with the original form given in (8.13) for posterior computation. In [20], an efficient MCMC sampling scheme is proposed based on the recognition that it is a scale mixture of normals.

8.6 Confounder Selection

The confounding variable selection problem is important when estimating the effect of an exposure or treatment on an outcome while deciding on which confounders to include for adjustment. A decision-theoretic approach that draws on the PenCR method to post-process the posterior for confounder selection and effect estimation was proposed in [33].

A linear model for the outcome is given by

$$\boldsymbol{y} = x\beta_x + \boldsymbol{U}\boldsymbol{\beta_u} + \varepsilon_y,\ \varepsilon_y \sim \mathcal{N}(0, \sigma_y^2\mathbb{I}) \qquad (8.15)$$

along with an exposure model

$$x = \boldsymbol{U}\gamma_u + \varepsilon_x,\ \varepsilon_x \sim \mathcal{N}(0, \sigma_x^2\mathbb{I}).$$

Interest focuses on inference for the variable x, while the variables in the matrix $\boldsymbol{U}$ are the potential confounders, as they appear in both the outcome and exposure models. We denote $\boldsymbol{\beta} = (\beta_x, \boldsymbol{\beta}_{\boldsymbol{u}}^T)^T$, $\gamma_u = (\gamma_1, \ldots, \gamma_p)^T$, and $\boldsymbol{W} = (x, \boldsymbol{U})$ which is $n \times (p+1)$ matrix. Let $\mathcal{C}_\alpha^\beta$ and $\mathcal{C}_\alpha^\gamma$ be $(1-\alpha) \times 100\%$ credible regions for $\boldsymbol{\beta}$ and $\boldsymbol{\gamma}$ respectively. The proposed estimator is the minimizer

$$\widetilde{\boldsymbol{\beta}} = \arg\min_{\boldsymbol{\beta}} \left\{ \epsilon \beta_x^2 + \| |\boldsymbol{\beta}_{\boldsymbol{u}}| + |\gamma_{\boldsymbol{u}}| \|_0 \right\},$$

subject to $\boldsymbol{\beta} \in \mathcal{C}_\alpha^\beta$ and $\boldsymbol{\gamma} \in \mathcal{C}_\alpha^\gamma$, where $|\cdot|$ denotes a vector after taking the absolute value on each component, and $\|\cdot\|_0$ denotes the L_0 norm of a vector. The value ϵ is fixed and small so that the exposure effect is left unpenalized, but the solution is slightly regularized and thus will be uniquely defined. We note that the sparsity criterion here counts the number of variables having an effect on either the outcome or the exposure model, or both. Hence, it is interpreted as the effective number of variables overall.

By using the approximation to the L_0 norm discussed previously in the context of the PenCR approach as in [7], the objective function is finally transformed as

$$\widetilde{\boldsymbol{\beta}} = \underset{\boldsymbol{\beta} \in \mathbb{R}^{p+1}}{\operatorname{argmin}} \left\{ (\boldsymbol{\beta} - \widehat{\boldsymbol{\beta}})^T \widehat{\boldsymbol{\Sigma}}^{-1} (\boldsymbol{\beta} - \widehat{\boldsymbol{\beta}}) + \lambda \sum_{j=1}^{p} \frac{|\beta_j|}{(|\widehat{\beta}_j| + |\widehat{\gamma}_j|)^2} \right\}, \tag{8.16}$$

where $\widehat{\boldsymbol{\beta}}$ and $\widehat{\boldsymbol{\Sigma}}$ are the posterior mean and covariance of $\boldsymbol{\beta}$ respectively. The final solution will be based on solving (8.16).

Simulation Example

Extensive simulations were performed in [33]. We list one example here. $\boldsymbol{Y} \sim N(\boldsymbol{W}\boldsymbol{\beta}, \boldsymbol{I})$, where $\boldsymbol{W}$ is an $n \times (p+1)$ design matrix with variable of interest in the first column and $p = 57$ potential confounders. The covariates for the ith observation, $\boldsymbol{W}_i \sim N(0, \boldsymbol{\Sigma})$ independently for $i = 1, \ldots, n$, where $\boldsymbol{\Sigma} = \{\sigma_{jk}\}$ is the covariance matrix with $\sigma_{jj} = 1$, $\sigma_{jk} = 0.7^{j+k-2}$ for $j \neq k = 0, \ldots, 7$ and the remaining elements zero. The regression coefficients $\beta_x = \beta_1 = \cdots = \beta_{14} = 0.1$ and $\beta_{15} = \cdots = \beta_{57} = 0$. In this setup, there are seven confounding variables correlated with $\boldsymbol{Y}$ and $\boldsymbol{X}$, seven explanatory variables correlated with $\boldsymbol{Y}$ only, and 43 variables correlated with neither $\boldsymbol{X}$ nor $\boldsymbol{Y}$. The PenCR method is applied on the data set using three priors on $\boldsymbol{\beta}$ and $\boldsymbol{\gamma}$: flat prior ($p(\beta) \propto 1$), normal prior, and Gamma prior. The simulated data is also fit using the true frequentist linear model that includes only the first 14 covariates, the full frequentist linear model that includes all 57 covariates, Bayesian model averaging (BMA) with $\beta_j = \eta_j \alpha_j$, $\eta_j \sim$ Bernoulli(0.5) and $\alpha_j \sim N(0, 10^2)$, Bayesian adjustment for confounding (BAC [32]) with $\omega = \infty$ using the R package `BEAU` and adaptive LASSO [36]. The results for $n = 100$ is given in Table 8.2. In general, the PenCR method with normal and Gamma prior attains similar results with the true model in bias, MSE, and coverage performance, outperforming other methods. The PenCR with a flat prior does not perform well in the $n = 100$ example, however, as shown in [33]. But as sample size increases, the performance will improve.

Data Example

An example is given in [33] to estimate the effect of mean fine particulate matter ($PM_{2.5}$) during the first trimester of pregnancy on birth weight in Mecklenburg County, North Carolina, USA. There are several potential confounders, including the mother's socioeconomic status, medical history, seasonality, and other weather variables. $PM_{2.5}$ levels were obtained from the U.S. Environmental Protection Agency's Fused Air Quality Predictions Using Downscaling database (`https://www.epa.gov/`). The birth and covariate data were obtained from three sources – the North Carolina Vital Statistics (Births) from the years 2003 through 2007, the State Center for Health Statistics (SCHS), and the

TABLE 8.2
Simulation results for confounder selection. Bias, MSE, and coverage are reported for the effect of interest, i.e., β_x. Coverage means 95% confidence or credible interval coverage. Standard errors are reported in the brackets.

$n = 100$	Bias	MSE	Coverage	AUC
True	-0.005 (0.008)	0.032 (0.002)	0.95	NA
Full	-0.013 (0.011)	0.065 (0.004)	0.95	NA
PenCR (Flat Priors)	0.051 (0.009)	0.044 (0.003)	0.76	0.624
PenCR (Empirical Bayes)	0.001 (0.007)	0.028 (0.002)	0.93	0.680
PenCR (Gamma Priors)	0.006 (0.007)	0.028 (0.002)	0.91	0.670
BMA	0.100 (0.006)	0.030 (0.002)	0.80	0.627
BAC ($\omega = \infty$)	0.050 (0.008)	0.033 (0.002)	0.91	0.701
Adaptive LASSO	0.071 (0.009)	0.043 (0.003)	0.71	0.571

Howard W. Odum Institute for Research in Social Science at University of North Carolina at Chapel Hill. Temperature and dew point data were from National Oceanic and Atmospheric Administration Climate Data Online (NOAACDO; `http://www.ncdc.noaa.gov/cdo-web`).

At-risk women aged 40 and over were considered for the analysis, limiting to single births that reached at least 37 weeks of gestation. The exposure variable is the first trimester mean value of $PM_{2.5}$. The potential confounder variables include principal components (PCs) of mean daily temperature throughout the pregnancy, mean daily dew point, and the interaction of temperature and dew point, along with indicators for a birth in spring, summer, or fall. It was anticipated that these weather-related and seasonal variables would have potentially large contributions to both the outcome and the particulate matter levels.

The linear confounder model was applied along with the PenCR processing approach as described. The results show that the point estimates for the $PM_{2.5}$ effects are similar to those obtained without the credible region selection method. However, the standard errors are 10% to 20% smaller than methods without the selection approach. This translated into results which failed to find a significant relationship due to the larger variance, while the PenCR approach reduced that variance and found results that agreed with previous findings. More details are available in [33].

8.7 Time-Varying Coefficients

A global-local shrinkage prior can be used in large over-parameterized models to allow for shrinkage. In a time-varying coefficient model, one approach is to directly model the time-varying parameter in a functional form to reduce the number of parameters. An alternative approach is to allow it to take on a potentially unique value at each time step. Unfortunately, in the second instance, the model is then severely over-parameterized and regularization is needed. Recently, [18] used global-local shrinkage in this model along with the PenCR variable selection method to obtain sparse estimates after fitting the model with a continuous prior.

In particular, a formulation of the time-varying coefficient regression model is given by:

$$y_t = \boldsymbol{\beta}_t' \boldsymbol{x}_t + \varepsilon_t, \ \boldsymbol{\beta}_t = \boldsymbol{\beta}_{t-1} + \boldsymbol{w}_t \tag{8.17}$$

where $\boldsymbol{\beta}_t$ is a p-dimensional dynamic (time-varying) vector of regression coefficients.

Assumptions on the innovation vector, $\boldsymbol{w}_t$, then determine the extent to which the coefficients vary. Following [18], it is assumed that the $\boldsymbol{w}_t$'s are Gaussian innovations with zero mean and variance-covariance matrix $\mathbf{V} = \text{diag}\{v_1, \ldots, v_p\}$. Each v_j, $(j = 1, \ldots, p)$ may be interpreted as the process innovation variance associated with the jth coefficient which controls the amount of time-variation in β_{jt}.

In this case, shrinkage on the v_j results in less variation in β_{jt}. Hence, a global-local shrinkage prior can be applied on these variance parameters. Here, the model may be simplified by inferring the components for which $v_j = 0$ as it implies a non-varying component.

The posterior distribution of $(\boldsymbol{\beta}, \mathbf{V})$ may be computed via MCMC [18]. The PenCR approach can then be applied after obtaining the posterior mean and covariance. As an alternative, [18] performed a thresholding of each MCMC draw as proposed by [29] to obtain a sparse solution.

8.8 Discussion

In this chapter, variable selection is viewed as a post-processing step after obtaining a full posterior distribution. In this way, we can separate, or decouple, the shrinkage step from the selection step. The shrinkage step is accomplished by the choice of prior, followed by the selection step which is accomplished by a potential choice of loss function. The main approach considered is based on the notion that all points lying within a posterior credible region are feasible values. The idea behind PenCR follows the Occam's razor principle – choosing the simplest model from the posterior credible region.

We discuss applications of the PenCR approach to various prior and model choices and suggest that the use of global-local shrinkage priors leads to good variable selection performance. In particular, elliptical contours – centered at the mean and determined by the covariance – can be used to define posterior credible regions. Computationally tractable processing steps can then be used to perform the variable selection. In complex models, the separation of the estimation and variable selection steps allows for two-stage methods to stabilize some pitfalls of single-stage methods. Here, it is worth emphasizing that the PenCR method is still valid even in settings where elliptical credible regions are used as there is an exact one-to-one correspondence between the contour radius and the coverage level. More specifically, by changing the scale of any given shape, we obtain a sequence of regions, and for any choice of coverage level, there is a unique scale value that yields a region containing the required amount of posterior probability.

In comparison to fully Bayesian approaches for model selection, the PenCR enjoys several advantages. Typical approaches require a prior over the space of possible models. Such a prior specification is not required by the PenCR method. In the aspect of posterior computation for fully Bayesian approaches such as SSVS with a spike-and-slab type prior, it is computationally costly to estimate the posterior probability of each model as it involves summarizing MCMC draws from the posterior distribution over the model space. Moreover, if p is very large, an exceptionally large number of MCMC runs is often required to achieve a satisfactory level of stability in these estimates. Such difficulties are avoided in the PenCR method as it does not require us to make inference over the model space. A common way to summarize these model posterior probabilities and choose a model is through marginal inclusion probabilities (see for example, [4]). Unfortunately, such approaches may not perform well when predictors are strongly correlated, as models visited by the MCMC algorithm would tend to include only one predictor from a correlated set at each step. Hence, if there are important and strongly correlated predictors, they may each not be present in the visited models with sufficient frequency to be identified as important. In contrast, the PenCR method avoids this issue as the solution is selected from the credible region of the joint distribution of the coefficients.

The PenCR also separates the regularization and shrinkage step from the selection step. This separation allows more sophisticated shrinkage priors to be assigned without complicating the downstream posterior exploration. Although this can be an advantage, the separation of the shrinkage and selection step is also a disadvantage for the PenCR method. More specifically, PenCR does not allow us to account for the uncertainty in the model selection step, as it only captures the uncertainty from the shrinkage step. Hence, unlike a fully Bayesian approach it does not assess the full uncertainty in the resultant solution.

We also note that the use of elliptical contours in general do not correspond to the highest posterior density regions. Hence, while they can be made to contain the correct posterior probability, they would not have the smallest volume among the regions containing that fixed probability. Consequently, the volume of the ellipsoidal regions may grow to be quite large relative to the optimal region, particularly in high dimensional settings. Whether or not this will affect the performance of the sequence of models created by the approach is an area of further investigation. In addition, the PenCR method, at its current development, may not perform well in cases where the posterior is multimodal. In such cases, the elliptical credible region would have to be large and possibly span across multiple modes to have an adequate coverage level.

Future work is needed to potentially address these current limitations. Direct extension of the PenCR approach to variable selection for more complex models can be done using the elliptical regions whenever we can analytically or numerically compute a posterior mean and covariance. However, developing computationally feasible alternatives to the elliptical regions that may match more closely with high density regions in these models would be useful. For multimodal posteriors, rather than a global region, it may be possible to construct a mixture of elliptical regions around each of the local modes, although the search for the sparse solution in this mixture distribution becomes a less tractable computationally problem.

Software

The `BayesPen` package implementing the PenCR method discussed in this chapter is available on GitHub at `https://rdrr.io/github/AnderWilson/BayesPen`. An R markdown with examples and code to reproduce results in this chapter can be found in the online Supplementary Material of this edited book.

Bibliography

[1] A. Armagan, M. Clyde, and D.B. Dunson. Generalized beta mixtures of Gaussians. In *Advances in neural information processing systems*, pages 523–531, 2011.

[2] A. Armagan, D.B. Dunson, and J. Lee. Generalized double Pareto shrinkage. *Statistica Sinica*, 23(1):119, 2013.

[3] R. Bai and M. Ghosh. Large-scale multiple hypothesis testing with the normal-beta prime prior. *Statistics*, 53(6):1210–1233, 2019.

[4] M.M. Barbieri and J.O. Berger. Optimal predictive model selection. *The Annals of Statistics*, 32(3):870–897, 2004.

[5] A. Bhadra, J. Datta, N.G. Polson, and B. Willard. The horseshoe+ estimator of ultra-sparse signals. *Bayesian Analysis*, 2016.

[6] A. Bhattacharya, D. Pati, N.S. Pillai, and D.B. Dunson. Dirichlet–laplace priors for optimal shrinkage. *Journal of the American Statistical Association*, 110(512):1479–1490, 2015.

[7] H.D. Bondell and B.J. Reich. Consistent high-dimensional bayesian variable selection via penalized credible regions. *Journal of the American Statistical Association*, 107(500):1610–1624, 2012.

[8] P.J. Brown, M. Vannucci, and T. Fearn. Multivariate bayesian variable selection and prediction. *Journal of the Royal Statistical Society: Series B (Statistical Methodology)*, 60(3):627–641, 1998.

[9] E. Candes, T. Tao, et al. The dantzig selector: Statistical estimation when p is much larger than n. *Annals of statistics*, 35(6):2313–2351, 2007.

[10] C.M. Carvalho, N.G. Polson, and J.G. Scott. Handling sparsity via the horseshoe. In *International Conference on Artificial Intelligence and Statistics*, pages 73–80, 2009.

[11] C.M. Carvalho, N.G. Polson, and J.G. Scott. The Horseshoe estimator for sparse signals. *Biometrika*, 97:465–480, 2010.

[12] B. Efron, T. Hastie, I. Johnstone, R. Tibshirani, et al. Least angle regression. *The Annals of statistics*, 32(2):407–499, 2004.

[13] E.I. George and R.E. McCulloch. Variable selection via Gibbs sampling. *Journal of the American Statistical Association*, 88(423):881–889, 1993.

[14] E.I. George and R.E. McCulloch. Approaches for Bayesian variable selection. *Statistica sinica*, 7(2):339–373, 1997.

[15] J.E. Griffin and P.J. Brown. Inference with normal-gamma prior distributions in regression problems. *Bayesian Analysis*, 5(1):171–188, 2010.

[16] P.R. Hahn and C.M. Carvalho. Decoupling shrinkage and selection in bayesian linear models: a posterior summary perspective. *Journal of the American Statistical Association*, 110(509):435–448, 2015.

[17] C. Hans, A. Dobra, and M. West. Shotgun stochastic search for large p regression. *Journal of the American Statistical Association*, 102(478):507–516, 2007.

[18] F. Huber, G. Koop, and L. Onorante. Inducing sparsity and shrinkage in time-varying parameter models. *Journal of Business & Economic Statistics*, pages 1–15, 2020.

[19] H. Ishwaran and J.S. Rao. Spike and slab variable selection: Frequentist and Bayesian strategies. *Annals of Statistics*, pages 730–773, 2005.

[20] S. Kundu, B.K. Mallick, and V. Baladandayuthapan. Efficient bayesian regularization for graphical model selection. *Bayesian Analysis*, 14(2):449–476, 2019.

[21] H. Lan, M. Chen, J.B. Flowers, B.S. Yandell, D.S. Stapleton, C.M. Mata, E.T. Mui, M.T. Flowers, K.L. Schueler, K.F. Manly, et al. Combined expression trait correlations and expression quantitative trait locus mapping. *PLoS Genet*, 2(1):e6, 2006.

[22] J. Lv and Y. Fan. A unified approach to model selection and sparse recovery using regularized least squares. *The Annals of Statistics*, pages 3498–3528, 2009.

[23] T.J. Mitchell and J.J. Beauchamp. Bayesian variable selection in linear regression. *Journal of the American Statistical Association*, 83(404):1023–1032, 1988.

[24] N.N. Narisetty and X. He. Bayesian variable selection with shrinking and diffusing priors. *The Annals of Statistics*, 42(2):789–817, 2014.

[25] J.T. Ormerod, C. You, S. Müller, et al. A variational bayes approach to variable selection. *Electronic Journal of Statistics*, 11(2):3549–3594, 2017.

[26] K. Perrakis and S. Mukherjee. The Alzheimer's Disease Neuroimaging Initiative. Scalable bayesian regression in high dimensions with multiple data sources. *Journal of Computational and Graphical Statistics*, 29(1):28–39, 2020.

[27] N.G. Polson, J.G. Scott, and J. Windle. Bayesian inference for logistic models using Pólya–gamma latent variables. *Journal of the American Statistical Association*, 108(504):1339–1349, 2013.

[28] N. G. Polson and J. G. Scott. Shrink globally, act locally: Sparse bayesian regularization and prediction. *Bayesian Statistics*, 9:501–538, 2010.

[29] P. Ray and A. Bhattacharya. Signal adaptive variable selector for the horseshoe prior. *arXiv preprint arXiv:1810.09004*, 2018.

[30] V. Ročková and E.I. George. Emvs: The em approach to bayesian variable selection. *Journal of the American Statistical Association*, 109(506):828–846, 2014.

[31] Y. Tian, H.D. Bondell, and A. Wilson. Bayesian variable selection for logistic regression. *Statistical Analysis and Data Mining: The ASA Data Science Journal*, 12(5):378–393, 2019.

[32] C. Wang, G. Parmigiani, and F. Dominici. Bayesian effect estimation accounting for adjustment uncertainty. *Biometrics*, 68(3):661–671, 2012.

[33] A. Wilson and B.J. Reich. Confounder selection via penalized credible regions. *Biometrics*, 70(4):852–861, 2014.

[34] Y. Zhang and H.D. Bondell. Variable selection via penalized credible regions with dirichlet–laplace global-local shrinkage priors. *Bayesian Analysis*, 13(3):823–844, 2018.

[35] Y.D. Zhang, B.P. Naughton, H.D. Bondell, and B.J. Reich. Bayesian regression using a prior on the model fit: The r2-d2 shrinkage prior. *Journal of the American Statistical Association*, pages 1–13, 2020.

[36] H. Zou. The adaptive lasso and its oracle properties. *Journal of the American statistical association*, 101(476):1418–1429, 2006.

[37] H. Zou and R. Li. One-step sparse estimates in nonconcave penalized likelihood models. *Annals of statistics*, 36(4):1509, 2008.

Part III

Extensions to Various Modeling Frameworks

9

Bayesian Model Averaging in Causal Inference

Joseph Antonelli

The University of Florida (USA)

Francesca Dominici

Harvard T.H. Chan School of Public Health (USA)

CONTENTS

Bayesian model averaging has traditionally been used in the context of prediction, however, it has recently been applied to a different context: estimation of the effect of a treatment on an outcome. To estimate treatment effects, prior distributions should prioritize variables that are associated with both the treatment and outcome variables, commonly referred to as confounders. In this chapter, we begin with a brief overview of basic topics in causal inference, and then discuss recent work aimed at integrating traditional Bayesian model averaging techniques within a causal inference framework. We discuss different approaches to prior distributions and we illustrate how, if done properly, informative prior distributions can lead to substantial improvements in finite samples. A brief vignette and all R code used to implement the models described in this section can be found in the Supplementary Materials.

DOI: 10.1201/9781003089018-9

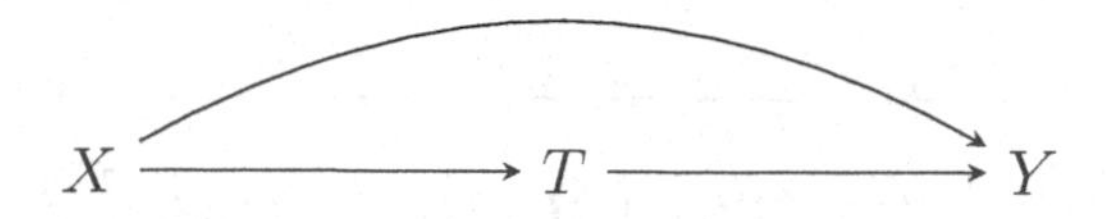

FIGURE 9.1
Causal diagram illustrating a situation in which the effect of T on Y is confounded by X.

9.1 Introduction to Causal Inference

Before we begin with the application of Bayesian model averaging to causal inference, we must first introduce certain fundamental ideas of causal inference. We do not intend to give a complete overview of the field of causal inference. Rather, we will introduce some of the basic terminology and assumptions that are standard in the estimation of causal effects, so that we can better understand how Bayesian model averaging can be applied in this context. For a more complete background to causal inference, we point readers to introductory textbooks in causal inference, such as [1] or [2].

Throughout this chapter, our interest will focus on estimating the causal effect of an exposure/treatment T on an outcome Y. A motivating example that we have encountered regularly in our own work is to estimate the causal effect of lowering the exposure to an environmental agent on a health outcome. The scientific goal is to understand how the risk for adverse health outcomes would change if we were able to intervene and lower the exposure to the environmental agent.

One of the biggest challenges to assessing causality from observational data is confounding. A confounder can be defined as a characteristic of the subjects that is a common cause of both the treatment and the outcome. Confounders (which we will denote by X) could mask or *confound* the relationship between T and Y which complicates causal attribution. An illustration of confounding by X can be found in Figure 9.1. For the example on environmental health, a potential confounder is poverty. It is possible that the poorest individuals are more likely to live in more polluted areas and at the same time are at higher risk of death or other adverse health outcomes due to inherent differences in access to healthcare.

The most common study design used to estimate causal effects are randomized controlled trials (RCTs). Randomization is considered the gold standard for assessing causality, especially in the context of efficacy and safety of new and existing drugs [3]. This is because randomization ensures that there are no systematic differences between the treated and untreated populations. In other words, randomization ensures that the two populations are as similar as possible with respect to all potential confounders, regardless of whether these potential confounders are measured. If the two groups are identical with respect to all the potential measured and unmeasured confounders, then we can use one group to infer what would have happened in the other group under the alternative treatment.

Unfortunately, randomization is often not possible, either because it is not ethical (such as in the context of exposure to environmental contaminants) or because it is hard to implement and/or data collection is costly and time consuming. Therefore the exposed and unexposed populations are potentially systematically different with respect to many confounders. If we fail to account for systematic differences between the treated and untreated populations, inference on causal effects will be biased.

9.1.1 Potential Outcomes, Estimands, and Identifying Assumptions

To define the quantities of interest in a causal analysis, it is important to first define potential outcomes. The potential outcome framework is rooted in the statistical work on randomized experiments by [4, 5] and [6], extended by [7–11] and subsequently by others to apply to observational studies. The potential outcomes are the values of the outcome that we would have observed under a particular treatment assignment. Suppose for now that the treatment of interest is a binary variable, and we are interested in the causal effect of $T = 1$ versus $T = 0$. In this scenario, there are two potential outcomes for subject i, which can be denoted by $Y_i(1)$ and $Y_i(0)$. These correspond to the outcomes we would have observed for subject i had they been exposed to the treatment or control condition, respectively. The individual treatment effect is therefore defined as $Y_i(1) - Y_i(0)$. Commonly interest lies in the average treatment effect, which is defined as $\text{ATE} = E[Y(1) - Y(0)]$. This represents the difference in the average outcome had everyone in the population been exposed to treatment versus if everyone in the population were given the control. Conditional treatment effects, such as $\text{CATE} = E[Y(1) - Y(0)|\boldsymbol{X}]$, are also of interest, however, we will focus on average treatment effects throughout this chapter.

The fundamental challenge is that we will never observe a potential outcome under a condition other than the one that actually occurred. [12] refers to this as *the fundamental problem of causal inference.* The missing potential outcome is typically named as a *counterfactual.* Thus, causal inference relies on the ability to predict the unobserved potential outcomes, or to identify features of the potential outcome distribution using the observed data distribution. Data alone are not enough to identify causal effects from the observed data, and we need to introduce several assumptions on the potential outcomes that permit causal analysis from observed data.

Assumption 1: Stable Unit Treatment Value Assumption (SUTVA). This assumption has two key components: the first is that there exists only one version of each treatment and therefore potential outcomes are well defined, while the second is that there is no interference. Interference occurs when the potential outcomes of one unit are affected by the treatment that was assigned to another unit. Importantly, SUTVA implies that $Y_i(T_i) = Y_i$. SUTVA is commonly violated in vaccine studies where the vaccination status of other individuals can affect a unit's risk of contracting a virus [13]. This assumption can also be violated in studies of air pollution regulations, because intervening at one location (e.g., a pollution source) likely affects health across many locations [14, 15]. Despite these potential difficulties, SUTVA is a widely used assumption in observational causal inference studies, and is realistic in many settings.

Our ability to predict the missing potential outcomes crucially depends on the treatment assignment mechanism which is the probabilistic rule, known or unknown, that determined $T = 1$ versus $T = 0$, and therefore determined if either $Y(1)$ or $Y(0)$ were observed or missing. The assignment mechanism is defined as the probability of getting the treatment conditional on $\boldsymbol{X}, Y(1), Y(0)$, e.g., $P(T|\boldsymbol{X}, Y(1), Y(0))$. This leads us to arguably the most controversial assumption required for many observational studies.

Assumption 2 (no unmeasured confounding): The assignment mechanism is unconfounded if: $P(T \mid \boldsymbol{X}, Y(1), Y(0)) = P(T \mid \boldsymbol{X})$. This assumption implies that if we can stratify the populations within subgroups that have the same values of the covariates (e.g., same age, gender, race, income), then within each of these strata, the treatment assignment (e.g., who gets the drug and who does not) is random. Another way to think of this assumption is that it implies that the treatment and potential outcomes are independent, conditional on the covariates, i.e., $Y(1), Y(0) \perp\!\!\!\perp T|\boldsymbol{X}$. The more covariates we have measured in our observed data, the more plausible this assumption becomes. However, this also complicates estimation by increasing the number of covariates in our respective models.

Assumption 3 (overlap or positivity): We define the propensity score for subject i as the probability of getting the treatment given the covariates [16] as $e(\boldsymbol{x}_i) = P(T_i = 1 \mid \boldsymbol{X} = \boldsymbol{x}_i)$. The assumption of overlap requires that all of the units have a propensity score that is strictly between 0 and 1, that is, they all have a positive chance of receiving either of the two treatments.

Now that we have described assumptions on the potential outcomes, we can show how the estimands of interest (such as the ATE) can be identified from the observed data. This will highlight different estimation strategies that can be used to estimate the treatment effects from the observed data, which will elucidate the different manners in which Bayesian model averaging can improve estimation of causal effects. We will focus here on identifying and therefore estimating $E[Y(t)]$ for $t = 0, 1$ and estimation of contrasts between these values follows immediately. First, we can see that the average potential outcome can be written as

$$\begin{aligned} E[Y(t)] &= E\{E[Y(t)|\boldsymbol{X}]\} \\ &= E\{E[Y(t)|T = t, \boldsymbol{X}]\} \quad \text{by unconfoundedness} \\ &= E\{E[Y|T = t, \boldsymbol{X}]\} \quad \text{by SUTVA} \end{aligned} \tag{9.1}$$

This implies that the average treatment effect can be identified if we are able to estimate the conditional mean of the outcome as a function of treatment status and the covariates, which is an observable quantity that can be estimated from the data. The average treatment effect can also be identified from the observed data using the propensity score, since

$$\begin{aligned} E\left\{\frac{\mathbb{I}\{T = t\}Y}{P(T = t|\boldsymbol{X})}\right\} &= E\left\{\frac{\mathbb{I}\{T = t\}Y(t)}{P(T = t|\boldsymbol{X})}\right\} \quad \text{by SUTVA} \\ &= E\left\{E\left[\frac{\mathbb{I}\{T = t\}Y(t)}{P(T = t|\boldsymbol{X})}\middle| Y(t), \boldsymbol{X}\right]\right\} \\ &= E\left\{\frac{Y(t)}{P(T = t|X)} P[T = t|\boldsymbol{X}]\right\} \quad \text{by unconfoundedness} \\ &= E[Y(t)] \end{aligned} \tag{9.2}$$

This suggests that the ATE can be identified from the observed data from either $E(Y|T, \boldsymbol{X})$ or $P(T = 1|\boldsymbol{X})$. Both of these are models in which Bayesian model averaging can be utilized to improve estimates of causal effects.

9.1.2 Estimation Strategies Using Outcome Regression, Propensity Scores, or Both

The previous section showed that the ATE can be identified from observational data using either $E(Y|T, \boldsymbol{X})$ or $P(T = 1|\boldsymbol{X})$. Now, we are in a position to discuss estimation strategies for the ATE, which will lead into our discussion about how these estimation strategies can utilize Bayesian model averaging to improve inference on causal estimates. First, let's suppose that we want to use the outcome model to estimate treatment effects, and we will see that this is the approach most commonly taken when using Bayesian model averaging in causal inference. A natural estimator of the ATE using the outcome model is

$$\begin{aligned} &\int_{\boldsymbol{x}} \Big(E[Y|T = 1, \boldsymbol{X} = \boldsymbol{x}] - E[Y|T = 0, \boldsymbol{X} = \boldsymbol{x}]\Big) f_{\boldsymbol{X}}(\boldsymbol{x}) d\boldsymbol{x} \\ \approx &\frac{1}{n}\sum_{i=1}^{n} \Big(E[Y|T = 1, \boldsymbol{X} = \boldsymbol{x}_i] - E[Y|T = 0, \boldsymbol{X} = \boldsymbol{x}_i]\Big). \end{aligned} \tag{9.3}$$

Note that this estimator requires a plug-in estimate of the outcome regression, and then integrates over the distribution of the covariates using the empirical distribution from our sample. We will discuss Bayesian estimation of these quantities in more detail in the following section, however, we can easily calculate this quantity for each posterior sample of the outcome regression model parameters, and take the posterior mean. We will see in Section 9.5 that inference on these quantities is more difficult than simply looking at the posterior distribution of this quantity.

Another estimator widely used in the causal inference literature is called the inverse probability weighted, or IPW, estimator. This estimator takes the following form:

$$\text{IPW} = \frac{1}{n}\sum_{i=1}^{n}\left[\frac{T_i Y_i}{e(\boldsymbol{x}_i)} - \frac{(1-T_i)Y_i}{1-e(\boldsymbol{x}_i)}\right], \tag{9.4}$$

where $e(\boldsymbol{x}_i)$ is the propensity score. This estimator differs from the previous one as it does not use the outcome model in any way. It relies solely on the correct specification of the propensity score model. One undesirable feature of both of these estimators is that they rely on correct specification of the outcome model (for the outcome model estimator) and the treatment model (for the IPW estimator). Substantial work has been done to improve these estimators by using so-called doubly robust estimators [17–23]. A doubly robust estimator of the ATE takes the form

$$\begin{aligned}\text{DR} = &\frac{1}{n}\sum_{i=1}^{n}\left[\frac{T_i Y_i}{e(\boldsymbol{x}_i)} - \frac{(T_i - e(\boldsymbol{x}_i))E(Y|T=1,\boldsymbol{X}=\boldsymbol{x}_i)}{e(\boldsymbol{x}_i)}\right]\\ &- \frac{1}{n}\sum_{i=1}^{n}\left[\frac{(1-T_i)Y_i}{1-e(\boldsymbol{x}_i)} + \frac{(T_i - e(\boldsymbol{x}_i))E(Y|T=0,\boldsymbol{X}=\boldsymbol{x}_i)}{1-e(\boldsymbol{x}_i)}\right]\end{aligned} \tag{9.5}$$

This estimator can be shown to be doubly robust in the sense that only one of the propensity score or outcome regression models needs to be correctly specified in order to obtain consistent estimates of the average treatment effect. This estimator is also advantageous for high-dimensional scenarios where the number of confounders is greater than the sample size as it leads to more desirable asymptotic properties and faster convergence rates [24], though this is beyond the scope of this chapter.

9.1.3 Why Use BMA for Causal Inference?

Now that we have presented some introductory concepts in causal inference, it is natural for us to ask how and why we utilize Bayesian model averaging (BMA) to improve estimates of causal effects. Whenever we refer to BMA, we are referring to Bayesian approaches that place prior distributions on the space of possible models defined by subsets of the covariate space, and which explore this space through Markov chain Monte Carlo (MCMC). We will discuss the exact formulation of these prior distributions in subsequent sections, and refer readers to [25] for an in-depth review. Many of the reasons for using BMA in this context are similar to the reasons that analysts use BMA for more traditional contexts such as accounting for model uncertainty in prediction. We will see that when the goal is estimating causal effects, additional subtleties must be addressed. An important motivation for leveraging BMA ideas in causal inference pertains to the no unmeasured confounding assumption, which is one of the most critical assumptions in studies of observational data in which the effect of T on Y is confounded by $\boldsymbol{X}$. By measuring more covariates in $\boldsymbol{X}$, we increase our chances that the no unmeasured confounding assumption is true, because it is more likely that we have measured all necessary confounders. While this is advantageous, it is also problematic as a very large number of covariates can make estimation of causal effects more difficult. We will be working under the assumption that the dimension of $\boldsymbol{X}$ is large. We will not explicitly discuss differences between settings when $p > n$ or $p < n$, but rather we will work under the premise that some form of dimension reduction is useful. Some might argue that if $p < n$ we should include all covariates and not perform model averaging, but this can lead to unstable estimates of the treatment effect if p is moderately large. BMA will be useful as it will allow us to avoid using models with too many covariates, which lead to less efficient estimates of causal effects. If done correctly, BMA will allow us to only utilize models that have removed unnecessary covariates, while including all covariates that are necessary for the no unmeasured confounding assumption to hold.

As we have seen in the previous sections, causal effects can be identified with some combination of either $P(T = 1|\boldsymbol{X})$, $E(Y|T, \boldsymbol{X})$, or both. We will explore the application of BMA on both of these models, and at times both of them simultaneously. For this reason, the space of models that we are considering are the 2^p models for $P(T = 1|\boldsymbol{X})$ and 2^p models for $E(Y|T, \boldsymbol{X})$ that come from deciding whether each of the p covariates is included or excluded in a particular model. Our goal will be to utilize MCMC to explore these large model spaces and ultimately place larger posterior probability on models that include all covariates in $\boldsymbol{X}$ that are necessary for the no unmeasured confounding assumption to hold, i.e., variables that are associated with both treatment and outcome. Note that the approaches considered in this chapter differ from those that first select a set of confounders, and then estimate causal effects conditional on this chosen set of variables [26, 27]. While these approaches show promise for estimating causal effects, we believe that BMA allows us to naturally account for uncertainty inherent to model/confounder selection and parameter estimation, which is crucial for finite sample performance of our estimators. Additionally, we must emphasize that the confounders $\boldsymbol{X}$ must all be pre-treatment variables. If any covariate is measured post-treatment and therefore could be affected by treatment, it should not be included when adjusting for confounding as it can distort the true treatment effect. This is important because Bayesian model averaging will not be able to separate between pre- and post-treatment variables, and will include post-treatment variables if they are correlated with the treatment or outcome.

9.2 Failure of Traditional Model Averaging for Causal Inference Problems

The first question that we should be asking is whether or not we can directly apply Bayesian model averaging to estimation of either $E(Y|T, \boldsymbol{X})$ or $P(T = 1|\boldsymbol{X})$. For simplicity, we will focus on the outcome model in this section, and we will further simplify this model by assuming it takes the following form:

$$E(Y|T = t, \boldsymbol{X} = \boldsymbol{x}_i) = \beta_0 + \beta_t t + \sum_{j=1}^{p} x_{ij}\beta_j. \tag{9.6}$$

This greatly simplifies the problem of estimating the ATE, as equation 9.3 simplifies to β_t, and therefore we can focus on how well BMA does at estimating this parameter. In one of the first papers on BMA, the authors treated all the regression parameters equally and therefore they allowed the parameter β_t to be shrunk towards zero [28]. It was later pointed out that when the goal is estimating β_t and accounting for model uncertainty due to confounder selection, it is important to force the variable T into the model [29]. For this reason, diffuse priors are placed on the intercept and parameter of interest β_t.

A straightforward way to utilize BMA in this model would be to assign independent spike-and-slab priors to each of the coefficients corresponding to the confounders x_j:

$$\begin{aligned} P(\beta_j|\gamma_j^y) &\sim (1-\gamma_j^y)\delta_0 + \gamma_j^y \mathcal{N}(0,\sigma_\beta^2) \\ p(\gamma_j^y|\theta) &= \theta^{\gamma_j^y}(1-\theta)^{1-\gamma_j^y} \\ \theta &\sim \text{Beta}(a_\theta, b_\theta). \end{aligned} \tag{9.7}$$

This prior distribution was first considered in [11] with a continuous version proposed in [6]. This distribution is a mixture between a point mass at zero and a normal distribution centered around zero. The binary variables γ_j^y indicate the importance of x_j for predicting the outcome. If x_j is an important predictor, then $P(\gamma_j^y = 1|\text{Data})$ will be close to 1. The posterior distribution for γ_j^y measures the association of x_j to the outcome, and is not based on the importance of x_j for confounding adjustment. If the true value of the regression coefficient β_j is small to moderate, then it is likely that this covariate will be excluded from the model with high probability. Unfortunately, if x_j is strongly associated with the treatment, excluding x_j from the outcome model, even when it only has a small association with the outcome, can lead to unacceptably high amounts of bias.

To see this, we will run a simple simulation study highlighting this negative feature of traditional BMA. We set the sample size to be $n = 200$, and draw $p = 10$ covariates from independent standard normal distributions. We set $\beta_t = 1$, and $\beta_j = 0$ for $j = 2, \ldots, p$. We vary $\beta_1 \in (0, 0.6)$, which allows for a range of associations between x_1 and Y that goes from no association to a strong association. We will generate the treatment from a Bernoulli distribution with probability given by $P(T = 1|\boldsymbol{X}) = \Phi(0.8x_1)$, implying that covariate 1 has a strong effect on the treatment assignment. We will run BMA as defined above across 1000 data sets for each value of β_1 considered and we will evaluate important metrics such as bias and 95% credible interval coverage of β_t, as well as $P(\gamma_1^y = 1|\text{Data})$, the posterior inclusion probability for covariate 1. We place independent prior distributions on each variable inclusion parameter such that $p(\gamma_j^y = 1) = p(\gamma_j^y = 0) = 0.5$ for all predictors in the data set.

Figure 9.2 shows the results for each of these three metrics as a function of β_1. It is clear from the figures why Bayesian model averaging, if not tailored to the specific problem of estimating β_t, fails in this setting. There is a range of values of β_t, roughly in the interval $(0.1, 0.4)$ for this simulation, in which the exclusion of x_1 in the model leads to non-negligible bias and decreased interval coverage. In this interval, BMA has low values of $P(\gamma_1^y = 1|\text{Data})$, which is leading to the degradation of estimation for β_t. This problem is not unique to this particular data generating process. For any sample size, there will be a range of coefficients for β_1 that lead to the sub-optimal performance here. This issue will occur for any covariates that have moderate to strong associations with the treatment, but won't be an issue if a particular covariate is only weakly associated with the treatment. The problems of this simulation can also be exacerbated if the number of covariates that have strong associations with the treatment and weak associations with the outcome grows. This example is intended to show that the standard prior distributions used for BMA will not lead to good performance for estimating treatment effects. We will see in the following section that informative prior distributions, which incorporate information about the treatment process, can be used to alleviate these issues and lead to improved finite sample performance.

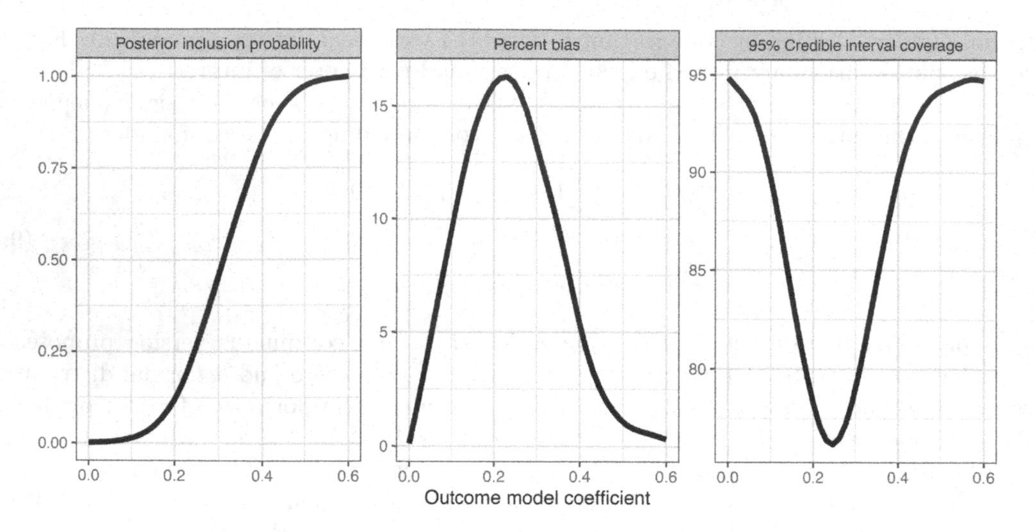

FIGURE 9.2
Results of the simulation study as a function of β_1. The left panel shows the posterior inclusion probability of covariate 1, the middle panel shows the percent bias for estimating β_t, and the right panel shows the percentage of time that the 95% credible interval contains β_t.

9.3 Prior Distributions Tailored Towards Causal Estimation

Before we introduce the prior distributions for model averaging in the context of causal inference, let us first define the treatment model, frequently referred to as the propensity score model. We will adopt a standard generalized linear model

$$g_t^{-1}(P(T = 1|\boldsymbol{X} = \boldsymbol{x}_i)) = \alpha_0 + \sum_{j=1}^{p} x_{ij}\alpha_j, \tag{9.8}$$

where $g_t(\cdot)$ is a link function such as the probit or logit link. This model will play a role in two distinct ways: the first is that predictions from this model can be used in their own right to estimate the average treatment effect, while the second is that this model can be used to develop informative prior distributions for the outcome model coefficients. In this section we focus on the latter of these two ideas, as the propensity score model coefficients can be used to determine which covariates to prioritize in the outcome model. First, it is important to understand the different types of variables in a causal analysis and their relative importance for obtaining unbiased and efficient estimates of treatment effects. In this context, there are four distinct types of variables based on whether they are correlated with the outcome, treatment, both, or neither. These variables and the names by which we will refer to them are summarized in Table 9.1. Confounders are the most critical variables in the estimation of causal effects as the exclusion of a single confounder can lead to biased results and invalid inferences. Predictors of the outcome are also important, however, excluding predictors of the outcome does not bias estimates of causal estimands. Predictors of the outcome can greatly improve the efficiency of treatment effect estimates, and therefore should be included into estimation whenever possible. Noise variables are no different in this context than in the context of predicting an outcome: inclusion of noise variables can

	$\beta_j \neq 0$	$\beta_j = 0$
$\alpha_j \neq 0$	Confounders	Instruments
$\alpha_j = 0$	Predictors	Noise

TABLE 9.1
Types of variables in a causal analysis

reduce efficiency of the model, but otherwise are not cause for concern. Lastly, instrumental variables are variables that are only associated with the treatment. Instrumental variables are interesting because in some cases, they can be used to estimate treatment effects when the unconfoundedness assumption does not hold. However, in the present setting where we are using an outcome model to estimate treatment effects, instrumental variables can be detrimental for a number of reasons. The first is that they reduce the efficiency of effect estimates because their correlation with the treatment makes it more difficult for the data to distinguish between the effect of the treatment and of the instrument. A second reason that instrumental variables can harm an analysis is if there is unmeasured confounding present, instrumental variables can actually inflate this bias when they are adjusted for [32]. Overall, our prior distributions should be used to favor the inclusion of covariates that are associated with both the treatment and outcome, as these are most critical to causal estimation. This amounts to assigning increased probability of inclusion into the outcome model for variables that are also associated with the treatment. This is a delicate issue, however, because if we overly favor covariates that are associated with the treatment, then we might unnecessarily include instrumental variables into the analysis.

9.3.1 Bayesian Adjustment for Confounding Prior

The first prior distribution for model averaging that was tailored towards effect estimation is called the Bayesian adjustment for confounding (BAC) prior. The BAC prior posits spike-and-slab priors for both the treatment and outcome models. We have already defined γ_j^y, which is a latent binary indicator of whether covariate j is used in the outcome model. We will specify analogous latent variables for the coefficients of the propensity score model, such that

$$P(\alpha_j|\gamma_j^t) \sim (1-\gamma_j^t)\delta_0 + \gamma_j^t\mathcal{N}(0,\sigma_\alpha^2).$$

The key difference between the BAC prior and the prior distribution defined in (9.7) is that we will link the latent variables in the outcome and treatment model a priori. Specifically, the BAC prior on the latent variables is such that the conditional odds of inclusion into the outcome model are defined by

$$\frac{P(\gamma_j^y=1|\gamma_j^t=1)}{P(\gamma_j^y=0|\gamma_j^t=1)} = \omega, \quad \frac{P(\gamma_j^y=1|\gamma_j^t=0)}{P(\gamma_j^y=0|\gamma_j^t=0)} = 1, \quad \text{for } j=1,\ldots,p$$

The tuning parameter $\omega \in [1,\infty)$ controls the degree of linkage between the treatment and outcome models. If $\omega = 1$, there is no linkage between the two models and our prior distributions revert back to traditional BMA. For large values of ω, our prior distribution will lead to variables being included with much higher probability in the outcome model if they are also associated with the treatment. For large values of ω, this can lead to the inclusion of instrumental variables, so to avoid this, the following conditional odds are imposed on the treatment model parameters

$$\frac{P(\gamma_j^t = 1 | \gamma_j^y = 0)}{P(\gamma_j^t = 0 | \gamma_j^y = 0)} = \frac{1}{\omega}, \quad \frac{P(\gamma_j^t = 1 | \gamma_j^y = 1)}{P(\gamma_j^t = 0 | \gamma_j^y = 1)} = 1, \quad \text{for } j = 1, \ldots, p.$$

This attempts to exclude variables in the outcome model that are only associated from the treatment. Overall, these two conditional odds imply the following joint prior distribution for (γ_j^t, γ_j^y):

$$P(\gamma_j^t = 0, \gamma_j^y = 0) = P(\gamma_j^t = 0, \gamma_j^y = 1) = P(\gamma_j^t = 1, \gamma_j^y = 1) = \frac{\omega}{3\omega + 1}$$

$$P(\gamma_j^t = 1, \gamma_j^y = 0) = \frac{1}{3\omega + 1}$$

Note that when $\omega = 1$, the prior probability of inclusion into either model is 0.5 for each covariate, which can potentially lead to an increase in false positives when p is large. To avoid this, one could specify that the prior probability of inclusion into the model decreases with p or specify a hyperprior distribution for the prior probability of inclusion into the model. However, in this context we are less worried about false positives and more worried about excluding important confounders, so we do not study this issue further. This prior distribution, or variations of it, have been used in a variety of contexts in causal inference such as heterogeneous treatment effects [33], missing data [34], and exposure-response curve estimation [35]. Similar ideas of linking exposure and outcome models have also been used in more difficult settings such as the estimation of causal effects for multivariate exposures [36].

The main difficulty with implementing this prior distribution is that the parameter ω must be chosen. There has been some work done aiming to choose this parameter in a data-driven way through cross-validation or the bootstrap [37]. These algorithms are intended to minimize the MSE of the resulting effect estimate as a function of ω, however, the authors found mixed results and determined that the algorithms were very sensitive to the data generating mechanism. Generally, we view this as a tuning parameter with which sensitivity analysis can be performed. Increasing ω can decrease the efficiency of the resulting estimates, but tends to have smaller bias, and therefore we recommend a large value of ω.

To illustrate how this prior distribution can improve estimates of treatment effects, we apply it to the same example as in Section 9.2 that we used to show how BMA can fail in certain settings. We apply the same model to these data, but we use the BAC prior with $\omega \in \{1, 25, 100, 1000, 50000\}$. The results can be found in Figure 9.3. It is clear that the BAC prior is an improvement on the traditional BMA prior ($\omega = 1$) that does not incorporate information from the treatment model. As ω increases, the bias decreases and the 95% credible intervals contain the true parameter a higher percentage of the time. The reason for this is the increase in the posterior inclusion probability of the important covariate, as seen in the left panel of Figure 9.3. In fact, the only model that is able to obtain unbiased estimates and nominal credible interval rates is the one with $\omega = 50000$, which effectively forces any covariate into the outcome model that is associated with the treatment. This has been seen in frequentist contexts for variable selection in causal inference that have shown that one must include all variables associated with *either* the treatment or outcome in order to obtain uniformly consistent results [26]. This does come at a cost, however, as we can see that even when the outcome model coefficient is zero, the posterior inclusion probability is close to one for $\omega = 50000$. We are able to reduce bias with very large ω values, but it will increase the variability in our estimates as well due to high posterior inclusion probabilities for covariates only associated with the treatment. We will encounter this same issue in the NHANES data analysis in Section 9.4.4.

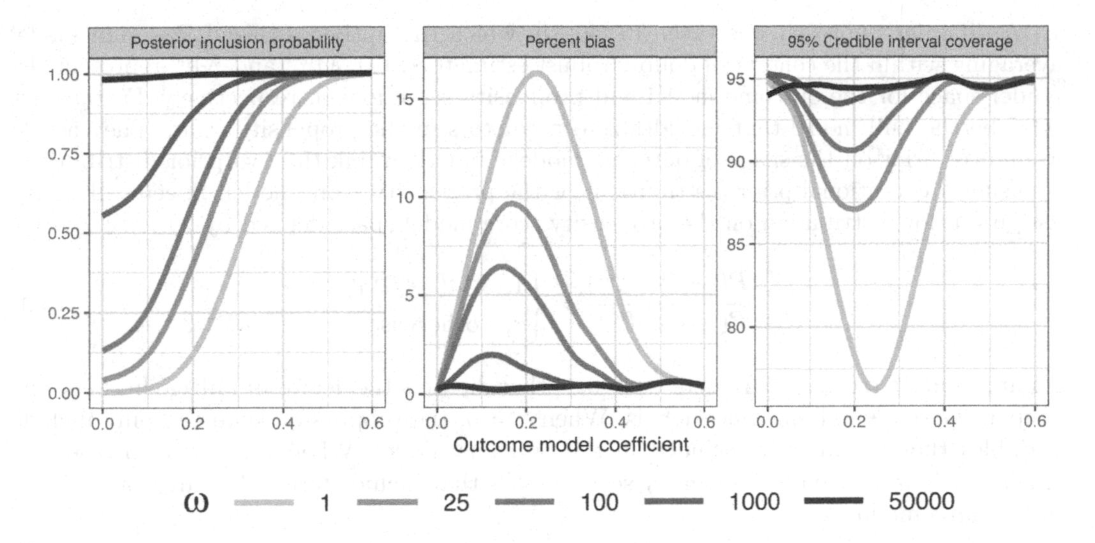

FIGURE 9.3
Results of the simulation study as a function of β_1. The left panel shows the posterior inclusion probability of covariate 1, the middle panel shows the percent bias for estimating β_t, and the right panel shows the percentage of time that the 95% credible interval contains β_t. The different lines denote the different values of ω considered.

9.3.2 Related Prior Distributions that Link Treatment and Outcome Models

Other approaches have built on similar ideas to construct prior distributions for the model space that improve effect estimation over naively applying traditional BMA that does not acknowledge the treatment model associations in any way. One approach extends the formulation in (9.7) through the introduction of a covariate-specific weight parameter w_j given by

$$p(\gamma_j^y|\theta) = \theta^{w_j \gamma_j^y}(1 - \theta^{w_j})^{1-\gamma_j^y}.$$

This covariate specific $w_j \in (0, 1)$ allows each covariate to have a different prior probability of entering into the model. If $w_j = 1$, then covariate j has prior probability of θ to enter into the model. If, however, $w_j < 1$, then covariate j has an increased probability of entering into the outcome model. The general idea is to set w_j to be small for covariates that have strong associations with the treatment. There are many ways one could go about setting these tuning parameters. One could make them inversely proportional to the covariate's association with the treatment, thereby increasing the prior inclusion probability for covariates with strong treatment associations. The approach taken in [38] is to first fit a frequentist estimate of the propensity score model using a standard approach such as the LASSO [39] that automatically performs variable selection. Then, for all covariates j that are deemed important by the propensity score model fit, w_j is set to δ, while all other weights are set to 1. This simplifies the problem to a single tuning parameter δ, however, the user must still choose a value for this parameter. In [38], the authors provide an automated way to select δ that maximizes the prior probability of inclusion for variables that are associated with the treatment while controlling the posterior probability of inclusion for instrumental variables, thereby balancing our desire to adjust for all confounders with our desire to eliminate instrumental variables.

A different approach was taken in [40], in which the authors utilized Bayesian model averaging within the context of doubly robust estimators. Let $\mathcal{M}^{om}$ and $\mathcal{M}^{ps}$ represent the model space for the outcome model and propensity score model, respectively. Further, let $\mathcal{M}_1^{ps}$ be the null model that includes zero predictors in the propensity score. They assign a uniform prior on the space of outcome models, but then link the two prior distributions by using a conditional prior distribution for the propensity score model. Specifically, they assign a prior distribution on the propensity score model space defined by

$$\frac{P(\mathcal{M}_i^{ps}|\mathcal{M}_j^{om})}{P(\mathcal{M}_1^{ps}|\mathcal{M}_j^{om})} = \begin{cases} 1, & \mathcal{M}_i^{ps} \subset \mathcal{M}_j^{om} \\ \tau, & \text{otherwise.} \end{cases} \tag{9.9}$$

Again, we have a tuning parameter $\tau \in [0, 1]$ that dictates the degree of linkage between the propensity score and outcome models. When $\tau = 0$, the propensity score can only include variables that are already included in the outcome model. When τ is between 0 and 1, smaller weight is given to propensity score models that include terms that are not included in the outcome model.

9.4 Bayesian Estimation of Treatment Effects

Now that we have discussed prior distributions on the model space of both the propensity score and outcome models, we can proceed with a discussion of how model averaging can be used to obtain improved treatment effect estimates. In this section, we will focus on constructing point estimates of treatment effects that utilize model averaging, and we will discuss issues with inference and constructing credible intervals in the following section. We will see that performing inference on many of the estimators discussed in this section is not as straightforward as simply looking at quantiles of the posterior distribution.

In terms of estimating treatment effects with Bayesian model averaging, the approaches that have been proposed can generally be put into three distinct groups, and we will discuss each of them separately. The first of which utilizes the propensity score, but only as a means for constructing informative prior distributions on the outcome model space. The second group of approaches also utilizes the propensity score, but they explicitly include the propensity score into the outcome model. Lastly, the third group of approaches can be seen as only approximately Bayesian, as they do not attempt to come up with a fully Bayesian estimate of the treatment effect. Instead, they use model averaging on either the propensity score or outcome model to improve estimates of those quantities, but then use the resulting estimates within traditional causal inference estimators that are not based on likelihoods and therefore are not fully Bayesian.

9.4.1 Outcome Model Based Estimation

This class of methods is based on the estimator of the treatment effect defined in (9.3). If we have a model for the outcome regression, then it is straightforward to estimate the ATE as we simply must integrate the model over the confounder distribution. The goal of these methods is to utilize Bayesian model averaging to improve estimation of the outcome model, which should in turn, lead to improved estimation of the treatment effect itself. These methods rely on factorizing the likelihood as $P(Y, T|\boldsymbol{X}, \boldsymbol{\alpha}, \boldsymbol{\beta}) = P(Y|T, \boldsymbol{X}, \boldsymbol{\beta})P(T|\boldsymbol{X}, \boldsymbol{\alpha})$. This implies that the likelihood separates into the outcome model and the propensity score model. If we were to use independent prior distributions for the parameters of these respective models, then the propensity score would play no role in the estimation of the outcome model, and it could be dropped for the purposes of treatment effect estimation.

We have seen, however, that informative prior distributions on the model space that link the treatment and outcome models can be used to improve estimation of the outcome model if we are interested in treatment effects. Due to this, these approaches aim to learn the joint posterior distribution of both the treatment and outcome model parameters. The posterior distribution of these two quantities are dependent, though this dependence is based solely on the prior distribution for the joint model space. Once we have posterior samples of both models, we can ignore the posterior distribution of the treatment model, and proceed with using the outcome model parameters to estimate treatment effects. Suppose that we have B posterior samples, and for $b = 1, \ldots, B$ we have $E(Y|T = t, \boldsymbol{X} = \boldsymbol{x}_i, \boldsymbol{\beta}^{(b)})$, posterior draws of the conditional outcome regression. A natural estimator of the average treatment effect can then be constructed as

$$\widehat{\text{ATE}} = \frac{1}{Bn} \sum_{b=1}^{B} \sum_{i=1}^{n} \Big(E[Y|T = 1, \boldsymbol{X} = \boldsymbol{x}_i, \boldsymbol{\beta}^{(b)}] - E[Y|T = 0, \boldsymbol{X} = \boldsymbol{x}_i, \boldsymbol{\beta}^{(b)}] \Big). \tag{9.10}$$

This is essentially the posterior mean of the quantity in (9.3), and therefore averages over all possible linear models for the outcome regression. Note that this quantity is simplified substantially if $E[Y|T = t, \boldsymbol{X} = \boldsymbol{x}_i, \boldsymbol{\beta}]$ is a linear model with no interactions between treatment and confounders, such as in (9.6). In this case, the treatment effect is simply β_t and therefore our estimator is simply the posterior mean of this quantity.

9.4.2 Incorporating the Propensity Score into the Outcome Model

The previous section dealt with approaches that only used the propensity score to help determine which covariates to prioritize in the outcome regression model. This section deals with approaches that include the propensity score into the outcome regression model itself, which has been commonly utilized in causal inference methodology and increases robustness to misspecification of the outcome model [41, 42]. These approaches also factorize the likelihood via $P(Y, T|\boldsymbol{X}, \boldsymbol{\alpha}, \boldsymbol{\beta}) = P(Y|T, \boldsymbol{X}, \boldsymbol{\beta})P(T|\boldsymbol{X}, \boldsymbol{\alpha})$, and utilize the same propensity score as defined in (9.8). The key difference is the formulation of the outcome model, which now takes the form

$$g_y^{-1}(E(Y|T = t, \boldsymbol{X} = \boldsymbol{x}_i)) = \beta_0 + \beta_t t + m(e(\boldsymbol{x}_i)) + \sum_{j=1}^{p} x_{ij}\beta_j, \tag{9.11}$$

where $g_y(\cdot)$ is a link function, and $e(\boldsymbol{x}_i) = g_t(\alpha_0 + \sum_{j=1}^{p} \alpha_j x_{ij})$ is the propensity score. The $m(\cdot)$ function describes how the propensity score is included in the outcome model. Possible choices include non-linear functions of the propensity score or dummy variables indicating membership into subclasses defined by quantiles of the propensity score distribution. The propensity score can potentially remove any confounding bias on its own, but the remaining covariate adjustment $\sum_{j=1}^{p} x_{ij}\beta_j$ can help to remove any residual confounding that is left. This closely resembles doubly robust approaches [20] as these two methods for confounding adjustment provide users with two chances to remove any confounding bias. In principle, model averaging could be used in either the propensity score model or the residual adjustment in the outcome model, or both. Model averaging on the propensity score was studied thoroughly in [41], where the authors also considered model averaging on the residual confounding adjustment in the outcome model.

One subtle issue with this model is referred to as model feedback, which is caused by the fact that the $\boldsymbol{\alpha}$ parameters appear in both the treatment and outcome components of the likelihood. This implies that the outcome will help to inform the parameters of the propensity score model. A key property of propensity scores that allows them to be used for confounding adjustment is called the balancing property, which states that

$$T \perp\!\!\!\perp \boldsymbol{X} | e(\boldsymbol{X}).$$

Utilizing information from the outcome when updating the propensity score can break this assumption, and it is not clear whether the propensity score will be a useful quantity for confounding adjustment. In fact, recent work has shown that a fully Bayesian analysis that allows the outcome to influence the propensity score can bias estimates of the treatment effect [43]. The same authors showed that this model feedback issue can be addressed in two distinct ways. The first is that if all covariates in the propensity score model are also included in the residual adjustment term $\sum_{j=1}^{p} x_{ij}\beta_j$, then the propensity score is not distorted and one can still obtain unbiased estimates of the causal effect. The second method is to "cut" the feedback, which amounts to an approximately Bayesian approach that only utilizes the treatment model when updating the propensity score model parameters, and ignores the information from the outcome component of the likelihood. One such approach in the current context can be found in [41], while an interesting discussion on cutting information across distinct Bayesian models can be found in [44]. Regardless of how the outcome model is ultimately fit, inference on the ATE proceeds in the same manner as the previous section and equation (9.10).

9.4.3 BMA Coupled with Traditional Frequentist Estimators

The final set of estimators that incorporate Bayesian model averaging do not attempt to provide fully Bayesian inference on treatment effects, but rather borrow ideas from Bayesian model averaging to improve frequentist estimates of treatment effects. The estimators defined in (9.4) or (9.5) could both incorporate Bayesian model averaging in a relatively straightforward manner. Both of these estimators rely on estimates of the propensity score and outcome regression: both of which we have already seen can be improved with Bayesian model averaging. Interest has particularly focused on the estimator defined in (9.5) as it is doubly robust, and only requires one of the two models to be correctly specified to obtain consistent estimates of treatment effects. Bayesian model averaging was used with this estimator in [40]. The authors used Bayesian model averaging with the prior distribution given in (9.9) to obtain $P(\mathcal{M}_i^{ps}, \mathcal{M}_j^{om}|\text{Data})$, where $\mathcal{M}_i^{ps}$ and $\mathcal{M}_j^{om}$ represent the model spaces for the treatment and outcome models respectively. For each chosen model defined by $\mathcal{M}_i^{ps}$ and $\mathcal{M}_j^{om}$, the authors calculate frequentist estimates of the propensity score and outcome regression functions. Given these estimates, the doubly robust estimator for this combination of models can be defined as $\widehat{\Delta}_{ij}^{DR}$. A model averaged doubly robust estimator can be defined as $\widehat{\Delta}_{DR}^{MA} = \sum_{ij} \widehat{w}_{ij} \widehat{\Delta}_{ij}^{DR}$, where the weights are given by $\widehat{w}_{ij} = P(\mathcal{M}_i^{ps}, \mathcal{M}_j^{om}|\text{Data})$. The authors showed that if either the true propensity score or outcome regression model lies within the class of models being considered, then this model averaged estimator is consistent for the ATE.

Some recent work has proposed fully Bayesian versions of these doubly robust estimators [45], though there has been substantial debate about whether an estimate of counterfactual outcomes can utilize the propensity score within the Bayesian framework [46]. [47] showed that a Bayesian analysis which honors the likelihood principle can not utilize the propensity score. Therefore, any approach targeting the estimators in (9.4) or (9.5) will likely only be approximately Bayesian, and inference will be carried out in a traditional frequentist manner. For a detailed discussion of this, and ways to perform inference that utilize posterior samples, see [48].

9.4.4 Analysis of Volatile Compounds on Cholesterol Levels

In this section we apply the aforementioned methodology to estimate the effects of volatile organic compounds on LDL cholesterol, HDL cholesterol, and triglyceride levels. The data to be examined come from The National Health and Nutrition Examination Survey (NHANES), which is a cross-sectional data source made publicly available by the Centers for Disease Control and Prevention (CDC). This data set was originally developed in [49], and has been subsequently analyzed in [36] and [48]. All data and analyses run in this section can be found along with a vignette that walks users through each step of the analyses in the online Supplementary Materials.

One interesting aspect of this analysis is that there are three outcomes and ten treatments, leading to thirty distinct analyses. The sample size in the data is $n = 179$ and there are $p = 82$ potential confounders to adjust for. Since the treatment variables are continuous measures of volatile organic compound exposure, the majority of the aforementioned approaches that incorporate the propensity score do not directly apply. For this reason, we will restrict attention to the BAC prior discussed in Section 9.3.1. We are fitting an outcome model defined in equation (9.6), and we are going to vary the prior distribution on the β_j coefficients corresponding to each potential confounder. Note that we are only considering a single volatile compound in each model, and will run separate models for each volatile compound. We will vary $\omega \in \{1, 25, 100, 1000, 50000\}$. When $\omega = 1$, the treatment model does not play any role in estimation of the causal effect and the prior probability of entry into the model for each covariate is 0.5. As we increase ω, covariates that are associated with the treatment are assigned higher prior probabilities of being nonzero in the outcome model, with $\omega = 50000$ corresponding to the case where we effectively force covariates into the outcome model if they are included in the treatment model.

The results are displayed in Figure 9.4, which shows posterior means and 95% credible intervals for the effect of each volatile compound on the three outcomes of interest. In many instances, the effect estimates and 95% credible intervals are similar across the range of ω values considered, however, there are a few exceptions where the results vary substantially by this tuning parameter. For all analyses examining the 7th volatile compound (VC 7), the credible interval widths greatly increase as ω increases. To understand this issue deeper, we can look at posterior inclusion probabilities for each covariate into the outcome model as a function of ω. Figure 9.5 shows the posterior inclusion probabilities for all covariates that had a posterior inclusion probability of at least 0.03 for one of the ω values in the analysis of VC 7 on HDL cholesterol levels. Covariates 8 and 15 have posterior inclusion probabilities near 1 for any value of ω, which indicates that these are strongly predictive of HDL cholesterol. Covariate 68 on the other hand has a posterior inclusion probability near 0 when $\omega = 1$, but an increasing posterior inclusion probability as ω increases. The 68th predictor has a marginal correlation of 0.83 with VC 7, which indicates a very strong relationship with the treatment variable. From a regression perspective it is clear that this will increase the posterior variance of β_t, the effect of interest, as it leads to a high degree of multicollinearity. From a causal inference perspective, we could be introducing an instrumental variable, which is known to reduce the efficiency of the causal effect of interest.

Another instance where the results vary across values of ω is the effect of VC 4 on HDL cholesterol. As in the previous example, this is a situation where a covariate is included in more MCMC iterations due to an association with the treatment and a large ω value. In this case, however, the widths of the credible interval only increase slightly, while the effect estimate doubles from 0.1 to 0.2 and the 95% credible interval no longer contains 0. It is possible that this covariate is only associated with the treatment and is being included into the outcome model unnecessarily as we increase ω. It is also possible that it has a weak association with the outcome and should be included in the model to obtain unbiased

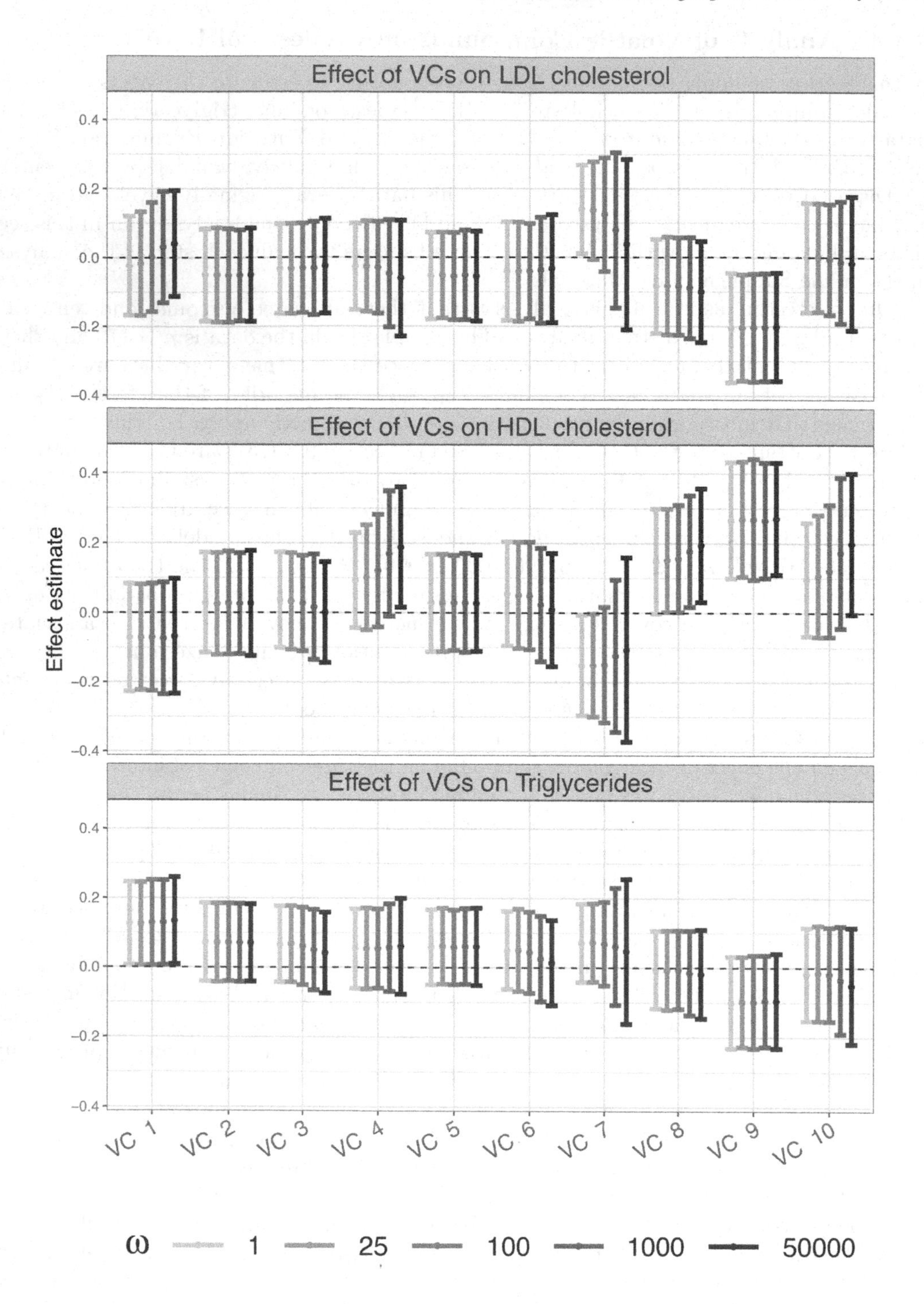

FIGURE 9.4
Results of the NHANES analysis using the Bayesian adjustment for confounding prior with varying levels of ω.

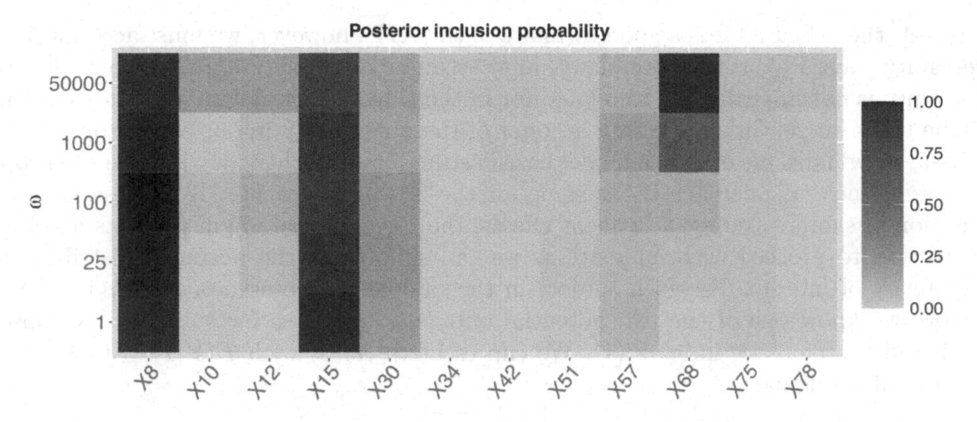

FIGURE 9.5
Posterior inclusion probabilities from the NHANES analysis estimating the effect of VC 7 on HDL cholesterol level.

estimates of the true effect of VC 4 on HDL cholesterol. In practice we can never know which of these two scenarios is occurring, but we favor larger values of ω to protect against the exclusion of an important confounding variable. As seen in Figure 9.3, only extremely large values of ω can fully protect against bias from an omitted confounder, though this will likely also lead to increased variance in our effect estimates.

9.5 Assessment of Uncertainty

Generally, one of the advantages of working within the Bayesian paradigm is that inference is automatic once we have posterior samples of all unknown parameters. Inference on any function of the unknown parameters can easily be accommodated with no additional effort, and no asymptotic approximations need be made. Unfortunately, this general principle does not hold for most applications of Bayesian model averaging to causal inference. It should be stressed that this is not unique to Bayesian model averaging, but rather all uses of Bayesian models for the propensity score or outcome regression models.

Before we discuss inference in greater detail, it is important to distinguish between sample average treatment effects and population average treatment effects. Throughout, we have been using the ATE as a motivating example, which is a population average causal effect as it is the expected value with respect to the population of interest. The sample average treatment effect (SATE) is of the form

$$\frac{1}{n}\sum_{i=1}^{n}\Big[Y_i(1) - Y_i(0)\Big],$$

while the population counterpart to this estimand is

$$E(Y(1) - Y(0)).$$

Sample average treatment effects treat potential outcomes as fixed values, while population average treatment effects treat potential outcomes as random variables drawn from a superpopulation. If we believe that our sample is a random sample from the superpopulation of interest, then the SATE is a good estimate of the ATE, however, we must account for the uncertainty caused by using a sample mean to estimate a population mean. This additional uncertainty is not immediately accounted for in the Bayesian paradigm and additional care must be taken to get an appropriate account of the uncertainty in our estimator.

First, we will discuss outcome model based estimators, whether they include the propensity score or not, as inference is the same for these two approaches. If interest lies in the estimation of sample average treatment effects, the Bayesian paradigm presents a straightforward way to proceed with inference as we can simply use the posterior distribution of the quantity of interest. For each subject in the sample, we observe one of $Y_i(1)$ or $Y_i(0)$, therefore we know one of the two potential outcomes. Suppose for subject i, we observe $T_i = 1$ and therefore we know $Y_i(1)$. We can estimate $Y_i(0)$ with $E(Y|T_i = 0, \boldsymbol{X}_i = \boldsymbol{x}_i)$. Therefore, if we define

$$\Delta_i = \begin{cases} Y_i - E(Y|T_i = 0, \boldsymbol{X} = \boldsymbol{x}_i), & T_i = 1 \\ E(Y|T_i = 1, \boldsymbol{X} = \boldsymbol{x}_i) - Y_i, & T_i = 0, \end{cases}$$

the sample average treatment effect can simply be estimated by $\Delta = \frac{1}{n}\sum_{i=1}^{n} \Delta_i$. Posterior inference on this quantity is straightforward because the only unknown quantity is $E(Y|T, \boldsymbol{X})$, which is a function of the unknown parameters and is easily accounted for through the posterior distribution of the outcome model. Inference can proceed by calculating Δ for every posterior sample and taking the relevant quantiles of the posterior distribution to construct credible intervals.

Inference is more complicated for population average treatment effect estimates. Suppose that we choose to follow (9.3) and estimate the ATE with

$$\frac{1}{n}\sum_{i=1}^{n} \Big(E[Y|T = 1, \boldsymbol{X} = \boldsymbol{x}_i] - E[Y|T = 0, \boldsymbol{X} = \boldsymbol{x}_i]\Big).$$

It may seem as though this is simply a function of $E(Y|T, \boldsymbol{X})$ and therefore inference can proceed directly from the posterior. However, this estimator is also averaging over the covariate distribution $\boldsymbol{x}_i$ that is observed in the sample to estimate a population average, and this uncertainty must be accounted for as well. Thankfully, a relatively straightforward modification allows us to use the posterior samples to perform inference on the ATE. We will construct weights ξ_i using the Bayesian bootstrap [50]. Define $u_0 = 0, u_n = 1$ and $u_1, \dots, u_{n-1}$ to be the order statistics from $n-1$ draws of a standard uniform distribution. Then the weights can be defined as $\xi_i = u_i - u_{i-1}$. We will construct one set of weights for each of the B posterior samples, and therefore we have $\xi_i^{(b)}$ for $i = 1, \dots, n$ and $b = 1, \dots, B$. For every posterior sample, we can calculate

$$\frac{1}{n}\sum_{i=1}^{n} \xi_i^{(b)} \Big(E[Y|T = 1, \boldsymbol{X} = \boldsymbol{x}_i, \boldsymbol{\beta}^{(b)}] - E[Y|T = 0, \boldsymbol{X} = \boldsymbol{x}_i, \boldsymbol{\beta}^{(b)}]\Big).$$

We can treat this as a posterior distribution for the ATE and can perform inference using the posterior quantiles as a credible interval. Effectively what this is doing is treating the covariates $\boldsymbol{x}_i$ as unknown variables and approximating their distribution by assigning weights $\xi_i^{(b)}$ to the empirical distribution of the sample.

Now we will divert attention to cases where Bayesian model averaging is used to improve estimates of the propensity score or outcome regression models, which are then used in the context of non-likelihood based estimators such as the inverse probability weighted or doubly robust estimator. For brevity we will only briefly mention approaches to inference in this context and point to important references exploring this problem in greater detail.

We have already discussed the estimator of [40], which combined Bayesian model averaging and the doubly robust estimator in a way that uses informative prior distributions on the model space for both the treatment and outcome models. This approach does not use the posterior distribution to account for uncertainty, and simply applies the standard nonparametric bootstrap [51] to the entire procedure to perform inference. While this approach works in some settings, it is not ideal because it does not readily extend to cases where the number of covariates is large relative to the sample size, and can be computationally intensive to calculate the posterior distribution of the propensity score and outcome models for each bootstrap sample. This approach was recently extended to nonparametric and high-dimensional models for the propensity and outcome models in a manner that does not require updating the posterior distribution for each bootstrap sample [48]. They showed that inference can be greatly improved in high-dimensional settings by accounting for parameter uncertainty in the propensity score and outcome regression models. This approach is also not fully Bayesian and does not attempt to construct a posterior distribution from which to perform inference. Instead they use Bayesian inference to account for a difficult source of uncertainty that is typically ignored in asymptotic approximations of frequentist estimators.

Lastly, recent work has focused on accounting for estimation in the propensity score when estimating causal effects [52]. There has been a fair amount of interest in estimating propensity scores within the Bayesian paradigm so that useful Bayesian ideas, such as Bayesian model averaging, can be used to improve causal estimation [41, 53]. Propensity scores can subsequently be used to estimate causal effects in many distinct fashions such as inverse weighting, matching, stratification, or outcome regression modeling. Most of these, however, are not likelihood based, and it is not clear how inference that accounts for uncertainty in propensity score estimation can proceed in the Bayesian paradigm. The different sources of uncertainty are delineated in [52], and the authors show how approximate Bayesian inference can proceed by separating the propensity score estimation stage and the estimation of causal effects.

9.6 Extensions to Shrinkage Priors and Nonlinear Regression Models

The discussion so far has centered around generalized linear models for both the propensity score and outcome regression model, where the two models are linked through prior distributions on the separate model spaces to encourage the inclusion of variables that are associated with both treatment and outcome. There are two key extensions to these ideas, which we briefly discuss here: first we will discuss continuous shrinkage priors, and then we discuss extending these models to semiparametric or nonparametric Bayesian models. The first extension no longer links the treatment and outcome models through informative prior distributions, but rather re-parameterizes the model in a way that avoids overly shrinking parameter estimates that correspond to confounders, thereby reducing bias in estimating treatment effects [54]. We restrict attention to continuous treatments, but extending to the binary setting is relatively straightforward. This approach is based on re-parameterizing the outcome and treatment models to

$$
\begin{aligned}
T &= \boldsymbol{X}\boldsymbol{\beta}_c + \epsilon \\
Y &= \beta_t(T - \boldsymbol{X}\boldsymbol{\beta}_c) + \boldsymbol{X}\boldsymbol{\beta}_d + \nu.
\end{aligned}
$$

Here $\boldsymbol{\beta}_c$ corresponds to the confounding effect of $\boldsymbol{X}$ in the sense that any covariate j that is a confounder will have $\beta_{c,j} \neq 0$. On the other hand, $\boldsymbol{\beta}_d$ corresponds to the direct effect of the covariates on the outcome. This means that any covariate with an effect on the outcome that is not through its effect on the treatment will have $\beta_{d,j} \neq 0$. Note that $\boldsymbol{\beta}_c$ now appears in both the treatment and outcome models. Therefore the amount of shrinkage of this parameter depends on both the treatment and outcome, thereby reducing shrinkage for covariates that are associated with both the treatment and outcome. By parameterizing the model in this way, we no longer need to use informative prior distributions, but can instead employ independent prior distributions for $\boldsymbol{\beta}_c$ and $\boldsymbol{\beta}_d$ and still achieve desirable amounts of shrinkage if interest is in the estimation of β_t. It was initially proposed to use shrinkage priors such as

$$\begin{aligned}\pi(\beta_{c,j}) &\sim \frac{1}{v}\log\left(1+\frac{4}{(\beta_{c,j}/v)^2}\right) \quad \text{for } j=1,\ldots,p\\ \pi(v) &\sim C^+(0,1),\end{aligned}$$

where $C^+(0,1)$ denotes the standard folded Cauchy distribution. Alternatively, we could employ Bayesian model averaging or independent spike-and-slab prior distributions for these parameters to eliminate unnecessary covariates completely from the model.

An additional extension to the ideas presented in earlier sections is to introduce non-linear functions of the covariates in both the propensity score and outcome regression models. We can posit the following models for the treatment and outcome:

$$\begin{aligned}g_y^{-1}(E(Y|T=t,\boldsymbol{X}=\boldsymbol{x}_i)) &= \beta_0 + f_t(t) + \sum_{j=1}^{p} f_j(x_{ij})\\ g_t^{-1}(P(T=1|\boldsymbol{X}=\boldsymbol{x}_i)) &= \alpha_0 + \sum_{j=1}^{p} h_j(x_{ij}).\end{aligned} \tag{9.12}$$

Now we need to specify prior distributions for the $f_j(\cdot)$ and $h_j(\cdot)$ functions in a way that both allows for non-linear relationships between $\boldsymbol{X}$ and T or Y, but also permits model averaging and linkage of the prior distributions in the two models. Fortunately, spike-and-slab priors have been adopted for many commonly used Bayesian semiparametric or nonparametric models, such as Gaussian processes [55]. These ideas have also been employed specifically in the context of causal inference [48]. If we do not want to assume additivity of the models in 9.12, then tree-based approaches that incorporate model averaging can also be used [56, 57]. This would additionally allow for the treatment effect to vary as a function of covariates, otherwise known as treatment effect heterogeneity.

As long as spike-and-slab prior distributions are used for both the treatment and outcome models, the informative prior distributions discussed in Section 9.3 can be applied to non-linear models. Nearly all work done that applies Bayesian model averaging in causal inference tailors the prior distribution on the variable inclusion parameters or model space in a way that reduces bias in finite samples. We simply need to have binary indicators of variable inclusions for each covariate, denoted by (γ_j^t, γ_j^y) for $j = 1, \ldots p$. These binary inclusion indicators have the exact same interpretation and purpose as in the linear settings described in previous sections, and therefore the informative prior distributions that link these two parameters, such as the BAC prior, can also be utilized here in a non-linear modeling setting.

9.7 Conclusion

In this chapter we have reviewed a variety of ways that Bayesian model averaging can be utilized to improve estimates of causal effects. Causal inference differs from traditional applications of Bayesian model averaging as we need to prioritize variables that are associated with both the treatment and outcome to obtain unbiased estimates of the effects of interest. Many causal estimators are functions of either the propensity score, the outcome model, or both. All of these models can be improved with Bayesian model averaging, but additional care must be given to the prior distributions on the model space to ensure that important confounders are included. Typically this is done by prioritizing variables in the outcome model that are also associated with the treatment. The extent to which variables are prioritized in the outcome model is an important tuning parameter that can greatly influence subsequent effect estimates as we have seen in both simulated data and the NHANES analysis. There is an inherent bias-variance trade-off with this decision: as we increase priority for variables associated with the treatment, our estimates tend to have smaller biases but larger variances. Finding the value of the tuning parameter that optimizes this trade-off is still an open question that merits further research.

There is an entirely different strategy to estimating causal effects in the Bayesian paradigm that could also benefit from Bayesian model averaging. This viewpoint explicitly models the joint distribution of the potential outcomes $(Y(1), Y(0))$. Of course, we only observe one of these two potential outcomes for any subject, and the other potential outcomes are treated as missing data. This missing data can be treated as an unknown parameter, and we aim to find the posterior distribution of this missing data. Once we have the posterior distribution of the missing counterfactuals, then we can sample from the posterior distribution of any causal contrast, such as the average treatment effect [10, 58]. For illustration, let's assume the following model for the joint distribution of the potential outcomes:

$$\begin{pmatrix} Y(1) \\ Y(0) \end{pmatrix} = \mathcal{N}\left(\begin{pmatrix} \boldsymbol{X}\boldsymbol{\beta}_1 \\ \boldsymbol{X}\boldsymbol{\beta}_0 \end{pmatrix}, \begin{pmatrix} \sigma_1^2 & \rho \\ \rho & \sigma_0^2 \end{pmatrix} \right).$$

Note that there is no information in the data to inform ρ, the correlation between the potential outcomes, because we never observe both of these quantities jointly. Typically sensitivity analysis is performed by seeing how inference varies as a function of ρ, or alternatively a prior distribution on ρ is placed to average over possible values of this parameter. Model averaging can be used to improve estimation of these models when the number of covariates in $\boldsymbol{X}$ is large. Many of the same ideas from before apply directly here, because we want to prioritize variables that are also associated with the treatment. All of the aforementioned prior distributions that were discussed in Section 9.3 can apply directly here to improve estimation of $\boldsymbol{\beta}_1$ and $\boldsymbol{\beta}_0$. However, further research is required to understand the exact manner in which Bayesian model averaging would improve estimates in this framework.

Bibliography

[1] G.W. Imbens and D.B. Rubin. *Causal inference in statistics, social, and biomedical sciences.* Cambridge University Press, 2015.

[2] M.D. Hernán and J.M. Robins. *Causal Inference: What If.* Boca Raton: Chapman & Hall/CRC, 2020.

[3] R. Collins, L. Bowman, M. Landray, and R. Peto. The magic of randomization versus the myth of real-world evidence. *New England Journal of Medicine*, 7:674–678, 2020.

[4] R.A. Fisher. The causes of human variability. *Eugenics Review*, 10:213–220, 1918.

[5] R.A. Fisher. *Statistical Methods for Research Workers.* London: London: Oliver & Boyd, 1925.

[6] J. Neyman. On the application of probability theory to agricultural experiments, section 9. *Roczniki Nauk Roiniczych*, X:1–51, 1923, 1990. reprinted in Statistical Science, 1990, 5, 463–485.

[7] D.B. Rubin. Estimating causal effects of treatments in randomized and nonrandomized studies. *Journal of Educational Psychology*, 66(5):688–701, 1974.

[8] D.B. Rubin. Inference and missing data. *Biometrika*, 63:581–592, 1976.

[9] D.B. Rubin. Assignment to treatment group on the basis of a covariate. *Journal of Educational Statistics*, 2(1):1–26, 1977.

[10] D.B. Rubin. Bayesian inference for causal effects: The role of randomization. *Ann. Statist.*, 6(1):34–58, 1978.

[11] D.B. Rubin. Formal mode of statistical inference for causal effects. *Journal of statistical planning and inference*, 25(3):279–292, 1990.

[12] P.W. Holland. Statistics and causal inference. *Journal of the American Statistical Association*, 81(396):945–960, 1986.

[13] M.G. Hudgens and M.E. Halloran. Toward causal inference with interference. *Journal of the American Statistical Association*, 103(482):832–842, 2008.

[14] G. Papadogeorgou, F. Mealli, and C.M. Zigler. Causal inference with interfering units for cluster and population level treatment allocation programs. *Biometrics*, 75(3):778–787, 2019.

[15] L. Forastiere, E.M. Airoldi, and F. Mealli. Identification and estimation of treatment and interference effects in observational studies on networks. *Journal of the American Statistical Association*, 2020.

[16] P.R. Rosenbaum and D.B. Rubin. The central role of the propensity score in observational studies for causal effects. *Biometrika*, 70(1):41–55, 1983.

[17] D.O. Scharfstein, A. Rotnitzky, and J.M. Robins. Adjusting for nonignorable drop-out using semiparametric nonresponse models. *Journal of the American Statistical Association*, 94(448):1096–1120, 1999.

[18] J.M. Robins. Robust estimation in sequentially ignorable missing data and causal inference models. *Proceedings of the American Statistical Association Section on Bayesian Statistical Science 1999*, 6–10, 2000.

[19] J.K. Lunceford and M. Davidian. Stratification and weighting via the propensity score in estimation of causal treatment effects: a comparative study. *Statistics in Medicine*, 23(19):2937–2960, 2004.

[20] H. Bang and J.M. Robins. Doubly robust estimation in missing data and causal inference models. *Biometrics*, 61(4):962–973, 2005.

[21] J.M. Robins and A. Rotnitzky. Semiparametric efficiency in multivariate regression models with missing data. *Journal of the American Statistical Association*, 90(429):122–129, 1995.

[22] J.M. Robins, A. Rotnitzky, and L.P. Zhao. Analysis of semiparametric regression models for repeated outcomes in the presence of missing data. *Journal of the American Statistical Association*, 90(429):106–121, 1995.

[23] J.M. Funk, D. Westreich, C. Wiesen, T. Stürmer, M. Brookhart, and M. Davidian. Doubly robust estimation of causal effects. *American journal of epidemiology*, 173:761–7, 03 2011.

[24] M.H. Farrell. Robust inference on average treatment effects with possibly more covariates than observations. *Journal of Econometrics*, 189(1):1–23, 2015.

[25] J.A. Hoeting, D. Madigan, A.E. Raftery, and C.T. Volinsky. Bayesian model averaging: a tutorial. *Statistical science*, 382–401, 1999.

[26] A. Belloni, V. Chernozhukov, and C. Hansen. Inference on treatment effects after selection among high-dimensional controls. *The Review of Economic Studies*, rdt044, 2013.

[27] A. Ertefaie, M. Asgharian, and D.A. Stephens. Variable selection in causal inference using a simultaneous penalization method. *Journal of Causal Inference*, 6(1), 2018.

[28] M. Clyde. Model uncertainty and health effect studies for particulate matter. *Environmetrics: The official journal of the International Environmetrics Society*, 11(6):745–763, 2000.

[29] D.C. Thomas, M. Jerrett, N. Kuenzli, T.A. Louis, F. Dominici, S. Zeger, J. Schwarz, R.T. Burnett, D. Krewski, and D. Bates. Bayesian model averaging in time-series studies of air pollution and mortality. *Journal of Toxicology and Environmental Health, Part A*, 70(3-4):311–315, 2007.

[30] T.J. Mitchell and J.J. Beauchamp. Bayesian variable selection in linear regression. *Journal of the american statistical association*, 83(404):1023–1032, 1988.

[31] E.I. George and R.E. McCulloch. Variable selection via gibbs sampling. *Journal of the American Statistical Association*, 88(423):881–889, 1993.

[32] J. Pearl. Invited commentary: understanding bias amplification. *American journal of epidemiology*, 174(11):1223–1227, 2011.

[33] C. Wang, F. Dominici, G. Parmigiani, and C. Zigler. Accounting for uncertainty in confounder and effect modifier selection when estimating average causal effects in generalized linear models: Accounting for uncertainty in confounder and effect modifier selection when estimating aces in glms. *Biometrics*, 71, 04 2015.

[34] J. Antonelli, C. Zigler, and F. Dominici. Guided bayesian imputation to adjust for confounding when combining heterogeneous data sources in comparative effectiveness research. *Biostatistics*, 18(3):553–568, 2017.

[35] G. Papadogeorgou, F. Dominici, *et al.*. A causal exposure response function with local adjustment for confounding: Estimating health effects of exposure to low levels of ambient fine particulate matter. *Annals of Applied Statistics*, 14(2):850–871, 2020.

[36] A. Wilson, C.M. Zigler, C.J. Patel, and F. Dominici. Model-averaged confounder adjustment for estimating multivariate exposure effects with linear regression. *Biometrics*, 74(3):1034–1044, 2018.

[37] G. Lefebvre, J. Atherton, and D. Talbot. The effect of the prior distribution in the bayesian adjustment for confounding algorithm. *Computational Statistics & Data Analysis*, 70:227–240, 2014.

[38] J. Antonelli, G. Parmigiani, and F. Dominici. High-dimensional confounding adjustment using continuous spike and slab priors. *Bayesian analysis*, 14(3):805, 2019.

[39] R. Tibshirani. Regression shrinkage and selection via the lasso. *Journal of the Royal Statistical Society: Series B (Statistical Methodology)*, 58(1):267–288, 1996.

[40] M. Cefalu, F. Dominici, N. Arvold, and G. Parmigiani. Model averaged double robust estimation. *Biometrics*, 2016.

[41] C.M. Zigler and F. Dominici. Uncertainty in propensity score estimation: Bayesian methods for variable selection and model-averaged causal effects. *Journal of the American Statistical Association*, 109(505):95–107, 2014.

[42] P.R. Hahn, J.S. Murray, and C.M. Carvalho. Bayesian regression tree models for causal inference: regularization, confounding, and heterogeneous effects. *Bayesian Analysis*, 2020.

[43] C.M. Zigler, K. Watts, R.W. Yeh, Y. Wang, B.A. Coull, and F. Dominici. Model feedback in bayesian propensity score estimation. *Biometrics*, 69(1):263–273, 2013.

[44] P.E. Jacob, L.M. Murray, C.C. Holmes, and C.P. Robert. Better together? statistical learning in models made of modules. *arXiv preprint arXiv:1708.08719*, 2017.

[45] O. Saarela, L.R. Belzile, and D.A. Stephens. A bayesian view of doubly robust causal inference. *Biometrika*, 103(3):667–681, 2016.

[46] J.M. Robins, M.A. Hernán, and L. Wasserman. On bayesian estimation of marginal structural models. *Biometrics*, 71(2):296, 2015.

[47] J.M. Robins and Y. Ritov. Toward a curse of dimensionality appropriate (coda) asymptotic theory for semi-parametric models. *Statistics in medicine*, 16(3):285–319, 1997.

[48] J. Antonelli, G. Papadogeorgou, and F. Dominici. Causal inference in high dimensions: A marriage between bayesian modeling and good frequentist properties. *Biometrics*, 2020.

[49] C.J. Patel, N. Pho, M. McDuffie, J. Easton-Marks, C. Kothari, I.S. Kohane, and P. Avillach. A database of human exposomes and phenomes from the us national health and nutrition examination survey. *Scientific data*, 3(1):1–10, 2016.

[50] D.B. Rubin. The bayesian bootstrap. *The annals of statistics*, pp. 130–134, 1981.

[51] B. Efron and R.J. Tibshirani. *An introduction to the bootstrap*. CRC press, 1994.

[52] S. Liao and C. Zigler. Uncertainty in the design stage of two-stage bayesian propensity score analysis. *arXiv preprint arXiv:1809.05038*, 2018.

[53] L. McCandless, P. Gustafson, and P. Austin. Bayesian propensity score analysis for observational data. *Statistics in Medicine*, 15:94–112, 2009.

[54] P.R. Hahn, C.M. Carvalho, D. Puelz, J. He, *et al.*. Regularization and confounding in linear regression for treatment effect estimation. *Bayesian Analysis*, 13(1):163–182, 2018.

[55] B.J. Reich, C.B. Storlie, and H.D. Bondell. Variable selection in bayesian smoothing spline anova models: Application to deterministic computer codes. *Technometrics*, 51(2):110–120, 2009.

[56] A.R. Linero. Bayesian regression trees for high-dimensional prediction and variable selection. *Journal of the American Statistical Association*, 113(522):626–636, 2018.

[57] A.R. Linero and Y. Yang. Bayesian regression tree ensembles that adapt to smoothness and sparsity. *Journal of the Royal Statistical Society: Series B (Statistical Methodology)*, 80(5):1087–1110, 2018.

[58] F. Mealli, B. Pacini, and D.B. Rubin. Statistical inference for causal effects. in *Modern Analysis of Customer Satisfaction Surveys*, Wiley, 2011.

10

Variable Selection for Hierarchically-Related Outcomes: Models and Algorithms

Hélène Ruffieux
University of Cambridge, UK

Leonardo Bottolo
University of Cambridge, UK

Sylvia Richardson
University of Cambridge, UK

CONTENTS

High-dimensional regression problems with multiple outcomes are increasingly common in real-world settings. For example, biomedical studies can involve from a handful to over twenty thousand responses, such as gene expression levels. We will focus on context-driven Bayesian variable selection approaches that can leverage complex dependence structures across the outcomes, while coherently conveying uncertainty. In particular, we will discuss the relative merits of diverse sparse priors in the context of hierarchically-related responses, and consider adaptations when the number of responses is large. We will also touch on aspects of accuracy and scalability of inference for such models, as well as algorithmic enhancements to best handle posterior multimodality. Genetic association problems involving molecular phenotypes – in particular on the regulation of gene expression – will provide us with a series of compelling case studies to illustrate the advantages of multiple-response hierarchical models tailored to large and complex datasets. We will conclude by highlighting some open statistical and computational challenges.

DOI: 10.1201/9781003089018-10

10.1 Introduction

Real-world situations multiply where large numbers of candidate predictors are considered for a series of related outcomes. They concern a variety of domains, including genetics, economy, epidemiology and social science. For example, psychiatric studies very often consider disorders that are not uniquely characterised by a single outcome measurement. Researchers typically attempt to identify risk factors for a collection of complementary instruments acting as surrogates for the disorder under consideration [1, 2]. Such a task could in principle be recast into a single-outcome regression set-up, either based on pooling rules, whereby all outcomes are combined into a single composite endpoint, or based on separate analyses of each outcome. However this can induce a substantial loss of information and, subsequently, of statistical power.

Principled variable selection approaches which can leverage shared association structures among multiple outcomes are therefore very much needed. These must involve flexible modelling strategies, that can accommodate the complexity and growing dimensions of present-day problems. The Bayesian framework is particularly suited as it permits representing all variables in a structured hierarchical fashion, thereby allowing a borrowing of information across outcomes associated with the same predictors, while incorporating prior knowledge and coherently conveying uncertainty.

While Bayesian variable selection for single-response regression problems has retained substantial attention over the past thirty years, the multiple-response case has been looked at relatively recently. Many concepts developed for the former setting naturally extend to the latter when a few response variables are considered, but new methodological and practical aspects arise when the number of responses largely exceeds a handful.

In this chapter, we will discuss the different paradigms entailed by these two types of problems - with a handful or a large number of outcomes q – and we will review different statistical approaches for these, all proposed over the past decade. We will see that conceptual and practical aspects are strongly interlinked and that it is therefore particularly critical to consider statistical modelling and inference algorithms hand in hand. We will also show that certain choices have broad implications on the type of posterior output obtained and on its interpretability. Collectively, the statistical questions involved in Bayesian variable selection for multiple-response regression problems span a large spectrum of aspects, which all closely depend on the primary aim of inference, as well as on the intrinsic characteristics of the data and applied problem tackled.

The chapter is organised as follows. We will first give an overview of the statistical and computational challenges encountered in multiple-response Bayesian variable selection, and describe different strategies for approaching them. We will then focus on presenting a selection of methods for a few or a large number of outcomes, as well as extensive case studies for them. The applications will all concern the domain of statistical genetics and genomics, as it provides compelling multiple-outcome examples, with large datasets involving dependent outcomes controlled by complex genetic association patterns. We will end the chapter by highlighting important open problems, that are likely to concentrate much research effort in the next decade.

10.2 Model Formulations, Computational Challenges and Tradeoffs

Consider the multivariate linear regression model,

$$\boldsymbol{Y} = \boldsymbol{X}_\gamma \boldsymbol{B}_\gamma + \boldsymbol{E}, \tag{10.1}$$

where $\boldsymbol{Y}$ is an $n \times q$ matrix of q responses for each of n independent samples, $\boldsymbol{X}_\gamma$ is an $n \times p_\gamma$ design matrix of p_γ predictors and $\boldsymbol{B}_\gamma$ is the corresponding $p_\gamma \times q$ matrix of regression coefficients. Without loss of generality, we assume the columns of $\boldsymbol{Y}$ and $\boldsymbol{X}_\gamma$ to be centred, so the intercept is set to zero. Moreover, $\boldsymbol{E}$ is an $n \times q$ error term which is often assigned a matrix-variate normal distribution, $\mathcal{MN}_{n\times q}(\boldsymbol{0}, \boldsymbol{I}_n, \boldsymbol{\Sigma})$, as defined by [3], meaning that the rows $\boldsymbol{E}_{i\cdot} = (\varepsilon_{i1}, \ldots, \varepsilon_{iq})$ are independent identically distributed as $\mathcal{N}_q(\boldsymbol{0}, \boldsymbol{\Sigma})$, with $\boldsymbol{\Sigma}$, a $q \times q$ symmetric positive definite matrix.

The multivariate framework (10.1) has been proposed by [4, 5] and is a generalisation of the univariate Gaussian regression framework first introduced by [6, 7]. It is described in the context of model choice, whereby a full $n \times p$ matrix of candidate predictors $\boldsymbol{X}$ is considered, along with a $p \times 1$ latent indicator parameter $\boldsymbol{\gamma}$, whose entry j takes value unity if candidate predictor $\boldsymbol{X}_j$ is included in the model and value zero otherwise, and where $p_\gamma = \sum_{j=1}^{p} \gamma_j$. This discrete model space framework offers a flexible ground for specifying diverse types of prior structures for the top-level model hierarchy — i.e., for prior and hyperprior distributions as well as for other model assumptions used for $\boldsymbol{B}_\gamma$, $\boldsymbol{\gamma}$ and $\boldsymbol{E}$. This hierarchy is typically specified in relation to the inference aim, and has impact on the size and complexity of the parameter spaces to be explored.

Here, we will focus primarily on variable selection tasks aimed at best leveraging shared dependence across the responses in regimes where the predictor space is high-dimensional ($p \gg n$). We will thus discuss how the model hierarchy can be adapted and twisted to best achieve these goals. In particular, we will describe different assumptions concerning three key components of multiple-response regression models, as sketched in Figure 10.1.

It is natural to first consider formulating a prior on the model space. This can be directly achieved via the prior specification for $\boldsymbol{\gamma}$; the Bernoulli distribution is a common choice, $\gamma_j \sim \text{Bern}(\omega)$, $j = 1, \ldots, p$, where ω is a probability of inclusion parameter, typically chosen small to induce sparsity. The posterior mean of γ_j then corresponds to the marginal posterior probability of inclusion (PPI) of variable $\boldsymbol{X}_j$, $\text{E}(\gamma_j \mid \boldsymbol{Y}) = \text{pr}(\gamma_j = 1 \mid \boldsymbol{Y})$, which is a useful posterior quantity for variable selection: PPIs are a direct measure of support for the presence or absence of individual associations, from which Bayesian false discovery rates can be obtained [9]. Different strategies to calculate PPIs using Markov chain Monte Carlo (MCMC) output are outlined in [10].

Placing a prior on ω, possibly with a further hyperprior structure in a fully Bayesian fashion, offers flexible avenues for controlling sparsity levels while borrowing information across variables. Since it is a conjugate to the Bernoulli distribution, the Beta distribution is natural choice. Moreover, the assumption that the indicators γ_j are independent conditional on ω but dependent marginally can be relaxed using a variable specific inclusion probability ω_j; when hyperpriors on ω_j are used, this modification also enables more control on the modelling of each predictor inclusion. [10] provide a general discussion on the implementation of Bayesian model selection.

Given this prior model space, several prior formulations can be considered for modelling the regression coefficients, with important distinctions depending on whether the rows of $\boldsymbol{B}$ are treated as independent or not.

The hierarchical mixture framework based on spike-and-slab distributions is a classical

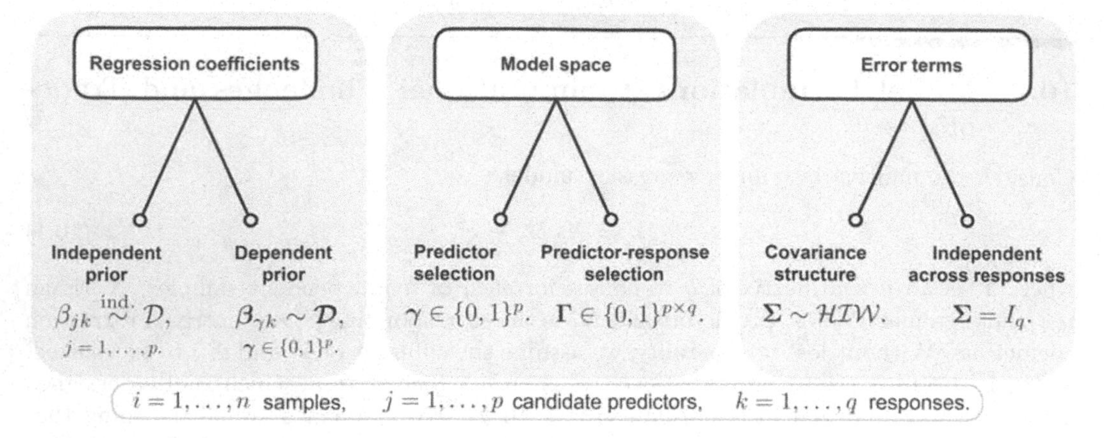

FIGURE 10.1
Key assumptions for the hierarchical structure of multiple-response models. Left: dependent or independent prior specification for the regression coefficients. Middle: predictor or predictor-response latent selection indicator. Right: direct modelling of the residual response dependence or conditionally independent responses (dependence encoded via the model hierarchy). $\mathcal{D}$ and $\boldsymbol{\mathcal{D}}$ stand for some generic univariate, respectively multivariate distribution and $\mathcal{HIW}$ stands for the hyper inverse Wishart distribution [8]. Each multiple-response model involves a specific combination of the three types of assumptions, some of which permitting distributional conjugacy (see main text).

setting where coefficients are conditionally independent,

$$\beta_{jk} \mid \gamma_j \sim \gamma_j f_\beta + (1 - \gamma_j)\delta_0 , \qquad j = 1, \ldots, p, \; k = 1, \ldots q, \tag{10.2}$$

with f_β an absolutely continuous density (typically a diffuse centred normal distribution) modelling the coefficients of the predictors included in the model ($\gamma_j = 1$) and δ_0 a point-mass at zero modelling the remaining coefficients ($\gamma_j = 0$). Spike-and-slab priors were introduced by [6, 11] and have become very popular since then.

The g-prior [12] is an important example of dependent priors for modelling the coefficients corresponding to the selected predictors using a subset-selection formulation. It uses the following multivariate Gaussian formulation for the kth column of $\boldsymbol{B}_\gamma$ corresponding to response k,

$$\boldsymbol{\beta}_{\gamma k} \mid \gamma, g, \sigma_k^2 \sim \mathcal{N}_{p_\gamma} \left(\boldsymbol{0}, g \left(\boldsymbol{X}_\gamma^T \boldsymbol{X}_\gamma \right)^{-1} \sigma_k^2 \right) , \tag{10.3}$$

where σ_k^2 is the response error variance, and $g > 0$ is a parameter controlling the expected sizes of effects and whose hyperprior specification has been the subject of considerable discussion [see 13, for a review]. The g-prior accommodates information about the predictor precision structure by directly involving the design matrix $\boldsymbol{X}_\gamma$. On one hand, this data dependence requires careful handling in presence of highly collinear predictors, as nearly singular $\boldsymbol{X}_\gamma^T \boldsymbol{X}_\gamma$ can cause instabilities. On the other hand, it can improve selection by explicitly accounting for the correlation between predictors and penalising the inclusion of spurious predictors highly correlated to the true predictor [14]. This prior is also employed for its conjugacy properties, which allow eliminating a determinant term from the analytical expression of the marginal likelihood, thus leading to cheaper computations.

More generally, distributional conjugacy with respect to regression coefficients and/or residual covariance substantially reduces the burden associated with the posterior exploration. One key assumption in model (10.1) is that a given predictor, if included in the model, is selected for every response variable; this simplification enables conjugacy when classical prior distributions, such as the inverse Wishart distribution, are employed for $\boldsymbol{\Sigma}$.

[15–17] all rely on this assumption to analytically integrate out the error covariance and regression coefficient matrices. They obtain faster MCMC mixing, as the model search is carried out solely over the space of the indicator vector and the remaining hyperparameters. Note that, in addition to variable selection, [17] consider covariance selection: they use an undirected decomposable graph formulation in which they assign a hyper-inverse Wishart distribution on $\boldsymbol{\Sigma} \mid \boldsymbol{G}$, where $\boldsymbol{G}$ is the decomposable graph underlying the response network structure [8]. This formulation induces sparsity both in the rows of the regression matrix and in the residual covariance matrix, which permits gaining insight into the set of responses concerned by the selected predictors.

In general, while selecting each predictor based on its relevance for either all or none of the responses (so-called "row sparsity") can be sensible in settings *with a few related outcomes*, this will be unreasonable or overly restrictive in other settings where the rows of $\boldsymbol{B}_\gamma$ are believed to be sparse. Examples include problems where the outcomes display block dependence structures arising from associations with different sets of predictors. In such cases, pairwise selection of predictors and their associated outcomes can be implemented using a variant of model (10.1), where a $p \times q$ matrix of latent indicators $\boldsymbol{\Gamma} = \{\gamma_{jk}\}$ defines the sparsity pattern of the regression matrix $\boldsymbol{B}$, i.e., with $\gamma_{jk} = 1$ if predictor j is associated with response k and $\gamma_{jk} = 0$ otherwise ("cell sparsity"). With no further assumption, conjugacy cannot be exploited, yet proposals have been made which use workarounds enabling tractable inference. [18, 19] introduce a sparse seemingly unrelated regression (SUR) framework [20, 21] and exploit a reparametrisation based on a decomposition of the covariance matrix to speed up MCMC computation. [22] couple a spike-and-slab specification on the precision matrix $\boldsymbol{\Omega} = \boldsymbol{\Sigma}^{-1}$ with a deterministic inference algorithm implementing expectation/conditional maximisation.

However, formulations which attempt a direct modelling of the residual response covariance matrix in a sparse regression set up all require exploring a large multimodal parameter space. This poses statistical and computational challenges as soon as the numbers of candidate predictors and responses are in the hundreds. Known issues pertain to unstable estimates, posterior exploration prone to entrapment in local modes, as well as large runtime and memory requirements. For instance, [17] indicate that inferring the covariance graph using their hyper-inverse Wishart framework scales as $O(q^2)$, which intrinsically limits its applications to problems with a modest number of responses. Envisioning reliable joint estimation of both regression and covariance matrices therefore appears out of reach in many applications where $q > 100$ and $p > 1\,000$. One important example – which will serve as illustration throughout this chapter – is that of molecular quantitative trait locus (QTL) studies, which infer associations between hundreds of thousands of SNPs (candidate predictors) and hundreds to tens of thousands of gene, protein or metabolite levels (responses).

To enable joint inference for large predictor and response spaces, an alternative line of research has been proposed, which circumvents modelling the covariance of the responses and permits borrowing information across them solely through the model hierarchy; this idea was initially put forth by [23] and developed further by [24]. By assuming $\boldsymbol{\Sigma} = \boldsymbol{I}_q$ in (10.1), these authors turn the modelling framework into a series of conditionally independent regressions, one per response,

$$\boldsymbol{y}_k = \boldsymbol{X}\boldsymbol{\beta}_k + \boldsymbol{\varepsilon}_k, \qquad \boldsymbol{\varepsilon}_k \sim \mathcal{N}_n\left(\boldsymbol{0}, \sigma_k^2 \boldsymbol{I}_n\right), \qquad k = 1, \ldots, q, \tag{10.4}$$

where $\boldsymbol{\beta}_k$ is a $p \times 1$ regression parameter and $\boldsymbol{\varepsilon}_k$ a $n \times 1$ error term. Each response $\boldsymbol{y}_k$ has a specific residual variance, σ_k^2, which is assigned an inverse Gamma prior. As we shall explain below, dependence across the parallel regression models is captured via the prior on the regression coefficients, namely, via parameters which are shared across all the responses. This simplifies computation and yet improves the detection of weak association effects shared across the outcomes compared to univariate response modelling [see, e.g., 23, 24].

This framework also easily accommodates response-specific selection and is therefore tailored to most applications in the large response setting, where it is assumed that each row of $\boldsymbol{B}_\gamma$ is sparse. For instance, in molecular QTL studies, a given SNP usually controls a subset of molecular outcomes, and this subset may vary depending on the SNP considered. Both independent and dependent priors on $\boldsymbol{B}$ can be naturally extended to induce "cell sparsity" by modelling β_{jk} conditionally on γ_{jk} rather than γ_j. For instance,

$$\beta_{jk} \mid \gamma_{jk}, \sigma_\beta^2, \sigma_k^2 \sim \gamma_{jk} \mathcal{N}\left(0, \sigma_\beta^2 \sigma_k^2\right) + (1 - \gamma_{jk})\,\delta_0, \qquad j = 1, \ldots, p,\ k = 1, \ldots, q,$$

is a multiple-response spike-and-slab prior adapted from (10.2) and

$$\boldsymbol{\beta}_{\gamma_k k} \mid \boldsymbol{\gamma}_k, g, \sigma_k^2 \sim \mathcal{N}_{p_{\gamma_k}}\left(\mathbf{0}, g\left(\boldsymbol{X}_{\gamma_k}^T \boldsymbol{X}_{\gamma_k}\right)^{-1} \sigma_k^2\right), \qquad k = 1, \ldots, q,$$

is a multiple-response g-prior adapted from (10.3), where $\boldsymbol{\gamma}_k$ denotes the kth column of $\boldsymbol{\Gamma}$. The scaling parameters σ_β^2, respectively g, which are typically assigned an inverse Gamma prior, are common to all responses and thus enable borrowing information across them. Importantly, further dependence between the responses can be encoded via the top-level prior hierarchy placed on γ_{jk} and, in particular, on ω_{jk} when using

$$\gamma_{jk} \mid \omega_{jk} \sim \text{Bernoulli}\left(\omega_{jk}\right). \tag{10.5}$$

The proposals of [23, 24], as well as more recent variants by [25–29], primarily differ in the prior structure assumed for ω_{jk} and its impact for modelling shared signals among the predictors and responses. [23, 27] use the simple formulation

$$\omega_{jk} \equiv \theta_j \sim \text{Beta}(a_j, b_j), \qquad a_j, b_j > 0, \tag{10.6}$$

whereas [24, 26] decouple the predictor and response contributions to the association probability by setting

$$\omega_{jk} \equiv \theta_j \times \zeta_k, \qquad 0 \leq \omega_{jk} \leq 1, \tag{10.7}$$

$$\theta_j \sim \text{Gamma}(c_j, d_j), \qquad \zeta_k \sim \text{Beta}(a_k, b_k), \qquad c_j, d_j, a_k, b_k > 0.$$

More recently, [28] proposed carrying over the decoupling within a probit formulation,

$$\omega_{jk} \equiv \Phi(\theta_j + \zeta_k), \quad \theta_j \mid \tau \sim \text{HS}(\tau), \quad \zeta_k \sim \mathcal{N}(n_0, t_0^2), \quad n_0 \in \mathbb{R}, t_0 > 0, \tag{10.8}$$

where HS$(\cdot)$ stands for the horseshoe distribution (details in Section 10.3.2). This probit-link formulation also lends itself to model extensions which leverage additional data that may provide insights into the propensity of predictors to be involved in associations. These data can be encoded as r *predictor-level* variables - gathered in a $p \times r$ matrix $\boldsymbol{V} = (\boldsymbol{V}_1, \ldots, \boldsymbol{V}_r)$ - which modulate the association probabilities, e.g., using a second-stage hierarchical regression extension:

$$\omega_{jk} \equiv \Phi(\theta_j + \zeta_k + \boldsymbol{V}_j^T \boldsymbol{\xi}), \qquad \xi_l \mid \rho_l \sim \rho_l \mathcal{N}(0, s^2) + (1 - \rho_l)\,\delta_0, \qquad l = 1, \ldots, r, \tag{10.9}$$

where ρ_l is assigned a Bernoulli prior distribution and ξ_l represents the contribution of predictor-level variable $\boldsymbol{V}_l$ to the predictor effects on the responses. This formulation is employed by [29] in the context of molecular QTL studies, whereby epigenetic annotations on SNPs are collected and used as a source of information on the functional potential of these variants. Of note, (10.9) assigns a spike-and-slab prior to the predictor-level covariate effects which enables modelling large panels of candidate covariates, a fraction of which may be inferred as relevant to the primary regression effects. More generally however, any link function h can in principle be used in the spike-and-slab framework for the primary effects (10.2):

$$\begin{aligned} \beta_{jk} \mid \gamma_j &\sim \gamma_j f_\beta + (1-\gamma_j)\delta_0, \quad \gamma_j \mid \omega_j \sim \text{Bernoulli}(\omega_j), \\ \omega_j &= h^{-1}(\mathbf{V}_j^T \boldsymbol{\xi}), \qquad\qquad\qquad j = 1, \ldots, p. \end{aligned}$$

A reparametrisation allows expressing the hyperparameters for the prior on ζ_k in (10.7), (10.8) and (10.9) in terms of the expected number of predictors associated with each response and its variance, which are appealing quantities for controlling sparsity levels [26, 28]. Crucially, in all four formulations (10.6), (10.7), (10.8) and (10.9), θ_j is an important parameter for capturing shared signals between the responses. This parameter is also involved in another important inference task, which is only present in problems comprising a relatively large number of responses: that of the modelling *hotspots*, namely, predictors which are associated with several responses. Such scenarios are of particular interest in molecular QTL studies, since hotspot SNPs, which control a possibly large number of molecular levels, may have key functional roles underlying certain diseases, but are nevertheless typically difficult to uncover due to relatively weak association effects [30]. By modulating the probability of association ω_{jk}, the predictor-specific parameter θ_j directly models the propensity of each predictor to be a hotspot. This representation, and the subsequent model hierarchy placed on θ_j, boosts the probability of association when there is evidence that many responses are associated with a given predictor (hotspot).

Collectively, the complete model hierarchy placed on the conditionally-independent regression framework (10.4) offers a particularly convenient setting for selecting pairs of associated predictors and responses, *as well as* hotspots. Modelling dependence via the prior structure on the association effects naturally serves variable selection as this set-up directly leverages strength across responses associated with the same predictors; [23, 24] and the more recent proposals [27–29] demonstrate *in silico* that this is sufficient to significantly improve selection performance over separate single-response regressions, even when the residual correlations are substantial. Moreover, this formulation yields directly interpretable measures of support for associations, via the posterior means of the hotspot propensity parameters θ_j (e.g., which genetic variant is a key regulator of a large set of genes?) and of the pairwise selection parameters γ_{jk} (e.g., which genetic variant affects which gene level?).

In presence of highly correlated responses, it is nevertheless essential to examine the potential sources of dependence and the impact of the conditional independence assumption on downstream inference, as will be discussed in Section 10.4. Moreover, this specification also requires careful considerations on aspects related to hyperparameter specification, hotspot shrinkage properties and Bayesian multiplicity adjustment [31] for the dimension of the responses. This constitutes the focus of the work in [28], which proposes a flexible, fully Bayesian global-local representation of the hotspots with the horseshoe prior (see Section 10.3.2).

Beside variations at the modelling level, the most recent proposals implement a shift from MCMC algorithms [14, 32] to more scalable approximate inference algorithms which can also robustly accommodate highly multimodal distributions: [27] propose an efficient block-variational algorithm and [28, 29] couple it with simulated annealing schemes. Simulated annealing introduces a so-called *temperature* parameter indexing a series of *heated* distributions which aim to sweep the local modes away and ease the progression to the global mode. This, in essence, corresponds to a transposition of tempering MCMC procedures, which the initial proposal of [26] already employs. Obviously, scalability is paramount to the uptake of the above approaches in practice. Applications in the large q regime would otherwise have to be recast into a series of smaller, tractable problems, at the risk of failing to capture dependence patterns which are informative for variable selection.

10.3 Illustrations on Published Case Studies

In this section, we will illustrate and exploit the advantages of the above hierarchical regression approaches for two active applied research fields in statistical genetics. Both problems concern expression quantitative trait locus (eQTL) studies, which aim to identify associations between hundreds of thousands of genetics variants, typically single nucleotide polymorphisms (SNPs), and thousands of traits corresponding to gene or transcript expression levels possibly measured in multiple conditions, for n individuals. SNPs can affect the expression of a nearby gene (*cis* eQTL) or they can control the expression of remote genes (*trans* eQTL), where the latter type of effects tend to be weaker than the former and hence more difficult to uncover. In particular, the detection of genetic *hotspots* – master SNPs regulating the expression of tens or possibly hundreds of genes in *trans* – is particularly challenging and a subject of considerable debate as they may shed light on important functional processes [33–35]. More generally, identifying the complex mechanisms of action of SNPs on gene expression is a prelude to disentangling the downstream consequences at the protein (the proteome), metabolite (the metabolome) level, and eventually on specific disease endpoints. In particular, thanks to the increased availability of samples collected in diverse tissues and cell types, much of the recent research efforts are on understanding condition-specific eQTL effects. These specialised analyses permit assessing the specificity or conservation of signals across given tissues and cell types, thereby helping refine our understanding of the genetic architecture and molecular circuitry underlying given diseases. Moreover, several studies have found that cell-type stimulation (e.g., via inflammatory proxies interferon-γ or lipopolysaccharide) tends to promote *trans* hotspot activity [36, 37] and hence offers an ideal setting for studying the genetic landscape of hotspots.

Hierarchical Bayesian methods which account collectively for all candidate predictor SNPs have been particularly helpful to enhance statistical power for eQTL mapping and identify small *trans* effects that would be missed by classical marginal "single-SNP" approaches [38–40]. A substantial performance gain is further obtained when also modelling jointly the multiple expression levels or multiple conditions that are controlled by overlapping sets of SNPs. With this in mind, we will next illustrate two multiple-response regression methods on such problems, namely for modelling (1) the covariance structure of gene levels across multiple tissues and (2) the hotspot activity under different cell-type stimuli. These studies give rise to different multiple regression set-ups, where the response matrix can represent either the measurements of several genes in a unique condition (tissue or cell type) or the measurements of a same gene in several conditions. Figure 10.2 provides a schematic representation of both scenarios and will serve as support for introducing the two modelling approaches.

10.3.1 Modelling eQTL Signals across Multiple Tissues

In recent years, the availability of 'omics data recorded on multiple tissues or time points has further increased the potential of genome-wide association studies with quantitative traits. In this framework, eQTL analysis is performed between a matrix of genotype values measured on a set of individuals and a multidimensional array, or tensor, describing the expression levels across tissues or time points for the same set of individuals. Multiple-tissue gene expression data can be very informative about the nature and the architecture of the genetic regulation, for instance, it can reveal whether the association between a genetic marker and a gene is conserved across tissues or is tissue-specific or, in a time-course analysis, how the regulation evolves over different time points. From a statistical

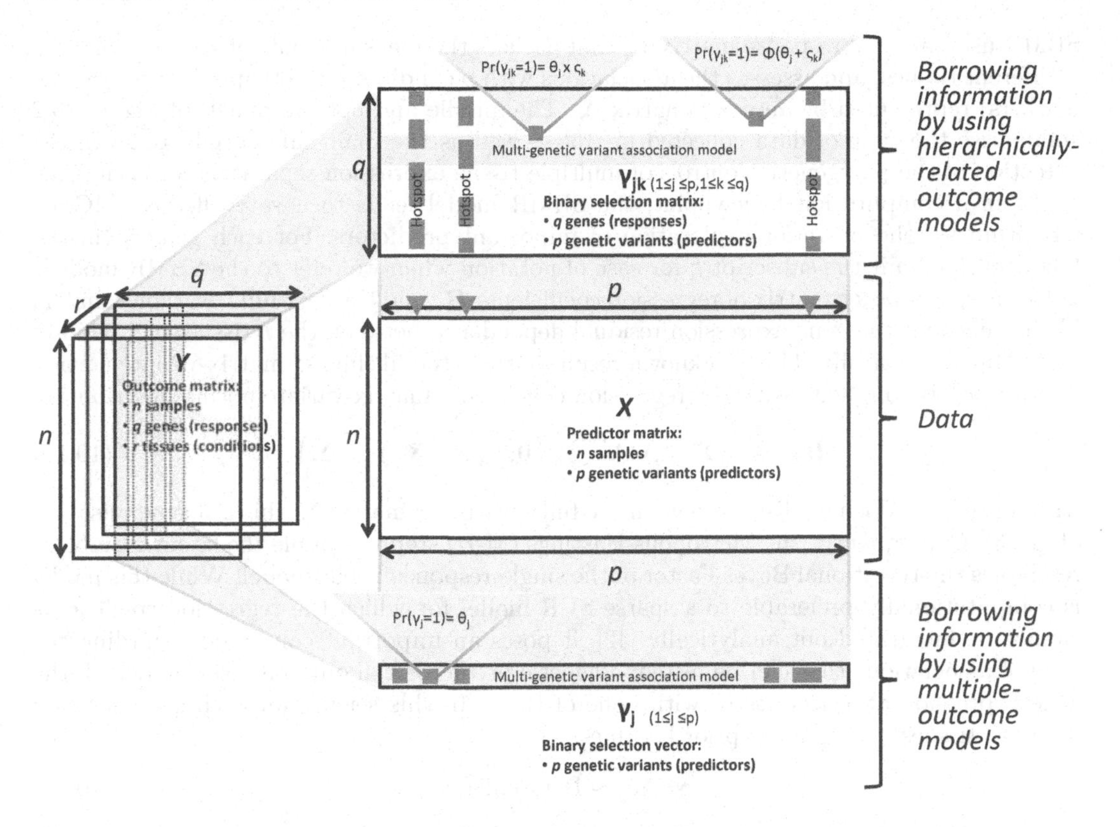

FIGURE 10.2
Schematic representation of the two modelling approaches for the real case studies presented in Section 10.3 (figure adapted from [41]). Middle panel: Data consist of a candidate predictor matrix of SNPs or genetic markers measured on a set of individuals (right) and a response multidimensional array, or tensor, describing the expression levels across conditions or time points for the same set of individuals (left). Bottom panel: For each gene, the genetic map for multiple-tissue gene levels is obtained by a multi-genetic marker association model (see Section 10.3.1). The model assumes that a genetic marker can only be associated with the gene levels in all tissues at the same time or with none of them. The level of sparsity for each multiple-tissue gene level is controlled by the prior expert elicitation of the number of expected associations and its variance. Top panel: Hierarchically-related outcome models are used to analyse all genes together, either tissue-by-tissue or pooling all tissues. The prior probability of genetic marker-gene association is decomposed into its marginal effects, i.e., the level of sparsity for each gene (similarly to the multi-genetic marker association model in the bottom panel) and the relative propensity of a predictor to influence several responses at the same time, using either a multiplicative model or an additive model within a probit formulation (see Section 10.3.2).

perspective, the joint analysis of multi-condition gene expression facilitates the discovery of small *trans* effects that would have gone unnoticed otherwise especially when they are conserved across tissues/time points. The advantage of the Bayesian paradigm, and in particular of Bayesian hierarchical models, is especially important in this scenario since it allows information sharing between different tissues or time points.

The first Bayesian solution to detect eQTL signals across multiple tissues has been proposed by [15]. Based on an evolutionary MCMC for the efficient exploration of the predictors' posterior model space [14, 32], Sparse Bayesian Multiple Regression (SBMR) can be seen as a multiple response regression model [4, 5] where a large number of genetic markers are tested together on their ability to jointly predict the variation of a same gene across multiple tissues. Hence, unlike in the standard single-tissue multiple regression setup where an $n \times q$ response matrix $\boldsymbol{Y}$ represents the expression levels of q genes for n individuals,

SBMR uses an $n \times r$ response matrix $\boldsymbol{Y}_k$ that gathers the measurements of a given gene k in r different tissues, and assesses the associations with p candidate genetic predictors (genetic markers) represented by an $n \times p$ matrix $\boldsymbol{X}$. The middle and bottom panels of Figure 10.2 summarise the type of data collected in [15] as well as the main inferential task, i.e., the detection of the polygenetic controls of multiple-tissue expression separately for each gene.

From a computational viewpoint, the SBMR model leads to a very effective MCMC algorithm for the posterior exploration of important predictors. For each gene k (in the following, we omit the subscript k for ease of notation when we refer to the SBMR model), both the sparse $p_\gamma \times r$ matrix of regression coefficients $\boldsymbol{B}_\gamma$ and the $r \times r$ full covariance matrix $\boldsymbol{\Sigma}$ that encodes the gene expression residual dependence between the r tissues can be integrated out analytically. This is a known result [3] since the likelihood matrix-variate normal distribution is conjugate with the regression coefficients matrix-variate normal distribution

$$\boldsymbol{B}_\gamma \mid \boldsymbol{\gamma}, g, \boldsymbol{\Sigma} \sim \mathcal{MN}_{p_\gamma \times r}\left(\boldsymbol{0}, g\left(\boldsymbol{X}_\gamma^T \boldsymbol{X}_\gamma\right)^{-1}, \boldsymbol{\Sigma}\right), \tag{10.10}$$

with an Inverse Wishart distribution on the full covariance matrix $\boldsymbol{\Sigma}$; this is a generalisation of (10.3). Consequently, the Metropolis-Hastings (M-H) step to sample the latent variable $\boldsymbol{\gamma}$ resembles the traditional Bayes Factor of the single-response linear model. While this model is computationally preferable to a sparse SUR model for which the regression coefficients cannot be integrated out analytically [42], it poses an important constraint regarding the type of associations that can be tested: a genetic predictor can only be associated with the gene expression of all tissues or with none of them. In this set-up, for each gene across r different tissues, the sparsity prior becomes

$$\gamma_j \mid \omega_j \sim \text{Bernoulli}\left(\omega_j\right) \tag{10.11}$$

with $\omega_j \sim \text{Beta}(a_j, b_j), a_j, b_j > 0$. Therefore, the SBMR model presented in [15] corresponds to the selection of an entire row of the latent binary matrix $\boldsymbol{\Gamma} = \{\gamma_{jl}\}$, $l = 1, \ldots, r$ ("row sparsity" model, see Section 10.2). In contrast, in a sparse SUR model, the selected predictors can be associated with any combination of the responses ("cell sparsity" model). This flexibility comes at a price, i.e., the regression coefficients cannot be integrated out analytically. In Figure 10.3 we show that, despite the "row sparsity" model assumption, it is possible to disentangle the role of each tissue in a post-precessing analysis. See [43] for an alternative solution of multiple-tissue gene expression problem when only one genetic marker is considered at-a-time, essentially performing selection of important responses rather than predictors.

The second main assumption of the SBMR model is the specification of a full covariance matrix $\boldsymbol{\Sigma}$ for the gene expression residual dependence between tissues, using an Inverse Wishart distribution. From a modelling perspective, the choice of a full covariance matrix is not as strong as the testing assumption described above: as discussed in [15], if the number of tissues analysed is relatively small (roughly $r < 10$), the posterior inference is not affected by the over-parametrisation of the model due to the estimation of all off-diagonal elements of $\boldsymbol{\Sigma}$. However, as shown in [44] for the same "row sparsity" model, the advantage of a sparse covariance matrix $\boldsymbol{\Sigma}$ is apparent when the number of tissues/conditions is larger, in the order of tens or hundreds of responses.

We complete the presentation of the SBMR model by discussing the hierarchical structure on the shrinkage factor g in (10.10) which can be interpreted as the quantity of information provided by the prior on the regression coefficients relative to the sample size; its value can influence the results of the variable selection procedure. To avoid arbitrary tuning, [15] let it adapt to the data by deriving the Jeffreys' prior

$$p(g) = \frac{1}{\log(1+m)(1+g)}, \qquad 0 \leq g \leq m, \tag{10.12}$$

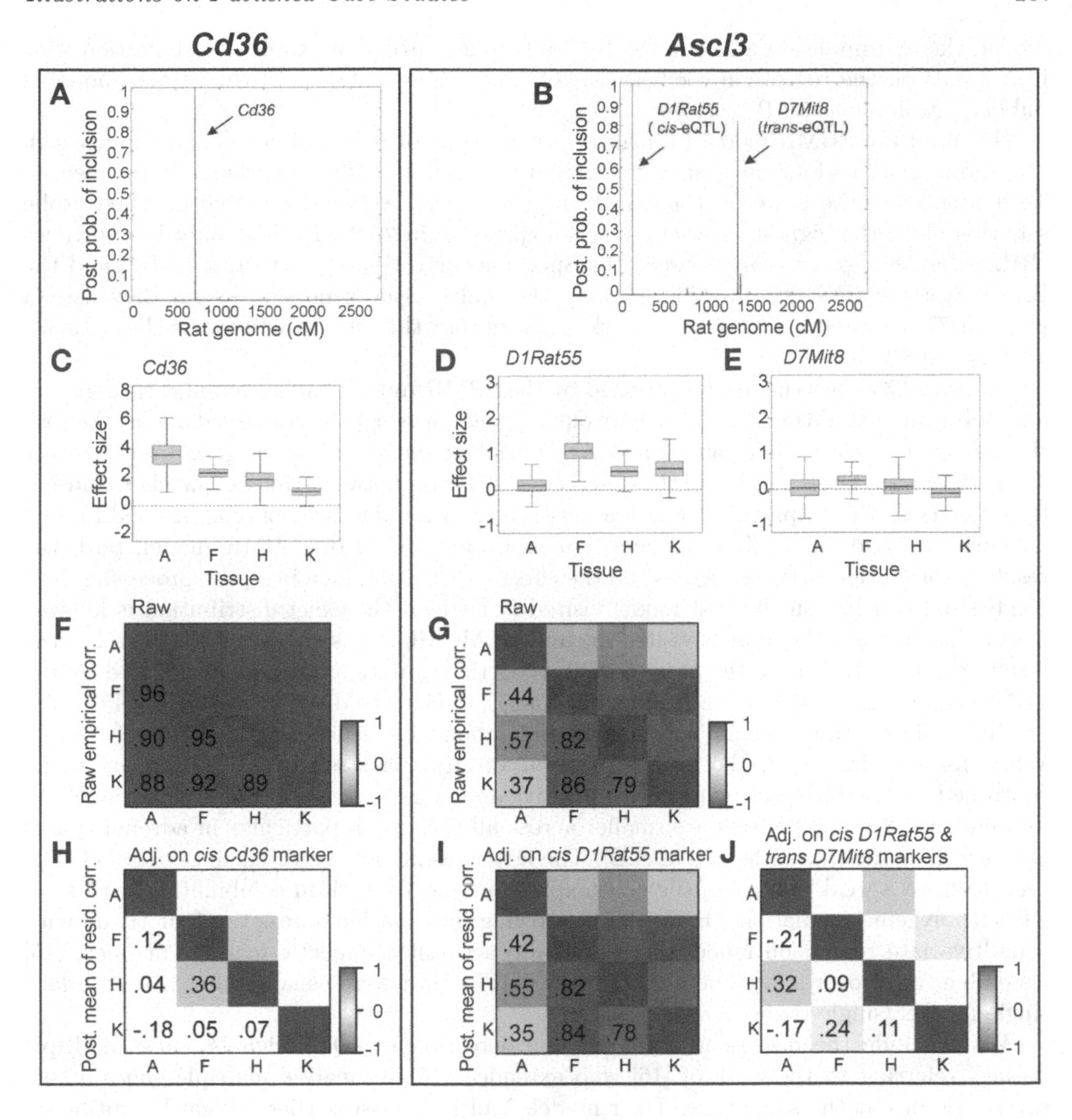

FIGURE 10.3
cis and *trans* eQTLs detected by the SBMR model for two genes *Cd36* (left panels) and *Ascl3* (right panels) across four different tissues: A, adrenal; F, fat; H, heart; K, kidney. (**A**, **B**): For each gene, the set of markers associated with high posterior probability of inclusion show monogenic control for *Cd36* gene (*Cd36* marker) and polygenic control for *Ascl3* gene (*cis* marker *D1Rat55* and *trans* marker *D7Mit8*). (**C**, **D**, **E**): Boxplots of the eQTLs posterior effect size for each tissue. While for gene *Cd36* the effect of *Cd36* marker is large and positive in all tissues, for gene *Ascl3* the *cis* effects are smaller across all tissues, in particular in adrenal, as well as the *trans* effects. (**F**, **G**): For each gene, the raw empirical correlation of gene expression across four tissues is reported. (**H**, **I**, **J**): The posterior mean of the residual correlation matrix is depicted for each gene. In panels **H** and **J**, posterior residual correlations were obtained conditionally on marker *Cd36* for gene *Cd36* and on both markers *D1Rat55* and *D7Mit8* for gene *Ascl3*, respectively. In panel **I**, the posterior residual correlations were obtained conditionally only on the *cis* eQTL. Despite the small effects of *trans* marker *D7Mit8*, it has a pivotal role in explaining the observed correlation between tissues. (Figure adapted from [15].)

with $m = \min(n, p^2)$, which is the default value assigned to the coefficient g if no prior distribution is specified [45]. The combination of (10.10) and (10.12) can be seen as a scale-normal mixture of g-priors. It is also an alternative formulation to the Zellner and Siow's prior density on the shrinkage factor g [13]. Note, however, that in the Zellner and Siow model, the marginalisation of g in (10.10) leads to a multivariate Cauchy distribution with heavy tails on the regression coefficients [26]. The implementation of the SBMR model is publicly available as an R package.

[15] used the SBMR model to identify genetic control points of gene expression, which are common across four rat tissues: fat, kidney, adrenal and left ventricle (hereafter heart). To demonstrate the power of the SBMR approach, they selected a subset of 2,000 probe sets that show the highest variation in gene expression in 29 rat Recombinant Inbred Lines (RILs) derived from a cross between the Spontaneously Hypertensive Rat (SHR) and the Brown Norway (BN) strains [46]. To map the multi-tissue gene expression, they used a panel of 770 non-redundant genetic markers. Note that the set of predictors is the same for all genes in all tissues.

Figure 10.3 shows the results obtained by the SBMR model in two exemplar cases of *cis*- and polygenic-regulation, (i.e., simultaneous *cis* and *trans* eQTL) conserved across the four tissues for gene *Cd36* (left panels) and *Ascl3* (right panels). While for gene *Cd36*, marker *Cd36* explains a large fraction of the observed correlation between tissues (panels F and H), for gene *Ascl3* the empirical correlation between tissues fades away only after conditioning on both *cis*- and *trans*-effect markers (panels G and J). In the SBMR model, both the residual correlation between tissues and the effect size are obtained in a post-processing step conditionally either on the best model visited (for which the exact distribution is known) or conditionally on the models visited during the MCMC. This simplifies considerably the downstream analysis once the pattern of the genetic regulation has been uncovered by the SBMR algorithm. Panels C, respectively D and E, depict the distribution of the effect size conditionally on the *cis*- and polygenic-control points for gene *Cd36*, respectively *Ascl3*. While for the first gene, the gene expression variation across all tissues appears to be controlled by the *cis*-effect marker with positive and large regression coefficients (panel C), for gene *Ascl3* the *cis* effects are smaller across all tissues, in particular in adrenal (panel D), as well as the *trans* effects (panel E). The latter results are expected since distal eQTLs tend to have a weak control of the gene expression and to work in combination with a *cis* effect (polygenic regulation). Hence, by borrowing information across the four tissue with a multivariate regression model that takes into account all genetic markers at once, [15] were able to uncover a large number of *trans* eQTLs that were usually missed by standard single-marker/single-tissue models.

We conclude the discussion regarding the detection of eQTL signals across multiple tissues, referring to the work of [16] who extended [15] to analyse multiple genes across several tissues at the same time. Their model, Multiple Tissues Hierarchical Evolutionary Stochastic Search, MT-HESS, can be seen as a combination of the SBMR approach and a hierarchical model based on a multiplicative decomposition of the selection probabilities outlined in (10.7). A similar decomposition is the basis of the scalable version for hotspot modelling presented in the following section. The schematic representation of the MT-HESS model, and the type and dimensions of the data analysed in [16] are presented in the middle and top panels of Figure 10.2.

10.3.2 Modelling eQTL Hotspots under Different Experimental Conditions

The material presented next focuses on modelling *hotspots*, i.e., predictors associated with several responses in large-p large-q regression problems. In this illustration, the response $\boldsymbol{Y}$ is an $n \times q$ matrix gathering the measurements for q gene expression levels in each of n individuals, and we will aim to estimate the associations with a (sparse) subset of p SNPs gathered in a $n \times p$ candidate predictor matrix $\boldsymbol{X}$ for the same individuals. This corresponds to a single-tissue eQTL regression setup whereby the gene levels are modelled jointly; see the top and middle panels of Figure 10.2. As outlined in Section 10.2, hierarchical approaches for this task need to fulfil two specific requirements: (i) they need to accommodate robust hyperprior specifications, that can flexibly infer the hotspot sizes while adjusting for the dimension of the response vector, and (ii) they need to implement scalable and accurate inference for large parameter spaces, as these often involve tens of thousands of parameters.

Our case study will exploit the fully Bayesian hotspot modelling approach ATLASQTL [28] to describe the landscape of hotspot SNPs that influence monocyte expression in humans. As introduced in (10.8), the method uses a flexible hierarchical regression framework that models the propensity of predictors to be hotspots using the horseshoe shrinkage prior, $\theta_j \mid \tau \sim \text{HS}(\tau)$. The horseshoe prior is a popular example of absolutely continuous global-local shrinkage priors that have an infinite spike at the origin and regularly-varying tails [44, 47]. This prior also enjoys theoretical guarantees, such as near-minimaxity in estimation [48]. It is usually formulated as follows:

$$\theta_j \mid \lambda_j, \tau \sim \mathcal{N}\left(0, \tau^2\lambda_j^2\right), \qquad \lambda_j \overset{\text{iid}}{\sim} \text{C}^+(0,1), \qquad j = 1, \ldots, p, \tag{10.13}$$

where $\text{C}^+(\cdot,\cdot)$ is the half-Cauchy distribution. In the context of the ATLASQTL model, the *local scale* λ_j permits a robust modelling of individual hotspot signals, which leaves their effects unshrunk, while the *global scale* τ shrinks noise globally to accommodate sparse settings.

Importantly, the response dimension needs to be accounted for as part of the prior specification for τ in order for the model to enjoy the same shrinkage properties as the horseshoe prior in the classical normal means model [49]. Indeed, it can be shown that by setting $\tau \sim \text{C}^+(0, q^{-1/2})$, the prior on the *shrinkage factor* controlling the regularisation of the hotspot effects reduces to the horseshoe-shaped shrinkage factor prior Beta$(1/2, 1/2)$ described in [49]. With this prior unbounded at 0 and 1, one expects *a priori*, either large hotspot effects or no effect. Moreover, one can show that the scaling factor $q^{-1/2}$ results in a response multiplicity adjustment that is critical to regularise θ_j and prevents spurious hotspot effects in weakly informative large-q settings [28]. This fully Bayesian specification on θ_j thus permits inferring the hotspot propensity variances from the data with no *ad-hoc* choice that may bias the estimation of the hotspot sizes.

Besides these modelling adaptations, ATLASQTL also tailors inference to the specificities of the parameter spaces to be explored, namely, to the size of the discrete search space created by the binary latent matrix $\Gamma = \{\gamma_{jk}\}$ (of dimension $2^{p\times q}$) and to its multimodality, further reinforced by the complex dependence structures typically observed among SNPs and gene products in molecular QTL problems. To bypass the notorious mixing issues encountered by classical MCMC approaches, [28] investigate deterministic alternatives to sampling-based approaches. They build on [27, 50], which provide positive evidence for accurate posterior exploration with variational inference in the context of genetic and QTL association studies, and derive a variational algorithm, which they augment with a simulated annealing scheme to enhance the exploration of multimodal posteriors. The procedure involves closed-form updates, with a so-called *temperature* parameter. High temperatures flatten the distribution of interest during the first iterations of the algorithm, thereby facilitating the search for the global optimum. The algorithm implements an efficient block coordinate ascent optimisation, where the variational parameters are updated in turn and by blocks for all the responses. ATLASQTL is publicly available as an R package.

[28] illustrate the advantages of ATLASQTL on an expression QTL (eQTL) analysis involving > 24K transcript levels measured in CD14$^+$ monocytes. The data consist of SNPs obtained from Illumina arrays, for 432 healthy European individuals. They also involve expression for four different monocytic conditions, before and after immune stimulation, namely, obtained by exposing the monocytes to the inflammation proxies interferon-γ (IFN-γ) or differing durations of lipopolysaccharide (LPS 2h or LPS 24h). Previous studies have

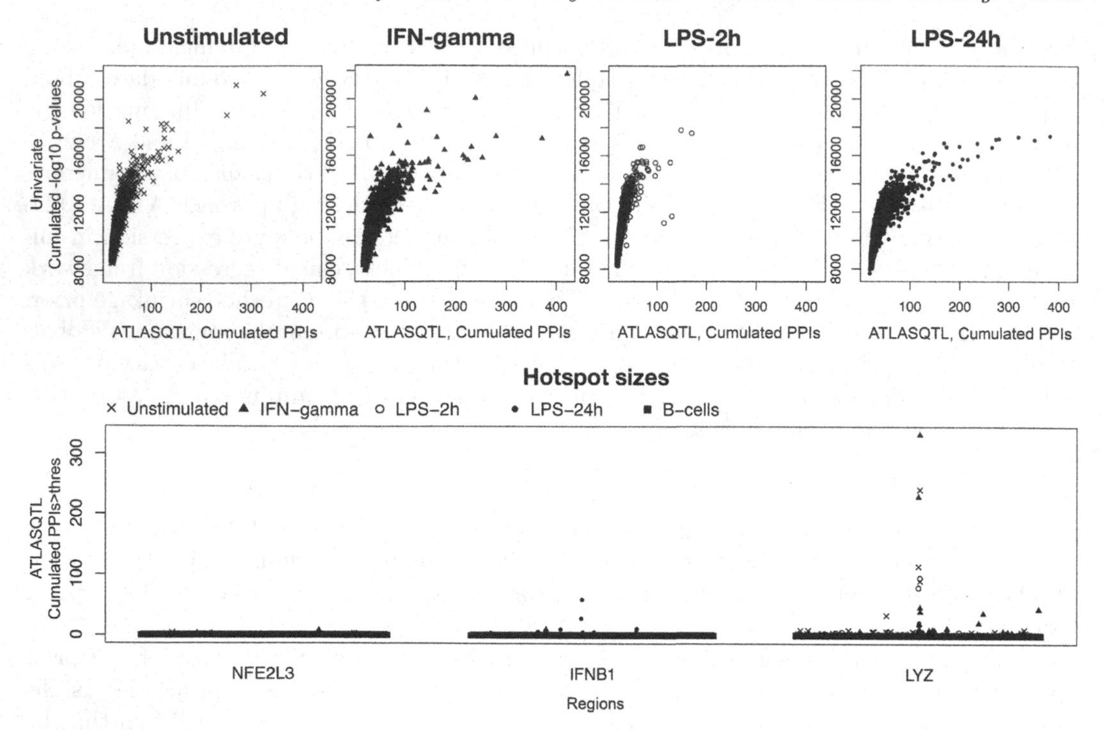

FIGURE 10.4
Hotspots from stimulated eQTL analyses, for the *NFEL2L3*, *IFNB1* and *LYZ* genomic regions with the four monocyte conditions and the B-cell negative controls. Top: for each condition, raw hotspot evidence for all three regions comprising *NFEL2L3*, *IFNB1* and *LYZ*. Scatterplots with $-\log_{10}$ *p*-values of univariate screening summed across responses ($-\sum_{k=1}^{q} \log_{10} p_{jk}$), versus variational PPIs (posterior probability of inclusion) obtained by ATLASQTL, summed across responses ($\sum_{k=1}^{q} \text{pr}(\gamma_{jk} = 1 \mid \boldsymbol{Y})$, see equation (10.5)). Bottom: Hotspot sizes, as estimated using a permutation-based FDR threshold of 20% on the PPIs. (Figure adapted from [28].)

suggested that gene stimulation may trigger substantial *trans*-regulatory activity (i.e., between a SNP and a distal gene product) [36, 37], which creates favourable conditions for the manifestation of hotspot SNPs.

Here, we will examine three genomic regions comprising genes thought to play a central role in the pathogenesis of immune disorders: *NFE2L3* on chromosome 7, *IFNB1* on chromosome 9, and *LYZ* on chromosome 12. Each region involves 1 500 SNPs and spans from 7.5 to 12 Mb. Step-by-step guidance for re-creating this analysis on synthetic data is available in the Supplementary Material under the form of an Rmarkdown script.

After quality control, [28] ran ATLASQTL for all four monocyte conditions, as well as for B cell levels, (five $n \times q$ response matrices $\boldsymbol{Y}^{\text{unstim}}$, $\boldsymbol{Y}^{\text{IFN}\gamma}$, $\boldsymbol{Y}^{\text{LPS2h}}$, $\boldsymbol{Y}^{\text{LPS24h}}$, $\boldsymbol{Y}^{\text{Bcells}}$, each composed of the $q > 24$K transcript levels measured in the corresponding condition, with $n = 413, 366, 260, 321$ or 275, respectively), and on each of the three regions ($n \times p$ candidate predictor matrices $\boldsymbol{X}^{NFE2L3}$, $\boldsymbol{X}^{IFNB1}$, $\boldsymbol{X}^{LYZ}$, each composed of SNPs located in the corresponding region, with $p = 1\,500$), resulting in 15 separate analyses. Figure 10.4 displays the evidence for hotspots as obtained by ATLASQTL versus a plain univariate screening, using MATRIXEQTL [51]. It compares the nominal $-\log_{10}$ *p*-values of MATRIXEQTL with the raw posterior probabilities of inclusion (PPIs) of ATLASQTL, both summed across responses: while the two approaches agree on the small or moderate evidence, ATLASQTL appears to boost the detection of hotspot effects.

Thanks to the computational efficiency of the variational algorithm, [28] could then obtain empirical Bayesian false discovery rates (FDR) by permutation analysis for each region and condition [following 27]. This allowed them to compute hotspot sizes by counting, for each SNP, the number of estimated PPIs (for its association with each response) that exceed the 20% FDR threshold. Figure 10.4 indicates an increased hotspot activity under stimulation with IFN-γ and LPS 24h. This activity is endorsed by the absence of hotspots in B cells which can be considered as valid negative controls [36]. The location and size of hotspots also varies greatly across the regions: the *NFE2L3* region is essentially inactive; its largest hotspot is of size 8 and appears under IFN-γ stimulation, confirming previous observations [37]. The *IFNB1* region displays more hotspots under LPS 24h stimulation in line with other existing work [37], but ATLASQTL finds that these hotspots have larger sizes. The top LPS 24h hotspot in the *IFNB1* region, rs3898946, is a known eQTL for genes *FOCAD* and *MLLT3* in the tibial artery, and for gene *PTPLAD2* in skin tissues [52].

The *LYZ* region has also been previously described for its hotspot activity [53] and is indeed very active in our analyses. While [37] mostly report stimuli-specific *trans*-regulatory effects, [28]'s top hit, rs6581889, located only 9 Kb downstream of the *LYZ* gene, is persistent across all four conditions: it is the largest hotspot in the unstimmulated condition with size 242, in the IFN-γ condition with size 333, and in the LPS 2h condition with size 96, and it is the second largest hotspot in the LPS 24h condition with size 18. This activity seems to have been largely triggered by the IFN-γ stimulation, but was also observed after 2 hours and 24 hours of LPS stimulation. Again, the B-cell data provide a good negative control, with no activity in the *LYZ* region: the largest number of transcripts associated with a given SNP is three, and the signal does not co-localise with any hotspot uncovered in monocytes. Finally, the study confirms the known associations of rs6581889 as a *cis* eQTL for *LYZ* and *YEATS4* in multiple tissues.

In summary, the above analysis revealed important monocyte-specific pleiotropic activity, while also confirming existing findings. It also paved the way for a more comprehensive case study [29], aimed at better understanding the functional mechanisms underlying this activity. This case study employed the EPISPOT method - which extends the ATLASQTL model according to (10.9) - to leverage a large panel of epigenetic marks annotating the SNP data. In particular, EPISPOT selected a monocyte DNase-I sensitivity site annotation from > 150 epigenetic annotations, which provided insights on mediation and cell-type specific effects for the *LYZ* hotspots.

10.4 Discussion

We have reviewed several approaches for modelling and borrowing information across large numbers of responses in a hierarchical fashion. We have also discussed the different steps towards timely and accurate inference, and how this involves marrying appropriate modelling choices with purpose-built inference algorithms. In our experience, the analysis of real datasets with a few hundred responses is feasible using MCMC inference – in particular when adaptive MCMC schemes that "visit" less frequently responses which are not linked with the predictors are used [24] – but they can take several hours to run. Conversely, deterministic alternatives, e.g., based on tailored annealed variational inference [28, 29], can prove particularly robust even in the presence of strong correlation structures, and typically run within minutes to a few hours on large molecular QTL data (in profiling studies for [28], this was about 30 times faster than MCMC inference). A number of important questions remain however on the modelling front, and we will focus on just two of them which we feel deserve particular attention.

First, the impact of the conditional independence assumption employed to bypass the modelling of large covariance matrices needs to be carefully considered. Using the hierarchically-related regression models presented in this chapter, the borrowing of information across outcomes is achieved exclusively via the model hierarchy, which suffices to improve the detection of weak association signals as shown in [23, 24, 27, 28]. However, when the outcomes are highly correlated for reasons other than shared predictor effects, ignoring their dependence structure can lead to anti-conservative inference [18]. This, and related aspects, have triggered important debate in the past decades in the applied genetics community, as several studies have shown how specifically accounting for the traits' dependence structure (by directly modelling it or by inferring latent factors along with the regression effects of interest) can clear away weak *trans* QTL signals, while not accounting for this structure can result in spurious genetic association effects due to confounding issues [54–58]. Finding the optimal balance between removing the unknown variation and retaining the signal of interest – and whether to attempt this by joint modelling or a two-step approach – remains an open question from both applied and methodological viewpoints.

Second, molecular and clinical datasets are growing in size and in diversity. In particular, protein, lipid or metabolite levels are now routinely collected along with genetic data for QTL analyses and, in some instances, with the intent of using them as molecular proxies for specific clinical endpoints. For these data, the hierarchically-related response models described in this chapter can often be employed without modification. Indeed, all assume Gaussian responses, which turns out to be a reasonable assumption for many molecular and continuous clinical trait settings (possibly after some variance-stabilising transformation). For a concrete example, a simpler variant of the ATLASQTL method by [28], namely the LOCUS method [27] based on formulation (10.6) and implemented using variational inference, has been successfully applied to map human protein abundance in plasma. The method enabled fully joint analysis of ≈ 300K SNPs and $100-1,000$ protein levels, which yielded replication rates of $> 80\%$ in independent-cohort data, as well as novel pQTL discoveries with potential biomedical implications [59]. Still, as new quantification techniques emerge, the need for more flexible link functions – e.g., capable to directly model count data in the case of RNAseq measurements or discrete clinical parameters – becomes more apparent. Moreover, when several types of 'omics data, such as gene expression, lipidomics and proteomics, are collected on the same samples in order to find common hotspots, the hierarchical model would need to be extended to allow for meaningful common parameters while enabling separate noise calibration.

Finally, [60] tried to address the problem of specifying a suitable link function for responses of diverse types, i.e., continuous, binary, categorical and count data, from studies involving high-throughput data, clinically relevant measurements and end-phenotypes. They proposed to perform Bayesian variable selection for combinations of continuous and/or discrete responses without any approximation by coupling Bayesian variable selection with a flexible link function provided by the Gaussian copula model [61]. Similarly to the SUR model, information is shared across responses by the Gaussian copula that models the dependence between the continuous observations and the latent variables of the discrete outcomes. Their approach, Bayesian Variable Selection for Gaussian Copula Regression (BVSGCR) allows the joint analysis of continuous (e.g., Gaussian) and discrete (e.g., binary, ordinal and nominal categorical and count) responses with missing values, as well as the specification of a sparse decomposable or non-decomposable graphical model for the conditional dependence structure between the responses. The latter instance of their method takes full advantage of the C++ computational efficiency of a recently proposed approach [62] to sample non-decomposable graphs. Despite this advantage, the overall number of responses that can be jointly analysed by BVSGCR model is limited, due to the high computational cost of sampling a graphical model and the associated sparse covariance matrix at each MCMC iteration [17, 18]. To make their model operational in the analysis of high-throughput data of diverse types, possibly with discrete clinical phenotypes, one may consider simplifying the dependence detectable by the graphical Gaussian copula model. For instance, this can be achieved by specifying a block diagonal graphical model with an accompanying suitable sparsity prior, where the dependence of the responses within each block is fully considered while blocks are regarded as independent but linked by a hierarchical structure, basically combining the advantages of the Gaussian copula model and the hierarchical-related regression model.

Software

All R packages mentioned in this chapter are available on GitHub at `https://github.com/lb664` and `https://github.com/hruffieux`. An Rmarkdown script emulating the multi-stimulus monocyte eQTL analysis can be found in the Supplementary Material.

Acknowledgements

This research was funded by the UK Medical Research Council programme MRC MC UU 00002/10 (HR, SR) and MR M0 13138/1, MR S0 2638X/1 (LB); the Lopez-Loreta Foundation (HR); the Engineering and Physical Sciences Research Council EP/R018561/1 (SR); the BHF-Turing Cardiovascular Data Science Awards 2017 & the Alan Turing Institute under the Engineering and Physical Sciences Research Council grant EP/N510129/1 (LB); the Alan Turing Institute Fellowship number TU/B/000092 (SR). This work was also supported by the NIHR Cambridge BRC. The views expressed are those of the author(s) and not necessarily those of the NHS, the NIHR or the Department of Health and Social Care.

Bibliography

[1] A. Teixeira-Pinto, J. Siddique, R. Gibbons, and S. L. Normand. Statistical approaches to modeling multiple outcomes in psychiatric studies. *Psychiatric Annals*, 39:729–735, 2009.

[2] G. I. Papakostas. Surrogate markers of treatment outcome in major depressive disorder. *International Journal of Neuropsychopharmacology*, 15:841–854, 2012.

[3] A. P. Dawid. Some matrix-variate distribution theory: notational considerations and a Bayesian application. *Biometrika*, 68:265–274, 1981.

[4] P. J. Brown, M. Vannucci, and T. Fearn. Multivariate Bayesian variable selection and prediction. *Journal of the Royal Statistical Society: Series B (Statistical Methodology)*, 60:627–641, 1998.

[5] P. J. Brown, M. Vannucci, and T. Fearn. Bayes model averaging with selection of regressors. *Journal of the Royal Statistical Society: Series B (Statistical Methodology)*, 64:519–536, 2002.

[6] E. I. George and R. E. McCulloch. Variable selection via Gibbs sampling. *Journal of the American Statistical Association*, 88:881–889, 1993.

[7] E. I. George and R. E. McCulloch. Approaches for Bayesian variable selection. *Statistica Sinica*, 7:339–373, 1997.

[8] A. P. Dawid and S. L. Lauritzen. Hyper Markov laws in the statistical analysis of decomposable graphical models. *The Annals of Statistics*, pages 1272–1317, 1993.

[9] M. A. Newton, A. Noueiry, D. Sarkar, and P. Ahlquist. Detecting differential gene expression with a semiparametric hierarchical mixture method. *Biostatistics*, 5:155–176, 2004.

[10] H. Chipman, E. I. George, R. E McCulloch, M. Clyde, D. P. Foster, and R. A. Stine. The practical implementation of Bayesian model selection. *Lecture Notes-Monograph Series*, pages 65–134, 2001.

[11] T. J. Mitchell and J. J. Beauchamp. Bayesian variable selection in linear regression. *Journal of the American Statistical Association*, 83:1023–1032, 1988.

[12] A. Zellner. On assessing prior distributions and Bayesian regression analysis with g-prior distributions. In *Studies in Bayesian Econometrics*, volume 6, pages 233–243. Elsevier, New York, United States, 1986. P. K. Goel and A. Zellner, editors.

[13] F. Liang, R. Paulo, G. Molina, M. A. Clyde, and J. O. Berger. Mixtures of g-priors for Bayesian variable selection. *Journal of the American Statistical Association*, 103: 410–423, 2008.

[14] L. Bottolo and S. Richardson. Evolutionary stochastic search for Bayesian model exploration. *Bayesian Analysis*, 5:583–618, 2010.

[15] E. Petretto, L. Bottolo, S. R. Langley, M. Heinig, C. McDermott-Roe, R. Sarwar, M. Pravenec, N. Hübner, T. J. Aitman, and S. A. Cook. New insights into the genetic control of gene expression using a Bayesian multi-tissue approach. *PLoS Computational Biology*, 6:e1000737, 2010.

[16] A. Lewin, H. Saadi, J. E. Peters, A. Moreno-Moral, J. C. Lee, K. G. C. Smith, E. Petretto, L. Bottolo, and S. Richardson. MT-HESS: an efficient Bayesian approach for simultaneous association detection in OMICS datasets, with application to eQTL mapping in multiple tissues. *Bioinformatics*, 32:523–532, 2015.

[17] A. Bhadra and B. K. Mallick. Joint high-dimensional Bayesian variable and covariance selection with an application to eQTL analysis. *Biometrics*, 69:447–457, 2013.

[18] L. Bottolo, M. Banterle, S. Richardson, M. Ala-Korpela, M. R. Jarvelin, and A. Lewin. A computationally efficient Bayesian Seemingly Unrelated Regressions model for high-dimensional Quantitative Trait Loci discovery. *Journal of the Royal Statistical Society: Series C (Applied Statistics)*, 2021. doi: https://doi.org/10.1101/467019. To Appear.

[19] Z. Zhao, M. Banterle, L. Bottolo, S. Richardson, A. Lewin, and M. Zucknick. BayesSUR: An R package for high-dimensional multivariate Bayesian variable and covariance selection in linear regression. *Journal of Statistical Software*, 2021. To appear.

[20] A. Zellner. An efficient method of estimating seemingly unrelated regressions and tests for aggregation bias. *Journal of the American Statistical Association*, 57:348–368, 1962.

[21] A. Zellner and T. Ando. A direct Monte Carlo approach for Bayesian analysis of the seemingly unrelated regression model. *Journal of Econometrics*, 159:33–45, 2010.

[22] S. K. Deshpande, V. Ročková, and E. I. George. Simultaneous variable and covariance selection with the multivariate spike-and-slab lasso. *Journal of Computational and Graphical Statistics*, pages 1–11, 2019.

[23] Z. Jia and S. Xu. Mapping quantitative trait loci for expression abundance. *Genetics*, 176:611–623, 2007.

[24] S. Richardson, L. Bottolo, and J. S. Rosenthal. Bayesian models for sparse regression analysis of high-dimensional data. In J. M. Bernardo, M. J. Bayarri, J. O. Berger, A. P. Dawid, D. Heckerman, A. F. M. Smith, and M. West, editors, *Bayesian Statistics*, volume 9, pages 539–569. Oxford University Press, New York, United States, 2010.

[25] M. P. Scott-Boyer, G. C. Imholte, A. Tayeb, A. Labbe, C. F. Deschepper, and R. Gottardo. An integrated hierarchical Bayesian model for multivariate eQTL mapping. *Statistical Applications in Genetics and Molecular Biology*, 11:1515–1544, 2012.

[26] L. Bottolo, E. Petretto, S. Blankenberg, F. Cambien, S. A. Cook, L. Tiret, and S. Richardson. Bayesian detection of expression quantitative trait loci hot spots. *Genetics*, 189:1449–1459, 2011.

[27] H. Ruffieux, A. C. Davison, J. Hager, and I. Irincheeva. Efficient inference for genetic association studies with multiple outcomes. *Biostatistics*, 18:618–636, 2017.

[28] H. Ruffieux, A. C. Davison, J. Hager, J. Inshaw, B. P. Fairfax, S. Richardson, and L. Bottolo. A global-local approach for detecting hotspots in multiple-response regression. *Annals of Applied Statistics*, 14:905–928, 2020.

[29] H. Ruffieux, B. Fairfax, I. Nassiri, E. Vigorito, C. Wallace, S. Richardson, and L. Bottolo. EPISPOT: an epigenome-driven approach for detecting and interpreting hotspots in molecular QTL studies. *The American Journal of Human Genetics*, 108:983–1000, 2021

[30] R. Breitling, Y. Li, B. M. Tesson, J. Fu, C. Wu, T. Wiltshire, A. Gerrits, L. V. Bystrykh, G. De Haan, and A. I. Su. Genetical genomics: spotlight on QTL hotspots. *PLOS Genetics*, 4:e1000232, 2008.

[31] J. G. Scott and J. O. Berger. Bayes and empirical-Bayes multiplicity adjustment in the variable-selection problem. *Annals of Statistics*, 38:2587–2619, 2010.

[32] L. Bottolo, M. Chadeau-Hyam, D.I. Hastie, S.R. Langley, E. Petretto, L. Tiret, D. Tregouet, and S. Richardson. ESS++: a C++ objected-oriented algorithm for Bayesian stochastic search model exploration. *Bioinformatics*, 27:587–588, 2011.

[33] H.J. Westra, M. J. Peters, T. Esko, H. Yaghootkar, C. Schurmann, J. Kettunen, M. W. Christiansen, B. P. Fairfax, K. Schramm, and J. E. Powell. Systematic identification of trans eQTLs as putative drivers of known disease associations. *Nature Genetics*, 45: 1238, 2013.

[34] B. Brynedal, J. Choi, T. Raj, R. Bjornson, B. E. Stranger, B. M. Neale, B. F. Voight, and C. Cotsapas. Large-scale trans-eQTLs affect hundreds of transcripts and mediate patterns of transcriptional co-regulation. *The American Journal of Human Genetics*, 100:581–591, 2017.

[35] C. Yao, R. Joehanes, A. D. Johnson, T. Huan, C. Liu, J. E. Freedman, P. J. Munson, D. E. Hill, M. Vidal, and D. Levy. Dynamic role of trans regulation of gene expression in relation to complex traits. *The American Journal of Human Genetics*, 100:571–580, 2017.

[36] B. P. Fairfax, S. Makino, J. Radhakrishnan, K. Plant, S. Leslie, A. Dilthey, P. Ellis, C. Langford, F. O. Vannberg, and J. C. Knight. Genetics of gene expression in primary immune cells identifies cell type–specific master regulators and roles of HLA alleles. *Nature Genetics*, 44:502, 2012.

[37] B. P. Fairfax, P. Humburg, S. Makino, V. Naranbhai, D. Wong, E. Lau, L. Jostins, K. Plant, R. Andrews, and C. McGee. Innate immune activity conditions the effect of regulatory variants upon monocyte gene expression. *Science*, 343:1246949, 2014.

[38] Y. Guan and M. Stephens. Bayesian variable selection regression for genome-wide association studies and other large-scale problems. *Annals of Applied Statistics*, 5: 1780–1815, 2011.

[39] J. Yang, T. Ferreira, A. P. Morris, S. E. Medland, P. A. F. Madden, A. C. Heath, N. G. Martin, G. W. Montgomery, M. N. Weedon, and R. J. Loos. Conditional and joint multiple-SNP analysis of GWAS summary statistics identifies additional variants influencing complex traits. *Nature Genetics*, 44:369, 2012.

[40] M. E. Goddard, K. E. Kemper, I. M. MacLeod, A. J. Chamberlain, and B. J. Hayes. Genetics of complex traits: prediction of phenotype, identification of causal polymorphisms and genetic architecture. *Proceedings of the Royal Society of London B: Biological Sciences*, 283, 2016.

[41] A. Lewin, L. Bottolo, and S. Richardson. Bayesian methods for gene expression analysis. In David Balding, Ida Moltke, and John Marioni, editors, *Handbook of Statistical Genomics, Fourth Edition*, pages 843–877. Wiley Online Library, 2019.

[42] H. Wang. Sparse seemingly unrelated regression modelling: Applications in finance and econometrics. *Computational Statistics & Data Analysis*, 54:2866–2877, 2010.

[43] T. Flutre, X. Wen, J. Pritchard, and M. Stephens. A statistical framework for joint eqtl analysis in multiple tissues. *PLoS Genetics*, 9:e1003486, 2013.

[44] A. Bhadra, J. Datta, N. G. Polson, and B. Willard. Default Bayesian analysis with global-local shrinkage priors. *Biometrika*, 103:955–969, 2016.

[45] C. Fernandez, E. Ley and M.F.J. Steel. Benchmark priors for Bayesian model averaging. *Journal of Econometrics*, 100:381–427, 2001.

[46] N. Hubner, C. A. Wallace, H. Zimdahl, E. Petretto, H. Schulz, F. Maciver, M. Mueller, O. Hummel, J. Monti, V. Zidek, et al. Integrated transcriptional profiling and linkage analysis for identification of genes underlying disease. *Nature Genetics*, 37:243–253, 2005.

[47] N. G. Polson and J. G. Scott. Shrink globally, act locally: sparse Bayesian regularization and prediction. In J. M. Bernardo, M. J. Bayarri, J. O. Berger, A. P. Dawid, D. Heckerman, A. F. M. Smith, and M. West, editors, *Bayesian Statistics*, volume 9, pages 501–538. Oxford University Press, New York, United States, 2010.

[48] S. van der Pas, B. Szabó, and A. van der Vaart. Adaptive posterior contraction rates for the horseshoe. *Electronic Journal of Statistics*, 11:3196–3225, 2017.

[49] C. M. Carvalho, N. G. Polson, and J. G. Scott. The horseshoe estimator for sparse signals. *Biometrika*, 97:465–480, 2010.

[50] P. Carbonetto and M. Stephens. Scalable variational inference for Bayesian variable selection in regression, and its accuracy in genetic association studies. *Bayesian Analysis*, 7:73–108, 2012.

[51] A. A. Shabalin. Matrix eQTL: ultra fast eQTL analysis via large matrix operations. *Bioinformatics*, 28:1353–1358, 2012.

[52] GTEx Consortium. The Genotype-Tissue Expression (GTEx) pilot analysis: multitissue gene regulation in humans. *Science*, 348:648–660, 2015.

[53] M. Rotival, T. Zeller, P. S. Wild, S. Maouche, S. Szymczak, A. Schillert, R. Castagné, A. Deiseroth, C. Proust, and J. Brocheton. Integrating genome-wide genetic variations and monocyte expression data reveals trans-regulated gene modules in humans. *PLoS Genetics*, 7:e1002367, 2011.

[54] H. M. Kang, C. Ye, and E. Eskin. Accurate discovery of expression quantitative trait loci under confounding from spurious and genuine regulatory hotspots. *Genetics*, 180: 1909–1925, 2008.

[55] O. Stegle, L. Parts, R. Durbin, and J. Winn. A Bayesian framework to account for complex non-genetic factors in gene expression levels greatly increases power in eQTL studies. *PLOS Computational Biology*, 6:e1000770, 2010.

[56] O. Stegle, L. Parts, M. Piipari, J. Winn, and R. Durbin. Using probabilistic estimation of expression residuals (PEER) to obtain increased power and interpretability of gene expression analyses. *Nature Protocols*, 7:500, 2012.

[57] N. Fusi, O. Stegle, and N. D. Lawrence. Joint modelling of confounding factors and prominent genetic regulators provides increased accuracy in genetical genomics studies. *PLOS Computational Biology*, 8:e1002330, 2012.

[58] J. W. J Joo, J. H. Sul, B. Han, C. Ye, and E. Eskin. Effectively identifying regulatory hotspots while capturing expression heterogeneity in gene expression studies. *Genome Biology*, 15:r61, 2014.

[59] H. Ruffieux, J. Carayol, R. Popescu, M.-E. Harper, R. Dent, W. H. M. Saris, A. Astrup, J. Hager, A. C. Davison, and A. Valsesia. A fully joint Bayesian quantitative trait locus mapping of human protein abundance in plasma. *PLOS Computational Biology*, 16: e1007882, 2020.

[60] A. Alexopoulos and L. Bottolo. Bayesian Variable Selection for Gaussian copula regression models. *Journal of Computational and Graphical Statistics*, pages 1–16, 2020.

[61] M. Pitt, D. Chan, and R. Kohn. Efficient Bayesian inference for Gaussian copula regression models. *Biometrika*, 93:537–554, 2006.

[62] A. Mohammadi and E. C. Wit. BDgraph: An R package for Bayesian structure learning in graphical models. *Journal of Statistical Software*, 89:1–30, 2019.

11

Bayesian Variable Selection in Spatial Regression Models

Brian J. Reich
North Carolina State University (USA)

Ana-Maria Staicu
North Carolina State University (USA)

CONTENTS

Spatially-referenced data pose challenges for Bayesian variable selection, including residual spatial correlation and cumbersome computing for large datasets. In this chapter, we discuss foundational approaches and recent advances in Bayesian variable selection for spatial regression models. We begin by introducing the canonical spatial regression model and discuss how Bayesian variable selection priors can be adapted to account for spatial correlation in the residuals. We then address the more challenging problem of allowing different regression relationships in different regions, including allowing the subset of covariates included in the model to vary across space. We introduce several models for this spatial variation, apply some of the models to a large spatial microbiome dataset.

11.1 Introduction

Spatial data naturally arise in many fields including environmental science, epidemiology, economics and neuroimaging. Due to advances in remote sensing and other forms of data collection, spatial data are increasingly large, both in terms of the number of observations sampled and the number of variables under study. For example, consider the data application studied in [12]. The response variable is ambient air pollution measured from a car driving through Oakland, CA. Air pollution is regressed onto dozens of covariates including the time

DOI: 10.1201/9781003089018-11

of day, type of road segment, distances from known emission sources, etc. The objectives of this regression are to identify the factors that lead to high air pollution and to use this fitted regression model for short-term forecasting.

A key assumption in spatial statistics is that observations from locations that are close together are likely to be correlated. This spatial correlation complicates regression, which typically assumes the observations are independent. In spatial linear regression, both the variables to be included in the mean model and the spatial covariance parameters must be estimated. Thus, performing variable selection within a hierarchical Bayesian model is a natural choice to simultaneously estimate the mean and covariance functions while accounting for uncertainty in both model components. In this chapter, we review methods that have been developed for Bayesian variable selection for spatial data.

Our review is divided into two types of Bayesian spatial variable selection. We begin in Section 11.2 by discussing extensions of standard Bayesian variable selection methods to deal with residual spatial correlation. In Sections 11.3 and 11.4 we review a second type of Bayesian spatial variable selection for models that allow the set of variables included in the model to be different across varying spatial locations. Bayesian spatial variable selection is illustrated with an example in Section 11.5.

11.2 Spatial Regression

In this section we introduce the standard spatial linear regression model; for a more detailed description of the model, with emphasis on Bayesian implementation, see [1] or [8]. Consider the response variable y_i collected at spatial location $\mathbf{s}_i \in \mathcal{R}^2$ (e.g., latitude/longitude) for $i \in \{1, ..., n\}$. The response is regressed onto covariates $\mathbf{x}_i = (x_{1i}, ..., x_{ip})^T$. The covariates can either be spatial variables, such as the elevation of location $\mathbf{s}_i$, or non-spatial variables such as the time of day the observation was taken. The spatial linear model is

$$y_i = \alpha + \sum_{j=1}^{p} x_{ij}\beta_j + \theta(\mathbf{s}_i) + \varepsilon_i \tag{11.1}$$

where α is the intercept, β_j is the coefficient for covariate j, $\theta(\mathbf{s}_i)$ is a spatial random effect and $\varepsilon_i \sim \text{Normal}(0, \sigma_e^2)$ independent over i. If the spatial term $\theta(\mathbf{s}_i)$ is removed the model becomes the standard multiple regression model. In (11.1), as in non-spatial regression, β_j is the expected increase in the response due to a unit increase in x_{ij} with all other variables held fixed. In the following, we assume the covariates are standardized to have mean zero and variance one to ensure the prior is comparable across the covariates.

The residuals in (11.1) are decomposed into spatial and non-spatial components. The spatial component $\theta(\mathbf{s})$ is typically assumed be a continuous function of $\mathbf{s}$ to account for unmeasured or unknown processes with a spatial pattern. For example, the spatial component might be induced by a missing covariate that has a strong spatial pattern, or by a spatial physical process such as shared proximity to a point source (e.g., a power plant). The so-called "nugget" term ε_i accounts for non-spatial variation such as measurement error or unmeasured or unknown local processes.

The model (11.1) naturally extends to non-Gaussian responses using a generalized spatial linear model [4]. In this setting, we specify a link function g and the model $g\{\text{E}(y_i)\} = \eta_i = \alpha + \sum_{j=1}^{p} x_{ij}\beta_j + \theta(\mathbf{s}_i)$ where β_j and $\theta(\mathbf{s}_i)$ are modeled as in the Gaussian response model. For example, if $y_i \in \{0, 1, 2, ...\}$ is a count, then a reasonable model might be to take g to be the natural logarithm function so that $y_i|\eta_i \sim \text{Poisson}\{\exp(\eta_i)\}$, where the y_i are independent given the η_i. Note that the conditional independence of the Poisson responses given the linear predictors induces a nugget effect, and thus the Gaussian nugget term ε_i has been omitted from the model; however, this term can be included as well to account for over-dispersion with respect to the Poisson model.

Returning to the Gaussian model in (11.1), spatial regression differs from non-spatial regression because the residuals are assumed to be correlated, with the correlation assumed to decay with the distance between observations. There are many parametric models for the covariance [8], but the simplest is the isotropic exponential model $\text{Cov}\{\theta(\mathbf{s}_i), \theta(\mathbf{s}_k)\} = \sigma_s^2 \exp\{-d(\mathbf{s}_i, \mathbf{s}_k)/\phi\}$, where σ_s^2 is the spatial variance, $d(\mathbf{s}_i, \mathbf{s}_k)$ is the distance between locations $\mathbf{s}_i$ and $\mathbf{s}_k$ and the parameter ϕ determines the range of spatial correlation. Under this model, the spatial terms $\theta(\mathbf{s}_i)$ are approximately independent (correlation 0.05) for two observations separated by distance 3ϕ, and thus ϕ controls the range of spatial correlation.

For analyses with many covariates, a common approach is to first perform variable selection under a working assumption of independent residuals, and then use the selected subset in a second-stage spatial analysis. However, this is inadvisable because estimation of the mean parameters can change dramatically between spatial and non-spatial fits, especially when the covariates themselves are spatially correlated [14, 22, 25]. Therefore, performing variable selection and estimating mean and covariance parameters simultaneously is preferred. Bayesian variable selection is ideally suited for this purpose because it naturally propagates uncertainties through the model including averaging over the selected covariates.

The regression coefficients can be modelled using any of the Bayesian variable selection methods for non-spatial data discussed in previous chapters. Since the β_j are not spatial processes, they can be given independent (over j) sparse priors, such as spike-and-slab or horseshoe priors. For example, [17] use stochastic search variable selection in a spatial logistic regression model. Therefore, the conceptual jump from non-spatial to spatial variable selection is rather small. The main hurdle to overcome is computational. The likelihood calculation involves manipulating the $n \times n$ covariance matrix, which can be slow for even moderate n. Fortunately, there is now a plethora of methods to deal with large spatial datasets such as approximating the spatial random effects with a low-rank basis expansion, approximating the covariance or inverse covariance matrix as a sparse matrix (i.e., with many zero entries), and divide-and-conquer methods that partition the spatial domain to subregions that are treated as independent (e.g., [13]).

11.3 Regression Coefficients as Spatial Processes

The regression model in Section 11.2 summarized the effect of each covariate with a single regression coefficient. More advanced models allow the regression coefficients to be spatial processes, i.e., the scalar quantity β_j becomes the spatial process $\beta_j(\mathbf{s})$. In this section we review models of this form and discuss the inference that can be performed.

11.3.1 Spatially-Varying Coefficient Model

The spatially-varying coefficient (SVC) model [9] extends the linear model in (11.1) to have spatially-varying regression coefficients. The model is

$$y_i = \alpha(\mathbf{s}_i) + \sum_{j=1}^{p} x_{ij}\beta_j(\mathbf{s}_i) + \epsilon_i \tag{11.2}$$

where $\alpha(\cdot)$ is a spatially-varying intercept, and $\beta_j(\cdot)$ captures the spatial-varying association of the covariate x_{ij}, and ϵ_i is independent error. The random effect $\theta(\mathbf{s})$ in (11.1) has been removed because it is redundent given the spatially-varying intercept $\alpha(\mathbf{s})$. The SVC model allows the relationships between the covariates and response to be different in different regions, perhaps due to the omission of an unknown spatial interaction variable. A classic motivating example is that the age of a home may have a negative relationship with its value in a suburban neighborhood, but a positive relationship in an historical neighborhood.

When the number of covariates is large, Bayesian variable selection could be used to select a subset of the covariates to include in the model. The excluded variable would have $\beta_j(\mathbf{s}) = 0$ for all $\mathbf{s}$ for covariate j. In some settings, it may be desirable to allow the subset of coefficients included in the model to vary with space, so that, for example, $\beta_j(\mathbf{s}_1) = 0$ but $\beta_j(\mathbf{s}_2) \neq 0$. A Bayesian spatial variable selection prior should therefore be sparse, i.e., place mass at (or near) zero, $\beta_j(\mathbf{s}) = 0$. However, for a spatial analysis we also desire a prior that encourages the covariate effect to vary smoothly over space so that $\beta_j(\mathbf{s}_1) \approx \beta_j(\mathbf{s}_2)$ if $\mathbf{s}_1 \approx \mathbf{s}_2$. To capture this prior belief requires an extension of univariate Bayesian variable selection priors to allow for spatial variation; this is the topic of Section 11.4.

Summarizing the posterior distribution for the SVC model is more complicated than the non-spatial case because we must not only determine which covariates are important, but also describe the areas where they are important. The posterior probabilities can be used in a Bayesian inference to detect the spatial subregions $\beta_j(\cdot)$ that are significant. For discrete mixture priors, one important quantity is the posterior inclusion probability $\pi_j(\mathbf{s}) = \text{Prob}\{\beta_j(\mathbf{s}) \neq 0|\mathbf{y}\}$, which can be estimated as the proportion of non-null draws from an MCMC chain, and that provides insight into the subregions of the spatial domain where the signal is non-zero. This can be viewed also as a hypothesis testing procedure about the spatial locations that are "significant", using frequentist language, and addressing it requires careful consideration of the multiple testing. Recent literature addresses this issue in terms of controlling the false discovery rate (either Bayesian or frequentist), in the context of spatially dependent hypotheses [6, 27, 33, 34]. For example, in Section 11.5 we follow [33] and declare a significant result for covariate j at $\mathbf{s}$ if $\pi_j(\mathbf{s}) > T$, where the threshold T is selected to control the Bayesian false discovery rate at α, i.e.,

$$\frac{\sum_{j=1}^{p}\sum_{i=1}^{n} I\{\pi_j(\mathbf{s}_i) > T\}\{1 - \pi_j(\mathbf{s}_i)\}}{\sum_{j=1}^{p}\sum_{i=1}^{n} I\{\pi_j(\mathbf{s}_i) > T\}} \approx \alpha.$$

11.3.2 Scalar-on-Image Regression

In classic geostatistical modeling, observations are made at a specific spatial location. In contrast, in scalar-on-image regression models, the response is a non-spatial scalar quantity and the covariate(s) is an entire spatial process. For example, the response for patient i might be their disease status and the predictor might be a 3D scan of their brain. Scalar-on-image regression was used by [11] to relate the white matter tissue damage to the cognitive disability in multiple sclerosis patients, and by [16] to help explain the relationship between alcoholism and brain activity assessed through electroencephalography. In this setting, the responses are independent and the challenge is to identify features of the image predictor that are predictive of disease status.

Let y_i be a scalar response (e.g., disease status) for observation i and $\mathbf{x}_i = \{x_i(\mathbf{s}_1), \dots, x_i(\mathbf{s}_q)\}$ be the image predictor (e.g., $x_i(\mathbf{s}_j)$ is a measure of brain activity in voxel j) defined at q spatial locations. The scalar-on-image regression model is

$$y_i = \alpha + \sum_{j=1}^{q} x_i(\mathbf{s}_j)\beta(\mathbf{s}_j) + \varepsilon_i, \tag{11.3}$$

where α is a scalar intercept, $\beta(\mathbf{s})$ is the spatially-varying regression coefficient defined over the same spatial structure as the image predictor, and ε_i is a Gaussian measurement error with variance σ_e^2. The framework can be extended to accommodate multiple image covariates or additional scalar covariates. We note that the intercept α is held constant because unlike the SVC model in (11.2), the responses are not associated with specific locations and therefore a spatially-varying intercept cannot be estimated.

While (11.3) has only a single covariate, its effect is a spatial process defined over a potentially large number locations, q. Therefore, a Bayesian spatial variable selection prior with sparsity and smoothness remains useful. In essence, we assume that the signal captured by the coefficient image $\beta(\mathbf{s})$ is organized in contiguous subregions and is smooth in every non-zero subregion. For example, in the brain imaging example above, it may be that brain activity only in a certain region is predictive of the disease, and so $\beta(\mathbf{s})$ would be non-zero in this region and zero outside this region.

11.4 Sparse Spatial Processes

In this section we review the literature on prior distributions for sparse spatial processes, i.e., processes that have spatial dependence and also mass at or near zero. Any of these processes could be applied as priors in the models introduced in Section 11.3, e.g., as the prior for $\beta_j(\mathbf{s})$ in (11.2) or $\beta(\mathbf{s})$ in (11.3). These methods are grouped by whether they are discrete and have prior mass exactly at $\beta_j(\mathbf{s}) = 0$ or have a continuous density function but with mass near zero. The essence of Bayesian variable selection for non-spatial problems is specifying priors for the regression coefficients that have both mass at (or near) zero, to remove (or shrink) the effect of irrelevant predictors, and heavy tails to avoid bias for the effects in important predictors. Many Bayesian spatial variable selection methods are motivated by extending these univariate prior distributions to multivariate distributions with similar properties.

Before proceeding with the review of spatial processes, we note that spatial methods are often classified by whether they assume the number of spatial locations under consideration is finite or uncountable [8]. Cases where the number of spatial locations is finite often arise when observations are assigned to areas, such as cancer rates in a geopolitical regions. Since geopolitical regions can be irregularly-shaped, there is no natural measure of distance between regions and so spatial dependence is often described in terms of adjacencies between regions. The foundational models for areal data are the conditional autoregression/Gaussian Markov random field model or simultaneous autoregressive models.

On the other hand, when there are potentially an uncountable number of spatial locations, such as temperature measurements, adjacency is poorly defined because measurements can be taken arbitrarily close to each other. For these point-referenced data, spatial dependence is typically described by the distance between locations and the leading model is the Gaussian process. While the choice to model data using areal or point-referenced methods is an important decision, the main concepts of Bayesian variable selection can be applied to either type of spatial data. Therefore, for simplicity, we assume a point-referenced data for the rest of the chapter. However, we emphasize that the methods can be applied to areal data, often by replacing a Gaussian process with say a conditionally autoregressive model.

11.4.1 Discrete Mixture Priors

Let $\beta(\mathbf{s})$ be the process value at spatial location $\mathbf{s} = (s_1, s_2)$. Assume the process is defined for the uncountably-many locations in the spatial domain $\mathcal{D}$ that is a subset of $\mathcal{R}^2$. At a particular location $\mathbf{s} \in \mathcal{D}$, define an arbitrary process as $z(\mathbf{s})$ and $z = \{z(\mathbf{s}); \mathbf{s} \in \mathcal{D}\}$ as the process defined over the entire domain. Correlation between the process at two location $\mathbf{s}_i$ and $\mathbf{s}_k$ is often defined by the distance between locations, $d(\mathbf{s}_i, \mathbf{s}_k)$. Most of the sparse spatial processes we describe below are based on a Gaussian process (GP). We use the notation $z \sim \text{GP}(\mu, \tau^2, \rho)$ to denote that z is a GP with mean $\text{E}\{z(\mathbf{s})\} = \mu$, variance $\text{Var}\{z(\mathbf{s})\} = \tau^2$ and correlation function $\text{Cor}\{z(\mathbf{s}_i), z(\mathbf{s}_k)\} = \rho\{d(\mathbf{s}_i, \mathbf{s}_k)\}$. For example, a simple correlation function is the exponential model $\rho(d) = \exp(-d/\phi)$ where the parameter ϕ controls the range of spatial dependence (see [8] for more flexible models). The GP model implies that the marginal distribution at each location is $z(\mathbf{s}) \sim \text{Normal}(\mu, \tau^2)$ and that the joint distribution at any n locations $\{z(\mathbf{s}_1), ..., z(\mathbf{s}_n)\}^T$ is multivariate normal with mean vector $(\mu, ..., \mu)^T$ and $n \times n$ covariance matrix with (i, k) element $\tau^2\rho\{d(\mathbf{s}_i, \mathbf{s}_k)\}$.

First, we review non-spatial discrete spike-and-slab mixture priors [10, 19] for a single random variable, β. As described in detail in previous chapters, a non-spatial discrete spike-and-slab mixture prior for Bayesian variable selection assumes that β follows a $\text{Normal}(0, \tau^2)$ distribution (and thus the corresponding covariate is included in the model) with probability π and $\beta = 0$ (and thus the corresponding covariate is excluded from the model) with probability $1 - \pi$. This prior can be written as

$$\beta = \gamma z, \tag{11.4}$$

where $\gamma \sim \text{Bernoulli}(\pi)$ independent of $z \sim \text{Normal}(0, \tau^2)$. The remainder of this section discusses several extensions that allow β to be a spatial process and retain the essential feature that the spatial process has prior probability of being exactly zero. This is accomplished by allowing the binary and/or Gaussian components of the discrete mixture prior to be spatial processes.

Global variable selection priors: The simplest model for spatial variable selection is to set all spatial locations to zero simultaneously. For example, [23] and [5] propose placing prior probability on three possibilities:

(1) The entire surface is zero, i.e., $\beta(\mathbf{s}) = 0$ for all $\mathbf{s}$
(2) The entire surface is a non-zero constant, i.e., $\beta(\mathbf{s}) = \beta_0$ for all $\mathbf{s}$
(3) The $\beta(\mathbf{s})$ is a spatial Gaussian process

These three types of behavior can be encoded in a single process as

$$\beta(\mathbf{s}) = \gamma_1\beta_0 + \gamma_1\gamma_2 z(\mathbf{s}) \tag{11.5}$$

where γ_1 and γ_2 have Bernoulli priors and $z \sim \text{GP}(0, \tau^2, \rho)$. Under (11.5), if $\gamma_1 = 0$ then we have case (1), if $\gamma_1 = 1$ and $\gamma_2 = 0$ we have case (2) and if $\gamma_1 = \gamma_2 = 1$ we have case (3). This approach is useful for global variable selection where the same set of covariates are included in the model for each spatial location, but does not permit local variable selection where the set of covariates included varies by spatial location.

Two-component mixture prior: Local variable selection can be achieved via a two-component mixtures prior that allows both the binary and continuous component of the mixture to be spatial processes. That is

$$\beta(\mathbf{s}) = \gamma(\mathbf{s})z(\mathbf{s}) \tag{11.6}$$

where $\gamma(\mathbf{s}) \in \{0, 1\}$ is a spatial binary process and $z \sim \text{GP}(\mu_z, \tau_z^2, \rho)$ is a spatial Gaussian process. Different models have been proposed for the binary spatial process [3, 11, 18, 29, 32], but arguably the simplest is the hard-threshold Gaussian process, $\gamma(\mathbf{s}) = I\{w(s) > 0\}$ where $w \sim \text{GP}(\mu_w, \tau_w^2, \rho)$ independent of z. Under this process the prior inclusion probability $\pi = \text{Prob}\{\gamma(\mathbf{s}) = 1\} = \Phi(\mu_w/\tau_w)$ where Φ is the standard normal distribution function.

Figure 11.1a plots a realization from the two-component mixture prior with $\mu_z = \mu_w = 0$, $\tau_z = \tau_w = 1$ and $\rho(d) = \exp\{-d/5\}$ for $\mathbf{s}$ arranged on a 100×100 square grid with grid spacing one. There are large regions with $w(\mathbf{s}) < 0$ and thus $\beta(\mathbf{s}) = 0$, and $\beta(\mathbf{s})$ is spatially smooth in the non-zero regions. To further understand this process, Figure 11.2a plots $\beta(\mathbf{s})$ by s_1 along the transect defined by $s_2 = 50$. As desired, $\beta(\mathbf{s})$ is exactly zero in spatial clusters of locations. Another feature that is apparent from Figure 11.2a is that sharp jumps from zero to large magnitude values are possible. For example, around $s_1 = 60$ the process jumps from 0 to 0.4 from one grid cell to the next. This feature is a result of the independence of the discrete and continuous components of the prior, and is useful for identifying sharp breaks in a spatial surface.

Trans-Gaussian priors: A second class of local variable selection priors uses a single Gaussian process $z \sim \text{GP}(0, 1, \rho)$ and achieves sparsity via a marginal transformation,

$$\beta(\mathbf{s}) = F\{z(\mathbf{s})\}, \tag{11.7}$$

where F is a transformation function that maps some values of $z(\mathbf{s})$ to $\beta(\mathbf{s}) = 0$. This is related to a Gaussian copula [21], which is often used to model dependence of a joint distribution while specifying the marginal distribution of each variable. For example, if G is the distribution function of a Gamma(a, b) distribution and $F(z) = G^{-1}\{\Phi(z)\}$, then $\beta(\mathbf{s})$ follows a Gamma(a, b) marginal distribution for each $\mathbf{s}$ and the process β inherits spatial dependence from the latent Gaussian process z.

There are several options for the transformation function. [35] select F so that the marginal distribution of $\beta(\mathbf{s})$ matches a spike-and-slab prior. [16] use the soft-thresholding function

$$F(z) = \tau \text{sign}(z)(|z| - \lambda)_+$$

where τ is the scale parameter, λ is the threshold and $(z)_+ = \max\{z, 0\}$. By construction, any $z(\mathbf{s}) \in [-\lambda, \lambda]$ gives $\beta(\mathbf{s}) = 0$, and thus the prior inclusion probability is $\pi = \text{Prob}\{\beta(\mathbf{s}) \neq 0\} = 2\Phi(-\lambda)$. This prior is referred to as the soft-thresholded Gaussian process (STGP) [16].

Figures 11.1b and 11.2b show a realization from this STGP with $\tau = \lambda = 1$ and $\rho(d) = \exp(-d/5)$. A key difference between the STGP and the two-component mixture is that because the STGP involves only a single GP, the realizations vary more smoothly across space than the two-component mixture prior. In fact, if both F and z are continuous functions, then β is also a continuous function.

Sparse basis expansion priors: Any smooth spatial process can be approximated using a finite basis expansion

$$\beta(\mathbf{s}) = \sum_{l=1}^{L} B_l(\mathbf{s}) b_l \tag{11.8}$$

where $B_1(\mathbf{s}), ..., B_L(\mathbf{s})$ are known basis functions and b_l are unknown coefficients that determine the shape of $\beta(\mathbf{s})$. For example, the spectral representation theorem [7] states that any stationary Gaussian process can be written in this form with $L = \infty$, the B_l set to certain trigonometric functions, and the b_l given independent Gaussian distributions. The Gaussian distribution can be replaced with a Bayesian variable selection prior with $\text{Prob}(b_l = 0) = \pi_l > 0$, but this is generally not a satisfying variable selection prior because $\beta(\mathbf{s}) = 0$ only if $b_l = 0$ for all l with $B_l(\mathbf{s}) \neq 0$. However, combining a sparse prior for b_l and sparse basis functions, i.e., functions with $B_l(\mathbf{s}) = 0$ for many $\mathbf{s}$, can give substantial prior

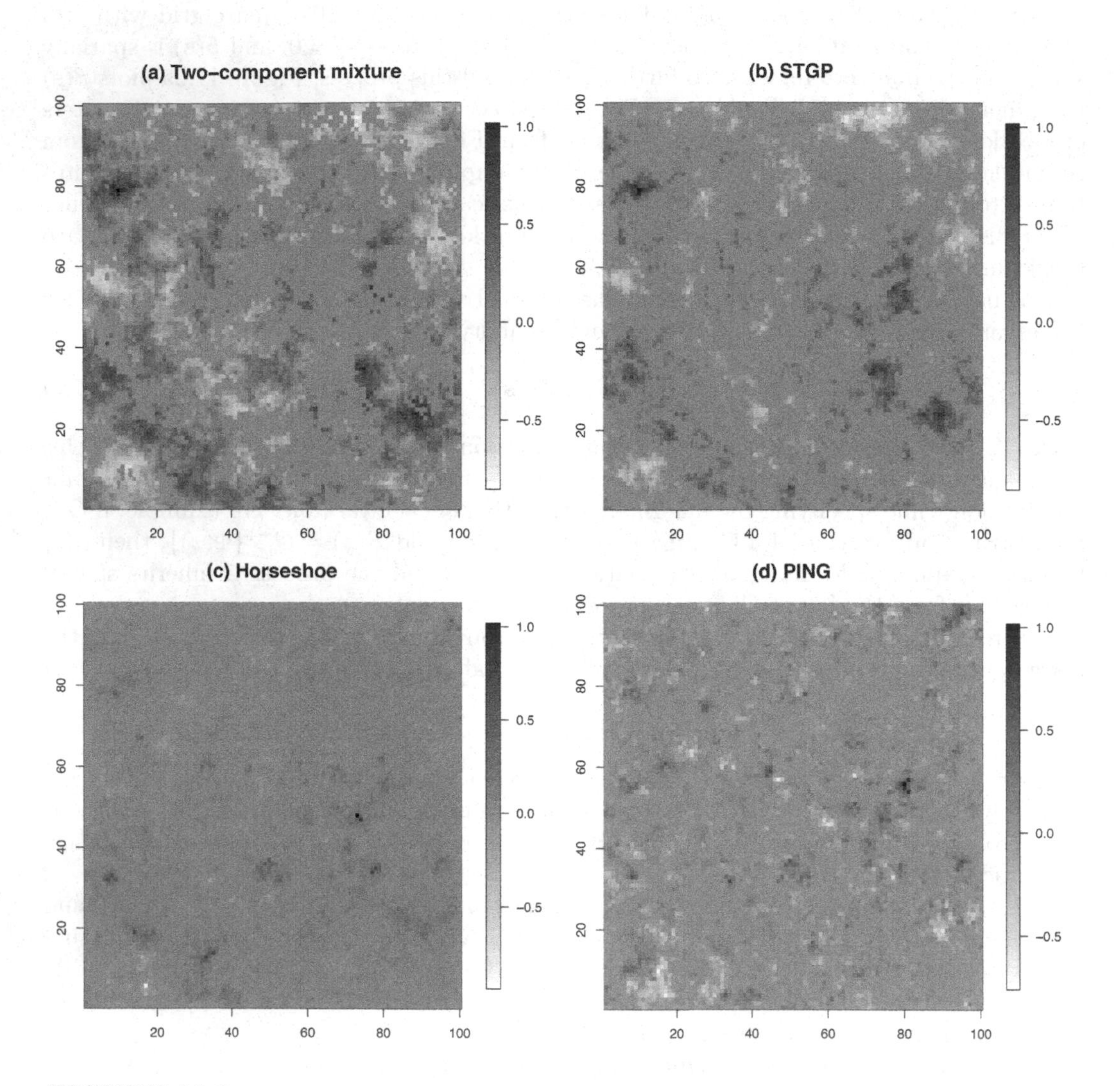

FIGURE 11.1
Realizations on a 100×100 regular grid of locations from (a) the two-component spatial mixture prior, (b) the soft-thresholded Gaussian process (STGP) prior, (c) the spatial horseshoe prior and (d) the product inverse Gaussian process (PING) with $q = 3$ components. All four realizations are scaled to $[-1,1]$ for comparison.

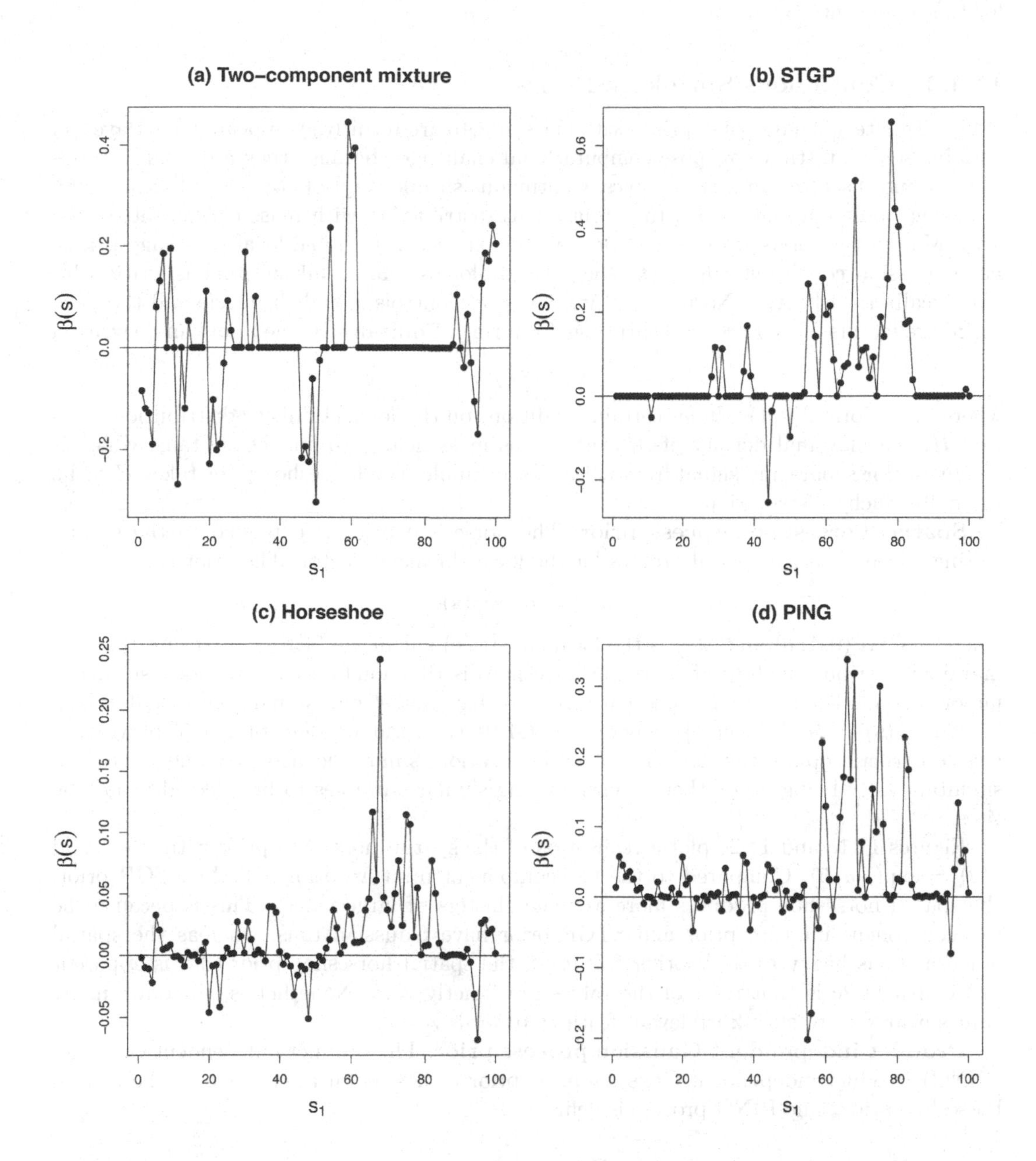

FIGURE 11.2
Realizations in Figure 11.1 plotted by s_1 for locations with $s_2 = 50$ for (a) the two-component spatial mixture prior, (b) the soft-thresholded Gaussian process (STGP) prior, (c) the spatial horseshoe prior and (d) the product inverse Gaussian process (PING) with $q = 3$ components. All four realizations are scaled to [-1,1] for comparison.

probability that $\beta(\mathbf{s}) = 0$ and thus an appropriate Bayesian spatial variable selection prior. For example, [20, 30, 36] combine sparse wavelet basis functions $B_l(\mathbf{s})$ with sparse priors for the coefficients b_k.

11.4.2 Continuous Shrinkage Priors

While discrete spike-and-slab priors with mass at zero are intuitively appealing for Bayesian variable selection, they can pose computational challenges because they are a mix of continuous and discrete random variables. Continuous shrinkage priors offer an alternative by replacing point mass at zero with a continuous distribution with mass concentrated near zero. Many continuous shrinkage priors can be expressed as global-local shrinkage priors. For regression coefficients $\beta_1, ..., \beta_p$, the general global-local shrinkage prior is written hierarchically as $\beta_j|\tau, \lambda_j \sim \text{Normal}(0, \tau^2\lambda_j^2)$ where τ^2 controls global shrinkage and the local shrinkage factors $\lambda_j \sim H$ control shrinkage for term j. Equivalently, the prior can be written

$$\beta_j = \tau\lambda_j z_j$$

where $z_j \sim \text{Normal}(0, 1)$. Under certain conditions on the local shrinkage distribution function H, the marginal density of β_j over λ_j has mass near zero and heavy tails, which is conducive for separating signal from noise. For example, the horseshoe prior takes H to be the half-Cauchy distribution.

Spatial horseshoe process prior: The horseshoe prior is extended to the spatial setting [15] by using a spatial process for the local shrinkage factor. The prior is

$$\beta(\mathbf{s}) = \tau\lambda(\mathbf{s})z(\mathbf{s}) \tag{11.9}$$

where $z \sim \text{GP}(0, 1, \rho)$ and $\lambda(\mathbf{s})$ is the local (in space) shrinkage factor. To ensure that the marginal distribution of $\beta(\mathbf{s})$ follows the horseshoe distribution for each $\mathbf{s}$, the local shrinkage factors are modeled using a Gaussian copula to have half-Cauchy marginal distributions. That is, $\lambda(\mathbf{s}) = H^{-1}[\Phi\{w(\mathbf{s})\}]$, where $w \sim \text{GP}(0, 1, \rho)$, independent of z. [15] prove that this construction puts mass on pairs of nearby locations simultaneously being near zero and simultaneously being large, thereby encouraging similar variables to be selected at nearby sites.

Figures 11.1c and 11.2c plot a realization of the spatial horseshoe prior with $\tau = 1$ and $\rho(d) = \exp(-d/5)$. Compared to the two-component mixture prior and the STGP prior, the spatial horseshoe produces more extreme clusters of large values. This is because the two-component mixture prior and STGP prior have Gaussian tails, whereas the spatial horseshoe has heavy tails. Another feature of the spatial horseshoe prior that is apparent in Figures 11.2c is that none of the values are exactly zero. Nonetheless, there are many values near zero to shrink irrelevant features towards zero.

Product independent Gaussian process prior: The product independent Gaussian (PING) Product independent Gaussian prior prior of [28] is an alternative to the spatial horseshoe prior. The PING process is defined as

$$\beta(\mathbf{s}) = \tau \prod_{l=1}^{q} z_l(\mathbf{s}) \tag{11.10}$$

where $z_1, ..., z_q$ are independent GPs with $z_1, ..., z_q \sim \text{GP}(0, 1, \rho)$. An advantage of this construction is that it does not involve complex functions such as the half-Cauchy distribution function and therefore the prior is conjugate in many applications. In addition to computational benefits, [28] show that as q increases the concentration of mass at zero and kurtosis of the marginal and joint distributions increase, and therefore with large q the PING prior serves as a spatial shrinkage prior. As shown in Figures 11.1d and 11.2d, its properties resemble the spatial horseshoe prior.

TABLE 11.1
Global Bayesian variable selection: Posterior inclusion probability (PIP), mean and 90% credible interval for the global variable selection analysis with spatial and non-spatial residual model.

	Non-spatial			Spatial		
Covariate	PIP	Mean	90% Interval	PIP	Mean	90% Interval
Latitude	0.42	−0.03	(−0.20, 0.09)	0.47	0.07	(0.00, 0.28)
Longitude	0.68	0.06	(0.00, 0.15)	0.31	0.00	(−0.16, 0.14)
Temperature	0.97	−0.27	(−0.46, −0.14)	0.72	−0.14	(−0.32, 0.00)
Precipitation	0.24	0.01	(0.00, 0.10)	0.16	0.00	(−0.01, 0.04)
NPP	0.43	−0.03	(−0.12, 0.00)	0.13	0.00	(−0.03, 0.00)
Elevation	0.89	−0.11	(−0.21, 0.00)	0.30	−0.02	(−0.14, 0.00)
House type	0.95	0.10	(0.00, 0.16)	0.76	0.06	(0.00, 0.13)
Bedrooms	0.26	0.01	(0.00, 0.09)	0.26	0.01	(0.00, 0.09)

11.5 Application to Microbial Fungi across US Households

To illustrate Bayesian spatial variable selection, we reanalyze data from a study of the factors that drive the household microbiome. The data originate from [2] and are analyzed using Bayesian methods in [24] and spatial methods in [31]. Data are collected at $n = 1,133$ households in the US. The response variable is the log number of fungal operational taxonomic units identified in the sample taken outside the home. This response is regressed onto $p = 8$ covariates: latitude, longitude, elevation, average temperature and precipitation, net primary productivity (NPP), the number of bedrooms in the home, and home type (a binary indicator of whether the house is a detached single-family home versus apartment and townhouse). The response and each covariate are standardized to have mean zero and variance one.

We begin by analyzing the data using a global model that assumes the covariates have constant effects as across space as in Section 11.2 and (11.1). We use the discrete mixture prior $\beta_j = \gamma_j z_j$ where $\gamma_j \sim \text{Bernoulli}(0.5)$ and $z_j \sim \text{Normal}(0, \sigma_z^2)$, all independent over j. For priors we select $\alpha, \log(\rho) \sim \text{Normal}(0, 10)$ and $\sigma_e^2, \sigma_s^2, \sigma_z^2 \sim \text{InvGamma}(0.1, 0.1)$. The results for this spatial model are compared to a non-spatial version with $\sigma_s^2 = 0$ in Table 11.1. The non-spatial model has four variables (longitude, temperature, elevation and house type) with inclusion probability over 0.5. Properly accounting for residual dependence reduces the inclusion probability for all four of these variables leaving only two (temperature and house type) with inclusion probability greater than 0.5. This highlights the importance of considering the correlation structure when performing variable selection.

Because the nature of the microbiome changes across the country due to climate and land-use variation, we allow for spatially-varying coefficients as in (11.2) with Bayesian variable selection priors. As shown below, the magnitude of the variation in the estimated effects and their inclusion probabilities suggest that a global model is likely insufficient. We begin with the STGP prior of [16]. For $j \in \{1, ..., p\}$, let $\beta_j(\mathbf{s}) = \text{sign}\{z_j(\mathbf{s})\}(|z_j(\mathbf{s})| - \lambda)_+$. Following [16], the latent processes are modelled using a finite basis expansion, $z_j(\mathbf{s}) = \sum_{l=1}^L B_l(\mathbf{s}) Z_{jl}$ where B_l are known basis functions and $Z_{jl} \sim \text{Normal}(0, \sigma_z^2)$. We take the basis functions to be the outer product of L_1 b-spline basis functions of longitude and $L_2 = L_1/2$ for b-spline basis function of latitude and normalize the basis functions so that $\sum_{l=1}^L B_l(\mathbf{s})^2 = 1$ for all $\mathbf{s}$. We show the results with $L_1 = 10$ and thus $L = 50$, but the results were similar for $L_1 = 15$ and $L_1 = 20$. Similarly, the intercept is $\alpha(\mathbf{s}) = \sum_{l=1}^{L} B_l(\mathbf{s})A_l$ with $A_l \sim \text{Normal}(0, \sigma_a^2)$. We set the threshold to $\lambda = 0.67\sigma_z$ so that the prior inclusion probability is approximately 0.5, and remaining priors are $\sigma_e^2, \sigma_a^2, \sigma_z^2 \sim \text{InvGamma}(0.1, 0.1)$.

The covariate effect with the most spatial variation is home type. Figure 11.3 maps the posterior mean of $\beta_j(\mathbf{s})$ and the posterior inclusion probability $\text{Prob}\{\beta_j(\mathbf{s}) \neq 0|\mathbf{y}\}$ for the STGP fit. The posterior mean and inclusion probability are the largest in the southeast, suggesting the fungi indigenous to these regions have stronger preference for single-family homes. Following the multiple testing procedure in [33], rejecting the local null hypothesis that $\beta_j(\mathbf{s}) = 0$ if the inclusion probability exceeds 0.65 controls the Bayesian false discovery rate at 0.2; most southeastern areas exceed this threshold. However, the null hypotheses cannot be rejected at any location if we control the Bayesian false discovery rate at 0.1.

As with any Bayesian analysis, a prior sensitivity analysis is warranted. The key hyperparameter in the STGP process is σ_z^2, which controls the prior variance of the latent spatial process. The fit in Figure 11.3 uses prior $\sigma_z^2 \sim \text{InvGamma}(\epsilon, \epsilon)$ for $\epsilon = 0.1$. We refit the model with $\epsilon = 0.01$ and $\epsilon = 1$ and partial results are plotted in Figure 11.4. We find that while the posterior of σ_z^2 is affected by ϵ, the resulting posterior inclusion probabilities are less sensitive to ϵ. In particular, the posterior mean (standard deviation) of σ_z^2 is 0.018 (0.012), 0.040 (0.015) and 0.124 (0.037) for $\epsilon = 0.01$, $\epsilon = 0.1$ and $\epsilon = 1$, but in all cases the posterior inclusion probability for home type in Figures 11.3 and 11.4 are large in the southeast and small for the rest of the country.

We compare the discrete-mixture STGP prior with the continuous-mixture spatial horseshoe process (SHP) prior in (11.9). The SHP prior is the product of two processes, $\beta_j(\mathbf{s}) = \lambda_j(\mathbf{s})z_j(s)$, where the local shrinkage factor $\lambda_j(\mathbf{s}) = H^{-1}[\Phi\{w_j(\mathbf{s})\}]$ is constructed to be a spatial process with half-Cauchy marginal distribution. We use the same finite basis expansion as in the STGP application, $z_j(\mathbf{s}) = \sum_{l=1}^{L} B_l(\mathbf{s})Z_{jl}$ and $w_j(\mathbf{s}) = \sum_{l=1}^{L} B_l(\mathbf{s})W_{jl}$, with $Z_{jl} \sim \text{Normal}(0, \sigma_z^2)$ and $W_{jl} \sim \text{Normal}(0, 1)$. The intercept and priors are the same as in the STGP application.

Figure 11.5 maps the estimated value of $\beta_j(\mathbf{s})$ and the location shrinkage factor $\lambda_j(\mathbf{s})$ for house type under the SHP prior. These estimates are the values of $\beta_j(\mathbf{s})$ and $\lambda_j(\mathbf{s})$ evaluated at the posterior mean of Z_{jl} and W_{jl}, which is more stable than the posterior mean of $\beta_j(\mathbf{s})$ and $\lambda_j(\mathbf{s})$ in areas with sparse data. The estimates of $\beta_j(\mathbf{s})$ under the STGP and SHP prior are both largest in the southeast. The estimated shrinkage factor $\lambda_j(\mathbf{s})$ is also the largest in the southeast, therefore applying less shrinkage in the southeast where the signal is strong and more shrinkage towards zero in other regions.

11.6 Discussion

In this chapter, we have reviewed recent advances in Bayesian variable selection for spatial regression models. While this area has advanced rapidly, there remain limitations that require future work. One example is to incorporate more realistic spatial correlation structures. We have considered only stationary correlation functions while allowing mean effects to vary spatially. A stationary correlation function is hard to justify for most large spatial datasets that cover diverse spatial domains. In particular, stationarity is clearly questionable for neuroimaging applications where distant brain regions can be functionally connected. Another area of future work is to connect the ideas of Bayesian spatial variable selection with spatial causal inference (see [26] for a recent review). A common scenario is a single treatment variable, but a large number of potential confounding variables. Carefully considering how to select confounding variables for the purpose of conducting valid spatial causal

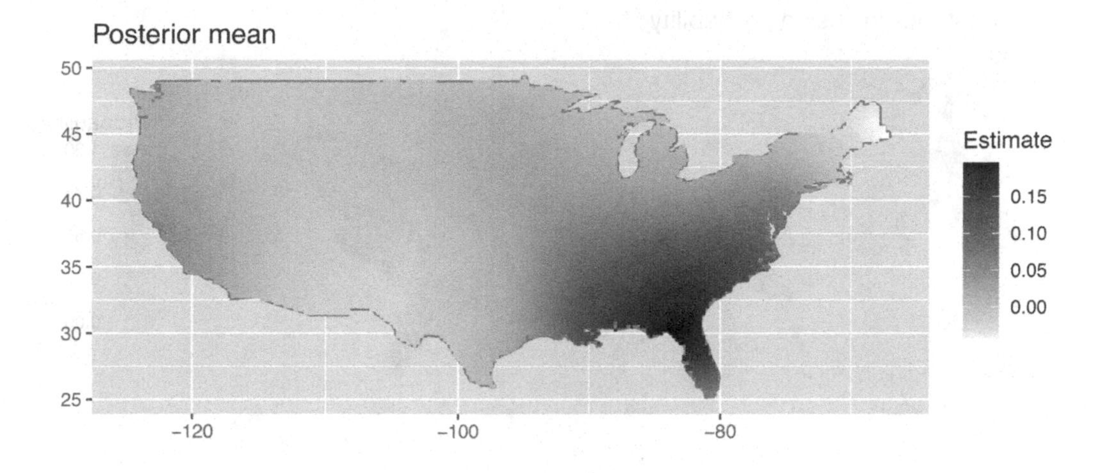

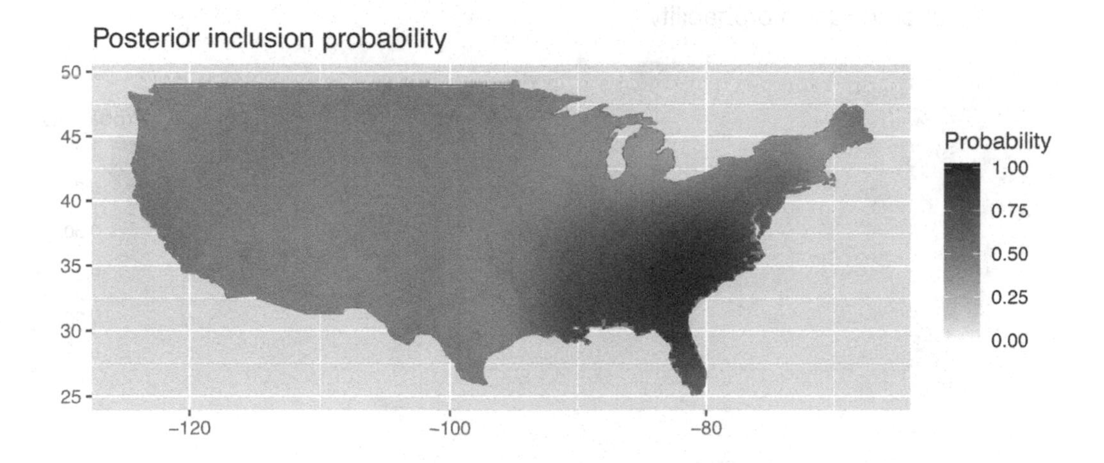

FIGURE 11.3
Posterior mean and inclusion probability for the spatially-varying effect $\beta_j(\mathbf{s})$ of home type on microbiome species richness under the soft-thresholded Gaussian process prior.

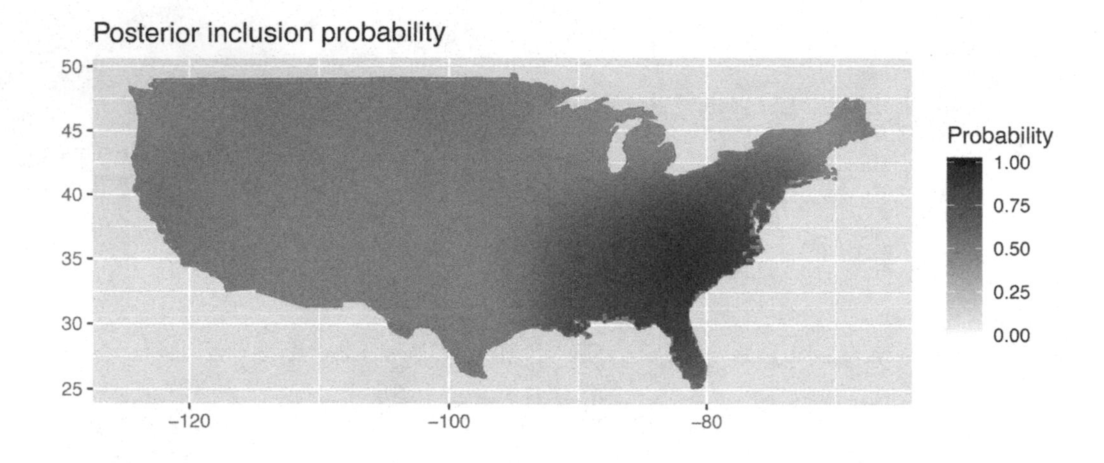

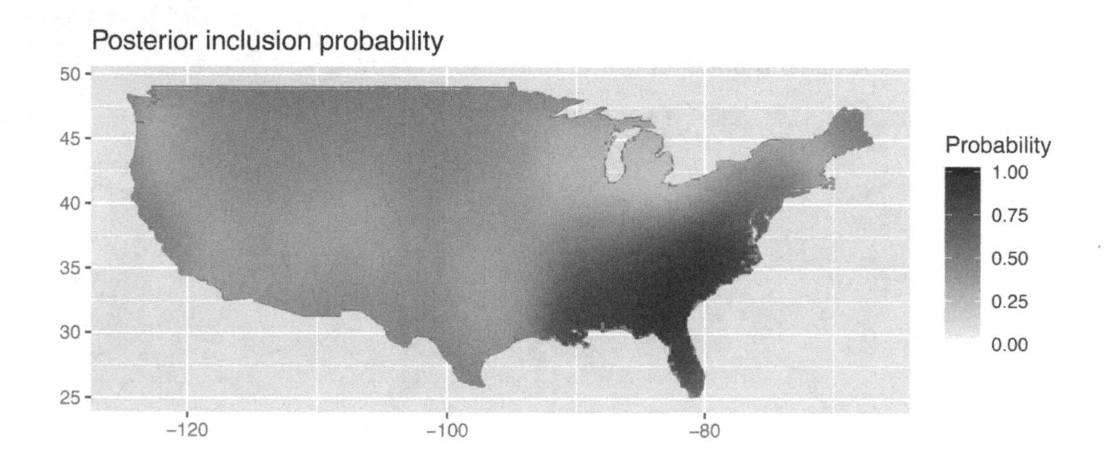

FIGURE 11.4
Posterior inclusion probability for the spatially-varying effect $\beta_j(\mathbf{s})$ of home type on microbiome species richness under the soft-thresholded Gaussian process prior assuming $\sigma_z^2 \sim \text{InvGamma}(0.01, 0.01)$ (top) versus $\sigma_z^2 \sim \text{InvGamma}(1, 1)$ (bottom).

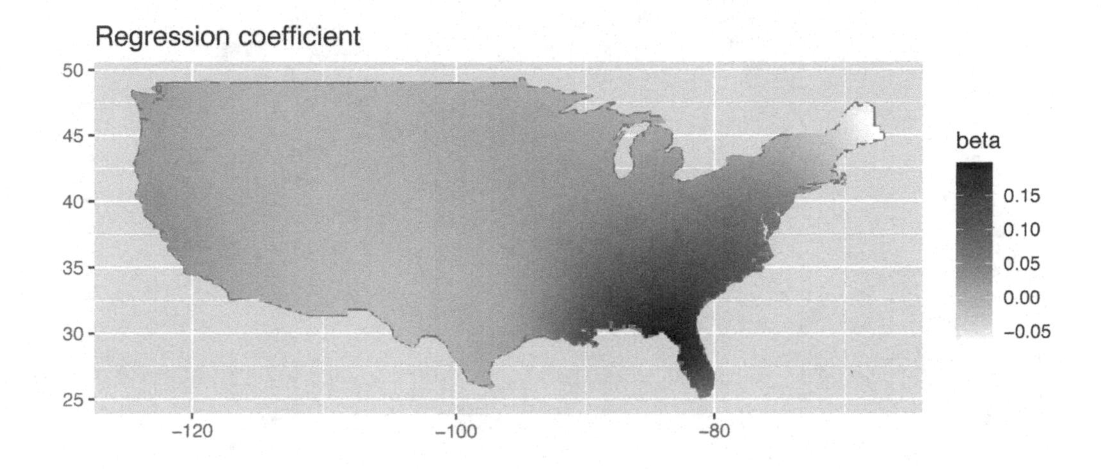

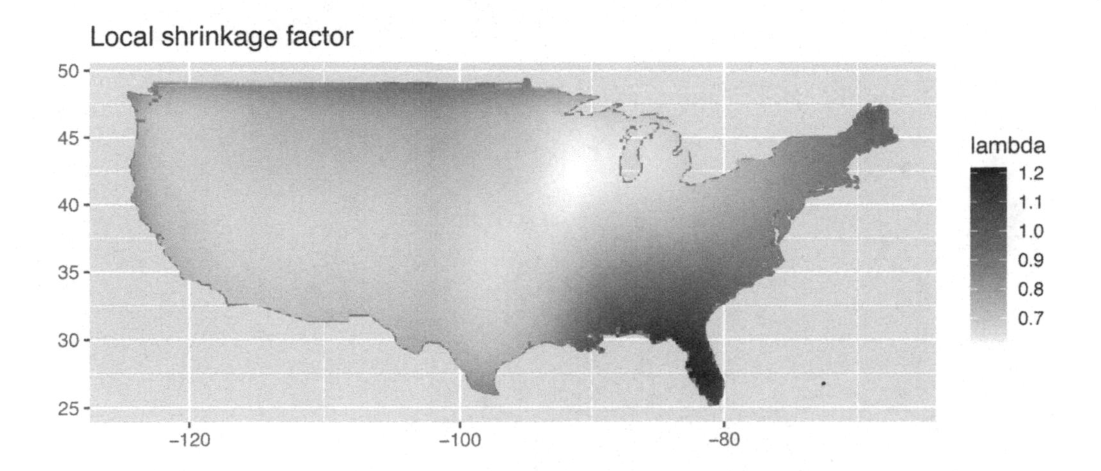

FIGURE 11.5
Estimated spatially-varying effect $\beta_j(\mathbf{s})$ and local shrinkage factor $\lambda_j(\mathbf{s})$ for home type on microbiome species richness under the spatial horseshoe prior.

inference would require new models and methods. Finally, while many Bayesian spatial variable selection methods can be implemented in general-purpose Bayesian software such as JAGS or STAN and the code to replicate the results in this chapter available as part of the online Supplementary Material of the book, rigorous software development for these methods would be a welcomed addition.

Bibliography

[1] S. Banerjee, B. P. Carlin, and A. E. Gelfand. *Hierarchical modeling and analysis for spatial data.* CRC press, 2014.

[2] A. Barberán, R. R. Dunn, B. J. Reich, K. Pacifici, E. B. Laber, H. L. Menninger, J. M. Morton, J. B. Henley, J. W. Leff, S. L. Miller, et al. The ecology of microscopic life in household dust. *Proceedings of the Royal Society B: Biological Sciences*, 282(1814):20151139, 2015.

[3] J. Choi and A. B. Lawson. Bayesian spatially dependent variable selection for small area health modeling. *Statistical Methods in Medical Research*, 27(1):234–249, 2018.

[4] P. J. Diggle, J. A. Tawn, and R. A. Moyeed. Model-based geostatistics. *Journal of the Royal Statistical Society: Series C (Applied Statistics)*, 47(3):299–350, 1998.

[5] A. O. Finley and S. Banerjee. Bayesian spatially varying coefficient models in the spBayes R package. *Environmental Modelling & Software*, 125:104608, 2020.

[6] J. P. French, S. R. Sain, et al. Spatio-temporal exceedance locations and confidence regions. *The Annals of Applied Statistics*, 7(3):1421–1449, 2013.

[7] M. Fuentes and B. J. Reich. Spectral domain. *Handbook of Spatial Statistics*, pages 57–77, 2010.

[8] A. E. Gelfand, P. Diggle, P. Guttorp, and M. Fuentes. *Handbook of Spatial Statistics.* CRC press, 2010.

[9] A. E. Gelfand, H.-J. Kim, C. Sirmans, and S. Banerjee. Spatial modeling with spatially varying coefficient processes. *Journal of the American Statistical Association*, 98(462):387–396, 2003.

[10] E. I. George and R. E. McCulloch. Approaches for Bayesian variable selection. *Statistica Sinica*, pages 339–373, 1997.

[11] J. Goldsmith, L. Huang, and C. M. Crainiceanu. Smooth scalar-on-image regression via spatial Bayesian variable selection. *Journal of Computational and Graphical Statistics*, 23(1):46–64, 2014.

[12] Y. Guan, M. C. Johnson, M. Katzfuss, E. Mannshardt, K. P. Messier, B. J. Reich, and J. J. Song. Fine-scale spatiotemporal air pollution analysis using mobile monitors on Google Street View vehicles. *Journal of the American Statistical Association*, pages 1–14, 2019.

[13] M. J. Heaton, A. Datta, A. O. Finley, R. Furrer, J. Guinness, R. Guhaniyogi, F. Gerber, R. B. Gramacy, D. Hammerling, M. Katzfuss, et al. A case study competition among methods for analyzing large spatial data. *Journal of Agricultural, Biological and Environmental Statistics*, 24(3):398–425, 2019.

[14] J. S. Hodges and B. J. Reich. Adding spatially-correlated errors can mess up the fixed effect you love. *The American Statistician*, 64(4):325–334, 2010.

[15] A.-T. Jhuang, M. Fuentes, J. L. Jones, G. Esteves, C. M. Fancher, M. Furman, and B. J. Reich. Spatial signal detection using continuous shrinkage priors. *Technometrics*, 0(0):1–21, 2019.

[16] J. Kang, B. J. Reich, and A.-M. Staicu. Scalar-on-image regression via the soft-thresholded Gaussian process. *Biometrika*, 105(1):165–184, 2018.

[17] J.-i. Kim, A. B. Lawson, S. McDermott, and C. M. Aelion. Variable selection for spatial random field predictors under a Bayesian mixed hierarchical spatial model. *Spatial and Spatio-temporal Epidemiology*, 1(1):95–102, 2009.

[18] F. Li, T. Zhang, Q. Wang, M. Z. Gonzalez, E. L. Maresh, J. A. Coan, et al. Spatial Bayesian variable selection and grouping for high-dimensional scalar-on-image regression. *The Annals of Applied Statistics*, 9(2):687–713, 2015.

[19] T. J. Mitchell and J. J. Beauchamp. Bayesian variable selection in linear regression. *Journal of the American Statistical Association*, 83(404):1023–1032, 1988.

[20] J. S. Morris, V. Baladandayuthapani, R. C. Herrick, P. Sanna, and H. Gutstein. Automated analysis of quantitative image data using isomorphic functional mixed models, with application to proteomics data. *The Annals of Applied Statistics*, 5(2A):894, 2011.

[21] R. B. Nelsen. *An Introduction to Copulas.* Springer, 2006.

[22] C. J. Paciorek. The importance of scale for spatial-confounding bias and precision of spatial regression estimators. *Statistical Science*, 25(1):107–125, 2010.

[23] B. J. Reich, M. Fuentes, A. H. Herring, and K. R. Evenson. Bayesian variable selection for multivariate spatially varying coefficient regression. *Biometrics*, 66(3):772–782, 2010.

[24] B. J. Reich and S. K. Ghosh. *Bayesian Statistical Methods.* CRC Press, 2019.

[25] B. J. Reich, J. S. Hodges, and V. Zadnik. Effects of residual smoothing on the posterior of the fixed effects in disease-mapping models. *Biometrics*, 62(4):1197–1206, 2006.

[26] B. J. Reich, S. Yang, Y. Guan, A. B. Giffin, M. J. Miller, and A. G. Rappold. A review of spatial causal inference methods for environmental and epidemiological applications. *arXiv preprint arXiv:2007.02714*, 2020.

[27] M. D. Risser, C. J. Paciorek, and D. A. Stone. Spatially dependent multiple testing under model misspecification, with application to detection of anthropogenic influence on extreme climate events. *Journal of the American Statistical Association*, 114(525):61–78, 2019.

[28] A. Roy, B. J. Reich, J. Guinness, R. T. Shinohara, and A.-M. Staicu. Spatial shrinkage via the product independent Gaussian process prior. *arXiv:1805.03240*, 2018.

[29] I. Scheel, E. Ferkingstad, A. Frigessi, O. Haug, M. Hinnerichsen, and E. Meze-Hausken. A Bayesian hierarchical model with spatial variable selection: The effect of weather on insurance claims. *Journal of the Royal Statistical Society: Series C (Applied Statistics)*, 62(1):85–100, 2013.

[30] Z. Shang and M. K. Clayton. An application of Bayesian variable selection to spatial concurrent linear models. *Environmental and Ecological Statistics*, 19(4):521–544, 2012.

[31] S. P. Singh, A.-M. Staicu, R. R. Dunn, N. Fierer, B. J. Reich, et al. A nonparametric spatial test to identify factors that shape a microbiome. *The Annals of Applied Statistics*, 13(4):2341–2362, 2019.

[32] M. Smith and L. Fahrmeir. Spatial Bayesian variable selection with application to functional magnetic resonance imaging. *Journal of the American Statistical Association*, 102(478):417–431, 2007.

[33] W. Sun, B. J. Reich, T. T. Cai, M. Guindani, and A. Schwartzman. False discovery control in large-scale spatial multiple testing. *Journal of the Royal Statistical Society. Series B, Statistical methodology*, 77(1):59, 2015.

[34] W. Tansey, O. Koyejo, R. A. Poldrack, and J. G. Scott. False discovery rate smoothing. *arXiv preprint arXiv:1411.6144*, 2014.

[35] L. F. B. Vock, B. J. Reich, M. Fuentes, and F. Dominici. Spatial variable selection methods for investigating acute health effects of fine particulate matter components. *Biometrics*, 71(1):167–177, 2015.

[36] H. Zhu, P. J. Brown, and J. S. Morris. Robust, adaptive functional regression in functional mixed model framework. *Journal of the American Statistical Association*, 106(495):1167–1179, 2011.

12

Effect Selection and Regularization in Structured Additive Distributional Regression

Paul Wiemann

Georg-August-Universität Göttingen (Germany)

Thomas Kneib

Georg-August-Universität Göttingen (Germany)

Helga Wagner

Johannes Kepler Universität Linz (Austria)

CONTENTS

Structured additive distributional regression provides a general framework for semiparametric regression where effects are assumed for all parameters characterizing the distribution of the response. This allows the analyst to focus on distributional aspects beyond the mean such as variability, skewness or other shape features. The predictors assigned to the parameters comprise structured additive combinations of various effect types such as non-linear effects of continuous covariates, spatial effects, random effects, varying coefficient terms, interaction surfaces, or high-dimensional vectors of linear effects. However, the resulting flexibility makes effect selection and regularization a challenge that can be conveniently tackled in a Bayesian framework by placing prior structures on blocks of regression coefficients relating to the different effects. We will review both effect selection priors based on

DOI: 10.1201/9781003089018-12

spike-and-slab structures and regularization priors enforcing shrinkage and smoothness of effect estimates from a conceptual as well as a computational point of view.

12.1 Introduction

Regression modeling in linear and generalized linear models captures the effects of covariates on the expected value of the response variable of interest. In contrast, structured additive distributional regression is a much more flexible modeling approach that allows modeling various types of effects on all parameters that characterize the distribution of a response variable, including location, scale, skewness, or other shape features. In this framework, not only the mean but each distributional parameter is linked to a structured additive predictor which may contain various types of effects, for example non-linear effects of continuous covariates, spatial effects, random effects, varying coefficient terms, interaction surfaces, or high-dimensional vectors of linear effects.

Specification of distributional regression models is a difficult task as the analyst has to decide for each predictor whether a covariate is included and how its effect is modeled. In our application presented in detail in Section 12.4, we use data on the undernutrition of children in India to analyze the stunting score, which indicates chronic undernutrition, using a distributional regression model. The stunting score is assumed to follow a normal distribution where both mean and variance are modeled via a structured additive predictor and potential covariates are the age of the child, age of the mother, the mother's body mass index, and an indicator for the district a child is living in. Many modeling decisions have to be made, for example whether the effects of age of mother and child are non-linear, linear, or irrelevant and whether an interaction between them should be included in the predictors for mean and variance.

Model selection could be accomplished by fitting all or a set of different models, however a more convenient strategy (at least for models including a small to moderate number of potential effects) is to specify the predictors with all possible effects in their most general form and determine the most adequate model during Markov chain Monte Carlo (MCMC) inference in the spirit of Bayesian variable selection. However, in contrast to the selection of single, scalar regression coefficients in variable selection, effect selection in structured additive predictors aims at the selection of more complex effects, typically captured by a vector of coefficients and hence requires specific prior distributions. Additionally, interest is not only in the binary decision of inclusion/exclusion of the effect of a covariate but also in which form this effect should be specified. Thus, for a continuous covariate like mother's age it is not only of interest whether there is an effect on a specific distributional parameter but also whether the effect can be specified as linear or a non-linear model is required.

To tackle these challenges, Bayesian effect selection relies on an appropriate decomposition of effects in components subject to selection together with a hierarchical prior on each component where a scalar parameter captures effect importance. Explicit effect selection is then feasible by placing a spike-and-slab prior on this "importance parameter" while implicit effect selection can be accomplished by a shrinkage prior.

Methods for Bayesian variable selection based on spike-and-slab priors introduced for linear regression in different variants by [10, 11, 13, 24] have been extended to the selection of functional effects modeled by Gaussian process priors with an appropriate covariance structure in [5, 30, 37] or by a linear combination of basis functions in [27, 32, 36]. Except for [26] who specify the spike-and-slab prior directly on the vector of basis function coefficients, all other approaches use a spike-and-slab prior on the scale of the multivariate coefficient vector to select the blocks of coefficients related to a functional effect. [32] proposed an approach that is not restricted to variable and function selection but allows for selection of any type of effects in a structured additive predictor in mean regression models. Effect selection for distributional parameters is considered only in [5] for double exponential regression models, where both the mean parameter and the dispersion parameter are linked to a structured additive predictor and [14] in general distributional regression models.

As mentioned above, implicit selection of regression coefficients can be obtained by priors which encourage heavy shrinkage of small effects and leave large effects almost unshrunken. These priors mimic the two-component spike-and-slab mixture prior by featuring a substantial mass close to zero combined with heavy tails [29]. Shrinkage priors for regression coefficients like the normal gamma [12] or the horseshoe prior [2] are obtained as scale mixtures of Gaussians. Shrinkage of the different types of effects in structured additive predictors requires shrinkage priors on the importance or the variance parameters.

Priors for different effect types must take the structure of the effect type into account but feature shrinkage goals that diverge from a null effect. The Bayesian P-spline specification [20], based on random walk priors, allows for a constant effect (null effect), a linear effect, or higher-order polynomials as the unpenalized effect. The functional horseshoe prior [34] enables a very flexible configuration of the target of the shrinkage procedure. [33] and [20] allow for adaptive regularization of continuous effects when the underlying function locally differs in complexity.

The rest of this chapter is structured as follows: Section 12.2 reviews the essential modelling components of structured additive distributional regression. In Section 12.3, we discuss how effect selection and effect shrinkage can be accomplished in these models. Section 12.4 illustrates Bayesian effect selection in two models for undernutrition in India. We highlight two particular priors for shrinkage of functional effects which diverge from the usual shrinkage towards a null effect in Section 12.5, including the recently introduced concept of shrinkage towards a functional subspace. Section 12.6 concludes the chapter. Code and data for reproducing our analyses is available as part of the online supplement of this edited volume.

12.2 Structured Additive Distributional Regression

12.2.1 Basic Model Structure

In structured additive distributional regression, we model the conditional distribution of the observed responses y_i, $i = 1, \dots, n$, given the covariate information $\boldsymbol{x}_i$ and the vector of model parameters $\boldsymbol{\theta}$ by means of a K-parametric probability density function (pdf)

$$f(y_i|\boldsymbol{x}_i, \boldsymbol{\theta}) = f(y_i|\boldsymbol{\nu}(\boldsymbol{x}_i, \boldsymbol{\theta})), \quad \boldsymbol{\nu}(\boldsymbol{x}_i, \boldsymbol{\theta}) = (\nu_1(\boldsymbol{x}_i, \boldsymbol{\theta}), \dots, \nu_K(\boldsymbol{x}_i, \boldsymbol{\theta}))^T$$

where the parameters $\boldsymbol{\nu}(\boldsymbol{x}_i, \boldsymbol{\theta})$ of the pdf depend deterministically on the covariate information $\boldsymbol{x}_i$ and the vector of model parameters $\boldsymbol{\theta}$ (including regression coefficients as well as hyperparameters, see the end of this section for a more precise definition). In particular, each distributional parameter $\nu_k(\boldsymbol{x}_i, \boldsymbol{\theta})$ is mapped to a structured additive predictor $\eta_k(\boldsymbol{x}_i, \boldsymbol{\theta})$ via a bijective response function h_k:

$$\nu_k(\boldsymbol{x}_i, \boldsymbol{\theta}) = h_k(\eta_k(\boldsymbol{x}_i, \boldsymbol{\theta})), \quad k = 1, \dots, K.$$

The purpose of this is to unambiguously map the domain of the predictor on the domain of the parameter and to avoid difficulties arising from restrictions on the parameter space. This is required since many distributional parameters are restricted (for example to the positive domain) and the additive predictor can, in general, take any value from the real numbers. Common choices for the response function are the identity function if the distributional parameter is unrestricted, the exponential function if the distributional parameter is restricted to the positive domain, and the expit function, $\text{expit}(\eta) = 1/(1+\exp(-\eta))$, for parameters limited to the unit interval.

The structured additive predictor $\eta_k(\boldsymbol{x}_i, \boldsymbol{\theta})$ is additively composed of J_k elements, that is,

$$\eta_k(\boldsymbol{x}_i, \boldsymbol{\theta}) = \sum_{j=1}^{J_k} f_{kj}(\boldsymbol{x}_i, \boldsymbol{\theta}).$$

Each of the effect components f_{kj}, $j = 1, \ldots, J_k$, is a function of the covariates $\boldsymbol{x}_i$ and the model parameters $\boldsymbol{\theta}$. Usually, the value of the function depends only on a sub-vector of the covariates with small dimension and a sub-vector of the model parameters but for notational convenience we use the complete vectors in the definition of the functions. Within this notation, we are able to represent different kinds of effect types such as linear effects, smooth effects of continuous covariates, spatial effects and many more. For that, we assume that f_{kj} can be represented by a basis function expansion with L_{kj} basis functions, i.e.,

$$f_{kj}(\boldsymbol{x}_i, \boldsymbol{\theta}) = \sum_{l=1}^{L_{kj}} \beta_{kjl} B_{kjl}(\boldsymbol{x}_i), \tag{12.1}$$

where $B_{kjl}(\boldsymbol{x}_i)$ denote the pre-specified basis functions and the basis coefficients β_{kjl} are elements of the vector of model parameters $\boldsymbol{\theta}$. As a major advantage, the resulting model is again linear in the basis coefficients such that each term in the structured additive predictor is associated with a design matrix $\boldsymbol{B}_{kj}$ obtained from evaluating the basis functions at the observed covariate values $\boldsymbol{x}_i$, $i = 1, \ldots, n$. However, the dimension of the coefficient vectors associated with the different effects is often of considerable dimension such that informative priors are assigned to them to achieve desirable properties such as smoothness.

A generic prior for the coefficient vector $\boldsymbol{\beta}_{kj} = (\beta_{kj1}, \ldots, \beta_{kjL_{kj}})^T$ that achieves such behavior is given by the (potentially degenerate) normal distribution with density

$$p(\boldsymbol{\beta}_{kj}|\lambda_{kj}) \propto \exp\left(-\frac{1}{2\lambda_{kj}^2}\boldsymbol{\beta}_{kj}^T \boldsymbol{K}_{kj} \boldsymbol{\beta}_{kj}\right) \tag{12.2}$$

where $\boldsymbol{K}_{kj}$ denotes an appropriate and pre-defined penalty matrix for the chosen effect type and λ_{kj}^2 can be thought of as a variance or local shrinkage parameter. These two factors together form the precision matrix $\boldsymbol{Q}_{kj} = \lambda_{kj}^{-2}\boldsymbol{K}_{kj}$ in the multivariate normal prior. The structure of $\boldsymbol{K}_{kj}$ depends on the considered effect type and additional information such as the neighbourhood-structure when modeling spatial dependency. For $\boldsymbol{K}_{kj} = \boldsymbol{0}$, Equation (12.2) is equivalent to a flat prior. For all other cases, the prior of the variance parameter λ_{kj}^2 depends on the chosen effect type and whether the effect is subject to effect selection or regularization. We leave this vague for now and just introduce $\boldsymbol{\psi}$ as the vector of hyper-parameters in the joint prior on λ_{kj}^2, $j = 1, \ldots, J_k$ and $k = 1, \ldots, K$.

For identification issues, we will typically restrict $\boldsymbol{\beta}_{kj} \in \mathbb{R}^{L_{kj}}$ by requiring that the linear constraint $\boldsymbol{A}_{kj}\boldsymbol{\beta}_{kj} = \boldsymbol{0}$ with a pre-specified constraint matrix $\boldsymbol{A}_{kj}$ is fulfilled. The resulting prior density is then given by

$$p(\boldsymbol{\beta}_{kj}|\lambda_{kj}) \propto \exp\left(-\frac{1}{2\lambda_{kj}^2}\boldsymbol{\beta}_{kj}^T \boldsymbol{K}_{kj}\boldsymbol{\beta}_{kj}\right)\mathbb{I}\left(\boldsymbol{A}_{kj}\boldsymbol{\beta}_{kj} = \boldsymbol{0}\right) \tag{12.3}$$

where $\mathbb{I}(\cdot)$ denotes the indicator function.

Now, that most model parameters are defined, $\boldsymbol{\theta}$ can be substantiated. The vector is created by stacking the so far discussed model parameters, that is,

$$\boldsymbol{\theta} = \left(\boldsymbol{\beta}_{11}^T, \ldots, \boldsymbol{\beta}_{1J_1}^T, \ldots, \boldsymbol{\beta}_{K1}^T, \ldots, \boldsymbol{\beta}_{KJ_K}^T, \lambda_{11}, \ldots, \lambda_{1J_1}, \ldots, \lambda_{K1}, \ldots, \lambda_{KJ_K}^T, \boldsymbol{\psi}^T\right)^T.$$

In the upcoming sections, we give an overview of concrete priors and provide references.

12.2.2 Predictor Components

In our applications below, we will consider four types of predictor components: linear effects, penalized splines, tensor product penalized splines and Markov random fields. However, the set of available predictor components is much broader, see [7, Ch. 9] and [19] for overviews.

For *linear effects*, the basis functions simply pick elements from the vector of covariates $\boldsymbol{x}_i$. In most cases, a flat prior with $\boldsymbol{K}_{kj} = \boldsymbol{0}$ is assigned to such linear effects, but informative priors leading to Bayesian versions of ridge regression or LASSO regularization are also possible, see for example [6].

Bayesian *penalized splines* [20] enable flexible estimation of non-linear effects of continuous covariates based on a polynomial spline representation in combination with a random walk prior. More precisely, the domain $[a, b]$ of the covariate of interest is decomposed into a moderately large number of intervals based on equidistant knots $a = \kappa_1 < \cdots < \kappa_m = b$. On these, we define the indicator basis functions

$$B_l^{(0)}(x) = \mathbb{I}(\kappa_l \leq x < \kappa_{l+1}) = \begin{cases} 1 & \kappa_l \leq x < \kappa_{l+1} \\ 0 & \text{otherwise} \end{cases} \qquad l = 1, \ldots, m-1$$

leading to a piecewise constant function representation. Piecewise polynomial approximations of higher degree $D > 0$ can then be achieved by the recursion

$$B_l^{(D)}(x) = \frac{x - \kappa_{l-D}}{\kappa_l - \kappa_{l-D}} B_{l-1}^{(D-1)}(x) + \frac{\kappa_{l+1} - x}{\kappa_{l+1} - \kappa_{l+1-D}} B_l^{(D-1)}(x).$$

Utilizing such Dth degree basis functions leads to the $L = m + D - 1$ dimensional function space of polynomial splines $f(x)$ with (i) $f(x)$ being a polynomial of degree D on each of the intervals $[\kappa_l, \kappa_{l+1})$ and (ii) $f(x)$ being $D - 1$ times continuously differentiable, where $D = 1$ implies continuity without being differentiable and $D = 0$ does not make any smoothness assumptions. An rth order random walk prior

$$\Delta_r \beta_l | \lambda^2 \sim \mathcal{N}(0, \lambda^2)$$

with rth order differences Δ_r and variance parameter λ^2 then allows to control the overall amount of smoothness by assigning appropriate hyperpriors to λ^2.

Tensor product splines extend this concept to bivariate effects $f(x_1, x_2)$ where the interaction surface for the two continuous covariates is constructed from the tensor product of two univariate penalized splines and a Kronecker sum of the penalty matrix to achieve smoothness of the surface along both covariate axes, see [19] for an in-depth discussion. *Markov random fields* allow for representing spatial effects for discrete spatial information such as administrative regions or districts where spatial smoothness is assumed across neighboring regions, see [31] for a detailed exposition.

12.2.3 Common Response Distributions

While we restrict ourselves to normally distributed responses in this chapter, we want to emphasize the generality of distributional regression. In particular, distributional regression allows to deal with various types of non-negative continuous distributions such as the gamma, Weibull or Box-Cox-Power exponential distribution avoiding the need of applying transformations to the response to match with the assumption of normally distributed error terms. Skewed distributions provide alternatives to the normal distribution (see for example [23]), while extended models for count data regression take aspects such as zero-inflation and/or overdispersion into account, (see for example [17]). Fractional responses, i.e., responses representing a single or multiple percentages can be modelled with the beta or the Dirichlet distribution (see [16] for an example of the latter) and mixed discrete-continuous distributions allow to supplement a continuous distribution for the response by discrete support points that are assigned a non-zero probability. Finally, multivariate response vectors can be considered, either based on multivariate extensions of standard distributions (see for example [17, 23]) or utilizing copulae to construct flexible forms of dependence in combination with arbitrary marginal distributions (see for example [15]).

12.2.4 Basic MCMC Algorithm

For Bayesian inference in structured additive distributional regression, we can rely on a generic Markov chain Monte Carlo simulation algorithm. In the following, we briefly review this algorithm for a basic model, ignoring priors on hyperparameters that will later be introduced to achieve effect selection and/or regularization.

Due to the non-conjugacy between the multivariate normal priors for the regression coefficients and the non-normal response distributions, usually no closed form full conditionals are available for the regression coefficients. On the other hand, blocks of coefficients corresponding to one functional effect in the additive predictor should usually be updated simultaneously due to potentially large within-block correlation. As a consequence, we require high-dimensional proposal densities for blocks of coefficients $\boldsymbol{\beta}_{kj}$. The basic idea is now to derive such a proposal by approximating the log of the full conditional with a normal distribution as suggested by asymptotic normal theory for posteriors under suitable regularity conditions. Note that we do not require strict asymptotic proofs here, since we are only constructing a proposal distribution. As long as this distribution is close enough to the full conditional, the acceptance probability will correct for any deviation between the approximation and the true full conditional while still achieving a large acceptance probability.

More precisely, we generate proposals based on the normal approximation $\boldsymbol{\beta}_{kj} \sim \mathcal{N}(\boldsymbol{\mu}_{kj}, \boldsymbol{P}_{kj}^{-1})$ with expectation

$$\boldsymbol{\mu}_{kj} = \boldsymbol{P}_{kj}^{-1} \boldsymbol{B}_{kj}^T \boldsymbol{W}_k (\tilde{\boldsymbol{y}}_k - \boldsymbol{\eta}_{k,-j})$$

and precision matrix

$$\boldsymbol{P}_{kj} = \boldsymbol{B}_{kj}^T \boldsymbol{W}_k \boldsymbol{B}_{kj} + \frac{1}{\lambda_{kj}^2} \boldsymbol{K}_{kj}$$

where $\boldsymbol{B}_{kj}$ is the design matrix of the effect of interest, $\boldsymbol{\eta}_{k,-j} = \boldsymbol{\eta}_k - \boldsymbol{B}_{kj}\boldsymbol{\beta}_{kj}$ is a partial predictor, and $\boldsymbol{W}_k$ and $\tilde{\boldsymbol{y}}_k$ are a diagonal matrix of working weights and a vector of working responses that resemble the quantities involved in the iteratively weighted least squares (IWLS) estimate in frequentist inference, see [18] for a detailed discussion. This proposal is applied iteratively in a loop over predictors and effects in the additive predictor and can easily be supplemented with additional steps for updating hyperparameters. The proposals can also directly be adjusted to incorporate linear constraints as defined by $\boldsymbol{A}_{kj}$ in Equation (12.3) when considering the decomposition of effects into different components (see Algorithm 2.6 in [31] for details).

As a major advantage, our sampler automatically adapts to the location and the curvature of (multivariate) full conditionals without the need for manual tuning. Furthermore, the acceptance probabilities of the sampler provide us with a simple yet effective way of quantifying the accuracy of the normal approximation where acceptance rates close to 100% indicate that the full conditional is very close to normality. Note that, in contrast to random walk proposals, high acceptance rates in our sampler are not a result of small step sizes but rather of a close approximation of the full conditional.

12.3 Effect Selection Priors

12.3.1 Challenges

Going beyond purely linear predictors and implementing effect selection in models with structured additive predictors leads to several additional challenges.

Combining regularization and selection. For most of the effects $f_{kj}(\boldsymbol{x}_i, \boldsymbol{\theta})$ it is desirable to combine regularized estimation, enforcing desirable properties such as smoothness or shrinkage, with selection in one prior hierarchy. In addition, it may be desirable to implement a local/global prior specification where overall regularization and selection are controlled by shared parameters that overcome the usual prior independence assumption. This can be achieved in the multivariate normal prior structure introduced above where the structure of the precision matrix $\boldsymbol{K}_{kj}$ determines the type of regularization while the variance parameter λ^2_{kj} controls both selection and the amount of regularization when assuming an appropriate prior hierarchy with common hyperparameters, for example, a prior determining the number of a priori expected effects in the model.

Selection of coefficient blocks. Based on the basis expansion from Equation (12.1), effect selection concerns the simultaneous (de-)selection of complete blocks of coefficients $\boldsymbol{\beta}_{kj}$ instead of scalar parameters. In particular for spatial or random effects, the coefficient blocks may be of considerably high dimension comprising several hundred parameters. Selection within effect blocks can be achieved by placing separate indicators on the individual basis functions while simultaneous selection leads to the use of one single, scalar selection parameter.

Hierarchical selection decisions. While the inclusion/exclusion of effects $f_{kj}(\boldsymbol{x}_i, \boldsymbol{\theta})$ certainly represents one question of considerable interest, it is often at least of similar importance to evaluate the possibility to reduce complex effects to simpler versions. For example, if $f_{kj}(\boldsymbol{x}_i, \boldsymbol{\theta})$ represents the non-linear, smooth effect of a continuous covariate, one may consider the hierarchical selection decisions of whether the covariate has an effect at all, whether this effect can be represented by a simple, parametric specification (for example a linear effect) or whether a completely non-linear model variant is indeed required. For interaction surfaces, the number of potential special cases increases considerably, where varying coefficient type interactions, non-linear main effects, parametric interactions, or parametric main effects form components of potential simpler model variants. In our application, we will illustrate such effect decompositions and apply selection indicators separately on the decomposed effects.

Computational considerations. Computational aspects are of considerable importance for effect selection in structured additive regression for several reasons. First of all, blocks of parameters that represent the different effects in the predictor, have to be handled together due to their inherent dependence to yield good mixing and convergence. In our framework based on basis functions, multivariate normal priors, and multivariate normal proposal densities, working with sparse matrix structures enable considerable efficiency gains, see [21]. Besides, the non-Gaussianity of most of the full conditionals can pose challenges when aiming at good mixing and avoiding convergence problems, in particular with stochastic search variable selection schemes based on binary inclusion/exclusion indicators, see [14, 32].

Hyperparameter elicitation. In any form of Bayesian variable selection, the determination of appropriate hyperparameter settings is a challenging task that requires considerable care. This is, even more, the case in structured additive regression models where various types of effects are additively combined. The respective effects can not easily be standardized to make them comparable. Thus, one can in general not devise one set of default hyperparameters but interpretable rules for hyperparameter elicitation are required and we will illustrate in the applications below how this can be achieved for our model class.

Further difficulties arise in the distributional regression setting:

Multiple regression predictors. Compared to usual mean regression specifications, distributional regression models rely on a multitude of regression predictors such that the sheer number of selection decisions grows considerably. As a consequence, questions concerning efficiency and computation are even more important.

Concurvity across predictors. Since the same types of covariates and covariate effects will often appear in multiple predictors, one may suspect that false positive and false negative inclusion probabilities of such effects will be dependent across the predictors. For example, if we erroneously exclude an effect from the predictor characterizing the location parameter of the distribution, this may affect the inclusion decision for the same covariate effect in the predictor for the scale parameter.

12.3.2 Spike-and-Slab Priors for Effect Selection

In principle, Bayesian effect selection in structured additive distributional regression can be accomplished similarly as in simple linear models, i.e., by introducing indicator variables $\delta_{kj} \in \{0, 1\}$ that indicate exclusion ($\delta_{kj} = 0$) vs. inclusion ($\delta_{kj} = 1$) of effect j in the kth predictor such that the predictor effectively reads

$$\eta_k(\boldsymbol{x}_i, \boldsymbol{\theta}) = \sum_{j=1}^{J_k} \delta_{kj} f_{kj}(\boldsymbol{x}_i, \boldsymbol{\theta})$$

complemented by a discrete Bernoulli prior, with potentially more hierarchical levels to make the prior better adapt to the required selection properties, for example

$$P(\delta_{kj} = 1) = \omega_{kj}, \quad P(\delta_{kj} = 0) = 1 - \omega_{kj}, \quad \omega_{kj} \sim \mathcal{B}eta(a_0, b_0). \tag{12.4}$$

An alternative is to place the indicators inside the prior of the parameters characterising the effect $f_{kj}(\boldsymbol{x}_i, \boldsymbol{\theta})$: for a linear effect with Gaussian prior

$$\beta_{kj} \sim \mathcal{N}(0, \delta_{kj}\lambda_{kj}^2), \quad \lambda_{kj}^2 \sim \mathcal{IG}(a_{kj}, b_{kj}),$$

the indicator decides whether the prior variance λ_{kj}^2 should be set to zero (such that the prior degenerates to a point mass on the prior mean) or whether the prior of λ_{kj}^2 should indeed be $\mathcal{IG}(a_{kj}, b_{kj})$. The prior on the effect β_{kj} thus is a mixture of a discrete spike at zero and a continuous slab component. Both approaches share the advantage that posterior probabilities for inclusion/exclusion of effects can be estimated and can also be shown to have good theoretical properties, but from an applied perspective they often suffer from considerable mixing problems in MCMC inference.

Spike-and-slab-type priors with continuous spikes replace the dichotomy between inclusion and exclusion of an effect by the difference between weak and heavy regularization. For example, in the setting above, one could specify

$$P(\delta_{kj} = 1) = \omega_{kj}, \quad P(\delta_{kj} = \xi) = 1 - \omega_{kj}, \quad \omega_{kj} \sim \mathcal{B}eta(a_0, b_0) \tag{12.5}$$

with ξ small yet positive such that plugging the indicator variable into

$$\beta_{kj} \sim \mathcal{N}(0, \delta_{kj}\lambda_{kj}^2), \quad \lambda_{kj}^2 \sim \mathcal{IG}(a_{kj}, b_{kj})$$

yields either a prior with rather small prior variance $\xi\lambda_{kj}^2$ (the spike, which induces heavy regularization) or a prior with the usual prior variance λ_{kj}^2 (the slab, which yields smaller regularization).

Many variants of spike-and-slab priors have been proposed for the selection of single effects in linear regression models. Spike-and-slab priors with a discrete spike were introduced in [11, 24] and a continuous spike in [10, 13], see [25] and [22] for an overview.

As discussed above, for a multivariate effect $\boldsymbol{\beta}_{kj}$ selection should be combined with regularization which would suggest to keep the structure of the prior on $\boldsymbol{\beta}$ and formulate a multivariate version of a spike-and-slab prior as

$$p(\boldsymbol{\beta}_{kj}) \propto \exp\left(-\frac{1}{2\delta_{kj}\lambda_{kj}^2}\boldsymbol{\beta}_{kj}^T\boldsymbol{K}_{kj}\boldsymbol{\beta}_{kj}\right)\mathbb{I}\left\{\boldsymbol{A}_{kj}\boldsymbol{\beta}_{kj} = \boldsymbol{0}\right\}$$

with the prior on the inclusion indicator δ_{kj} as in Equation (12.5). However, as shown in [32, Online Appendix] such a prior is not useful for selection of a block of coefficients from a computational point of view, as overconditioning increases with the dimension of $\boldsymbol{\beta}_{kj}$ and results in severe convergence and mixing problems of MCMC inference. A viable approach results by moving from the centered parameterization of the effects to their non-centered parameterization, given as

$$f_{kj}(\boldsymbol{x}_i, \boldsymbol{\theta}) = \lambda_{kj}\sum_{l=1}^{L_{kj}}\tilde{\beta}_{kjl}B_{kjl}(\boldsymbol{x}_i), \tag{12.6}$$

where the effects $\boldsymbol{\beta}_{kj}$ and $\tilde{\boldsymbol{\beta}}_{kj}$ are related by

$$\boldsymbol{\beta}_{kj} = \lambda_{kj}\tilde{\boldsymbol{\beta}}_{kj}$$

and the prior density for $\tilde{\boldsymbol{\beta}}$ is specified as

$$p(\tilde{\boldsymbol{\beta}}_{kj}) \propto \exp\left(-\frac{1}{2}\tilde{\boldsymbol{\beta}}_{kj}^T\boldsymbol{K}_{kj}\tilde{\boldsymbol{\beta}}_{kj}\right)\mathbb{I}\left\{\boldsymbol{A}_{kj}\tilde{\boldsymbol{\beta}}_{kj} = \boldsymbol{0}\right\}.$$

In the non-centered parameterization, effect selection can be accomplished by placing a spike-and-slab prior on the squared importance parameter λ_{kj}^2 rather than on the coefficient vector $\boldsymbol{\beta}_{kj}$.

Though the two parameterizations seem to be equivalent, this is only the case if the prior precision matrix $\boldsymbol{K}_{kj}$ is not rank deficient. If, for example, a linear effect is not penalized by the prior, λ_{kj}^2 approaching zero implies a linear effect in the centered parameterization (see

Equation (12.1)) but an effect equal to zero in the rescaled version (see Equation (12.6), for more details see [14]). Hence in the non-centered parameterization the non-penalized effects, can additionally be included in the structured additive predictor which allows separate selection of penalized and non-penalized parts of the effects.

In the non-centered parameterization, effect selection can be accomplished by specifying a spike-and-slab prior on the scale parameter λ_{kj}^2 as

$$\begin{aligned} \lambda_{kj}^2 | \delta_{kj}, \psi_{kj}^2 &\sim \mathcal{G}\left(\frac{1}{2}, \frac{1}{2\delta_{kj}\psi_{kj}^2}\right) \\ \psi_{kj}^2 &\sim \mathcal{IG}(a_{kj}, b_{kj}) \end{aligned} \tag{12.7}$$

with the prior on the indicators δ_{kj} as in Equation (12.5). After marginalizing over the parameter ψ_{kj}^2, the spike-and-slab prior on λ_{kj}^2 is mixture of two scaled beta prime distributions with shape parameters $p = 1/2$ and $q = a_{kj}$ and scale parameters $c = 2b_{kj}$ for $\delta_{kj} = 1$, and $c = 2\xi b_{kj}$ otherwise. The scaled beta prime distribution is defined via its probability density function for $x > 0$ as $p(x|p, q, c) \propto (x/c)^{p-1}(1 + x/c)^{-p-q}$ with $p, q, c > 0$. Hence the resulting prior on $\boldsymbol{\beta}$ is called normal beta prime spike-and-slab (NBPSS) prior (as proposed in [14]).

A similar prior for general effect selection in generalized additive models was proposed in [32] on the effects resulting from a mixed model decomposition with basically the same spike-and-slab prior specification on the squared importance parameter (though formulated as a normal spike-and-slab prior on λ_{kj}^2), but a bimodal prior with modes at $+1$ and -1 on the standardized effects in the mixed model decomposition. The main disadvantage of this specification is that the mixed model representation destroys sparsity properties which are usually present in the design matrices (for example band structures for P-splines) and hence requires higher computation times, see [14] for a thorough comparison of the two priors.

Practical application of effect selection via the NBPSS prior requires to choose values for the three parameters in the Beta prime spike-and-slab prior (the parameters a_{kj} and b_{kj} and the scale parameter ξ) as well as the parameters a_0 and b_0 of the Beta prior on the inclusion probability ω. Based on theoretical and computational considerations, [14] recommend as default choice a flat prior on the inclusion indicators, i.e., $a_0 = b_0 = 1$, and a value of $a_{kj} = 5$ which guarantees the existence of moments but yet heavy tails (to allow for moves between spike-and-slab component in MCMC). No general recommendations can be given for the two remaining parameters b_{kj} (which determines the mean of the slab component) and the parameter ξ (which is the ratio of the means of spike-and-slab component) and hence they propose to determine these parameters from probability statements on inclusion and exclusion of effects. Details on how these parameters can be determined in an application are given in Section 12.4.2.

A further issue that arises for selection of effects in structured additive predictors is how structural relationships between predictors can be taken into account in selection decisions. For hierarchically related predictors according to the inheritance principle (see [4]), higher order terms should be included in the model only if lower order terms are, with strong or weak heredity depending on whether every or at least one associated lower-order predictor is included.

Relations among predictors like heredity can easily be taken into account by a hierarchical specification of the dependence structure of the indicator variables. For example, for two covariates with inclusion indicators δ_{kj} and $\delta_{kj'}$, the prior distribution on the inclusion indicator $\delta_{k,jj'}$ for the interaction effect would be specified conditional the values of δ_{kj} and $\delta_{kj'}$ by $P(\delta_{k,jj'} = 1|\delta_{kj'}, \delta_{kj})$. The heredity principle restricts one or more of these conditional inclusion probabilities to zero. Under weak heredity the interaction effect is excluded only if both main effects are excluded and hence

$$P(\delta_{k,jj'} = 1 | \delta_{kj} = 0, \delta_{kj'} = 0) = 0,$$

whereas under strong heredity the interaction effect is excluded whenever any of the main effects is excluded which implies

$$P(\delta_{k,jj'} = 1 | \delta_{kj} = 0, \delta_{kj'} = 0) = P(\delta_{k,jj'} = 1 | \delta_{kj} = 1, \delta_{kj'} = 0) = \\ = P(\delta_{k,jj'} = 1 | \delta_{kj} = 0, \delta_{kj'} = 1) = 0.$$

Another form of structural selection is sparse group selection, see [3], where effects, which are subject to selection, are partitioned into groups. Selection decisions are first made on the group level and then on single effects within a selected group. For the effect of a continuous covariate, an alternative to separate selection of the linear and the non-linear effect via independent indicators, is group selection via a joint indicator on both effects and within group selection via effect-specific indicators.

For effects represented by a vector of coefficients, interest might also be in selection of single effects. For example, one could be interested to identify the subjects with a non-negligible subject-specific deviation caused by subject-specific random intercepts in the linear predictor. For independent subject-specific effects such a selection is feasible by putting spike-and-slab priors on the elements of the diagonal covariance structure matrix $\boldsymbol{K}^{-1}$, see [9]. Selection of the dependence structure of a coefficient vector (which is required to guarantee the positive semi-definiteness of their covariance matrix during selection) can be accomplished, for example, by putting spike-and-slab priors on the elements of the Cholesky decomposition of $\boldsymbol{K}^{-1}$, see [8].

Finally, as a structured additive predictor can contain many different effects, a goal might be to control the overall complexity of the predictors. This cannot be achieved by priors that are independent across effects but requires to specify the prior similarly as in [29], as a product of a global parameter shared by the effects in one predictor and an effect-specific parameter to achieve local shrinkage.

12.3.3 Regularization Priors for Effect Selection

Besides spike-and-slab priors, priors based on the idea of regularization or shrinkage can be used for effect selection. The main difference to spike-and-slab priors is that such priors do not feature a binary selection variable. Instead, the marginal prior on the coefficients is constructed such that it features a distinct spike in the origin and heavy tails. This design allows us to implicitly distinguish noise and signals (corresponding to uncertain and strong effects, respectively) and to manage them differently. Uncertain effects are shrunken towards a zero effect while strong effects should be kept almost untouched.

The LASSO penalty [35] is one of the most prominent examples of shrinkage approaches. Having its foundations in the frequentist world, the LASSO method was adapted to the Bayesian framework [28]. However, the Bayesian version does not share the implicit variable selection property of shrinking regression coefficients to exact zero that made its frequentist counterpart so popular.

More recently, the concept of global-local shrinkage has emerged, having the horseshoe prior [2] as one of the most prominent examples. Here, the goal is not to shrink distinct regression coefficients towards zero but to shrink a vector of regression coefficients towards a sparse solution. In the Bayesian context, sparsity needs to be interpreted differently. It is sufficient that a large share of coefficients has its density mass close to zero.

This concept of sparsity can also be used for effect selection using shrinkage priors. An effect is considered de-selected if the target of the shrinkage procedure is within the corresponding credibility interval. To give some examples, for a regression coefficient that could be 0, for a Bayesian P-spline that could be a constant effect (when using first order differences) or a linear effect (when using second order differences). Another option to perform effect selection can be based on an assessment of the magnitude of shrinkage. For example, with the horseshoe prior one can construct a so called shrinkage coefficient within the interval $(0, 1)$. This coefficient is derived from the variance on the regression coefficient and can be interpreted as the magnitude of shrinkage. Consequently, a value of zero implies no shrinkage and a value of 1 full shrinkage. Therefore, effect selection can be based on a thresholding procedure considering, for example, values larger than 0.5 as sufficient for effect de-selection. We return to the concept of regularization and shrinkage in Section 12.5 in which we highlight two particular concepts of regularization for functional effects.

12.4 Application: Childhood Undernutrition in India

In the following, we illustrate the application of the spike-and-slab type effect selection priors discussed in Section 12.3.2 for two different situations, reflecting some of the challenges arising in the context of structured additive distributional regression. In the first setting, we consider an additive main effects location-scale model where both the mean and the standard deviation of the normally distributed response variable are related to an additive predictor consisting of potentially non-linear effects of the continuous covariates and a spatial effect represented by a Gaussian Markov random field. In addition, the non-linear effects are decomposed into linear effects and the orthogonal deviation such that the effect selection priors enable us to decide not only whether to include a variable at all, but also whether the effect can be safely reduced to a linear effect or whether a more general, non-linear modeling form is indeed required.

In the second setting, we investigate the decomposition of an interaction surface modeled as the tensor product of two univariate penalized splines. More specifically, we decompose the interaction surface into various additive components (including linear and non-linear main effects but also interaction effects such as a parametric, linear interaction and varying coefficient type effects) and place separate selection indicators on the additive components. This allows us to study whether a complex interaction is indeed required or whether the interaction can be reduced to a simpler form. To limit computational complexity, the second setting will only consider a mean regression specification.

12.4.1 Data

For our analyses, we rely on a data set provided along with the R package `gamboostLSS` on childhood malnutrition in India, containing a subsample of 4,000 observations from the Indian Demographic and Health Survey conducted in 1998/99, see `www.dhsprogramme.com` for details. Despite considerable progress over the past decades, childhood malnutrition is still among the most urgent public health challenges in developing and transition countries, affecting not only the growth of children directly but also several long term socio-economic and health-related outcomes indirectly.

The nutritional status of children is usually quantified based on Z-scores that relate the observed nutritional status (as reflected by different anthropometric characteristics) to the nutritional status derived from a reference population. More precisely, such a Z-score is

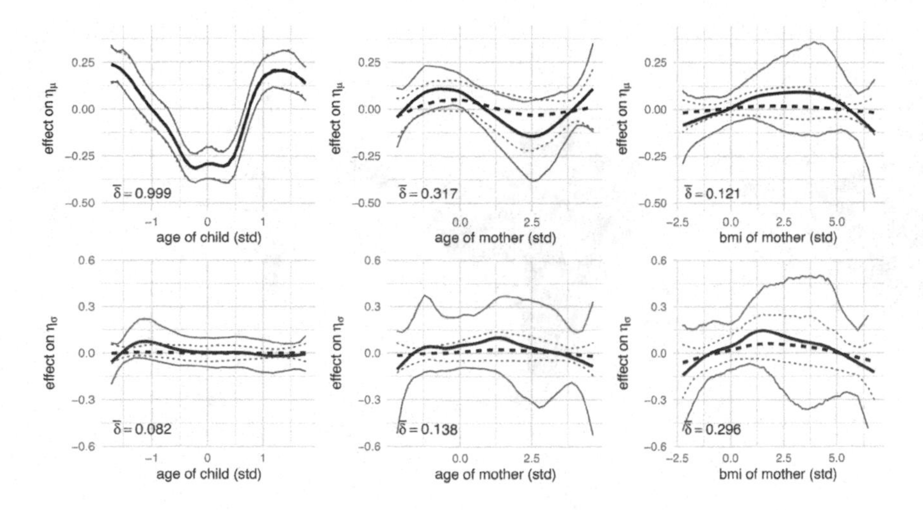

FIGURE 12.1
Main effects model: Estimated non-linear deviation effects, where the thick line represents posterior means while the thin lines indicate 95% pointwise posterior credibility intervals. Solid lines refer to the results without effect selection while dashed lines indicates results with effect selection priors. The estimated inclusion probability is included in the lower left of the figures.

commonly defined as

$$Z_i = \frac{\text{AC}_i - \mu_{\text{AC}}}{\sigma_{\text{AC}}}$$

where AC_i represents the anthropometric characteristic for child i while μ_{AC} and σ_{AC} refer to the location and scale measures in the reference population (stratified with respect to age and gender). In the following, we will focus on chronic malnutrition (also referred to as stunting), utilizing (insufficient) height for age as the anthropometric characteristic of interest.

As explanatory variables, we consider the age of the child in months (cage), the age of the mother in years (mage), the body mass index of the mother (mbmi) and the area of residence (area). Note that our analyses are of a purely illustrative nature while more thorough and more in-depth distributional analyses can be found, for example, in [16] (however, not including effect selection based on appropriate prior structures).

12.4.2 A Main Effects Location-Scale Model

In our first illustration, we assume

$$\text{stunting}_i \sim \mathcal{N}(\mu_i, \sigma_i^2)$$

where both the mean $\mu_i = \eta_\mu(\boldsymbol{x}_i)$ and the variance $\sigma_i^2 = \exp(\eta_\sigma(\boldsymbol{x}_i))$ of the normal distribution are equipped with additive predictors

$$\begin{aligned}\eta_\bullet(\boldsymbol{x}_i) &= \alpha_{\bullet 0} + \beta_{\bullet 1}\text{cage}_i + \beta_{\bullet 2}\text{mage}_i + \beta_{\bullet 3}\text{mbmi}_i \\ &\quad + f_{\bullet 1}(\text{cage}_i) + f_{\bullet 2}(\text{mage}_i) + f_{\bullet 3}(\text{mbmi}_i) + f_{\bullet\text{spat}}(s_i).\end{aligned}$$

In this specification, the index $\bullet$ refers to either the mean μ or the variance σ^2, $\alpha_{\bullet 0}$ is the overall intercept of the predictor equation that is assigned a flat prior such that the

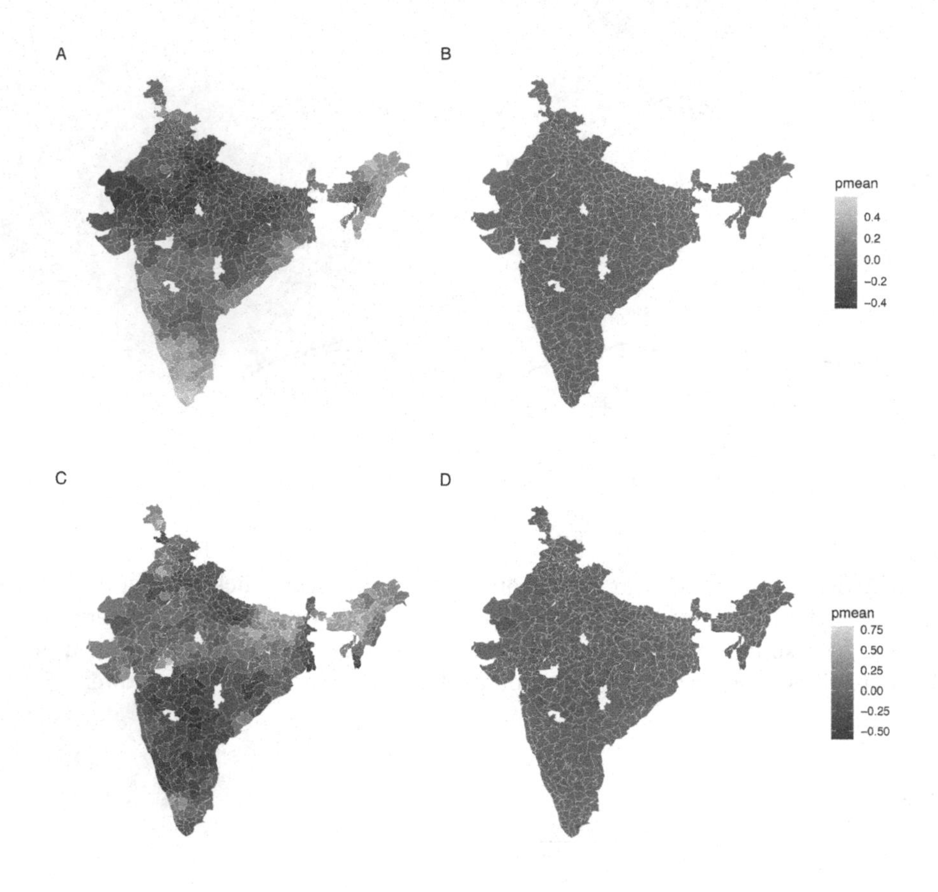

FIGURE 12.2
Main effects model: Estimated spatial effects, where the first row shows effects on the expectation μ while the second row shows effects on the predictor η_σ of the variance σ^2. The left column refers to the results without effect selection while the right column comprises the differences between the results without and with effect selection priors.

intercept is not subject to selection, $\beta_{\bullet 1}, \beta_{\bullet 2}, \beta_{\bullet 3}$ are the linear effects of the three continuous covariates cage, mage, and mbmi, $f_{\bullet 1}, f_{\bullet 2}, f_{\bullet 3}$ are the non-linear deviations from the linear effects, modeled as cubic penalized splines with 20 inner knots and second order random walk prior subject to a linear constraint that makes the effects orthogonal to the linear effects (see [19] for details), and $f_{\bullet \text{spat}}$ is a spatial effect based on the administrative district s_i that child i is living in, modeled as a Gaussian Markov random field.

All covariates and the response variable have been standardized prior to the analysis to facilitate prior elicitation and numerical stability. We estimated two models based on the specification above: Model M_1 where flat priors are assigned to the parametric effects while multivariate normal priors of the form Equation (12.2) are assigned to the penalized splines and the Markov random field supplemented by inverse gamma hyperpriors for the variance parameters λ^2 with hyperparameters $a = b = 0.001$, and model M_2 where (multivariate) normal priors are assumed for all parameters (except for the intercept) supplemented by NBPSS priors for the corresponding variance priors to achieve effect selection.

For the hyperparameters of the NBPSS prior, we rely on default values for the parameters of the Beta distribution of the prior inclusion probability ω ($a_0 = b_0 = 1$) and the parameter

a of the inverse gamma prior on ψ^2 ($a = 5$, corresponding to heavy tails but existing first two moments) while the two remaining parameters b (for the inverse gamma prior of ψ^2) and ξ (the multiplier for the variance ψ^2 in the spike-and-slab specification) are elicited from prior assumptions on the effect sizes. Following the simulation-based evidence provided in [14], we considered the two probability statements

$$P\left(\sup_{\boldsymbol{x}} |f(\boldsymbol{x})| \leq c \,\middle|\, \delta = 1\right) = \zeta \tag{12.8}$$

and

$$P\left(\sup_{\boldsymbol{x}} |f(\boldsymbol{x})| \leq c \,\middle|\, \delta = \xi\right) = 1 - \zeta \tag{12.9}$$

where Equation (12.8) refers to the probability that the supremum norm of an effect is smaller than a pre-specified level c for all design points $\boldsymbol{x}$ given selection ($\delta = 1$), such that ζ and c should be small while Equation (12.9) is the probability of not exceeding the threshold c given non-selection ($\delta = \xi$) such that the probability is reversed to $1 - \zeta$. We considered $c = 0.1$ and $\zeta = 0.1$ in all our analyses.

The resulting estimates are shown in Figures 12.1 (non-linear deviation effects) and 12.2 (spatial effects) and Table 12.1 (linear effects). Obviously, there is basically no difference between the spatial effects estimated with or without effect selection priors and indeed the estimated inclusion probability is virtually equal to one for both the effects on the mean and the variance. This matches well with earlier findings that identified considerable spatial variation in undernutrition patterns, which in our case is further supported by the limited complexity of our model. In contrast, there are stronger differences for the estimated non-linear effects (note that the linear effects have been removed by a linear constraint as discussed above). While the age of the child is a strong predictor for the mean that remains virtually untouched by employing a selection prior, all other effects are rather small when being estimated without effect selection. Consequently, we find that the effects are shrunken closely to zero when augmenting the selection prior, leading also to small posterior inclusion probabilities. We also find significantly reduced posterior uncertainty when relying on effect selection priors. This has been previously observed in [14] who showed that this indeed represents an efficiency gain that does not impair the coverage properties of the posterior credibility intervals.

		additive model		with effect selection		
	variable	pmean	95% CI	pmean	95% CI	I
μ	Intercept	-0.04	[-0.15, 0.07]	-0.01	[-0.09, 0.06]	-
	age of child	-0.37	[-0.42, -0.31]	-0.37	[-0.42, -0.32]	1
	age of mother	0.06	[-0.01, 0.14]	0.02	[-0.02, 0.09]	0.45
	bmi of mother	0.14	[0.05, 0.19]	0.14	[0.10, 0.18]	1
σ	Intercept	-0.46	[-0.69, -0.29]	-0.37	[-0.48, -0.28]	-
	age of child	0.09	[0.03, 0.16]	0.06	[0.01, 0.12]	0.31
	age of mother	0.02	[-0.10, 0.15]	0.01	[-0.02, 0.07]	0.24
	bmi of mother	0.02	[-0.11, 0.15]	0.03	[-0.03, 0.10]	0.45

TABLE 12.1
Table of linear effects, posterior means with 95% credible intervals, on both predictors of the additive model with and without effect selection. All covariates have been standardized. The last column denotes the selection probability.

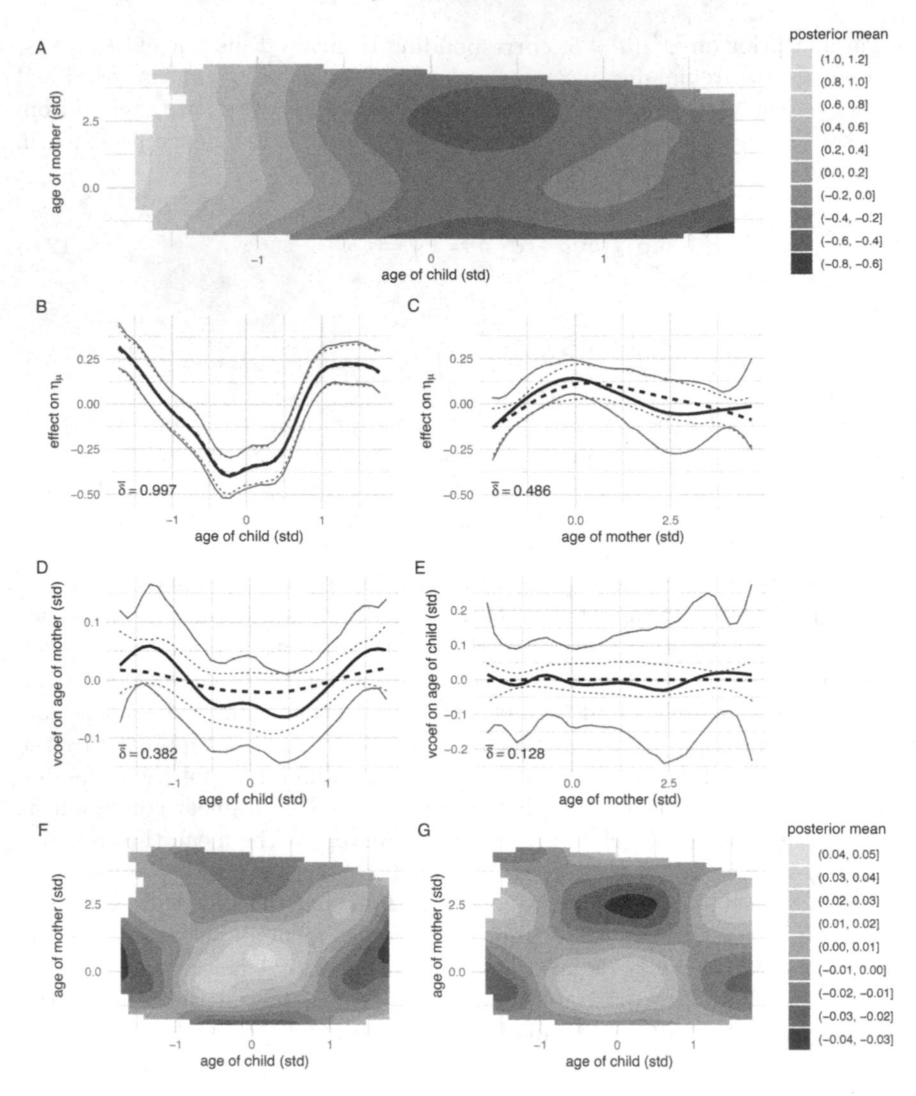

FIGURE 12.3
Interaction surface: Estimated complete interaction effect without selection prior (panel A) and estimated effect decomposition with non-linear main effects (panels B and C), varying coefficient terms (panels D and E) and remaining interaction effect without selection prior (panel F) and with selection prior (panel G). In panels B to E, the thick line represents posterior means while the thin lines indicate 95% pointwise posterior credibility intervals. Solid lines refer to the results without effect selection while dashed lines indicates results with effect selection priors.

12.4.3 Decomposing an Interaction Surface

One of the main limitations in the analysis presented in the previous section is the restriction to a purely additive main effects model. In the following, we consider the analysis of a potentially complex interaction along the example of the age of the child and the age of the mother. To reduce computational complexity, we assumed

$$\text{stunting}_i \sim \mathcal{N}(\mu_i, \sigma^2)$$

where only the mean μ_i is related to a regression predictor while the error variance σ^2 is treated as a constant with an inverse gamma prior, more precisely $\sigma^2 \sim \mathcal{IG}(0.001, 0.001)$. For the mean, we consider an interaction surface of the form

$$\eta_{\mu i} = f(\text{cage}_i, \text{mage}_i)$$

where $f(\text{cage}_i, \text{mage}_i)$ is modeled as the tensor product of two univariate cubic penalized splines with second order random walk prior and 20 inner knots.

While the resulting interaction can be cast into the general framework presented in Section 12.2, it is often quite difficult to interpret the resulting estimate. Panel A in Figure 12.3 shows an image plot of the estimated interaction effect when a global variance parameter λ^2 with inverse gamma prior and hyperparameters $a = b = 0.001$ is considered. While there seems to be some indication of an interaction, it is hard to derive a sensible interpretation and it would be more convenient to decompose the interaction surface into some simplified effects plus the deviating interaction effect.

Utilizing the modular construction of tensor product interaction effects in [19], we can achieve such a decomposition by assigning linear constraints to the interaction surface to remove certain simpler cases while adding the removed parts explicitly as separate effects in the model. To illustrate this point, we considered the decomposed predictor

$$\begin{aligned}\eta_{\mu i} &= \alpha + \beta_1 \text{cage}_i + \beta_2 \text{mage}_i + \beta_3 \text{cage}_i \text{mage}_i + f_1(\text{cage}_i) + f_2(\text{mage}_i)\\ &\quad + \text{mage}_i f_3(\text{cage}_i) + \text{cage}_i f_4(\text{mage}_i) + f_5(\text{cage}_i, \text{mage}_i)\end{aligned}$$

where $\alpha, \beta_1, \beta_2, \beta_3$ are coefficients representing the intercept, linear effects for the age of the child and the age of the mother, and a parametric, linear interaction, f_1, f_2 are non-linear main effects that can be represented as univariate penalized splines (subject to a linear constraint to remove the linear main effects), f_3, f_4 are varying coefficient interactions also modeled based on univariate penalized splines (subject to a linear constraint to remove the parametric interaction term), and f_5 is the remaining interaction surface, after removing all other effects by linear constraints.

This decomposition already facilitates effect selection by visually inspecting the estimates for the different model components. We further substantiate this by placing spike-and-slab effect selection priors on all components of the decomposed model specification (except for the intercept), applying the same specifications and prior elicitation strategies as in the previous section. It turns out (see Figure 12.3 and Table 12.2) that the linear and non-linear main effect of the age of the child are the main component of the interaction surface while all other components of the effect decomposition are shrunken closely to zero when applying the effect selection prior (with the effect of the age of the mother as an exception that shows a small remaining effect).

		decomposed model		with effect selection		
	variable	pmean	95% CI	pmean	95% CI	I
μ	Intercept	-0.04	[-0.15, 0.04]	-0.03	[-0.12, 0.05]	-
	age of child	-0.37	[-0.47, -0.27]	-0.38	[-0.44, -0.32]	1
	age of mother	0.04	[-0.04, 0.12]	0.01	[-0.02, 0.05]	0.19
	age[child]*age[mother]	0.01	[-0.06, 0.08]	0.00	[-0.02, 0.02]	0.17

TABLE 12.2
Table of linear effects, posterior means with 95% credible intervals, on the mean effect of the decomposed model with and without effect selection. All covariates have been standardized. The last column denotes selection probability.

12.5 Other Regularization Priors for Functional Effects

In this section, we highlight two specific concepts of regularization or shrinkage of smooth effects based on regression splines: locally adaptive regularization (Section 12.5.1) and shrinkage towards a functional subspace (Section 12.5.2). The former has benefits when fitting a spline to a function of locally varying complexity while the latter shrinks the estimated function towards functions from a predefined family.

Although the introduced works limit their scope towards the mean effect of normally distributed responses with constant variance or to distributions from the exponential family, we find them nonetheless of high interest for the readership of this book. Both methods fit into the model structure defined in Section 12.2.1 and the regularization is changed by modifying the prior on $\boldsymbol{\beta}_{kj}$. However, shrinkage operates fundamentally different compared to the spike-and-slab approach. Unlike the spike-and-slab approach, shrinkage priors do not use discrete variables to decide on weak or strong regularization. Nonetheless, shrinkage priors are typically designed to have a large (potentially infinite) spike in the origin and heavy tails and, therefore, achieve a similar behavior with respect to the regularized parameters [2, 29].

Since we restrict the presentation in this part of the chapter to normally distributed responses with an additive predictor only on the mean effect, we suppress in the following the index for the notation of the distributional parameter, i.e., k, and furthermore denote the constant error variance with σ^2.

12.5.1 Locally Adaptive Regularization

Bayesian P-splines are based on B-spline basis functions with a moderately large number of equidistant knots. To avoid a wiggly functional estimate and to prevent overfitting, the spline coefficients are regularized with random walk priors which allow for flexible but smooth approximations of the underlying function. However, when the underlying function has locally varying complexity, using one global variance parameter is not sufficient. Locally adaptive regularization overcomes this issue by employing multiple variance parameters that can adapt to local phenomena. That means, for example, that the function is shrunken towards a constant effect for small values of the covariate but can be almost not regularized for large values of the covariate and thus adapt the pattern in the data well.

Examples for locally adaptive priors for Bayesian P-spline can, for example, be found in [20] or [33]. Both authors modify the original random walk prior to achieving local adaptivity. Recall, the Bayesian P-spline prior on the spline coefficients $\boldsymbol{\beta}_j$ is defined as

$$p(\boldsymbol{\beta}_j|\lambda_j^2) \propto \exp\left(-\frac{1}{2\lambda_j^2}\boldsymbol{\beta}_j^T \boldsymbol{K}_j \boldsymbol{\beta}_j\right)$$

with penalty matrix $\boldsymbol{K}_j$ constructed from the crossproduct of a difference matrix and an inverse gamma prior on λ_j^2. Equivalently, the prior can be expressed as a random walk of first or second order such that

$$\begin{aligned} \beta_{jl}|\beta_{jl-1} &= \beta_{jl-1} + u_{jl}, \quad l = 2, \dots, L && \text{(first order) or} \\ \beta_{jl}|\beta_{jl-1}, \beta_{jl-2} &= 2\beta_{jl-1} - \beta_{jl-1} + u_{jl}, \quad l = 3, \dots, L && \text{(second order)} \end{aligned}$$

with independent increments $u_{jl} \sim \mathcal{N}(0, \lambda_j^2)$ and flat priors on β_{j1} or β_{j1} and β_{j2}, respectively.

[20] suggest to replace the prior on the increments by $u_{jl} \sim \mathcal{N}(0, \lambda_j^2/\delta_{jl})$ with

$\delta_{jl} \sim \mathcal{G}(0.5, 0.5)$. On the one hand, this prior allows the variances in the random walk to adjust to local phenomenae in the underlying function and, on the other hand, allows for a better fit when the underlying function is highly oscillating. The latter is due to heavier tails in the marginal prior on $\beta_{jl}|\beta_{jl-1}$ (after marginalizing out δ_{jl} and λ_j^2).

The Normal-Exponential-Gamma prior on the differences of B-spline coefficients (NEG prior) is introduced by [33]. The increments are split into M subsequent blocks of sizes $S_1, S_2, \ldots, S_M$. Within each block, the variance is constant, such that

$$(u_{j2}, \ldots, u_{jL})^T \sim \mathcal{N}(0, \text{diag}(\lambda_{j1}^2 \boldsymbol{I}_{S_1}, \ldots, \lambda_{jM}^2 \boldsymbol{I}_{S_M})).$$

The prior on the variance parameters are independent Exponential-Gamma priors, in particular

$$\lambda_{jm}^2 | z_{jm} \sim \mathcal{E}xp(z_{jm}) \text{ for } m = 1, \ldots, M$$
$$z_{jm} | a_z, b_z \sim \mathcal{G}(a_z, b_z) \text{ for } m = 1, \ldots, M$$

where $a_z > 0$ and $b_z > 0$ are predefined hyperparameters. The number of variance blocks M as well as their sizes $S_1, \ldots, S_M$ can either be predefined or estimated during the inference procedure. For that and within an MCMC algorithm, the number of variance blocks is estimated via a reversible jump algorithm and in each MCMC iteration, a Metropolis-Hastings step determines the size of the variance blocks. The advantage of the second approach is the piecewise constant variance on the increments but comes at the cost of a more complex estimation procedure.

Making these regularization approaches available to smooth effects in additive predictors of all distributional parameters should be conceptually straightforward. However, future research has yet to substantiate this claim. Besides, the NEG prior is to the best of our knowledge not implemented in standard software for distributional regression.

12.5.2 Shrinkage towards a Functional Subspace

Shin et al. [34] introduce the functional horseshoe prior (fHS), a novel approach for shrinkage of functional smooth effects of a covariate in additive models. Instead of using the random walk prior on the spline coefficients and therefore shrinking towards a constant or a linear effect, the prior is altered such that the spline-based function gets shrunken towards a predefined functional subspace. This subspace is comprised of a family of parametric functions of the covariate. Examples for this subspace could be all linear functions, all quadratic functions, or even more complex functions like all trigonometric polynomials of a certain order.

The fundamental idea of this prior is to create a projection matrix to project the estimated function evaluated at the observed values of the covariate into the orthogonal complement of the used subspace. Then, a squared norm of the projected vector should be minimized while a good fit to the data is kept. This trade-off is done via a specific prior on the variance parameter.

The functional horseshoe prior is defined by altering the precision matrix in the prior of the spline coefficients and assigning a different prior to the variance parameter λ_j^2. Let $\boldsymbol{Z}_j$ be a basis of the functional subspace that is the target of the shrinking. For example, this could be $\boldsymbol{Z}_j = (\boldsymbol{1}, \boldsymbol{x}_j)$ for a linear effect or $\boldsymbol{Z}_j = (\boldsymbol{1}, \boldsymbol{x}_j, \boldsymbol{x}_j^2)$ for a quadratic effect. Here $\boldsymbol{x}_j = (x_{1j}, \ldots, x_{nj})^T$ denotes the vector of the values of the j-th covariate for all observations and the square operator is meant to be applied element-wise. Then, $\boldsymbol{P}_j = \boldsymbol{Z}_j(\boldsymbol{Z}_j^T \boldsymbol{Z}_j)^{-1} \boldsymbol{Z}_j^T$ is the projection matrix into this subspace and $\bar{\boldsymbol{P}}_j = \boldsymbol{I} - \boldsymbol{P}_j$ is the projection matrix into the orthogonal complement. The prior density of the spline coefficients and its hyperparameter is defined as

$$p(\boldsymbol{\beta}_j|\lambda_j, \sigma^2) \propto \exp\left(-\frac{1}{2\lambda_j^2\sigma^2}\boldsymbol{\beta}_j^T \boldsymbol{B}_j^T \bar{\boldsymbol{P}}_j \boldsymbol{B}_j \boldsymbol{\beta}_j\right)$$

and

$$p(\lambda_j|a_j, b_j) \propto \frac{(\lambda_j^2)^{b_j - \frac{1}{2}}}{(1+\lambda_j^2)^{a_j+b_j}} \mathbb{I}\left\{\lambda_j \in (0, \infty)\right\}$$

with $a_j, b_j > 0$. The latter implies a Beta-prime prior on λ_j^2. Further defining $\boldsymbol{K}_j = \boldsymbol{B}_j^T \bar{\boldsymbol{P}}_j \boldsymbol{B}_j$ yields a structure similar to Equation (12.2). Besides the differently targeted shrinking, the fHS prior also conditions on σ^2. In the spike-and-slab paradigm for linear effects, George and McCulloch [11] study the effect of conditioning on σ^2 and find advantages for both specifications. One particular advantage is the analytical simplification when conditioning on σ^2.

This advantage also exists in the present context: considering the simplified model with just one covariate effect (that is $J = 1$) and provided that $\text{span}(\boldsymbol{Z}_j) \subsetneq \text{span}(\boldsymbol{B}_j)$, it can be shown that the posterior mean of the spline function is given by

$$E(\boldsymbol{B}_j\boldsymbol{\beta}_j|\boldsymbol{y}, \lambda_j) = \left(\frac{1}{1+\lambda_j^2}\boldsymbol{P}_j + \left(1 - \frac{1}{1+\lambda_j^2}\right)\boldsymbol{P}_j^B\right)\boldsymbol{y} \tag{12.10}$$

where $\boldsymbol{P}_j^B = \boldsymbol{B}_j(\boldsymbol{B}_j^T\boldsymbol{B}_j)^{-1}\boldsymbol{B}_j^T$ is the projection matrix into the space spanned by the basis function of the spline. This formulation gives insight into the strength of the shrinkage as the result of the shrinking process is, in expectation, a weighted average of the spline solution and the parametric solution with the shrinkage weight $\kappa_j = 1/(1+\lambda_j^2)$. The closer κ_j is to 1, the stronger is the shrinkage towards the parametric solution while $\kappa_j \to 0$ allows the solution to be the unpenalized spline solution.

[34] show that the prior from above implies a beta prior on the shrinkage weight. In particular, the implied prior on κ_j is $\mathcal{B}eta(a_j, b_j)$. This prior has the typical horseshoe shape when $a_j = b_j = 0.5$ and thus justifies the prior's name. The authors suggest setting $a_j = 0.5$ and choosing b_j according to the number of basis functions as well as the number of observations.

We find the functional horseshoe prior very promising since it allows the user to diverge from the usual shrinkage target, that is, the null effect. Instead, parametric functions or even semi-parametric functions can be specified as the goal of the shrinkage procedure, which might make shrinkage a suitable procedure in more areas of application. However, it seems that integrating the fHS prior in the class of distributional regression is not straightforward since the scale parameter of the error variance σ^2 is part of the prior specification while the class of distributional regression features models with non-constant error variance. This makes the interpretation of the shrinkage weight as shown in Equation (12.10) more difficult. Besides, distributional regression links structured additive predictors to all distributional parameters and is not limited to normally distributed responses. The fHS prior needs to be adapted to those situations and further research is required to investigate the properties of the fHS prior in this context.

We demonstrate here briefly the functional shrinkage with a simulated data example. For that, we consider two simple scenarios each with 50 observations and one covariate. The two scenarios differ only in the underlying data-generating function. In the first scenario, the underlying function is a third-degree polynomial ($f_{s1}(x) = 1 - x + x^2 + 2x^3$) and in the second scenario the function corresponds to the sine function ($f_{s2}(x) = 1 + \sin(2\pi x)$).

In both scenarios, we generate the response, conditioned on the covariate x, independently according to $y|x \sim \mathcal{N}(f(x), 0.5^2)$ where f is either f_{s1} or f_{s2}. The covariate is uniformly distributed between -1 and 1.

To contrast different approaches, we estimate three models based on a B-spline of third order with 10 basis functions with different priors (a non-informative prior, the functional horseshoe prior, and the Bayesian P-spline prior) on the spline coefficients. Thereby, we define the functional subspace that is target of the shrinkage by $\boldsymbol{Z} = [\boldsymbol{1}, \boldsymbol{x}, \boldsymbol{x}^2, \boldsymbol{x}^3]$. Thus, we expect to see strong shrinkage in the first scenario and weak or no shrinkage in the second scenario. Besides, we show the estimated effect of the best fitting parametric model (in terms of least squares) when restricting to functions within the null space (cubic polynomials).

The posterior means of the estimated functions are shown in Figure 12.4. We observe that the spline with fHS prior resembles a cubic function for the first scenario. In the second scenario, the function estimated with fHS diverges from the null space and is able to detect the underlying pattern. This is also reflected in the posterior mean of the shrinkage coefficient $\bar{\kappa} = 0.97$ and $\bar{\kappa} = 0.14$ in scenario one and two, respectively. Interpreting these numbers as indicators for effect selection as describe in Section 12.3.3, we conclude that a cubic polynomial is sufficient to model the effect of the covariate in scenario one. In scenario two, we have $\bar{\kappa} = 0.14 < 0.5$ which implies that the effect is selected and thus a more complex or different relationship between the response and the covariate seem realistic.

The P-spline and B-spline seem to detect the underlying function, although the B-spline is quite wiggly in the first scenario and the P-spline is not pushed towards a cubic polynomial. The parametric estimate fails as anticipated when misspecified. For simplicity, we employed STAN [1] for the estimation of the P-spline and the spline with fHS prior.

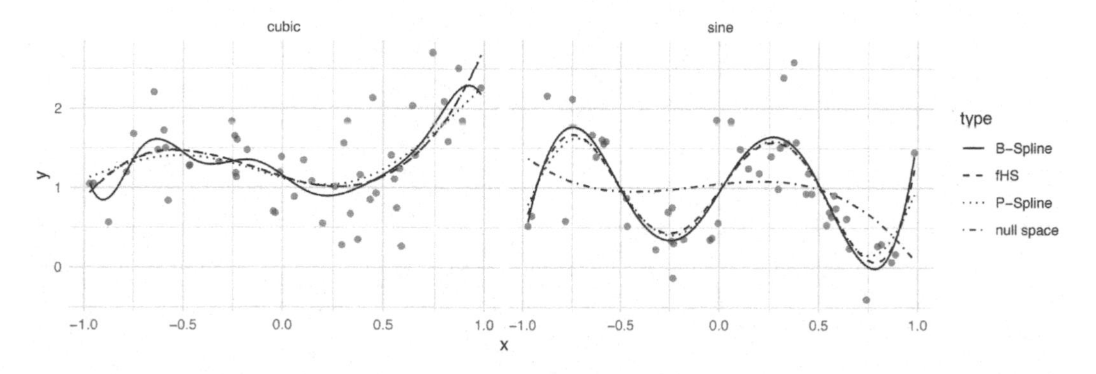

FIGURE 12.4
Posterior mean estimates of the functions based on B-splines with different priors on the spline coefficients as well as the parametric solution (null space). The left panel displays the first scenario while the right panel shows the second scenario.

12.6 Summary and Discussion

In this chapter, we have reviewed and investigated some approaches for effect selection and regularization in the broad class of structured additive distributional regression. Our focus has been on spike-and-slab-type prior structures and in an application to childhood undernutrition, we empirically demonstrated the ability of a certain class of spike-and-slab priors to assist researchers in complex model choice and effect selection decisions. This also has the advantage of effectively controlling potential overfitting. While already standard inverse gamma priors automatically adapt to the required complexity of nonparametric effects in structured additive regression, the total number of effects rapidly grows in distributional regression settings with multiple predictors such that overfitting may still be an issue. The ability to automatically deselect some effects based on spike-and-slab effect selection priors provides an effective means to combat this.

Although the adaptation of alternative prior structures, such as the ones discussed in Section 12.5 for achieving adaptive smoothing or smoothing towards a pre-specified subspace should conceptually be easy to transfer to a distributional context, several challenges have to be faced when turning this into practice. On the one hand, these challenges often relate to numerical efficiency and computational aspects such as achieving desirable convergence and mixing properties. Furthermore, many regularization priors explicitly include the error variance as quantification of uncertainty in their prior formulation, which is, in general, not possible in distributional regression models. As a consequence, we firmly believe that there still is considerable room for improving the specification of informative priors in distributional regression models to foster desirable properties of the model estimates and to assist applied scientists in setting up sensible models.

Bibliography

[1] B. Carpenter, A. Gelman, M. D. Hoffman, D. Lee, B. Goodrich, M. Betancourt, M. Brubaker, J. Guo, P. Li, and A. Riddell. Stan: a probabilistic programming language. *Journal of Statistical Software*, 76(1):1–32, 2017.

[2] C. M. Carvalho, N. G. Polson, and J. G. Scott. The horseshoe estimator for sparse signals. *Biometrika*, 97(2):465–480, 2010.

[3] R.-B. Chen, C.-H. Chu, S. Yuan, and Y. N. Wu. Bayesian sparse group selection. *Journal of Computational and Graphical Statistics*, 25(3):665–683, 2016.

[4] H. Chipman. Bayesian variable selection with related predictors. *The Canadian Journal of Statistics*, 107(1):17–36, 1996.

[5] R. Cottet, R. J. Kohn, and D. J. Nott. Variable selection and model averaging in semiparametric overdispersed generalized linear models. *Journal of the American Statistical Association*, 103(482):661–671, 2008.

[6] L. Fahrmeir, T. Kneib, and S. Konrath. Bayesian regularization in structured additive regression: a unifying perspective on shrinkage, smoothing and predictor selection. *Statistics and Computing*, 20:203–219, 2010.

[7] L. Fahrmeir, T. Kneib, S. Lang, and B. Marx. *Regression*. Springer, Berlin, Heidelberg, 2013.

[8] S. Frühwirth-Schnatter and R. Tüchler. Bayesian parsimonious covariance estimation for hierarchical linear mixed models. *Statistics and Computing*, 18:1–13, 2008.

[9] S. Frühwirth-Schnatter and H. Wagner. Bayesian variable selection for random intercept modeling of Gaussian and non-Gaussian data. In J. M. Bernardo, M. J. Bayarri, J. O. Berger, A. P. Dawid, D. Heckerman, A. F. M. Smith, and M. West, editors, *Bayesian Statistics 9*, pages 165–200. Oxford, 2011.

[10] E. I. George and R. E. McCulloch. Variable selection via Gibbs sampling. *Journal of the American Statistical Association*, 88:881–889, 1993.

[11] E. I. George and R E. McCulloch. Approaches to Bayesian variable selection. *Statistica Sinica*, 7:339–373, 1997.

[12] P. J. Griffin and J. E. Brown. Inference with normal-gamma prior distributions in regression problems. *Bayesian Analysis*, 5(1):171–188, 2010.

[13] H. Ishwaran and J. S. Rao. Detecting differentially expressed genes in microarrays using Bayesian model selection. *Journal of the American Statistical Association*, 98:438–455, 2003.

[14] N. Klein, M. Carlan, T. Kneib, S. Lang, and H. Wagner. Bayesian effect selection in structured additive distributional regression models. *Bayesian Analysis*, 16(2):545–573, 2021.

[15] N. Klein and T. Kneib. Simultaneous inference in structured additive conditional copula regression models: a unifying Bayesian approach. *Statistics and Computing*, 26(4):841–860, 2016.

[16] N. Klein, T. Kneib, S. Klasen, and S. Lang. Bayesian structured additive distributional regression for multivariate responses. *Journal of the Royal Statistical Society. Series C (Applied Statistics)*, 64(4):569–591, 2015.

[17] N. Klein, T. Kneib, and S. Lang. Bayesian generalized additive models for location, scale, and shape for zero-inflated and overdispersed count data. *Journal of the American Statistical Association*, 110(509):405–419, 2015.

[18] N. Klein, T. Kneib, S. Lang, and A. Sohn. Bayesian structured additive distributional regression with an application to regional income inequality in Germany. *The Annals of Applied Statistics*, 9(2):1024–1052, 2015.

[19] T. Kneib, N. Klein, S. Lang, and N. Umlauf. Modular regression - a Lego system for building structured additive distributional regression models with tensor product interactions. *TEST*, 28(1):1–39, 2019.

[20] S. Lang and A. Brezger. Bayesian P-splines. *Journal of Computational and Graphical Statistics*, 13(1):183–212, 2004.

[21] S. Lang, N. Umlauf, P. Wechselberger, K. Harttgen, and T. Kneib. Multilevel structured additive regression. *Statistics and Computing*, 24(2):223–238, 2014.

[22] G. Malsiner-Walli and H. Wagner. Comparing spike and slab priors for Bayesian variable selection. *Austrian Journal of Statistics*, 40:241–264, 2011.

[23] P. Michaelis, N. Klein, and T. Kneib. Bayesian multivariate distributional regression with skewed responses and skewed random effects. *Journal of Computational and Graphical Statistics*, 27(3):602–611, 2018.

[24] T. J. Mitchell and J. J. Beauchamp. Bayesian variable selection in linear regression. *Journal of the American Statistical Association*, 83:1023–1032, 1988.

[25] R. B. O'Hara and M. J. Sillanpää. A review of Bayesian variable selection methods: What, how and which. *Bayesian Analysis*, 4:85–118, 2009.

[26] A. Panagiotelis and M. S. Smith. Bayesian density forecasting of intraday electricity prices using multivariate skew t distributions. *International Journal of Forecasting*, 24(4):710–727, 2008.

[27] A. Panagiotelis and M. S. Smith. Bayesian identification, selection and estimation of functions in high-dimensional additive models. *Journal of Econometrics*, 143:291–316, 2008.

[28] T. Park and G. Casella. The Bayesian Lasso. *Journal of the American Statistical Association*, 103(482):681–686, 2008.

[29] N. G. Polson and J. G. Scott. Shrink globally, act locally: sparse Bayesian regularization and prediction. In J. M. Bernardo, M. J. Bayarri, J. O. Berger, A. P. Dawid, D. Heckerman, A. F. M. Smith, and M. West, editors, *Bayesian Statistics 9*. Oxford, 2010.

[30] B. J. Reich, C. B. Storlie, and H. D. Bondell. Variable selection in Bayesian smoothing spline ANOVA models: application to deterministic computer codes. *Technometrics*, 51(2):110–120, 2009.

[31] H. Rue and L. Held. *Gaussian Markov random fields.* Chapman & Hall / CRC, New York, 1 edition, 2005.

[32] F. Scheipl, L. Fahrmeir, and T. Kneib. Spike-and-slab priors for function selection in structured additive regression models. *Journal of the American Statistical Association*, 107(500):1518–1532, 2012.

[33] F. Scheipl and T. Kneib. Locally adaptive Bayesian P-splines with a Normal-Exponential-Gamma prior. *Computational Statistics & Data Analysis*, 53(10):3533–3552, 2009.

[34] M. Shin, A. Bhattachrya, and V. E. Johnson. Functional horseshoe priors for subspace shrinkage. *Journal of the American Statistical Association*, 115(532):1784–1797, 2020.

[35] R. Tibshirani. Regression shrinkage and selection bia the Lasso. *Journal of the Royal Statistical Society: Series B (Methodological)*, 58(1):267–288, 1996.

[36] P. Yau, R. Kohn, and S. Wood. Bayesian variable selection and model averaging in high-dimensional multinomial nonparametric regression. *Journal of Computational and Graphical Statistics*, 12(1):23–54, 2003.

[37] H. Zhu, M. Vannucci, and D. D. Cox. A Bayesian hierarchical model for classification with selection of functional predictors. *Biometrics*, 66(2):463–473, 2010.

13

Sparse Bayesian State-Space and Time-Varying Parameter Models

Sylvia Frühwirth-Schnatter
Vienna University of Economics and Business (Austria)

Peter Knaus
Vienna University of Economics and Business (Austria)

CONTENTS

In this chapter, we review variance selection for time-varying parameter (TVP) models for univariate and multivariate time series within a Bayesian framework. We show how both continuous as well as discrete spike-and-slab shrinkage priors can be transferred from variable selection for regression models to variance selection for TVP models by using a non-centered parametrization. We discuss efficient MCMC estimation and provide an application to US inflation modeling.

DOI: 10.1201/9781003089018-13

13.1 Introduction

Time-varying parameter (TVP) models and, more generally, state space models are widely used in time series analysis to deal with model coefficients that change over time. This ability to capture gradual changes is one of state space models greatest advantages. The flipside of this high degree of flexibility, however, is that they run the risk of overfitting with a growing number of coefficients, as many of them might, in reality, be constant over the entire observation period. This will be exemplified in the present chapter with an economic application. We will model US inflation through a TVP Phillips curve, where, out of 18 potentially time-varying coefficients, only a single one actually changes over time. We will show that allowing static coefficients to be time-varying leads to a considerable loss of statistical efficiency, both in uncertainty quantification for the parameters and forecasting future time series observations. We will also show that substantial statistical efficiency can be gained by applying a Bayesian estimation strategy that is able to single out parameters that are indeed constant or even insignificant.

Identifying constant coefficients in a TVP model amounts to a *variance selection* problem, involving a decision on whether the variances of the shocks driving the dynamics of a time-varying parameter are equal to zero. Variance selection in latent variable models is known to be a non-regular problem within the framework of classical statistical hypothesis testing [25]. The introduction of shrinkage priors for the variances of a TVP model within a Bayesian framework has proven to be a very useful strategy which is capable of automatically reducing time-varying coefficients to static ones if the model overfits.

In pioneering work, [18] reformulated the *variance selection* problem for state space models as a *variable selection* problem in the so-called non-centered parametrization of the TVP model. This insight established a general strategy for extending shrinkage priors from standard regression analysis to this more general framework. For variance selection in "sparse" state space and TVP models, [18] employed discrete spike-and-slab priors, [2] relied on the Bayesian Lasso prior, [5] applied the normal-gamma prior of [23] and [6] introduced the triple gamma prior, which is related to the normal-gamma-gamma prior [24] and contains the horseshoe prior [9] as a special case.

The present chapter reviews this literature, starting in Section 13.2 with univariate time-varying parameter models. In particular, we will demonstrate that the commonly used inverse gamma prior on the process variances prevents variance selection. Using a ridge prior in the non-centered TVP model instead of an inverse gamma prior provides a simple, yet useful alternative. The ridge prior can be translated into a gamma prior for the variances and leads to more reliable uncertainty quantification in parameter estimation and forecasting for sparse state space models. Starting from the ridge prior, continuous shrinkage priors for variance selection are discussed in Section 13.3, whereas Section 13.4 discusses discrete spike-and-slab priors. In both sections, we also review strategies for efficient Markov chain Monte Carlo (MCMC) estimation, which is even more challenging for state space models than for standard regression models. Section 13.5 discusses extensions to multivariate time series, including TVP Bayesian vector autoregressive models and TVP Cholesky stochastic volatility models, shows how to compare various shrinkage priors through log predictive density scores and addresses the issues of classifying coefficients into dynamic or constant ones. Section 13.6 concludes with a brief discussion.

13.2 Univariate Time-Varying Parameter Models

13.2.1 Motivation and Model Definition

In this section, we consider time-varying parameter (TVP) models for a univariate time series y_t. For $t = 1, \ldots, T$, we have that

$$\begin{aligned} \boldsymbol{\beta}_t &= \boldsymbol{\beta}_{t-1} + \mathbf{w}_t, \qquad & \mathbf{w}_t &\sim \mathcal{N}_p\left(\mathbf{0}, \mathbf{Q}\right), \\ y_t &= \mathbf{x}_t \boldsymbol{\beta}_t + \varepsilon_t, \qquad & \varepsilon_t &\sim \mathcal{N}\left(0, \sigma^2\right), \end{aligned} \tag{13.1}$$

where $\boldsymbol{\beta}_t = (\beta_{1t}, \ldots, \beta_{pt})^\top$ is a latent state variable and the covariance matrix $\mathbf{Q} = \text{Diag}\left(\theta_1, \ldots, \theta_p\right)$ of the innovations $\mathbf{w}_t$ is diagonal. $\mathbf{x}_t = (x_{1t}, \ldots, x_{pt})$ is a p-dimensional row vector containing the explanatory variables at time t. The variables x_{jt} can be exogenous (i.e., determined outside the model) control variables and/or be equal to lagged values of y_t. Usually, one of the variables, say x_{1t}, corresponds to the intercept, but an intercept need not be present. In Section 13.5, this approach is extended to multivariate time series $\mathbf{y}_t$.

To fully specify the model, a distribution has to be defined for the initial value $\boldsymbol{\beta}_0$ of the state process, with a typical choice being a normal distribution, e.g., $\boldsymbol{\beta}_0 \sim \mathcal{N}_p\left(\boldsymbol{\beta}, \mathbf{Q}\right)$, with initial expectation $\boldsymbol{\beta} = (\beta_1, \ldots, \beta_p)^\top$. An alternative choice is to assume a diffuse prior with fixed initial expectation and a very uninformative prior covariance matrix, e.g., $\boldsymbol{\beta}_0 \sim \mathcal{N}_p\left(\mathbf{0}, 10^5 \cdot I_p\right)$ where I_p is the p-dimensional identity matrix. However, such a choice is not recommended for TVP models where overfitting presents a concern.

The goal is to recover the unobserved state process $\boldsymbol{\beta}_0, \ldots, \boldsymbol{\beta}_T$ given the observed time series $\mathbf{y} = (y_1, \ldots, y_T)$. If $\boldsymbol{\beta}$, $\mathbf{Q}$ and σ^2 were known, this is easily achieved by the famous Kalman filter and smoother [30]. For illustration, a time series y_t is generated from model (13.1) with $T = 200$, $p = 3$, $x_{1t} = 1$, $x_{jt} \sim \mathcal{N}(0, 1)$, $j = 2, 3$, $\sigma^2 = 1$, $(\beta_1, \beta_2, \beta_3) = (1, -0.5, 0)$ and $(\theta_1, \theta_2, \theta_3) = (0.02, 0, 0)$. The paths of the hidden process $\boldsymbol{\beta}_t$ are reconstructed using the Kalman filter and smoother based on the true values of $\boldsymbol{\beta}$, θ_1 and σ^2 and very small values for $\theta_2 = \theta_3 = 10^{-6}$ and compared to the true paths in the left-hand side of Figure 13.1. Since the marginal posterior of $\boldsymbol{\beta}_t|\mathbf{y}$ is a Gaussian distribution for each t, point-wise credible regions for $\boldsymbol{\beta}_t$ are easily obtained which are very helpful for uncertainty quantification. Although the TVP model used for estimation overfits, the Kalman smoother is rather accurate in recovering the true paths and clearly indicates that the last coefficients are constant, *assuming* that θ_2 and θ_3 are very close to 0.

However, in real-world applications, the variances θ_j are unknown and estimated from the observed time series, together with the entire path $\mathbf{z} = (\boldsymbol{\beta}_0, \ldots, \boldsymbol{\beta}_T)$. As evident from the Kalman filter, the variances $\mathbf{Q}$ of the innovations $\mathbf{w}_t$ play an important role in quantifying the loss from propagating the filtering density $\boldsymbol{\beta}_{t-1}|\mathbf{y}^{t-1} \sim \mathcal{N}_p\left(\mathbf{m}_{t-1|t-1}, \mathbf{P}_{t-1|t-1}\right)$, given $\mathbf{y}^{t-1} = (y_1, \ldots, y_{t-1})$, into the future to forecast $\boldsymbol{\beta}_t$:

$$\boldsymbol{\beta}_t|\mathbf{y}^{t-1} \sim \mathcal{N}_p\left(\mathbf{m}_{t-1|t-1}, \mathbf{P}_{t-1|t-1} + \mathbf{Q}\right).$$

A comparably minor change of $\mathbf{Q}$ can have a strong effect on uncertainty quantification. For instance, assuming $\theta_2 = \theta_3 = 0.001$ (instead of 10^{-6}) for the simulated data has a huge effect on the recovered paths, as shown in the right-hand side of Figure 13.1. Not only are the credible intervals much broader, we can also no longer be sure if the two coefficients β_{2t} and β_{3t} are time-varying or constant.

In a maximum likelihood framework, the Kalman filter is used to compute the likelihood function, which is maximized to obtain estimates of $\theta_1, \ldots, \theta_p$, σ^2, and $\beta_1, \ldots, \beta_p$ (if the

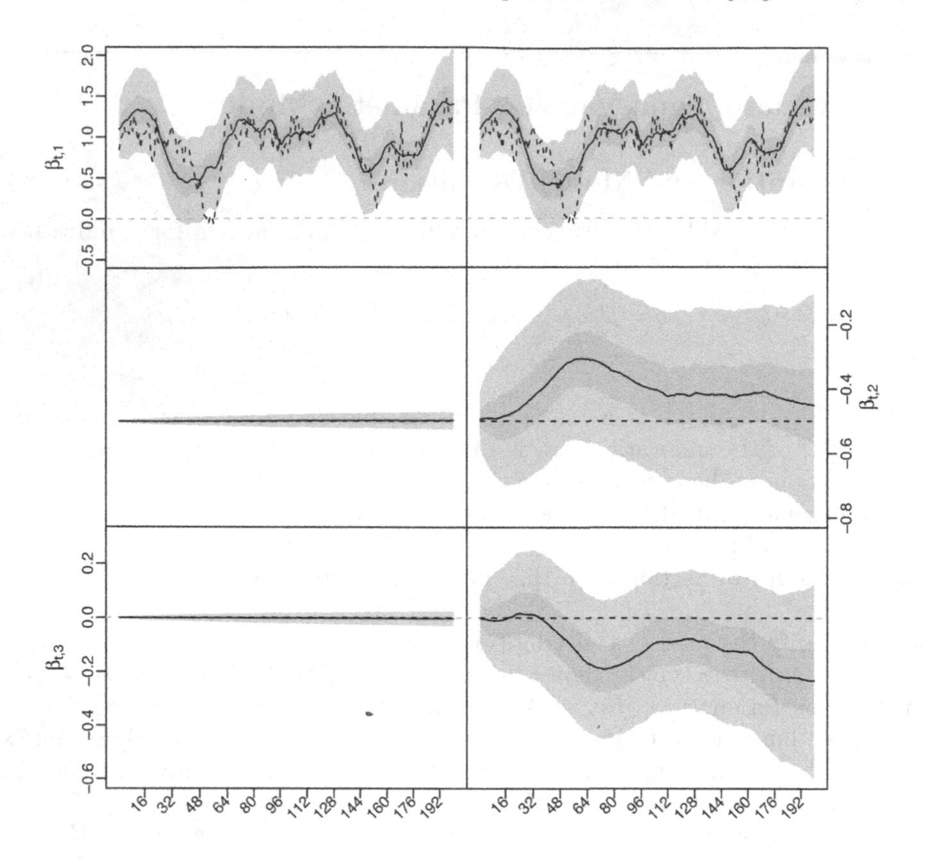

FIGURE 13.1
Recovery of the hidden parameters of a TVP model for simulated data using the Kalman filter and smoother in an overfitting TVP model with $\theta_2 = \theta_3 = 10^{-6}$ (left-hand side) and $\theta_2 = \theta_3 = 0.001$ (right-hand side). The gray shaded regions represent pointwise 95% and 50% credible intervals, respectively, while the black solid line represents the pointwise median. The black dashed line represents the true hidden parameter values.

initial means are unknown). Reconstructing $\mathbf{z} = (\boldsymbol{\beta}_0, \ldots, \boldsymbol{\beta}_T)$ then operates conditional on these estimates, see e.g., [25].

For Bayesian inference, priors are chosen for $\theta_1, \ldots, \theta_p$, σ^2, and $\beta_1, \ldots, \beta_p$. Given time series observations $\mathbf{y} = (y_1, \ldots, y_T)$, the joint posterior distribution $p(\mathbf{z}, \boldsymbol{\beta}, \mathbf{Q}, \sigma^2|\mathbf{y})$ is the object of interest from which marginal posteriors $p(\boldsymbol{\beta}_t|\mathbf{y})$ are derived for each t. These can be used for uncertainty quantification as in Figure 13.1, while also taking uncertainty in the model parameters into account. Different algorithms have been developed to sample from the joint posterior $p(\mathbf{z}, \boldsymbol{\beta}, \mathbf{Q}, \sigma^2|\mathbf{y})$, in particular two-block Gibbs samplers that alternate between drawing from $p(\mathbf{z}|\boldsymbol{\beta}, \mathbf{Q}, \sigma^2, \mathbf{y})$ using forward-filtering, backward-sampling (FFBS) [7, 14] and drawing from $p(\boldsymbol{\beta}, \mathbf{Q}, \sigma^2|\mathbf{z}, \mathbf{y})$.

Both maximum likelihood (ML) and Bayesian inference work well for TVP models where all state variables β_{jt} are dynamic. If one of the variances θ_j is equal to 0, ML estimation leads to a non-regular testing problem, since the true value lies on the boundary of the parameter space [25]. As opposed to this, Bayesian inference is able to deal with such sparse TVP models and, more generally, sparse state space models. The two main challenges from the Bayesian perspective are the choice of an appropriate prior for the variances θ_j and computational challenges with regards to efficient MCMC estimation.

13.2.2 The Inverse Gamma Versus the Ridge Prior

A popular prior choice for the process variance θ_j is the inverse gamma distribution,

$$\theta_j \sim \mathcal{IG}(s_0, S_0), \tag{13.2}$$

which is often applied with very small hyperparameters, e.g., $s_0 = S_0 = 0.001$ [42]. Given the latent process $(\beta_{j0}, \ldots, \beta_{jT})$, this prior is conditionally conjugate in the so-called centered parametrization (13.1), since the density $p(\beta_{j0}, \ldots, \beta_{jT}|\theta_j)$ is the kernel of an inverse gamma distribution. Hence, prior (13.2) leads to an inverse gamma posterior distribution $p(\theta_j|\beta_{j0}, \ldots, \beta_{jT})$. However, this prior performs poorly when dealing with a sparse TVP model, as it is bounded away from zero, making it incapable of inducing strong shrinkage [18].

The effect of choosing a specific prior becomes more apparent when we rewrite model (13.1) in the non-centered parametrization introduced in [18]:

$$\begin{aligned} \tilde{\boldsymbol{\beta}}_t &= \tilde{\boldsymbol{\beta}}_{t-1} + \tilde{\mathbf{w}}_t, \qquad \tilde{\mathbf{w}}_t \sim \mathcal{N}_p(\mathbf{0}, I_p), \\ y_t &= \mathbf{x}_t\boldsymbol{\beta} + \mathbf{x}_t \text{Diag}\left(\sqrt{\theta}_1, \ldots, \sqrt{\theta}_p\right)\tilde{\boldsymbol{\beta}}_t + \varepsilon_t, \quad \varepsilon_t \sim \mathcal{N}\left(0, \sigma^2\right), \end{aligned} \tag{13.3}$$

with initial distribution $\tilde{\boldsymbol{\beta}}_0 \sim \mathcal{N}_p(\mathbf{0}, I_p)$. An affine transformation connects the two parametrizations:

$$\beta_{jt} = \beta_j + \sqrt{\theta}_j\tilde{\beta}_{jt}, \quad t = 0, \ldots, T, \quad j = 1, \ldots, p. \tag{13.4}$$

Evidently, both representations are equivalent, and we can specify a prior either on the variances θ_j in (13.1) or on the scale parameters $\sqrt{\theta}_j$ in (13.3). Since the conjugate prior for $\sqrt{\theta}_j$ in the non-centered parametrization (13.3) is the normal distribution, the scale parameter $\sqrt{\theta}_j$ is assumed to be Gaussian:

$$\sqrt{\theta}_j|\sigma^2 \sim \mathcal{N}\left(0, \sigma^2\tau\right) \quad \Leftrightarrow \quad \theta_j|\sigma^2 \sim \mathcal{G}\left(\frac{1}{2}, \frac{1}{2\tau\sigma^2}\right). \tag{13.5}$$

Here, $\sqrt{\theta}_j \in \mathbb{R}$ is allowed to take on both positive and negative values. This implies that $\theta_j = (\sqrt{\theta}_j)^2$ follows a re-scaled χ^2_1-distribution. [16] introduced such a shrinkage prior (with fixed scale parameter τ) for the process variance in a univariate TVP model (that is $p = 1$), and [18] extended this idea to state space models with $p > 1$. Alternatively, it can be assumed that the prior scale is independent of σ^2, i.e.,

$$\sqrt{\theta}_j \sim \mathcal{N}(0, \tau) \quad \Leftrightarrow \quad \theta_j \sim \mathcal{G}\left(\frac{1}{2}, \frac{1}{2\tau}\right). \tag{13.6}$$

As shown by [37], such a prior has certain advantages compared to (13.5) and allows for the introduction of stochastic volatility in model (13.1), see [33] and Section 13.5.1.

From the viewpoint of variable selection, prior (13.6) is a ridge prior in a standard regression model, conditional on the hidden path $\mathbf{z} = (\tilde{\boldsymbol{\beta}}_0, \ldots, \tilde{\boldsymbol{\beta}}_T)$. Many variable selection priors have been introduced for standard regression models (albeit with known rather than latent regressors), see [4] for a recent review. Given the non-centered parametrization (13.3), any of these priors can be, in principle, applied in the context of sparse TVP and state space

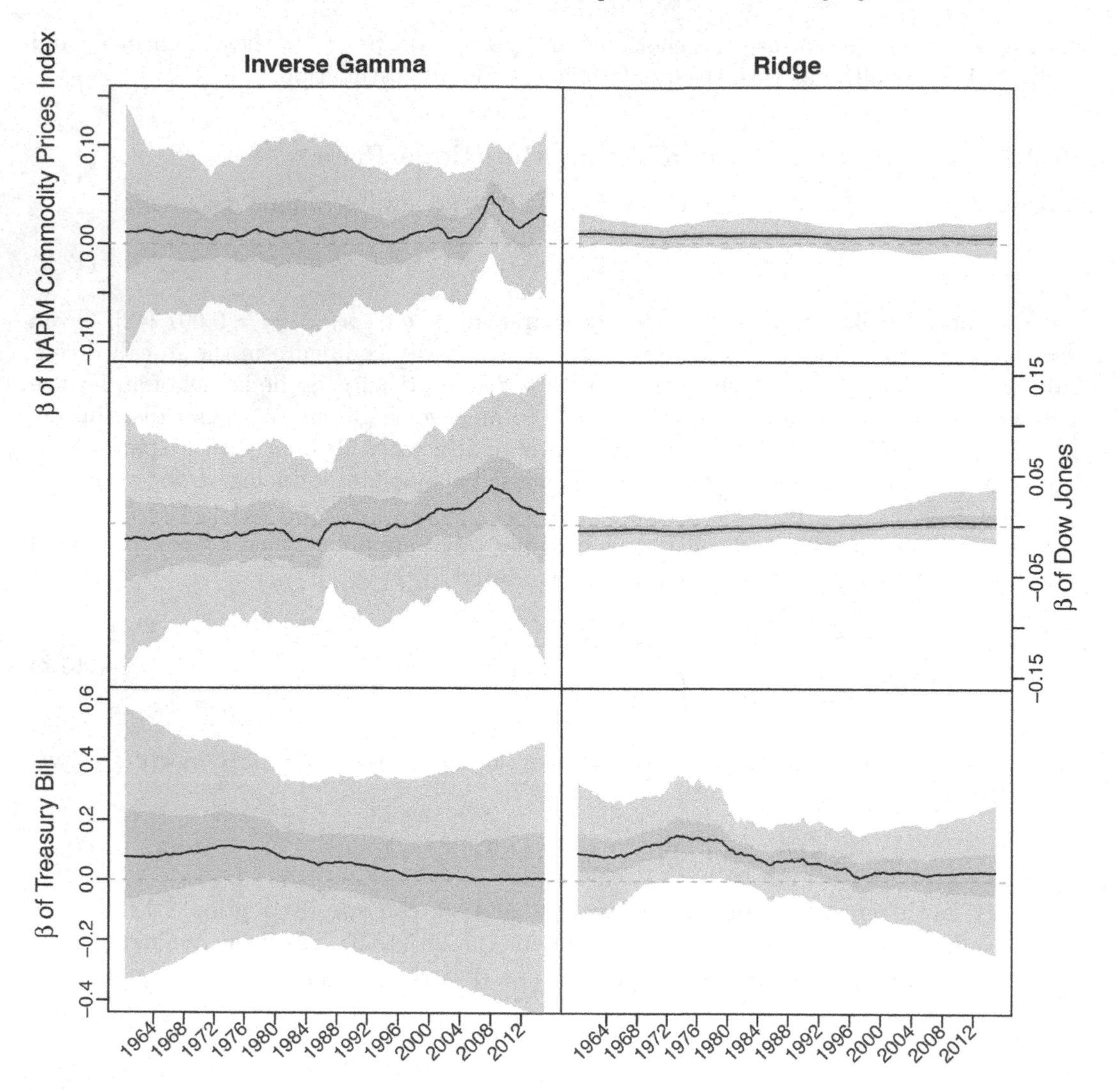

FIGURE 13.2
Recovery of the time-varying parameters for the inflation data under the inverse gamma $\theta_j \sim \mathcal{IG}(0.001, 0.001)$ and under the ridge prior $\theta_j \sim \mathcal{G}(0.5, 10)$. The gray shaded regions represent pointwise 95% and 50% credible intervals, respectively, while the black line represents the pointwise median.

models for variance selection. And, indeed, the literature has seen an increasing number of papers following this lead [2, 5, 6, 16, 18].

Shrinking θ_j toward the boundary value is achieved by shrinking $\sqrt{\theta}_j$ toward 0 (which is an interior point of the parameter space in the non-centered parametrization). For a sparse state space model, prior (13.6) substitutes the inverse gamma prior (13.2) with a gamma prior. This change in the prior specification is negligible for truly dynamic models, where the posterior distribution $p(\mathbf{z}, \boldsymbol{\beta}, \mathbf{Q}, \sigma^2|\mathbf{y})$ is fairly robust to prior choices $p(\theta_j)$, but has a considerable effect on uncertainty quantification for the unknown path $\mathbf{z}$ for a sparse state space model. This is illustrated in Figure 13.2, where the gamma prior $\theta_j \sim \mathcal{G}(0.5, 10)$ is

Algorithm 1 MCMC sampling in the non-centered parametrization of a TVP model under the ridge prior.

(a) sample the latent variables $\mathbf{z} = (\tilde{\beta}_0, \ldots, \tilde{\beta}_T)$ conditional on the model parameters $\boldsymbol{\alpha} = (\beta_1, \ldots, \beta_p, \sqrt{\theta}_1, \ldots, \sqrt{\theta}_p)$ and σ^2 from $\mathbf{z}|\boldsymbol{\alpha}, \sigma^2, \mathbf{y}$, using, e.g., FFBS;

(b) sample $(\boldsymbol{\alpha}, \sigma^2)$ conditional on $\mathbf{z}$:

  (b-1) sample σ^2, respectively, from the inverse gamma density $\sigma^2|\mathbf{z}, \mathbf{y}$ or $\sigma^2|\boldsymbol{\alpha}, \mathbf{z}, \mathbf{y}$ depending on whether the ridge priors' scale depends on σ^2 or not;

  (b-2) sample $\boldsymbol{\alpha}$ from the multivariate Gaussian $\boldsymbol{\alpha}|\sigma^2, \mathbf{z}, \mathbf{y}$.

compared to the inverse gamma prior $\theta_j \sim \mathcal{IG}(0.001, 0.001)$ for the inflation data that will be discussed in detail in Section 13.3.3.

13.2.3 Gibbs Sampling in the Non-Centered Parametrization

A two-block Gibbs sampler is available to sample the latent variables $\mathbf{z} = (\tilde{\beta}_0, \ldots, \tilde{\beta}_T)$ and the model parameters $\boldsymbol{\alpha} = (\beta_1, \ldots, \beta_p, \theta_1, \ldots, \theta_p)$ and σ^2 in the non-centered parametrization, see Algorithm 1. In step (b), if the prior scale in (13.5) depends on σ^2 and, similarly, $\beta_j|\sigma^2 \sim \mathcal{N}(0, \sigma^2\tau)$, then, conditional on $\mathbf{z}$, the non-centered parametrization (13.3) is a standard Bayesian regression model for $\boldsymbol{\alpha}$ with a conjugate prior.

13.3 Continuous Shrinkage Priors for Sparse TVP Models

13.3.1 From the Ridge Prior to Continuous Shrinkage Priors

The ridge prior (13.6) for $\sqrt{\theta}_j$ can be rewritten in the following way,

$$\sqrt{\theta}_j|\psi_j^2 \sim \mathcal{N}(0, \tau\psi_j^2) \quad \Leftrightarrow \quad \theta_j|\psi_j^2 \sim \mathcal{G}\left(\frac{1}{2}, \frac{1}{2\tau\psi_j^2}\right), \tag{13.7}$$

where $\psi_j^2 = 1$ is a fixed scale parameter and τ controls the global level of shrinkage of θ_j, since $\mathrm{E}[\theta_j|\tau] = \tau$. In a sparse state space model, we expect that only a fraction of the coefficients are indeed dynamic, while the remaining coefficients are (nearly) constant. This prior perception should be reflected in the choice of the prior distribution of the unknown variances $\theta_1, \ldots, \theta_p$. In this section, we discuss how to incorporate this information through continuous shrinkage priors. In Section 13.4, we discuss mixture priors, also called spike-and-slab priors, in the context of variance selection.

Under the ridge prior (13.7), $\psi_j^2 \sim \delta_1$ follows a point mass prior on 1, which does not allow for any local adaptation. Continuous shrinkage priors take the form of global-local shrinkage priors in the sense of [43], where ψ_j^2 follows a prior $p(\psi_j^2)$ that encourages many small values, representing coefficients that are nearly constant, while at the same time some of the ψ_j^2's are allowed to take on larger values to represent coefficients that are indeed time-varying.

For univariate sparse state space and TVP models, [2] introduced the Bayesian Lasso prior [41], where ψ_j^2 follows an exponential distribution:

$$\sqrt{\theta}_j|\psi_j^2 \sim \mathcal{N}(0, \tau\psi_j^2), \qquad \psi_j^2 \sim \mathcal{E}xp(1). \tag{13.8}$$

This prior is extended by [5] to the normal-gamma prior [23], where the exponential prior for $p(\psi_j^2)$ is generalized to a gamma prior:

$$\sqrt{\theta}_j|\psi_j^2 \sim \mathcal{N}\left(0, \tau\psi_j^2\right), \qquad \psi_j^2|a^\xi \sim \mathcal{G}\left(a^\xi, a^\xi\right). \tag{13.9}$$

For both priors, τ acts as a global shrinkage parameter in a similar manner as for the ridge prior (13.7), however each innovation variance θ_j is mixed over its *own* (local) scale parameter ψ_j^2, each of which follows an independent exponential (13.8) or a gamma distribution (13.9). Hence, the ψ_j^2's play the role of local (component specific) shrinkage parameters. (13.9) obviously reduces to the Bayesian Lasso prior for $a^\xi = 1$, but encourages more prior shrinkage toward small values and, at the same time, more extreme values than the Bayesian Lasso prior for $a^\xi < 1$.

The normal-gamma prior (13.9) for $\sqrt{\theta}_j$ can be represented in the following way as a "double gamma" on θ_j [5]:

$$\theta_j|\xi_j^2 \sim \mathcal{G}\left(\frac{1}{2}, \frac{1}{2\xi_j^2}\right), \quad \xi_j^2|a^\xi \sim \mathcal{G}\left(a^\xi, \frac{a^\xi \kappa_B^2}{2}\right), \tag{13.10}$$

where $\kappa_B^2 = 2/\tau$. [6] proposed an extension of the double gamma prior (13.10) to a triple gamma prior, where another layer is added to the hierarchy:

$$\theta_j|\xi_j^2 \sim \mathcal{G}\left(\frac{1}{2}, \frac{1}{2\xi_j^2}\right), \quad \xi_j^2|a^\xi, \kappa_j^2 \sim \mathcal{G}\left(a^\xi, \frac{a^\xi \kappa_j^2}{2}\right), \quad \kappa_j^2|c^\xi \sim \mathcal{G}\left(c^\xi, \frac{c^\xi}{\kappa_B^2}\right). \tag{13.11}$$

The main difference to the double gamma prior is that the prior scale of the ξ_j^2's is not identical, as each local parameter ξ_j^2 depends on yet another local scale parameter κ_j^2. A similar prior is applied to the initial expectations β_j:

$$\beta_j|\lambda_j \sim \mathcal{N}\left(0, \lambda_j\right), \quad \lambda_j|\tau_j^2 \sim \mathcal{G}\left(a^\tau, \tau_j^2\right), \quad \tau_j^2 \sim \mathcal{G}\left(c^\tau, \frac{2c^\tau}{a^\tau \lambda_B^2}\right). \tag{13.12}$$

[6] show that the triple gamma prior (13.11) can be represented as a global-local shrinkage prior in the sense of [44], with the local shrinkage parameter ψ_j^2 arising from an $\mathrm{F}\left(2a^\xi, 2c^\xi\right)$ distribution:

$$\sqrt{\theta}_j|\psi_j^2 \sim \mathcal{N}\left(0, \tau\psi_j^2\right), \quad \psi_j^2|a^\xi, c^\xi \sim \mathrm{F}\left(2a^\xi, 2c^\xi\right), \tag{13.13}$$

with global shrinkage parameter $\tau = 2/\kappa_B^2$. An interesting special case of the triple gamma is the horseshoe prior [9] which results for $a^\xi = c^\xi = 1/2$, since $\psi_j^2 \sim \mathrm{F}(1,1)$ implies that $\psi_j \sim t_1$. [6] show that many other well-known shrinkage priors introduced in a regression context are special cases of the triple gamma, which itself can be regarded as an application of the normal-gamma-gamma prior [24] to variance selection in the non-centered parametrization (13.3).

Among other representations, the triple gamma prior has a representation as a generalized beta mixture prior introduced by [1] for variable selection in regression models:

$$\sqrt{\theta}_j|\rho_j \sim \mathcal{N}\left(0, 1/\rho_j - 1\right), \quad \rho_j|a^\xi, c^\xi, \phi^\xi \sim \mathcal{TPB}\left(a^\xi, c^\xi, \phi^\xi\right), \tag{13.14}$$

where $\phi^\xi = 2c^\xi/(\kappa_B^2 a^\xi) = \tau c^\xi/a^\xi$ and $\mathcal{TPB}\left(a^\xi, c^\xi, \phi^\xi\right)$ is the three-parameter beta distribution. This relationship makes it possible to investigate the shrinkage profile $p(\rho_j)$ of the triple gamma prior. Figure 13.3 contrasts a triple gamma prior with $a^\xi = c^\xi = 0.1$ with a

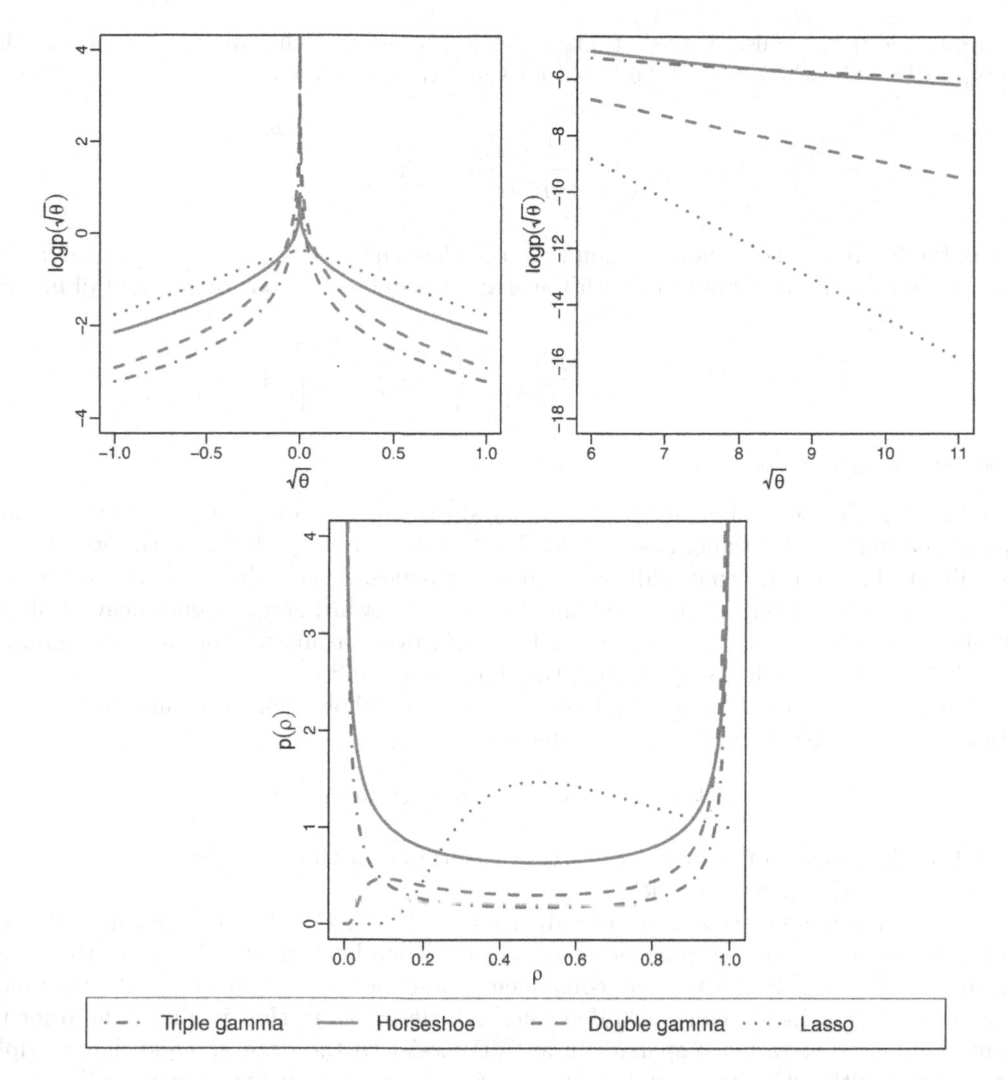

FIGURE 13.3
Spike (top left-hand side) and tail (top right-hand side) of the marginal prior $p(\sqrt{\theta_j})$ and corresponding shrinkage profiles $p(\rho_j)$ (bottom) under the triple gamma prior with $a^\xi = c^\xi = 0.1$ in comparison to the horseshoe prior, the double gamma prior with $a^\xi = 0.1$ and the Lasso prior. $\tau = 1$ ($\kappa_B^2 = 2$) for all prior specifications.

few of its special or limiting cases, showing the behavior around the origin, in the tails, as well as the shrinkage profiles.

The graphical representation of the triple gamma prior in Figure 13.3 is based on [6] who prove the following closed form expression for the marginal prior $p(\sqrt{\theta_j}|\phi^\xi, a^\xi, c^\xi)$:

$$p(\sqrt{\theta_j}|\phi^\xi, a^\xi, c^\xi) = \frac{\Gamma(c^\xi + \frac{1}{2})}{\sqrt{2\pi\phi^\xi} B(a^\xi, c^\xi)} U\left(c^\xi + \frac{1}{2}, \frac{3}{2} - a^\xi, \frac{\theta_j}{2\phi^\xi}\right), \tag{13.15}$$

where $U(a, b, z) = \int_0^\infty e^{-zt} t^{a-1}(1+t)^{b-a-1} dt$ is the confluent hyper-geometric function of the second kind.

The parameter a^ξ and c^ξ control, respectively, the behavior of this shrinkage prior at the origin and in the tails. [6] prove that the triple gamma prior has an infinite spike at the origin, if $a^\xi \leq 0.5$, where for $a^\xi < 0.5$ and for small values of $\sqrt{\theta}_j$:

$$p(\sqrt{\theta}_j|\phi^\xi, a^\xi, c^\xi) = \frac{\Gamma(\frac{1}{2} - a^\xi)}{\sqrt{\pi}(2\phi^\xi)^{a^\xi} B(a^\xi, c^\xi)} \left(\frac{1}{\sqrt{\theta}_j}\right)^{1-2a^\xi} + O(1).$$

Hence, the infinite spike is more pronounced, the closer a^ξ is to 0. As $\sqrt{\theta}_j \to \infty$, the triple gamma prior has polynomial tails, with the shape parameter c^ξ controlling the tail index:

$$p(\sqrt{\theta}_j|\phi^\xi, a^\xi, c^\xi) = \frac{\Gamma(c^\xi + \frac{1}{2})(2\phi^\xi)^{c^\xi}}{\sqrt{\pi} B(a^\xi, c^\xi)} \left(\frac{1}{\sqrt{\theta}_j}\right)^{2c^\xi+1} \left[1 + O\left(\frac{1}{\theta_j}\right)\right].$$

Choosing the hyperparameters

A challenging question is how to choose the parameters a^ξ, c^ξ and ϕ^ξ of the triple gamma prior in the context of variance selection for TVP models. In high-dimensional settings it is appealing to have a prior that addresses two major issues: first, high concentration around the origin to favor strong shrinkage of small variances toward zero; second, heavy tails to introduce robustness to large variances and to avoid over-shrinkage. For the triple gamma prior, both issues are addressed through the choice of a^ξ and c^ξ.

a^ξ and c^ξ can be fixed, as for the Lasso and the horseshoe prior, or estimated from the data under a suitable prior. [6], e.g., assume that

$$2a^\xi \sim \mathcal{B}eta\left(\alpha_{a^\xi}, \beta_{a^\xi}\right), \qquad 2c^\xi \sim \mathcal{B}eta\left(\alpha_{c^\xi}, \beta_{c^\xi}\right), \tag{13.16}$$

restricting the support of a^ξ and c^ξ to $(0, 0.5)$, ensuring that the triple gamma prior is more aggressive than the horseshoe prior.

Ideally, one should place a hyperprior distribution on the global shrinkage parameter ϕ^ξ. Such a hierarchical triple gamma prior introduces dependence among the local shrinkage parameters $\xi_1^2, \ldots, \xi_p^2$ in (13.11) and, consequently, among $\theta_1, \ldots, \theta_p$ in the joint (marginal) prior $p(\theta_1, \ldots, \theta_p)$. Introducing such dependence is desirable in that it allows the prior to adapt the degree of variance sparsity in a TVP model to the data at hand. For a triple gamma prior with arbitrary a^ξ and (finite) c^ξ, [6] assume the following prior on ϕ^ξ:

$$\phi^\xi | a^\xi, c^\xi \sim \mathcal{BP}\left(c^\xi, a^\xi\right). \tag{13.17}$$

Prior (13.17) reduces to $\phi^\xi | a^\xi \sim \mathrm{F}\left(2a^\xi, 2a^\xi\right)$ for $a^\xi = c^\xi$. Hence, for the horseshoe prior, $\phi^\xi \sim \mathrm{F}(1, 1)$ and the global shrinkage parameter $\tau = \sqrt{\phi^\xi}$ follows a Cauchy prior as in [4, 8]. As shown by [6], under this hyperprior, the triple gamma prior exhibits behavior similar to Bayesian Model Averaging (BMA), with a uniform prior on an appropriately defined model size, see Section 13.5.5.

For infinite c^ξ, hierarchical versions of the Lasso and the double gamma prior in TVP models are based on a gamma prior for the global shrinkage parameter κ_B^2, $\kappa_B^2 \sim \mathcal{G}\left(d_1, d_2\right)$ [2, 5]. This leads to a heavy-tailed extension of both priors, where each marginal density $p(\sqrt{\theta}_j|d_1, d_2)$ follows a triple gamma prior with the same parameter a^ξ (being equal to one for the Bayesian Lasso) and tail index $c^\xi = d_1$. In this light, very small values of d_1 had to be applied in these papers to ensure heavy tails of $p(\sqrt{\theta}_j|d_1, d_2)$.

13.3.2 Efficient MCMC Inference

The two-block Gibbs sampler outlined in Section 13.2.1 can be extended to perform MCMC inference for continuous shrinkage priors by exploiting the normal scale mixture representation underlying any global-local shrinkage prior.

Assume, for illustration, that we want to apply a normal-gamma prior for the initial expectations β_j and a double gamma prior for θ_j:

$$\begin{aligned} \beta_j|\lambda_j &\sim \mathcal{N}(0, \lambda_j), & \lambda_j|a^\tau, \lambda_B^2 &\sim \mathcal{G}\left(a^\tau, \frac{a^\tau \lambda_B^2}{2}\right), \\ \theta_j|\xi_j^2 &\sim \mathcal{G}\left(\frac{1}{2}, \frac{1}{2\xi_j^2}\right), & \xi_j^2|a^\xi, \kappa_B^2 &\sim \mathcal{G}\left(a^\xi, \frac{a^\xi \kappa_B^2}{2}\right), \end{aligned} \tag{13.18}$$

with fixed global shrinkage parameters a^τ, λ_B^2, a^ξ and κ_B^2. In this case, we can run a three-block Gibbs sampler to draw (a) the latent state process from $p(\mathbf{z}|\boldsymbol{\alpha}, \sigma^2, \mathbf{y})$, (b) the model parameter $\boldsymbol{\alpha} = (\beta_1, \dots, \beta_p, \sqrt{\theta}_1, \dots, \sqrt{\theta}_p)$ from $p(\boldsymbol{\alpha}, \sigma^2|\boldsymbol{\lambda}, \boldsymbol{\xi}, \mathbf{z}, \mathbf{y})$ conditional on knowing the local scale parameters $\boldsymbol{\lambda} = (\lambda_1, \dots, \lambda_p)$ and $\boldsymbol{\xi} = (\xi_1^2, \dots, \xi_p^2)$, and (c) the local scale parameters from $p(\lambda_j|\beta_j, \lambda_B^2)$ and $p(\xi_j^2|\theta_j, \kappa_B^2)$ for $j = 1, \dots, p$.

Let us consider step (c) in more detail, since sampling the local shrinkage parameter from $\xi_j^2|\theta_j, \kappa_B^2$ (and similarly from $\lambda_j|\lambda_B^2, \beta_j$) is less standard. The double gamma prior $\theta_j|\xi_j^2$ in (13.18) leads to a density for ξ_j^2 given θ_j which is the kernel of an inverse gamma density. In combination with the gamma prior for $\xi_j^2|a^\xi, \kappa_B^2$ also appearing in (13.18), this leads to a posterior distribution arising from a generalized inverse Gaussian (GIG) distribution: $\xi_j^2|\theta_j, a^\xi, \kappa_B^2 \sim \mathcal{GIG}\left(a^\xi - 1/2, a^\xi \kappa_B^2, \theta_j\right)$. A very stable generator from the GIG distribution is implemented in the R-package `GIGrvg` [26].

To center or to non-center?

In step (a) and (b) of the three–block sampler described above, we have the option to either work with the centered parametrization (13.1) or the non-centered parametrization (13.3). Regardless of the parametrization, sampling the state process is straightforward, using either FFBS [7, 14] or a one-block sampler such as "all without a loop" (AWOL) [5, 31].

In the centered parametrization, the conditional posterior $\theta_j|\beta_{j0}, \dots, \beta_{jT}, \beta_j$ is again a GIG distribution, since the gamma prior for θ_j in (13.18) is combined with the density $p(\beta_{j0}, \dots, \beta_{jT}|\theta_j, \beta_j)$, which is the kernel of an inverse gamma density. However, like many MCMC schemes which alternate between sampling from the full conditionals of the latent states and the model parameters, the resulting sampler suffers from slow convergence and poor mixing if some of the true process variances are small or even zero.

As shown by [18], MCMC estimation based on the non-centered parametrization proves to be useful, in particular if the process variances are close to zero. Using the representation of the double gamma prior for θ_j as a conditionally normal prior, $\sqrt{\theta}_j|\xi_j^2 \sim \mathcal{N}\left(0, \xi_j^2\right)$, we obtain a joint Gaussian prior for $\boldsymbol{\alpha} = (\beta_1, \dots, \beta_p, \sqrt{\theta}_1, \dots, \sqrt{\theta}_p)$, where the local shrinkage parameters $\boldsymbol{\lambda}$ and $\boldsymbol{\xi}$ change the prior scale in a dynamic fashion during MCMC sampling. Hence, in the non-centered parametrization (13.3), conditional on $\boldsymbol{\lambda}, \boldsymbol{\xi}$ and the latent process $\mathbf{z}$, we are dealing with a Bayesian regression model under a non-conjugate analysis and sampling from $p(\boldsymbol{\alpha}, \sigma^2|\boldsymbol{\lambda}, \boldsymbol{\xi}, \mathbf{z}, \mathbf{y})$ can be implemented as in Algorithm 1.

[16] discusses the relationship between the various parametrizations for a simple TVP model with $p = 1$ and the computational efficiency of the resulting MCMC samplers, see also [40]. For TVP models with $p > 1$, MCMC estimation in the centered parametrization is preferable for all coefficients that are actually time-varying, whereas the non-centered

parametrization is preferable for (nearly) constant coefficients. For practical time series analysis, both types of coefficients are likely to be present and choosing a computationally efficient parametrization in advance is not possible.

As shown by [5] in the context of TVP models, these two data augmentation schemes can be combined through the *ancillarity-sufficiency interweaving strategy* (ASIS) introduced by [51] to obtain an efficient sampler combining the "best of both worlds". ASIS provides a principled way of interweaving the centered and the non-centered parametrization of a TVP model by re-sampling certain parameters conditional on the latent variables in the alternative parametrization of the model. More specifically, [5] sample β_j and $\sqrt{\theta}_j$ in the non-centered parametrization from the joint conditionally Gaussian distribution and interweave into the centered parametrization to resample θ_j from the conditional GIG distribution (and β_j from yet another conditionally Gaussian distribution). This leads to an MCMC sampling scheme which increases posterior sampling efficiency considerably compared to sticking with either of the two parametrizations throughout sampling, while the additional computational cost of the interweaving step is minor. ASIS was extended by [6] to the more general triple gamma prior.

MCMC sampling is extended by additional steps for hierarchical versions of the triple gamma prior, by sampling all unknown global shrinkage parameters a^τ, c^τ, λ_B^2, a^ξ, c^ξ and κ_B^2 from the appropriate conditional posterior distributions. A full description of these algorithms can be found in [5] for the double gamma prior and in [6] for the more general triple gamma prior.

The `shrinkTVP` *package*

The R package **`shrinkTVP`** [33] offers efficient implementations of MCMC algorithms for TVP models with continuous shrinkage priors, specifically the triple gamma prior and its many special and limiting cases. It is designed to provide an easy entry point for fitting TVP models with shrinkage priors, while also giving more experienced users the option to adapt the model to their needs. The computationally demanding portions are written in C++ and then interfaced with R, combining the speed of compiled code with the ease-of-use of interpreted code.

13.3.3 Application to US Inflation Modelling

In our application we model quarterly US inflation (1964:Q1–2015:Q4) as a generalized Phillips curve with time-varying parameters in the spirit of [34]. This means that inflation at time t is modeled as

$$\begin{aligned} \boldsymbol{\beta}_t &= \boldsymbol{\beta}_{t-1} + \mathbf{w}_t, \qquad && \mathbf{w}_t \sim \mathcal{N}_p\left(\mathbf{0}, \mathbf{Q}\right), \\ y_t &= \mathbf{x}_{t-1}\boldsymbol{\beta}_t + \varepsilon_t, && \varepsilon_t \sim \mathcal{N}\left(0, \sigma_t^2\right), \end{aligned}$$

where y_t is inflation at time t, $\mathbf{x}_{t-1}$ is a set of $p = 18$ predictors including an intercept, exogenous variables from the previous time period and y_{t-1} to y_{t-r}, a series of lagged observations of inflation. For the application at hand we assume that $r = 3$. The exogenous predictors included are broad and represent many different potential determinants of inflation. Table 13.1 offers an overview of the data, the sources used and the transformations applied to achieve (approximate) stationarity. For this application we assume the error variance σ_t^2 follows a stochastic volatility specification as in Section 13.5.1.

Three different priors are placed on the expected initial values $\beta_1, \dots, \beta_p$ and on the variances of the innovations $\theta_1, \dots, \theta_p$, namely the ridge prior, as defined in equation (13.6), the Lasso prior, as defined in equation (13.8), and the triple gamma prior, as defined in equation (13.13). In the case of the Lasso prior, the global shrinkage parameters λ_B^2 and κ_B^2 are learned from the data under a gamma prior, specifically $\lambda_B^2 \sim \mathcal{G}\left(0.001, 0.001\right)$ and

TABLE 13.1
US inflation data description and sources

Mnemonic	Description	Database name	Source	Tc
inf	Consumer Price Index	CPI	PHIL	4
unemp	Unemployment rate	RUC	PHIL	1
cons	Real Personal Consumption Expenditures	RCON	PHIL	4
dom_inv	Real Gross Private Domestic Investment	RINVRESID	PHIL	4
gdp	Real GDP	ROUTPUT	PHIL	4
hstarts	Housing Starts	HSTARTS	PHIL	3
emp	Non-farm Payroll Employment	EMPLOY	PHIL	4
pmi	ISM Manuf.: PMI Composite Index	NAPM	FRED	2
treas	3m Treasury Bill: Secondary Market	TB3MS	FRED	1
spread	Spread 10-year T-Bond yield/3m T-Bill	TB3MS - GS10	FRED	1
dow	Dow Jones Industrial Average	UDJIAD1	BCB	4
m1	M1 Money Stock	M118Q2	PHIL	4
exp	Expected Changes in Inflation Rates	-	UoM	1
napmpri	NAPM Commodity Prices Index	NAPMPRI	FRED	2
napmsdi	NAPM Vendor Deliveries Index	NAPMSDI	FRED	2

Notes: Tc refers to the transformation applied to the data. Let z_{it} be the original time series and x_{it} be the transformed time series, then 1 - no transformation, 2 - first difference, $x_{it} = z_{it} - z_{i,t-1}$, 3 - logarithm, $x_{it} = \log z_{it}$, 4 - first difference of logarithm $x_{it} = 100(\log z_{it} - \log z_{i,t-1})$. Sources are the Federal Reserve Bank of Philadelphia (PHIL), the Federal Reserve Bank of St. Louis (FRED), the University of Michigan (UoM) and the Banco Central do Brasil (BCB).

$\kappa_B^2 \sim \mathcal{G}(0.001, 0.001)$. In the triple gamma case, the hyperparameters are also learned from the data, under the priors defined in equations (13.16) and (13.17), with hyperparameter values $\alpha_{a^\xi} = \alpha_{a^\tau} = 5$ and $\beta_{a^\xi} = \beta_{a^\tau} = 10$.

Figure 13.4 shows how the three prior setups recovered the same states that were already presented in Figure 13.2. While all three are noticeably smaller in scale than the states recovered under the inverse gamma prior, they still differ in this regard as a consequence of the degree of shrinkage imposed, with the triple gamma prior imposing the most, followed by the Lasso prior and the ridge prior, in that order. This can be seen in the parameter for the Dow Jones - the median is virtually zero under the triple gamma prior while displaying much more movement under the other two priors. The parameter of the commodity prices index turns out to be significant, but practically constant under the triple Gamma prior, while the two other priors also assign considerable posterior mass to negative values. In the case of the parameter for the treasury bill, the most pronounced movement comes from the state estimated under the triple gamma prior, indicating that truly time-varying states are more likely to be picked up in such a sparse environment if the non time-varying parameters are effectively shrunken towards fixed ones.

Another way to examine the effect that various levels of shrinkage have on the inference that follows is to look at the model implied predictions. Figure 13.5 plots the posterior predictive density of the three different models and contrasts these with the true levels of inflation. Two things are noteworthy: first, the stronger the shrinkage imposed by the prior, the less closely the median follows the true observation. This can be seen as shrinkage preventing the model from overfitting. Second, the error variance appears to be larger for the models with more shrinkage, as the spurious time variation in some parameters is dampened, leaving more of the variance to be soaked up by the error term. That this is beneficial for prediction can be seen in Section 13.5.4.

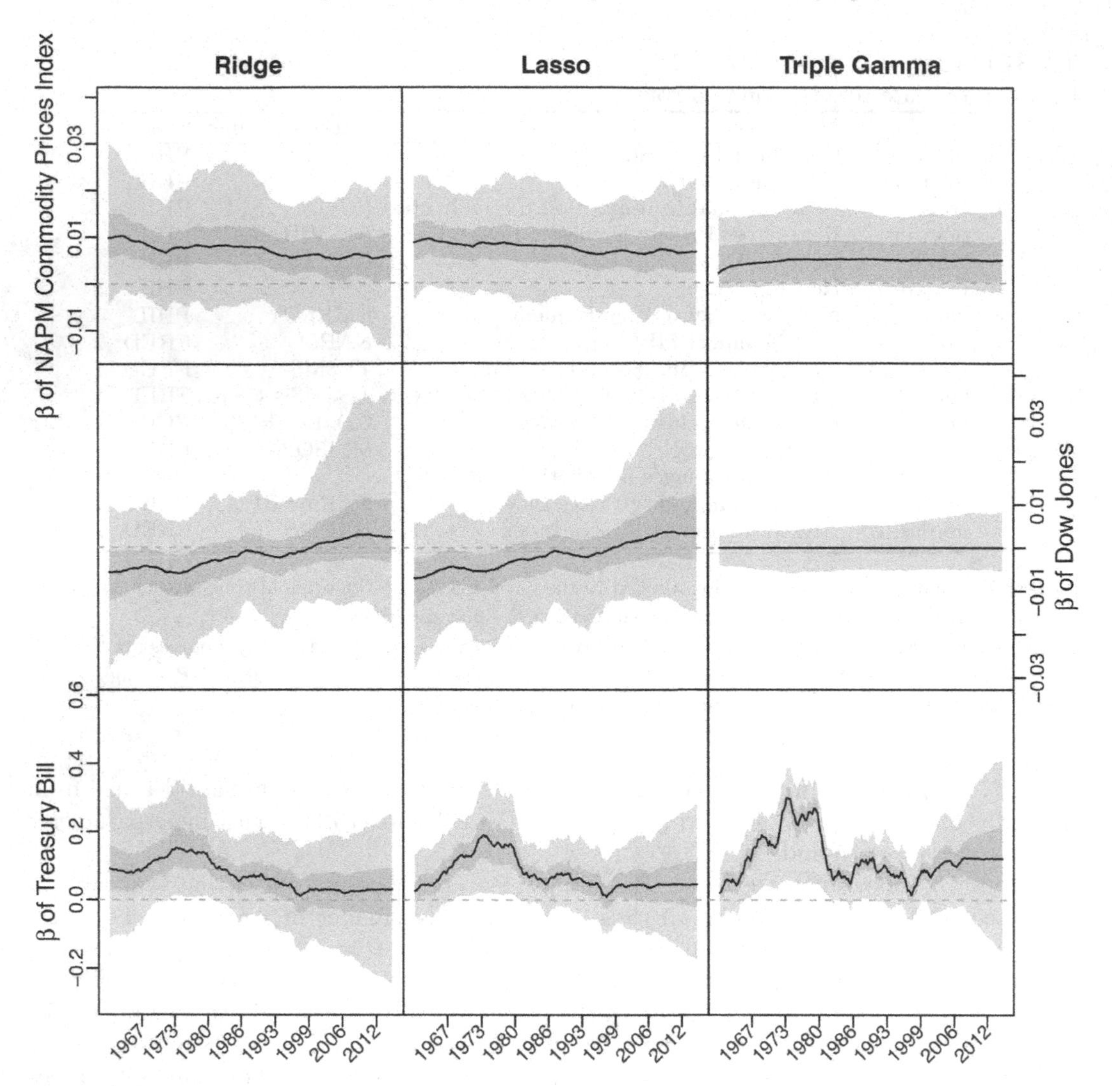

FIGURE 13.4
Recovery of the time-varying parameters for the inflation data under the ridge prior, Lasso prior and triple gamma prior. The gray shaded regions represent pointwise 95% and 50% credible intervals, respectively, while the black line represents the pointwise median.

13.4 Spike-and-Slab Priors for Sparse TVP Models

13.4.1 From the Ridge prior to Spike-and-Slab Priors

A spike-and-slab prior is a finite mixture distribution with two components, where one component (the *spike*) has much stronger global shrinkage than the second component (the *slab*). Such mixture shrinkage priors were introduced by [20, 21] for variable selection for regression models and aim to identify zero and non-zero regression effects. However, they are useful far beyond this problem and allow, for instance, parsimonious covariance modelling

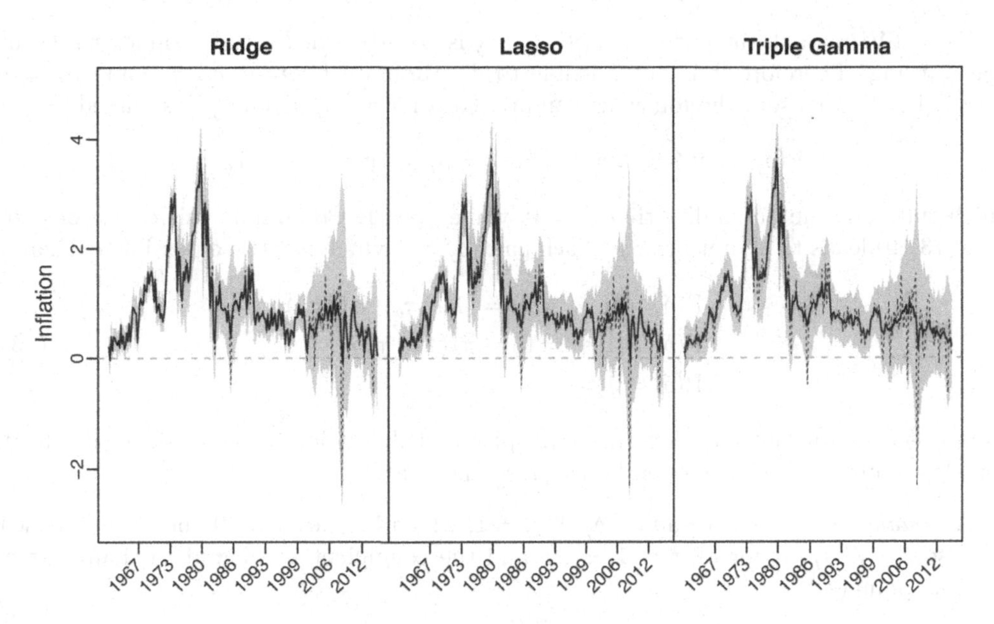

FIGURE 13.5
Predicting the levels of inflation under the ridge prior, Lasso prior and triple gamma prior. The gray shaded regions represent pointwise 95% and 50% credible intervals, respectively, while the solid black line represents the pointwise median. The dashed black line indicates the actual level of inflation.

for longitudinal data [48], covariance selection in random effects models [17] and robust random effects estimation [19].

Discrete spike-and-slab priors were introduced in state space modeling by [18] to achieve shrinkage of time-varying state variables toward fixed components. In TVP models, such a prior is introduced for the variance θ_j and reads $\theta_j \sim (1-\pi_\gamma)\delta_0 + \pi_\gamma p_{\text{slab}}(\theta_j)$, with the spike being a point measure at 0 and $p_{\text{slab}}(\theta_j)$ being the distribution in the slab. [18] introduced the following prior for the scale parameter $\sqrt{\theta}_j$ in the non-centered parametrization (13.3), with a ridge prior in the slab:

$$\sqrt{\theta}_j|\sigma^2 \sim (1-\pi_\gamma)\delta_0 + \pi_\gamma \mathcal{N}\left(0, \sigma^2 B_\gamma\right).$$

With γ_j being a binary indicator that separates the spike from the slab, π_γ controls the prior occurrence of dynamic coefficients:

$$\mathrm{P}(\gamma_j = 1|\pi_\gamma) = \pi_\gamma. \tag{13.19}$$

Again, this prior can be seen as an extension of the ridge prior, this time with a binary local scale parameter $\psi_j^2 = \gamma_j$ taking either the value 0 or 1: $\sqrt{\theta}_j|\psi_j^2 = \gamma_j \sim \mathcal{N}\left(0, \sigma^2 B_\gamma \gamma_j\right)$. A discrete spike-and-slab prior is also applied to the initial expectation β_j:

$$\beta_j|\sigma^2 \sim (1-\pi_\delta)\delta_0 + \pi_\delta \mathcal{N}\left(0, \sigma^2 B_\delta\right),$$

with a corresponding binary indicator δ_j to separate the spike from the slab. The dependence of the prior scale on the error variance σ^2 in both priors $p(\beta_j|\sigma^2)$ and $p(\sqrt{\theta}_j|\sigma^2)$ allows sampling the indicators γ_j and δ_j without conditioning on any model parameters, see Section 13.4.2.

For a TVP model, the initial expectation β_j is not identified if the parameter is actually time-varying. Therefore it is not possible to discriminate between $\delta_j = 0$ and $\delta_j = 1$, if $\gamma_j = 1$. For this reason, the following conditional prior for δ_j given γ_j is assumed:

$$\mathrm{P}(\delta_j = 1|\gamma_j = 0, \pi_\delta) = \pi_\delta, \quad \mathrm{P}(\delta_j = 1|\gamma_j = 1) = 1,$$

which rules out the possibility that $\delta_j = 0$, while $\gamma_j = 1$. Combining this conditional prior with (13.19) leads to a joint prior for each pair (δ_j, γ_j) which has three possible realizations:

$$\begin{aligned} \mathrm{P}(\delta_j = 0, \gamma_j = 0) &= (1 - \pi_\delta)(1 - \pi_\gamma), \\ \mathrm{P}(\delta_j = 1, \gamma_j = 0) &= \pi_\delta(1 - \pi_\gamma), \\ \mathrm{P}(\delta_j = 1, \gamma_j = 1) &= \pi_\gamma. \end{aligned} \tag{13.20}$$

As opposed to continuous priors, discrete spike-and-slab priors allow explicit classification of the variables in a TVP model, based on δ_j and γ_j:

(1) A *dynamic* coefficient results if $\delta_j = \gamma_j = 1$, which implies $\beta_j \neq 0$ and $\theta_j \neq 0$, in which case $\beta_{jt} \neq \beta_{j,t-1}$ for all $t = 1, \ldots, T$ and the coefficient is allowed to change at each time point.

(2) A *fixed, non-zero* coefficient results if $\gamma_j = 0$ but $\delta_j = 1$, which implies $\beta_j \neq 0$ while $\theta_j = 0$, in which case $\beta_{jt} = \beta_j$ for all $t = 1, \ldots, T$ and the coefficient is significant, but fixed.

(3) A *zero* coefficient results if $\delta_j = \gamma_j = 0$, which implies $\beta_j = 0$ and $\theta_j = 0$, in which case $\beta_{jt} = 0$ for all $t = 1, \ldots, T$ and the coefficient is insignificant.

The probabilities given in (13.20) are the prior probabilities for classifying coefficients into these three categories. Based on this prior, in a fully Bayesian inference, the joint posterior distribution of $p(\boldsymbol{\delta}, \boldsymbol{\gamma}|\mathbf{y})$ of all indicators $\boldsymbol{\delta} = (\delta_1, \ldots, \delta_p)$ and $\boldsymbol{\gamma} = (\gamma_1, \ldots, \gamma_p)$ is derived and can be used for posterior classification, e.g., by deriving the model most often visited, or the median probability model.

Choosing hyperparameters for discrete spike-and-slab priors

First, the prior probabilities π_γ and π_δ to observe a dynamic or a constant parameter, respectively, have to be chosen. As for standard variable selection, the strategy to fix π_γ and π_δ is very informative on the model sizes. The numbers p_d, p_f and p_0 of dynamic, constant and zero coefficients, respectively, are given by

$$p_d = \sum_{j=1}^{p} \gamma_j, \quad p_f = \sum_{j=1}^{p} \delta_j(1 - \gamma_j), \quad p_0 = \sum_{j=1}^{p} (1 - \delta_j)(1 - \gamma_j).$$

Hence, apriori, $p_d|\pi_\gamma \sim \mathcal{B}in\,(p, \pi_\gamma)$, $p_f|\pi_\delta, p_d \sim \mathcal{B}in\,(p - p_d, \pi_\delta)$, while p_0 given p_d and p_f is deterministic, $p_0 = p - (p_d + p_f)$.

Alternatively, a hyperprior can be assumed for both probabilities in order to learn the desired degree of sparsity from the data. Such a hierarchial prior allows more adaptation to the required level of sparsity and assumes that the prior probabilities π_δ and π_γ are unknown, each following a beta distribution:

$$\pi_\delta \sim \mathcal{B}eta\left(a_0^\delta, b_0^\delta\right), \qquad \pi_\gamma \sim \mathcal{B}eta\left(a_0^\gamma, b_0^\gamma\right). \tag{13.21}$$

Choosing $a_0^\delta = b_0^\delta = 1$ and $a_0^\gamma = b_0^\gamma = 1$ implies that the prior on p_d is uniform on $\{0, \ldots, p\}$, while $p_f|p_d$ is uniform on $\{0, \ldots, p - p_d\}$.

Algorithm 2 Model space MCMC under a discrete spike-and slab prior with a conjugate Gaussian slab.

(a) Sample indicators $\boldsymbol{\gamma} = (\gamma_1, \ldots, \gamma_p)$ and $\boldsymbol{\delta} = (\delta_1, \ldots, \delta_p)$ from $p(\boldsymbol{\delta}, \boldsymbol{\gamma}|\mathbf{z}, \mathbf{y})$ conditional on the latent variables $\mathbf{z} = (\tilde{\boldsymbol{\beta}}_0, \ldots, \tilde{\boldsymbol{\beta}}_T)$;

(b) sample the model parameters $\boldsymbol{\beta}_{\delta,\gamma}$ and σ^2 conditional on $\mathbf{z}$ and $(\boldsymbol{\delta}, \boldsymbol{\gamma})$:

(b-1) sample σ^2 from the inverse gamma density $\sigma^2|\boldsymbol{\delta}, \boldsymbol{\gamma}, \mathbf{z}, \mathbf{y}$

(b-2) sample $\boldsymbol{\beta}_{\delta,\gamma}$ from the multivariate Gaussian $\boldsymbol{\beta}_{\delta,\gamma}|\sigma^2, \mathbf{z}, \mathbf{y}$.

(c) sample $\mathbf{z}$ conditional on $\boldsymbol{\vartheta} = (\beta_1, \ldots, \beta_p, \sqrt{\theta}_1, \ldots, \sqrt{\theta}_p, \sigma^2)$ from $\mathbf{z}|\boldsymbol{\vartheta}, \mathbf{y}$, using FFBS or AWOL.

Second, the prior in the slab has to be specified. For a discrete spike-and-slab prior, all θ_js with $\gamma_j = 0$ and all β_js with $\delta_j = 0$ are switched off in the non-centered model (13.3). Hence, a prior has to be chosen for the parameter $\boldsymbol{\beta}_{\delta,\gamma}$ collecting all remaining non-zero β_js and $\sqrt{\theta}_j$s. Under a Gaussian slab distribution, such a prior reads

$$\boldsymbol{\beta}_{\delta,\gamma}|\sigma^2 \sim \mathcal{N}_k\left(\mathbf{0}, \sigma^2 \tau I_k\right), \tag{13.22}$$

where $k = p_f + 2p_d$. However, as for variable selection in regression models, the choice of τ is influential in a higher-dimensional setting. A certain robustness is achieved by choosing a hierarchial Student-t slab, where

$$\begin{aligned} \beta_j|\delta_j = 1 &\sim \mathcal{N}\left(0, \sigma^2\lambda^2/\tau_j^2\right), & \tau_j^2 &\sim \mathcal{G}\left(a^\tau, a^\tau\right), \\ \sqrt{\theta}_j|\gamma_j = 1 &\sim \mathcal{N}\left(0, \sigma^2\kappa^2/\xi_j^2\right), & \xi_j^2 &\sim \mathcal{G}\left(a^\xi, a^\xi\right), \end{aligned}$$

with hyperpriors $\lambda^2 \sim \mathcal{G}\left(a^\lambda, a^\lambda\right)$ and $\kappa^2 \sim \mathcal{G}\left(a^\kappa, a^\kappa\right)$ with small degrees of freedom, e.g., $a^\tau = a^\xi = a^\lambda = a^\kappa = 0.5$.

Alternatively, [18] consider a fractional prior which is commonly used in model selection, as it adapts the prior scale automatically in a way that guarantees model consistency [39]. For TVP models, [18] defined a fractional prior for $\boldsymbol{\beta}_{\delta,\gamma}$ conditional on the latent process $\mathbf{z}$ as $p(\boldsymbol{\beta}_{\delta,\gamma}|b, \cdot) \propto f(\mathbf{y}|\boldsymbol{\beta}_{\delta,\gamma}, \sigma^2, \mathbf{z})^b$. This prior can be interpreted as the posterior of a non-informative prior combined with a small fraction b of the complete data likelihood $f(\mathbf{y}|\boldsymbol{\beta}_{\delta,\gamma}, \sigma^2, \mathbf{z})$.

13.4.2 Model Space MCMC

MCMC inference under discrete spike-and-slab priors is challenging, since the sampler is operating in a very high-dimensional model space. Each of the p covariates defines three types of coefficients, hence the sampler needs to navigate through 3^p possible models. The various steps of model space MCMC are summarized in Algorithm 2 for the conjugate slab distribution (13.22).

Naturally, the most challenging part is Step (a). If p is not too large, then Step (a) can be implemented as a full enumeration Gibbs step by computing the marginal likelihood $f(\mathbf{y}|\boldsymbol{\delta}, \boldsymbol{\gamma}, \mathbf{z})$ for all 3^p possible combinations of indicators, as illustrated by [18] for unobserved component state space models. Note that, conditional on the latent process $\mathbf{z}$, $f(\mathbf{y}|\boldsymbol{\delta}, \boldsymbol{\gamma}, \mathbf{z})$ is the marginal likelihood of a constrained version of regression model (13.3) under the conjugate prior (13.22) and therefore has a simple closed form. To derive the posterior $f(\boldsymbol{\delta}, \boldsymbol{\gamma}|\mathbf{z}, \mathbf{y}) \propto p(\mathbf{y}|\boldsymbol{\delta}, \boldsymbol{\gamma}, \mathbf{z}) f(\boldsymbol{\delta}, \boldsymbol{\gamma})$, these marginal likelihoods are combined with the prior

$p(\boldsymbol{\delta}, \boldsymbol{\gamma})$ for all models, which is available in closed form even under the hierarchical prior (13.21).

In cases where such a full enumeration Gibbs step becomes unfeasible because p is simply too large, Step (a) can be implemented as a single move sampler: loop randomly over all pairs of indicators $(\delta_j, \gamma_j), j = 1, \dots, p$, and propose to move from the current model $s = (\delta_j, \gamma_j)$ to a new model $s^{\text{new}} = (\delta_j^{\text{new}}, \gamma_j^{\text{new}})$ with probability $q_{s \to s^{\text{new}}}$. Accept $(\boldsymbol{\delta}, \boldsymbol{\gamma})^{\text{new}}$ with probability $\min(1, \alpha)$ where

$$\alpha = \frac{f(\mathbf{y}|(\boldsymbol{\delta}, \boldsymbol{\gamma})^{\text{new}}, \mathbf{z}) p((\boldsymbol{\delta}, \boldsymbol{\gamma})^{\text{new}})}{f(\mathbf{y}|\boldsymbol{\delta}, \boldsymbol{\gamma}, \mathbf{z}) p(\boldsymbol{\delta}, \boldsymbol{\gamma})} \times \frac{q_{s^{\text{new}} \to s}}{q_{s \to s^{\text{new}}}}.$$

The art here is to design sensible moves. One strategy is to move with equal probability to one of the two alternative categories. For instance, if currently $\delta_j = \gamma_j = 1$ defines a dynamic coeffcient, then propose, respectively, with probability 0.5 to either move to a fixed coeffcient, where $\delta_j^{\text{new}} = 0$ (while $\gamma_j^{\text{new}} = \gamma_j = 1$) or to a zero coeffcient, where $\delta_j^{\text{new}} = \gamma_j^{\text{new}} = 0$. In general, moves involving a change from a fixed to a dynamic coefficient are not easily accepted. Given that $\gamma_j = 0$, the current latent path $\mathbf{z}_j = (\tilde{\beta}_{j0}, \dots, \tilde{\beta}_{jT})$ was sampled from the prior $p(\mathbf{z}_j)$ which can be very different from the smoothed posterior $p(\mathbf{z}_j|\gamma_j^{\text{new}} = 1, \mathbf{y})$, in particular if T is large.

Having updated the vector of indicators $(\boldsymbol{\delta}, \boldsymbol{\gamma})$, a modified version of Algorithm 1 is applied in Step (b) and (c) of Algorithm 2 to sample the unconstrained model parameters $\boldsymbol{\beta}_{\delta,\gamma}, \sigma^2|\mathbf{z}, \mathbf{y}$ and $\mathbf{z}|\boldsymbol{\vartheta}, \mathbf{y}$ in the restricted version of the non-centered parametrization. In particular, the sampling order is interchanged to obtain a valid sampler, since $(\boldsymbol{\delta}, \boldsymbol{\gamma})$ are updated without conditioning on the parameter $\boldsymbol{\vartheta} = (\beta_1, \dots, \beta_p, \sqrt{\theta}_1, \dots, \sqrt{\theta}_p, \sigma^2)$.

13.4.3 Application to US Inflation Modelling

The analysis in Section 13.3.3 is extended, using discrete spike-and-slab priors for β_j and $\sqrt{\theta}_j$ with following slab distributions: (1) Gaussian with $\tau = 1$, (2) fractional priors with $b = 10^{-4}$ and (3) hierarchial Student-t with $a^{\tau} = a^{\xi} = a^{\lambda} = a^{\kappa} = 0.5$. The hierarchical prior $\sigma^2|C_0 \sim \mathcal{IG}(0.5, C_0)$, $C_0 \sim \mathcal{G}(5, 10/3)$ is assumed for the (homoscedastic) variance σ^2. The prior of π_δ and π_γ is chosen as in (13.21) with $a_0^\delta = b_0^\delta = a_0^\gamma = 1$ and $b_0^\gamma = 2$.

Model space MCMC sampling was run for 100.000 iteration after a burn-in of 10.000. Under the hierarchial Student-t slab, the sampler exhibits acceptance rates of around 20% for all classes of moves. This indicates relatively good performance, given that the latent variables $\mathbf{z}$ are unobserved and imputed under the old indicators. For Gaussian and fractional slabs, the average acceptance rate of moves between fixed and dynamic components was less than 5%. To verify convergence, the sampler was run twice, starting either from a full TVP model with all γ_js equal to 1 or from a standard regression model with all γ_js equal to 0. Under the Student-t slab, we found high concordance between the models sampled by both chains after burn-in. Under Gaussian and fractional slabs, however, the two chains were sampling totally different models, depending on the starting value.

The time-varying parameters recovered under the hierarchical Student-t slab are shown in Figure 13.6. We see a similar discrimination between a dynamic path (treasury bills), a constant path (commodity prices index) and a zero path (Dow Jones) as we saw in Figure 13.4 under the triple gamma prior. A more formal discrimination based on the sampled indicators δ_j and γ_j will be performed in Section 13.5.5.

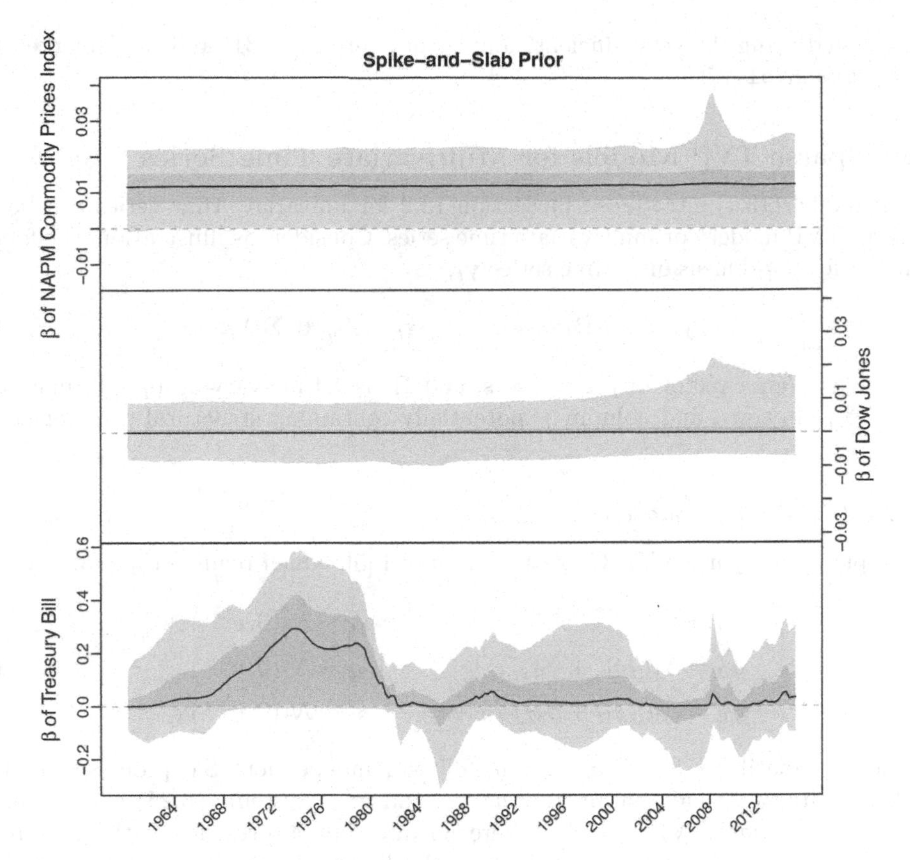

FIGURE 13.6
Recovery of the time-varying parameters for the inflation data under the discrete spike-hierarchical-Student-t-slab prior. The gray shaded regions represent pointwise 95% and 50% credible intervals, respectively, while the black line represents the pointwise median.

13.5 Extensions

13.5.1 Including Stochastic Volatility

Assuming a homoscedastic error variance σ^2 in the observation equation of the TVP model (13.1) may create spurious time-variation in the coefficients, as discussed by [47]. To be robust against conditional heteroscedasticity, σ_t^2 is often assumed to be time-varying over $t = 1, \ldots, T$:

$$\begin{aligned} \boldsymbol{\beta}_t &= \boldsymbol{\beta}_{t-1} + \mathbf{w}_t, \qquad && \mathbf{w}_t \sim \mathcal{N}_p(\mathbf{0}, \mathbf{Q}), \\ y_t &= \mathbf{x}_t \boldsymbol{\beta}_t + \varepsilon_t, \qquad && \varepsilon_t \sim \mathcal{N}\left(0, \sigma_t^2\right). \end{aligned}$$

For TVP models, it is common to assume a stochastic volatility (SV) specification [28], where the log volatility $h_t = \log \sigma_t^2$ follows an AR(1) process:

$$h_t | h_{t-1}, \mu, \phi, \sigma_\eta^2 \sim \mathcal{N}\left(\mu + \phi(h_{t-1} - \mu), \sigma_\eta^2\right). \tag{13.23}$$

The unknown model parameters μ, ϕ, and σ_η^2 in (13.23) and the entire latent volatility process $\{h_0, h_1, \ldots, h_T\}$ are added to the set of unknown variables. MCMC estimation is easily extended using the very efficient sampler developed by [31] and implemented in the R-package `stochvol` [27].

13.5.2 Sparse TVP Models for Multivariate Time Series

The TVP model (13.1) introduced in Section 13.2 for univariate time series can be easily extended to TVP models for multivariate time series. Consider, as illustration, the following TVP model for a q-dimensional time series $\mathbf{y}_t$,

$$\mathbf{y}_t \quad = \quad \mathbf{B}_t\mathbf{x}_t + \boldsymbol{\varepsilon}_t, \qquad \boldsymbol{\varepsilon}_t \sim \mathcal{N}_q\left(\mathbf{0}, \boldsymbol{\Sigma}_t\right), \tag{13.24}$$

where $\mathbf{x}_t$ is a *column* vector of p regressors, and $\mathbf{B}_t$ is a time-varying $(q \times p)$ matrix with coefficient $\beta_{ij,t}$ in row i and column j, potentially containing structural zeros or constant values.

Sparse TVP Cholesky SV models

One example is the sparse TVP Cholesky SV model [5], which reads for $q = 3$:

$$\begin{aligned} y_{1t} &= \varepsilon_{1t}, & \varepsilon_{1t} &\sim \mathcal{N}\left(0, \mathrm{e}^{h_{1t}}\right), \\ y_{2t} &= \beta_{21,t}y_{1t} + \varepsilon_{2t}, & \varepsilon_{2t} &\sim \mathcal{N}\left(0, \mathrm{e}^{h_{2t}}\right), \\ y_{3t} &= \beta_{31,t}y_{1t} + \beta_{32,t}y_{2t} + \varepsilon_{3t}, & \varepsilon_{3t} &\sim \mathcal{N}\left(0, \mathrm{e}^{h_{3t}}\right), \end{aligned} \tag{13.25}$$

where the log volatilities h_{it}, $i = 1, \ldots, q$, follow q independent SV processes as defined in (13.23), with row specific parameters μ_i, ϕ_i, and $\sigma^2_{\eta,i}$. System (13.25) consists of three independent univariate TVP models, where no intercept is present. In the first row, no regressors are present either and only the log volatility h_{1t} has to be estimated. In the i-th equation, $i-1$ regressors are present and $i-1$ time-varying regression coefficients $\beta_{ij,t}$ as well as the time-varying volatility h_{it} need to be estimated. System (13.25) can be written as

$$\mathbf{y}_t \sim \mathcal{N}_q\left(\mathbf{B}_t\mathbf{x}_t, \mathbf{D}_t\right),$$

where $\mathbf{B}_t$ is a $q \times q$ matrix with time-varying coefficients $\beta_{ij,t}$, which are 0 for $j \geq i$. $\mathbf{D}_t = \mathrm{Diag}\left(\mathrm{e}^{h_{1t}}, \ldots, \mathrm{e}^{h_{qt}}\right)$ is a diagonal matrix and the q-dimensional vector $\mathbf{x}_t = (y_{1t}, \ldots, y_{qt})^\top$ is equal to $\mathbf{y}_t$.

It is possible to show that this system is equivalent to the assumption of a dynamic covariance matrix, $\mathbf{y}_t \sim \mathcal{N}_q\left(\mathbf{0}, \boldsymbol{\Sigma}_t\right)$, where $\boldsymbol{\Sigma}_t = \mathbf{A}_t\mathbf{D}_t\mathbf{A}_t^\top$ and the dynamic Cholesky factor $\mathbf{A}_t$ is lower triangular with ones on the main diagonal and related to $\mathbf{B}_t$ through $\mathbf{A}_t = (I_q - \mathbf{B}_t)^{-1}$.

Both in (13.25) as well as in the more general system (13.24), the unconstrained time-varying coefficients $\beta_{ij,t}$ are assumed to follow independent random walks as in the univariate case:

$$\beta_{ij,t} = \beta_{ij,t-1} + \omega_{ij,t}, \quad \omega_{ij,t} \sim \mathcal{N}\left(0, \theta_{ij}\right), \tag{13.26}$$

with initial values $\beta_{ij,0} \sim \mathcal{N}\left(\beta_{ij}, \theta_{ij}\right)$. Each of the time-varying coefficients $\beta_{ij,t}$ is potentially constant, with the corresponding process variance θ_{ij} being 0. A constant coefficient $\beta_{ij,t} = \beta_{ij}$ is potentially insignificant, in which case $\beta_{ij} = 0$. Hence, as for the univariate case, discrete spike-and-slab priors as introduced in Section 13.4 or continuous shrinkage priors as introduced in Section 13.3 are imposed on the fixed regression coefficients β_{ij}, as well as the process variances θ_{ij}. This defines a sparse multivariate TVP model for identifying which of these scenarios holds for each coefficient $\beta_{ij,t}$.

It is advantageous to introduce (hierarchical) shrinkage priors which are independent row-wise. For instance, [5], introduce a hierarchical double gamma prior for θ_{ij} and a hierarchical normal-gamma prior for β_{ij} for each row i of the TVP Cholesky SV model. Alternatively, independent discrete spike-and-slab priors with row-specific inclusion probabilities can be specified. Any of these choices leads to prior independence across the q rows of the system (13.24) and both model space MCMC as well as boosted MCMC can be applied row-wise to perform posterior inference.

Sparse TVP-VAR-SV models

Another important example are time-varying parameter vector autoregressive models of order r with stochastic volatility (TVP-VAR-SV), where the q-dimensional time series $\mathbf{y}_t$ is assumed to follow

$$\mathbf{y}_t = \mathbf{c}_t + \mathbf{\Phi}_{1,t}\mathbf{y}_{t-1} + \dots \mathbf{\Phi}_{r,t}\mathbf{y}_{t-r} + \boldsymbol{\varepsilon}_t, \qquad \boldsymbol{\varepsilon}_t \sim \mathcal{N}_q\left(\mathbf{0}, \mathbf{\Sigma}_t\right), \tag{13.27}$$

where $\mathbf{c}_t$ is the q-dimensional time-varying intercept, $\mathbf{\Phi}_{j,t}$, for $j = 1, \dots, r$ is a $q \times q$ matrix of time-varying coefficients, and $\mathbf{\Sigma}_t$ is the time-varying variance covariance matrix of the error term. Since the influential paper of [45], this model has become a benchmark for analyzing relationships between macroeconomic variables that evolve over time, see [11–13, 35, 38], among many others.

Since all q equations share the same predictor $\mathbf{x}_t = (1, \mathbf{y}_{t-1}^\top, \dots, \mathbf{y}_{t-r}^\top)^\top$ (a vector of length $p = qr + 1$), the TVP-VAR-SV model can be written in a compact notation exactly as in (13.24) with matrix

$$\mathbf{B}_t = \left(\mathbf{c}_t \ \mathbf{\Phi}_{1,t} \ \cdots \ \mathbf{\Phi}_{r,t}\right).$$

All coefficients $\beta_{ij,t}$ in $\mathbf{B}_t$ follow independent random walks as in (13.26) with initial expectation β_{ij} and process variance θ_{ij}. Due to the high dimensional nature of the time-varying matrix $\mathbf{B}_t$, shrinkage priors are instrumental for efficient inference, even for moderately sized systems. For instance, [6] introduce independent hierarchical triple gamma priors for β_{ij} and θ_{ij} in each row $i = 1, \dots, q$ of the TVP-VAR-SV model and demonstrate considerable efficiency gain compared to other shrinkage priors, such as the Lasso.

Since $\mathbf{\Sigma}_t$ is typically a full covariance matrix, the rows of the system (13.27) are not independent, as the various components in $\boldsymbol{\varepsilon}_t$ are correlated. Following [17], [6] use the Cholesky decomposition $\mathbf{\Sigma}_t = \mathbf{A}_t\mathbf{D}_t\mathbf{A}_t^\top$ to represent the TVP-VAR-SV model as a triangular system with independent errors $\boldsymbol{\eta}_t \sim \mathcal{N}_q\left(\mathbf{0}, \mathbf{D}_t\right)$. $\mathbf{A}_t$ is lower triangular with ones on the main diagonal and the unconstrained elements $a_{ij,t}$ in the i-th row and j-th column of $\mathbf{A}_t$ again follow random walks, with their own set of shrinkages priors on the corresponding variances and initial expectations.

The TVP-VAR-SV model then has a representation as a system of q univariate TVP models, e.g., for $q = 3$:

$$\begin{aligned}
y_{1t} &= \mathbf{x}_t\boldsymbol{\beta}_t^1 + \eta_{1t}, & \eta_{1t} &\sim \mathcal{N}\left(0, \sigma_{1t}^2\right), \\
y_{2t} &= \mathbf{x}_t\boldsymbol{\beta}_t^2 + a_{21,t}\eta_{1t} + \eta_{2t}, & \eta_{2t} &\sim \mathcal{N}\left(0, \sigma_{2t}^2\right), \\
y_{3t} &= \mathbf{x}_t\boldsymbol{\beta}_t^3 + a_{31,t}\eta_{1t} + a_{32,t}\eta_{2t} + \eta_{3t}, & \eta_{3t} &\sim \mathcal{N}\left(0, \sigma_{3t}^2\right),
\end{aligned}$$

where $\boldsymbol{\beta}_t^i$ is the ith row of $\mathbf{B}_t$. For $i > 1$, the ith equation is a univariate TVP model with the residuals $\eta_{1t}, \dots, \eta_{i-1,t}$ of the preceding $i - 1$ equations serving as explanatory variables. Nevertheless, the time-varying parameters $\boldsymbol{\beta}_t^i$ in each row can be estimated equation by equation [6].

It should be noted that both models might be sensitive to the ordering of the variables of the multivariate outcome $\mathbf{y}_t$, see [32] for a thorough discussion.

13.5.3 Non-Gaussian Outcomes

While the discussion of this chapter is centered around Gaussian time series, all methods can be extended to non-Gaussian time series, as demonstrated in [18], who also considered time series of small counts based on the Poisson distribution. The main idea is to augment auxiliary latent variables $\boldsymbol{\omega}$ such that conditional on $\boldsymbol{\omega}$ a Gaussian TVP model results. Variable and variance selection is then performed conditional on $\boldsymbol{\omega}$, while an additional step in the MCMC scheme imputes $\boldsymbol{\omega}$ given the remaining variables.

Examples include the representation of student-t errors as scale mixtures of Gaussians and binary time series, where the representation $d_t = \mathbb{I}(y_t > 0)$ leads to the conditionally Gaussian state space model (13.1). A similar strategy is pursued in [19, 50] for non-Gaussian random effects models and in [49] for dynamic survival models, see also [3] for a recent review on regularisation in complex and deep models.

13.5.4 Log Predictive Scores for Comparing Shrinkage Priors

Log predictive density scores (LPDS) are a widely used scoring rule to compare models; see, e.g., [22]. As shown by [5], log predictive density scores are also a useful means of evaluating and comparing different shrinkage priors for TVP models. It is common in this framework to use the first t_0 time series observations $\mathbf{y}^{\text{tr}} = (\mathbf{y}_1, \dots, \mathbf{y}_{t_0})$ as a "training sample", while evaluation is performed for the remaining observations $\mathbf{y}_{t_0+1}, \dots, \mathbf{y}_T$.

For univariate time series y_t, LPDS is defined as:

$$\text{LPDS} = \log p(y_{t_0+1}, \dots, y_T | \mathbf{y}^{\text{tr}}) = \sum_{t=t_0+1}^{T} \text{LPDS}_t^\star, \quad \text{LPDS}_t^\star = \log\, p(y_t | \mathbf{y}^{t-1}).$$

For each point in time, $\text{LPDS}_t^\star$ analyzes the performance separately for each y_t and is obtained by evaluating the one-step ahead predictive density $p(y_t|\mathbf{y}^{t-1})$ given observations $\mathbf{y}^{t-1} = (y_1, \dots, y_{t-1})$ up to $t-1$ at the *observed* value y_t. LPDS is an aggregated measure of performance for the entire time series. As shown by [15] in the context of selecting time-varying and fixed components for a basic structural state space model, LPDS can be interpreted as a log marginal likelihood based on the training sample prior $p(\boldsymbol{\vartheta}|\mathbf{y}^{\text{tr}})$, since

$$p(y_{t_0+1}, \dots, y_T | \mathbf{y}^{\text{tr}}) = \int p(y_{t_0+1}, \dots, y_T | \mathbf{y}^{\text{tr}}, \boldsymbol{\vartheta}) p(\boldsymbol{\vartheta}|\mathbf{y}^{\text{tr}}) d\,\boldsymbol{\vartheta},$$

where $\boldsymbol{\vartheta} = (\beta_1, \dots, \beta_p, \sqrt{\theta}_1, \dots, \sqrt{\theta}_p, \sigma^2)$ summarises the unknown model parameters. Hence, log predictive density scores provide a coherent foundation for comparing the predictive power of different types of shrinkage priors.

Determining $\text{LPDS}_t^\star$ for each $t = t_0+1, \dots, T$ can be challenging computationally. In [5], a Gaussian mixture approximation, called the *conditionally optimal Kalman mixture approximation*, is introduced to determine $p(y_t|\mathbf{y}^{t-1})$ independently for each t, based on M draws $\boldsymbol{\vartheta}^{(m)}, m = 1, \dots, M$ from the posterior distribution $p(\boldsymbol{\vartheta}|\mathbf{y}^{t-1})$.

The whole concept can be extended to multivariate time series by defining

$$\text{LPDS} = \log p(\mathbf{y}_{t_0+1}, \dots, \mathbf{y}_T | \mathbf{y}^{\text{tr}}) = \sum_{t=t_0+1}^{T} \text{LPDS}_t^\star, \quad \text{LPDS}_t^\star = \log\, p(\mathbf{y}_t | \mathbf{y}^{t-1}).$$

In a triangular system such as the TVP Cholesky SV model and the TVP-VAR-SV model discussed in Section 13.5.2, errors are uncorrelated and we can exploit that

$$\text{LPDS}_t^\star = \sum_{i=1}^{q} \text{LPDS}_{it}^\star, \quad \text{LPDS}_{it}^\star = \log\, p(y_{it} | y_{1t}, \dots, y_{i-1,t}, \mathbf{y}^{t-1}).$$

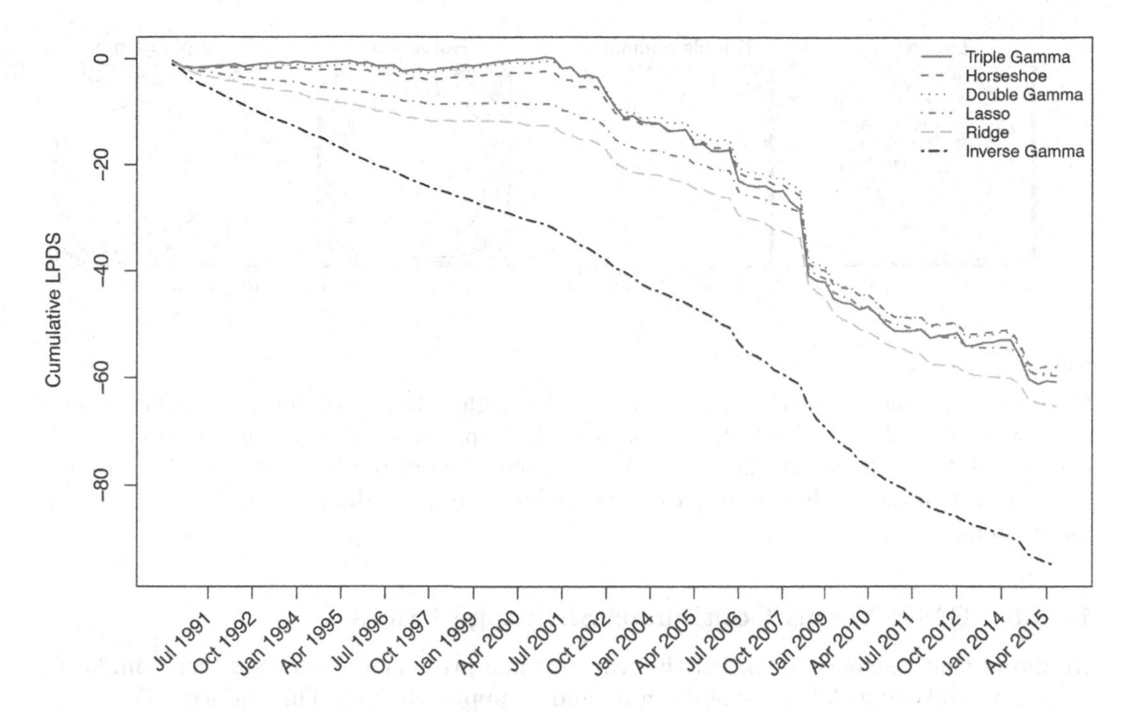

FIGURE 13.7
Cumulative LPDSs for the last 100 quarters of the inflation dataset introduced in Section 13.3.3, for six different continuous (shrinkage) priors.

Since we condition on *observed* values $y_{1t}, \dots, y_{i-1,t}$ in equation i, $\text{LPDS}^{\star}_{it}$ can be determined independently for each t and for each equation i. This allows one to fully exploit the computational power of modern parallel computing facilities.

Application to US inflation modelling

To demonstrate the benefit that shrinkage provides with regards to out-of-sample prediction, we calculate one-step ahead LPDSs for the last 100 time points of the inflation dataset introduced in Section 13.3.3 and compute the cumulative sum. Six different prior choices are considered here: (1) the triple gamma prior, (2) the horseshoe prior, (3) the double gamma prior, (4) the Lasso prior, (5) the ridge prior and, finally, (6) the inverse gamma prior. Figure 13.7 displays the results, with higher numbers equating to better out-of-sample prediction. It is immediately obvious that the inverse gamma prior does not appear to be competitive in this regard. While it displays a high degree of in-sample fit (as evidenced by Figure 13.5, Section 13.3.3), the forecasting performance severely lags behind the other prior choices. Similarly, if not quite as drastically, the ridge prior does not forecast as well as the more strongly regularized approaches. The three priors with the most shrinkage, the triple gamma, the horseshoe and the double gamma, all perform comparably, while the Lasso prior initially lags behind, only to gain ground during the subprime mortgage crisis between 2007 and 2009.

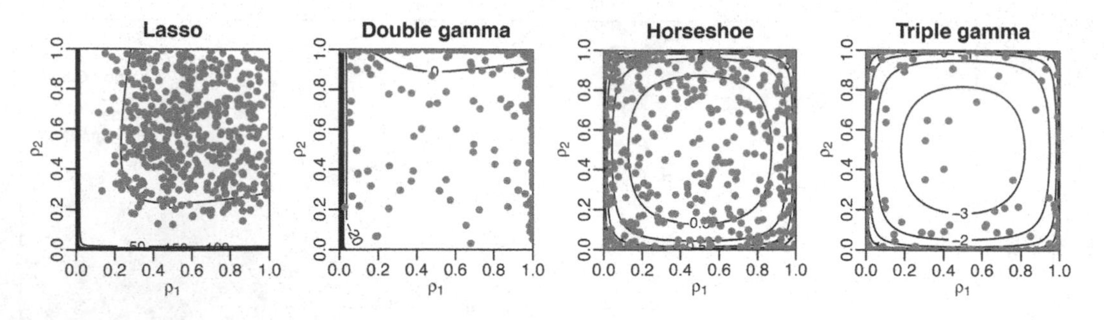

FIGURE 13.8
Bivariate shrinkage profile $p(\rho_1, \rho_2)$ for (from left to right) the Lasso prior, the double gamma prior with $a^\xi = 0.1$, the horseshoe prior, and the triple gamma prior with $a^\xi = c^\xi = 0.1$, with $\tau = 1$ ($\kappa_B^2 = 2$) for all the priors. The contour plots of the bivariate shrinkage profile are shown, together with 500 samples from the bivariate prior distribution of the shrinkage parameters.

13.5.5 BMA Versus Continuous Shrinkage Priors

An interesting insight of [6] is that the triple gamma prior shows behaviour very similar to a discrete spike-and-slab prior as both a^ξ and c^ξ approach zero. This induces BMA-type behaviour on the joint shrinkage profile $p(\rho_1, \ldots, \rho_p)$, with an infinite spike at all corner solutions, where some ρ_j are very close to one, whereas the remaining ones are very close to zero. For illustration, Figure 13.8 compares bivariate shrinkage profiles of various continuous shrinkage priors. The BMA-type behavior of the triple gamma becomes evident through the large amount of mass placed in the four corners, with the overlayed 500 samples from the prior following suit and clustering in those areas.

Following [8], a natural way to perform variable selection in the continuous shrinkage prior framework is through thresholding. Specifically, when $(1 - \rho_j) > 0.5$, or $\rho_j < 0.5$, the variable is included, otherwise it is not. Notice that thresholding implies a prior on the model dimension p_d defined as

$$p_d = \sum_{j=1}^{p} \mathbb{I}\{\rho_j < 0.5\} \sim \mathcal{B}in\,(p, \pi_\gamma)\,, \quad \pi_\gamma = \mathrm{P}(\rho_j < 0.5),$$

where $\rho_j \sim \mathcal{TPB}\left(a^\xi, c^\xi, \phi^\xi\right)$, see (13.14). The choice of the global shrinkage parameter ϕ^ξ strongly impacts the prior on p_d. For a symmetric triple gamma prior with $a^\xi = c^\xi$ and $\phi^\xi = 1$ fixed, for instance, $\pi_\gamma = 0.5$ and we obtain $p_d \sim \mathcal{B}in\,(p, 0.5)$, regardless of a^ξ. This leads to similar problems as with fixing $\pi_\gamma = 0.5$ for a discrete spike-and-slab prior. Placing a hyperprior on ϕ^ξ as discussed in Section 13.3.1 is as vital for variance selection through a continuous shrinkage prior as making π_γ random is for a discrete spike-and-slab prior. [8] show that the hyperprior for ϕ^ξ defined in (13.17) leads to a uniform prior distribution on the model dimension p_d, since $\pi_\gamma \sim \mathcal{U}\,[0, 1]$ is uniformly distributed.

Application to US inflation modelling

For illustration, we compare discrete spike-and-slab priors and hierarchical continuous shrinkage priors with regard to classification of the time-varying parameters for the inflation data set introduced in Section 13.3.3. The posterior probabilities of each coefficient

TABLE 13.2
Classifying the coefficients for the inflation data in Table 13.1 into zero coefficients (z), constant coefficients (f) and time-varying coefficients (d) under a discrete spike-and-slab prior with hierarchical Student-t slab.

β_{jt}	P(z|**y**)	P(f|**y**)	P(d|**y**)	β_{jt}	P(z|**y**)	P(f|**y**)	P(d|**y**)
intercept	0.30	0.41	0.29	emp	0.38	0.42	0.20
y_{t-1}	0.39	0.39	0.22	pmi	0.57	0.35	0.08
y_{t-2}	0.36	0.39	0.25	treas	0.03	0.11	**0.86**
y_{t-3}	0.28	0.35	0.37	spread	0.37	0.41	0.22
unemp	0.46	0.36	0.18	dow	**0.61**	0.30	0.09
cons	0.28	0.39	0.33	m1	0.39	0.44	0.17
dom_inv	0.58	0.34	0.08	exp	0.27	0.45	0.27
gdp	0.45	0.39	0.16	napmpri	0.05	**0.78**	0.17
hstarts	0.41	0.43	0.16	napmsdi	0.60	0.34	0.06

to be either zero, fixed or dynamic are estimated from the M posterior draws of $(\delta_j^{(m)}, \gamma_j^{(m)})$:

$$\mathrm{P}(\beta_{jt}\ \text{dynamic}|\mathbf{y}) = \frac{1}{M}\sum_{m=1}^{M}\gamma_j^{(m)}, \quad \mathrm{P}(\beta_{jt}\ \text{fixed}|\mathbf{y}) = \frac{1}{M}\sum_{m=1}^{M}\delta_j^{(m)}(1-\gamma_j^{(m)}),$$

and $\mathrm{P}(\beta_{jt}\ \text{zero}|\mathbf{y}) = 1 - \mathrm{P}(\beta_{jt}\ \text{dynamic}|\mathbf{y}) - \mathrm{P}(\beta_{jt}\ \text{fixed}|\mathbf{y})$. The indicators $(\delta_j^{(m)}, \gamma_j^{(m)})$ are an immediate outcome of the model space MCMC sampler for the discrete spike-and-slab prior and are derived for continuous shrinkage priors using thresholding as explained above.

According to this procedure, none of the coefficients is classified other than zero for the Lasso prior, which is not surprising in light of Figure 13.4. Somewhat unexpectedly, the same classification results for the triple gamma prior, for which a clear visual distinction can be made in Figure 13.4 between the relatively dynamic coefficient of treasury bills and the other two coefficients which are shrunken toward a fixed coefficient.

As opposed to this, the discrete spike-and-slab prior shows more power to discriminate between the different types of coefficients for this specific data set. The corresponding classification probabilities are reported for each coefficient in Table 13.2 and match the behavior of the recovered time-varying coefficients in Figure 13.6. More specifically, treasury bills is clearly classified as dynamic, the commodity prices index is classified as having a positive, but fixed effect on inflation, and the Dow Jones is clearly classified as insignificant.

13.6 Discussion

This chapter illustrates the importance of variance selection for TVP models. If the true model underlying a time series is sparse, with many coefficients being constant or even zero, then a full-fledged TVP model might quickly overfit. To avoid loss of statistical efficiency in parameter estimation and forecasting that goes hand-in-hand with the application of an overfitting model, we generally recommend to substitute the popular inverse gamma prior for the process variances by suitable shrinkage priors. As demonstrated in this chapter, shrinkage priors are indeed able to automatically reduce time-varying coefficients to constant or even insignificant ones.

Within the class of continuous shrinkage priors, flexible priors such as hierarchical versions of the double gamma, the triple gamma or the horseshoe prior typically turn out to be preferable to less flexible priors such as the hierarchical Lasso. These priors often show a comparable behavior in terms of model comparison through log predictive density scores and they beat the inverse gamma prior by far. This was illustrated with an application to US inflation modelling using a TVP Phillips curve.

Discrete spike-and-slab priors are an attractive alternative to continuous shrinkage priors as they allow explicit classification of the time-varying coefficients into dynamic, constant and zero ones. For continuous shrinkage priors, such a classification can be achieved only indirectly through thresholding and the appropriate choice of the truncation level is still an open issue for TVP models. However, convergence problems with model space MCMC algorithms are common with discrete spike-and-slab priors and the sampler might get stuck in different parts of the huge model space, depending on where the algorithm is intialized. In our illustrative application, discrete spike-and-slab priors were more successful in classifying obviously time-varying coefficients than any continuous shrinkage prior, but only in combination with a Student-t slab distribution. For other slab distributions, in particular Gaussian ones, severe convergence problems with trans-dimensional MCMC estimation were encountered.

A key limitation of any of the approaches reviewed in this chapter is that they can only differentiate between parameters that are constantly time-varying or not time-varying at all. One could think of many scenarios in which a parameter may be required to be time-varying over a stretch of time and be constant elsewhere. The design of suitable *dynamic shrinkage priors* that are able to handle such a situation is cutting-edge research in the area of state space and TVP models. Very promising approaches toward dynamic shrinkage priors were put forward by a number of authors, including [10, 29, 36, 46].

A script to replicate select results from this chapter and instructions on how to download software routines in `R` is made available as part of the online supplement of this edited volume.

Bibliography

[1] A. Armagan, D.B. Dunson, and C. Merlise. Generalized beta mixtures of Gaussians. In *Advances in Neural Information Processing Systems*, pages 523–531, 2011.

[2] M. Belmonte, G. Koop, and D. Korobolis. Hierarchical shrinkage in time-varying parameter models. *Journal of Forecasting*, 33:80–94, 2014.

[3] A. Bhadra, J. Datta, Y. Li, and N. Polson. Horseshoe regularisation for machine learning in complex and deep models. *International Statistical Review*, 34:405–427, 2019.

[4] A. Bhadra, J. Datta, N.G. Polson, and B. Willard. Lasso meets horseshoe: A survey. *Statistical Science*, 34:405–427, 2019.

[5] A. Bitto and S. Frühwirth-Schnatter. Achieving shrinkage in a time-varying parameter model framework. *Journal of Econometrics*, 210:75–97, 2019.

[6] A. Cadonna, S. Frühwirth-Schnatter, and P. Knaus. Triple the gamma – A unifying shrinkage prior for variance and variable selection in sparse state space and TVP models. *Econometrics*, 8:20, 2020.

[7] C. K. Carter and R. Kohn. On Gibbs sampling for state space models. *Biometrika*, 81:541–553, 1994.

[8] C.M. Carvalho, N.G. Polson, and J.G. Scott. Handling sparsity via the horseshoe. *Journal of Machine Learning Research W&CP*, 5:73–80, 2009.

[9] C.M. Carvalho, N.G. Polson, and J.G. Scott. The horseshoe estimator for sparse signals. *Biometrika*, 97:465–480, 2010.

[10] A. Cassese, W. Zhu, M. Guindani, and M. Vannucci. A Bayesian nonparametric spiked process prior for dynamic model selection. *Bayesian Analysis*, 14:553–572, 2019.

[11] J.C.C. Chan and E. Eisenstat. Bayesian model comparison for time-varying parameter VARs with stochastic volatilty. *Journal of Applied Econometrics*, 218:1–24, 2016.

[12] E. Eisenstat, J.C.C. Chan, and R.W. Strachan. Stochastic model specification search for time-varying parameter VARs. *SSRN Electronic Journal 01/2014; DOI: 10.2139/ssrn.2403560*, 2014.

[13] M. Feldkircher, F. Huber, and G. Kastner. Sophisticated and small versus simple and sizeable: When does it pay off to introduce drifting coefficients in Bayesian VARs, 2017. ArXiv: 1711.00564.

[14] S. Frühwirth-Schnatter. Data augmentation and dynamic linear models. *Journal of Time Series Analysis*, 15:183–202, 1994.

[15] S. Frühwirth-Schnatter. Bayesian model discrimination and Bayes factors for linear Gaussian state space models. *Journal of the Royal Statistical Society, Ser. B*, 57:237–246, 1995.

[16] S. Frühwirth-Schnatter. Computationally efficient Bayesian parameter estimation for state space models based on reparameterizations. In Andrew Harvey, Siem Jan Koopman, and Neil Shephard, editors, *State Space and Unobserved Component Models: Theory and Applications*, pages 123–151. Cambridge University Press, Cambridge, 2004.

[17] S. Frühwirth-Schnatter and R. Tüchler. Bayesian parsimonious covariance estimation for hierarchical linear mixed models. *Statistics and Computing*, 18:1–13, 2008.

[18] S. Frühwirth-Schnatter and H. Wagner. Stochastic model specification search for Gaussian and partially non-Gaussian state space models. *Journal of Econometrics*, 154:85–100, 2010.

[19] S. Frühwirth-Schnatter and H. Wagner. Bayesian variable selection for random intercept modeling of Gaussian and non-Gaussian data. In J. M. Bernardo, M. J. Bayarri, J. O. Berger, A. P. Dawid, D. Heckerman, A. F. M. Smith, and M. West, editors, *Bayesian Statistics 9*, pages 165–200. Oxford University Press, Oxford (UK), 2011.

[20] E. I. George and R. McCulloch. Variable selection via Gibbs sampling. *Journal of the American Statistical Association*, 88:881–889, 1993.

[21] E. I. George and R. McCulloch. Approaches for Bayesian variable selection. *Statistica Sinica*, 7:339–373, 1997.

[22] T. Gneiting and A. Raftery. Strictly proper scoring rules, prediction, and estimation. *Journal of the American Statistical Association*, 102:359–378, 2007.

[23] J. E. Griffin and P. J. Brown. Inference with normal-gamma prior distributions in regression problems. *Bayesian Analysis*, 5:171–188, 2010.

[24] J. E. Griffin and P. J. Brown. Hierarchical shrinkage priors for regression models. *Bayesian Analysis*, 12:135–159, 2017.

[25] A. C. Harvey. *Forecasting, Structural Time Series Models and the Kalman Filter*. Cambridge University Press, Cambridge, 1989.

[26] W. Hörmann and J. Leydold. GIGrvg: Random variate generator for the GIG distribution. R package version 0.4, url: `http://CRAN.R-project.org/package=GIGrvg`. 2015.

[27] D. Hosszejni and G. Kastner. Modeling univariate and multivariate stochastic volatility in R with stochvol and factorstochvol. *Journal of Statistical Software*, 2021. (available as arXiv report 1906.12123).

[28] E. Jacquier, N. G. Polson, and P. E. Rossi. Bayesian analysis of stochastic volatility models. *Journal of Business & Economic Statistics*, 12:371–417, 1994.

[29] M. Kalli and J. E. Griffin. Time-varying sparsity in dynamic regression models. *Journal of Econometrics*, 178:779–793, 2014.

[30] R. E. Kalman. A new approach to linear filtering and prediction problems. *Transactions ASME Journal of Basic Engeneering*, 82:35–45, 1960.

[31] G. Kastner and S. Frühwirth-Schnatter. Ancillarity-sufficiency interweaving strategy (ASIS) for boosting MCMC estimation of stochastic volatility models. *Computational Statistics and Data Analysis*, 76:408–423, 2014.

[32] L. Kilian and H. Lütkepohl. *Structural Vector Autoregressive Analysis*. Themes in Modern Econometrics. Cambridge University Press, Cambridge, 2017.

[33] P. Knaus, A. Bitto-Nemling, A. Cadonna, and S. Frühwirth-Schnatter. Shrinkage in the time-varying parameter model framework using the R package shrinkTVP. *Journal of Statistical Software*, 2021. conditionally accepted (available as arXiv report 1907.07065).

[34] G. Koop and D. Korobilis. Forecasting inflation using dynamic model averaging. *International Economic Review*, 53:867–886, 2012.

[35] G. Koop and D. Korobilis. Large time-varying parameter VARs. *Journal of Econometrics*, 177:185–198, 2013.

[36] D. R. Kowal, D. S. Matteson, and D. Ruppert. Dynamic shrinkage processes. *Journal of the Royal Statistical Society, Ser. B*, 81:781–804, 2019.

[37] G. E. Moran, V. Ročková, and E. I. George. Variance prior forms for high-dimensional Bayesian variable selection. *Bayesian Analysis*, 14:1091–1119, 2019.

[38] J. Nakajima. Time-varying parameter VAR model with stochastic volatility: An overview of methodology and empirical applications. *Monetary and Economic Studies*, 29:107–142, 2011.

[39] A. O'Hagan. Fractional Bayes factors for model comparison. *Journal of the Royal Statistical Society, Ser. B*, 57:99–138, 1995.

[40] O. Papaspiliopoulos, G. Roberts, and M. Sköld. A general framework for the parameterization of hierarchical models. *Statistical Science*, 22:59–73, 2007.

[41] T. Park and G. Casella. The Bayesian Lasso. *Journal of the American Statistical Association*, 103:681–686, 2008.

[42] G. Petris, S. Petrone, and P. Campagnoli. *Dynamic Linear Models with R*. Springer, New York, 2009.

[43] N.G. Polson and J.G. Scott. Shrink globally, act locally: Sparse Bayesian regularization and prediction. In J. M. Bernardo, M. J. Bayarri, J. O. Berger, P. Dawid, D. Heckerman, A. F. M. Smith, and M. West, editors, *Bayesian Statistics 9*, pages 501–538. Oxford University Press, Oxford, 2011.

[44] N.G. Polson and J.G Scott. Local shrinkage rules, Lévy processes, and regularized regression. *Journal of the Royal Statistical Society, Ser. B*, 74:287–311, 2012.

[45] G.E. Primiceri. Time varying structural vector autoregressions and monetary policy. *Review of Economic Studies*, 72:821–852, 2005.

[46] V. Ročková and K. McAlinn. Dynamic variable selection with spike-and-slab process priors. *Bayesian Analysis*, page forthcoming, 2020.

[47] C.A. Sims. Macroeconomics and reality. *Econometrica*, 48:1–48, 1980.

[48] M. Smith and R. Kohn. Parsimonious covariance matrix estimation for longitudinal data. *Journal of the American Statistical Association*, 97:1141–1153, 2002.

[49] H. Wagner. Bayesian estimation and stochastic model specification search for dynamic survival models. *Statistics and Computing*, 21:231–246, 2011.

[50] H. Wagner and C. Duller. Bayesian model selection for logistic regression models with random intercept. *Computational Statistics & Data Analysis*, 56:1256–1274, 2012.

[51] Y. Yu and X.-L. Meng. To center or not to center: that is not the question - an ancillarity-suffiency interweaving strategy (ASIS) for boosting MCMC efficiency. *Journal of Computational and Graphical Statistics*, 20:531–615, 2011.

14

Bayesian Estimation of Single and Multiple Graphs

Christine B. Peterson

The University of Texas MD Anderson Cancer Center (USA)

Francesco C. Stingo

University of Florence (Italy)

CONTENTS

In this chapter, we discuss Bayesian approaches for the inference of both single and multiple networks. To begin, we provide an overview of Bayesian graphical modeling approaches to learn directed and undirected networks, from either Gaussian or non-Gaussian data. We then provide an in-depth description of Bayesian models for the joint estimation of multiple undirected networks, where either the inclusion of edges or the edge values themselves are related across settings. These approaches highlight some of the advantages of taking a Bayesian approach, as hierarchical priors enable sharing of information across groups, automatic learning of parameters, and posterior estimates of uncertainty regarding individual networks and their similarity. We include simulation studies comparing the performance of Bayesian and frequentist methods for the inference of multiple networks, as well as a case study on brain structural connectivity networks during the progression of Alzheimer's disease.

DOI: 10.1201/9781003089018-14

14.1 Introduction

In statistical analysis, graphical models are applied to learn the dependence among a set of random variables, where each variable corresponds to a node in the graph, and the edges correspond to dependence relations. When the samples are drawn from a single population, the goal of inference is to learn a single graph structure G from the observations. Each edge can be represented by a binary indicator, and mixture priors can be used for model selection. However, in settings with heterogeneous populations, where network structures may differ, the assumption of a single graph structure is overly restrictive. In this setting, if the samples are drawn from known groups, a sensible goal is instead to estimate group-specific graph structures G_k, for $k = 1, \ldots, K$. This challenge is common in settings where there are known experimental or treatment groups, and the interest lies in understanding the network structures within each group and identifying the shared vs. group-specific relationships. In more complex settings where group membership is not known or the presence of an edge may depend on a continuous covariate, clustering or regression-based approaches can be used to learn cluster- or subject-specific graphs.

In this chapter, we first review methods for estimating single networks, which may be either directed or undirected. We also discuss Bayesian models for network inference from non-Gaussian data, which has been a popular topic in research years. We then move on to the primary focus of the chapter, which is approaches which enable the joint estimation of multiple undirected graphs in the Bayesian framework, relying on hierarchical priors to link edge selection or edge values across settings. Bayesian approaches for network inference offer several key advantages over frequentist alternatives: firstly, one can obtain posterior estimates of uncertainty regarding the inclusion of edges in the graph. This enables us to attach a measure of confidence to a selected network or edge. Another important advantage is that parameters can be learned automatically within a fully Bayesian approach. This is particularly important in the multi-network setting, where existing penalized approaches require the selection of more than one penalty parameter [8, 18], which can be challenging in practice. Finally, Bayesian approaches allow additional summaries of interest; in particular, the multiple network approaches discussed in this chapter enable posterior estimates of the extent of shared structure or shared edge values across networks.

14.2 Bayesian Approaches for Single Graph Estimation

In this section, we begin with some background on graphical models, then discuss Bayesian priors for inference of both undirected and directed networks. We also comment on recent methods that address the challenges of network inference for non-Gaussian data.

14.2.1 Background on Graphical Models

Mathematically, a graph G is represented by the pair (V, E), where V corresponds to a set of vertices, and E to a set of edges. In an *undirected* graph, an edge $(i, j) \in E$ if and only if $(j, i) \in E$. In a *directed* graph, each edge has an associated direction, and the edge (i, j) is often represented visually as an arrow. Directed graphs that contain no paths leading from a node back to itself are referred to as *directed acyclic graphs*, or DAGs. Finally, *chain graphs* and *reciprocal graphs* may include both directed and undirected edges. These graph structures can be used to represent the conditional dependence relations among a set of random variables. The choice of a specific type of graph is problem-dependent; for example, if the observed random variables can be naturally divided in multiple groups, and these groups follow a natural ordering, chain graphs are the most appropriate model, with groups represented by chain components. Similarly, if the observed random variables follow a natural ordering and it is reasonable to assume lack of feedback loops, DAGs become the most appropriate choice. In this chapter we will mostly focus on undirected graphs; these models assume that all observed random variables are on an equal footing.

In many Bayesian graphical modeling approaches, the presence or absence of an edge is represented using a latent indicator g_{ij}, where $g_{ij} = 1$ if $(i,j) \in E$, and 0 otherwise. When the data arise from a multivariate normal distribution $\mathcal{N}(\mathbf{0}, \mathbf{\Sigma})$, there is a convenient link between the inverse of the covariance matrix $\mathbf{\Sigma}^{-1} = \mathbf{\Omega}$, referred to as the precision or concentration matrix, and the structure of an undirected graph: specifically, the precision matrix entry ω_{ij} will be exactly 0 if and only if $(i,j) \notin E$, reflecting that the random variables i and j are conditionally independent given the remaining observed variables [10]. Estimation of a sparse version of $\mathbf{\Omega}$ therefore corresponds to learning an undirected graph structure. We illustrate this concept schematically in Figure 14.1. Models in this framework are known as *Gaussian graphical models.*

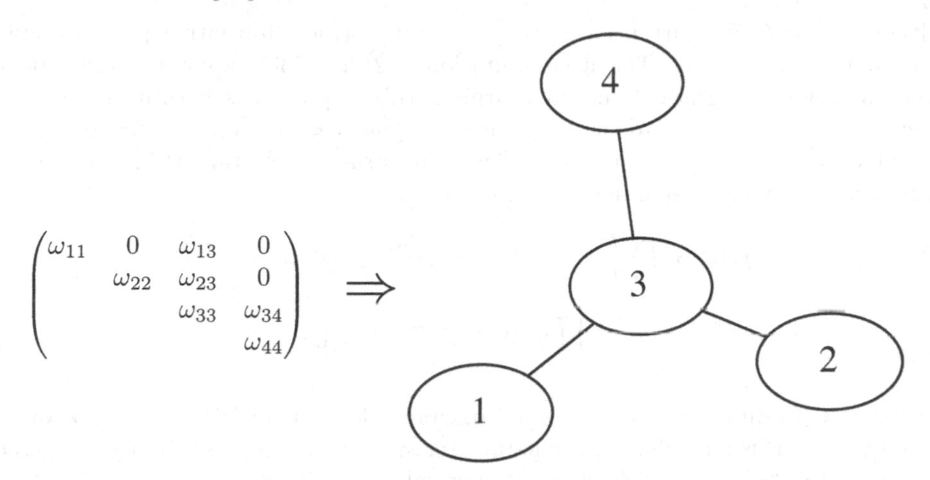

FIGURE 14.1
Schematic illustration showing the relation between edge selection and elements of the precision matrix in an undirected graph. In this illustration, the non-zero off-diagonal elements of the precision matrix $\mathbf{\Omega}$ at left correspond to the edges depicted in the network at right. For example, the non-zero precision matrix element ω_{13} is represented by the edge between nodes 1 and 3 in the graph at right.

14.2.2 Bayesian Priors for Undirected Networks

In Bayesian methods for the inference of Gaussian graphical models, a key modeling choice is how to formulate an appropriate prior on the precision matrix $\mathbf{\Omega}$ and the graph structure G. One possible prior on $\mathbf{\Omega}$ is the G-Wishart [43], which is the conjugate prior for the precision matrix of a multivariate normal distribution, restricted to the space P_G of symmetric positive definite matrices with exact zeros corresponding to the graph structure G. For $\delta > 2$ and D positive definite, $W_G(\delta, D)$ is a G-Wishart distribution on the space

P_G with density

$$P(\mathbf{\Omega}|G,\delta,D) = \frac{1}{I_G(\delta,D)}|\mathbf{\Omega}|^{(\delta-2)/2}\exp\left(-\frac{1}{2}tr(\mathbf{\Omega}^T D)\right). \tag{14.1}$$

Generating posterior samples in this framework poses a number of computational challenges, however; in particular, not only is the posterior normalizing constant intractable, but also that of the prior.

We now review some approaches developed for the inference of single undirected graphs in the Bayesian framework. Working with a G-Wishart prior, [11] proposed a joint search over the space of graphs and precision matrices, using a reversible jump algorithm to handle the changing dimension of the parameter space. [54] proposed a more efficient approach, using the exchange algorithm to avoid evaluation of the prior normalization constant, while [26] introduced a birth-death algorithm that offers improved convergence and computing time. One way to simplify the problem is to focus on the space of decomposable graphs, where it is possible to compute the normalizing constant analytically [9, 17]. Building on results in this framework, [47] proposed an updating scheme based on graph-theoretic results for decomposable graphs, with a generalization to approximate inference for unrestricted graphs.

Although the G-Wishart prior is the conjugate prior, alternative prior formulations may be more amenable to efficient computation. [52] and [35] explored the use of double exponential priors for graph structure learning; this approach is known as the *Bayesian graphical lasso*, an analogue to the frequentist graphical lasso [15]. Inspired by continuous spike-and-slab priors in the context of Bayesian variable selection [16], [53] proposed a mixture prior on the precision matrix elements,

$$p(\mathbf{\Omega}) \propto \prod_{i<j}\left\{(1-\pi)\mathcal{N}(\omega_{ij}|0,\nu_0^2) + \pi\mathcal{N}(\omega_{ij}|0,\nu_1^2)\right\}\cdot$$
$$\prod_i \text{Exp}(\omega_{ii}|\lambda/2)\cdot \mathbf{1}_{\mathbf{\Omega}\in M^+}, \tag{14.2}$$

where the first product reflects that the off-diagonal elements of $\mathbf{\Omega}$ follow a two-component mixture prior. In this formulation, the parameter $\nu_0 > 0$ should be set to a small value, and the parameter $\nu_1 > 0$ should be set to a larger value, resulting in a mixture of two normal densities, one concentrated close to 0, and one more diffuse. The second product term expresses that the diagonal elements of $\mathbf{\Omega}$ follow exponential densities, and the indicator function $\mathbf{1}$ represents that $\mathbf{\Omega}$ is constrained to the space of positive definite symmetric matrices M^+. As described in [53], this formulation enables efficient block Gibbs updates to the columns of $\mathbf{\Omega}$. [24] proposed a further gain in computational efficiency by avoiding MCMC altogether, instead relying on an expectation conditional maximization algorithm to estimate the posterior mode.

14.2.3 Bayesian Priors for Directed Networks

When there is a known ordering of the variables, for example, where there are established upstream factors, such as genetic sequence variation or microRNA abundances, that influence downstream gene or protein expression, the interest lies in inferring a set of directed edges. Given an *a priori* ordering of the variables, the joint likelihood can be factorized into a product of conditional distributions, which can be framed as regression problems. Working in this framework, [48] proposed an approach for the inference of regulatory networks, using mixture priors to learn the structure of a DAG. [29] integrated data from multiple platforms along with clinical outcomes in a directed graphical model, relying on Bayes factors for network selection. [30] allowed for non-linear effects through the incorporation of penalized splines, with structure learning via a mixture prior on the spline coefficients. Finally, [1] proposed the use of non-local priors [19] to enable faster learning of a directed graph structure.

14.2.4 Bayesian Network Inference for Non-Gaussian Data

As is the case for many statistical modeling approaches, much of the literature on graphical model inference in the Bayesian framework has focused on the Gaussian setting, due to its convenient theoretical and computational properties. Moving beyond the assumption of normality, there have been a number of proposals in recent years aimed at addressing settings with non-Gaussian data. [13, 14] proposed robust graphical modeling approaches that can accommodate outliers or heavy tails in the observed variables, using alternative and Dirichlet t-distributions. Building on this framework, [7] developed a flexible clustering approach using hierarchical completely normalized random measures to share information on deviations from normality across observations.

These methods enhance the applicability of graphical models for non-Gaussian continuous data, but are not suitable for count-based observations, which are obtained in many real-world applications such as RNA sequencing experiments or microbiome studies. Recent work addressing the challenge of count data includes [33], which proposed a hierarchical model where the observed compositional count data are modeled using a multinomial-Dirichlet distribution, with parameters that depend on a latent Gaussian layer. Selection of both the network structure and relevant covariates takes place at the latent layer, which captures the underlying absolute abundances. Latent Gaussian layers have also been proposed in Bayesian methods for network inference from Poisson data [51] and binary data [23, 49].

Finally, the most flexible class of approaches allow for variables arising from unknown or mixed distributions. For settings with non-normal or mixed discrete and continuous data, [40] proposed the use of a Gaussian copula model to learn the dependence structure, while [4] relied on Gaussian scale mixtures.

Given this background on approaches for the estimation of a single network, either directed or undirected, in the Bayesian framework, we proceed in the following sections to describe in more detail methods for the inference of multiple graphs.

14.3 Multiple Graphs with Shared Structure

As discussed in Section 17.1, in this chapter we are interested in not only the inference of single networks, but also network inference for problems where the data arise from multiple heterogeneous populations. In this section, we assume that there are K such groups, and that each sample in our data is drawn from a known group k, for $k = 1, \ldots, K$. We discuss approaches for the inference of multiple undirected graphs in the Bayesian framework, representing the dependence structures within each of these groups, using hierarchical models that flexibly encourage shared structure. We focus in particular on the modeling approach described in [37], which proposed the use of a Markov random field (MRF) prior, specifically an Ising model, to link the inclusion of edges across multiple networks. We describe this method in detail in the following subsections.

14.3.1 Likelihood

We let $\mathbf{X}_k$ represent the $n_k \times p$ matrix of observed data for the kth group, for groups $k = 1, \ldots, K$. In a practical setting, the K groups could be defined by experimental condition, disease status, or treatment group. We assume that the same p random variables are observed for each group, but allow the sample size n_k to be group-specific. The observed data for the ith subject in group k follow the multivariate normal likelihood $\mathbf{X}_{ik} \sim \mathcal{N}(\boldsymbol{\mu}_k, \boldsymbol{\Sigma}_k)$. Since the primary interest lies in inference of the covariance structure, rather than the mean, the data are assumed to be column-centered, so that $\boldsymbol{\mu}_k = \mathbf{0}$. Note that the precision matrix $\boldsymbol{\Omega}_k$ is group-specific. Following the Gaussian graphical modeling framework, the precision matrix within each group $\boldsymbol{\Omega}_k$ is assumed to be sparse, and the sparsity pattern is described using a group-specific graph structure G_k, where the binary indicator $g_{k,ij}$ denotes the inclusion of edge (i,j) in graph k.

14.3.2 Prior Formulation

The key innovation of [37] lies in the formulation of a Markov random field (MRF) prior to link the edge inclusion indicators across groups, thereby encouraging the joint selection of common edges. MRF priors have previously been applied to link the selection of variables [21, 38]. Here the prior serves a different purpose: to link the selection of edges across graphs. Consider the vector $\boldsymbol{g}_{ij} = (g_{1,ij}, \ldots, g_{K,ij})^T$, which represents the inclusion of edge (i,j) across the K groups. The joint MRF prior on $\boldsymbol{g}_{ij}$ is written

$$p(\boldsymbol{g}_{ij}|\nu_{ij}, \boldsymbol{\Theta}) \propto \exp\{\nu_{ij}\mathbf{1}^T\boldsymbol{g}_{ij} + \boldsymbol{g}_{ij}^T\boldsymbol{\Theta}\boldsymbol{g}_{ij}\},\ i = 1, \ldots, p;\ j = 1, \ldots, p, \tag{14.3}$$

where ν_{ij} is a scalar parameter influencing the prior probability of edge (i,j) across all graphs, and $\boldsymbol{\Theta}$ is a $K \times K$ symmetric matrix which reflects cross-group similarity of the networks. The entry θ_{km} in this matrix reflects the pairwise edge sharing between graphs G_k and G_m. This model is illustrated schematically in Figure 14.2, which shows an example configuration for $K = 3$ groups.

The joint prior for all graphs $G_1, \ldots, G_K$ is obtained by taking the product over all edges (i,j) of the density given in equation (14.3):

$$p(G_1, \ldots, G_K|\boldsymbol{\nu}, \boldsymbol{\Theta}) = \prod_{i<j} p(\boldsymbol{g}_{ij}|\nu_{ij}, \boldsymbol{\Theta}), \tag{14.4}$$

where $\boldsymbol{\nu} = \{\nu_{ij}|1 \leq i < j \leq p\}$. The parameter $\boldsymbol{\nu}$ can be used to control the prior probability of edge inclusion, without consideration of edge sharing, while $\boldsymbol{\Theta}$ flexibly captures shared structure across groups.

A key feature of the model proposed in [37] is that the cross-group relationships are not pre-specified. Instead, they are inferred within the Bayesian hierarchical modeling framework through the choice of an appropriate prior on the elements of the matrix $\boldsymbol{\Theta}$. Specifically, [37] propose the use of a mixture prior on θ_{km}:

$$p(\theta_{km}|\gamma_{km}) = (1 - \gamma_{km}) \cdot \delta_0 + \gamma_{km} \cdot \text{Gamma}(\alpha, \beta), \tag{14.5}$$

where δ_0 represents a Dirac delta function, which corresponds to the "spike" portion of the mixture, while the "slab" portion of the mixture is taken to be a gamma distribution, which has support on the interval $(0, \infty)$. The latent indicator γ_{km} represents whether graphs k and m are related, and follows a Bernoulli prior with mean w. The joint prior on $\boldsymbol{\Theta}$ is defined as the product over the element-wise prior of equation (14.5); the joint prior on the indicators of cross-group relatedness $\boldsymbol{\gamma}$ is similarly defined as a product over the independent Bernoulli densities. This formulation allows for great flexibility regarding the

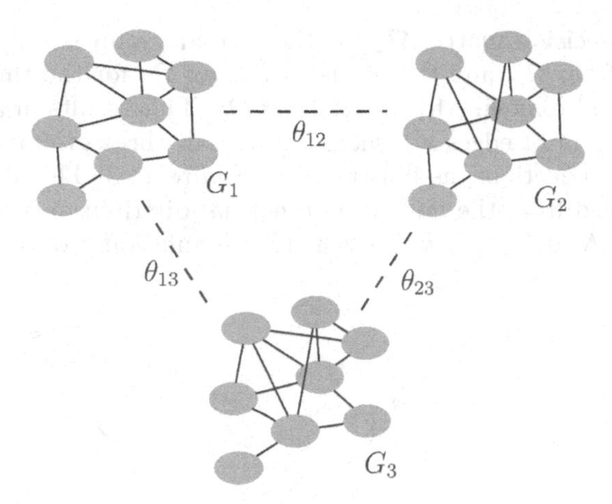

FIGURE 14.2
Schematic illustration of model, for an example setting with $K = 3$ groups. G_1, G_2, and G_3 represent the graph structures for sample groups 1, 2, and 3, respectively. The parameter θ_{12} captures the similarity in structure between graphs G_1 and G_2, θ_{13} captures the similarity between graphs G_1 and G_3, and θ_{23} captures the similarity between graphs G_2 and G_3.

extent of shared structure across groups, and also for inference on cross-group relatedness through posterior summaries of $\boldsymbol{\gamma}$ and $\boldsymbol{\Theta}$.

We now discuss the remaining elements of the prior specification. The parameter $\boldsymbol{\nu}$ influences the overall level of sparsity, and the prior on specific elements ν_{ij} can be tailored to reflect prior information regarding the inclusion of an edge (i, j). In the absence of edge-specific prior information, the prior on $\boldsymbol{\nu}$ can be formulated to put weight towards smaller values, to encourage overall graph sparsity. To achieve this, a Beta(a,b) prior is imposed on a transformation q_{ij} of ν_{ij}, specifically $q_{ij} = \frac{\exp(\nu_{ij})}{1+\exp(\nu_{ij})}$; see [37] for more detail on the choice of prior for this parameter. Finally, the prior on the precision matrix $\boldsymbol{\Omega}_k$ for each group is taken to be a G-Wishart. As discussed in Section 14.2, the G-Wishart is the conjugate prior for the precision matrix of a multivariate normal which follows a given sparsity pattern, described by a graph structure G. The formulation of this prior is given in equation (14.1).

Given this prior specification, posterior inference can be performed by generating Markov Chain Monte Carlo (MCMC) samples from the joint posterior. The MCMC scheme described in [37] allows for a reasonably efficient search over the space of possible graphs, precision matrices, and remaining model parameters. As discussed below, updating the group-specific graphs and precision matrices is the most expensive step, and subsequent work has enabled improved scalability in p through the adoption of a different prior on the precision matrices.

14.3.3 Simulation and Case Studies

Here we focus on a subset of the simulation results in [37] which highlight the benefits of the Bayesian joint estimation approach in comparison to alternative methods for graph structure learning. We also briefly summarize the case study on protein networks across cancer subtypes. In their simulation, [37] consider a setting where the number of nodes p equals 20, the sample size n per group is 50, and the number of sample groups K is 3. In the simulation set-up, the precision matrix $\boldsymbol{\Omega}_1$ for the first group is a band matrix with non-zero entries in the first two diagonals outside the main diagonal, corresponding to an

AR(2) graph. The precision matrix $\mathbf{\Omega}_2$ for the second group is obtained by adding and removing five edges from $\mathbf{\Omega}_1$, and the precision matrix $\mathbf{\Omega}_3$ for the third group is similarly obtained by adding and removing five edges from $\mathbf{\Omega}_2$. This results in a set of three graphs, where a large proportion of edges are shared across all three groups, but all graphs have some group-specific connections, as illustrated in Figure 14.3. The observed data for each group are then sampled from the multivariate normal distribution with the corresponding precision matrix, i.e., $\mathcal{N}(\mathbf{0}, \mathbf{\Sigma}_k)$ for $k = 1, 2, 3$. This simulation process is repeated 25 times.

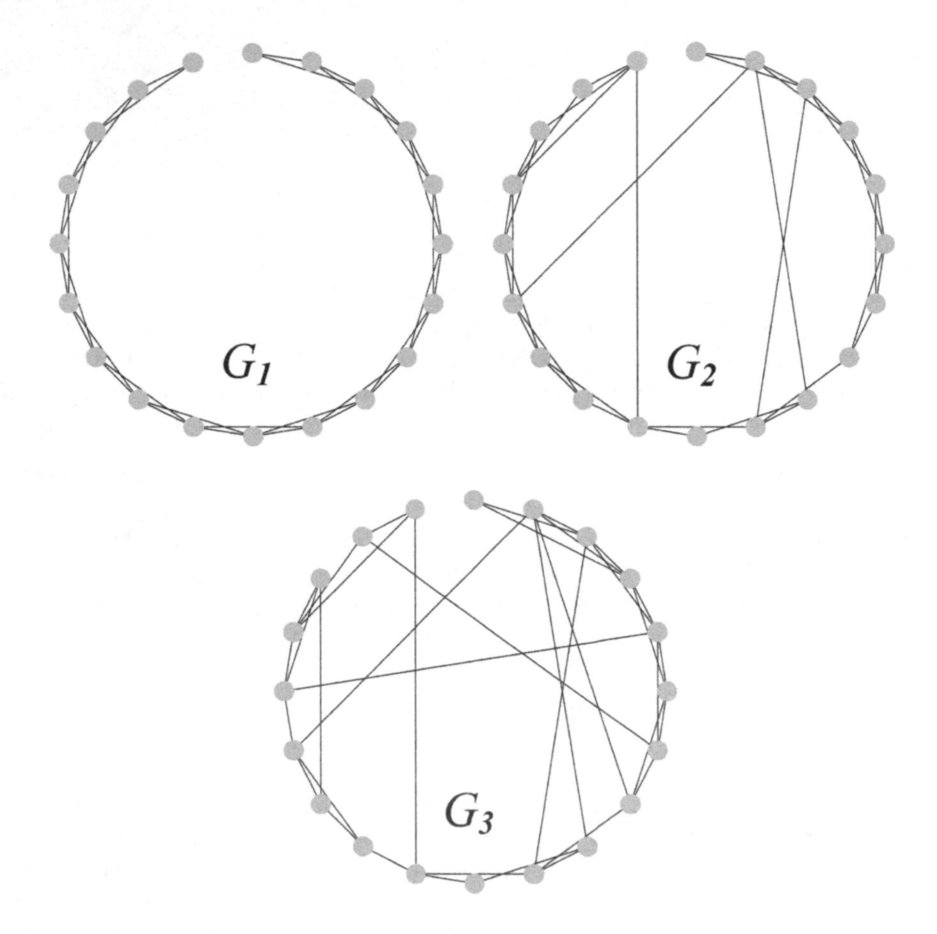

FIGURE 14.3
Graph structures G_1, G_2, and G_3 for the simulation described in Section 14.3.3.

The following methods are included in the performance comparison: the fused and joint graphical lasso [8], separate estimation using G-Wishart priors [54], and the joint Bayesian method of [37]. The within-group and cross-group penalty parameters for the fused and joint graphical lasso were chosen using a grid search with the Akaike information criterion as the objective. The prior probability of edge inclusion for [54] was taken as 0.2. The joint Bayesian method was applied with the parameter setting $\alpha = 2$, $\beta = 5$, $w = 0.9$, $a = 1$, $b = 4$, $\delta = 3$ and $D = I_p$; see [37] for a detailed justification of this hyperparameter setting. For both Bayesian methods, we ran 30,000 total iterations, and discarded the first 10,000 as the burn-in. Edges with a posterior probability of inclusion (PPI) > 0.5 are considered to be selected. Performance results and standard errors reflect averages over 25 simulated data sets.

	TPR	FPR	AUC
Fused graphical lasso	0.93 (0.03)	0.52 (0.10)	0.91 (0.01)
Group graphical lasso	0.93 (0.03)	0.55 (0.07)	0.88 (0.02)
Separate Bayesian	0.52 (0.03)	0.010 (0.006)	0.91 (0.01)
Joint Bayesian	**0.58** (0.04)	**0.008** (0.004)	**0.97** (0.01)

TABLE 14.1
True positive rate (TPR), false positive rate (FPR), and area under the receiver operating characteristic curve (AUC) for methods compared in the simulation study described in Section 14.3.3. Results are means over 25 simulated data sets, with standard error reported in parentheses.

We report the comparative performance results in Table 14.1, with an illustration of the receiver operating characteristic (ROC) curve for edge selection in Figure 14.4. As shown in Table 14.1, the Bayesian methods result in sparser networks, with a lower true positive rate and much lower false positive rate than the penalized methods considered. This is partly due to the fact that standard approaches for penalty parameter selection tend to result in insufficient sparsity [8]. The area under the ROC curve (AUC) provides an alternative metric that allows us to compare graph selection accuracy for varying values of the penalty parameters. The ROC curves are built by varying a threshold for the marginal posterior edge probability, for the Bayesian methods, and by varying the values of the penalty parameters, for the penalized likelihood approaches. The AUC values reported in Table 14.1 and the corresponding ROC curve provided in Figure 14.4 demonstrate that the joint Bayesian method achieves better selection performance across the range of model sizes. Notably, the joint Bayesian method improves sensitivity for edge selection over the separate Bayesian approach, demonstrating the benefit of joint estimation to enhance detection of shared structure.

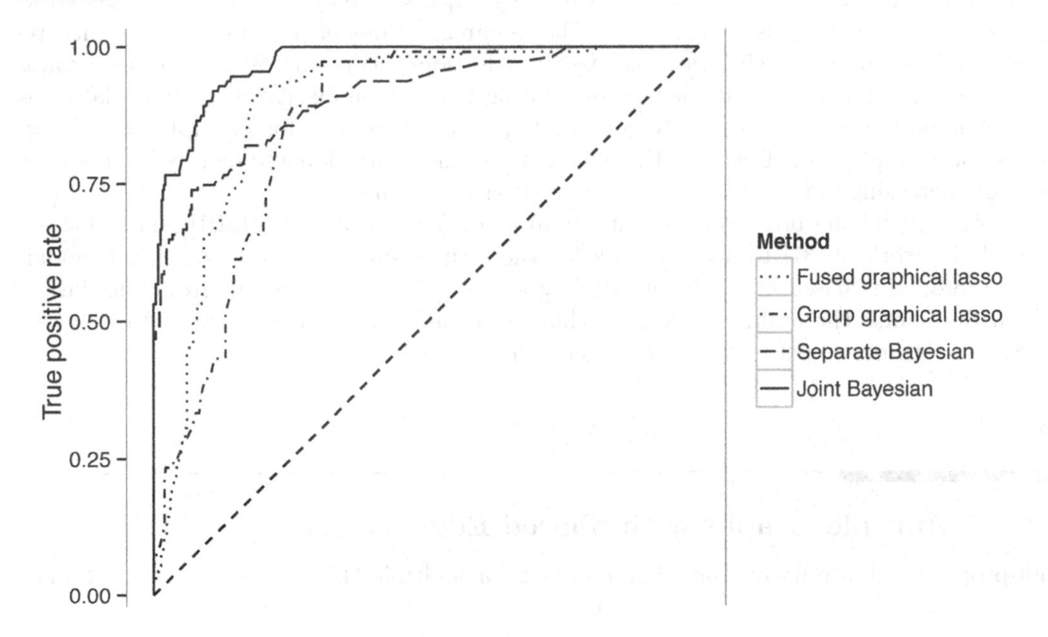

FIGURE 14.4
Receiver operating characteristic (ROC) curves for edge selection corresponding to the simulation study described in Section 14.3.3.

[37] include a case study examining the application of their joint Bayesian method to the inference of protein networks in cancer. In this application, they consider protein abundance data collected from subjects with various subtypes of acute myeloid leukemia (AML) [20], focusing on 18 proteins involved in apoptosis and cell cycle regulation, which are critical pathways in cancer pathobiology. There are 4 subtypes of AML analyzed, with varying sample sizes across the groups ($n_1 = 17, n_2 = 34, n_3 = 68$, and $n_4 = 59$). The resulting graphs include between 17 and 26 edges, with 9 shared across all 4 sample groups. This setting highlights a potential benefit of the joint Bayesian modeling framework, where information on common structure can be shared across groups, without allowing groups with larger sample sizes to dominate the results, as could potentially happen in pooled inference.

14.3.4 Related Work

We now briefly discuss approaches that build on the work of [37] to enable applications to larger number of variables p collected from a single platform or to data collected across multiple platforms. Although [37] demonstrate good performance of graph structure learning, the method is not scalable beyond a moderate number of variables ($p \approx 20 - 30$). The computational bottleneck is the joint updates of the graph and precision matrix within each group, which is based on a Metropolis-Hastings scheme proposed for single graph sampling under the G-Wishart prior [54]. By retaining the MRF prior across groups, but replacing the G-Wishart prior as in equation (14.1) within each group with a mixture prior as in equation (14.2), [46] were able to rely on a more efficient Gibbs sampling approach to update the graphs and precision matrices within each group, enabling scalability to larger problems ($p \approx 50 - 100$). Scalability in p could potentially be further improved by focusing on point estimation, as in [24].

The joint estimation of [37] assumes that p covariates are measured in each sample group. [45] consider a more challenging multi-omic study setting, where there are data from S different platforms collected across K sample groups, where p_s variables are observed for each subject for platforms $s = 1, \ldots, S$. The hierarchical model proposed by [45] incorporates a Markov random field prior relating the selection of edges across the sample groups, with a second Markov random field prior relating network similarity across the platforms. This formulation allows insight into coordinated network changes for multiple related data types, for example, for a disease setting where both gene expression and metabolite networks become increasingly disrupted for worsening severity of illness.

Motivated by the brain connectivity analysis of functional MRI (fMRI) data, [55] extended the work of [37] to a framework for the estimation of dynamic graphical models. This approach is based on a hidden Markov model (HMM) that learns from the data at which time point the brain connectivity changes, creating partitions of the data. For each partition a graphical network structure is estimated.

14.4 Multiple Graphs with Shared Edge Values

[36] proposed an alternative Bayesian approach for multiple Gaussian undirected graphical models based on a prior distribution that encourages similarity of edge values across groups, rather than similarity of the binary adjacency matrices that define the inclusion of edges in the networks, as originally proposed by [37]. This approach defines a joint framework that learns from the data the precision matrices within each group, and the cross-group similarity in terms of edge values. Akin to the fused graphical lasso of [8], similarity is

then defined in terms of the elements of the precision matrices; unlike alternative methods in the frequentist framework [39, 44], the approach of [36] does not require a first step to learn the cross-group similarities, but jointly learns both the within-group and cross-group associations from the data in a single step.

While the approach presented in Section 14.3 is appropriate when graphs may share edges regardless of the sign and magnitude of the corresponding connection (i.e., the corresponding value in the precision matrices), the method presented in this section assumes a deeper similarity, since similarity is defined not only in terms of presence/absence, but also the sign and magnitude of the edges.

14.4.1 Likelihood

We use the same notation as in Section 14.3.1. Within each group, observations are assumed to be independent and arising from a multivariate normal distribution, such that the distribution of each row of the observed matrix $\mathbf{X}_k$ of group k is distributed according to a multivariate normal distribution $\mathcal{N}(\boldsymbol{\mu}_k, \boldsymbol{\Sigma}_k)$. Note that the interest does not lie in the mean values, and we assume that data are centered by group, so that $\boldsymbol{\mu}_k = \mathbf{0}_k$ for $k = 1, \dots, K$. The main goal is to learn the group-specific precision matrices defined as the inverse of the covariance matrices $\boldsymbol{\Omega}_k \equiv (\omega_{k,ij}) = \boldsymbol{\Sigma}_k^{-1}, k = 1, \dots K$.

Graphical modeling within the Bayesian framework consists both in model selection (i.e., selection of the network structures) and in learning of the model parameters. These two steps, and corresponding parameters G_k and $\boldsymbol{\Omega}_k$, are interrelated, and similarity in the precision matrices will result in similarity in the graphs. The Bayesian approaches described in Section 14.3 encourage network similarity using prior distributions that link the graph structures G_k, whereas the approach proposed by [36] is based on a prior that links the groups through the parameters $\boldsymbol{\Omega}_k$, and similarity is defined in terms of the strength of the connections rather than only the inclusion/exclusion of edges. Methodologically, prior distributions on multiple precision matrices, and companion computational strategies, are harder to specify than priors on adjacency matrices, as precision matrices are constrained to be positive definite.

14.4.2 Prior Formulation

In the context of Gaussian undirected graphs, a range of prior distributions for the precision matrix $\boldsymbol{\Omega}$ have been proposed in the literature. First attempts required the graph to belong to the space of decomposable graphs [9, 17]; this assumption may limit the applicability of these types of methods but enables fast computational strategies based on marginal likelihoods of decomposable graphs that are analytically available. Methods that do not require restriction of the model space include the approach proposed by [52] based on shrinkage priors, and approaches based on conjugate priors [11, 43, 54], which, besides being limited in computational scalability, are difficult to extend to the multiple precision matrices case.

[36] constructed a novel prior distribution for multiple precision matrices building upon the stochastic search structure learning (SSSL) model proposed by [53]. This approach is based on a prior distribution for each element of the precision matrix: a mixture prior with two normal components on the off-diagonal entries and an exponential prior on the diagonal elements, as detailed in equation (14.2). The mixture prior is designed to perform graph selection, does not require any constraint on the graph structure, and enables implementation of efficient algorithms. In the context of multiple graphs, the SSSL framework can be extended to a joint model that borrows strength across groups through a joint prior on the precision matrices $\boldsymbol{\Omega}_1, \dots, \boldsymbol{\Omega}_K$.

The shrinkage prior for K precision matrices introduced by [36] is defined as

$$p(\boldsymbol{\Omega}_1 \ldots, \boldsymbol{\Omega}_K | \{\boldsymbol{\Theta}_{ij} : i < j\}) \propto \prod_{i<j} \mathcal{N}_K(\boldsymbol{\omega}'_{ij} | \mathbf{0}, \boldsymbol{\Theta}_{ij}) \prod_i \prod_k \text{Exp}(\omega_{k,ii} | \lambda/2) \mathbf{1}_{\boldsymbol{\Omega}_1 \ldots, \boldsymbol{\Omega}_K \in M^+}, \tag{14.6}$$

where $\boldsymbol{\omega}_{ij} = (\omega_{1,ij}, \ldots, \omega_{K,ij})$ are the elements corresponding to edge (i,j) across the K groups. The indicator function ensures that each precision matrix belong to M^+, the cone of positive definite symmetric matrices.

The key part of this prior that effectively links the K groups is the normal distribution on the off-diagonal elements. The covariance matrix $\boldsymbol{\Theta}_{ij}$ drives the amount of information shared across groups. A prior on $\boldsymbol{\Theta}_{ij}$ can be defined by exploiting the decomposition

$$\boldsymbol{\Theta}_{ij} = \text{diag}(\boldsymbol{\nu}_{ij}) \cdot \boldsymbol{\Phi} \cdot \text{diag}(\boldsymbol{\nu}_{ij}),$$

defined by a K-vector of edge-specific standard deviations $\boldsymbol{\nu}_{ij}$, and a $K \times K$ overall correlation matrix $\boldsymbol{\Phi}$ shared across all edges. With $\text{diag}(\boldsymbol{\nu}_{ij})$ we indicate a diagonal matrix with vector $\boldsymbol{\nu}_{ij}$ on the diagonal.

Edge selection is performed through a two-component continuous mixture prior such that each element of the vector $\boldsymbol{\nu}_{ij}$ is set to either a small (edge not selected) or large (edge selected) value [53]. Specifically, $\nu_{k,ij} = v_1$ if $g_{k,ij} = 1$, and $\nu_{k,ij} = v_0$ if $g_{k,ij} = 0$; $v_1 > 0$ and $v_0 > 0$ are set to large and small values, respectively.

Following [3], a joint uniform prior on the correlation matrix $\boldsymbol{\Phi}$ can be specified as follows:

$$p(\boldsymbol{\Phi}) \propto 1 \cdot \mathbf{1}_{\boldsymbol{\Phi} \in \mathcal{R}^K}, \tag{14.7}$$

where $\mathcal{R}^K$ is the set of $K \times K$ correlation matrices. Large positive values of the off-diagonal elements of $\boldsymbol{\Phi}$ encourage the inclusion of the same edges across groups, and, in the opposite case of $\boldsymbol{\Phi}$ being equal to an identity matrix, the K precision matrices are independent, and the entire method essentially reduces to a separate analysis of the K groups. In all middle ground cases, two groups k and l are related if the (k,l) element of $\boldsymbol{\Phi}$ assumes relatively larger values a posteriori.

A prior on the graphs $G_1, \ldots, G_K$ concludes the model specification. A simple option, adopted by [36], is a set of independent Bernoulli priors:

$$p(G_1, \ldots, G_K) \propto \prod_{k=1}^{K} \prod_{i<j} \{\pi^{g_{k,ij}} (1-\pi)^{1-g_{k,ij}}\}. \tag{14.8}$$

This distribution is defined only up to its normalizing constant. In order to simplify computations, the normalizing constant of prior (14.8) can be set to be proportional to the normalizing constant of prior (14.6) [53]; these normalizing constants will cancel out and therefore will not be part of the joint prior on $(\boldsymbol{\Omega}_k, G_k)$. As discussed in [53], π is not exactly equivalent to the probability of edge inclusion a priori; but it is relatively easily to set this parameter to values that achieve a targeted level of sparsity.

This prior formulation leads to computationally efficient algorithms [36] that automatically sample from the cone of positive definite precision matrices that corresponds to a given graph.

In simulation studies based on data generated to mimic the neuroimaging data described in the next section (Section 14.4.3), this method performed well in terms of structure learning and precision matrix estimation in comparison to alternative penalized likelihood and Bayesian approaches [8, 37, 53]: in general, Bayesian methods selected sparser graphs, leading to low false positive rates, whereas penalized likelihood methods selected denser graphs, leading to higher true and false positive rates. The linked precision matrix method of [36] achieves the best overall graph selection performance in terms of Matthews correlation coefficient (MCC) and AUC, and also the best accuracy of the precision matrix estimation in terms of the Frobenius loss.

14.4.3 Analysis of Neuroimaging Data

[36] analyzed neuroimaging data from the Australian Imaging, Biomarkers and Lifestyle (AIBL) study of ageing [12] with the goal of understanding changes in the structural connectivity networks during the progression of Alzheimer's disease. All individuals in this study are age 60 or older; their brain activity was measured using magnetic resonance imaging (MRI) and positron emission tomography (PET) brain imaging. Structural connectivity networks were inferred based on measurements of cortical thickness, a proxy of the atrophy of the cortical grey matter and thus of the neurodegenerative condition of the brain [41], for four groups of patients: high performing healthy control (hpHC, $n =$ 143 subjects), healthy control (HC, $n = 145$), mild cognitive impairment (MCI, $n = 148$), and Alzheimer's disease (AD, $n = 148$). This analysis focused on $p = 100$ brain regions of interest (ROI), based on a parcellation of the brain obtained using the Neuromorphometrics atlas (http://www.neuromorphometrics.com). The random variables of interest are the mean cortical thickness in each ROI. See [36] for a description of the data processing, as well as a discussion and guidelines for setting the hyperparameters of the model.

The MCMC algorithm was run multiple times for 20,000 iterations, after a burn-in period of 5,000 iterations; convergence checks did not point to any potential issues. A relevant summary of the posterior distributions obtained pulling together the multiple MCMC chains are the posterior probabilities of inclusion (PPI) for each edge. PPI for the hpHC and AD groups are summarized in Figure 14.5: the within group distribution of the PPIs is summarized by the histograms on the upper left and lower right, similarity between groups can be assessed looking at the scatter plot of the PPIs on the upper right and at the proportions of PPIs falling in each quadrant, in the lower left plot.

Overall, the resulting networks are quite sparse since many of the PPIs take very small values. Many edges are strongly supported across the two groups, highlighting the relevance of joint inference on the networks. From the PPIs, it is possible to estimate the structural connectivity networks, for example selecting all edges with PPI > 0.5 (often referred to as the median probability model).

The brain can be divided in five lobes: frontal, temporal, parietal, occipital and limbic cortex. For all disease groups, the vast majority of the edges identified by this analysis connect ROI belonging to the same lobe. Figure 14.6 displays the within-lobe estimated networks for two illustrative disease groups, hpHC and AD, for the frontal and occipital lobes; for each graph, the left side represents the left brain hemisphere, and the right side represents the right brain hemisphere. The edges are coded as follows: the lightweight dashed ones are shared by all 4 groups, the dotted ones are unique to a given group, and blue the solid ones are the edges shared by at least two groups. Confirming what was reported by [25], the method identifies many connections between the corresponding regions in the right and left hemispheres of the brain.

A key feature of this model is the ability of summarize the pairwise group similarity in terms of network structure. Specifically, the estimated submatrix of $\boldsymbol{\Phi}$ for the hpHC and AD groups:

$$\begin{pmatrix} & \text{hpHC} & \text{AD} \\ \text{hpHC} & 1.000 & \\ \text{AD} & 0.865 & 1.000 \end{pmatrix}$$

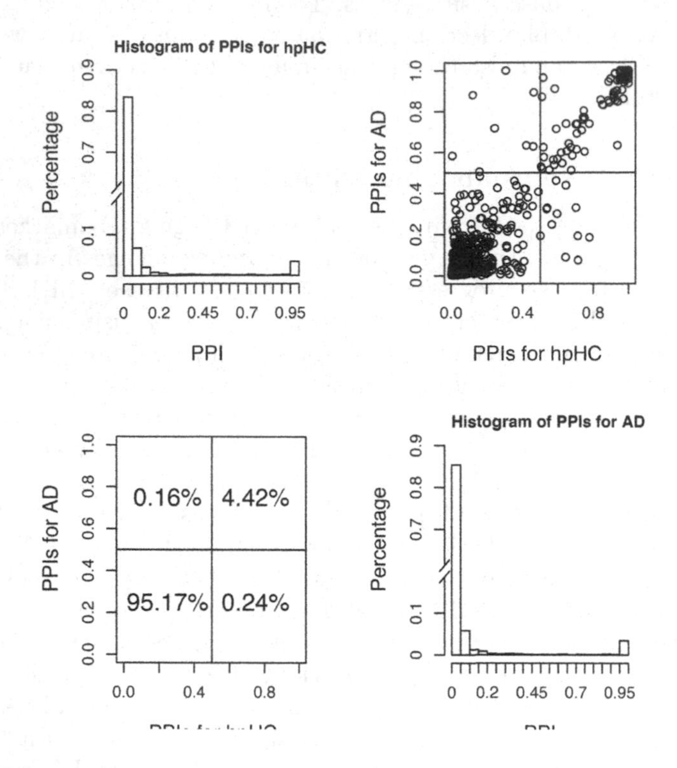

FIGURE 14.5
PPIs across two groups of subjects: hpHC (high-performing healthy controls) and AD (Alzheimer's disease). Plots on the upper left and lower right show histograms of the PPIs for the individual groups. The break in the y-axis is included to allow better visualization of the small PPIs. The upper right plot shows a scatter plot of the PPIs, and the lower-left plot shows the percents of PPIs falling in each quadrant for this pair of groups.

shows that, as expected, the hpHC and AD patients differ the most in terms of network structure, assessed in comparison to the remaining cross-group similarity parameters reported in [36].

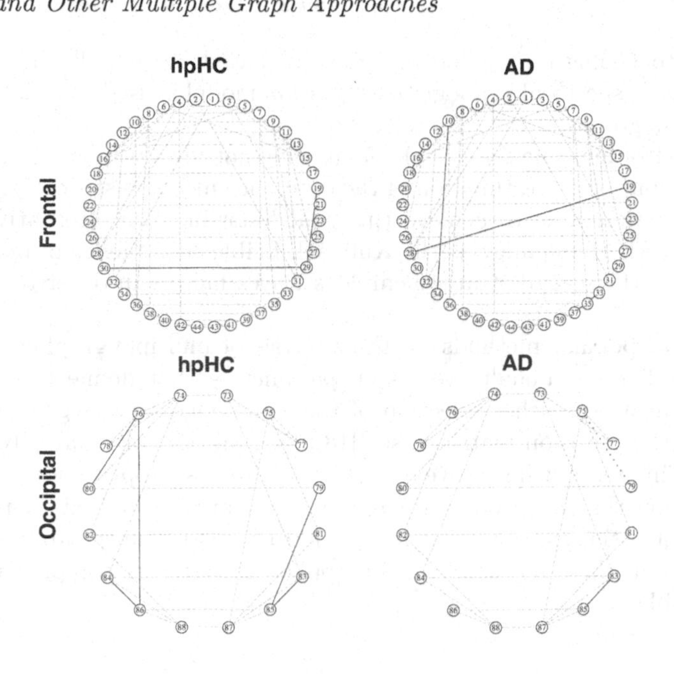

FIGURE 14.6
Illustrative subnetworks corresponding to the frontal (top) and occipital lobes (bottom), for 2 groups of subjects: hpHC (high-performing healthy controls) and AD (Alzheimer's disease)

14.5 Multiple DAGs and Other Multiple Graph Approaches

In this section, we briefly summarize methods that can be used for the joint analysis of multiple graphical models that are alternatives to the one previously discussed, with a particular emphasis on multiple DAGs. For example, DAGs are preferred in genetic analyses where directed pathways are of particular interest; additionally they represent the natural graphical framework to perform causal reasoning [34].

[57] proposed an approach for multiple directed graphs that can be implemented when the number of groups is 2. Specifically, one DAG is learnt for the first baseline group whereas the DAG for the second group is defined as the differential DAG. An alternative method for the analysis of multiple DAGs was proposed by [32]; this method is not limited to two groups and exploits integer linear programming for the exact estimation of multiple DAGs.

In the context of DAGs, a challenging aspect of model selection is the existence of Markov-equivalent DAGs, namely graphs which share the same set of independencies and cannot be distinguished through data alone [2]. [6] tackle this problem by considering a representative graph for each Markov equivalence class, named the essential graph. Specifically, they develop a methodology for Bayesian structural learning over multiple essential graphs, one for each of several groups, through novel prior on the graphs' skeletons; this approach relies on objective priors with regard to model parameters.

In a related paper, [28] proposed a model that goes beyond multiple DAGs and allows for the selection of cycles. The model space is then defined as the set of multiple directed cyclic graphs, and a Bayesian model is defined for the joint analysis of multiple related groups.

In the context of undirected graphs, [50] proposed a Bayesian approach for multiple groups related to Chung-Lu random graphs and based on a multiplicative prior on the adjacency matrices; specifically, a logistic regression model is used to link the edge inclusion probability across groups.

In the context of multiple graphs, the focus may not always be on the estimation of the graph structures but on the estimation of the difference in graph structures between groups; for example, [56] were able to achieve this goal following two alternative strategies: the first defines network discrepancy as the Kullback-Leibler divergence of posterior predictive distributions, and the second strategy employs Bayes factors in order to achieve the same goal.

Another set of popular methods for the analysis of multiple graphical models is based on penalized likelihoods. Penalty terms on parameters that define the group differences are defined to encourage either selection of the same edges across groups or similarity of the elements of the precision matrices; see [18], [58] and [5]. Alternatively, [8] were able to perform similar inference using convex penalization terms. Approaches based on penalized likelihoods assume that all groups are related, and are not suited to perform inference beyond the point estimation, making it very hard to assess uncertainty of the estimates. On the other hand, these methods can be applied to datasets comprised by a very large number of variables.

14.6 Related Topics

In the previous sections, we assumed that the conditional independence among variables is common to all observations in the same subgroup. In many applications, sample variability can not be completely explained by a discrete covariate used to define the subgroups.

For example, in many cancer genomic studies, the interest lies in the effect of clinically relevant biomarkers such as cancer stage, subtypes, or other prognostic factors, on the biological networks that drive the disease. These biomarkers can help explain the heterogeneity in the structure of the biological networks and thereby the heterogeneity among the cancer patients. The approach proposed by [31] introduces a subject-specific graph G_i to explicitly account for sampling heterogeneity, More specifically, [31] proposed a framework, termed graphical regression (GR), that leverages covariates, either discrete or continuous, in modeling subject-level graphs G_i. The main idea of GR is to express the subject-level parameters such as the graph G_i, that cannot be estimated by definition given the sample size of 1, as functions of covariates. These functions are characterized by parameters shared across all subjects and are therefore estimable. In the model proposed by [31] the response variables follow a DAG model specific for each subject, and covariates determine the structure of the subject-specific DAG through a function characterized by population-level parameters.

In other settings, there may not be covariates that can explain the heterogeneity among the subjects; for example, [42] proposed an infinite mixture of Gaussian graphical models based on Dirichlet process mixtures, that clusters observations into homogeneous subgroups, with each subgroup having its own conditional independence structure. The number of groups is learnt from the data, and this approach can be applied to any setting that does not include groups comprised by a very small number of subjects. The statistical model is rather complex and the companion algorithm does not scale well with the number of variables included in the graphical model.

14.7 Discussion

We have presented Bayesian methods for the estimation of single and multiple networks, with an in-depth discussion of Bayesian approaches based on mixture priors that can effectively infer the network structure of a finite number of subgroups. We focused on two approaches for undirected Gaussian graphical models: the first one [37] assumes that the subgroups may be similar in terms of the network structure, whereas the second one [36] assumes that the subgroups may be similar in terms of the elements of the precision matrices. Note that similarity of the precision matrices does imply similarity of the underlying graph, but not vice-versa. We also briefly described, in Section 14.5, methods that go beyond the undirected graphs and also, in Section 14.6, methods that do not assume a pre-specified subgrouping structure.

For the analysis of single networks, alternative approaches for the estimation of precision matrices based on horseshoe prior have been recently proposed. In the context of high-dimensional multivariate normal data [22] proposed a graphical horseshoe estimator that can be easily obtained through a Gibbs sampler.

In addition to the modeling aspects that have been extensively discussed here, another relevant area of research concerns computational strategies for these type of models, e.g., [24]. An alternative computational approach for multiple GGMs is proposed by [50]; posterior inference is obtained by means of a sequential Monte Carlo (SMC) algorithm which uses tempering techniques. As noted by the authors, this algorithm scales well in the number of groups, and has equivalent performance to methods based on G-Wishart priors.

Although methods based on penalized likelihood are computationally more efficient and scale better than the discussed Bayesian approaches in terms of the number of variables that can be included in the model, it is particularly important in the context of high-dimensional data to remark that Bayesian approaches provide measures of uncertainty, such as posterior model and edge inclusion probabilities, that clearly describe the confidence that we can have on a given selected model; this aspect is even more relevant for studies based on datasets of relatively small size, given that in this context many models may fit the data equally well. Finally, we want to mention that Bayesian modeling of single and, particularly, multiple networks remains an active area of research; current and future research topics of interest include computational approaches that may enable the analysis of larger networks, networks that change in structure with respect to covariates [31], and models for discrete or non-Gaussian random variables; see [27] for a recent review of Bayesian models for complex networks.

Acknowledgements

Figures 14.2–14.4 and Table 14.1 (Peterson et al. [37]) reprinted by permission of the American Statistical Association, www.amstat.org.

Bibliography

[1] D. Altomare, G. Consonni, and L. La Rocca. Objective Bayesian search of Gaussian directed acyclic graphical models for ordered variables with non-local priors. *Biometrics*, 69(2):478–487, 2013.

[2] S.A. Andersson, D. Madigan, and M.D. Perlman. A characterization of Markov equivalence classes for acyclic digraphs. *Annals of Statistics*, 25(2):505–541, 1997.

[3] J. Barnard, R. McCulloch, and X. Meng. Modeling covariance matrices in terms of standard deviations and correlations, with application to shrinkage. *Statistica Sinica*, 10(4):1281–1311, 2000.

[4] A. Bhadra, A. Rao, and V. Baladandayuthapani. Inferring network structure in non-normal and mixed discrete-continuous genomic data. *Biometrics*, 74(1):185–195, 2018.

[5] T. Cai, H. Li, W. Liu, and J. Xie. Joint estimation of multiple high-dimensional precision matrices. *Stat. Sinica*, 38:2118–2144, 2015.

[6] F. Castelletti, L. La Rocca, S. Peluso, F. C Stingo, and G. Consonni. Bayesian learning of multiple directed networks from observational data. *Statistics in Medicine*, 39(30):4745–4766, 2020.

[7] A. Cremaschi, R. Argiento, K. Shoemaker, C. Peterson, and M. Vannucci. Hierarchical normalized completely random measures for robust graphical modeling. *Bayesian Analysis*, 14(4):1271, 2019.

[8] P. Danaher, P. Wang, and D.M. Witten. The joint graphical lasso for inverse covariance estimation across multiple classes. *Journal of the Royal Statistical Society: Series B (Statistical Methodology)*, 76:373–397, 2014.

[9] A.P. Dawid and S.L. Lauritzen. Hyper markov laws in the statistical analysis of decomposable graphical models. *The Annals of Statistics*, pages 1272–1317, 1993.

[10] A.P. Dempster. Covariance selection. *Biometrics*, 28:157–175, 1972.

[11] A. Dobra, A. Lenkoski, and A. Rodriguez. Bayesian inference for general Gaussian graphical models with application to multivariate lattice data. *Journal of the American Statistical Association*, 106(496):1418–1433, 2011.

[12] K.A. Ellis, A.I. Bush, D. Darby, D. De Fazio, J. Foster, P. Hudson, et al. The Australian Imaging, Biomarkers and Lifestyle (AIBL) study of aging: methodology and baseline characteristics of 1112 individuals recruited for a longitudinal study of Alzheimer's disease. *Int. Psychogeriatr.*, 21(4):672–687, 2009.

[13] M. Finegold and M. Drton. Robust graphical modeling of gene networks using classical and alternative t-distributions. *The Annals of Applied Statistics*, pages 1057–1080, 2011.

[14] M. Finegold and M. Drton. Robust Bayesian graphical modeling using dirichlet t-distributions. *Bayesian Analysis*, 9(3):521–550, 2014.

[15] J. Friedman, T. Hastie, and R. Tibshirani. Sparse inverse covariance estimation with the graphical lasso. *Biostatistics*, 9(3):432–441, 2008.

[16] E.I. George and R.E. McCulloch. Variable selection via Gibbs sampling. *Journal of the American Statistical Association*, 88(423):881–889, 1993.

[17] P. Giudici and P.J. Green. Decomposable graphical Gaussian model determination. *Biometrika*, 86(4):785–801, 1999.

[18] J. Guo, E. Levina, G. Michailidis, and J. Zhu. Joint estimation of multiple graphical models. *Biometrika*, 98(1):1–15, 2011.

[19] V.E. Johnson and D. Rossell. On the use of non-local prior densities in Bayesian hypothesis tests. *Journal of the Royal Statistical Society: Series B (Statistical Methodology)*, 72(2):143–170, 2010.

[20] S.M. Kornblau, R. Tibes, Y.H. Qiu, W. Chen, H.M. Kantarjian, M. Andreeff, K.R. Coombes, and G.B. Mills. Functional proteomic profiling of AML predicts response and survival. *Blood*, 113(1):154–164, 2009.

[21] F. Li and N.R. Zhang. Bayesian variable selection in structured high-dimensional covariate spaces with applications in genomics. *Journal of the American statistical association*, 105(491):1202–1214, 2010.

[22] Y. Li, B.A. Craig, and A. Bhadra. The graphical horseshoe estimator for inverse covariance matrices. *Journal of Computational and Graphical Statistics*, 28(3):747–757, 2019.

[23] Z.R. Li, T.H. McComick, and S.J. Clark. Using Bayesian latent Gaussian graphical models to infer symptom associations in verbal autopsies. *Bayesian Analysis*, 15(3):781–807, 2020.

[24] Z.R. Li and T.H. McCormick. An expectation conditional maximization approach for Gaussian graphical models. *Journal of Computational and Graphical Statistics*, pages 1–11, 2019.

[25] A. Mechelli, K.J. Friston, R.S. Frackowiak, and C.J. Price. Structural covariance in the human cortex. *J. Neurosci.*, 25(36):8303–8310, 2005.

[26] A. Mohammadi and E.C. Wit. Bayesian structure learning in sparse Gaussian graphical models. *Bayesian Analysis*, 10(1):109–138, 2015.

[27] Y. Ni, V. Baladandayuthapani, M. Vannucci, and F.C. Stingo. Bayesian graphical models for modern biological applications. *Statistical Methods and Applications*, (Under revision), 2021.

[28] Y. Ni, P. Müller, Y. Zhu, and Y. Ji. Heterogeneous reciprocal graphical models. *Biometrics*, 74(2):606–615, 2018.

[29] Y. Ni, F.C. Stingo, and V. Baladandayuthapani. Integrative Bayesian network analysis of genomic data. *Cancer Informatics*, 13:CIN–S13786, 2014.

[30] Y. Ni, F.C. Stingo, and V. Baladandayuthapani. Bayesian nonlinear model selection for gene regulatory networks. *Biometrics*, 71(3):585–595, 2015.

[31] Y. Ni, F.C. Stingo, and V. Baladandayuthapani. Bayesian graphical regression. *Journal of the American Statistical Association*, 114(525):184–197, 2019.

[32] C.J. Oates, J.Q. Smith, S. Mukherjee, and J. Cussens. Exact estimation of multiple directed acyclic graphs. *Statistics and Computing*, 26(4):797–811, 2016.

[33] N. Osborne, C.B. Peterson, and M. Vannucci. Latent network estimation and variable selection for compositional data via variational EM. *Journal of Computational and Graphical Statistics*, in press, 2021.

[34] J. Pearl. *Causality: Models, reasoning, and inference.* Cambridge University Press, 2000.

[35] C. Peterson, M. Vannucci, C. Karakas, W. Choi, L. Ma, and M. Maletić-Savatić. Inferring metabolic networks using the bayesian adaptive graphical lasso with informative priors. *Statistics and its Interface*, 6(4):547, 2013.

[36] C.B. Peterson, N. Osborne, F.C. Stingo, P. Bourgeat, J.D. Doecke, and M. Vannucci. Bayesian modeling of multiple structural connectivity networks during the progression of Alzheimer's disease. *Biometrics*, 76(4), 1120–1132, 2020.

[37] C.B. Peterson, F.C. Stingo, and M. Vannucci. Bayesian inference of multiple Gaussian graphical models. *Journal of the American Statistical Association*, 110(509):159–174, 2015.

[38] C.B. Peterson, F.C. Stingo, and M. Vannucci. Joint Bayesian variable and graph selection for regression models with network-structured predictors. *Statistics in medicine*, 35(7):1017–1031, 2016.

[39] E. Pierson, the GTEx Consortium, D. Koller, A. Battle, and S. Mostafavi. Sharing and specificity of co-expression networks across 35 human tissues. *PLoS Computational Biology*, 11(5), 2015.

[40] M. Pitt, D. Chan, and R. Kohn. Efficient Bayesian inference for Gaussian copula regression models. *Biometrika*, 93(3):537–554, 2006.

[41] O. Querbes, F. Aubry, J. Pariente, J. Lotterie, J.-F. Démonet, V. Duret, et al. Early diagnosis of Alzheimer's disease using cortical thickness: impact of cognitive reserve. *Brain*, 132(8):2036–2047, 2009.

[42] A. Rodriguez, A. Lenkoski, and A. Dobra. Sparse covariance estimation in heterogeneous samples. *Electronic Journal of Statistics*, 5:981–1014, 2011.

[43] A. Roverato. Hyper inverse Wishart distribution for non-decomposable graphs and its application to Bayesian inference for Gaussian graphical models. *Scandinavian Journal of Statistics*, 29(3):391–411, 2002.

[44] T. Saegusa and A. Shojaie. Joint estimation of precision matrices in heterogeneous populations. *Electronic Journal of Statistics*, 10(1):1341–1392, 2016.

[45] E. Shaddox, C.B. Peterson, F.C. Stingo, N.A. Hanania, C. Cruickshank-Quinn, K. Kechris, R. Bowler, and M. Vannucci. Bayesian inference of networks across multiple sample groups and data types. *Biostatistics*, 21:561–576, 2020.

[46] E. Shaddox, F.C. Stingo, C.B. Peterson, S. Jacobson, C. Cruickshank-Quinn, K. Kechris, R. Bowler, and M. Vannucci. A Bayesian approach for learning gene networks underlying disease severity in COPD. *Statistics in Biosciences*, 10(1):59–85, 2018.

[47] F. Stingo and G.M. Marchetti. Efficient local updates for undirected graphical models. *Statistics and Computing*, 25(1):159–171, 2015.

[48] F.C. Stingo, Y.A. Chen, M. Vannucci, M. Barrier, and P.E. Mirkes. A Bayesian graphical modeling approach to microRNA regulatory network inference. *The Annals of Applied Statistics*, 4(4):2024, 2010.

[49] A. Talhouk, A. Doucet, and K. Murphy. Efficient Bayesian inference for multivariate probit models with sparse inverse correlation matrices. *Journal of Computational and Graphical Statistics*, 21(3):739–757, 2012.

[50] L.S. Tan, A. Jasra, M. De Iorio, and T. Ebbels. Bayesian inference for multiple Gaussian graphical models with application to metabolic association networks. *The Annals of Applied Statistics*, pages 2222–2251, 2017.

[51] G. Vinci, V. Ventura, M. A. Smith, and R.E. Kass. Adjusted regularization in latent graphical models: Application to multiple-neuron spike count data. *Annals of Applied Statistics*, 12(2):1068–1095, 2018.

[52] H. Wang. Bayesian graphical lasso models and efficient posterior computation. *Bayesian Analysis*, 7(4):867–886, 2012.

[53] H. Wang. Scaling it up: Stochastic search structure learning in graphical models. *Bayesian Analysis*, 10(2):351–377, 2015.

[54] H. Wang and S.Z. Li. Efficient Gaussian graphical model determination under *G*-Wishart prior distributions. *Electronic Journal of Statistics*, 6:168–198, 2012.

[55] R. Warnick, M. Guindani, E. Erhardt, E. Allen, V. Calhoun, and M. Vannucci. A Bayesian approach for estimating dynamic functional network connectivity in fMRI data. *Journal of the American Statistical Association*, 113(521):134–151, 2018.

[56] D.R. Williams, P. Rast, L. Pericchi, and J. Mulder. Comparing gaussian graphical models with the posterior predictive distribution and bayesian model selection, Feb 2019.

[57] M. Yajima, D. Telesca, Y. Ji, and P. Müller. Detecting differential patterns of interaction in molecular pathways. *Biostatistics*, 16(2):240–251, 12 2014.

[58] Y. Zhu, X. Shen, and W. Pan. Structural pursuit over multiple undirected graphs. *J. Am. Stat. Assoc.*, 109(508):1683–1696, 2014.

Part IV

Other Approaches to Bayesian Variable Selection

15

Bayes Factors Based on g-Priors for Variable Selection

Gonzalo García-Donato

Universidad de Castilla-La Mancha (Spain)

Mark F. J. Steel

University of Warwick (U.K.)

CONTENTS

Variable selection can be naturally seen as a model selection problem where the entertained models differ in which subset of variables explains the outcome of interest. Posterior model probabilities are a simple function of Bayes factors, a key inferential tool in Bayesian analysis. This approach to variable selection automatically provides sparse answers along with probabilistic assessments regarding their credibility. This methodology is, however, not exempt from difficulties including prior elicitation and numerical challenges related with its practical implementation. Particularly in the context of linear and generalized linear models, the so-called g-priors are often used due to their appealing properties formally described in [2]. In this chapter we review the implementation of variable selection through Bayes factors in linear and generalized linear models, using g-priors. Emphasis is placed on providing: i) practical guides for implementation, including documentation for the use of `R` packages and ii) the analysis of real examples which illustrate the enormous potential of this approach to variable selection.

15.1 Bayes Factors

Variable selection with Bayes factors is a methodology based on the *significance tests* by Sir Harold Jeffreys. These were introduced in a series of papers published during the first

DOI: 10.1201/9781003089018-15

decades of the 20th century and culminated in his famous book "Theory of Probability" [19], the first edition of which dates back to 1939. The reader is referred to [12] for an interesting account of the history of Bayes factors, and to [33] for a modern revision of Jeffreys' influential book.

Jeffreys' significance tests provide a solution to testing precise null hypotheses (i.e., that a certain parameter takes a particular value, such as zero) through the evidence that data give to each of the tested hypotheses. A key consideration is that hypotheses define statistical models and hence such evidence can be measured utilizing the relative support that each model receives from the data. The number that contains that relative evidence is what we call today the Bayes factor.

In the simple case with only two hypotheses, the data $\boldsymbol{y}$ follows a certain distribution under the null hypothesis $M_0 : \boldsymbol{y} \sim f_0(\boldsymbol{y} \mid \boldsymbol{\theta}_0)$, while under the alternative $M_1 : \boldsymbol{y} \sim f_1(\boldsymbol{y} \mid \boldsymbol{\theta}_1)$.

To decide which of these models/hypotheses provides a more appropriate representation for the underlying data generating process of $\boldsymbol{y}$ we obtain the Bayes factor of M_1 to M_0 as:

$$B_1 = \frac{m_1(\boldsymbol{y})}{m_0(\boldsymbol{y})}, \qquad m_\gamma(\boldsymbol{y}) = \int f_\gamma(\boldsymbol{y} \mid \boldsymbol{\theta}_\gamma)\, p_\gamma(\boldsymbol{\theta}_\gamma)\, d\boldsymbol{\theta}_\gamma,\ \gamma = 0, 1. \tag{15.1}$$

Above, m_1 and m_0 are the prior predictive marginals (often called marginal likelihoods) and $p_1(\cdot)$ and $p_0(\cdot)$ are the priors for the parameters within each model.

The Bayes factor B_1 is a measure of evidence in favor of M_1 and against M_0 provided by the data under the chosen prior distributions. The larger B_1, the stronger is the evidence supporting M_1. Several authors have provided rules to interpret B_1 [19, 21] but it is common to use them through their relation with the posterior probabilities of the models being compared. It can easily be seen that posterior odds ratios between models are equal to the prior odds multiplied by the appropriate Bayes factor:

$$\frac{p(M_1 \mid \boldsymbol{y})}{p(M_0 \mid \boldsymbol{y})} = \frac{p(M_1)}{p(M_0)} B_1, \tag{15.2}$$

where $p(M_\gamma)$ is the prior probability assigned to M_γ. Thus, conditionally on either M_0 or M_1 being the true model, the posterior probability of M_1 is

$$p(M_1 \mid \boldsymbol{y}) = \frac{p(M_1)B_1}{p(M_1)B_1 + p(M_0)}.$$

In order to make an explicit model selection we have to choose a threshold for the posterior probability for M_1. If $p(M_1 \mid \boldsymbol{y})$ is larger than this threshold we choose M_1 and otherwise we choose M_0. Such a threshold could, of course, be guided by a utility or loss function in a decision-theoretic setting (see e.g., [6]).

An important characteristic of this approach to model selection is that both models are given full consideration (different from methodologies where the selection is based on consideration of only the largest model). The advantages of this approach are nicely reviewed in [5] and here we want to emphasize two that we find particularly relevant. Firstly, the approach is automatically parsimonious (choosing the simplest model for a similar fit). This is essentially because, in logarithmic scale, Bayes factors can be approximated by a goodness-of-fit term minus a penalty for complexity that provides the mentioned automatic 'protection' to simpler models (for more details and related references on this issue see eg. [20]). Secondly, the method comes accompanied with a measure of uncertainty regarding the model selection exercise since it is based on the full posterior probability distribution over all considered models. This gives an accurate idea of the remaining uncertainty regarding which model to use. In this chapter we show, through real applications, the potential of the approach based on Bayes factors to variable selection highlighting the richness of the obtained inference.

The probability distribution over models provided by the formal Bayesian approach sketched above also allows us to formally include the uncertainty regarding models in our inference and decision-making by averaging over models with the posterior model distribution. This so-called Bayesian model averaging is the natural Bayesian response to uncertainty and was already described in [23] and used in e.g., [32] and [13]. This is a natural step to fully incorporate the model uncertainty, and is available in the packages used here. Key posterior quantities mentioned in the chapter, such as the posterior probability of inclusion of a regressor are derived by averaging over models.

In the situation where more than two models are entertained, the index γ takes values in a set $\mathcal{M}$ (called model space) and the posterior probability of any of the competing models is

$$p(M_\gamma \mid \boldsymbol{y}) \propto p(M_\gamma)B_\gamma,$$

where B_γ is the ratio of the marginal likelihood $m_\gamma(\boldsymbol{y})$ to the marginal likelihood of a fixed model (say, without loss of generality M_0). A multiple model selection problem naturally arises in *variable selection*. In this situation, the proposed models share a common functional form (e.g., a normal linear regression model) but differ in which explanatory variables, from a given set, are included to explain the response. The focus of this chapter will be on variable selection in the context of normal linear models (Section 15.2) and generalized linear models (Section 15.3).

Although very sound and cogent, the implementation of this methodology has two main challenges that we next describe. The first difficulty is a conceptual issue, namely that the prior used is going to have an important effect on the results. In contrast to the situation where we formulate a prior on the parameters of a single uncontested model, we do not have the luxury of priors that are "non-informative" in the sense that their effect is easily swamped by the data as we collect more observations. In addition, the prior needs to be proper on model-specific parameters. Indeed, any arbitrary constant in $p_\gamma(\boldsymbol{\theta}_\gamma)$ will affect the marginal likelihood $m_\gamma(\boldsymbol{y})$ in (15.1). Thus, if this constant emanating from an improper prior does not multiply the marginal likelihoods of all possible models, it clearly follows from (15.2) that posterior model probabilities are not determined. In this chapter, we pay special attention to the family of priors named g-priors. Strongly inspired by the work of Jeffreys to implement his significance tests, g-priors were introduced by [38, 39] and they have been the topic of renewed research interest over the last fifteen years or so. These types of priors, which some authors have also called *conventional* [1, 5], are introduced in Section 16.4.

A second challenge for the practical implementation is computational. In some cases (in particular, the linear Gaussian model with the class of g-priors mentioned in Subsection 16.4) the integral in (15.1) can be solved analytically, but in many other cases it does not admit an explicit solution and we need to resort to a simulated or approximated answer. In addition, the number of models in the model space can, for many applications be very large indeed, thus precluding an exhaustive enumeration of all possible models. As an example, the genetic example in Subsection 16.4 has a model space with 2^{4088} models, which is far larger than what can be dealt with exhaustively (typically, we can deal with model spaces up to size 2^{30} or so if we use complete enumeration). We discuss these numerical issues in Section 15.2.2. Fortunately, these methods are easily accesible to practitioners with specific `R` [30] packages. Here, we illustrate the use of the freely available packages `BayesVarSel` and `glmBfp`. All our examples are run on a Macbook pro laptop with 2.6 GHz Intel Core i5 processor without parallel computation, clearly indicating that the analysis of practically relevant problems is readily accessible.

15.2 Variable Selection in the Gaussian Linear Model

Consider a random sample $\boldsymbol{y} = (y_1, \ldots, y_n)^T$ with components y_i being independent, normally distributed with unknown variance σ^2. In the regression setup, the mean μ_i of y_i is assumed to be a linear combination of a subset of p possible explanatory variables $\{X_1, \ldots, X_p\}$, but it is uncertain which is the relevant subset. This situation implicitly defines 2^p entertained regression models which can be expressed by making use of a vector parameter $\boldsymbol{\gamma} = (\gamma_1, \ldots, \gamma_p)^T$ where $\gamma_j \in \{0, 1\}$ and $\gamma_j = 1$ indicates that x_j is included in the model. Hence the model space is $\mathcal{M} = \{0, 1\}^p$, where each $\boldsymbol{\gamma} \in \mathcal{M}$ assumes that

$$\mu_i = \alpha + \sum_{j=1}^{p} \gamma_j \, x_{ij} \beta_j, \ \forall i = 1, 2, \ldots, n,$$

with x_{ij} denoting the ith observation of variable x_j, i.e., the (i, j)th element of the full $(n \times p)$ covariate matrix $\boldsymbol{X}$. Using the notation in the introduction, now $\boldsymbol{\theta}_\gamma = (\alpha, \boldsymbol{\beta}_\gamma, \sigma)$ and the entertained models are

$$M_\gamma : f_\gamma(\boldsymbol{y} \mid \alpha, \boldsymbol{\beta}_\gamma, \sigma) = \mathcal{N}_n(\boldsymbol{y} \mid \alpha \mathbf{1}_n + \boldsymbol{X}_\gamma \boldsymbol{\beta}_\gamma, \sigma^2 I_n),$$

where $\boldsymbol{X}_\gamma$ is a $(n \times p_\gamma)$-dimensional ($p_\gamma = \sum_{j=1}^{p} \gamma_j$) data matrix formed using the included variables in M_γ (abusing notation, $\boldsymbol{\beta}_\gamma$ is empty when $\boldsymbol{\gamma}$ is the null vector) and $\mathbf{1}_n$ represents a n-dimensional unitary column vector. The covariates are standardized by subtracting their means, which makes them orthogonal to the intercept and renders the interpretation of the intercept common to all models.

Assuming that one of the models in $\mathcal{M}$ is the true model, the posterior probability of any model γ^* is

$$p(M_{\gamma^*} \mid \boldsymbol{y}) = \frac{B_{\gamma^*}(\boldsymbol{y}) p(M_{\gamma^*})}{\sum_\gamma B_\gamma(\boldsymbol{y}) p(M_\gamma)}, \tag{15.3}$$

where $p(M_\gamma)$ is the prior probability of M_γ and B_γ is the Bayes factor of M_γ with respect to a fixed model, say M_0 (without any loss of generality) and hence $B_\gamma = m_\gamma / m_0$ and $B_0 = 1$.

15.2.1 Objective Prior Specifications

Priors for the within model parameters: the g-priors

The prior on the model parameters assigns posterior point mass at zero for those regression coefficients that are not included in M_γ, which automatically induces sparsity. Without loss of generality, the prior distribution can be expressed as

$$p_\gamma(\boldsymbol{\beta}_\gamma, \alpha, \sigma) = p_\gamma(\boldsymbol{\beta}_\gamma \mid \alpha, \sigma) p_\gamma(\alpha, \sigma).$$

The common parameters are assumed to be equal for all models and for $p(\alpha, \sigma)$ a very commonly used objective prior is assumed

$$p_\gamma(\alpha, \sigma) = p(\alpha, \sigma) \propto \sigma^{-1}. \tag{15.4}$$

In the g-priors, the model-specific parameters have the following distribution specified conditionally on a hyperparameter $g > 0$ (which is the reason for the name of these priors):

$$p_\gamma(\boldsymbol{\beta}_\gamma \mid \alpha, \sigma, g) = \mathcal{N}_{p_\gamma}(\boldsymbol{\beta}_\gamma \mid \mathbf{0}_{p_\gamma}, g\sigma^2 (\boldsymbol{X}_\gamma^T \boldsymbol{X}_\gamma)^{-1}). \tag{15.5}$$

Proposal	Reference	Name
Fixed g		
$g = n$	[21, 38]	Unit information prior
$g = p^2$	[15]	Risk inflation criterion prior
$g = \max\{n, p^2\}$	[13]	Benchmark prior
$g = \log(n)$	[13]	Hannan-Quinn
Random g		
$g \sim IGa(1/2, n/2)$	[19, 39, 40]	Cauchy prior
$g \mid a \sim p(g) \propto (1+g)^{-a/2}$	[27]	hyper-g
$g \mid a \sim p(g) \propto (1+g/n)^{-a/2}$	[27]	hyper-g/n
$g \sim p(g) \propto (1+g)^{-3/2}$, $g > \frac{1+n}{p_\gamma+1} - 1$	[2]	Robust prior

TABLE 15.1
Specific proposals for the hyperparameter g in the literature.

This prior structure already appeared in [5, 13] and is now the most commonly used prior in the context of the normal linear model.

Without g, the prior covariance matrix above coincides with the expected Fisher information matrix corresponding to $\boldsymbol{\beta}_\gamma$ obtained from the model M_γ. The parameter g has the role of scaling the resulting matrix, for example such that the prior reflects a similar amount of information as one observation (this corresponds to a fixed value $g = n$ and leads to log Bayes factors that behave asymptotically like the BIC, see [13]). Several authors have argued in favor of treating g as an unknown hyperparameter for which a hyperprior needs to be assigned. There are theoretical reasons for the introduction of this extra layer of variability that relate to information consistency, which implies that the posterior probability tends to one for a model with arbitrarily large sampling evidence in its favour. In addition, from a practical perspective, treating g as random provides a prior for $\boldsymbol{\beta}_\gamma$ with flatter tails, hence accommodating regressors with a moderate impact on the response. In the section devoted to the sensitivity analysis, we will see a manifestation of this effect in practice. In Table 15.1 we have collected the most popular proposals for g. There are subtle conceptual differences that have lead different authors to propose these specific choices and the reader is referred to the original references for more details. Hence, the g-priors have, manifestly, been proposed based on constructive arguments. Nevertheless, much later [3] showed that these priors can also be derived using a mathematical formal rule based on the "distance" between competing models.

The ensuing methodology based on g-priors is endorsed by many appealing theoretical properties. A number of these have a frequentist flavor like consistency (ability to select the true model when the sample size grows to infinity or when the evidence in favour of a model becomes overwhelming) but others are specifically related with the desiderata for objective priors for testing and model selection. Within the latter category, we emphasize that g-priors are predictive matching (reporting inconclusive evidence when the sample size is extremely small). Finally, these priors produce Bayes factors that are invariant to affine transformations of the covariates. The reader is referred to [2] for a comprehensive discussion of these properties.

Priors over the model space $\mathcal{M}$

For priors over the model space $\mathcal{M}$, a very popular starting point is

$$p(M_\gamma \mid \zeta) = \zeta^{p_\gamma}(1-\zeta)^{p-p_\gamma}, \tag{15.6}$$

where p_γ is the number of covariates in M_γ, and the hyperparameter $\zeta \in (0,1)$ has the interpretation of the common prior probability that a given variable is included (independently of all others).

For the specific assignment in (15.6), some of the most popular default choices for ζ are

- Fixed $\zeta = 1/2$, which assigns equal prior probability to each model, i.e $p(M_\gamma) = 1/2^p$;
- Random $\zeta \sim \mathcal{U}(0,1)$, giving equal probability to each possible number of covariates or model size.

Of course many other choices for ζ – both fixed and random– have been considered in the literature. In general, fixed values of ζ have been shown to perform poorly in controlling for multiplicity (the occurrence of spurious explanatory variables as a consequence of performing a large number of tests) and can lead to rather informative priors. This issue can be avoided by using random distributions for ζ as, for instance, the second proposal above that has been studied in [35]. Additionally, [25] consider the use of $\zeta \sim \mathcal{B}eta(1,b)$ which results in a binomial-beta prior for the number of covariates in the model or the model size, p_γ:

$$p(p_\gamma = w \mid b) = \frac{b}{\Gamma(b+p+1)} \binom{p}{w} \Gamma(1+w)\Gamma(b+p-w), \; w = 0, 1, \ldots, p.$$

Notice that for $b = 1$ this reduces to the uniform prior on ζ and also on p_γ. As [25] highlight, this setting is useful to incorporate prior information about the mean model size, say $w^\star$. This would translate into $b = (p - w^\star)/w^\star$.

In variable selection, applications with a large number of explanatory variables p are becoming very common. In these situations, depending on the context and the prior information, it is typically a good idea to use a prior which implies a multiplicity correction or a prior which induces sparsity along the lines suggested in [8]. Additionally, in such contexts, we usually have to face situations where $p > n$ (or even $p \gg n$) and the set $\mathcal{M}$ contains models with $p_\gamma + 1 > n$ that are hence rank deficient (in the Gaussian setting these models are not estimable). Normally, these models are given zero prior probability (they are discarded) and the prior assignment in (15.6) applies only to those models for which $p_\gamma < n$ with a proportionality sign. A different treatment for these singular models in the normal case is given in [4]. These authors argue that while full rank models may contain decisive information concerning which covariates are related with the response, the rank deficient ones are not informative but will add uncertainty reflecting the fact that p is large compared to n. In this regard, [4] observe that rank deficient models are "copies" (reparameterizations) of the saturated model with $p_\gamma + 1 = n$ with the same marginal likelihood (cf. (15.7) with $SSE_\gamma = 0$ and $p_\gamma + 1 = n$). This justifies the use of unitary Bayes factors for all rank deficient models and thus avoids the need to assign zero prior probability to these models. In practical terms, both approaches are expected to provide similar results, unless n is very small. We have observed this agreement in the second of our applications which concerns a high dimensional study with $n = 71$ and $p = 4088$.

15.2.2 Numerical Issues

As already mentioned in Subsection 15.1 there are two main computational challenges in solving a model uncertainty problem through Bayes factors. Firstly, the integral in (15.1) and, secondly, the sum in the denominator of (15.3) which involves many terms if p is moderate or large.

Fortunately, in normal models, g-priors combine easily with the likelihood, and conditionally on g lead to closed forms for $m_\gamma(\boldsymbol{y})$. Hence, at most, a univariate integral needs to be computed when g is taken to be random. Thus Bayes factors have a very manageable expression:

$$B_\gamma(\boldsymbol{y}) = \int \left(1 + g\, Q_\gamma\right)^{-(n-1)/2} (1+g)^{(n-p_\gamma-1)/2}\, p(g)\, dg. \tag{15.7}$$

where Q_γ, is the ratio of the sum of squared errors of model M_γ to that of the null model M_0. Interestingly, there have been recent proposals for prior distributions, which despite assuming a hyperprior on g induce closed form marginals using special mathematical functions. This characteristic is shared by the robust prior of [2], the prior of [28] and the hyper-g in [27].

The second problem, related with the magnitude of the number of models in $\mathcal{M}$ (i.e., 2^p), could be a much more difficult one. If p is small (say, p in the twenties at most) exhaustive enumeration is possible but if p gets larger, exact approaches quickly become infeasible. Interesting exceptions include the recent work by [9, 37] who have developed, for certain particular problems (where $n = p$), exact algorithms that may handle problems with even very large p. For the general variable selection case, however, it is hard to imagine an exact solution and we will have to rely on some sort of approximation to the posterior distribution. This question has been studied in [17] who considered a simple Gibbs scheme that was suggested by [18]. The algoritm can be efficiently implemented in the following way. Start taking an initial model $\boldsymbol{\gamma}_{(0)} = (\gamma_{1(0)}, \gamma_{2(0)}, \ldots, \gamma_{p(0)})$ with Bayes factor $B_{\gamma_{(0)}}$ then repeat, for $i = 1, \ldots, N$ (N is the number of iterations) the following $p+1$ steps:

- Step j : $1 \leq j \leq p$. Define the model $\boldsymbol{\gamma}_* = (\gamma_{1(i-1)}, \ldots, 1 - \gamma_{j(i-1)}, \ldots, \gamma_{p(i-1)})$ and compute B_{γ_*} and $r = B_{\gamma_*} p(M_{\gamma_*})/(B_{\gamma_*} p(M_{\gamma_*}) + B_{\gamma_{(i-1)}} p(M_{\gamma_{(i-1)}}))$. With probability r re-define $\boldsymbol{\gamma}_{(i-1)} = \boldsymbol{\gamma}_*$ and $B_{\gamma_{(i-1)}} = B_{\gamma_*}$.
- Final step. Define and save $\boldsymbol{\gamma}_{(i)} = \boldsymbol{\gamma}_{(i-1)}$.

The result is $\{\boldsymbol{\gamma}_{(1)}, \boldsymbol{\gamma}_{(2)}, \ldots, \boldsymbol{\gamma}_{(N)}\}$, a sample from the posterior distribution (15.3) which is the only ingredient needed to obtain summaries solely based on the frequency of visits. For instance, the vector of posterior inclusion probabilities is obtained as the sample mean and so on.

Despite the apparent simplicity of this strategy, [17] show that this approach is potentially more precise than heuristic searching methods looking for 'good' models with estimates based on renormalization (i.e with weights defined by the analytic expression of posterior probabilities, cf. (15.3)). They show that the latter methods could be biased by the searching procedure.

15.2.3 `BayesVarSel` and Applications

There are several `R` packages that make it straightforward to implement variable selection based on g-priors. These have been considered in some detail in [14] who concluded that the results are comparable across the packages, although there are still important differences in cover and focus.

The results in this section are obtained with the `R` package `BayesVarSel` ([16]). This package was first released in December 2012 and has been periodically maintained and updated with new functionalities since then. `BayesVarSel` can be downloaded from the official `R` repository `https://cran.r-project.org` or from the github repository `https://github.com/comodin19/BayesVarSel`. The version used in this chapter is `2.2.0`. The code for running the examples is provided as supplementary material.

The package was conceived as a suite of tools to solve the variable selection problem for Gaussian linear models based on g-priors. It comes armed with many possibilities to summarize the posterior distributions and is very flexible regarding the choices of g (in particular, it incorporates the proposals in Table 15.1) and $p(M_\gamma)$. In our applications, we use the default choice in the package that corresponds to using $\zeta \sim \mathcal{U}(0,1)$ for $p(M_\gamma \mid \zeta)$ and the Robust prior [2] for $p(g)$. This configuration of prior inputs is the one that we ultimately recommend in general variable selection procedures.

Example of a moderate p (enumeration is feasible)

OBICE ([41]) was a study conducted in Spain during the years 2007-2008 to determine the association between diet, physical activity and obesity in children under 15 years of age. This study has a case-control design and the collected data come from a questionnaire completed by pediatricians. The survey collected a lot of information, some of which is redundant, and here we are considering $p = 15$ variables that provide a complete description of the aspects considered in the study (see Table 15.2). Data from the OBICE study is distributed with `BayesVarSel`. The model space contains $2^{15} = 32,768$ models, a size that allows us to compute posterior probabilities exactly (through exhaustive enumeration) in a few seconds. To avoid using imputation methods we include here only the children without any missing value leading to a sample size of $n = 996$ (84% of the number of recruited children). The response variable, y_i, is the Body Mass Index (BMI) and the age of the child is a variable that is always included (since it is known to influence BMI).

The possibility of reporting the degree of uncertainty regarding the variable selection problem in several informative ways is an important advantage of the methodology based on Bayes factors illustrated here. For instance, the model that in this study is most probable a posteriori (indicated as HPM in Table 15.2) contains (apart from the fixed one) 9 variables and has a posterior probability of 0.06. The model that follows in probability is the full model with a probability of 0.04. Interestingly, the smaller dimensionality of the HPM indicates that the information in some of the explanatory variables is really contained in others (e.g., the explanatory power of eating fruit and vegetables seem to be contained in the habit of having five meals per day). The individual importance of the variables can be assessed using the posterior inclusion probabilities. These are the aggregated probabilities of all models that contain a certain variable and are presented in Table 15.2. As expected, the variables included in the HPM are assigned large posterior inclusion probabilities (at least "strong evidence" according to the classification in [16]). None of the others are clearly ruled out, so the study doesn't have enough information to clarify whether these have an important role in the explanation of the body mass index. For instance, the sex of the child has an inclusion probability of 0.54 (quite similar to its prior probability) so there is no conclusive evidence whether this variable has any impact on obesity.

Additionally, we can explore the joint effect of variables in relation to their role in explaining the response (for a detailed study on this and related concepts the reader is referred to [24]). The information for such effect is contained in the probability that a certain variable is included, given that another is not, leading to a $p \times p$ matrix represented in Figure 15.1. To ease the interpretation, the row on top of the plot represents the marginal inclusion probabilities and the interrelations worth mentioning are when this probability differs substantially from the conditional probabilities in the table. In the great majority of cases there is barely any change (meaning that not including any other variable has no effect) but nevertheless several interesting facts arise. First is that we clearly see that either not considering the weight or height at birth diminishes the inclusion probability of the other (which makes sense since the response variable depends on both weight and height). More interesting is what happens with the dietary habit of having afternoon snacks. The importance of this variable increases substantially if the information contained in "having or not 5 meals per day" is not considered, allowing to conclude that this last variable has

Variable	Inc. Prob	HPM
Weight at birth	0.99	Y
Height at birth	0.99	Y
Sex	0.54	
The father is obese	1.00	Y
The mother is obese	1.00	Y
The child (regularly):		
...has 5 daily meals	1.00	Y
...eats vegetables	0.37	
...eats fruit	0.33	
...consumes afternoon snacks	0.83	Y
...was breastfed	0.42	
...practices sports	0.88	Y
Daily hours the child:		
...watches TV	1.00	Y
...plays with electronic devices	0.39	
...sleeps	0.42	
Daily candy consumption	0.99	Y

TABLE 15.2
Explanatory variables considered from the OBICE study. The dependent variable is Body Mass Index and the table contains posterior inclusion probabilities and an indicator of which variables appear in the highest posterior probability model (HPM).

a similar role as afternoon snacks. In other words, the two variables are substitutes (even though both appear in the HPM).

Example of a high dimensional setting

In [7], the relation of the production of riboflavin in *Bacillus subtilis* to the expression level of $p = 4088$ genes is studied. The dataset is distributed with the `R` package `hdi` [11], and consists of $n = 71$ samples.

In [7], the authors are primarily interested in comparing the results among different frequentist statistical methodologies for high dimensional variable selection based on controlling false positive statements (type I error) and p-values. In this regard, the authors obtain results that vary "to a certain extent" over the different methods. The LASSO method selects 30 genes and the other methods either select no gene; select only the gene called `YXLD_at` or select three genes (`LYSC_at`, `YOAB_at` and `YXLD_at`). These disparate results provide an idea of the inherent difficulties in these high dimensional problems.

The model space for this problem contains (many) rank deficient models that were assigned zero prior probability. In this case, the posterior probability of rank deficient models is negligible so it does not make any difference whether these models are a priori ruled out or not. With respect to the computation, we used Gibbs sampling with 50,000 iterations which took slightly more than 2 hours.

The genes with an estimated inclusion probability larger than 0.1 are collected in Table 15.3. The influence of the gene `YOAB_at` on the production of riboflavin is clear and is strongly endorsed by the data, leading to an inclusion probability of 0.97. A main difference with the results in the previous example is that several covariates in the HPM (also shown in Table 15.3) have very small posterior inclusion probabilities and hence the interpretation of the results is more subtle. The HPM has an estimated posterior probability of 0.01 and the model that follows, with a probability of 0.002, contains four genes of which only `YOAB_at` is

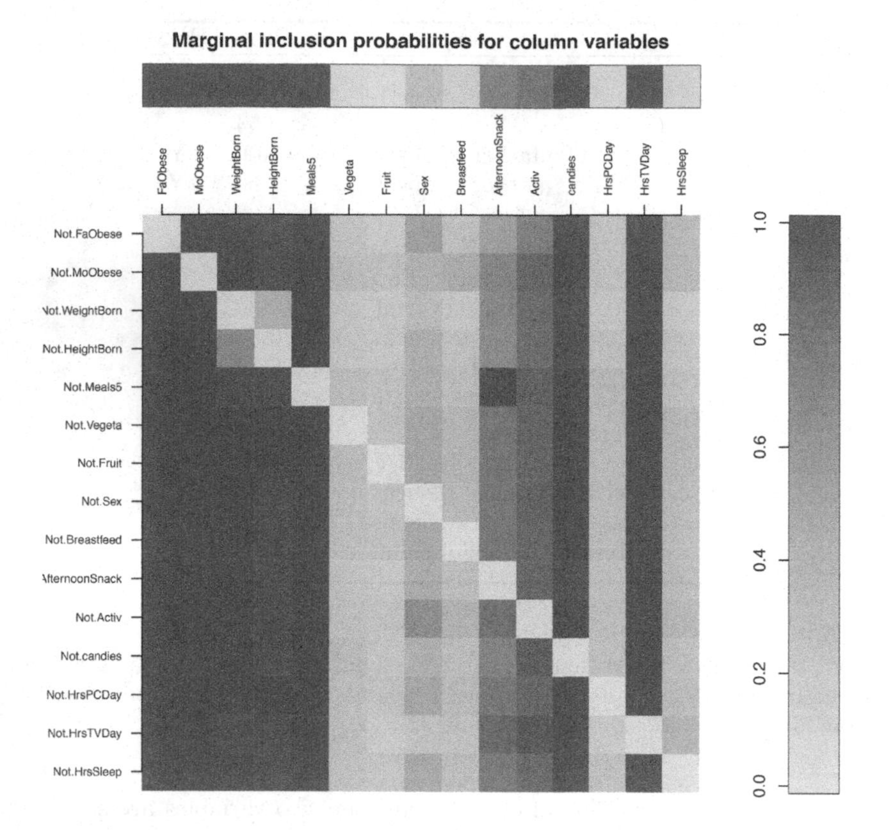

FIGURE 15.1
From the OBICE study, matrix of probabilities of inclusion probability of the column variable given that the variable in the row not included.

in the HPM. The third best model, which has a similar posterior probability, also proposes four genes, two of which are not included in the HPM. These results suggest a situation with multiple modes and hence several joint configurations of genes could provide a sensible explanation for the riboflavin detection. On the other hand, the results clearly indicate the great majority of genes are unimportant, with 97% of genes having an inclusion probability below 0.005.

Our interest in this experiment is mainly for illustrative purposes and with the above comments we wanted to highlight the richness of the posterior distribution to provide insight in the nature of the influence of the explanatory variables that goes far beyond a single model selected.

A final note is about the confidence in the numerical method used. As seems customary in MCMC methods, we ran in parallel (thus not requiring extra computational time) two other chains with randomly chosen initial values. The results were very similar to those described suggesting an efficient exploration of this large model space providing a reliable approximation to the posterior distribution.

Gene	Inc. Prob	HPM
YOAB_at	0.97	Y
YXLE_at	0.43	Y
ARGF_at	0.41	
YXLD_at	0.40	
CARB_at	0.21	
YFII_at	0.18	Y
YISU_at	0.17	
ARGB_at	0.14	Y
YHDZ_at	0.11	
YHEA_at	0.03	Y
YLXQ_at	0.08	Y

TABLE 15.3
Posterior inclusion probabilities of genes with a value larger than 0.1 and an indicator of which genes appear in the highest posterior probability model (HPM) in the riboflavin dataset.

15.2.4 Sensitivity to Prior Inputs

In the context of the two previous applications, we now conduct a sensitivity study to asses the impact of the particular choice of $p(g)$ within the g-prior family and that of $p(M_\gamma)$ in (15.6). Recall that in our applications we used the Robust prior and $\zeta \sim \mathcal{U}(0,1)$. Here we also computed the posterior distribution using the Cauchy, hyper-g/n and unit information priors for g (cf. Table 15.1) and using the fixed assignment $\zeta = 1/2$ on the model space.

Results for the OBICE study, in the form of posterior inclusion probabilities, are collected in Table 15.2.4. These are quite insensitive to the choice of $p(g)$ and $p(M_\gamma)$ and the main conclusions about the importance of entertained covariates remain unchanged. The largest differences are observed between the unit information prior (with a fixed $g = n$) and the rest (with random g, with $p(g)$ a function of n). The extra layer assumed in $p(g)$ provides flatter tails to the prior on $\boldsymbol{\beta}_\gamma$ hence producing methods that are more liberal (more easily allowing for the presence of signals). This essentially explains the increment in the evidence reported towards declaring influential covariates, with the unit information prior being the most conservative, followed by Cauchy, then robust and finally hyper-g/n. We can also see this effect through the comparison of the posterior model size, $p_\gamma \mid \boldsymbol{y}$, (summarized in this same table with its mean and standard deviation) which clearly shows this same ordering.

In this dataset, which has a moderate number of potential regressors, $p = 15$, the choice of $\zeta \sim \mathcal{U}(0,1)$ or $\zeta = 1/2$ has barely any impact on the results although we see $\zeta = 1/2$ behaving slightly more conservatively. This is a direct consequence of the intrinsic tendency of $\zeta \sim \mathcal{U}(0,1)$ to favor models of dimensions that are shared by fewer models (so downweighting models with dimensions around $p/2$, of which there are many) and it so happens that in this problem many of the interesting models have a dimension which is larger than $p/2$ (in the opposite case, we would observe that $\zeta \sim \mathcal{U}(0,1)$ favors simpler models).

The results in the high dimensional application with the riboflavin data were barely sensitive to the prior for the regression parameter. In particular, with the Cauchy, hyper-g/n and Unit Information priors we obtained posterior inclusion probabilities that were very similar to those shown in Table 15.3. Furthermore, the HPM found with these priors coincide. Nevertheless, the results dramatically change if instead of $\zeta \sim \mathcal{U}(0,1)$ we use the prior with a fixed probability $\zeta = 1/2$. In this case, if singular models are considered (with a unitary Bayes factor) then the posterior distribution tends to essentially mimic the prior distribution with all genes having posterior inclusion probabilities very close to

Variable	Robust	Cauchy	hyper-g/n	Unit Inf.
$\zeta \sim \mathcal{U}(0,1)$				
Weight at birth	0.99	0.97	0.99	0.81
Height at birth	0.99	0.96	0.99	0.77
Sex	0.54	0.38	0.62	0.16
The father is obese	1.00	1.00	1.00	1.00
The mother is obese	1.00	1.00	1.00	1.00
...has 5 daily meals	1.00	1.00	1.00	0.99
...eats vegetables	0.37	0.21	0.45	0.06
...eats fruit	0.33	0.18	0.41	0.05
...afternoon snacks.	0.83	0.70	0.86	0.40
...was breastfed	0.42	0.25	0.50	0.07
...practices sports	0.88	0.82	0.90	0.65
...watches TV	1.00	1.00	1.00	1.00
...plays electronic	0.39	0.22	0.47	0.06
...sleeps	0.42	0.25	0.51	0.07
Daily candy consumption	0.99	0.97	0.99	0.92
$E(p_\gamma \mid \boldsymbol{y})$	13.1	11.9	13.7	10
$SD(p_\gamma \mid \boldsymbol{y})$	1.8	1.6	1.9	1.4
$\zeta = 1/2$				
Weight at birth	0.98	0.96	0.98	0.82
Height at birth	0.97	0.94	0.97	0.78
Sex	0.32	0.25	0.35	0.13
The father is obese	1.00	1.00	1.00	1.00
The mother is obese	1.00	1.00	1.00	1.00
...has 5 daily meals	1.00	0.99	1.00	0.99
...eats vegetables	0.15	0.10	0.17	0.04
...eats fruit	0.12	0.09	0.14	0.03
...afternoon snacks.	0.67	0.58	0.69	0.36
...was breastfed	0.19	0.13	0.21	0.06
...practices sports	0.80	0.76	0.81	0.64
...watches TV	1.00	1.00	1.00	1.00
...plays electronic	0.16	0.11	0.18	0.04
...sleeps	0.19	0.13	0.22	0.06
Daily candy consumption	0.97	0.96	0.97	0.92
$E(p_\gamma \mid \boldsymbol{y})$	11.5	11	11.7	9.9
$SD(p_\gamma \mid \boldsymbol{y})$	1.2	1.1	1.2	1.1

TABLE 15.4

Posterior inclusion probabilities and summaries of the posterior distribution of the model size, p_γ, for different prior inputs.

0.5. The results are not more satisfactory if, still using $\zeta = 1/2$, singular models are given zero probability or if, for instance, only models with a number of regressors up to certain fixed value are given non-null prior probability. Any of these assignments leads to a posterior distribution that strongly concentrates on the largest possible dimension (showing a clear dependence on the prior) and without identifying any sensible model (all posterior inclusion probabilities again being approximately 0.5). In this problem, with such a large p, multiplicity correction is of crucial importance and none of these prior assignments, based on fixed $\zeta = 1/2$, provides any such control, letting the posterior distribution concentrate where there are more models leading to useless results. A different path would be a a prior inducing strong sparsity, strongly favoring models with few regressors through for instance the proposal in [25] with a mean model size $w^\star \ll p$. Although the original motivation for using such prior is sparsity, it seems to work well in practice even though it does not explicitly address the issue of multiplicity. Furthermore, and unlike for $\zeta \sim \mathcal{U}(0,1)$, the prior input $w^\star$ has to be specified and the results depend critically on its assumed value.

15.3 Variable Selection for Non-Gaussian Data

Consider a random sample $\boldsymbol{y} = (y_1, \ldots, y_n)^T$ with components y_i being independent and with a distribution in the exponential family ([29]):

$$f(y_i \mid \theta_i, \phi) = \exp\left\{\frac{y_i\theta_i - b(\theta_i)}{\phi/\omega_i} + c(y_i, \phi/\omega_i)\right\},$$

where $b(\cdot)$ and $c(\cdot)$ are functions that specify a distribution for the random variables. The mean of y_i is $\mu_i = b'(\theta_i)$. This defines a wide family of models denoted by generalized linear models (GLMs). In our covariate selection context, each $\boldsymbol{\gamma} \in \mathcal{M}$ assumes that

$$h(\mu_i) = \alpha + \sum_{j=1}^{p} \gamma_j\, x_{ij}\beta_j, \ \forall i = 1, 2, \ldots, n$$

where $h(\cdot)$ is the link function. In the special case of a linear Gaussian model, the link function is the identity function.

The implementation of g-priors in GLMs is less advanced and in particular there is no consensus about which proposal for the prior covariance matrix in (15.8) (see the Appendix) best generalizes the arguments in the linear model case. There are various such matrices that can be inspired by the information in $\boldsymbol{y}$ about $\boldsymbol{\beta}_\gamma$ and the expected Fisher information matrix (which would be the obvious candidate) cannot be used directly since it depends on the parameters $\boldsymbol{\beta}_\gamma$ themselves. We refer the reader to [34] and [26] (and references therein) for a detailed exposition of the different possibilities. The proposal in [34] is a natural extension of g-priors to non-Gaussian models. The authors propose using the expected information matrix for $\boldsymbol{\beta}_\gamma$ evaluated at $(\alpha, \boldsymbol{\beta}_\gamma) = (0, \mathbf{0})$ (imposing a similar prior scheme as in the normal linear model in which $\boldsymbol{\beta}_\gamma$ does not depend on α).

Another source of concern is the form assumed for the prior for the common parameters. The resulting joint prior is improper and hence it is not guaranteed that the prior predictive marginal exists (or equivalently that the posterior is proper) for all models in $\mathcal{M}$. As far as we know, there is no general result that ensures that this condition holds and, given the substantial mathematical differences within the GLM class, conditions under which posterior propriety holds must be checked on a case-by-case basis. In our applications the response y_i is Bernoulli, with the intercept α as the only common parameter. As a theoretical contribution of this chapter, the Appendix shows that for this likelihood and under very mild conditions, the use of a constant prior in (15.4) for the common parameters and in (15.8) for $\boldsymbol{\beta}_\gamma$ with fixed g ensures the existence of the posterior for the usual link functions.

The discussion about the prior over the model space, $p(M_\gamma)$, in Section 16.4 remains valid in GLM variable selection as it is fully independent of the statistical models entertained. Finally, the expression for the Bayes factor $B_\gamma(\boldsymbol{y})$ does not have a closed form and it has to be computed with numerical methods, usually based on Laplace integration. This idea was already proposed in [31].

15.3.1 `glmBfp` and Applications

The package **`glmBfp`** is distributed as accompanying software for the proposals in [34]. Its main command is **`glmBayesMfp`** which provides a user-friendly interface since its usage is similar to the base command **`glm`** to fit GLMs. Unfortunately, **`glmBfp`** seems to be in an early phase of development (current version is `0.0.60` and at the time of finishing the writing of this chapter is available through the archive repository of `R` packages) and its ability to explore the results is limited. We expect that more functionalities will be incorporated in the near future. Concerning the prior inputs, in our examples we use the unit information prior for g (see Table 15.1). For $p(M_\gamma)$ we use $\zeta \sim \mathcal{U}(0,1)$ as in [35], assuming that the prior probability for rank deficient models is zero. The code is provided as supplementary material.

Variable selection in logit models with moderate p

As a first example we will consider again the OBICE study ([41]) about child obesity but now the dependent variable is the indicator of whether the child was classified as obese ($y_i = 1$) or not ($y_i = 0$) by the pediatrician. In this situation, y_i follows a Bernoulli distribution (a member of the exponential family) with the probability of success as the mean μ_i. It is well known that in this model $\phi = 1$ and $\omega_i = 1$ and we will employ the logit link function $h(\mu_i) = \log\{\mu_i/(1-\mu_i)\}$ which corresponds to the canonical link function.

The potential explanatory variables are the same as in Subsection 15.2.3 plus age (which now is not fixed). This makes a total of $p = 16$ variables and exhaustive enumeration is still feasible (taking less than 5 minutes to run). Summaries of the posterior distribution in the form of inclusion probabilities and the model which has the highest posterior probability are displayed in Table 15.5.

The results are along the lines of those obtained in Table 15.2 but with interesting differences. In general, the posterior distribution points to simpler models with fewer explanatory variables influencing the response. In this approach, where the obesity condition is to be explained (and not the BMI as before), the weight and height at birth lose their explanatory capacity while the family genetics remain key variables. Among the habits, afternoon snacks and the practice of sports are no longer important determinants for being an obese child, a role that is now assumed by consuming five daily meals, the intensity of watching TV and daily candy intake.

Variable selection in probit models with large p

In our last applied example, we analyze the arthritis data in [36] which concern $n = 31$ patients with rheumatoid arthritis and osteoarthritis and $p = 755$ gene expression measurements. This dataset has been used to study the influence of the prior assignments in [22] in a context similar to ours. The response variable is a binary variable (the indicator of having the disease) and the link function is the probit function.

Variable	Inc. Prob	HPM
Weight at birth	0.11	
Height at birth	0.06	
Sex	0.10	
The father is obese	1.00	Y
The mother is obese	1.00	Y
Age	0.12	
The child (regularly):		
...has 5 daily meals	0.87	Y
...eats vegetables	0.03	
...eats fruit	0.07	
...consumes afternoon snacks	0.16	
...was breastfed	0.03	
...practices sports	0.18	
Daily hours the child:		
...watches TV	0.86	Y
...plays with electronic devices	0.03	
...sleeps	0.03	
Daily candy consumption	1.00	Y

TABLE 15.5
The dependent variable is the indicator of obesity. The table contains posterior inclusion probabilities and indicates which variables belong to the highest posterior probability model (HPM).

We simulate 10^6 draws from the posterior distribution taking approximately 47 minutes. We also run two other independent chains as a check on the reliability of the results. We find that the great majority of genes do not have any impact on the classification of the disease and only variables $V290$ and $V258$ have some role as determinants: their inclusion probabilities are 0.31 and 0.29, while all the others are smaller than 0.1. It is also relevant that the highest posterior probability model (with a probability of 0.13) is the one with only $V290$ followed very closely by the one with only $V258$ (probability of 0.12). These results are in agreement with [22] who also found these genes to be the most relevant.

15.4 Conclusion

Variable selection in regression models is a pervasive problem that occurs in a very wide variety of applied fields. This chapter focuses on principled Bayesian methods based on Bayes factors in the context of g-prior structures. Through empirical examples, we illustrate the ease of implementation of these methods using freely available R packages for both normal linear models and generalized linear models. We analyse applications with the number of possible covariates ranging from 15 to 4088 and show that reliable inference can be obtained quite rapidly with standard computing equipment. A rich tapestry of possible questions can then be answered and the results are easily interpretable. We also highlight that prior assumptions often have an important effect on the results, and we recommend robustifying the prior structures through priors on hyperparameters. In the Appendix we prove posterior propriety for commonly used generalized linear models.

Acknowledgments

The work of the first author has been partially funded by grant PID2019-104790GB-I00 from the Spanish Ministerio de Ciencia e Innovación and by grant SBPLY/17/180501/000491 from the Consejería de Educación, Cultura y Deportes de la Junta de Comunidades de Castilla-La Mancha.

Appendix

Theorem 15.4.1. *Consider the problem of variable selection within the generalized linear model where y_i follows a Bernoulli distribution. Suppose that i) the link function $h(\cdot)$ is either the probit, the logit or the log-log function; ii) not all observed y_i are equal and iii) the matrix of covariates in the full model $\boldsymbol{X}$ is of full rank. If the prior assumed for M_γ is*

$$p_\gamma(\boldsymbol{\beta}_\gamma, \alpha) \propto \mathcal{N}_{p_\gamma}(\boldsymbol{\beta}_\gamma \mid \mathbf{0}_{p_\gamma}, g\Sigma_\gamma), \tag{15.8}$$

where g is a fixed scalar and Σ_γ is a positive definite matrix, then the marginal $m_\gamma(\boldsymbol{y})$ exists for every model in $\mathcal{M}$.

Proof

We have to show that the integral

$$\int\int f(\boldsymbol{y} \mid \alpha, \boldsymbol{\beta}) \mathcal{N}_p(\boldsymbol{\beta} \mid \mathbf{0}_p, g\Sigma)\, d\alpha\, d\boldsymbol{\beta}$$

is finite (for simplicity the subscript γ has been removed). We partially base our proof on [10]: given the identity in their (4.4) the above integral can be expressed as:

$$\int\int\int 1\{\alpha\boldsymbol{\iota}^\star + \boldsymbol{X}^\star\boldsymbol{\beta} \leq \boldsymbol{u}\} \mathcal{N}_p(\boldsymbol{\beta} \mid \mathbf{0}_p, g\Sigma)\, d\alpha\, d\boldsymbol{\beta}\, d\boldsymbol{F}(\boldsymbol{u}),$$

where $\boldsymbol{F}(\boldsymbol{u}) = (h^{-1}(u_1), \ldots, h^{-1}(u_n))^T$ (i.e., component by component, the inverse of the link function); $\boldsymbol{\iota}^\star = (z_1, \ldots, z_n)^T$ where $z_i = 1$ if $y_i = 0$ and $z_i = -1$ if $y_i = 1$; $\boldsymbol{X}^\star$ is the matrix with rows $z_i(x_{i1}, \ldots, x_{ip})$ (i.e., $\boldsymbol{X}$ with row i multiplied by $z_i, i = 1, \ldots, n$). Obviously, the above integral equals:

$$\int\int\int 1\{\alpha\boldsymbol{\iota}^\star \leq \boldsymbol{u} - X^\star\boldsymbol{\beta}\} \mathcal{N}_p(\boldsymbol{\beta} \mid \mathbf{0}_p, g\Sigma)\, d\alpha\, d\boldsymbol{\beta}\, d\boldsymbol{F}(\boldsymbol{u}),$$

and now we apply Lemma 4.1 in [10] to bound the integral over α, resulting in the following upper bound (up to a constant)

$$\int\int ||\boldsymbol{u} - X^\star\boldsymbol{\beta}||\, \mathcal{N}_p(\boldsymbol{\beta} \mid \mathbf{0}_p, g\Sigma) d\boldsymbol{\beta}\, d\boldsymbol{F}(\boldsymbol{u}), \tag{15.9}$$

where $||\cdot||$ represents the Euclidean norm and the conditions in Lemma 4.1 apply given our conditions ii) and iii). Finally, we use the triangle inequality to bound (15.9) by the sum

$$\int ||\boldsymbol{u}||\, d\boldsymbol{F}(\boldsymbol{u}) + \int ||X^\star\boldsymbol{\beta}||\, \mathcal{N}_p(\boldsymbol{\beta} \mid \mathbf{0}_p, g\Sigma) d\boldsymbol{\beta},$$

and both integrals exist because the first moment of $d\boldsymbol{F}(\cdot)$ (for the cases assumed in i)) and that of $\mathcal{N}_p(\boldsymbol{\beta} \mid \mathbf{0}_p, g\Sigma)$ exist.

Bibliography

[1] M. J. Bayarri and G. García-Donato. Extending conventional priors for testing general hypotheses in linear models. *Biometrika*, 94(1):135–152, 2007.

[2] M.J. Bayarri, J.O. Berger, A. Forte, and G. García-Donato. Criteria for Bayesian model choice with application to variable selection. *The Annals of Statistics*, 40:1550–1577, 2012.

[3] M.J. Bayarri and G. García-Donato. Generalization of Jeffreys divergence-based priors for Bayesian hypothesis testing. *Journal of the Royal Statistical Society: Series B*, 70(5):981–1003, 2008.

[4] J. O. Berger, G. García-Donato, M.A. Martinez-Beneito, and V. Peña. Bayesian variable selection in high dimensional problems without assumptions on prior model probabilities. arXiv:1607.02993, July 2016.

[5] J.O. Berger and L.R. Pericchi. *Objective Bayesian Methods for Model Selection: Introduction and Comparison*, volume 38 of *Lecture Notes–Monograph Series*, pages 135–207. Institute of Mathematical Statistics, Beachwood, OH, 2001.

[6] J.M. Bernardo and A.F.M. Smith. *Bayesian Theory.* Chichester: Wiley, 1994.

[7] P. Buhlman, M. Kalisch, and L. Meier. High-dimensional statistics with a view towards applications in biology. *Annual Review of Statistics and its Applications*, 1:255–278, 2014.

[8] I. Castillo, J. Schmidt-Hieber, and A. van der Vaart. Bayesian linear regression with sparse priors. *Annals of Statistics*, 43:1986–2018, 2015.

[9] I. Castillo and A. van der Vaart. Needles and straw in a haystack: Posterior concentration for possibly sparse sequences. *Annals of Statistics*, 40(4):2069–2101, 2012.

[10] M.H. Chen and Q. Shao. Propriety of posterior distribution for dichotomous quantal responses. *Proceedings of the American mathematical society*, 129(1):293–302, 200.

[11] R. Dezeure, P. Bühlman, L. Meier, and N. Meinshausen. High-dimensional inference: confidence intervals, p-values and R-software hdi. *Statistical Science*, pages 533–558, 2015.

[12] A. Etz and E.-J. Wagenmakers. J.B.S. Haldane's contribution to the Bayes factor hypothesis test. *Statistical Science*, 32(2):313–329, 2017.

[13] C. Fernández, E. Ley, and M.F. Steel. Benchmark priors for Bayesian model averaging. *Journal of Econometrics*, 100:381–427, 2001.

[14] A. Forte, G. García-Donato, and M.F. Steel. Methods and tools for Bayesian variable selection and model averaging in normal linear regression. *International Statistical Review*, 86(2):237–258, 2018.

[15] D.P. Foster and E. I. George. The Risk Inflation Criterion for Multiple Regression. *The Annals of Statistics*, 22:381–427, 1994.

[16] G. García-Donato and A. Forte. Bayesian Testing, Variable Selection and Model Averaging in Linear Models using R with BayesVarSel. *The R Journal*, 10(1):155–174, 2018.

[17] G. Garcia-Donato and M.A. Martinez-Beneito. On Sampling strategies in Bayesian variable selection problems with large model spaces. *Journal of the American Statistical Association*, 108(501):340–352, 2013.

[18] E. George and R. McCulloch. Approaches for Bayesian variable selection. *Statistica Sinica*, 7:339–373, 1997.

[19] H. Jeffreys. *Theory of Probability.* Oxford University Press, 3rd edition, 1961.

[20] R.E. Kass and A.E. Raftery. Bayes factors. *Journal of the American Statistical Association*, 90(430):773–795, 1995.

[21] R.E. Kass and L. Wasserman. A reference Bayesian test for nested hypotheses and its relationship to the schwarz criterion. *Journal of the American Statistical Association*, 90(431):928–934, 1995.

[22] D. Lamnisos, J.E. Griffin, and M.F.J. Steel. Cross-validation prior choice in Bayesian probit regression with many covariates. *Statistics and Computing*, 22(2):359–373, 2012.

[23] E.E. Leamer. *Specification Searches: Ad Hoc Inference with Nonexperimental Data.* New York: Wiley, 1978.

[24] E. Ley and M.F. Steel. Jointness in Bayesian variable selection with applications to growth regression. *Journal of Macroeconomics*, 29(3):476 – 493, 2007. Special Issue on the Empirics of Growth Nonlinearities.

[25] E. Ley and M.F. Steel. On the effect of prior assumptions in Bayesian model averaging with applications to growth regression. *Journal of Applied Econometrics*, 24(4):651–674, 2009.

[26] Y. Li and M. Clyde. Mixtures of g-priors in generalized linear models. *Journal of the American Statistical Association*, 113:1828–1845, 2018.

[27] F. Liang, R. Paulo, G. Molina, M.A. Clyde, and J.O. Berger. Mixtures of g-priors for Bayesian variable selection. *Journal of the American Statistical Association*, 103(481):410–423, 2008.

[28] Y. Maruyama and E.I. George. Fully Bayes factors with a generalized g-prior. *The Annals of Statistics*, 39(5):2740–2765, 2011.

[29] P. McCullagh and J.A. Nelder. *Generalized Linear Models.* Chapman and Hall, 1989.

[30] R Core Team. *R: A Language and Environment for Statistical Computing.* R Foundation for Statistical Computing, Vienna, Austria, 2015.

[31] A. E. Raftery. Approximate Bayes factors and accounting for model uncertainty in generalised linear models. *Biometrika*, 83:251–266, 1996.

[32] A.E. Raftery, D. Madigan, and J. Hoeting. Bayesian model averaging for linear regression models. *Journal of the American Statistical Association*, 92:179–191, 1997.

[33] C.P. Robert, N. Chopin, and J. Rousseau. Harold Jeffreys' theory of probability revisited. *Statistical Science*, 24(2):141–172, 2009.

[34] D. Sabanes and L. Held. Hyper-g priors for generalized linear models. *Bayesian Analysis*, 6:387–410, 2011.

[35] J.G. Scott and J.O. Berger. Bayes and empirical-Bayes multiplicity adjustment in the variable-selection problem. *The Annals of Statistics*, 38(5):2587–2619, 2010.

[36] N. Sha, M.G. Vanucci, P.J. Tadesse, I. Brown, N. Dragoni, T.C. Davies, A. Roberts, M. Contestabile, M. Salmon, C. Buckley, and F. Falciani. Bayesian variable selection in multinomial probit models to identify molecular signatures of disease stage. *Biometrics*, 60:812–819, 2004.

[37] T. van Erven and B. Szabo. Fast exact Bayesian inference for sparse signals in the normal sequence model. *Bayesian Analysis*, 16:forthcoming, 2021.

[38] A. Zellner. On assessing prior distributions and Bayesian regression analysis with g-prior distributions. In A. Zellner, editor, *Bayesian Inference and Decision techniques: Essays in Honor of Bruno de Finetti*, pages 389–399. Edward Elgar Publishing Limited, 1986.

[39] A. Zellner and A. Siow. Posterior odds ratio for selected regression hypotheses. In J. M. Bernardo, M.H. DeGroot, D.V. Lindley, and Adrian F. M. Smith, editors, *Bayesian Statistics 1*, pages 585–603. Valencia: University Press, 1980.

[40] A. Zellner and A. Siow. *Basic Issues in Econometrics.* Chicago: University of Chicago Press, 1984.

[41] O. Zurriaga, J. Perez-Panades, J. Izquiero, M. Gil, Y. Anes, C. Quiñones, M. Margolles, A. Lopez-Maside, A. T. Vega-Alonso, and M.T. Miralles. Factors associated with childhood obesity in spain. the OBICE study: a case–control study based on sentinel networks. *Public Health Nutrition*, 14(6):1105–113, 2011.

16

Balancing Sparsity and Power: Likelihoods, Priors, and Misspecification

David Rossell
Pompeu Fabra University (Spain)

Francisco Javier Rubio
University College London (UK)

CONTENTS

Occam's razor is a classical argument for Bayesian model selection (BMS). Penalizing complexity leads to choosing a parsimonious model, whereas in frequentist tests one cannot prove the simpler model. This argument is correct but care is needed. Good model selection requires balancing parsimony and power, which depends on three elements: the priors, model, and whether it approximates well the data-generating mechanism (misspecification).

Standard finite-dimensional BMS does not penalize complexity sufficiently, whereas high-dimensional formulations impose stronger penalties that ensure consistency, but may reduce power. Regarding misspecification, BMS still targets a reasonable problem when (inevitably) assuming the wrong model, and the effect on false positives vanishes asymptotically, but power is exponentially reduced. To increase power one may enrich the model, but having more parameters can reduce power. We discuss these issues, practical strategies to improve sparsity-power tradeoffs, and offer our views on a long-standing debate on BMS sensitivity to prior dispersion.

DOI: 10.1201/9781003089018-16

16.1 Introduction

In many fields it is important not only to identify what covariates are truly associated to an outcome $\boldsymbol{y}$, but to quantify uncertainty in the existence of such association. Bayesian model selection (BMS) is a powerful tool for this task. We focus on structural learning problems where one seeks to test hypotheses or, more generally, choose what model represents best the true mechanism that generated the data. This problem is related to, but should be distinguished from, estimating regression parameters or predicting $\boldsymbol{y}$. For example, in prediction problems inference is typically robust to adding a few extra spurious parameters, since their point estimates are often close to zero, whereas in structure recovery one wishes to avoid false positives. Similarly, for prediction purposes failing to add variables that have small effects is often not very consequential, given that their contribution to the expected value of $\boldsymbol{y}$ is small, whereas in structure recovery one wishes to avoid false negatives.

BMS is appealingly general. Given a likelihood and priors for each model one obtains posterior model probabilities that guide model choice and assess model uncertainty. BMS is distinct from Bayesian strategies based on shrinkage priors, reviewed elsewhere in this handbook, see also [3, 18, 40]. The latter quantify uncertainty on parameters rather than on models, hence we view them as more suitable for estimation/prediction. Also, BMS is more general than variable selection where one considers the independent inclusion/exclusion of p variables. Within BMS it is easy to introduce groups of variables and hierarchical constraints, e.g., associated to non-linear effects or covariate interactions. Incorporating such structure in penalized likelihood and shrinkage prior methods can be challenging and computationally costly. For example, analyzing the salary data from Section 16.10.1 via BMS took seconds, whereas the hierarchical LASSO failed to return a solution after 2 hours. We also illustrate how, even without group or hierarchical constraints, BMS can attain competitive run times, see also Table 16.1 for a colon cancer dataset with $p = 10,172$ variables where the run time are similar to penalized likelihood methods. Sections 16.2, 16.3 and 16.4 introduce the specific model and prior formulation we consider, and discuss how to interpret the BMS solution under model misspecification. Section 16.5 discusses the robustness of the BMS solution to prior parameters.

Despite these appealing properties, there are important questions related to the frequentist validity of BMS. A first issue refers to the validity of $\pi(\gamma \mid \boldsymbol{y})$ to assess selection uncertainty (Section 16.6). For example, given a posterior probability $\pi(\gamma \mid \boldsymbol{y}) = 0.9$ for some model γ, is there any relation between this figure and the frequentist probability $P_{F_0}(\gamma^* = \gamma)$ of choosing the right model, and the type I or type II errors of individual parameters? It turns out that said validity holds asymptotically, at rates that depend on how fast $\pi(\gamma \mid \boldsymbol{y})$ concentrates on γ^*, which in turn depend on Bayes factors. It is natural to then ask how quickly does concentration occur, and what is the effect of misspecification. Interestingly, misspecified rates have the same functional form as in the well-specified case, but their coefficients do change (Section 16.7). Although misspecified Bayes factors to detect true signals grow exponentially in n, they are often exponentially slower than in the well-specified case [20, 33–35]. For example, Figure 16.1 (top right panel) shows that assuming Normal errors may significantly reduce the frequentist probability of choosing the right set of variables, when errors truly follow an asymmetric Laplace distribution that features heavy tails. Similar results hold in high-dimensional settings (Section 16.8). Briefly, there are three strategies to ensure that BMS consistently selects the right model γ^* as $n \to \infty$: penalizing complexity via $\pi(\gamma)$, via diffuse priors $\pi(\boldsymbol{\beta}_\gamma \mid \boldsymbol{y})$, or using non-local priors (NLPs, [20, 31]). Penalizing complexity leads to good asymptotic properties, but can significantly reduce finite n power. For example, Figure 16.1 (top left panel) illustrates a simulation where so-called Complexity priors perform an excellent job at discarding spurious parameters, but incur a significant cost in terms of power relative to other formulations. One then encounters an important issue: on the one hand both large p and misspecification often reduce power, but on the other if p is large then the sample size may not be large enough to consider more flexible models that could help reduce misspecification. In Section 16.9, we discuss simple strategies to improve power. One may learn whether enriching the likelihood via flexible terms is likely to increase or reduce power, and set non-local or empirical Bayes priors to penalize complexity adaptively.

We conclude this chapter by illustrating the practical implications of our discussion. Section 16.10 presents a salary survey example where p is moderate but one considers interactions. There, BMS is faster and attains significantly higher power than penalized likelihood alternatives. We also present high-dimensional continuous and survival regression examples that illustrate sparsity-power trade-offs where, under simple default priors, BMS attains highly-competitive performance. Section 16.11 offers some final thoughts. We remark that this chapter is a selective review of a wide literature. We summarize some of its main findings, with a certain focus on our work, and keep technical details to a minimum. R code for our examples is available online.

16.2 BMS in Regression Models

We introduce analogous notation to other chapters, adapted to our more general BMS setting. Let $\mathbf{y} = (y_1, \ldots, y_n)^T$ where $y_i \in \mathbb{R}$ is the response for individual $i = 1, \ldots, n$, $\boldsymbol{x}_i \in \mathbb{R}^p$ be the covariates and $\boldsymbol{X}$ the corresponding $n \times p$ matrix. We assume that $\boldsymbol{x}_i$ includes the intercept, if one wishes to force its inclusion (or that of any other term) one may set its prior inclusion probability to 1. We wish to consider several models, distinguished by what covariates have an effect on $\boldsymbol{y}$ and potentially also by the form of such effect, e.g., linear versus non-linear. We denote a specific model by $\gamma \in \Gamma$, by Γ the set of considered models, by $\boldsymbol{\beta}_\gamma$ regression parameters associated to model γ and by ϕ_γ any nuisance parameters (e.g., a dispersion parameter). To ease notation $\boldsymbol{\beta}_\gamma$ includes the coefficients associated to the intercept term and any other covariates forced into the model, denoted as α elsewhere in this book. Although the results we discuss hold more generally, we focus on regression models within the exponential family (e.g., Gaussian, logistic or Poisson regression). In the particular case of Gaussian regression, then $\phi = \sigma^2$ is the error variance, as denoted elsewhere in this book. Let $p_\gamma = \dim(\boldsymbol{\beta}_\gamma)$ and $\ell_\gamma(\boldsymbol{\beta}_\gamma, \phi_\gamma; \boldsymbol{y})$ the log-likelihood function under γ. A critical notion in this chapter is that we adopt a realistic viewpoint where the data are generated by some unknown probability distribution F_0, which is in general outside any of the considered models (model misspecification).

We are interested in situations where each model is defined by a predictor $\boldsymbol{X}_\gamma \boldsymbol{\beta}_\gamma$. We remark that $\boldsymbol{X}_\gamma \boldsymbol{\beta}_\gamma$ need not be a linear predictor, since $\boldsymbol{X}_\gamma$ may contain hierarchical interactions or basis expansions corresponding to semi-parametric and nonparametric regressions, for example. In canonical variable selection one considers the inclusion/exclusion of each individual column in $\boldsymbol{X}$. Here we consider a slightly more general version where $\boldsymbol{X} = (\boldsymbol{X}_1, \ldots, \boldsymbol{X}_J)$ contains J groups of variables, where $\boldsymbol{X}_j$ is an $n \times p_j$ matrix and p_j is the number of variables in group j. Let $\boldsymbol{\beta} = (\boldsymbol{\beta}_1, \ldots, \boldsymbol{\beta}_J)$ where $\boldsymbol{\beta}_j \in \mathbb{R}^{p_j}$ are the regression parameters for group j. Then each model is defined by a vector of indicators $\gamma = (\gamma_1, \ldots, \gamma_J)$, where $\gamma_j = \mathrm{I}(\boldsymbol{\beta}_j \neq 0)$, $\boldsymbol{X}_\gamma = \{\boldsymbol{X}_j\}_{\gamma_j=1}$ are the columns selected by γ and $\boldsymbol{\beta}_\gamma = \{\boldsymbol{\beta}_j\}_{\gamma_j=1}$ their regression coefficients. In particular, if each group contains $p_j = 1$

variable, then $J = p$ and one recovers the canonical setting. We denote the number of groups selected by γ by $s(\gamma) = \sum_{j=1}^{J} \gamma_j$.

Given prior densities $\pi(\boldsymbol{\beta}_\gamma, \phi_\gamma \mid \gamma)$ and prior model probabilities $\pi(\gamma)$, Bayes theorem gives posterior model probabilities

$$\pi(\gamma \mid \boldsymbol{y}) = \frac{\pi(\gamma) f(\boldsymbol{y} \mid \gamma)}{f(\boldsymbol{y})} = \left(1 + \sum_{\gamma' \neq \gamma} B_{\gamma'\gamma} \frac{\pi(\gamma')}{\pi(\gamma)}\right)^{-1}, \tag{16.1}$$

where $B_{\gamma'\gamma} = f(\boldsymbol{y} \mid \gamma')/f(\boldsymbol{y} \mid \gamma)$ is the Bayes factor between γ' and γ, and

$$f(\boldsymbol{y} \mid \gamma) = \int e^{\ell_\gamma(\boldsymbol{\beta}_\gamma, \phi_\gamma; \boldsymbol{y})} \pi(\boldsymbol{\beta}_\gamma, \phi_\gamma \mid \gamma) d\boldsymbol{\beta}_\gamma d\phi_\gamma,$$

the integrated (or marginal) likelihood. Common BMS strategies are choosing the model with highest $\pi(\gamma \mid \boldsymbol{y})$, variables with high marginal posterior probabilities $\pi(\gamma_j \neq 0 \mid \boldsymbol{y})$ or, in prediction problems, either weight models based on $\pi(\gamma \mid \boldsymbol{y})$ or find a single sparse model giving similar predictions to the weighted prediction [19, 29]. Either way $\pi(\gamma \mid \boldsymbol{y})$ plays a critical role, hence the importance to understand its behavior.

Standard regression models (e.g., continuous, exponential family or survival outcomes) make important assumptions on the relationship between $\boldsymbol{y}$ and $\boldsymbol{X}$. Suppose that these assumptions are met, that is, $(\boldsymbol{y}, \boldsymbol{X})$ truly arise from a probability distribution F_0 included in the family defined by ℓ_{γ^*}, for some $\gamma^* \in \Gamma$. Then, BMS assigns $\pi(\gamma^* \mid \boldsymbol{y})$ converging to 1 at well-known rates reviewed below (under mild regularity conditions) [10, 22, 31]. Even in this ideal case one should set priors with care, else $\pi(\gamma \mid \boldsymbol{y})$ may concentrate on γ^* slowly as n grows and, in high-dimensional problems where p grows faster than n, may fail to concentrate on γ^* [21].

Model misspecification is conceptually an even more important concern. Models are approximations to reality, one does not expect any model to represent F_0 perfectly. One may misspecify the mean $E(\boldsymbol{y} \mid \boldsymbol{X})$ by ignoring non-linear terms, representing them imperfectly with finite basis for computational convenience, or by omitting covariates that were not recorded. One may also miss asymmetries or thick tails in the errors or their covariance, for example. In generalized linear models one may choose a poor link function, and in survival models one may misspecify the hazard function or make undue assumptions on the censoring (e.g., being conditionally independent of $\boldsymbol{y}$, given $\boldsymbol{X}$). It is natural to then ask whether (16.1) remains meaningful, in the sense of addressing a (reasonable) well-specified problem. In Section 16.3, we explain that the answer is yes, BMS asymptotically selects the model γ^* with smallest dimension p_{γ^*} minimizing the expected loss implied by the log-likelihood. The interpretation is often simple, e.g., in linear regression BMS selects any covariate that reduces mean squared error under F_0, or equivalently drops covariates that are conditionally uncorrelated with $\boldsymbol{y}$ given $\boldsymbol{X}_{\gamma^*}$. Similar interpretations apply to generalized linear and survival models.

These issues concern the specification of the likelihood. BMS requires completing the probability model via prior distributions on parameters and models. For concreteness in Section 16.4 we introduce simple priors, these could be replaced by more complex choices, but they would have no material effect on our discussion.

16.3 Interpreting BMS Under Misspecification

The effects of misspecification on parameter estimation have been well-studied. Under general parametric models and suitable regularity conditions, the maximum likelihood estimator (MLE) $(\widehat{\boldsymbol{\beta}}_\gamma, \widehat{\phi}_\gamma)$ converges in probability to $(\boldsymbol{\beta}^*_\gamma, \phi^*_\gamma)$ maximizing

$$L_\gamma(\boldsymbol{\beta}_\gamma, \phi_\gamma) = \int \ell_\gamma(\boldsymbol{\beta}_\gamma, \phi_\gamma \mid \boldsymbol{y}, \boldsymbol{X}) dF_0(\boldsymbol{y}, \boldsymbol{X}), \tag{16.2}$$

the expected log-likelihood under $(\boldsymbol{y}, \boldsymbol{X}) \sim F_0$, and is asymptotically normally-distributed (e.g., see [44] and [42], Chapter 5). The effect of misspecification is reflected on a sandwich MLE covariance matrix which, intuitively, measures the discrepancy between F_0 and the assumed model. The same observations apply to Bayesian inference, since $\pi(\boldsymbol{\beta}_\gamma, \phi_\gamma \mid \boldsymbol{y})$ converges to the MLE sampling distribution as $n \to \infty$ under mild conditions ([42], Chapter 10).

These issues may appear technical but are actually quite simple. Suppose that one posits a linear regression model

$$\boldsymbol{y} = \boldsymbol{X}_\gamma \boldsymbol{\beta}_\gamma + \varepsilon, \tag{16.3}$$

with $\varepsilon \sim \mathcal{N}(0, \phi_\gamma I)$, where $\phi_\gamma = \sigma^2_\gamma$ is the variance, but truly $\boldsymbol{y} \mid \boldsymbol{X} \sim F_0(\boldsymbol{y} \mid \boldsymbol{X})$ and for simplicity we assume $\boldsymbol{X}_\gamma$ to be non-random. Then, as $n \to \infty$, the MLE $\widehat{\boldsymbol{\beta}}_\gamma = (\boldsymbol{X}_\gamma^T \boldsymbol{X}_\gamma)^{-1} \boldsymbol{X}_\gamma^T \boldsymbol{y}$ satisfies

$$V_\gamma^{-1/2}(\widehat{\boldsymbol{\beta}}_\gamma - \boldsymbol{\beta}^*_\gamma) \xrightarrow{D} \mathcal{N}(0, I), \tag{16.4}$$

with sandwich covariance $V_\gamma = (\boldsymbol{X}_\gamma^T \boldsymbol{X}_\gamma)^{-1} \boldsymbol{X}_\gamma^T \text{Cov}_{F_0}(\boldsymbol{y}) \boldsymbol{X} (\boldsymbol{X}_\gamma^T \boldsymbol{X}_\gamma)^{-1}$. The result holds in high-dimensions where p_γ grows with n, as long as $(\boldsymbol{X}_\gamma^T \boldsymbol{X}_\gamma)/n$ and $\boldsymbol{X}_\gamma^T \text{Cov}_{F_0}(\boldsymbol{y}) \boldsymbol{X}_\gamma / n$ converge in probability to strictly positive-definite matrices. If F_0 truly satisfies (16.3) for some $(\boldsymbol{\beta}^*_\gamma, \phi^*_\gamma)$, we obtain the usual $V_\gamma = \phi^*_\gamma (\boldsymbol{X}_\gamma^T \boldsymbol{X}_\gamma)^{-1}$. More generally, F_0 may be such that the mean of $\boldsymbol{y}$ is non-linear in $\boldsymbol{X}_\gamma$, the errors are non-Normal, correlated or heteroscedastic. Then, V_γ is typically inflated (larger determinant) relative to $\phi^*_\gamma (\boldsymbol{X}_\gamma^T \boldsymbol{X}_\gamma)^{-1}$, and such increased uncertainty affects the Bayes factor rates (Section 16.7).

We remark that for $\boldsymbol{X}_\gamma^T \boldsymbol{X}_\gamma$ to be invertible one must have $p_\gamma \leq n$, *i.e.*, even when $p \gg n$, for model selection purposes one typically restricts attention to models $\gamma \in \Gamma$ with at most $p_\gamma \leq n$ parameters. Under mild conditions, BMS based on (16.3) selects $\gamma^* \in \Gamma$ with smallest dimension p_{γ^*} minimizing the least-squares criterion under F_0. Specifically, let

$$M^* = \min_{\gamma \in \Gamma} E_{F_0}[(\boldsymbol{y} - \boldsymbol{X}_\gamma \boldsymbol{\beta}^*_\gamma)^T (\boldsymbol{y} - \boldsymbol{X} \boldsymbol{\beta}^*_\gamma)],$$

be the lowest MSE achievable across all models, $\Gamma^* = \{\gamma \in \Gamma : (\boldsymbol{y} - \boldsymbol{X}_\gamma \boldsymbol{\beta}^*_\gamma)^T (\boldsymbol{y} - \boldsymbol{X} \boldsymbol{\beta}^*_\gamma) = M^*\}$ the models achieving that MSE, and $\gamma^* = \arg\min_{\gamma \in \Gamma^*} p_\gamma$ that with smallest dimension. Then, under suitable conditions,

$$\pi(\gamma^* \mid \boldsymbol{y}) \xrightarrow{P} 1,$$

as $n \to \infty$. That is, the asymptotic solution γ^* drops any variables that do not improve MSE under F_0 or, equivalently, that are conditionally uncorrelated with $\boldsymbol{y}$ given $\boldsymbol{X}_{\gamma^*}$. Identifying conditional uncorrelations is a common goal when using regression models for explanatory (as opposed to predictive) purposes. Hence BMS targets a sensible structural learning problem concerning the data-generating F_0. For simplicity, we assumed that γ^* is unique, otherwise the results remain valid by defining γ^* as the union of all smallest optimal models.

The intuition from linear regression extends to more complex regression and survival models. We offer a brief discussion and refer the reader to [31, 33, 34] for details. Let M^* be the minimum of the loss function defined by the negative expected log-likelihood in (16.2), and Γ^* and γ^* be as before. Then, $\pi(\gamma^* \mid \boldsymbol{y}) \xrightarrow{P} 1$, as $n \to \infty$, under very general conditions, including high-dimensional p_γ and nonparametric settings beyond our scope [45, 46]. The interpretation of γ^* remains accessible, e.g., in logistic regression γ^* is the smallest model minimizing the logistic loss, and discards covariates that are conditionally uncorrelated with $\boldsymbol{y}$ according to that loss, given $\boldsymbol{X}_{\gamma^*}$. As another example consider survival analysis, where the response variable $y_i = \min\{o_i, c_i\}$ is the minimum between the survival time $o_i \in \mathbb{R}_+$ and the right-censoring time $c_i \in \mathbb{R}_+$. Here the censoring mechanism also plays a role in defining γ^*, since the behavior of BMS depends on the true F_0 generating (o_i, c_i). More specifically, it is possible to show that under censoring and misspecification, the asymptotic solution γ^* includes covariates that help predict survival times and/or censoring times, and discards other covariates [31].

16.4 Priors

Our discussion holds for a wide class of priors $\pi(\boldsymbol{\beta}_\gamma, \phi_\gamma \mid \gamma)$ and $\pi(\gamma)$. Such prior richness is an asset of BMS, but for asymptotic results the main relevant aspects are the prior mass $\pi(\gamma^*)$ assigned to the asymptotic solution γ^*, and around $(\boldsymbol{\beta}_\gamma^*, \phi_\gamma^*)$ under various γ's. This observation is an interesting contrast to estimation problems, where the tails of $\pi(\boldsymbol{\beta}_\gamma, \phi_\gamma)$ can also affect asymptotic rates. For example, in linear regression setting Laplace or thicker tails attains better estimation rates than Gaussian tails [6], such an issue does not occur in BMS (up to low order terms, [31]). For our purposes, it suffices to consider two prior families

$$\pi_N(\boldsymbol{\beta}_\gamma \mid \phi_\gamma, \gamma) = \prod_{\gamma_j=1} \mathcal{N}\left(\boldsymbol{\beta}_j; 0, \frac{\phi_\gamma g_N}{p_j}(\boldsymbol{X}_j^T \boldsymbol{X}_j/n)^{-1}\right), \tag{16.5}$$

$$\pi_M(\boldsymbol{\beta}_\gamma \mid \phi_\gamma, \gamma) = \prod_{\gamma_j=1} \frac{\boldsymbol{\beta}_j^T \boldsymbol{X}_j^T \boldsymbol{X}_j \boldsymbol{\beta}_j/(np_j)}{\phi_\gamma g_M/(p_j+2)} \mathcal{N}\left(\boldsymbol{\beta}_j; 0, \frac{\phi_\gamma g_M}{p_j+2}(\boldsymbol{X}_j^T \boldsymbol{X}_j/n)^{-1}\right), \tag{16.6}$$

where $\boldsymbol{X}_j$ is the submatrix of $\boldsymbol{X}$ corresponding to the jth variable group, $p_j = \dim(\boldsymbol{\beta}_j)$ and $g_N, g_M > 0$ are given dispersion parameters. If the model contains an unknown dispersion parameter ϕ_γ, we set an inverse gamma prior $\pi(\phi_\gamma \mid \gamma) = \text{IG}(\phi_\gamma; a_\phi/2, b_\phi/2)$ for known $a_\phi, b_\phi > 0$. Contrary to popular belief, Bayes factors are fairly robust to (g_N, g_L), provided they are constants not depending on n, and simple considerations lead to default $g_N = g_L = 1$ (Section 16.5).

The prior π_N can be replaced by other families. An important characteristic is that, although γ states that all elements in $\boldsymbol{\beta}_\gamma$ are non-zero, π_N assigns positive prior density to $\boldsymbol{\beta}_\gamma$ containing zeroes. Any prior with this property is a so-called local prior, which [20] argued leads to incoherent prior beliefs. Under general conditions, local priors are slower at discarding spurious coefficients relative to non-local priors ([35], Proposition 1). The density π_M is a group moment (MOM) prior density that generalizes the MOM prior in [20] and the product moment (pMOM) prior in [21]. It is a NLP that assigns vanishing density if and only if a group $\boldsymbol{\beta}_j = 0$ for any j. Other NLPs are possible [20, 34, 36], but for concreteness we focus on (16.6).

Our discussion largely applies to any model prior $\pi(\gamma)$, its main feature of interest being the implied penalty on model size. For concreteness, we lay out a structure that is sensible both when one wishes to include/exclude individual columns in $\boldsymbol{X}$ and when one wishes to consider joint inclusion/exclusion of groups of columns in $\boldsymbol{X}$. It is common to decompose $\pi(\gamma)$ by first setting a prior probability on the model size, and then assigning the same probability to all models of a given size. Model size can be measured by the number of model parameters $p_\gamma = \dim(\boldsymbol{\beta}_\gamma)$ or, alternatively, the number of active groups $s(\gamma) = \sum_{j=1}^{J} \mathrm{I}(\gamma_j = 1)$ within the total J groups. We focus on the latter, but note that in canonical settings with no groups then $p_\gamma = s(\gamma)$ and both measures of model size become equivalent. We consider

$$\pi(\gamma) = \pi(s(\gamma)) \binom{J}{s(\gamma)}^{-1}, \tag{16.7}$$

where popular high-dimensional choices include the Beta-Binomial prior [38] where $\pi(s(\gamma)) = 1/(J+1)$, and the Complexity prior [6] where $\pi(s(\gamma))$ decreases near-exponentially in $s(\gamma)$. See also [47] for using a Poisson distribution for $\pi(s(\gamma))$. Complexity and Poisson priors penalize model size strongly, which leads to better asymptotic properties in high-dimensions, at the cost of reduced finite-n power, see Section 16.8. Any desired model restrictions are easily added to (16.7) by truncating the prior support, e.g., setting $\pi(\gamma) = 0$ to any model with $p_\gamma > n$, or that violates a hierarchical restriction (e.g., if it includes interactions but not the corresponding main effects).

16.5 Prior Elicitation and Robustness

A common criticism to BMS (that is not well-grounded, in our view) is that this framework might be overly sensitive to prior parameters. Its root is the so-called Lindley's paradox [25], namely that under an infinite prior dispersion (e.g., $g_N = \infty$ in (16.5)) the Bayes factor between two models with $p_\gamma < p_{\gamma'}$ completely favors the smaller one, i.e., $B_{\gamma\gamma'} = \infty$. We neither view this issue as a paradox nor a particularly problematic issue. Many modern statistical methods rely on tuning parameters, BMS and penalized likelihood methods are no exception. For example, LASSO can exclude all variables or include $\min\{n, p\}$ variables, as the regularization parameter changes. The relevant question is not that BMS results are extreme under extreme prior dispersion values, rather how severe is the sensitivity and are there reasonable defaults.

Regarding sensitivity, as discussed below simple prior elicitation leads to (g_N, g_M) in (16.5)-(16.6) to constants that do not depend on n, a simple default being $g_N = g_M = 1$. From a theoretical perspective, Sections 16.7-16.8 show that if (g_N, g_M) are constant then the leading terms in Bayes factors are either polynomial or exponential in n. That is, the effect of any constant (g_N, g_M) washes off as n grows. From a practical point of view, Section 16.10 shows examples where inference is robust to (g_N, g_M) ranging around the default $g_N = g_M = 1$ by a factor of 100, whereas the results obtained for different values of n are significantly different. That is, posterior inference is largely driven by the data, rather than prior parameters. It is true that both rates and practical performance are affected if one lets (g_N, g_M) depend on n, e.g., grow with n, but these imply an unreasonable (to us) belief that all $\boldsymbol{X}_\gamma$ have an extremely high predictive power on $\boldsymbol{y}$, see Section 16.10.1 for an example.

We first review the unit information prior (UIP), a classical default motivating the BIC [37], and refer the reader to [9] for a deeper review of alternative defaults. The UIP sets its precision to match the likelihood precision provided by one observation, e.g., in linear regression it has prior precision matrix $\boldsymbol{X}_\gamma^T \boldsymbol{X}_\gamma / n\phi_\gamma$. The UIP can also be interpreted as the prior belief that the expected contribution of the linear predictor $\boldsymbol{X}_\gamma \boldsymbol{\beta}_\gamma$ relative to the total dispersion $n\phi$ satisfies $E(\boldsymbol{\beta}_\gamma^T \boldsymbol{X}_\gamma^T \boldsymbol{X}_\gamma \boldsymbol{\beta}_\gamma / [n\phi]) = p_\gamma$. This seems reasonable when there are no groups in $\boldsymbol{X}_\gamma$, as then $p_\gamma = s(\gamma)$, the number of variables in γ. If $\boldsymbol{X}_\gamma$ contains groups however, such as non-linear basis expansions, it seems more reasonable to assume that the predictive power is given by $s(\gamma) < p_\gamma$. In summary, we suggest default prior parameters such that

$$E\left(\frac{\boldsymbol{\beta}_\gamma^T \boldsymbol{X}_\gamma^T \boldsymbol{X}_\gamma \boldsymbol{\beta}_\gamma}{n\phi}\right) = E\left(\sum_{\gamma_j=1} \frac{\boldsymbol{\beta}_j^T \boldsymbol{X}_j^T \boldsymbol{X}_j \boldsymbol{\beta}_j}{n\phi}\right) = s(\gamma).$$

In (16.5)-(16.6) this rule gives $g_L = g_N = 1$. In our experience, defaults along these lines result in fairly competitive BMS. Of course, these are not the absolute best possible values for any given application, but they help define their order of magnitude. While different analysts may not agree exactly on (g_L, g_N) there is a fairly narrow range of values that are reasonable, and BMS is usually robust within that range, see Section 16.10.1 for an example.

Note also that prior parameters offer an opportunity to improve power to detect practically-relevant effects, a recurrent theme in this chapter. Practical relevance in linear regression can be measured by a covariate's contribution to the signal-to-noise ratio, given by $\boldsymbol{\beta}_\gamma^T \boldsymbol{X}_\gamma^T \boldsymbol{X}_\gamma \boldsymbol{\beta}_\gamma / (n\phi)$ and hence directly related to our discussion. Similarly in logistic regression $\boldsymbol{\beta}_\gamma$ defines odds-ratios, and in survival models it defines hazard ratios or time accelerations. By setting prior parameters such that $\pi(\boldsymbol{\beta}_\gamma, \phi_\gamma \mid \gamma)$ assigns high prior probability to effect sizes that one wishes to detect, say hazard ratios > 1.1 in absolute value, one can improve the finite-n power to detect such effects. See [32] (Section 3) and [34] (Section 3) for further discussion.

16.6 Validity of Model Selection Uncertainty

BMS is not only a tool to choose a model $\widehat{\gamma}$, but also to measure uncertainty in the selection. For instance, one may assess the joint selection uncertainty via $\pi(\widehat{\gamma} \mid \boldsymbol{y})$, and for variable group j via marginal probabilities $\pi(\gamma_j = \widehat{\gamma}_j \mid \boldsymbol{y})$. A natural question is whether these measures are valid from a frequentist point of view. Suppose that $\pi(\widehat{\gamma} \mid \boldsymbol{y}) = 0.9$. Does that provide any guarantee on the frequentist correct model selection probability $P_{F_0}(\widehat{\gamma} = \gamma^*)$, or on the type I or type II errors of individual variable groups? In a slight abuse of notation, the results in this section and in Section 16.7 hold both when P_{F_0} and E_{F_0} are probabilities and expectations under $(\boldsymbol{y}, \boldsymbol{X}) \sim F_0$ when the design matrix $\boldsymbol{X}$ is considered to be random, or those under $\boldsymbol{y} \mid \boldsymbol{X} \sim F_0(\boldsymbol{y} \mid \boldsymbol{X})$ when $\boldsymbol{X}$ is non-random. In Section 16.8 the results are stated conditional on $\boldsymbol{X}$, that is F_0 denotes $\boldsymbol{y} \mid \boldsymbol{X} \sim F_0(\boldsymbol{y} \mid \boldsymbol{X})$.

While it is hard to provide guarantees that hold for all n, asymptotically BMS provides valid uncertainty quantification. In full generality ([31], Proposition 1 and Corollary 1), if one takes $\widehat{\gamma} = \arg\max_\gamma \pi(\gamma \mid \boldsymbol{y})$ to be the posterior mode, then

$$P_{F_0}(\widehat{\gamma} \neq \gamma^*) \leq 2E_{F_0}\left[\pi(\gamma \neq \gamma^* \mid \boldsymbol{y})\right],$$

also family-wise type I and II error rates are $\leq P_{F_0}(\widehat{\gamma} \neq \gamma^*)$. Note that $E_{F_0}\left[\pi(\gamma \neq \gamma^* \mid \boldsymbol{y})\right]$

converging to 0 as $n \to \infty$ is, by definition of L_1 convergence, equivalent to $\pi(\gamma \neq \gamma^*) \xrightarrow{L_1} 0$. A similar result holds when selection is based on marginal posterior probabilities, that is $\widehat{\gamma}_j = 1$ if $\pi(\gamma_j = 1 \mid \boldsymbol{y}) > t$ for some threshold t (and $\widehat{\gamma}_j = 0$ otherwise). Then, type I and II error probabilities satisfy

$$P_{F_0}(\widehat{\gamma}_j = 1 \mid \gamma_j^* = 0) \leq \frac{1}{t} E_{F_0}[\pi(\gamma_j = 1 \mid \boldsymbol{y})],$$
$$P_{F_0}(\widehat{\gamma}_j = 0 \mid \gamma_j^* = 1) \leq \frac{1}{1-t} E_{F_0}[\pi(\gamma_j = 0 \mid \boldsymbol{y})].$$

That is, if one can show that $\pi(\gamma^* \mid \boldsymbol{y})$ converges to 1 at a certain rate (in the L_1 norm under F_0), then the frequentist correct selection probability $P_{F_0}(\widehat{\gamma} = \gamma^*)$ converges to 1 at that rate or faster. By guaranteeing such concentration (either for finite n or as $n \to \infty$), one also ensures correct frequentist selection and a form of control on type I-II errors. Posterior concentration is discussed in Sections 16.7- 16.8. To be precise, Section 16.7 gives rates for convergence in probability, which is weaker than the L_1 convergence required above. Section 16.8 discusses the latter for high-dimensional linear regression.

We omit details, but similar results apply to re-normalized L_0 selection criteria such as BIC and EBIC ([31], Section 3). By renormalized we mean that each model is scored via pseudo-posterior probabilities $\tilde{\pi}(\gamma) = \exp\{-\text{BIC}_\gamma/2\} / \sum_{\gamma' \in \Gamma} \exp\{-\text{BIC}_{\gamma'}/2\}$, where BIC_γ can be suitably-replaced by any other L_0 criterion. Then, the frequentist validity of $\tilde{\pi}(\gamma)$ can be established by studying $E_{F_0}[\tilde{\pi}(\gamma)]$. These findings are interesting in that L_1 penalties and shrinkage priors are often motivated as computationally-convenient alternatives to BMS and L_0 criteria, but quantifying uncertainty for the former is a hard problem. In contrast, BMS and L_0 criteria offer an asymptotically-valid strategy to assess model selection uncertainty.

16.7 Finite-Dimensional Results

This section reviews results on the frequentist behavior of Bayes factors when p is fixed and $n \to \infty$. These results hold for very general models [11, 20, 22, 43] and under misspecification [33, 34], and also provide helpful intuition for high-dimensional settings where p grows with n (Section 16.8). Although we state these results asymptotically, for certain model classes it is possible to obtain finite-n bounds, e.g., see [31] for (possibly non-linear) regression with Gaussian errors.

In general terms, under correct model specification, [11] and [20] have shown that local priors accumulate evidence in favor of a true alternative model at an exponential rate, while the rate in favour of true null models is accumulated at a polynomial rate. In order to illustrate this behavior, let us focus on the Gaussian regression (16.3), with priors (16.5)–(16.6). To characterize Bayes factors $B_{\gamma\gamma^*}$, we need to consider separately the case where $\gamma^* \subset \gamma$ (overfitted models) and the case where $\gamma^* \not\subset \gamma$ (non-overfitted models) separately.

For $\gamma^* \subset \gamma$, it holds that $B_{\gamma\gamma^*} = \text{O}_\text{p}(a_n)$, where $a_n = (ng_N)^{\frac{p_{\gamma^*} - p_\gamma}{2}}$ under π_N (16.5), and $a_n = (ng_M)^{3(p_{\gamma^*} - p_\gamma)/2}$ under π_M (16.6). In contrast, for $\gamma^* \not\subset \gamma$,

$$\log(B_{\gamma\gamma^*}) = -n[L_{\gamma^*}(\boldsymbol{\beta}^*_{\gamma^*}, \phi^*_{\gamma^*}) - L_\gamma(\boldsymbol{\beta}^*_\gamma, \phi^*_\gamma)] + \log(b_n) + \text{O}_\text{p}(1).$$

where $b_n = (ng_N)^{\frac{p_{\gamma^*} - p_\gamma}{2}}$ under π_N, and $b_n = (ng_M^3)^{(p_{\gamma^*} - p_\gamma)/2}$ under π_M. Recall that, as defined in (16.2), L_{γ^*} and L_γ denote the expected log-likelihood under F_0, $(\boldsymbol{\beta}^*_\gamma, \phi^*_\gamma)$ is the KL-optimal parameter value under γ, and $(\boldsymbol{\beta}^*_{\gamma^*}, \phi^*_{\gamma^*})$ that under γ^*. Thus, we can observe that the use of NLPs lead to an acceleration of the Bayes factor rates, compared to local alternative priors, which alleviates the imbalance of the accumulation of evidence.

Under model misspecification (e.g., misspecifying the tails of the errors, or the functional dependence on the covariates, or missing covariates), the asymptotic rates to discard overfitted models $\gamma \supset \gamma^*$ are unaffected, up to constant terms (though finite n performance can suffer, [33, 34]). These results remain valid in survival analysis contexts where the response variable may be censored [34]. In contrast, the rate to discard non-overfitted models is essentially exponential in n with a coefficient $L_{\gamma^*}(\boldsymbol{\beta}^*_{\gamma^*}, \phi^*_{\gamma^*}) - L_{\gamma}(\boldsymbol{\beta}^*_{\gamma}, \phi^*_{\gamma}) > 0$. This coefficient determines the drop of predictive ability in γ relative to γ^*, and is affected by misspecification and by censoring, typically resulting in an exponential power drop in the asymptotic power.

16.8 High-Dimensional Results

In Section 16.7, we distinguished models based on their size and whether they contained γ^* or not. Let $S_l = \{\gamma : \gamma^* \subset \gamma, p_\gamma = l\}$ be the overfitted models of size l that contain γ^* plus some spurious parameters. Similarly, let $S_l^c = \{\gamma : \gamma^* \not\subset \gamma, p_\gamma = l\}$ be the size l non-overfitted (or non-spurious) models. Clearly,

$$1 - \pi(\gamma^* \mid \boldsymbol{y}) = \sum_{l=p_{\gamma^*}+1}^{\bar{p}} \sum_{\gamma \in S_l} \pi(\gamma \mid \boldsymbol{y}) + \sum_{l=0}^{\bar{p}} \sum_{\gamma \in S_l^c} \pi(\gamma \mid \boldsymbol{y}), \tag{16.8}$$

where $\bar{p} = \max_{\gamma \in \Gamma} p_\gamma$ is the size of the largest model under consideration. Although not strictly necessary, one may assume $\bar{p} \ll n$, that is although maybe $p \gg n$ and $|\Gamma| \gg n$, one is not willing to select a model with $p_\gamma \gg n$ parameters. For precision, $a_n \ll b_n$ denotes $\lim_{n\to\infty} a_n/b_n = 0$ for two deterministic sequences $a_n, b_n > 0$.

Section 16.7 discussed rates for Bayes factors, and hence also for $\pi(\gamma^* \mid \boldsymbol{y})$, when p is fixed. The main elements were whether the prior on parameters is a local or a non-local prior, its dispersion ((g_N, g_M) in (16.5)-(16.6)), the prior on model size $\pi(s(\gamma))$, and the signal strength measured by a non-centrality parameter

$$\lambda_{\gamma,\gamma^*} = n[L_{\gamma^*}(\boldsymbol{\beta}^*_{\gamma^*}, \phi^*_{\gamma^*}) - L_{\gamma}(\boldsymbol{\beta}^*_{\gamma}, \phi^*_{\gamma})].$$

Roughly speaking, in (possibly non-linear) Gaussian regression $\lambda_{\gamma,\gamma^*}$ is proportional to n times the magnitude of $\boldsymbol{\beta}^*_{\gamma^*}$ (relative to ϕ^*_{γ} and correlations in $\boldsymbol{X}$), see below.

As it turns out, the strategies for $\pi(\gamma^* \mid \boldsymbol{y})$ to converge in high dimensions are again NLPs, diffuse priors (growing g) and sparse $\pi(s(\gamma))$. Briefly, [21] and [39] showed such consistency for strategies using NLPs on the parameters. [15, 28] proposed the use of diffuse LPs in high-dimensional regression, and [45] also proposed diffuse LPs in more general models. [6] showed that, by combining so-called (sparse) Complexity priors $\pi(s(\gamma))$ on the models with (local) Laplace priors $\pi(\boldsymbol{\beta}_\gamma \mid \gamma)$ on the parameters, one discards regression models that overshoot the dimension of p_{γ^*}. Using similar priors [7] proved consistency for symmetric nonparametric errors, [16] for general structured linear models under misspecified sub-Gaussian errors, and [30] for regression trees. These three strategies can be combined, e.g., by using both diffuse and Complexity priors in regression [46], or by using diffuse NLPs [39]. As an important practical issue, however, penalties from diffuse priors and Complexity priors $\pi(s(\gamma))$ do not depend on the data. Should γ^* be less sparse than anticipated, there can be a significant loss of power, see Section 16.9. Also, setting growing g may be problematic from a foundational point of view, in that it does not lead to reasonable prior beliefs (Section 16.5). Of course, in practice one could simply view g as a tuning parameter to be learnt from the data, akin to regularization parameters in penalized likelihood methods. We personally prefer to consider a framework where g is elicited a priori or at least constrained to be in a range reflecting beliefs about relevant problem features. See for example Figure 16.2 (right), fully described in Section 16.10.1, showing the prior expected value of the R^2 coefficient as a function of g. Values outside $g \in [0.1, 10]$ imply the belief that the proportion of variance in $\boldsymbol{y}$ explained by covariates is either lower or higher than one may deem reasonable in most applications.

The rest of this section summarizes high-dimensional regression results from [31]. The regression may be non-linear, and F_0 may contain omitted variables and heteroscedastic and correlated errors, although it is assumed to have Gaussian tails (thicker tails require adjusting certain Bayes factor bounds, but similar intuition applies). The strategy to study $\pi(\gamma^* \mid \boldsymbol{y})$ is as follows. The fact that p grows with n poses challenges, even if one could prove convergence in probability for each term in (16.8), their sum may not converge. The added terms can exhibit complex dependencies, and their number grows (very quickly) with n. A simpler strategy is to bound (16.8) in expectation under F_0, *i.e.*, study its L_1 convergence to 0. By bounding Bayes factor tails, it is possible to obtain rates for (16.8) similar to those for fixed p settings in Section 16.7, up to lower-order terms [31]. The latter terms are often logarithmic, but for simplicity we express them via constants $\alpha < 1$, $w < 1$ to be thought of as being close to 1. Specifically, $E_{F_0}[1 - \pi(\gamma^* \mid \boldsymbol{y})] \ll$

$$\sum_{l=p_{\gamma^*}+1}^{\bar{p}} \sum_{\gamma \in S_l} \frac{\pi(\gamma)}{\pi(\gamma^*)} (gn)^{\frac{-\alpha(p_\gamma - p_{\gamma^*})}{2}} + \sum_{l=0}^{\bar{p}} \sum_{\gamma \in S_l^c} \frac{\pi(\gamma)}{\pi(\gamma^*)} \frac{(gn)^{\frac{-\alpha(p_\gamma - p_{\gamma^*})}{2}}}{e^{w\lambda_{\gamma\gamma^*}^{\alpha}/2}}. \tag{16.9}$$

The first term in (16.9) measures the speed at which one discards spurious parameters. The rate applies to local priors such as π_N in (16.5). For NLPs this first term vanishes faster, e.g., π_M in (16.6) attains $\alpha < 3$. The second term relates to discarding models missing parameters from γ^*, *i.e.*, the power to detect truly active parameters. Provided that g and $\pi(\gamma)$ do not favor γ over γ^* too strongly, the leading factor is exponential $\lambda_{\gamma\gamma^*} > 0$. Here $\lambda_{\gamma\gamma^*}$ is the difference between mean squared error associated to $\boldsymbol{X}_\gamma \boldsymbol{\beta}_\gamma^*$ under F_0 and that of $\boldsymbol{X}_{\gamma^*} \boldsymbol{\beta}_{\gamma^*}$. To gain intuition, if $\boldsymbol{X}_{\gamma^*}^T \boldsymbol{X}_\gamma = I$ then $\lambda_{\gamma^*\gamma} = ||\boldsymbol{X}_{\gamma^*} \boldsymbol{\beta}_{\gamma^*}^*||_2^2 / \phi^*$, and more generally

$$\lambda_{\gamma^*\gamma} \geq n v_{\gamma^*\gamma} (\boldsymbol{\beta}_\gamma^*)^T \boldsymbol{\beta}_\gamma^* / \phi^*,$$

where $v_{\gamma^*\gamma}$ is the smallest non-zero eigenvalue of the residual covariance after regressing $\boldsymbol{X}_{\gamma^*}$ on $\boldsymbol{X}_\gamma$. By suitably adjusting $\lambda_{\gamma\gamma^*}$, (16.9) remains valid under misspecification, e.g., when $E_{F_0}(\boldsymbol{y} \mid \boldsymbol{X})$ is non-linear in ways different than allowed by the model, omitting covariates and heteroscedastic correlated errors. The expression also applies to re-normalized L_0 penalties, by replacing $\pi(\gamma)$ and gn by the associated penalty on model dimension, e.g., $g = 1$ and $\pi(\gamma)/\pi^*(\gamma^*) = 1$ for the BIC. For common priors and L_0 penalties (16.9) can be simplified. Under suitable conditions, $E_{F_0}[1 - \pi(\gamma^* \mid \boldsymbol{y})] \ll$

$$\frac{(p_{\gamma^*} + 1)(p - p_{\gamma^*})^{1-\alpha}}{(gn)^{\alpha/2} p^c} + e^{-w\underline{\lambda}^{\alpha'}/2 + (p_{\gamma^*} - 1)\log(p)} + e^{-\gamma\bar{\lambda}^{\alpha'}/2 + (\bar{p} - p_{\gamma^*} + 1)\log(p - p_{\gamma^*})}, \tag{16.10}$$

where $(\underline{\lambda}, \bar{\lambda})$ are lower and upper bounds on $\lambda_{\gamma\gamma^*}$ for non-overfitted models with $l < p_{\gamma^*}$ (sparser than γ^*) and $l > p_{\gamma^*}$ (less sparse than γ^*) respectively. The rate under a Beta-Binomial prior is obtained by plugging in $c = 0$, for the Complexity prior $c > 0$ is the parameter in $\pi(s(\gamma)) = e^{-cs(\gamma)}$, and for the EBIC [8] one takes $c = 0$, $g = 1$, for example. For the EBIC and certain priors the term $(p - p_{\gamma^*})^{1-\alpha}$ can be replaced by tighter logarithmic term in p.

We avoid a technical discussion and rather point out that Expression (16.10) provides useful intuition. It expresses how fast $\pi(\gamma^* \mid \boldsymbol{y})$ converges to 0, and how much posterior mass is assigned to overfitted and non-overfitted models, as a relatively simple function of n, p, the optimal model dimension p_{γ^*} (sparsity of the data-generating truth), and prior parameters such as g and c featuring in $\pi(\boldsymbol{\beta}_\gamma \mid \gamma)$ and $\pi(\gamma)$. The first term in (16.10) (overfitted $\gamma \in S$) is asymptotically larger than the other two terms, hence to obtain optimal rates one should focus on ensuring sparsity. Overfitted models receive vanishing probability under a Beta-Binomial prior and the EBIC when p is up to polynomial in gn (if $p_{\gamma^*} \ll (gn)^{\alpha/2}$), whereas diffuse priors and Complexity priors allow for arbitrarily large p. Hence these priors help obtain optimal asymptotic rates. However, these strategies can reduce finite n power. This is seen in the second term in (16.9) where large g and sparse $\pi(\gamma)$ favor small models ($p_\gamma < p_{\gamma^*}$) over γ^*. The power drop can be surprising even in simple examples, as illustrated next.

16.9 Balancing Sparsity and Power

The Bayesian approach has two elements to balance sparsity and power: the prior and the likelihood. Ideally, these are guided by subject matter expertise, see Section 16.5 for simple prior elicitation ideas. Nevertheless, as useful guidelines to aid these choices, we discuss (frequentist) type I error and power considerations. Ideally one seeks priors that encourage parsimony (prevent false positives), while not reducing power too strongly. Similarly, a flexible likelihood avoids power losses due to gross model misspecification, but too much flexibility (many parameters) also reduces power.

Theory provides useful insights on these important principles, but trusting asymptotic optimality results can be misleading, particularly in high dimensions. Of particular relevance, diffuse and Complexity priors [6, 28] underlie much of the state-of-the-art BMS theory. These priors excel at discarding spurious models, but enforcing sparsity a priori, regardless of the data, can lead to important power losses.

We discuss three natural strategies that can lead to significant practical gains. First, one may forgo asymptotic optimality to instead combine model priors that induce a mild complexity penalty (e.g., Beta-Binomial) with NLPs that (roughly speaking) penalize only truly spurious parameters. As an illustration consider a simplest Gaussian linear regression with $\boldsymbol{X}^T\boldsymbol{X} = I$, $n = 110$, $p = 100$ and $p_{\gamma^*} = 20$ truly active covariates with $\boldsymbol{\beta}^* = (\mathbf{b}, \mathbf{b}, \mathbf{b}, \mathbf{b}, 0, \ldots, 0)$ where $\mathbf{b} = (0.25, 0.5, 0.75, 1, 1.5)$ and $\phi^* = 1$, an example from [31]. We combined the Normal prior (16.5) (default $g_N = 1$) with a Beta-Binomial$(1,1)$ and a Complexity prior ($c = 1$). We also used the MOM prior (16.6) (default $g_M = 1$) plus a Beta-Binomial$(1,1)$. Using theory from Section 16.8, one can show that the Normal-Complexity prior has better asymptotic rates for $\pi(\gamma^* \mid \boldsymbol{y})$ than the other two, but Figure 16.1 (top left) shows a severe lack of power. The average $\pi(\gamma_j = 1 \mid \boldsymbol{y})$ is negligible even for large $\boldsymbol{\beta}_j^*$. The Normal-Beta Binomial prior improved power but results in large inclusion probabilities for spurious $\boldsymbol{\beta}_j^* = 0$. The MOM-Beta Binomial offered a compromise, it penalized truly zero or near-zero $\boldsymbol{\beta}_j^*$ and attained high power for the remaining (more practically relevant) active covariates.

A second strategy is to set the prior via empirical Bayes. Let g be the prior dispersion and ω a parameter determining $\pi(\gamma)$, e.g., $\omega = 1$ might indicate a Beta-Binomial prior and $\omega = 2$ a Complexity prior, or alternatively one could let $\omega = \pi(\gamma_j = 1)$ be the marginal

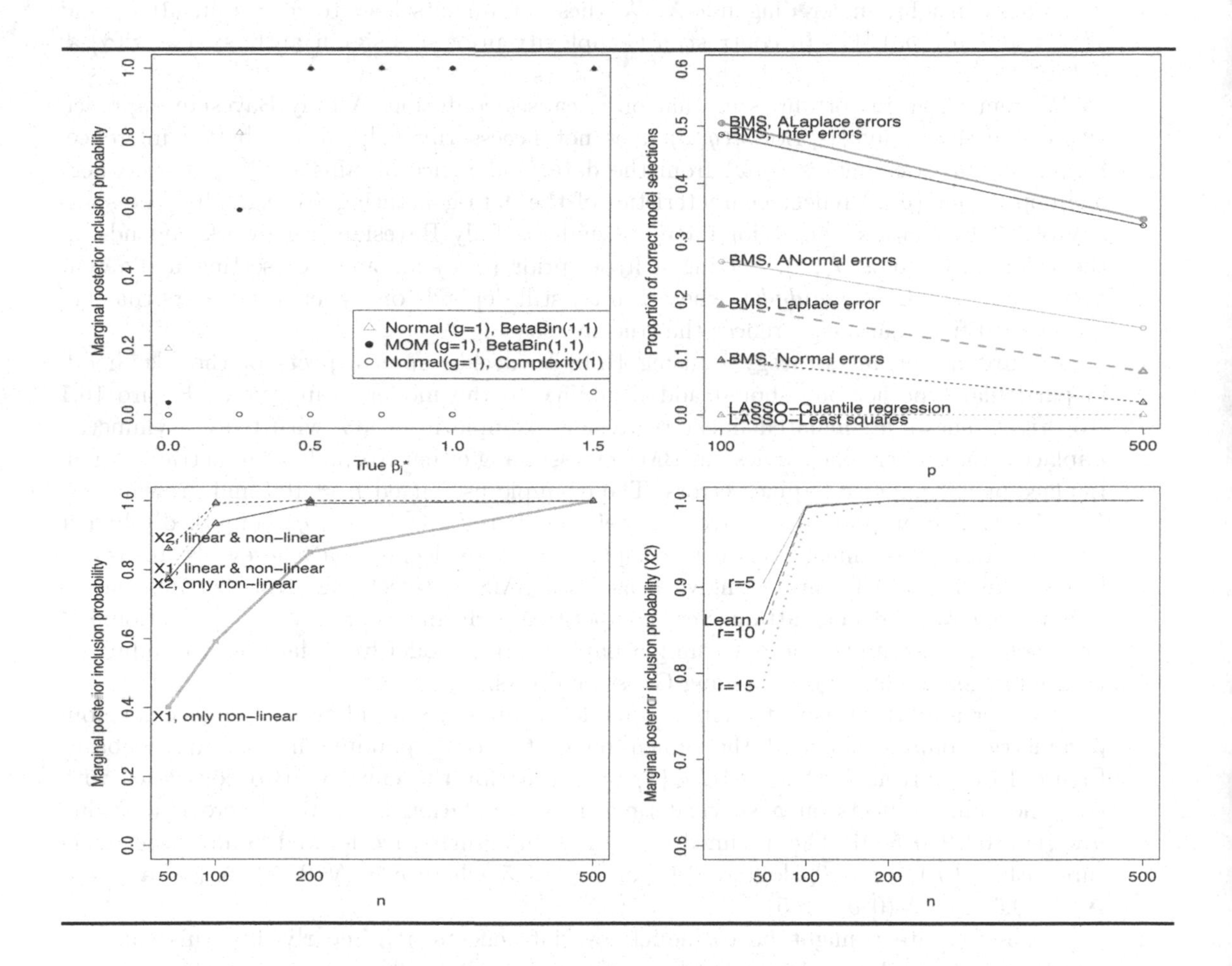

FIGURE 16.1
Balancing sparsity and power in simulations. Top left: mean $P(\gamma_j = 1 \mid \boldsymbol{y})$ versus $\boldsymbol{\beta}_j^*$ in well-specified linear regression with $n = 110$, $p = 100$, $p_{\gamma^*} = 20$ and uncorrelated covariates. Top right: correct model selections under several error assumptions in linear regression with true asymmetric Laplace errors, $n = 100$, $p \in \{100, 500\}$. Bottom: linear + non-linear effect decomposition of 2 truly non-linear covariates on survival significantly increases power (left); so does a spline basis with small dimension $r \in \{5, 10, 15\}$ (right)

prior inclusion probability. Standard empirical Bayes sets

$$(\widehat{g},\widehat{\omega}) = \arg\max_{g,\omega} f(\boldsymbol{y} \mid g,\omega) = \arg\max_{g,\omega} \sum_{\gamma\in\Gamma} f(\boldsymbol{y} \mid g,\gamma)\pi(\gamma \mid \omega),$$

see [17] for details and an alternative that jointly maximizes (γ, g, ω), and [24] to set only g. To gain insight, under diagonal $\boldsymbol{X}^T\boldsymbol{X}$ these arguments lead to $g \approx 1$ in (16.5) and $\pi(\gamma) \approx p_\gamma/(p - p_\gamma)$ [17]. In contrast, a Complexity prior sets significantly sparser $\pi(\gamma) \approx e^{-p_\gamma(c+\log p)}$.

We remark an important issue that often causes confusion. A fully Bayesian approach where one sets a hyper-prior $\pi(g,\omega)$ does not necessarily help obtain better inference. Empirical Bayes estimates $(\hat{g},\hat{\omega})$ from the data, and hence obtains $\pi(\gamma \mid \boldsymbol{y},\hat{g},\hat{\omega})$, where one hopes that $(\hat{g},\hat{\omega})$ reflect characteristics of the data-generating F_0, e.g., the true sparsity of γ^*. In contrast, posterior inference under a fully Bayesian framework depends on the prior only via $\pi(\boldsymbol{\beta}_\gamma,\gamma)$. Setting a hyper-prior (g,ω) amounts to setting a different $\pi(\boldsymbol{\beta}_\gamma,\gamma) = \int \pi(\boldsymbol{\beta}_\gamma,\gamma,g,\omega)dgd\omega$, where $\pi(g,\omega)$ still depends on hyper-parameters that are not learned from data, e.g., reflect the true sparsity of γ^*.

A third important strategy is using BMS to decide upon aspects of the likelihood, in particular whether one should add flexibility to the model assumptions. Figure 16.1 (top right) shows a simulated linear regression example from [33] with truly asymmetric Laplace errors, where one carries out BMS by assuming either Normal, asymmetric Normal, Laplace or asymmetric Laplace errors. The example uses fixed $n = 100$ and growing $p \in \{100, 500\}$. The proportion of correct model selections $P_{F_0}(\widehat{\gamma} = \gamma^* \mid \boldsymbol{y})$ is markedly higher when (correctly) assuming asymmetric Laplace errors. A deeper analysis reveals this is due to gains in power [33]. Interestingly, if one uses BMS to select the error distribution (as well as the covariates) one attains very competitive performance. See also [33] (Section 6.5) for a genomics example where learning a Laplace error model from data led to finding an additional gene, relative to assuming Gaussian errors.

Another important aspect related to model flexibility is deciding on the need for nonparametric components, and the dimension of the corresponding nonparametric basis. Figure 16.1 (bottom left) shows $P_{F_0}(\gamma_j = 1 \mid \boldsymbol{y})$ for the effect of two covariates with truly non-linear effects on a survival time, in a simulation from [34] where $p = 2$ and $n \in \{50, 100, 200, 500\}$. The assumed model has an additive accelerated failure time structure with $\log(y_i) = x_{i1} + 0.5\log(|x_{i2}|) + \epsilon_i$ and $c_i = 0.5$, where $\boldsymbol{x}_i \sim \mathcal{N}(0, \mathbf{A})$, $\mathbf{A}_{11} = \mathbf{A}_{22} = 1$, $\mathbf{A}_{12} = 0.5$, $\epsilon_i \sim \mathcal{N}(0, \sigma = 0.5)$.

A naive strategy might be to model covariate effects non-linearly, but this causes a significant power drop relative to decomposing covariate effects into a single-coefficient linear term plus a group of coefficients capturing deviations from linearity, e.g., $P(\gamma_1 = 1 \mid \boldsymbol{y})$ roughly doubles from 0.4 to 0.8. Intuitively, the linear component captures part of the covariate effects with a single parameter, the non-linear component captures said effects better but also incurs a larger BIC-type dimensionality penalty (Section 16.7). This example used a spline basis with dimension $r \in \{5, 10, 15\}$ learned via BMS. Note that a truly semi-parametric approach would consider large r (typically growing with n). This example illustrates that such large r it not only computationally inconvenient, but can also lead to a drop in power (Figure 16.1, bottom right). The power drop could potentially be addressed by using large r but reducing the effective number of parameters, e.g., via P-splines, but obtaining integrated likelihoods would remain computationally inconvenient in BMS settings where one seeks to consider many models. In our opinion, for BMS purposes where one mainly seeks to determine if a covariate has an effect, it is more convenient to consider finite basis with moderate r, and to decompose effects into linear plus deviation-from-linear components, to improve power and computational speed. Again, we remark

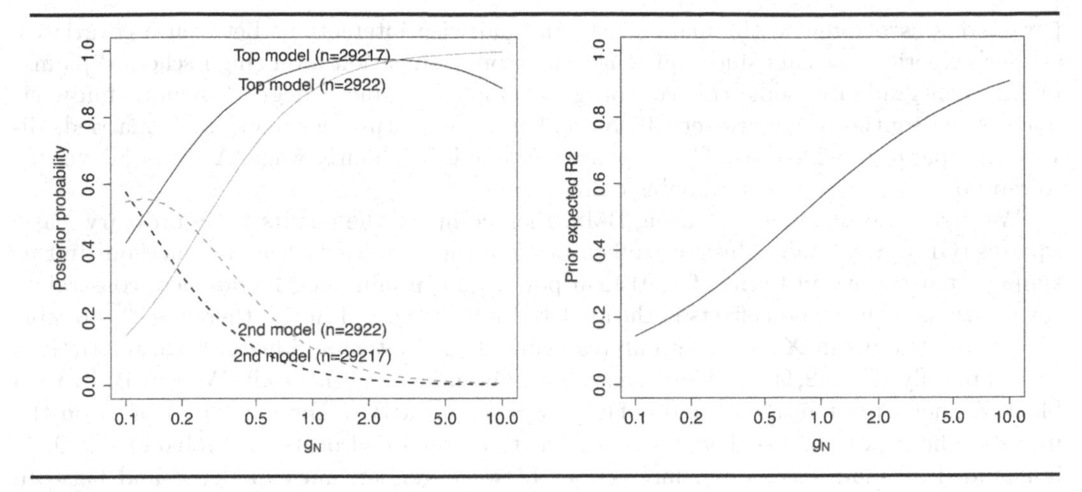

FIGURE 16.2
Salary data. Left: posterior probability of the top two models as a function of $g_N \in [0.1, 10]$ when using the full data ($n = 29217$) and a randomly selected subset ($n = 2922$). Right: prior expected R^2 under the block-Zellner prior for $g_N \in [0.1, 10]$

that these observations are in contrast to estimation/prediction problems, where one would typically consider large r.

16.10 Examples

We illustrate the use of BMS in several practical settings. We first consider a classical exercise that considers main effects and interactions for the association between covariates and salary. Despite its simplicity, it helps illustrate the flexibility of BMS to deal with group and hierarchical constraints, as well as study prior robustness issues discussed above. It also illustrates the difficulties encountered by state-of-the-art penalized likelihood methods in assessing significance, in incorporating hierarchical restrictions and sensitivity to regularization parameters. We next consider a high-dimensional linear regression example, where BMS attains competitive performance in terms of predictive ability and run times, but returns significantly simpler solutions than penalized likelihood methods. The example also illustrates differences between using local and non-local priors, as well as diffuse priors with large prior dispersion, that link to the theory presented earlier. Finally, we illustrate the generality of BMS via a survival analysis where, by considering non-linear covariate effects, one obtains refined patient prognosis results.

16.10.1 Salary

We analyze a dataset used in [27] based on a 2012 survey of the USA Census Bureau and the Bureau of Labor Statistics, recording the hourly salaries for white non-hispanic full-time workers within ages 25–54. There are $n = 29,217$ individuals and we consider $p = 67$

parameters associated to the main effects and pairwise interactions between 5 covariates: gender, experience (years since finishing education), education level (high school diploma, high school graduate, some college, college graduate, advanced degree), region (midwest, north east, southern and western USA) and marital status (never married, married, divorced, separated, widowed). The response variable is log-hourly wage. A R markdown file to reproduce the analysis is available online.

We first analyze these data using BMS, then compare the results to an ordinary least-squares (OLS) and finally illustrate difficulties in using penalized likelihood methods in this simple setting, both in terms of statistical power and in enforcing hierarchical constraints (interactions require main effects in the model). Interestingly, although there are 2^{67} possible subsets of columns in $\boldsymbol{X}$, due to group (categorical covariates) and hierarchical restrictions there are only $|\Gamma| = 2,900$ models, hence one can enumerate them all. We run BMS with block Zellner's prior in (16.5) and default $g_N = 1$ and a Beta-Binomial$(1,1)$ prior on the models. The task took less than 1 second. The top model had posterior probability > 0.95, it included all main effects and interactions between gender and experience and between education and experience. All other interactions were discarded. The second top model had 0.041 posterior probability and selected the same covariates plus the interaction between gender and marital status. As predicted by theory, when n is large posterior probabilities concentrate on few models.

As a benchmark on what results may be reasonable to obtain in these data, since n is large we can rely on ordinary least-squares (OLS) under the full model including all $p = 67$ regression parameters to provide fairly accurate estimates, confidence intervals and p-values. We remark that OLS is a useful reference but it does not lead to a clear model selection rule. For instance, a popular strategy is to apply stepwise methods where one drops terms with non-significant p-values one at a time, ensuring that main effects are not discarded before their corresponding interactions. From a statistical point of view, a caveat is that it is not clear how to measure type I error for such multi-step procedures. Using OLS, all main effects were again significant (p-value < 0.0001) and so were the interactions gender versus experience, education versus experience and gender versus marital status (p-values <0.0001, <0.0001, and 0.0002 respectively). Relative to BMS, the OLS p-values were more liberal in also suggesting interactions of gender versus education, and education versus region (p-value=0.021, 0.0047). This is no surprise, it is common for Bayesian tests to be more conservative than using a standard 0.05 thresholds on p-values, which [2] argued can help ameliorate the so-called reproducibility crisis in scientific research. From a predictive point of view least-squares was very similar to Bayesian model averaging, the correlation between the predicted $\boldsymbol{y}$ was 0.994, but the latter shrunk towards zero some of the parameter estimates.

Next we assessed sensitivity of the BMS results to the prior dispersion and to the sample size. Figure 16.2 (black lines) displays the posterior probability of the top two models described above for $g_N = 0.1, ..., 10$. Results were remarkably stable, most g_N returned the same top model and, for small g_N then the second top model receives higher posterior probability. Recall that these two models selected similar covariates and in combination receive a posterior probability near 1, for all these g_N. One might argue that with such large n it is hardly surprising that posterior inference is robust to g_N. To study this issue, we repeated the analysis by randomly selecting 10% of the observations, obtaining a new sample size $n = 2,922$. The top model (0.880 posterior probability) now only included main effects for gender, education and experience, and the second top model (0.108 posterior probability) added the main effect for region. It is not surprising that with smaller n one selects less covariates, as discussed the power to detect truly non-zero effects is exponential in n. In fact, a similar phenomenon was observed when applying OLS. In our opinion, the main message is that reducing n by a factor of 10 had a large effect on the BMS solution, whereas changing g_N by a factor of 100 did not.

One may wonder whether it is reasonable to consider $g_N \notin [0.1, 10]$. We do not believe so. Figure 16.2 (right) shows the prior expectation of the percentage of variance in $\boldsymbol{y}$ explained by $\boldsymbol{X}_\gamma$, the R^2 coefficient

$$R^2 = \frac{\boldsymbol{\beta}_\gamma^T \boldsymbol{X}_\gamma^T \boldsymbol{X}_\gamma \boldsymbol{\beta}_\gamma / n}{\phi + \boldsymbol{\beta}_\gamma^T \boldsymbol{X}_\gamma^T \boldsymbol{X}_\gamma \boldsymbol{\beta}_\gamma / n} = \left(1 + \frac{\phi n}{\boldsymbol{\beta}_\gamma^T \boldsymbol{X}_\gamma^T \boldsymbol{X}_\gamma \boldsymbol{\beta}_\gamma}\right)^{-1},$$

for $g_N \in [0.1, 10]$. The prior expected R^2 ranges from $< 15\%$ up to $> 90\%$, the default $g_N = 1$ roughly corresponding to $E(R^2) = 0.5$ (the variance explained by $\boldsymbol{X}_\gamma$ being of the same magnitude as the error variance ϕ).

Finally we analyze these data with penalized likelihood methods. We focus on group LASSO [1, 48] and attempt to control type I errors using LASSO post-selection inference [23] and Stability selection ([26], R package stabs). Given that group LASSO does not incorporate hierarchical constraints we attempted to use hierarchical LASSO ([4], R package hierNet), but in these data it failed to return a solution after 2 hours of computation. For group LASSO we used R package grplasso and the cross-validation wrapper function available at https://github.com/SoyeonKimStat/Grouplasso. We set the regularization parameter λ to two popular values returned by the software, $\widehat{\lambda}$ minimizing the cross-validated squared prediction error, and $\widehat{\lambda}$ plus 1 standard error. The former value returned non-zero estimates for all 67 parameters, whereas the latter returned only 10 non-zero estimates (main effects for female, education, experience and region) and selected no interaction terms. Neither solution $\widehat{\boldsymbol{\beta}}_{\widehat{\lambda}}$ seems satisfactory given that OLS p-values suggested clearly significant interactions. It is also unappealing that the solution is so sensitive to $\widehat{\lambda}$. To assess the statistical significance associated to $\widehat{\boldsymbol{\beta}}_{\widehat{\lambda}}$ we used LASSO post-selection inference. The method requires solving an optimization problem that took > 10 minutes to run and unfortunately ultimately failed to converge. As an alternative we considered stability selection, setting the per-family error rate to 0.05 and trying two different default values for a cutoff parameter. In both cases only a single coefficient (main effect of high-school education) was selected. Relative to the OLS p-values, these findings suggest a serious lack of power.

Altogether these examples illustrate that, even in a setting where n is large, standard penalized likelihood methods can face practical difficulties in sensitivity to tuning parameters and in balancing the prevention of false positives with statistical power. In this example, BMS provided a simple solution at low computational cost, and attained a better sparsity versus power trade-off.

16.10.2 Colon Cancer

We reproduce the colon cancer example in [35], the R code and data are available as a supplement to that article. TGFB is a gene playing an important role in colon cancer progression, and it is important to understand its relation to other genes. [5] found 172 genes related to TGFB in mice data, and verified their association to cancer progression in data with $n = 262$ patients. We applied BMS to study which of these $p = 172$ genes are associated to TGFB in humans, and which are conditionally uncorrelated given the identified genes. We then repeated the exercise after adding 10,000 extra genes selected randomly from the distinct 18,178 genes measured in the experiment.

We illustrate the effect of different prior formulations by using the MOM prior in (16.6) with default $g_M = 1/3$, the Benchmark prior (BP, [14]) setting a diffuse $g_N = p^2/n$ in (16.5), and the unit information prior (UIP) setting $g_N = 1$. We also report results for three

	$p = 172$		$p = 10,172$		
	$\bar{p}$	$\widehat{R}^2$	$\bar{p}$	$\widehat{R}^2$	CPU time
MOM	4.3	0.566	6.5	0.617	1m 52s
UIP	6.7	0.563	11.9	0.614	3m 10s
BP	4.2	0.562	3.0	0.586	1m 23s
SCAD	29	0.565	81	0.535	16.7s
LASSO	42	0.586	159	0.570	23.7s
ALASSO	24	0.569	10	0.536	2m 49s

TABLE 16.1
TGFB data with $p = 172$ or $10,172$ genes. $\bar{p}$: posterior mean $E(p_\gamma \mid \boldsymbol{y})$ (MOM, UIP, BP) or selected number of predictors (SCAD, LASSO, ALASSO). $\widehat{R}^2$ coefficient is between $(y_i, \widehat{y}_i)$ (leave-one-out cross-validation). CPU time on Linux OpenSUSE 13.1, 64 bits, 2.6GHz processor, 31.4Gb RAM for 1,000 Gibbs iterations (MOM,iMOM,BP) or 3×10^6 model updates (HG)

penalized likelihood methods, LASSO [41], SCAD [13] and adaptive LASSO (ALASSO, [49]).

Table 16.1 shows the posterior expected model size (for BMS) and the number of selected variables (for penalized likelihood), and the leave-one-out cross-validated $\widehat{R}^2$ coefficient between predictions (obtained via Bayesian model averaging, for Bayesian methods) and observations. The idea is that methods achieving a comparatively low $\widehat{R}^2$ are either dropping important variables, or including too many spurious variables.

For $p = 172$ all methods had a similar cross-validated $\widehat{R}^2$ but the three BMS methods selected significantly smaller models. For $p = 10,172$ there were larger differences between methods. The models selected by MOM achieved highest $\widehat{R}^2$ and had a parsimonious posterior expected model size of 6.5 genes. The UIP achieved a similar $\widehat{R}^2$, despite including a few more genes, suggesting that these were spurious, matching our theoretical discussion that local priors do not enforce sparsity sufficiently. Interestingly the BP, a local prior that sets a large dispersion g_L to improve sparsity, returned a smaller model at the cost of a reduced $\widehat{R}^2$. This observation again aligns with theoretical results that diffuse priors may cause a drop in power. The remaining methods selected substantially more predictors, at no gain in $\widehat{R}^2$.

The genes selected by MOM are known to be related to various cancer types, TGFB regulators and genes that alleviate cancer symptoms, see [35] for further discussion. These findings suggest that combining NLPs with a Beta-Binomial prior, *i.e.*, setting a mild prior penalty on model size, was effective at finding a parsimonious set of predictors that had competitive predictive power. We also note that all BMS computation times were highly competitive.

16.10.3 Survival Analysis of Serum Free Light Chain Data

Some studies suggest that the serum free light chain (FLC), a component of immunoglobulin production, may be a biomarker for immune dysregulation and plasma cell disorders [12]. The `flchain` dataset from the R package `survival` contains a stratified random sample with $n = 7,874$ subjects from a study of the relationship between FLC and survival time. The response variable is the time from enrolment until death (in years). We considered 5 covariates: gender, FLC kappa portion (kappa), FLC lambda portion (lambda), serum creatinine, and whether the patient was diagnosed with monoclonal gammapothy (yes/no).

We excluded the variable age from our analysis as it is left-truncated and the sample contains only advanced age patients. We scaled all continuous covariates to have zero mean and unit variance. For simplicity, we focus the analysis on the $n = 6,524$ individuals with no missing values. The data were heavily censored, $n_o = 1,962$ individuals died and $n_c = 4,562$ were alive at the end of the study follow-up. Our purpose is to illustrate the interest in being able to seamlessly incorporate non-linear effects into Bayesian model selection.

For illustration we used a Normal accelerated failure times model. We first applied BMS considering only linear effects, using Zellner's prior. The top model had 0.69 posterior probability and included only FLC kappa and FLC lambda (marginal inclusion probabilities 1 and 0.833 respectively). As a purely descriptive external validation, both variables had a p-values < 0.01 for a maximum likelihood fit under the full model. We next extended the exercise to consider non-linear effects for the 3 continuous variables. This analysis again selected linear effects for FLC kappa and FLC lambda (posterior inclusion probabilities of 1 and 0.9997), but also included a non-linear effect for FLC kappa and for creatinine and marginal evidence for gender (posterior inclusion probabilities 1, 0.9998 and 0.524).

To illustrate the differences between the selected linear and non-linear models, we compared their predicted 0.75 survival quantiles. The left panel in Figure 16.3 shows that the difference between these predictions is practically relevant. The right panel shows that the non-linear model predicts higher survival for low values of FLC kappa, and lower survival for high values of kappa. If one were to unduly assume linearity, one would under-estimate the impact of these covariates on survival.

As an important remark, we chose the 0.75 quantile over the perhaps more conventional median survival given that $> 50\%$ of the observations were censored. That is, predicting median survival would require an extrapolation beyond the study's follow-up time that can be sensitive to model misspecification (survival times having log-normal tails). As a token, a non-negligible fraction of predicted median survival times are > 100 years, such predictions are obviously not reasonable. This is an interesting contrast to BMS, which as discussed selects the covariates that show predictive power on the average over the observed censoring times, and hence does not depend on such extrapolations.

16.11 Discussion

We summarized several fundamental issues that have an important impact both on the theory and practice of BMS. From a theoretical viewpoint, for large enough n, BMS is equipped with strong properties to select the right model, to portray the Bayesian uncertainty via posterior probabilities, and to bound the frequentist probabilities of choosing the right model, type I and II errors. Methodologically, BMS is appealingly general. One can easily incorporate group and hierarchical restrictions and compare non-nested models, for example. It is also interesting that, even when all models are misspecified, BMS targets a reasonable problem that asymptotically returns the best smallest model (minimizing the log-likelihood associated loss). Despite these nice properties, for BMS to perform well in practice it is necessary to strike a balance between sparsity and power. This is the recurrent tension between type I and II errors in statistical inference, and interacts with potential power drops caused by misspecification and choosing overly sparse priors (large prior dispersion, or too large probability on small models). We took a pragmatic view and discussed simple strategies to set priors and to enrich the likelihood by learning from the data whether one should add non-linear and/or nonparametric terms. The beauty is that deciding upon

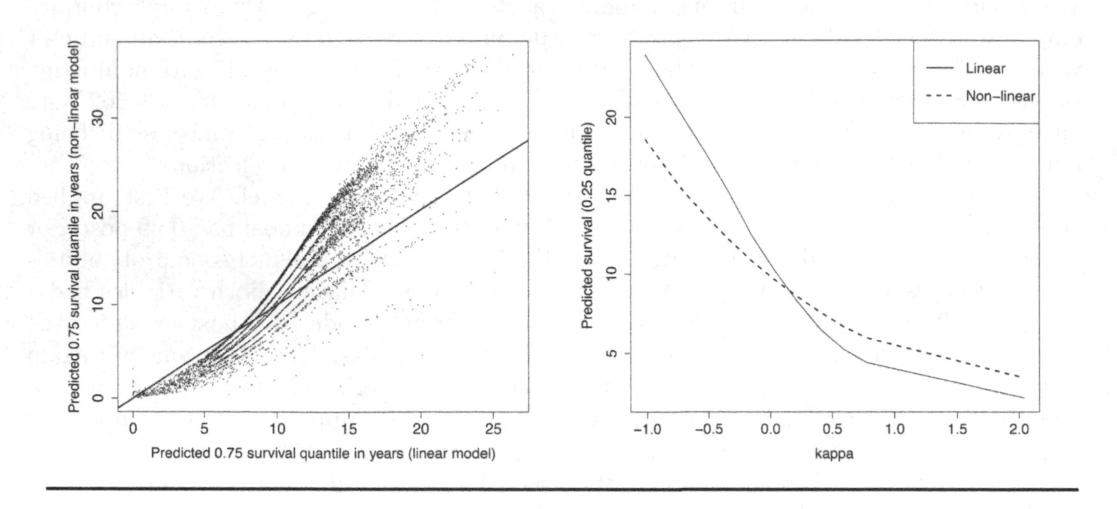

FIGURE 16.3
Serum free light chain data. Left: predicted 0.75 survival time by the selected linear and non-linear models. Right: average estimated 0.75 survival time versus FLC kappa.

the addition of such terms can be seamlessly integrated within the BMS framework. In our experience, simple ideas along the discussed lines are fairly effective in practice.

As argued throughout, we believe that there is a misconception in the literature on BMS being overly sensitive to prior dispersion parameters. Theory shows that the dependence on the sample size n is of a larger order of magnitude than on the prior dispersion, this was supported empirically by the salary data and by many other datasets we have analyzed during our career. Further, within Bayesian inference, prior parameters have a natural interpretation that severely restrict the range of reasonable values in most applications. This is in contrast to frequentist or machine learning methods where tuning parameters often have no subject-matter interpretation. We do not imply that the simple defaults suggested here are the ultimate solution, there is definitely value in studying how to set better defaults. We merely remark that these defaults attain a pretty competitive performance, relative to existing methods.

We avoided discussing computational issues. BMS can face computational bottlenecks, either due to the integrated likelihoods $f(\boldsymbol{y} \mid \gamma)$ being costly or to exploring a large model space. In our experience Markov chain Monte Carlo methods and simple heuristics lead to fairly effective model exploration, but approximating $f(\boldsymbol{y} \mid \gamma)$ quickly can be more challenging. Partial solutions are reducing the complexity of these models, e.g., replacing a nonparametric term by a finite-dimensional approximation, but more research in this area is needed. We did show however examples where BMS attained competitive run times relative to popular penalized likelihood methods, which at the very least proves that one should not rule out BMS a priori as being computationally infeasible.

Bibliography

[1] S. Bakin. *Adaptive regression and model selection in data mining problems.* PhD thesis, The Australian National University, Canberra, Australia, 5 1999.

[2] D.J. Benjamin, J.O. Berger, M. Johannesson, B.A. Nosek, E.-J. Wagenmakers, R. Berk, K.A. Bollen, B. Brembs, L. Brown, C. Camerer, et al. Redefine statistical significance. *Nature Human Behaviour*, 2(1):6, 2018.

[3] A. Bhattacharya, D. Pati, N.S. Pillai, D.B. Dunson, and B. David. Bayesian shrinkage. *Journal of the American Statistical Association*, 110(512):1479–1490, 2015.

[4] J. Bien, J. Taylor, and R. Tibshirani. A LASSO for hierarchical interactions. *Annals of Statistics*, 41(3):1111–1141, 2013.

[5] A. Calon, E. Espinet, S. Palomo-Ponce, D.V.F. Tauriello, M. Iglesias, M.V. Céspedes, M. Sevillano, C. Nadal, P. Jung, X.H.-F. Zhang, D. Byrom, A. Riera, D. Rossell, R. Mangues, J. Massague, E. Sancho, and E. Batlle. Dependency of colorectal cancer on a tgf-beta-driven programme in stromal cells for metastasis initiation. *Cancer Cell*, 22(5):571–584, 2012.

[6] I. Castillo, J. Schmidt-Hieber, and A.W. van der Vaart. Bayesian linear regression with sparse priors. *The Annals of Statistics*, 43(5):1986–2018, 2015.

[7] M. Chae, L. Lin, and D.B. Dunson. Bayesian sparse linear regression with unknown symmetric error. *arXiv*, 1608.02143:1–34, 2016.

[8] J. Chen and Z. Chen. Extended Bayesian information criteria for model selection with large model spaces. *Biometrika*, 95(3):759–771, 2008.

[9] G. Consonni, D. Fouskakis, B. Liseo, and I. Ntzoufras. Prior distributions for objective Bayesian analysis. *Bayesian Analysis*, 13(2):627–679, 2018.

[10] A.P. Dawid. Prequential analysis, stochastic complexity and Bayesian inference. In *Bayesian statistics 4*, pages 109–125, Oxford, 1992. Oxford University Press.

[11] A.P. Dawid. The trouble with Bayes factors. Technical report, University College London, 1999.

[12] A. Dispenzieri, J.A. Katzmann, R.A. Kyle, D.R. Larson, T.M. Therneau, C.L. Colby, R.J. Clark, G.P. Mead, S. Kumar, and L J. Melton-III. Use of nonclonal serum immunoglobulin free light chains to predict overall survival in the general population. In *Mayo Clinic Proceedings*, volume 87, pages 517–523, 2012.

[13] J. Fan and R. Li. Variable selection via nonconcave penalized likelihood and its oracle properties. *Journal of the American Statistical Association*, 96:1348–1360, 2001.

[14] C. Fernández, E. Ley, and M.F.J. Steel. Benchmark priors for Bayesian model averaging. *Journal of Econometrics*, 100:381–427, 2001.

[15] D.P. Foster and E.I. George. The risk inflation criterion for multiple regression. *The Annals of Statistics*, 22(4):1947–1975, 1994.

[16] C. Gao, A.W. van der Vaart, and H.H. Zhou. A general framework for Bayes structured linear models. *arXiv*, 1506.02174:1–44, 2015.

[17] E.I. George and D.P. Foster. Calibration and empirical Bayes variable selection. *Biometrika*, 87(4):731–747, 2000.

[18] J. Griffin and P. Brown. Hierarchical shrinkage priors for regression models. *Bayesian Analysis*, 12(1):135–159, 2017.

[19] P.R. Hahn and C.M. Carvalho. Decoupling shrinkage and selection in Bayesian linear models: a posterior summary perspective. *Journal of the American Statistical Association*, 110(509):435–448, 2015.

[20] V.E. Johnson and D. Rossell. On the use of non-local prior densities for default Bayesian hypothesis tests. *Journal of the Royal Statistical Society B*, 72:143–170, 2010.

[21] V.E. Johnson and D. Rossell. Bayesian model selection in high-dimensional settings. *Journal of the American Statistical Association*, 24(498):649–660, 2012.

[22] B.J.K. Kleijn. On the frequentist validity of Bayesian limits. *arXiv*, 1611.08444:1–55, 2017.

[23] J.D. Lee, D.L. Sun, Y. Sun, and J.E. Taylor. Exact post-selection inference, with application to the lasso. *The Annals of Statistics*, 44(3):907–927, 2016.

[24] F. Liang, R. Paulo, G. Molina, M.A. Clyde, and J.O. Berger. Mixtures of g-priors for Bayesian variable selection. *Journal of the American Statistical Association*, 103:410–423, 2008.

[25] D.V. Lindley. A statistical paradox. *Biometrika*, 44:187–192, 1957.

[26] N. Meinshausen and P. Bühlman. Stability selection. *Journal of the Royal Statistical Society B*, 72(4):417–473, 2010.

[27] C.B. Mulligan and Y. Rubinstein. Selection, investment, and women's relative wages over time. *The Quarterly Journal of Economics*, 123(3):1061–1110, 2008.

[28] N.N. Narisetty and X. He. Bayesian variable selection with shrinking and diffusing priors. *The Annals of Statistics*, 42(2):789–817, 2014.

[29] J. Piironen, M. Paasiniemi, and A. Vehtari. Projective inference in high-dimensional problems: Prediction and feature selection. *Electronic Journal of Statistics*, 14(1):2155–2197, 2020.

[30] V. Ročková and S. van der Pas. Posterior concentration for Bayesian regression trees and their ensembles. *arXiv*, 1708.08734:1–40, 2017.

[31] D. Rossell. Concentration of posterior model probabilities and normalized l0 criteria. *Bayesian Analysis*, (in press):1–27, 2021.

[32] D. Rossell and P. Müller. Sequential stopping for high-throughput experiments. *Biostatistics*, 14(1):75–86, 2013.

[33] D. Rossell and F.J. Rubio. Tractable Bayesian variable selection: beyond normality. *Journal of the American Statistical Association*, 113(524):1742–1758, 2018.

[34] D. Rossell and F.J. Rubio. Additive Bayesian variable selection under censoring and misspecification. *arXiv*, 1907.13563:1–57, 2019.

[35] D. Rossell and D. Telesca. Non-local priors for high-dimensional estimation. *Journal of the American Statistical Association*, 112:254–265, 2017.

[36] D. Rossell, D. Telesca, and V.E. Johnson. High-dimensional Bayesian classifiers using non-local priors. In *Statistical Models for Data Analysis XV*, pages 305–314. Springer, 2013.

[37] G. Schwarz. Estimating the dimension of a model. *Annals of Statistics*, 6:461–464, 1978.

[38] J.G. Scott and J.O. Berger. Bayes and empirical Bayes multiplicity adjustment in the variable selection problem. *The Annals of Statistics*, 38(5):2587–2619, 2010.

[39] M. Shin, A. Bhattacharya, and V.E. Johnson. Scalable Bayesian variable selection using nonlocal prior densities in ultrahigh-dimensional settings. *Statistica Sinica*, 28(2):1053–1078, 2018.

[40] Q. Song and F. Liang. Nearly optimal Bayesian shrinkage for high dimensional regression. *arXiv*, 1712.08964:1–45, 2017.

[41] R. Tibshirani. Regression shrinkage and selection via the Lasso. *Journal of the Royal Statistical Society, B*, 58:267–288, 1996.

[42] A.W. van der Vaart. *Asymptotic statistics.* Cambridge University Press, New York, 1998.

[43] S.G. Walker. Modern Bayesian asymptotics. *Statistical Science*, 19(1):111–117, 2004.

[44] H. White. Maximum likelihood estimation of misspecified models. *Econometrica: Journal of the Econometric Society*, pages 1–25, 1982.

[45] Y. Yang and D. Pati. Bayesian model selection consistency and oracle inequality with intractable marginal likelihood. *arXiv*, 1701.00311:1–38, 2017.

[46] Y. Yang, M.J. Wainwright, and M.I. Jordan. On the computational complexity of high-dimensional Bayesian variable selection. *The Annals of Statistics*, 44(6):2497–2532, 2016.

[47] Z. Yang and A. Womack. Revisiting high dimensional Bayesian model selection for gaussian regression. *arXiv*, 1905.06224:1–16, 2019.

[48] M. Yuan and Y. Lin. Model selection and estimation in regression with grouped variables. *Journal of the Royal Statistical Society: Series B (Statistical Methodology)*, 68(1):49–67, 2006.

[49] H. Zhou. The adaptive LASSO and its oracle properties. *Journal of the American Statistical Association*, 101(476):1418–1429, 2006.

17

Variable Selection and Interaction Detection with Bayesian Additive Regression Trees

Carlos M. Carvalho

The University of Texas at Austin (USA)

Edward I. George

The University of Pennsylvania (USA)

P. Richard Hahn

Arizona State University (USA)

Robert E. McCulloch

Arizona State University (USA)

CONTENTS

Bayesian Additive Regression Trees (BART) has emerged as a highly effective Bayesian approach to ensemble modeling with many binary trees. The BART Markov Chain Monte Carlo (MCMC) algorithm provides effective stochastic search in a complex model space and Bayesian uncertainty. As is the case with many modern approaches, the overall complexity of the model makes interpretation difficult. In practice, investigators often wish to know what predictor variables are important or, more generally, which roles variables play in the model. In this chapter we review some approaches for understanding how variables enter the BART model. We present simple ways to find out which variables are most important, which pairs

DOI: 10.1201/9781003089018-17

of variables interact in the model, and which subsets of variables allow us to approximate the full information inference according to a user defined metric. In all cases, our approaches are based on post processing the output from basic BART modeling, naturally capturing the uncertainty by the usual MCMC variation in a straightforward way.

17.1 Introduction

Modern statistical methods have, to a remarkable extent, advanced our ability to uncover complex, high dimensional relationships. In particular, for directed problems in which we predict Y from $\mathbf{x} = (x_1, \ldots, x_p)$, methods based on ensembles of trees have performed amazingly well in practice. Although the inner workings of these "black box" models are necessarily complex and somewhat intimidating, these ensembles are particularly well suited for "model-free" variable selection because of their flexible nonparametric nature. In contrast to popular parametric approaches such as normal linear regression, where variable selection is tantamount to submodel selection, the unrestricted nature of tree ensembles allows for a richer set of possibilities for discovering the potential relatedness of Y and a subset of the predictors.

Bayesian Additive Regression Trees (BART, [4]) is a Bayesian approach to tree ensemble modeling which has proven to be remarkably effective for prediction and inference, often with minimal tuning. In this chapter, we show how the output of BART is particularly congenial for the related problems of variable selection and interaction detection. The essential idea is to simply assess relative frequencies with which variables and their interactions appear in the tree components of the ensemble. We also present a utility driven approach for finding subsets of variables which allow us to approximate the full BART information inference according to a user defined metric. In all cases, our approaches are based on post processing the output from basic BART modeling, naturally capturing the uncertainty by the usual MCMC variation in a straightforward way.

17.2 BART Overview

We begin with a review of the essentials of BART as introduced by [4], hereafter CGM10. In a nutshell, BART is a fully Bayesian approach for modeling the regression Y on $\mathbf{x} = (x_1, \ldots, x_p)$ with a flexible sum-of-trees model of the form

$$Y = \sum_{j=1}^{m} g(\mathbf{x}; \mathcal{T}_j, \mathcal{M}_j) + \epsilon, \qquad \epsilon \sim \mathcal{N}(0, \sigma^2). \tag{17.1}$$

Here, each $\mathcal{T}_j$ is a recursive binary regression tree associated with a set $\mathcal{M}_j$ of terminal node constants μ_{ij}, for which $g(\mathbf{x}; \mathcal{T}_j, \mathcal{M}_j)$ is the step function which assigns $\mu_{ij} \in \mathcal{M}_j$ to $\mathbf{x}$ according to the sequence of splitting rules in $\mathcal{T}_j$. These splitting rules are binary splits of the predictor space of the form $\{\mathbf{x} \in A\}$ vs $\{\mathbf{x} \notin A\}$ where A is a subset of the range of $\mathbf{x}$. When the number of trees $m = 1$, (17.1) reduces to the single tree model used by [3], hereafter CGM98, for Bayesian CART.

For each value of $\mathbf{x}$, under (17.1), $E(Y \mid \mathbf{x})$ is equal to the sum of all the terminal node μ_{ij}'s assigned to $\mathbf{x}$ by the $g(x; \mathcal{T}_j, \mathcal{M}_j)$'s. Thus, the sum-of-trees function is flexibly capable of approximating a wide class of functions from R^n to R, especially when the number of trees m is large. Note also that the sum-of-trees representation is simply the sum of many simple multidimensional step functions from R^n to R, namely the $g(\mathbf{x}; \mathcal{T}_j, \mathcal{M}_j)$, rendering it much more manageable than basis expansions with more complicated elements such as multidimensional wavelets or multidimensional splines.

The BART model specification is completed by introducing a prior distribution over all the parameters of the sum-of-trees model, namely $(\mathcal{T}_1, \mathcal{M}_1), \ldots, (\mathcal{T}_m, \mathcal{M}_m)$ and σ. Note that $(\mathcal{T}_1, \mathcal{M}_1), \ldots, (\mathcal{T}_m, \mathcal{M}_m)$ entail all the bottom node parameters as well as the tree structures and splitting rules, a very large number of parameters, especially when m is large. To cope with this parameter explosion, we use a "regularization" prior that effectively constrains the fit by keeping each of the individual tree effects from being unduly influential. Without such a regularizing influence, large tree components would overwhelm the rich structure of (17.1), thereby limiting its scope of fine structure approximation.

17.2.1 Specification of the BART Regularization Prior

To simplify the specification of this regularization prior, we restrict attention to symmetric independence priors of the form

$$p((\mathcal{T}_1, \mathcal{M}_1), \ldots, (\mathcal{T}_m, \mathcal{M}_m), \sigma) = \left[\prod_j \left(\prod_i p(\mu_{ij} \mid \mathcal{T}_j) \right) p(\mathcal{T}_j) \right] p(\sigma), \qquad (17.2)$$

where $\mu_{ij} \in \mathcal{M}_j$, thereby reducing prior specification to the choice of prior forms for $p(\mathcal{T}_j), p(\mu_{ij} \mid \mathcal{T}_j)$ and $p(\sigma)$. To simplify matters further we use identical prior forms for every $p(\mathcal{T}_j)$ and for every $p(\mu_{ij} \mid \mathcal{T}_j)$. As detailed below, each of these prior forms are controlled by just a few interpretable hyperparameters that can be calibrated to yield surprisingly effective default specifications for regularization of the sum-of-trees model.

For $p(\mathcal{T}_j)$, we use the sequential tree-generating process which is specified by three aspects: (i) the probability that a node at depth d $(= 0, 1, 2, \ldots)$ is non-terminal, given by

$$\alpha(1 + d)^{-\beta}, \qquad \alpha \in (0, 1), \beta \in [0, \infty), \qquad (17.3)$$

(ii) the distribution on the splitting variable assignments at each interior node, and (iii) the distribution on the splitting rule assignment in each interior node, conditional on the selected splitting variable. As default choices, CGGM 98 and CGM10 recommend $\alpha = 0.95$ and $\beta = 2$ for (1), and a uniform priors on each set of possibilities for (ii) and (iii). Interesting alternative enhancements of these choice for $p(\mathcal{T}_j)$ have been proposed by [7–9].

For $p(\mu_{ij} \mid \mathcal{T}_j)$, we use the conjugate normal distribution $\mathcal{N}(\mu_\mu, \sigma_\mu^2)$ which allows μ_{ij} to be marginalized out, vastly simplifying MCMC posterior calculations. To guide the specification of the hyperparameters μ_μ and σ_μ, we note that under (17.1), it is highly probable that $E(Y \mid \mathbf{x})$ lies between y_{min} and y_{max}, the minimum and maximum of the observed values of Y in the data, and that the prior distribution of $E(Y \mid \mathbf{x})$ is $\mathcal{N}(m\,\mu_\mu, m\,\sigma_\mu^2)$, (because $E(Y \mid \mathbf{x})$ is the sum of m independent μ_{ij}'s under the sum-of-trees model). Based on these facts, we use the informal empirical Bayes strategy of choosing μ_μ and σ_μ so that $\mathcal{N}(m\,\mu_\mu, m\,\sigma_\mu^2)$ assigns substantial probability to the interval (y_{min}, y_{max}). This is conveniently done by choosing μ_μ and σ_μ so that $m\,\mu_\mu - k\,\sqrt{m}\,\sigma_\mu = y_{min}$ and $m\,\mu_\mu + k\,\sqrt{m}\,\sigma_\mu = y_{max}$ for some preselected value of k such as 1, 2 or 3. For example, $k = 2$ would yield a 95% prior probability that $E(Y \mid \mathbf{x})$ is in the interval (y_{min}, y_{max}). The goal of this specification strategy for μ_μ and σ_μ is to ensure that the implicit prior for $E(Y \mid \mathbf{x})$ is in the right "ballpark" in the sense of assigning substantial probability to the entire region of plausible values of $E(Y \mid \mathbf{x})$ while avoiding overconcentration and overdispersion of the prior with respect to the likelihood. As long as this goal is met, BART seems to be very robust to variations of these specifications.

For $p(\sigma)$, we also use a conjugate prior, here the inverse chi-square distribution $\sigma^2 \sim \nu\lambda/\chi^2_\nu$. Here again, we use an informal empirical Bayes approach to guide the specification of the hyperparameters ν and λ, in this case to assign substantial probability to the entire region of plausible values of σ while avoiding overconcentration and overdispersion of the prior. Essentially, we calibrate the prior df ν and scale λ with a "rough data-based overestimate" $\hat{\sigma}$ of σ. Two natural choices for $\hat{\sigma}$ are (i) a "naive" specification, the sample standard deviation of Y, or (ii) a "linear model" specification, the residual standard deviation from a least squares linear regression of Y on all the predictors. We then pick a value of ν between 3 and 10 to get an appropriate shape, and a value of λ so that the qth quantile of the prior on σ is located at $\hat{\sigma}$, that is $P(\sigma < \hat{\sigma}) = q$. We consider large values of q such as 0.75, 0.90 or 0.99 to center the distribution below $\hat{\sigma}$.

17.2.2 Posterior Calculation and Information Extraction

Combining the regulation prior with the likelihood, $L((\mathcal{T}_1, \mathcal{M}_1), \ldots, (\mathcal{T}_m, \mathcal{M}_m), \sigma \,|\, \mathbf{y})$ induces a posterior distribution

$$p((\mathcal{T}_1, \mathcal{M}_1), \ldots, (\mathcal{T}_m, \mathcal{M}_m), \sigma \,|\, \mathbf{y}) \tag{17.4}$$

over the full sum-of-trees model parameter space. Here $\mathbf{y}$ is the observed $n \times 1$ vector of Y values in the data which are assumed to be independently realized. Note also that here and below we suppress explicit dependence on $\mathbf{x}$ as we assume $\mathbf{x}$ to be fixed throughout. Although analytically intractable, the following backfitting MCMC algorithm can be used to very effectively simulate samples from this posterior.

This algorithm is a Gibbs sampler at the outer level. Let $\mathcal{T}_{(j)}$ be the set of all trees in the sum *except* $\mathcal{T}_j$, and similarly define $\mathcal{M}_{(j)}$, so that $\mathcal{T}_{(j)}$ will be a set of $m-1$ trees, and $\mathcal{M}_{(j)}$ the associated terminal node parameters. A Gibbs sampling strategy for sampling from (17.4) is obtained by m successive draws of $(\mathcal{T}_j, \mathcal{M}_j)$ conditionally on $(\mathcal{T}_{(j)}, \mathcal{M}_{(j)}, \sigma)$:

$$(\mathcal{T}_j, \mathcal{M}_j) \,|\, T_{(j)}, \mathcal{M}_{(j)}, \sigma, \mathbf{y}, \tag{17.5}$$

$j = 1, \ldots, m$, followed by a draw of σ from the full conditional:

$$\sigma \,|\, \mathcal{T}_1, \ldots \mathcal{T}_m, \mathcal{M}_1, \ldots, \mathcal{M}_m, \mathbf{y}. \tag{17.6}$$

The draw of σ in (17.6) is simply a draw from an inverse gamma distribution, which can be straightforwardly obtained by routine methods. More subtle is the implementation of the m draws of $(\mathcal{T}_j, \mathcal{M}_j)$ in (17.5). This can be done by taking advantage of the following simplifying reduction. First, observe that the conditional distribution $p(\mathcal{T}_j, \mathcal{M}_j \,|\, \mathcal{T}_{(j)}, \mathcal{M}_{(j)}, \sigma, \mathbf{y})$ depends on $(\mathcal{T}_{(j)}, \mathcal{M}_{(j)}, \mathbf{y})$ only through $\mathbf{R}_j = (r_{j1}, \ldots, r_{jn})'$, the $n \times 1$ vector of partial residuals

$$r_{ji} \equiv y_i - \sum_{k \neq j} g(\mathbf{x}_i; T_k, \mathcal{M}_k), \tag{17.7}$$

obtained from a fit that excludes the jth tree. Thus, the m draws of $(\mathcal{T}_j, \mathcal{M}_j)$ given $(\mathcal{T}_{(j)}, M_{(j)}, \sigma, \mathbf{y})$ in (17.5) are equivalent to m draws from

$$(\mathcal{T}_j, \mathcal{M}_j) \,|\, \sigma, \mathbf{R}_j, \tag{17.8}$$

$j = 1, \ldots, m$. Because we have used a conjugate prior for $\mathcal{M}_j$, $p(\mathcal{T}_j \,|\, \sigma, \mathbf{R}_j)$ can be obtained in closed form, up to a normalizing constant. This allows us to carry out each draw from (17.8) in two successive steps as

$$\mathcal{T}_j \mid \sigma, \mathbf{R}_j \tag{17.9}$$

followed by

$$\mathcal{M}_j \mid \mathcal{T}_j, \sigma, \mathbf{R}_j. \tag{17.10}$$

The draw of $\mathcal{T}_j$ in (17.9), although somewhat elaborate, can be obtained with the Metropolis-Hastings (MH) algorithm of CGM98. The draw of $\mathcal{M}_j$ in (17.10) is simply a set of independent draws of the terminal node μ_{ij}'s from normal distributions. The draw of $\mathcal{M}_j$ enables the calculation of the subsequent residual $\mathbf{R}_{j+1}$, which is then used for the draw of $\mathcal{T}_{j+1}$.

We initialize the chain with m simple single node trees, and then repeat iterations until satisfactory convergence is obtained. Fortunately, this backfitting MCMC algorithm appears to mix very well, as we have found that different restarts give remarkably similar results even in difficult problems. At each iteration, each tree may increase or decrease the number of terminal nodes by one, or change one or two splitting rules. The sum-of-trees model, with its abundance of unidentified parameters, allows the "fit" to glide freely from one tree to another. Because each move makes only small incremental changes to the fit, we can imagine the algorithm as analogous to sculpting a complex figure by adding and subtracting small dabs of clay.

For inference based on our MCMC sample, we rely on the fact our backfitting algorithm is ergodic. Thus, the induced sequence of sum-of-trees functions

$$f^*(\cdot) = \sum_{j=1}^{m} g(\cdot\,; \mathcal{T}_j^*, \mathcal{M}_j^*), \tag{17.11}$$

from the sequence of draws $(\mathcal{T}_1^*, \mathcal{M}_1^*), \ldots, (\mathcal{T}_m^*, \mathcal{M}_m^*)$, is converging to $p(f \mid \mathbf{y})$, the posterior distribution of the "true" $f(\cdot)$. Thus, by running the algorithm long enough after a suitable burn-in period, the sequence of f^* draws, say $f_1^*, \ldots, f_K^*$, may be regarded as an approximate, dependent sample of size K from $p(f \mid \mathbf{y})$. Bayesian inferential quantities of interest can then be approximated with this sample as follows.

To estimate $f(\mathbf{x})$ or predict Y at a particular $\mathbf{x}$, in-sample or out-of-sample, a natural choice is the average of the after burn-in sample $f_1^*, \ldots, f_K^*$,

$$\frac{1}{K} \sum_{k=1}^{K} f_k^*(\mathbf{x}), \tag{17.12}$$

which approximates the posterior mean $E(f(\mathbf{x}) \mid \mathbf{y})$. Posterior uncertainty about $f(\mathbf{x})$ may be gauged by the variation of $f_1^*(\mathbf{x}), \ldots, f_K^*(\mathbf{x})$. For example, a natural and convenient $(1-\alpha)\%$ posterior interval for $f(\mathbf{x})$ is obtained as the interval between the upper and lower $\alpha/2$ quantiles of $f_1^*(\mathbf{x}), \ldots, f_K^*(\mathbf{x})$.

17.3 Model-Free Variable Selection with BART

We now describe and illustrate how variable selection may be accomplished by direct inspection of the basic BART output. The essential idea to select those components of $\mathbf{x}$ that tend to be used most often in the MCMC sequence of fitted sum-of-trees models, $f_1^*, \ldots, f_K^*$. To measure this tendency, we simply record for each model f_k^*, the proportion of splitting rules using each component of $\mathbf{x}$, and then average these proportions across the whole sequence of simulated models. Those components with the largest average usage proportions are then selected. The attribution of classical statistical significance levels to these proportions can be obtained with the permutation based methods proposed by [1].

While it makes intuitive sense that relevant predictors will tend to be used frequently in the fitted sum-of-trees ensemble, it turns out that when the number of trees m is large, the redundancy of the overly rich basis of trees will also allow for many irrelevant predictors to be mixed in with the relevant ones. Fortunately, as m is decreased, this redundancy is diminished, thereby encouraging BART to strongly favor the more relevant predictors for its fit. Thus, a direct BART approach to variable selection can be accomplished by observing what happens to the $\mathbf{x}$ component usage frequencies in a sequence of MCMC samples $f_1^*, \ldots, f_K^*$ as the number of trees m is set smaller and smaller.

More precisely, for each simulated sum-of-trees model f_k^*, let z_{ik} be the proportion of all splitting rules that use the ith component of $\mathbf{x}$. Then

$$v_i \equiv \frac{1}{K} \sum_{k=1}^{K} z_{ik} \tag{17.13}$$

is the average usage per splitting rule of the ith component of $\mathbf{x}$. As m is set smaller and smaller, the sum-of-trees model will tend to more strongly favor inclusion of those $\mathbf{x}$ components which improve prediction of the Y values and exclusion of those $\mathbf{x}$ components that do not. In effect, smaller m will create a bottleneck that forces the $\mathbf{x}$ components to compete for entry into the sum-of-trees model. As we illustrate below, the $\mathbf{x}$ components with larger v_i's will be those that provide the most information for predicting Y.

17.3.1 Variable Selection with the Boston Housing Data

We illustrate this approach with the well known Boston housing data which is available in many data science environments (library `MASS` in R). There are $n = 506$ observations, each of which corresponds to a neighborhood in the Boston area. The response of interest is $y =$ the median price of a house in thousands of dollars. For each neighborhood, thirteen characteristics were recorded. We refer the reader to the R documentation for details, but a few of the variables which will be important are (i) `lstat`: lower status percentage of the population, (ii) `rm`: average number of rooms per dwelling, and (iii) `nox`: nitrogen oxides concentration (parts per 10 million).

For this Boston housing data, we ran BART for 10,000 burn-in iterations and then 10,000 post burn-in iterations. We used the `BART` package implementation in the R library with all BART hyperparameters left at their default values, with the exception of the number of trees m, which we reset from 200 to 20.

Figure 17.1 displays the posterior distributions of percent usage for each variable. The red points are the posterior mean estimates (averages over MCMC draws) and the vertical (blue) lines are the 90% posterior intervals (with 5% and 95% quantiles from the MCMC draws as endpoints). Although there is still substantial uncertainty, three variables stand out from the rest. With reasonably high posterior probability, the variables `nox`, `rm` and `lstat` are most important in the sense of being used most often for the fit of the sum-of-trees model.

The posterior estimates in Figure 17.1 are based on a BART ensemble with $m = 20$ trees, which is considerably smaller than the $m = 200$ default in the `BART` package. For predictive accuracy, the larger number of trees often works best. BART is designed to let a variable enter with "a little bit of fit" and the amount of fit gets smaller as we add trees. With a large number of trees, a variable may be used in several tree splitting rules without

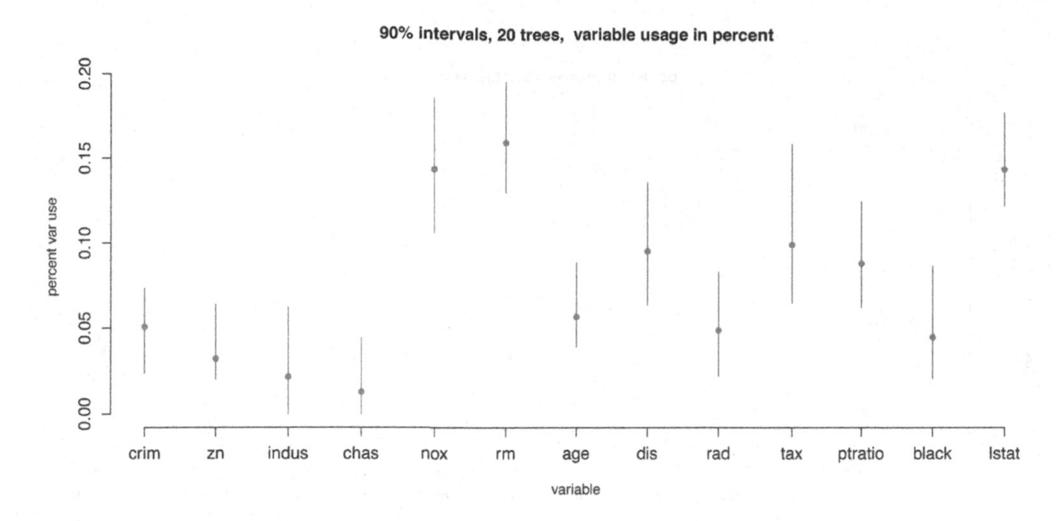

FIGURE 17.1
Posterior means and 90% credibility intervals for variable usage percentages for a model with $m =$ 20 trees.

really having much of an effect. However, for variable selection, fewer trees may give a clearer picture of the most relevant variables by narrowing the focus to those variables that make meaningful contributions to the fit.

For comparison of these two choices of m, Figure 17.2 plots the posterior means of the variable percent usages from our previous BART run with $m = 20$ trees and a BART run (same number burn-in and kept draws) with the default $m = 200$ trees. We can see that it would be much harder to pick out a promising variable subset from the $m = 200$ tree run.

While the percent usage approach provides an attractive simple approach to variable selection, the cost of using fewer trees may also degrade the fit. Figure 17.3 displays an in sample scatterplot matrix of the response y (= median house price), the BART fits with $m = 200$ trees, the BART fits with $m = 20$ trees, and the fits from a linear model for the Boston housing data. In this case we see that the BART default fit is very good and slightly better than the $m = 200$ tree fit. However both are much better better that the linear fit which is dramatically worse. Note also that although we have used $m = 20$ as a typical case of smaller m to illustrate the variable selection potential of BART, it may be useful to try other values of m, both larger and smaller, for exploration.

In practice, as in the Boston example, it is often straightforward to find a number of trees m for which BART identifies the more important variables but still gives a fit similar to the default. It is noteworthy that the percent usage approach can be effective without the explicit use of an additional variable selection prior by relying on just a simple measure of variable usage, as opposed to how much its usage improves the loss (as in CART). [7] proposes such an alternative approach which works very nicely by adding prior information about sparsity to BART. However, it requires another layer of prior comprehension and specification.

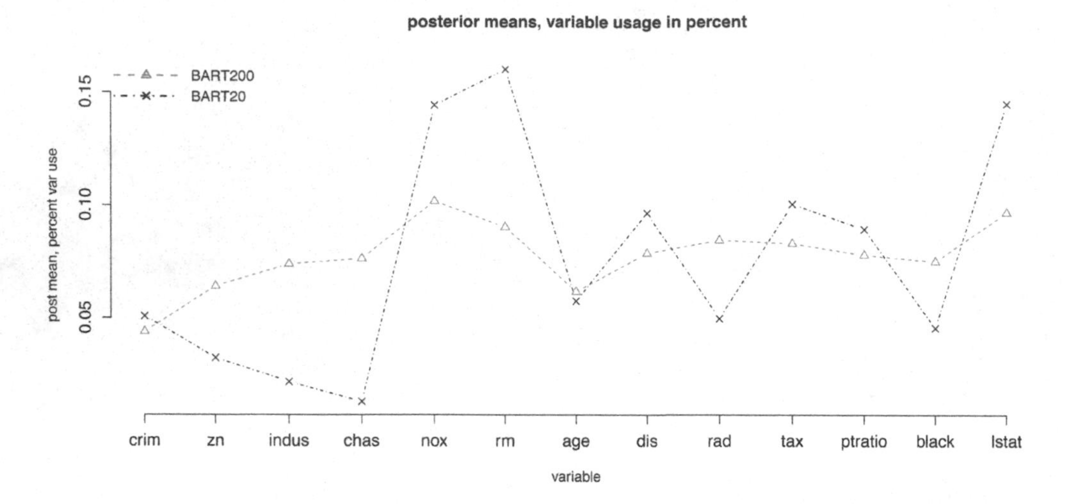

FIGURE 17.2
Comparison of the posterior means for variable usage percentages for a model with m =20 trees and a model with m = 200 trees. With a smaller number of trees, the more important variables are included much more frequently.

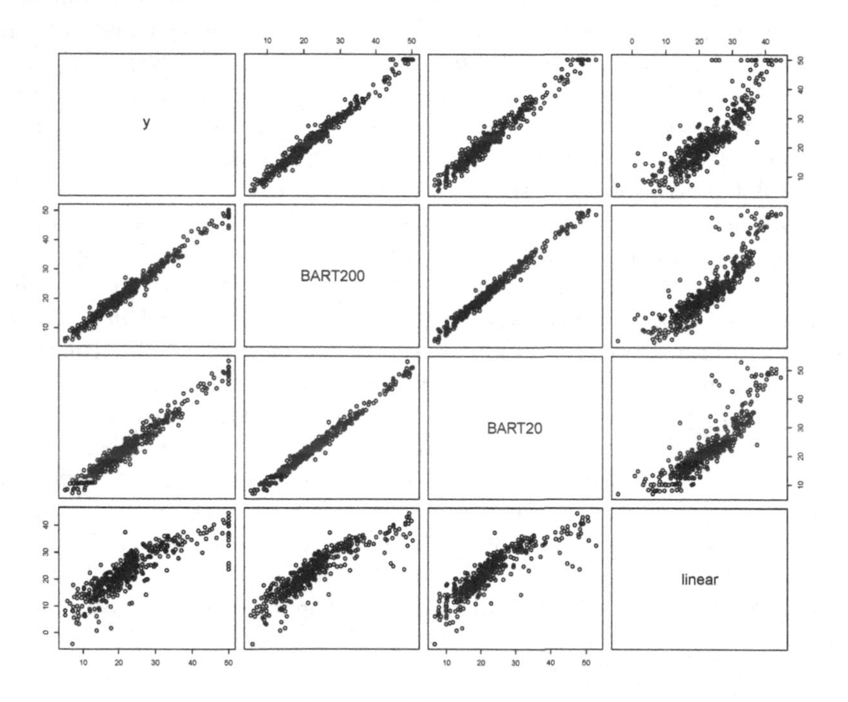

FIGURE 17.3
In sample scatterplot matrix of the response y (= median house price), the BART fits with 200 trees, the BART fits with 20 trees, and the fits from a linear model for the Boston housing data.

17.4 Model-Free Interaction Detection with BART

In a regression of y on $\mathbf{x}$, an interaction between two predictors x_i and x_j is said to exist when the effect of x_i on y depends on the value of x_j, and vice-versa. Such an interaction is manifested in a sum-of-trees model when the x_i and x_j components of $\mathbf{x}$ tend to both be used for splitting rules in common trees. Note that this notion of an interaction is much more general than the commonly used approach of measuring an interaction by the coefficient of the product $x_i * x_j$ in a linear model.

Analogously to the previously described percentage usage approach for variable selection, interaction detection can be similarly carried out by identifying those interactions which occur most frequently over the MCMC sequence of fitted sum-of-trees models $f_1^*, \ldots, f_K^*$. More precisely, for each simulated sum-of-trees model f_k^*, let z_{ijk} be the proportion of trees in f_k^* where both x_i and x_j are used for splitting rules. Then

$$v_{ij} \equiv \frac{1}{K} \sum_{k=1}^{K} z_{ijk} \tag{17.14}$$

is the average proportion of trees with x_i,x_j interactions across $f_1^*, \ldots, f_K^*$. Those interactions with the largest average proportions are deemed most important. This approach can be straightforwardly extended to three-way and higher order interaction detection. (Note that improved measures of interaction may also be considered. For example, it would be more relevant to count only those trees where x_i and x_j are used in the same path from the root to the terminal node. However, this improvement is apt to be slight because most of the BART ensemble trees containing both x_i and x_j will already satisfy this property, a consequence of the tendency of the regularization prior to keep most trees small).

In contrast to our BART variable selection approach, it turns out to be less critical to use a small number of trees m for our interaction detection approach. Because the number of potential interactions grows rapidly with the number of variables, the chances of accumulation for irrelevant interaction becomes much smaller as well. Nevertheless, it is still illuminating to carry out the procedure with both large and small values of m.

17.4.1 Variable Selection and Interaction Detection with the Friedman Simulation Setup

Let us first illustrate our BART approaches to both variable selection and interaction detection with the well-known Friedman simulation setup. We here simulate $n = 500$ observations from the basic model

$$y = f(\mathbf{x}) + \sigma Z, \;\; Z \sim \mathcal{N}(0, 1),$$

with $\mathbf{x}$ ten-dimensional and

$$f(x_1, x_2, \ldots, x_{10}) = 10 \sin(\pi x_1 x_2) + 20 (x_3 - .5)^2 + x_4 + x_5.$$

The x_i are iid uniform on $(0, 1)$ and $\sigma = 1$.

This simulation setup was devised by [5] to study the efficacy of non-linear regression techniques. However, the setup also turns out to be perfect for illustrating both variable selection and interaction discovery. Only the first five of the ten $\mathbf{x}$ components matter, and only one pairwise interaction is present: namely the interaction between x_1 and x_2. In a real application it would be of tremendous interest to know that only these two variables interact, especially without having further knowledge of the functional form.

Results for one simulated data set are displayed in Figure (17.4). Panel (a) provides the variable selection results with $m = 20$ and $m = 200$ trees. This panel corresponds closely to Figure 5 of [4]. For each variable, we plot the posterior mean of the *percentage of rules* (across all m trees) which use that variable. With $m = 20$, we very clearly identify the first five variables as being important.

Panel (b) provides the interaction detection results with $m = 20$ and $m = 200$ trees. With ten variables, there are 10*9/2=45 possible variable pairs. For each pair, we plot the posterior mean of the *percentage of trees* (out of m trees) which use both of the variables in their splitting rules. We normalize the $m = 20$ and $m = 200$ results by dividing by each set of 45 posterior means by the maximum. Thus, the largest value displayed in each case is one. With both $m = 20$ and $m = 200$ we clearly identify the first pair (x_1 and x_2) as being of interest. With two variables involved, a pair is less likely to come in inconsequentially, so that the identification of interesting pairs is less sensitive to the choice of m than in the case of variable selection.

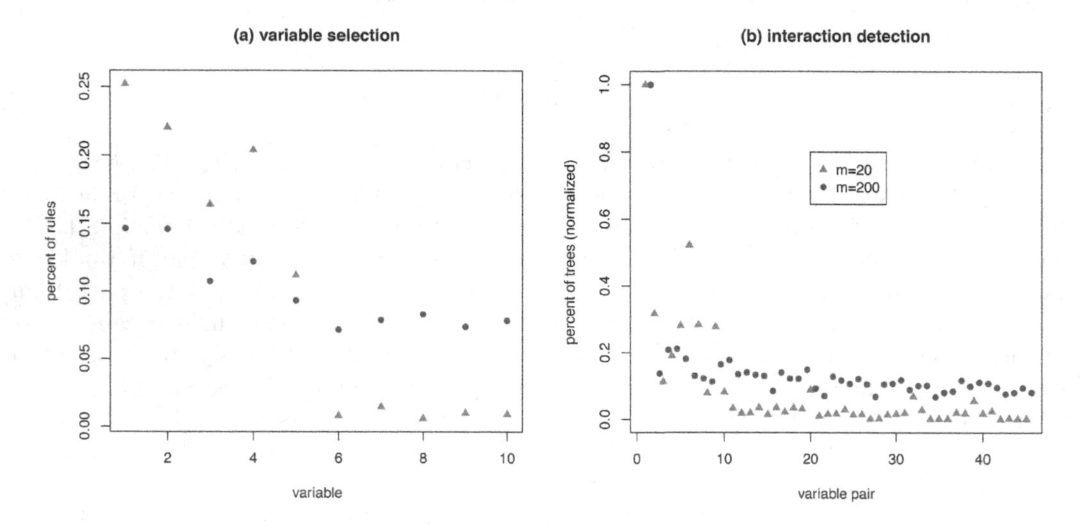

FIGURE 17.4
Variable usage percentages for the Friedman data with $m = 20$ and $m = 200$ trees. In panel (a) we correctly identify the first five variables as being important. In panel (b) we correctly identify the first interaction, which corresponds to variables x_1 and x_2.

17.4.2 Interaction Detection with the Boston Housing Data

For our second example, we return to the Boston housing data for which we illustrated variable selection in Section 17.3. Recall that each of the $n = 506$ observations corresponds to neighborhood. The response is the median house price in the neighborhood. There are 13 explanatory variables measuring characteristics of the neighborhoods. We did a preliminary variable selection (using the approach illustrated in the previous section) and tossed out three of the variables. Fitted values (from BART) with and without the three variables are very similar.

Figure (18.10) displays the results of the interaction detection with $m = 20$ and $m = 200$ trees. The format is the same as in panel (b) of Figure (17.4). Several pairs of interest are identified. Our real data has more interesting structure than our simulated data! We will investigate the pair `dis` and `lstat` simply because these variables are more easily understood. `dis` is the "weighted distances to five Boston employment centres". `lstat` is the "lower status perecentage of the population".

In Figure (17.6) we attempt to graphically see the interaction between `dis` and `lstat` suggested by Figure (18.10). In panel (a) we plot `dis` vs. `lstat`. Four subsets of points are identified depending on whether `dis` and `lstat` are "low" or "high". In the (b) panel we plot the fitted values from the BART run with $m = 200$. Before fitting we subtracted off the average response so the vertical axis is actually the amount the median value for a neighborhood is above the average. The four boxplots correspond to the four data subsets indicated in panel (a).

So, for example, the first boxplot displays the fitted prices when both `dis` and `lstat` are low. The observations included here correspond to those highlighted in the bottom left corner of panel (a). The label "`dL_lL`" indicates that `dis` is Low and `lstat` is Low. Similarly, the third boxplot is labelled "`dH_lL`", indicating that `dis` is High and `lstat` is Low.

The first pair of boxplots indicate the effect of increasing `lstat` when `dis` is low. The second pair of boxplots indicate the effect of increasing `lstat` when `dis` is high. Clearly, the boxplots indicate a strong interaction. For low `dis`, the effect of a the change in `lstat` is much more pronounced. A nice neighborhood close to the city center is highly desirable whereas a bad neighborhood close to the city center may be very bad.

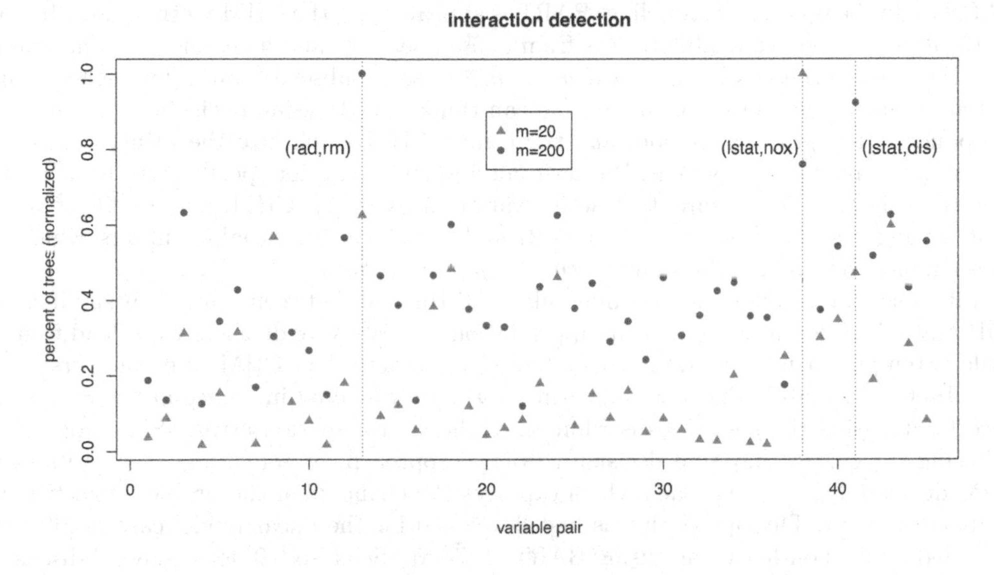

FIGURE 17.5
Interaction detection for the Boston housing data with $m = 20$ and $m = 200$ trees. Here, ten explanatory variables give rise to 50 possible pairwise interactions.

17.5 A Utility Based Approach to Variable Selection using BART Inference

In this section we discuss an approach to variable selection developed by [2], henceforth CHM. While the CHM approach may be used with any model, CHM emphasize the use of

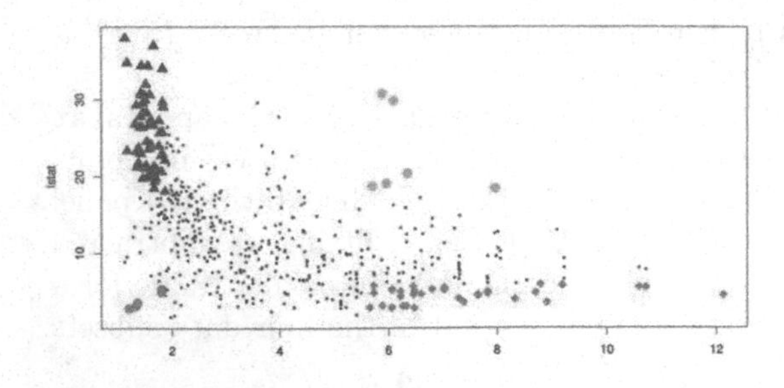

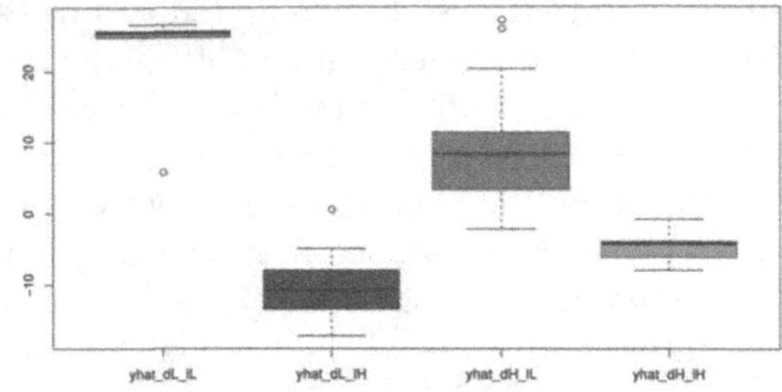

FIGURE 17.6
In the left panel we identify four subsets of our data by whether each of `dis` and `lstat` are low or high. In the right panel the boxplots display the fitted values (median house values) for the observations in the four subsets. The average of the dependent variable was subtracted off so that the vertical axis is the amount the median value of a neighborhood is above average. The first pair of boxplots both have low values of `dis`. The first box has low values of `lstat` and the second box has high values of `lstat`. The second pair of boxplots again compare low and high `lstat` but now `dis` is high.

BART to do the non-linear modeling. BART is appealing for the CHM methodology because of it's ability to get reasonable results from default settings and assessment of uncertainty.

CHM cast variable selection as a *decision* to use a subset of variables rather than a search for the "true" subset. Generally we can think of a Bayesian decision as having three components, the prior, likelihood, and the utility. CHM emphasize the utility component rather than the prior component. Rather than developing a prior specification that believes that there is some small subset of active variables as in [7], CHM run BART at its default setting (as discussed in Section 17.2) and then look for variable subsets which can approximate the simple BART inference *as a practical matter.*

Infamously, the ubiquitous p-value fails to distinguish between "practical significance" and "statistical significance". Recent appreciation of this severe deficiency has lead to many futile attempts to fix a fundamentally flawed approach. The CHM approach first looks for subsets of variables that can approximate the unrestricted inference and then assesses uncertainty using the posterior distribution of the approximation error. While not strictly following the correct Bayesian decision theoretic approach, by remaining true to its spirit, CHM develop a simple approach which captures the elements of the problem practitioners really care about. This approach was first developed for the linear model case in [6] and is extended to the non-linear case using BART in CHM. See also [10] for a general discussion of the "projection strategy" which looks for simplifying structure suggested by unrestricted inferences.

Let S denote a subset of the variables. Let $\hat{f}(\mathbf{x})$ be the the posterior mean of $f(\mathbf{x})$ obtained from BART. Let X_P denote a set of $\mathbf{x}$ values at which we want to learn $E(Y\,|\,\mathbf{x}) = f(\mathbf{x})$. The CHM approach has three basic steps:

1. Use BART to learn $\hat{f}$ such that $\hat{f}(\mathbf{x}) \approx E(Y \mid \mathbf{x})$ for $\mathbf{x}$ values in the training data. Standard default choices for BART parameters may be used.

2. Find a subset S of the variables for which there is a nonlinear function $\gamma_S(\mathbf{x})$ such that γ_S only depends on the subset of variables indicated by S and $\gamma_S(\mathbf{x}) \approx \hat{f}(\mathbf{x})$, for $\mathbf{x} \in X_P$.

3. Use the BART draws to assess our uncertainty about the approximation error.

Note the choice of X_P above. To select variables, we need to specify how we going to use the model in practice! Often a simple default choice is to let X_P be the values observed in the training data. It is however, not uncommon for an application to consider a much richer set of $\mathbf{x}$ values. This consideration, which seems fundamental, is clearly missing from most variable selection approaches.

17.5.1 Step 1: BART Inference

Figure 17.7 displays results from step 1 for the Boston housing data. BART was run at the default settings for 20,000 MCMC iterations of which the last 10,000 post burn-in iterations were kept. The full data set of $n = 506$ observations were used for training. Figure 17.7 reproduces some of the information from Figure 17.3 in Section 17.3, but here the intervals for $f(\mathbf{x})$ are included. The crucial point here is that we are just running standard BART.

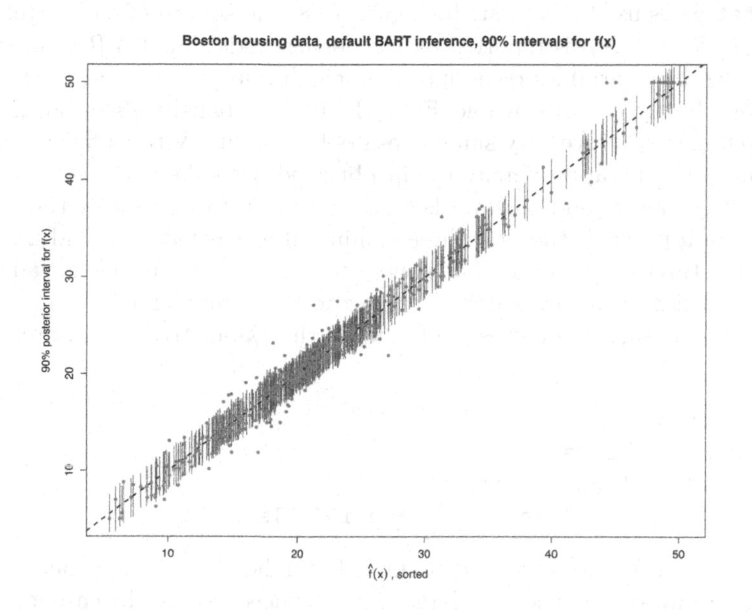

FIGURE 17.7
Sorted by the values of the posterior means $\hat{f}(\mathbf{x})$, points are plotted at $(\hat{f}(\mathbf{x}), y)$ for all $(\mathbf{x}, y)$ in the training data. The vertical lines at each point are the 90% posterior intervals for $f(\mathbf{x})$. The dashed line indicating perfect prediction has intercept 0 and slope 1.

17.5.2 Step 2: Subset Search

We now tackle the problem of finding subsets S and non-linear functions $\gamma_S(\mathbf{x})$ which depend on $\mathbf{x}$ only through the subset of variables indicated by S, such that γ_S approximates $\hat{f}$ for $\mathbf{x} \in X_P$. Again, $\hat{f}$ is obtained from step 1.

To search for subsets, we use the time honored forward selection and backwards selection approaches. When the number of variables is small (as in our Boston example) we can simply try all possible subsets. The tricky part is finding γ_S given S. We do not want to rerun BART at each candidate subset since this would be too time consuming. In addition, we are seeking an arbitrarily good approximation to the function $\hat{f}$. While BART is engineered to deal with a noisy y, this problem has different characteristics so that simple default BART may not be effective.

The key idea in CHM is to use a large decision tree for γ_S. Given S, we fit a large tree to the data $(\mathbf{x}_S, \hat{f}(\mathbf{x}))$, where $\mathbf{x}_S$ is the subset of variables indicated by S. When fitting a decision tree to data, we have to worry about the bias variance trade-off when choosing the tree size. Here, since we are simply seeking an approximation to already smoothed data, we find that a simple default large tree size works very well. When we have sufficient information in $\mathbf{x}_S$ we cannot overfit. If $\mathbf{x}_S$ does not capture the relevant information in $\mathbf{x}$, our big tree could, in principle, overfit, but we have not found this to be a problem (see Figure 17.9). Informally, CHM refer to this approach as "fit-the-fit" as we use the big tree to explain the fit $\hat{f}(\mathbf{x})$ rather than y.

Figure 17.8 shows the results from forward search (top panel), backward search (middle panel) and all subsets search (bottom panel). So, for example, if we try to fit-the-fit with just one of the variables x_j, $j = 1, 2, \ldots, p$, by fitting a big tree to the data $(x_j, \hat{f}(\mathbf{x}))$ we find that `lstat` gives us the best fit. In Figure 17.8, the square of the correlation between $\gamma_S(\mathbf{x})$ and $\hat{f}(\mathbf{x})$ for $\mathbf{x} \in X_P$ is reported on the vertical axis (labeled R-squared). In the top panel we can see how variables come into our search one by one, and in the middle panel we can see how they go out one by one. From the bottom panel we see that in this example, all three search methods give very similar results for the fit. With as few as three variables, (`lstat`, `rm`, `nox`) we can approximate the fit obtained with the entire $\mathbf{x}$ very well.

Interestingly, forward and backwards search do not provide exactly the same subsets of variables. In the top panel, the first three coming in are `lstat`, `rm`, and `nox`. These three variables are the three left in the middle panel. But, the fourth variable is `rad` in the forward search and dis in the backward search. Indeed, `rad` is the first variable out in the backward search. The variable subsets of sizes 1-5 found by the exhaustive search are:

```
1: "lstat"
2: "rm"     "lstat"
3: "nox"    "rm"     "lstat"
4: "nox"    "rm"     "rad"    "lstat"
5  "nox"      "rm"        "dis"      "ptratio" "lstat"
```

Note that while both the forward and backward searches have the property that a smaller set of variables is always a subset of a larger set, this need not be the case for the all subsets search. We see that, for all subsets search, the best subset of size four has `rad` and not `dis`, while the best subset of size 5 has `dis` and not `rad`. These two variables, `rad` and `dis`, have a strong non-linear relationship, which explains the different results from the different searches. CHM are careful not to claim they know the "truth" about `rad` and `dis`. They only know that particular subsets approximate well.

Given found subsets S_j of size $j = 1, 2, \ldots, p$, CHM sometimes find that we can improve the approximation by rerunning BART using the training data and the indicated subsets. Figure 17.9 plots $\hat{f}(\mathbf{x})$ vs the approximating function found by the big tree and re-running BART for $j = 1, 2, 3, 4$. BART fits are in the left column and big tree fits are in the right column. We see that after three or four variables have entered, we have an excellent fit and that the results from both refitting BART and the big tree are sufficiently comparable to give us faith in our search mechanism. It does not seem that the big tree is overfitting for small j.

17.5.3 Step 3: Uncertainty Assessment

In step 2, we only used the posterior means from the BART inference to provide $\hat{f}$. In this section we use the BART MCMC draws to assess our uncertainty about the approximation error.

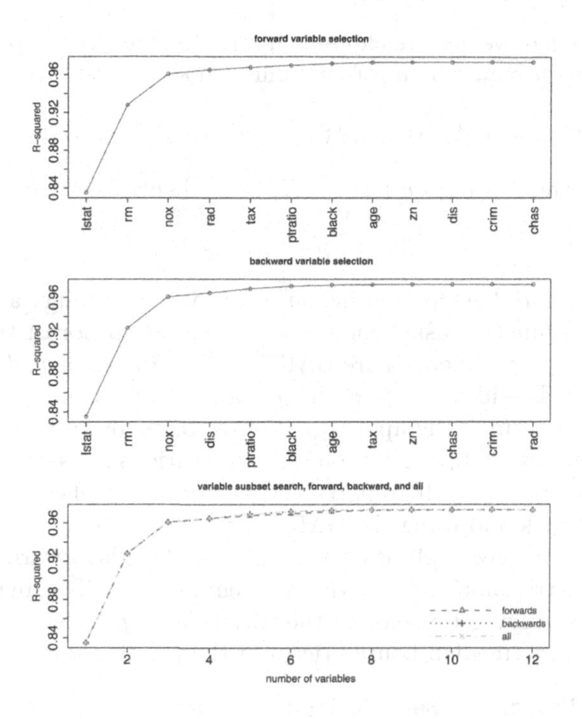

FIGURE 17.8
Search for functions which approximate $\hat{f}(\mathbf{x}) \approx E(Y \mid \mathbf{x})$ obtained from initial BART run using all the variables. Approximating functions use subsets of the variables. In each panel the vertical axis is the square of the correlation between $\hat{f}(\mathbf{x})$ and the approximation. The top panel shows results from the forward search in which variables enter one at a time. The middle panel shows results from the backward search in which variables exit one at a time. The plot on the bottom panel plots the fit from forwards, backwards, and all subsets search.

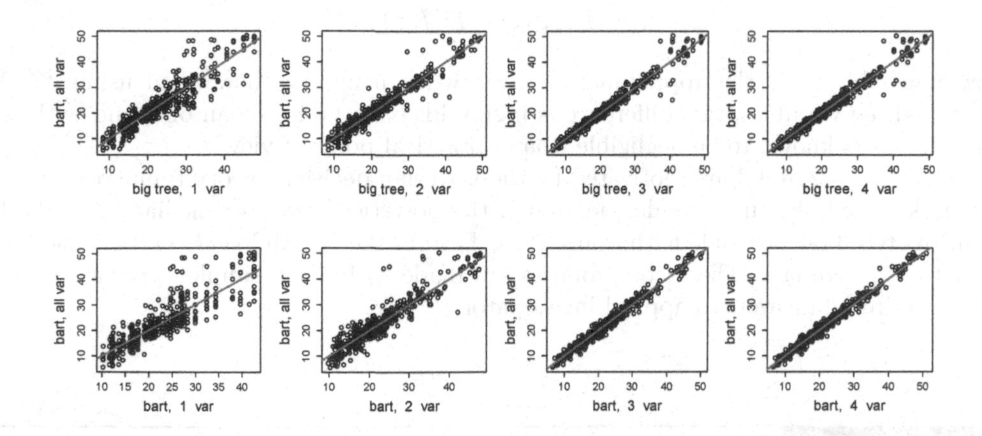

FIGURE 17.9
Comparing approximations to $\hat{f}(\mathbf{x})$ using both big trees and rerunning BART on selected subsets. In each graph the posterior mean estimates $\hat{f}(\mathbf{x})$ is on the vertical axis. The four columns correspond to using the first 1, 2, 3 and 4 variables, respectively, as found by the all subsets search. In both the all subsets and forward search the first four variables variables enter in the order `lstat`, `rm`, `nox`, and `rad`. In the bottom, row of plots, the horizontal axis is the posterior mean estimate of $f(\mathbf{x})$ obtained by rerunning BART using only the subset of variables. In the top row is the big-tree fit to $\hat{f}(\mathbf{x})$.

Crucially, at this stage we need to choose a metric to quantify the practical consequences of a certain level of approximation error. We must choose a distance

$$D(f,g) \equiv D((f(x_1), f(x_2), \ldots, f(x_p)), (g(x_1), g(x_2), \ldots, g(x_p)), \mathbf{x} \in X_P)$$

to capture how different f is from g for $\mathbf{x} \in X_P$. We then report the posterior distribution of

$$D(f, \gamma_S),$$

where f is the random variable representing our posterior uncertainty about $f(\mathbf{x}) = E(Y \mid \mathbf{x})$ and γ_S is a candidate function using the subset S. The posterior distribution is estimated by the set of values $D(f_d, \gamma)$ where f_d are BART MCMC draws of f, $d = 1, 2, \ldots, N$. In our Step 1, N was 10,000. The idea is that each f_d is a plausible candidate for a good function and we want to see how small the approximation error tends to be. We are reporting the posterior distribution of the approximation error for various subsets of the variables in $\mathbf{x}$.

The left panel of Figure 17.10 reports the posterior for the subsets found using all possible subsets and γ_S found using the BART refit. For D here, we have used root mean squared error, though in any application it would be straightforward to replace this with a more meaningful metric motivated by the problem at hand. To gauge the information in Figure 17.10, it is necessary get a sense of the variation of y, for example from the usual R summary of the y values (median house prices) in the training data:

```
Min. 1st Qu.  Median    Mean 3rd Qu.    Max.
5.00   17.02   21.20   22.53   25.00   50.00
```

Clearly, with only the three variables (`lstat`, `nox`, `rm`) the approximation error is highly likely to be close to 2. This may be deemed sufficiently accurate as a practical matter given the range of y. With seven variables, the error is highly likely to be comparable to that with all the variables.

The left panel of Figure 17.10 reports the posterior distribution of

$$D(f, \gamma_S) - D(f, \hat{f}).$$

How much bigger is the approximation error when using γ_S than when using $\hat{f}$? With our top three variables, this difference is highly likely to be less than one, and with seven variables it is known to be negligible from a practical point of view!

While steps 2 and 3 are motivated by the Bayesian decision theory framework, they do not stick to it fully. In particular, in step 2, the posterior mean (or median) $\hat{f}$ is the holy grail. In step 3 our attitude is that any draw f_d from the MCMC *could* be the function we want to approximate. These compromises are made to build a simple approach that will give meaningful answers to applied investigators.

17.6 Conclusion

Modern statistical methodology provides us with remarkably effective tools for uncovering potentially non-linear relationships between an outcome y and a complex, high dimensional predictor $\mathbf{x}$. Understanding the nature of an uncovered relationship is inherently difficult given the necessarily complex mathematical representation of the relationship.

A fundamental aspect of a relationship which investigators quite commonly wish to learn about is variable selection. Out of all the variables in $\mathbf{x}$, which are *the important*

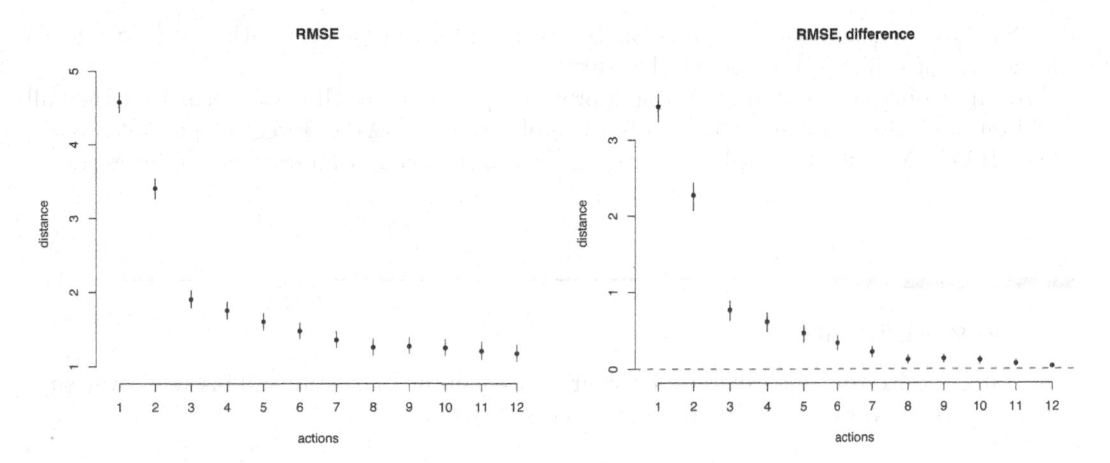

FIGURE 17.10
Posterior distribution of the approximation error. Approximations are found by running BART with the subsets found by the all subsets search. The left panel shows the posterior distribution of the approximation error. The right panel shows the posterior distribution of the between the error using the subset and the error using all of the variables.

ones? The development of tools for variable selection in the context of a linear model has been an important area of research for many years. The development of methods for variable selection for modern non-linear modeling is a crucial issue for applied use. Remarkably, key `R` packages such as `rpart` (for classification and regression trees) and `randomForest` include "variable importance measures".

Any approach to variable selection involves some difficult choices given the complexity of the problem. In this chapter we describe some relative simple, but effective, approaches using Bayesian Additive Regression trees (BART). BART has several advantages. Some key aspects of BART are the relatively simple default choices for the prior and the Markov Chain Monte Carlo (MCMC) stochastic search and uncertainty representation. All of the methods described in this chapter allow for use of the MCMC draws to develop a representation of the uncertainty (see Figures 17.1 and 17.10).

Section 17.3 describes an extremely simple yet effective approach to variable selection using BART inference. We simply count the number of times a particular component of $\mathbf{x}$ is used for a splitting rule in a BART ensemble of trees. This may be done for each MCMC draw. A drawback is that it may be necessary to use a smaller number of trees m in the BART ensemble than might be optimal for predictive inference. However, it will often be a simple matter to find a value of m for which BART identifies the more important variables without drastically compromising the fit (Figure 17.3). In practice this may be much easier than altering the basic model to incorporate prior choices about variable sparsity.

Going beyond the problem of variable selection, Section 17.4 suggests how BART output may be used to detect interactions of pairs of variables. The key simple idea is that a single tree can capture interactions by using different variables for its splitting rules. Rather than count how often a variable is used for a splitting rule, we count how often pairs of variables are used for splitting rules for the same tree. This is a much more flexible notion of interaction than the common strategy of including a product of the variables in the model.

Section 17.5 presents a completely different approach. Given the initial fit from a comprehensive application of BART, we search for an approximation to that fit which uses only a subset of the variables. A simple yet effective search algorithm is presented. Given a potentially useful subset, we assess the subset and our uncertainty through the posterior distribution of the approximation error. By focusing on the approximation error, we focus on the practical importance of variable subsets.

An appealing aspect shared by the approaches presented in this chapter is that they all build on analyzing the output from a basic application of BART. Effective post processing of the BART MCMC draws allows us to uncover variables and interactions of interest.

Acknowledgments

We thank the Editors and referees for their many helpful suggestions. This work was supported by NSF Grants DMS-1916245 and DMS-1916233.

Bibliography

[1] J. Bleich, A. Kapelner, E. I. George, and S.T. Jensen. Variable selection for BART: An application to gene regulation. *Ann. Appl. Stat.*, 8(3):1750–1781, 09 2014.

[2] C. Carvalho, P.R. Hahn, and R. E. McCulloch. Fitting the fit, variable selection using surrogate models and decision analysis, a brief introduction and tutorial. *arxiv*, 2020.

[3] E.I. George, H.A. Chipman, and R.E. Mcculloch. Bayesian cart model search. *Journal of the American Statistical Association*, 93(443):935–948, 1998.

[4] H.A. Chipman, E.I. George, and R.E. McCulloch. BART: Bayesian additive regression trees. *The Annals of Applied Statistics*, 4(1):266–298, 2010.

[5] J. H. Friedman. Multivariate adaptive regression splines (Disc: P67-141). *The Annals of Statistics*, 19:1–67, 1991.

[6] P.R. Hahn and C. M. Carvalho. Decoupling shrinkage and selection in Bayesian linear models: a posterior summary perspective. *Journal of the American Statistical Association*, 110(509):435–448, 2015.

[7] A. Linero. Bayesian regression trees for high dimensional prediction and variable selection. *Journal of the American Statistical Association*, 113(522):626–36, 2018.

[8] V. Ročková and E. Saha. On theory for BART. volume 89 of *Proceedings of Machine Learning Research*, pages 2839–2848. PMLR, 16–18 Apr 2019.

[9] V. Ročková and S. van der Pas. Posterior concentration for Bayesian regression trees and forests. *Ann. Statist.*, 48(4):2108–2131, 08 2020.

[10] S. Woody, C.M. Carvalho, and J. S. Murray. Model interpretation through lower-dimensional posterior summarization. *Journal of Computational and Graphical Statistics*, 0(0):1–9, 2020.

18

Variable Selection for Bayesian Decision Tree Ensembles

Antonio R. Linero

University of Texas at Austin (USA)

Junliang Du

Florida State University (USA)

CONTENTS

Abstract: We consider nonparametric variable selection using Bayesian decision tree ensembles, primarily from a fully-Bayesian perspective. The variable selection problem for Bayesian decision trees seems, on the surface, to be very simple – if a variable has been used as a branch in a decision tree, then it is in the model, and otherwise it is not. We show that the situation is more subtle than this, with poorly chosen priors resulting in overly-dense models. We review methods that do not use the model itself to determine variable importance and fully-Bayes approaches which derive model inclusion probabilities from the model itself. Fully-Bayesian approaches rely on sparsity-inducing priors to perform well; such priors include the Dirichlet prior, Spike-and-Forest priors, and a new class of Gibbs priors which we introduce. We illustrate these approaches on simulated and benchmark datasets. We also discuss extensions of these approaches when there exists additional structure, such as graphical or grouping structure, on the predictors.

DOI: 10.1201/9781003089018-18

18.1 Introduction

The majority of research on Bayesian variable selection has focused on (generalized) linear models. Arguably, however, this problem is more critical in nonparametric and semiparametric problems due to the curse of dimensionality, which makes nonparametric problems of even moderate dimension highly difficult. A striking example of this is given in Figure 18.1, which illustrates that failing to perform variable selection in Bayesian additive regression tree (BART) models can lead to a disastrous decrease in prediction performance (see Section18.1.1 for a description of this problem).

This chapter focuses on the *Bayesian additive regression trees* (BART) model. We will consider the *semiparametric regression* model

$$y_i = f(\boldsymbol{x}_i) + \epsilon_i, \qquad \epsilon_i \sim \mathcal{N}(0, \sigma^2),$$

where $f(\cdot)$ is modeled in a "black-box" fashion as a sum of regression trees. All of the main points discussed here carry over to more general problems without modification, and BART models have been developed for survival analysis [1, 4, 31], loglinear models [23], density estimation [17], and multivariate or mixed-response models [19], among many other settings.

The main appeal of BART is twofold:

1. it is very easy to use, with minimal prior specification required and a variety of fast software packages available; and

2. it works extremely well in practice, especially when used in problems where no massive feature engineering is required (e.g., typical causal inference problems, but not in computer vision or natural language processing problems).

Starting from the initial proposal of [6], our goal is to cover the possible ways to use BART as a variable selection tool. We also outline some newer contributions which improve on recently proposed "fully-Bayesian" variable selection procedures, and show how to move beyond simple sparsity to other problems. Our focus will be on variable selection procedures which incorporate variable selection into a fully-Bayesian framework using *sparsity inducing* priors.

18.1.1 Running Example

To illustrate concepts throughout, we will use the following simulation setting. We consider $y_i = f(\boldsymbol{x}_i) + \epsilon_i$ where $\epsilon_i \sim \mathcal{N}(0, \sigma^2)$ and $f(\boldsymbol{x})$ has the form

$$f(\boldsymbol{x}) = 10\sin(\pi x_1 x_2) + 20(x_3 - 0.5)^2 + 10x_4 + 5x_5. \tag{18.1}$$

Note that $f(\boldsymbol{x})$ does not depend on x_j for $j > 5$. Estimating $f(\boldsymbol{x})$ of the form (18.1) is an interesting prediction task, as it consists of both linear and non-linear terms, as well as a nonlinear interaction. Our interest lies in recovering the relevant set of covariates $\mathcal{S} = \{1, 2, \ldots, 5\}$ of $f(\boldsymbol{x})$. The term $(x_3 - 0.5)^2$ is particularly challenging for linear models, as there is no linear trend in X_3 when $X_3 \sim \mathcal{U}(0, 1)$. We refer to this as the "Friedman example" because it was introduced by [13]; this example is also commonly used in the BART literature as a test case [6, 18].

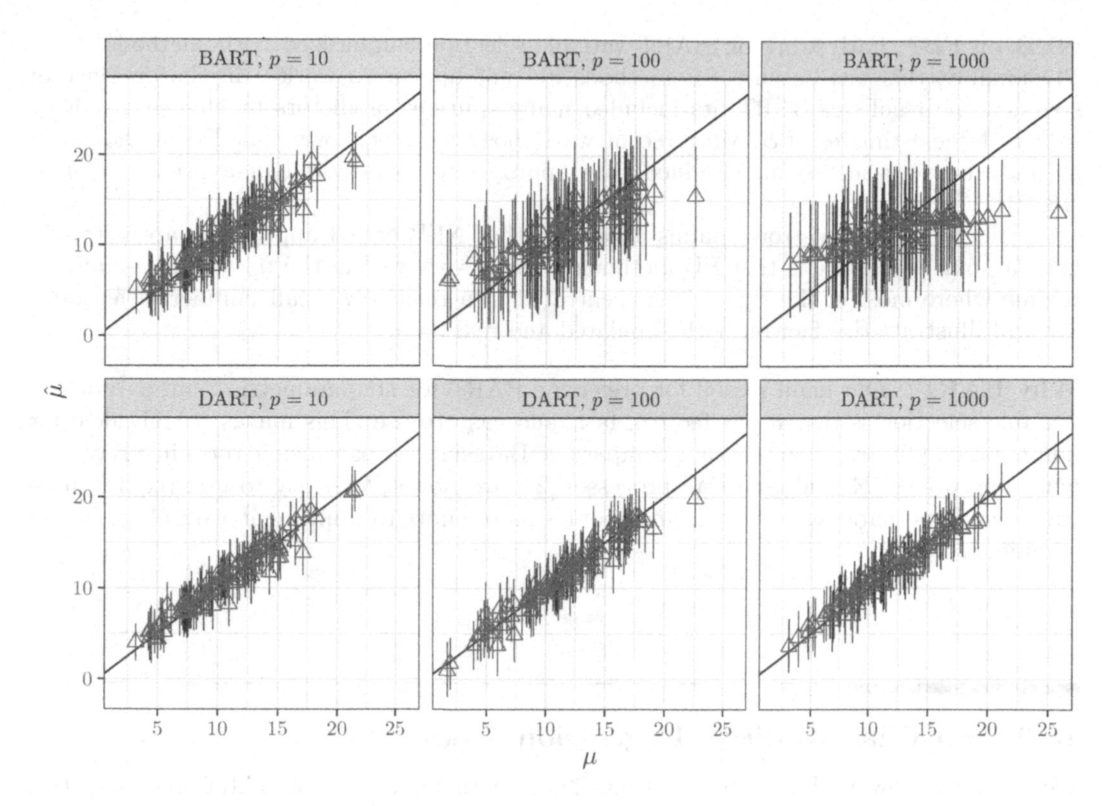

FIGURE 18.1
Predictive performance of BART on the Friedman function given in Section 18.1.1 as the dimension of the predictor p varies. The x-axis gives the true value of $f(\boldsymbol{x})$ for 100 held-out values of $\boldsymbol{x}$ while the y-axis gives the predicted value $f(\boldsymbol{x})$ from the model. The top row gives results for the default BART model [6]; the bottom row gives results using the modified BART model of [18] (and 95% credible intervals).

18.1.2 Possible Strategies

Our formal goal is to recover the set of relevant covariates of $f(\cdot)$, defined to be $\mathcal{S} = \{j : f(\boldsymbol{x}) \text{ is non-constant as a function of } x_j\}$. Unlike Bayesian GLMs, determination of $\mathcal{S}$ cannot be summarized in terms of the sparsity of a regression coefficient $\boldsymbol{\beta}$.

We divide variable selection methods for BART into two types of approaches: fully-Bayes (FB) and non-fully-Bayes (NFB) methods.

Non-fully-Bayesian methods An NFB method takes a model fit (or many different model fits) and post-processes the results to obtain measures of variable importance or conduct hypothesis tests. NFB methods are not justified by the Bayesian calculus – instead, we informally examine variable importance metrics derived from the fits, or use techniques like frequentist hypothesis testing [3] based on these metrics.

Fully-Bayesian methods FB methods use the Bayesian hierarchical model itself to estimate $\mathcal{S}$. For example, in a linear model, we might use a spike-and-slab prior to impose sparsity in a regression coefficient $\boldsymbol{\beta}$ in order to estimate the probability that a given coefficient β_j is non-zero. For nonparametric models, the simplest way to perform FB variable selection is to compute the proportion of the time that $f(\boldsymbol{x})$ depends on x_j given some posterior samples of $f(\boldsymbol{x})$.

NFB or FB? Early work on BART variable selection emphasized NFB methods [3, 6]. Presumably, this was because FB methods did not appear to work – in the presence of sparsity, the results of BART fits included many spurious predictors by chance, resulting in very large estimates of $\mathcal{S}$. More recent work, however, has shown that FB methods can actually be very successful, provided that appropriate sparsity-inducing priors are used [18, 21].

While we make no strong claims about whether NFB or FB approaches are better in general, our experience is that FB methods perform very well in the *needles-in-a-haystack* regime where most of the signal is concentrated in a relatively small number of features. We will illustrate this fact on both simulated and real data.

Why BART? Our main reason for preferring BART for nonparametric/semi-parametric variable selection is that it is effective, fast, and easy to use. This makes BART ideal for non-experts. Our experience is that comparable Bayesian nonparametric tools like Bayesian neural networks [24] and Gaussian processes [27] are slower, less easy to use, require more extensive hyperparameter tuning, and require more effort to implement variable selection techniques.

18.2 Bayesian Additive Regression Trees

We briefly review BART in this section. Those with knowledge of BART may skip this section after referring to Table 18.1 for our notation.

18.2.1 Decision Trees and their Priors

Our basic building block is the *decision tree*. Figure 18.2 shows a decision tree fit to the salaries of professional Major League Baseball (MLB) players during a season, measured in thousands of dollars. To obtain a prediction for a player we start from the top of the tree and apply the decision rules iteratively until we reach the bottom of the tree, and then read off the prediction. For example, if a player scored more than 46 runs, was a free agent, was not eligible for salary arbitration, and batted in more than 82 runs, then the tree predicts a salary of approximately \$3,038,000 dollars.

We now introduce our notation (summarized in Table 18.1). A *binary decision tree* $\mathcal{T}$ consists of a collection of nodes $\{\eta_o\}$ where o is a finite binary string consisting of the symbols L (left) and R (right). In Figure 18.2, for example, η_{RLRR} has prediction \$3,038,000. All trees have a root node $\eta_\emptyset$. Given a node $\eta_o \in \mathcal{T}$, its left and right children are given by η_{oL} and η_{oR} respectively. The nodes η_{oL} and η_{oR} are simultaneously either present or absent in $\mathcal{T}$; if they are present, then their *parent* η_o must also be present. A node η_o consists of a *splitting coordinate* j_o (i.e., the variable used to construct the decision rule) and a *cutpoint* C_o (the value of x_{j_o} used to make the decision). For example, at the root of the tree in Figure 18.2, the splitting coordinate corresponds to "Runs" and the cutpoint corresponds to 46. Let $\mathcal{B} \subseteq \mathcal{T}$ consist of all *branch nodes* (also called *non-terminal* nodes), which are nodes $\eta_o \in \mathcal{T}$ such that $\eta_{oL}, \eta_{oR} \in \mathcal{T}$. The collection of *leaf nodes* $\mathcal{L}$ (or *terminal* nodes) consists of all $\eta_o \in \mathcal{T}$ which are not branches. We denote the predictions made by the tree as $M = \{\mu_o : \eta_o \in \mathcal{T}\}$. If $\boldsymbol{x}$ is associated to leaf node $\ell \in \mathcal{L}$, we write $\boldsymbol{x} \rightsquigarrow \ell$. We let $g(\boldsymbol{x}; \mathcal{T}, M)$ denote the prediction for $\boldsymbol{x}$, i.e., $g(\boldsymbol{x}; \mathcal{T}, M) = \mu_o$ if-and-only-if $\boldsymbol{x} \rightsquigarrow \eta_o$ and $\eta_o \in \mathcal{L}$.

Bayesian inference requires a prior on $(\mathcal{T}, M)$. Let ϑ be a collection of hyperparameters.

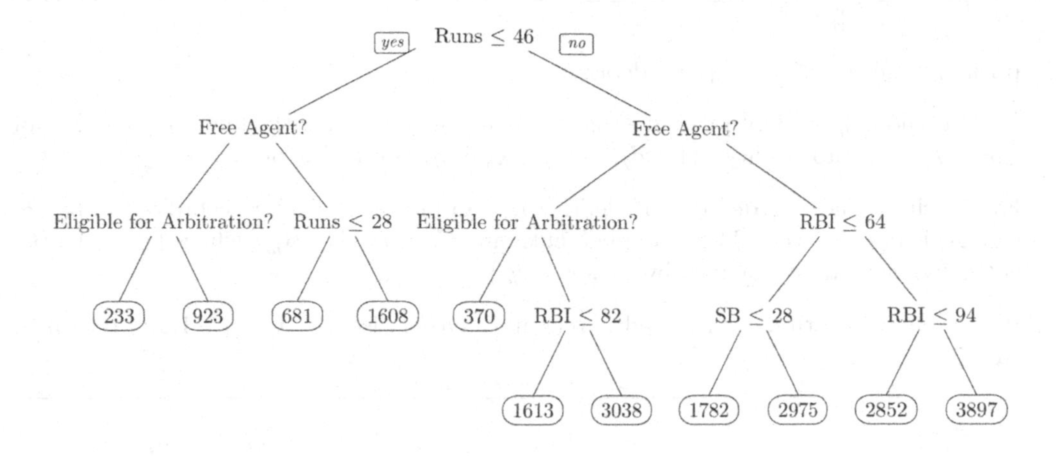

FIGURE 18.2
Example of a decision tree for predicting the salary of baseball players from features of that player. Bottom nodes represent the predicted salary in thousands of dollars.

$\mathcal{T}$:	a decision tree	$\mathcal{L}$:	collection of leaves
M:	collection of predictions	$\mathcal{B}$:	collection of branches
η_o:	a node of the decision tree	o:	a string of L's and R's
j_o:	coordinate of η_o's splitting rule	ϑ:	hyperparameters of $p_{\mathcal{T}}$
μ_o:	prediction at a leaf node	$\boldsymbol{x} \rightsquigarrow \ell$:	$\boldsymbol{x}$ associated to leaf ℓ
C_o :	cutpoint for η_o	g:	$g(\boldsymbol{x}; \mathcal{T}, M)$ returns μ_o if $x \rightsquigarrow \eta_o$
σ_μ:	variance of μ_o's	(β, α):	splitting probability $\alpha(1+d)^{-\beta}$
s:	$j_o \sim Categorical(s)$		

TABLE 18.1
Summary of our notation for decision trees.

We write $p_{\mathcal{T}}(\mathcal{T} \mid \vartheta)$ for the prior on $\mathcal{T}$, which we take to be a *branching process prior*. One possible algorithm for sampling from our chosen prior (*not* the posterior) is given in Algorithm 1. The hyperparameter of this prior is $\vartheta = (s, \alpha, \beta)$.

BART implementations vary somewhat at Step 3, and we have chosen this specification to simplify some of the full conditionals we will discuss when we consider priors on s. See [18] for a discussion of other possibilities and the role they play in inference. The parameters α and β describe the overall shape of the tree, with α denoting the probability that $\mathcal{T}$ has at least one branch and β controlling how deep the trees tend to grow.

Our prior on M will typically vary according to the type of model under consideration. For semiparametric regression, $[Y_i \mid \boldsymbol{X}_i = \boldsymbol{x}] \sim \mathcal{N}(\mu_o, \sigma^2)$ if $\boldsymbol{x} \rightsquigarrow \eta_o$ in which case the prior $\mu_o \stackrel{\text{iid}}{\sim} \mathcal{N}(0, \sigma_\mu^2)$ has the benefit of *conditional conjugacy* and also allows for the marginal likelihood of $\mathcal{T}$ to be evaluated in closed form. Similarly, if Y_i is categorical taking values in $\{1, \ldots, C\}$ then we might set $[Y_i \mid \boldsymbol{X}_i = \boldsymbol{x}] \sim \text{Categorical}(\mu_o)$ where $\mu_o \sim \mathcal{D}(a_1, \ldots, a_C)$. More details on these types of models are given by [5, 9].

Algorithm 1 Algorithm for sampling from the prior $p_{\mathcal{T}}(\mathcal{T} \mid \vartheta)$.

Input: $\vartheta = (s, \alpha, \beta)$

1. Initialize the tree $\mathcal{T} = \{\eta_\emptyset\}$ and depth $d = 0$.
2. For each node η_o of depth d, make that node a branch node, with children $\eta_{oL} \in \mathcal{T}$ and $\eta_{oR} \in \mathcal{T}$, with probability $\alpha(1 + d)^{-\beta}$; otherwise, η_o is a leaf node.
3. For each branch node η_o of depth d, sample a splitting coordinate $j_o \sim \text{Categorical}(s_1, \ldots, s_p)$. Then sample a cutpoint $C_o \sim \mathcal{U}(A_{j_o}, B_{j_o})$ where $\prod_{j=1}^{p}[A_j, B_j]$ is the hyper-rectangle defined by $\{\boldsymbol{x} : \boldsymbol{x} \rightsquigarrow \eta_o\}$.
4. If all nodes of depth d are leaf nodes, return $\mathcal{T}$. Otherwise, set $d \leftarrow d + 1$ and return to step 2.

18.2.2 The BART Model

Rather than modeling an unknown function $f(\boldsymbol{x})$ using a single decision tree, BART models $f(\boldsymbol{x})$ as a *sum* of decision trees, i.e.,

$$f(\boldsymbol{x}) = \sum_{k=1}^{m} g(\boldsymbol{x}; \mathcal{T}_k, \mathcal{M}_k).$$

The $\mathcal{T}_k$'s are encouraged by the prior to be *shallow* in that the maximal depth of each tree is relatively small (typically 2 or less). The BART model was motivated by the technique of *boosting weak learners* in machine learning [12]. By decomposing $f(\boldsymbol{x})$ in terms of shallow decision trees, the BART model expresses a preference for *low order interactions*. BART is most successful in problems where the assumption that interactions are low-order is reasonable.

Default prior for $\mathcal{T}_1, \ldots, \mathcal{T}_m$: Our prior for the trees assumes independence conditional on ϑ, i.e., $\mathcal{T}_k \stackrel{\text{iid}}{\sim} p_{\mathcal{T}}(\mathcal{T} \mid \vartheta)$ with $p_{\mathcal{T}}$ being the prior given by the generative model described by Algorithm 1. The values $\alpha = 0.95$ and $\beta = 2$ are commonly used to ensure that $\mathcal{T}_k$ is unlikely to only have a root while keeping the trees shallow.

Default prior for $\mathcal{M}_1, \ldots, \mathcal{M}_m$: Given $\mathcal{T}_1, \ldots, \mathcal{T}_m$, we specify independent Gaussian priors for the leaf node parameters, with $\mu_{mo} \stackrel{\text{iid}}{\sim} \mathcal{N}(0, \sigma_\mu^2)$ (different choices may be more appropriate for non-Gaussian likelihoods [1, 19, 23]). The parameter σ_μ^2 controls the amount of signal expected in the data. The original recommendation is to simply fix $\sigma_\mu = 0.5/(k\sqrt{m})$ for some $k \in \{1, 2, 3\}$ after scaling the response to lie in the interval $[-0.5, 0.5]$ [6], although it is certainly possible to endow σ_μ with a hyperprior.

Default Prior for σ^2: The prior for σ^2 is selected in an informal data-based fashion. A common approach [6] is to impose first that BART accounts for more signal than a linear regression model and second that not too much mass is placed near 0; the second requirement is sensible, as models which allow for $\sigma \approx 0$ may encourage overfitting. We set $\sigma^{-2} \sim \mathcal{G}(1.5, \beta)$ with β chosen so that $P(\sigma < \widehat{\sigma}) = 0.9$ where $\widehat{\sigma}$ is an estimate of σ based on a linear model.

Choice of s: the original BART approach [6] takes $s = (p^{-1}, \ldots, p^{-1})$, the idea being that selecting each predictor with equal probability is "non-informative." It is now understood that s plays a role in variable selection for both FB and NFB methods. We examine the choice of s in Sections 18.3 and 18.4.

18.3 Variable Importance Scores

Heuristically, we might expect an "important" variable to occur frequently in the tree ensemble. Let the variable importance of predictor j be $v_j = \mathbb{E}(m_j/B \mid \text{Data})$ where m_j is the number of times predictor j is used in a splitting rule and B denotes the number of branches, i.e., v_j is the posterior expected proportion of rules that use predictor j.

In Figure 18.3 we see variable importance scores for the Friedman example with $p \in \{10, 250\}$, $n = 250$, and $m \in \{20, 200\}$. As a general rule, we see that taking M smaller makes the gap between the relevant and irrelevant predictors larger. This was noted by [6] to occur because there is more competition for splitting rules among the predictors when M is small.

Figure 18.3 reveals problems with using raw variable importance scores. First, it is unclear how to interpret them in isolation to decide if a variable has a sufficiently high importance. The scale of the scores is also difficult to interpret; if we have $v_1 = 2v_2$, does this mean that x_1 is "twice as important" as x_2? The fact that the scores themselves depend on m make claims of this nature dubious.

The original BART implementation takes $s = (p^{-1}, \dots, p^{-1})$, which has undesirable effects on the behavior of the v_j's. Better variable selection is obtained by taking the number of trees m small, so that predictors have to "compete" for a small number of branches. This creates the following trade-off: on the one hand, we should make the number of trees small so as to avoid including irrelevant predictors, but on the other we should use a large number of trees to obtain a flexible model. It is unclear how to balance these two goals, particularly when one is interested in both prediction accuracy and variable selection at the same time. Even when the number of trees is restricted, there is still a tendency to include extraneous variables in the model with this approach. Nevertheless, the m_j's are actually very useful in performing variable selection; as we will see, the trick is to use them to learn s rather than having s fixed.

18.3.1 Empirical Bayes and Variable Importance Scores

To improve on the raw variable importances, we now give the informal analysis of variable importance scores a probabilistic interpretation as the first step of an expectation-maximization algorithm for computing an empirical Bayes estimate of s. We consider maximizing the marginal likelihood of the data $\mathfrak{m}_s(\text{Data}) = \mathbb{E}_s \mathfrak{p}(\text{Data} \mid \theta)\, p(\theta \mid s)$. We can optimize $\mathfrak{m}_s(\text{Data})$ over s using the expectation-maximization algorithm (EM) algorithm [8]

$$\begin{aligned} s^{(t+1)} &= \arg\max_s \mathbb{E}_{s^{(t)}} \left[\log \mathfrak{p}(\text{Data} \mid \theta) + \log p(\theta \mid s) \mid \text{Data}\right] \\ &= \arg\max_s \mathbb{E}_{s^{(t)}} \left[\log p(\theta \mid s) \mid \text{Data}\right] \end{aligned} \tag{18.2}$$

In the case of the $p_{\mathcal{T}}$ described in Section 18.2.1, $\log p(\theta \mid s) = \text{constant} + \sum_j m_j \log s_j$, so that the maximum of (18.2) is given by

$$s_j^{(t+1)} = \frac{\mathbb{E}_{s^{(t)}}(m_j \mid \text{Data})}{\mathbb{E}_{s^{(t)}}(B \mid \text{Data})}$$

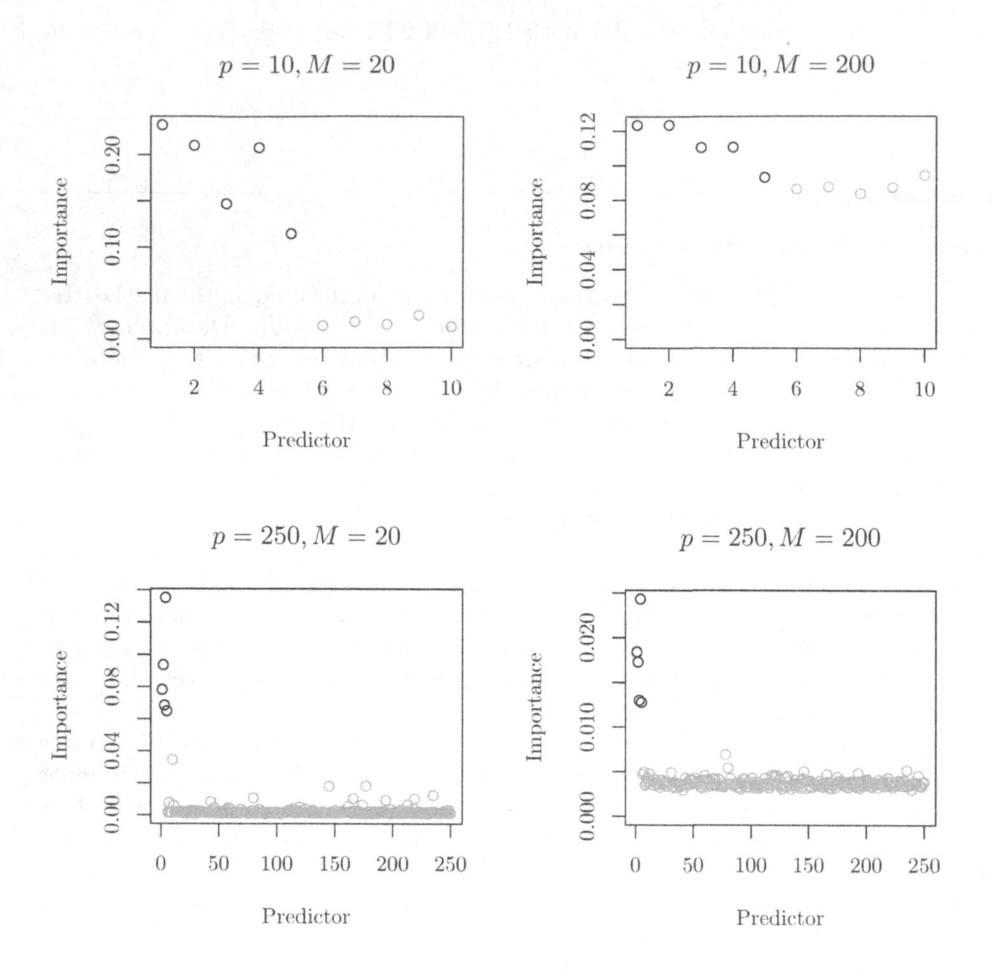

FIGURE 18.3
Variable importance scores for the Friedman test function for varying numbers of predictors and trees. Relevant predictors are colored black while irrelevant predictors are colored gray.

where we recall that m_j is the number of splitting rules based on predictor j and $B = \sum_j m_j$ is the number of branches in the ensemble. This update for s_j is similar to computing variable importance score v_j, except that v_j is the expectation of a ratio ($\mathbb{E}(m_j/B \mid \text{Data})$) rather than a ratio of expectations. Starting from $s^{(0)} = (p^{-1}, \dots, p^{-1})$ we see that ranking the variables using $s^{(1)}$ is similar to ranking them using the v_j's.

It is natural to ask if there is any improvement in using $s^{(K)}$ for some $K \geq 1$ rather than $s^{(1)}$. The left panel of Figure 18.4 revisits the Friedman example of Section 18.1.1 ($n = 250, p = 250, \sigma = 1$), showing the trajectory of $s_j^{(t)}$ as t varies. We see that, in the case of the Friedman simulation, the empirical Bayes algorithm results in the relevant predictors $j = 1, \dots, 5$ being assigned high importance, while all irrelevant predictors are assigned low importance. Relative to the v_j's, the $s_j^{(K)}$'s make it very easy to identify the relevant predictors without any ambiguity, provided that K is taken sufficiently large.

Going one step further, we might want an estimate of $\mathcal{S}$, which suggests that we would like the s to be *sparse*. One way to impose sparsity is to add an additional *sparsity inducing*

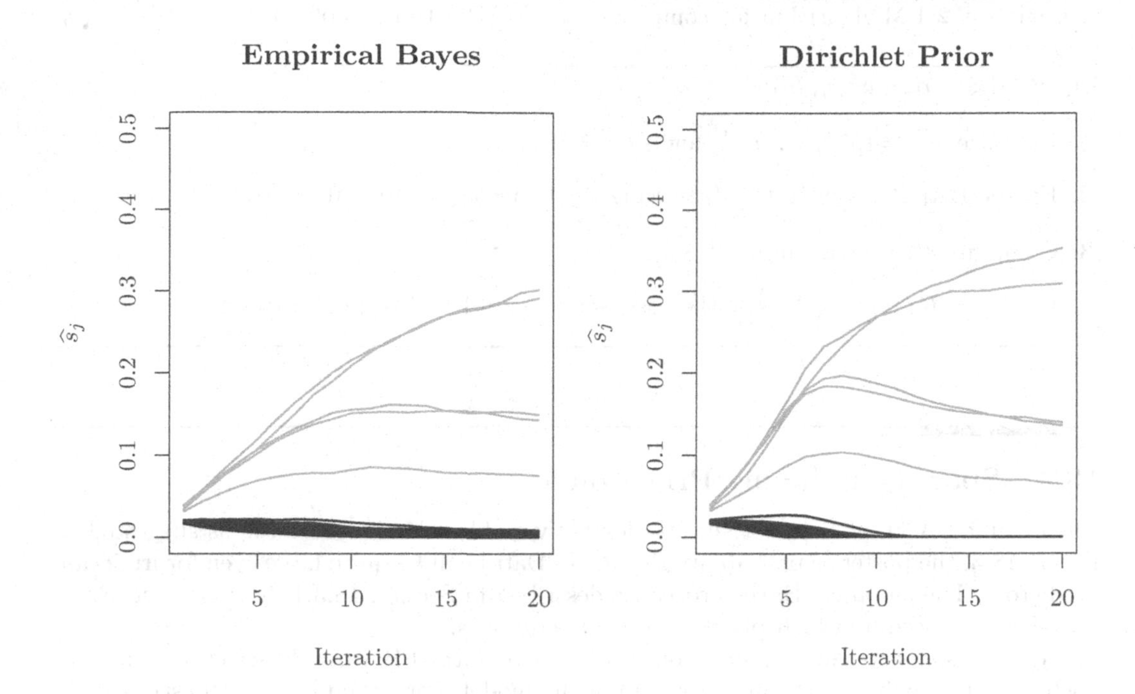

FIGURE 18.4
Left: estimate of s_j at each iteration of 20 iterations of the EM algorithm. Right: estimate of s_j using the EM algorithm to compute the MAP estimate with (18.3) and $\eta = 0.003$.

penalty into (18.2). The right panel of Figure 18.4 considers the penalty $(\eta - 1)\sum_j \log s_j$. This choice of penalty corresponds to the maximum a-posteriori (MAP) estimator of s when $s \sim \mathcal{D}(\eta, \ldots, \eta)$. Under this penalty the associated update for s from the EM algorithm is

$$s_j^{(t+1)} \leftarrow \frac{\max\{\mathbb{E}_{s^{(t)}}(m_j \mid \text{Data}) + \eta - 1, 0\}}{\sum_{j'} \max\{\mathbb{E}_{s^{(t)}}(m_{j'} \mid \text{Data}) + \eta - 1, 0\}}. \tag{18.3}$$

Note that $\eta = 1$ corresponds to the original EM algorithm. The Bayesian EM algorithm is given in Algorithm 2. For $\eta < 1$ this corresponds to a *thresholding* rule, where $s_j^{(t+1)} \leftarrow 0$ if $\mathbb{E}_{s^{(t)}}(m_j \mid \text{Data}) \le 1 - \eta$; in words, this says we eliminate predictor j if it occurs in fewer than $1 - \eta$ branches on average. We see in Figure 18.4 that after roughly 10 iterations, all of the irrelevant predictors have been completely eliminated from the model. The empirical Bayes approach is particularly easy to implement using the package `bartMachine` in R.

There are three main difficulties with this EM algorithm. First, Algorithm 2 finds only a local optimum value for s; sensitivity to local minima can be somewhat reduced by slowly reducing η from 1 to its target value in Algorithm 2. The second difficulty is that the empirical Bayes approach does not give any meaningful uncertainty quantification; if we use a sparsity inducing penalty we will only have a point estimate of $\mathcal{S}$ rather than a posterior distribution. Finally, the EM algorithm requires fitting the model a potentially large number of times. We remark that fitting the model multiple times is also quite common for NFB methods: the permutation testing approach of [3] also requires the model to be refit many times. Hence, this EM algorithm is not overly burdensome relative to other approaches.

Algorithm 2 EM algorithm for computing the MAP estimator of s under a $\mathcal{D}(\eta, \ldots, \eta)$ prior.

Input: (Data, $\alpha, \beta, \sigma_\mu^2, \eta, K$)

1. Initialize $s^{(0)} = (p^{-1}, \ldots, p^{-1})$ and set $t = 0$.
2. Fit the BART model to the data using the hyperparameter values $(\alpha, \beta, \sigma_\mu^2, s^{(t)})$.
3. Compute $s^{(t+1)}$ according to (18.3).
4. If $t + 1 = K$, return $s^{(t+1)}$. Otherwise, set $t \leftarrow t + 1$ and return to step 2.

18.4 Sparsity Inducing Priors on s

The default BART prior is not suitable for FB variable selection because, as suggested by Figure 18.3, the posterior probability $\Pr(j \in \mathcal{S} \mid \text{Data})$ can be quite large even for irrelevant predictors. The empirical Bayes procedure described in Section 18.3.1, however, suggests a fully-Bayes procedure which places a prior directly on s.

Given a sparsity-inducing prior on s, we can perform FB variable selection using the posterior probabilities that predictors are in the model. For example, we can estimate the set of relevant covariates as $\widehat{\mathcal{S}}_\omega = \{j : \Pr(j \in \mathcal{S} \mid \text{Data}) \geq \omega\}$; in particular, $\widehat{\mathcal{S}}_{0.5}$ corresponds to the *median probability model*. We will see that, if the prior on s is selected appropriately, the median probability model becomes a very attractive estimate of $\mathcal{S}$.

18.4.1 The Uniform Prior on s

To build intuition, we first discuss a poor choice of prior on s: the uniform prior $s \sim \mathcal{D}(1, \ldots, 1)$. Like the default prior $s = (p^{-1}, \ldots, p^{-1})$, one might be drawn to this prior as an "uninformative" choice. Far from capturing any intuitive sense of being uninformative, this prior on s encodes *dogmatic* information on the number of predictors included in the model when the number of predictors p is large [18].

Theorem 18.4.1. *Let Q denote the number of predictors used in a BART ensemble constructed according to Algorithm 1. Then, conditional on the number of branches B, $\mathbb{E}(Q \mid B) \geq B + C_{1B}/p$ and $\operatorname{Var}(Q \mid B) \leq C_{2B}/p$ for the settings $s \sim \mathcal{D}(1, \ldots, 1)$ and $s = (p^{-1}, \ldots, p^{-1})$, where C_{1B} and C_{2B} are constants depending only on B.*

Theorem 18.4.1 makes it clear that both the default prior and the uniform prior are completely at odds with a prior belief of sparsity. For large values of p, the ensemble becomes completely saturated with predictors –for B branches, we will have B unique predictors, and each predictor will be used only once.

18.4.2 The Dirichlet Prior

The problem with the uniform prior is that it does not encourage s to be *nearly sparse.* Note that if s has $D \ll p$ non-zero entries then our ensemble will use at-most D variables. One way to make s nearly-sparse is to use the *sparsity inducing Dirichlet prior* $s \sim \mathcal{D}(\xi/p, \ldots, \xi/p)$. This is illustrated in Figure 18.5, which shows samples of the $\mathcal{D}(\xi/3, \xi/3, \xi/3)$ distribution for different values of ξ. We see that the uniform prior ($\xi = 3$) places too much mass on the

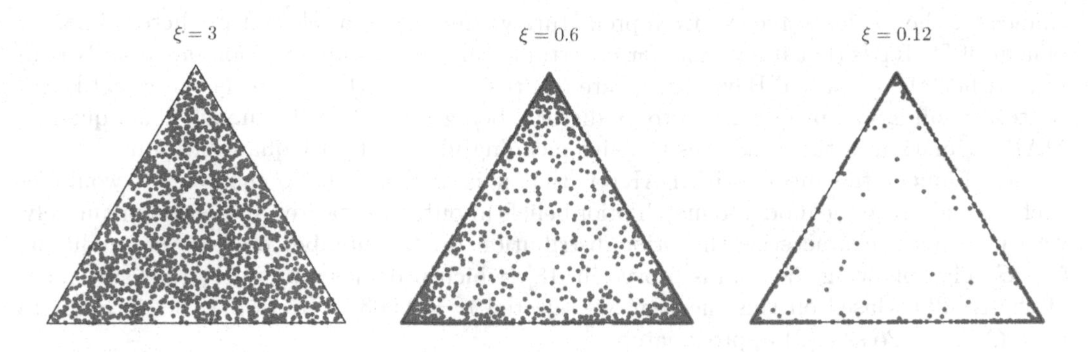

FIGURE 18.5
Samples from a $\mathcal{D}(\xi/3, \xi/3, \xi/3)$ prior on $\mathbb{S}_3$ for different values of ξ. Here, the vertices of the simplex correspond to the points $(1, 0, 0)$, $(0, 1, 0)$, and $(0, 0, 1)$, the edges correspond to 2-sparse vectors, and the interior corresponds to dense vectors. The midpoint of the simplex corresponds to the point $(1/3, 1/3, 1/3)$.

interior of the simplex $\mathbb{S}_p = \{s : \sum_{j=1}^{p} s_j = 1,\ s_j \geq 0\}$, whereas sparsity in s occurs on the edges and vertices of $\mathbb{S}_p$. Smaller values of ξ are seen to lead to more concentration near the edges and vertices, which correspond to models with fewer than p non-zero entries of s.

Another point in favor of the sparsity-inducing Dirichlet prior is given by the results of Section 18.3.1. We have already seen that using a penalty motivated by the sparsity-inducing Dirichlet prior can lead to effective variable selection. The remaining issue is whether it is practical to use this prior.

Computations using the Dirichlet prior are surprisingly straight-forward due to the conjugacy of the Dirichlet distribution with multinomial sampling [18]. Under the choice of $p_{\mathcal{T}}$ given in this chapter we have the full-conditional $s \sim \mathcal{D}(\xi/p + m_1, \ldots, \xi/p + m_p)$, and hence the update for s can be added to commonly-used BART algorithms to automatically perform variable selection, as in Algorithm 3 (for choices of $p_{\mathcal{T}}$ for which this is not the correct full conditional, the $\mathcal{D}(\xi/p + m_1, \ldots, \xi/p + m_p)$ is a very accurate Metropolis-Hastings proposal). To distinguish this procedure from the vanilla BART model, we will refer to this model as the DART model.

Algorithm 3 An iteration of MCMC to fit the DART model

Input: Data, $\{\mathcal{T}_k, M_k : k = 1, \ldots, m\}$, σ_μ, α, β, s, ξ

1. Update $(\mathcal{T}_1, M_1), \ldots, (\mathcal{T}_m, M_m)$ using the Bayesian backfitting algorithm described by [16].

2. Update $s \sim \mathcal{D}(\xi/p + m_1, \ldots, \xi/p + m_p)$ where m_j is the total number of splits on predictor j across all trees in the ensemble.

3. Return to step 1.

The fully-Bayes approach using DART has several advantages over the empirical Bayes approach. First, and perhaps most importantly, it does not require fitting the model multiple times. The stochastic nature of MCMC can also make the occurrence of multiple modes less severe. For example, examining the update (18.3), we see that if $\mathbb{E}_{s^{(t)}}(m_j \mid \theta) \leq 1 - \eta$ then $s_j^{(t+1)} = 0$ and variable j will be eliminated permanently; because of this, there is no chance for the model to recover from prematurely eliminating a relevant predictor. Another benefit of DART is that it accounts for uncertainty in the variable selection process. By contrast, when the empirical Bayes procedure returns the estimate $s_j = 0$ the empirical Bayes posterior will assign probability 0 to predictor j being present in the model. Consequently, MAP estimation in this case does not give meaningful uncertainty quantification.

One immediate concern with DART is that it is unclear how to choose ξ. It would be useful to calibrate the model to match prior beliefs about the expected size of $\mathcal{S}$. Fortunately, we can exactly characterize the prior distribution on the number of relevant predictors $Q \in \mathcal{S}$. The following theorem is proved in [18] (which provides a more rigorous statement of the result). Based on this theorem, we selected $\eta = 0.003$ in Section 18.3.1 in order to have $Q - 1 \sim \text{Poisson}(2)$ approximately.

Theorem 18.4.2 (Heuristic). *Suppose that* $s \sim \mathcal{D}(\xi/p, \ldots, \xi/p)$ *and let* B *denote the number of branches in the ensemble. Then, conditional on* B*, we have* $Q-1$ *is approximately* $\text{Poisson}(\theta)$ *for large* p *and* B *where* $\theta = p \times \left[1 - \frac{\{\xi(1-p^{-1})\}^{(B)}}{\xi^{(B)}}\right] = \xi\{\psi(\xi+B) - \psi(\xi)\} + C_{\xi,B}/p$ *where* $\psi(x) = \frac{d}{dx}\log\Gamma(x)$ *and* $C_{\xi,B}$ *is bounded as a function of* p.

While Theorem 18.4.2 is useful when our beliefs about Q can be modeled using a Poisson distribution, it is actually quite informative about the scale of Q – it roughly implies that Q will vary within a range of $\theta \pm k\sqrt{\theta}$ for some k. A more flexible prior can be obtained by placing a hyperprior on ξ. The prior $\xi/(\xi + p) \sim \text{Beta}(0.5, 1)$ has been suggested [18], and we use that as our default in the remainder of this chapter. An approach which gives similar results to DART, but allows for *direct* specification of the prior on Q, is given in Section 18.4.4.

Figure 18.6 shows how DART is capable of performing fully-Bayesian variable selection for the Friedman function in a way that default BART cannot. The top two panels show the posterior distribution of s_j for each of $p = 100$ predictors, as well as the posterior probability of each variable being present in the model. We see that the relevant predictors have by far the largest values of s_j and are included with probability 1. The median probability model for DART also selects the true model with $\widehat{\mathcal{S}} = \{1, 2, 3, 4, 5\}$. Using the default BART prior does not work in this regard, regardless of the number of trees used; while we do select the relevant predictors, there are many false positives for both $m = 20$ and $m = 200$.

18.4.3 The Spike-and-Forest Prior

As an alternative to the Dirichlet prior, we consider the *Spike-and-Forest* prior [30]. In words, the Spike-and-Forest prior first imposes that no more than that $D \le p$ of the predictors are relevant, with the distribution $p_D(d)$ chosen to reflect prior knowledge about the number of relevant predictors. Then, like the original BART prior, each of the predictors is used to construct splitting rules with equal probability. The model is given by

$$
\begin{aligned}
D &\sim p_D(d), \\
[\gamma \mid D] &\sim \binom{p}{D}^{-1} I(\sum_j \gamma_j = D, \gamma \in \{0,1\}^P), \\
s &= (\gamma_1/D, \ldots, \gamma_P/D).
\end{aligned}
\tag{18.4}
$$

This prior is more in line with traditional Bayesian variable selection approaches than DART, and it is much easier to reason about how predictors enter the model.

Benefits: The Spike-and-Forest priors has several advantages over DART. First, specification of a prior on D is easier to understand intuitively; note, however, that D *does not*

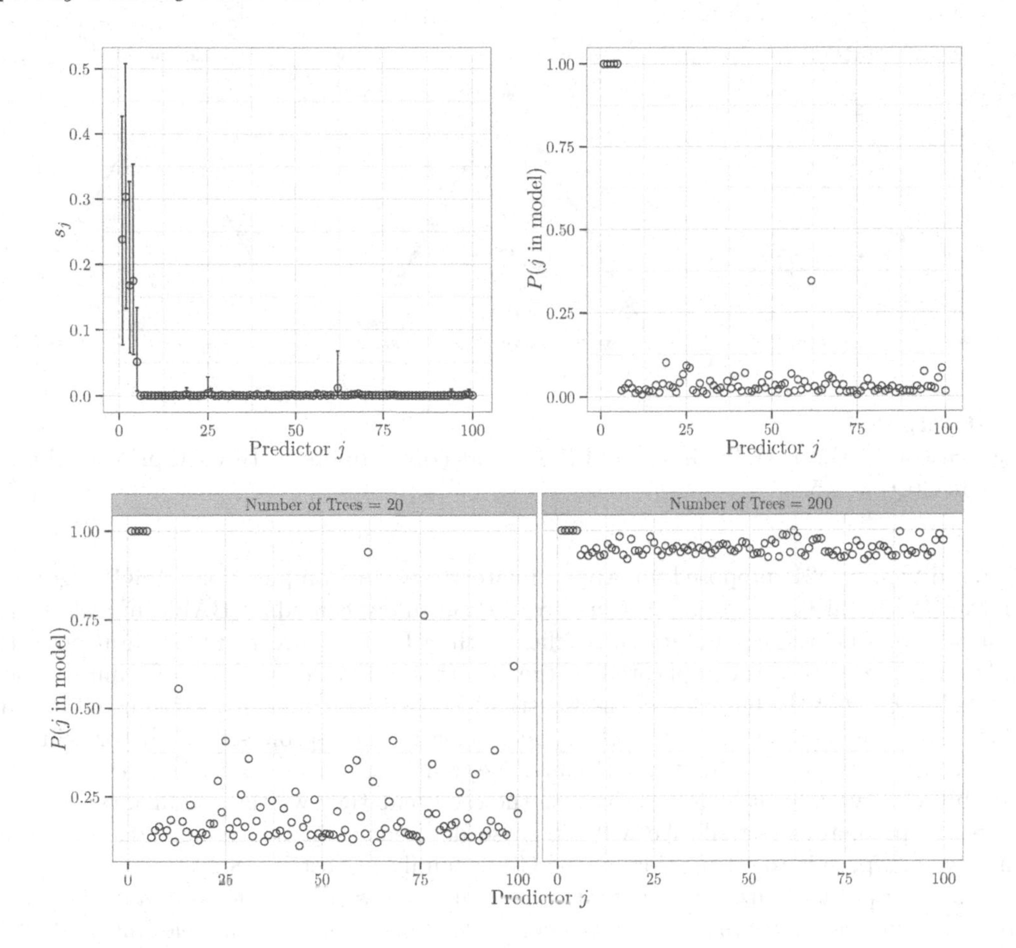

FIGURE 18.6
Top left: posterior distribution for the splitting probabilities s_j for the DART model with 200 trees. Top right: posterior probability that a given predictor is included in the model. Bottom: posterior probabilities for the BART model of a given predictor being included in the model for $m = 20$ (left) and $m = 200$ (right).

represent the number of relevant predictors, as it is possible for the ensemble to omit predictor j even if $\gamma_j = 1$. Second, the Spike-and-Forest prior has better asymptotics as the tree ensemble size diverges: as the number of branches $B \to \infty$, the number of relevant predictors Q converges in probability to $D < \infty$, whereas Q grows at a rate $Q \sim \xi \log(1 + B/\xi)$ for DART. This implies that, while the prior on Q in DART is *mostly* unaffected by the choice of the number of trees m, it is not completely free of it. Lastly, DART has the undesirable feature that the order statistics of the weights $s_{(1)} < s_{(2)} < \cdots < s_{(p)}$ decays very quickly, with $\mathbb{E}(s_{(j)}) \approx O\left(\left(\frac{\xi}{\xi+1}\right)^j\right)$. Hence, the DART prior will encourage most of the mass to concentrate on a small number of relevant predictors, rather than spreading the mass evenly among all relevant predictors. The Spike-and-Forest prior does not have this drawback, as it explicitly gives all relevant predictors the same splitting probability of $s_j = 1/D$.

Drawbacks: In our experience, the main drawback of the Spike-and-Forest prior is that it is difficult to implement in a way that allows for efficient MCMC to be performed. Because

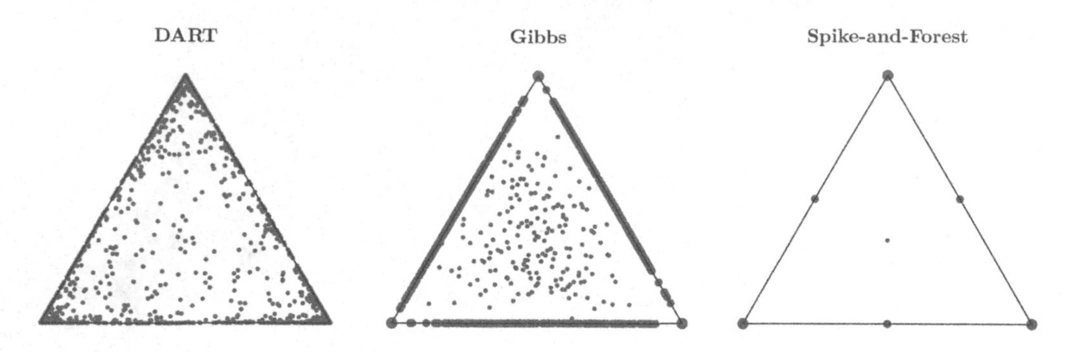

FIGURE 18.7
Comparison of the DART, Gibbs, and Spike-and-Forest priors, with each prior calibrated so that $\mathbb{E}(Q) = 1.5$.

of this difficulty, [21] proposed an Approximate Bayesian Computation (ABC) algorithm for the Spike-and-Forest prior; this involves fitting many candidate BART models to the data, which is far less computationally efficient than DART. Direct implementation of the Spike-and-Forest prior is complicated by the fact that active components of s must appear in roughly B/D of the branches of the forest, which makes it difficult for the usual MCMC schemes to add new predictors to the model since a newly-proposed predictor will only appear in $1 \ll B/D$ of the branches. Because these MCMC schemes must leave the posterior distribution invariant, it is necessarily also the case that they will have difficulty *removing* irrelevant predictors as well. We will see these deficiencies in the Spike-and-Forest prior when we compare it to other priors in the Gibbs family in Section 18.4.4.

A second potential drawback is that, by not allowing s_j to vary for relevant predictors, the model will be unable to adapt to varying levels of importance in the relevant predictors. The authors have not encountered any examples where this is likely to be a large problem, however, and the main constraint on the use of the Spike-and-Forest prior remains its computational difficulties.

Remark: The prior initially described by [30] and later used by [21] to study consistent model selection does not, strictly speaking, correspond to (18.4). Packaged with the prior on (γ, D) is a prior on the tree topologies $\mathcal{T}_1, \ldots, \mathcal{T}_m$, which is uniform over a class of trees defining "admissible partitions." The reason for this non-standard prior is to make the posterior more amenable to theoretical analysis. We believe that, from a practical perspective, nothing serious is lost by our casting of the Spike-and-Forest in terms of s.

18.4.4 Finite Gibbs Priors

At one extreme, the DART prior has the advantage of computational tractability, while the Spike-and-Forest prior enjoys stronger theoretical properties and more intuitive prior elicitation. Both methods also sit at the extremes of how they treat s –a-priori, DART loads most of the mass on a small number of the relevant predictors, while the Spike-and-Forest prior splits the mass evenly among all relevant predictors. An ideal method would sit in the middle of these two extremes, with the intuitive appeal and strong theoretical properties of the Spike-and-Forest prior and the efficient computation of DART.

The *Gibbs priors* we discuss now give a compromise between the DART and Spike-and-Forest priors. For intuition, we start with the following hierarchical model that might be called a *Spike-and-DART* prior:

$$D \sim p_D(d),$$
$$\gamma \sim \binom{p}{D}^{-1} I(\sum_j \gamma_j = D), \tag{18.5}$$
$$s \sim \mathcal{D}(\gamma_1\eta, \gamma_2\eta, \ldots, \gamma_p\eta),$$

where $p_D(d)$ is a mass function supported on $\{1, \ldots, p\}$. This model is equivalent to DART when $D \equiv p$ (with $\eta = \xi/p$) and equivalent to the Spike-and-Forest when $\eta \to \infty$. Hence this model interpolates between the Spike-and-Forest and DART priors depending on the choice of the hyperparameters η and $p_D(d)$. A comparison between the DART, Spike-and-DART, and Spike-and-Forest priors is given in Figure 18.7.

The Spike-and-DART prior has the benefits of the DART and Spike-and-Forest priors without their drawbacks. Prior specification in terms of $p_D(d)$ gives Spike-and-DART the interpretability of Spike-and-Forest. Additionally, one might conjecture that computations for Spike-and-DART are more like DART than Spike-and-Forest. To see why, recall that the MCMC for the Spike-and-Forest prior is difficult because each predictor should appear in $O(B/D)$ of the splitting rules, which makes it difficult to introduce new predictors without making large modifications to the ensemble. By allowing the components of s to vary by predictor, however, we bypass this issue – a predictor appearing only once in the ensemble is not implausible because it might have a small s_j.

Directly working with (18.5) turns out also to make it difficult to derive efficient MCMC schemes, but it points us in the correct direction. Following [22], the trick is to *marginalize over* (s, γ, D). The prior obtained in this fashion can be described in terms of a *Gibbs prior* on the partition of the branches $\mathcal{B} = \{b_1, \ldots, b_B\}$, where two branches belong to the same class if they split on the same predictor. Let $\mathcal{C}$ be a partition of $\mathcal{B}$ such that $b \sim b'$ if $j_b = j_{b'}$. A Gibbs prior on $\mathcal{C}$ is any prior of the form

$$\mathfrak{f}(\mathcal{C}) = V_B(Q) \prod_{c \in \mathcal{C}} \frac{\Gamma(\eta + |c|)}{\Gamma(\eta)}$$

where $Q = |\mathcal{C}|$ is the number of predictors used. To obtain a prior directly on the splitting rules, we can then assign a unique predictor to each one of the partition classes. Let $j^\star = (j_1^\star, \ldots, j_Q^\star)$ denote the unique predictors assigned to each subset of branches. Then a Gibbs prior for the splitting rules is given by

$$\mathcal{C} \sim \mathfrak{f}(\mathcal{C}),$$
$$\mathfrak{f}(j_1^\star, \ldots, j_Q^\star \mid \mathcal{C}) = \frac{(P - Q)!}{P!}.$$

Conveniently, this model can also be characterized by a *Pölya Urn* scheme [2, 7], and it can be shown that the probability of $j_b = j$ given $\mathcal{J}_b = \{j_{b'} : b' \neq b\}$ is given by

$$p(j_b = j \mid \mathcal{J}_b) = \begin{cases} \frac{V_B(Q_b)}{V_{B-1}(Q_b)}(\eta + m_j^{(-b)}) & \text{if } m_j^{(-b)} \neq 0, \\ \frac{V_B(Q_b+1)}{(P-Q^{(-b)})V_{B-1}(Q_b)}\eta & \text{otherwise,} \end{cases} \tag{18.6}$$

where $m_j^{(-b)}$ is the number of splitting rules other than j_b which use predictor j and Q_b is the number of unique values of j in $\mathcal{J}_b$. This is reminiscent of the *Chinese restaurant process* [25] commonly encountered in Bayesian nonparametrics.

The Spike-and-DART prior can be characterized through a (finite) Gibbs prior, with $\eta = \eta$ and

$$V_B(t) = \sum_{d=t}^{p} \frac{d!}{(d-t)!} \frac{\Gamma(\eta d)}{\Gamma(\eta d + B)} p_D(d). \tag{18.7}$$

A proof of this statement is given by [22]. This also implies that the DART prior is a type of Gibbs prior with $\eta = \xi/p$ and $V_B(t) = \frac{p!}{(p-t)!} \frac{\Gamma(\xi)}{\Gamma(\xi+B)}$. The Spike-and-Forest model can also be implemented within this framework by taking $\eta \to \infty$, the limiting expression of (18.6) being easily computed in closed form.

Computation with Gibbs priors is almost as convenient as it is with DART, and has the additional benefit that specification of a hyperprior on η appears to be unneeded. At any point of the MCMC algorithm described by [16] where j_b would be updated by Metropolis-Hastings, we simply sample j_b according to (18.6) rather than sampling according to s; by sampling from the conditional (18.6), no changes to the acceptance ratio are required. For computational efficiency, it is best to cache and reuse any evaluations of $V_B(t)$. Provided that caching is used, this algorithm is no more computationally expensive than the usual DART algorithm.

Applying the Gibbs prior to the Friedman example

Figure 18.8 compares the Spike-and-Forest and Gibbs (Spike-and-DART) priors. We use the default setting $\eta = 1$ for the Gibbs prior. We use a *truncated zeta* distribution for $p_D(d)$ for the Spike-and-Forest and Gibbs priors, with $p_D(d) = 1/(d^\zeta H_{\zeta,p})$ and $H_{\zeta,p}$ the generalized harmonic number $\sum_{d=1}^{p} d^{-\zeta}$. As a default, we set $\zeta = 1$ to encourage sparsity. We consider the high-dimensional setting with $p = 1000$ and $n = 250$ observations.

The results show the fundamental difficulty of correctly implementing the Spike-and-Forest prior. The Spike-and-Forest prior never eliminates all of the predictors and, once a predictor enters the model, it is very difficult to remove it. The false positives seen in Figure 18.8 (corresponding to posterior probabilities exceeding 0.5 for the median probability model) are most likely due to the failure of the MCMC. By contrast, the Gibbs prior has no false positives and mixes reasonably well on the number of predictors. It also has no trouble removing predictors once they enter the model.

Table 18.2 compares performance of the DART, Gibbs, and Spike-and-Forest prior on the Friedman example for different values of p and σ. The median probability model is used to make the final model selection. Methods are compared according to their precision (ratio of true positives to positives), recall (ratio of true positives to number of active predictors), and F_1 score (harmonic mean of the precision and recall). We see that, particularly for small p, the Gibbs prior outperforms DART, and never performs worse. The Spike-and-Forest prior performs poorly overall, and we conjecture that this is due to poor MCMC mixing.

18.5 An Illustration: The WIPP Dataset

We consider a dataset obtained from a computer model for two phase fluid flow used in an application for the Waste Isolation Pilot Plant [32]. There are $p = 31$ input variables and $n = 300$ simulations were run. The response is the "cumulative brine flow in m^3 into the waste repository at 10,000 years assuming there was a drilling intrusion at 1,000 years." We refer to this throughout as the WIPP dataset. Our goal is to determine which of the model

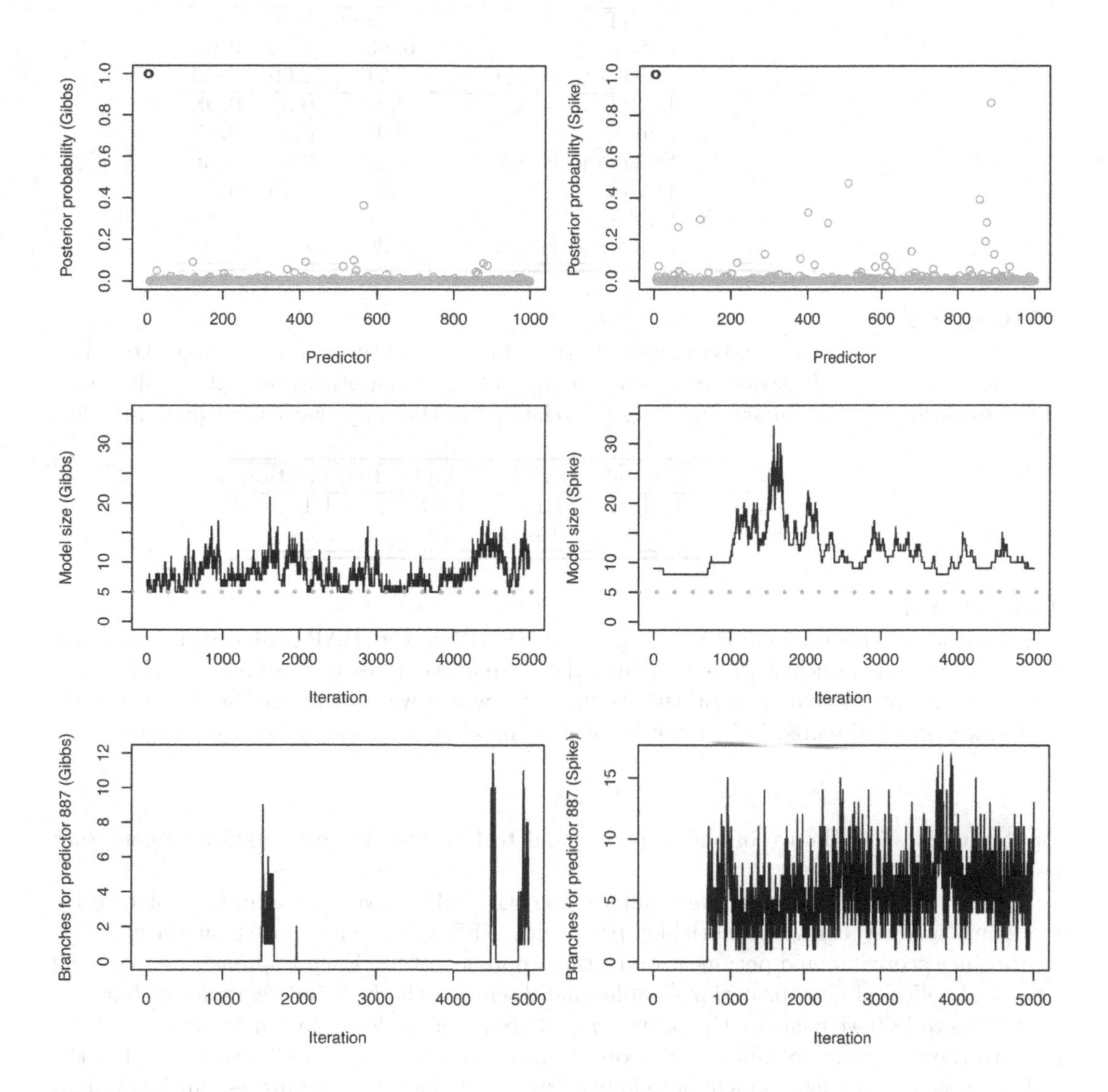

FIGURE 18.8
Top row: posterior probability of each predictor being included in the model using the Gibbs (left) and Spike-and-Forest (right) priors. Middle row: Trace plot of the model size (defined to be the number of unique covariates which appear in the ensemble) when fitting the model by MCMC for the Gibbs (left) and Spike-and-Forest (right) models. Bottom: traceplot of m_{887} (the number of splits on the 887th predictor) for the Gibbs (left) and Spike-and-Forest (right) models.

p	σ	Method	Precision	Recall	F_1
7	3	DART	0.75	1.00	0.85
		Gibbs	**0.96**	1.00	**0.98**
		Spike-and-Forest	0.71	1.00	0.83
	5	DART	0.71	1.00	0.83
		Gibbs	**0.89**	1.00	**0.94**
		Spike-and-Forest	0.71	1.00	0.83
1000	3	DART	0.97	0.99	**0.98**
		Gibbs	0.95	0.99	**0.97**
		Spike-and-Forest	0.22	0.98	0.36
	5	DART	0.95	0.75	**0.84**
		Gibbs	0.91	0.82	**0.85**
		Spike-and-Forest	0.30	**0.84**	0.43

TABLE 18.2
Results of the simulation study; standard errors for all quantities are less than 0.01. The precision, recall, and F_1 scores represent averages of these quantities over 200 replications of the experiment. Particularly good results relative to other approaches are given in bold.

Method	BART	DART	Permutation
RMSE	1.10	1.00	1.14
Time	16s	18s	230s

TABLE 18.3
Comparison of the original BART algorithm (BART), the BART algorithm with the sparsity-inducing Dirichlet (DART), and the permutation testing approach. RMSE denotes root-mean-squared error on the testing data, which was normalized so that the best-performing method scored 1.00. Time denotes the number of seconds taken to fit each model.

inputs are most predictive of the response, and to leverage this information to construct better prediction models.

Using the WIPP dataset, we compare the FB (fully-Bayesian) approach of directly modeling sparsity using the Dirichlet prior with NFB approaches based on the variable importance scores [6] and permutation testing approach of [3]. Each approach uses $m = 50$ trees and collected 6000 posterior samples and discarded the first 1000 samples to burn-in.

In Figure 18.9 we compare the posterior probability of model inclusion with the Dirichlet prior to the raw variable importance scores v_j described in Section 18.3. We argue that the fully-Bayesian approach is much more informative, as it clearly distinguishes which variables are associated from those which are not. It also gives a simple decision rule for including variables – we select a variable if it is more-likely-than-not in the ensemble. By contrast, there is no obvious rule for deciding a cutoff for selection using the v_j's.

Next, we compare the variables selected by the "Global-Max" approach of [3], which is based on a nonparametric permutation test using the variable importance scores, with the results obtained from the Dirichlet prior. Despite the Dirichlet prior unambiguously including 11 variables, the Global-Max procedure (which does not rank variables strictly by their importance) selects only `BPCOMP`, `BHPERM`, `WMICDFLG`, and `HALPOR`.

While we cannot fully assess the efficacy of the variable selection on this dataset, a reasonable benchmark for comparison is held-out predictive accuracy. We divided the dataset into training and testing sets with 80% of the data included in the training set. We then fit

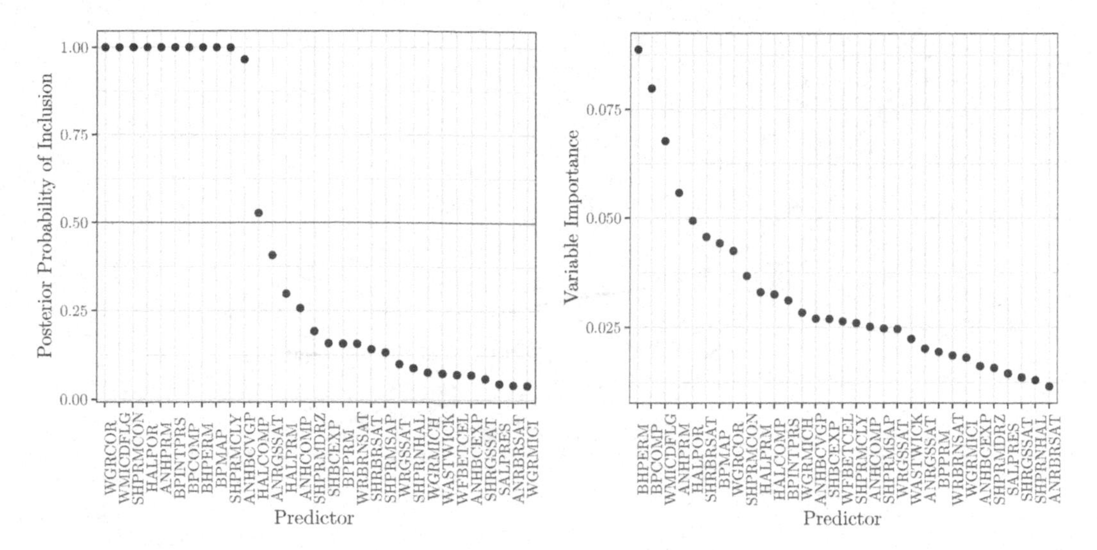

FIGURE 18.9
Comparison of the posterior inclusion probability for each predictor in the WIPP dataset using the sparsity-inducing Dirichlet prior with the original variable importance scores proposed by [6].

(i) the BART model with no modifications, (ii) the BART model with the default sparsity-inducing Dirichlet prior, and (iii) a model which includes only the variables selected from the approach of [3]. Table 18.3 compares these three approaches according to root mean-squared error $\sqrt{\sum_{i \in \text{test}} \{Y_i - \widehat{f}(X_i)\}^2 / n_{\text{test}}}$ and the total amount of time (in seconds) required to fit each method. DART is the best performing method by a wide margin, and the computation time for DART is roughly equivalent to BART. By contrast, the permutation method takes more than 10 times as long to fit and actually performs worse than if we had not performed the variable selection to begin with.

In the case of the WIPP dataset, it seems that fully-Bayesian approaches are clearly preferred. We remark that the computational gap between the methods only increase as the number of predictors increase, as the permutation testing approaches involve permuting the data many times for each individual predictor.

18.6 Extensions

The same basic strategy of penalizing the entries of s can be extended to a variety of other settings.

18.6.1 Interaction Detection

One of the proposed reasons for the strong empirical performance of the BART prior is that it encourages $f(\boldsymbol{x})$ to consist of low-order interactions; draws from the BART prior can be represented as $f(\boldsymbol{x}) = \sum_{d=1}^{D} f_d(\boldsymbol{x}^{(d)})$ where $f_d(\boldsymbol{x}^{(d)})$ represents an interaction between $Q_d \ll p$ of the predictors.

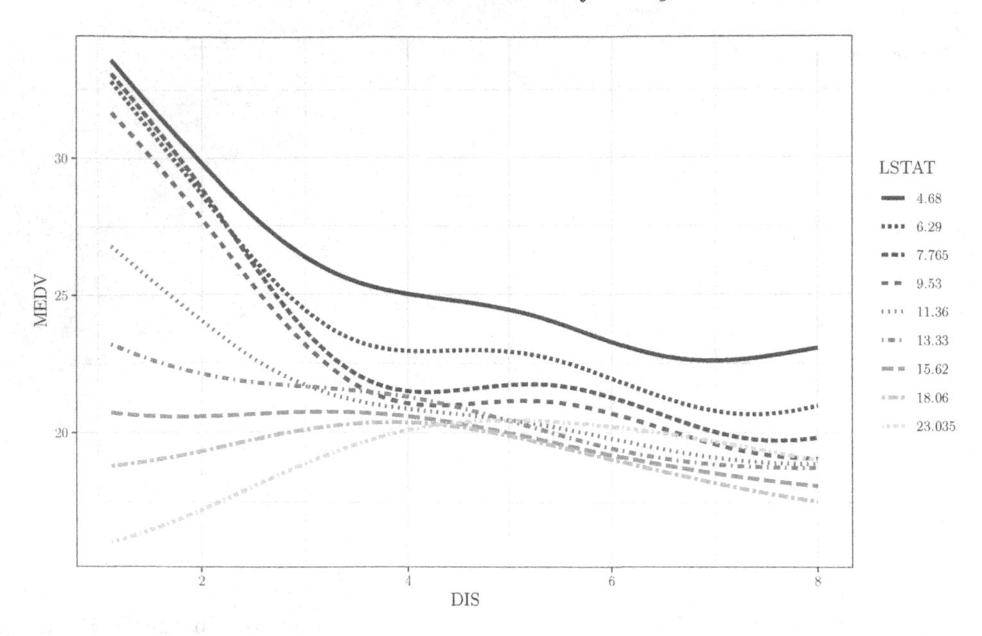

FIGURE 18.10
Relationship between DIS and MEDV found by fitting the DPF model to the Boston housing dataset; the relationship clearly changes as LSTAT is varied.

We focus on FB methods for interaction detection. Just as the original BART prior does not induce sparsity in predictors, the class of Gibbs priors tends to overstate the number of interactions in the data [11]. To address this, [11] introduce the *Dirichlet process forest* (DPF).

The problem with using DART for interaction detection is that there is no penalty which discourages variables from interacting with each other once they are selected. The DPF corrects this problem by clustering the trees into groups $Z_1, \ldots, Z_m \in \{1, \ldots, K\}$ such that each cluster has its own splitting proportion $s^{(k)} \sim \mathcal{D}(\xi w_1, \ldots, \xi w_p)$. In essence, we divide the trees up into "teams" which work together to each learn a low-order interaction. The clustering structure of the DPF is given by

$$\pi \sim \mathcal{D}(\omega/K, \ldots, \omega/K) \qquad \text{and} \qquad [Z_1, \ldots, Z_m \mid \pi] \sim \text{Categorical}(\pi).$$

This is essentially a finite approximation to a Dirichlet process prior [33].

An interesting feature of the DPF model is that it captures several commonly used models as special cases for limiting values of the hyperparameters (ξ, ω). When $\xi \to 0$ and ω is fixed, the DPF converges to a *sparse additive model* (SPAM) [28] while as $\omega \to 0$ with ξ fixed the model converges to DART.

We apply the DPF to the famous Boston housing dataset [14], a common benchmark dataset in the machine learning literature in which the goal is to predict the median housing price in land tracts of Boston from features such as their distance to the city center (DIS) and proportion of lower status individuals living in an area (LSTAT). The DPF strongly preferred models in which there was an interaction between DIS and LSTAT, and a spline fit of this interaction is presented in Figure 18.10 – we see that, as LSTAT increases, the relationship changes from monotonically decreasing to increasing. We find this result interesting because (i) there is strong evidence that it is present, with the DPF always including the interaction across many different splits of the training data and (ii) it is not found by other nonparametric approaches which have used this dataset [26]. We remark that NFB approaches based on defining appropriate interaction importance measures also pick up this interaction, but have the usual disadvantage of not providing a clear decision rule regarding whether an interaction is present or not.

18.6.2 Structure in Predictors

BART can also accommodate incorporation of additional structure in the predictors to aide in variable selection [10]. We consider incorporating *grouping* structure in the columns of $\boldsymbol{X}$. This structure can be represented by a sparse matrix $\mathcal{G} \in \{0,1\}^{G \times p}$ with predictor j a member of group g if $\mathcal{G}_{gj} = 1$. Exploitation of grouping structure in predictors is commonly performed in high-dimensional linear models [15, 29, 34], but there seems to be limited attention to this problem in Bayesian nonparametric settings.

A particularly important instance where grouping structure is useful occurs when one wishes to include categorical predictors in the BART model. This is usually done by converting variables with C categories into C separate binary variables. In this case, the C binary variables form a natural group, and we should prefer that they are all either relevant or irrelevant.

To incorporate grouping structure into the prior, we specify $s = W\pi$ where $\pi \in \mathbb{S}_G$ is a mixing probability over the groups and $W = (w_1, \dots, w_G) \in \mathbb{R}^{p \times G}$ is a matrix of group-specific probabilities $w_g \in \mathbb{S}_p$. To incorporate the grouping structure, we impose the restriction that W has the same sparsity pattern as $\mathcal{G}$. A computationally convenient prior for this problem is to set $\pi \sim \mathcal{D}(\xi\lambda_1, \dots, \xi\lambda_G)$ and $w_g \sim \mathcal{D}(\psi q_{g1}, \dots, \psi q_{gp})$, (with $q_{gj} \equiv 0$ if $\mathcal{G}_{gj} = 0$) which can be shown to be conditionally conjugate for the choice of $p_{\mathcal{T}}$ used in this chapter. Additional details on the use of this are given by [10], who apply this methodology to find active gene pathways in a breast cancer dataset.

In [10], the authors show the impact of leveraging grouping structure for the Friedman dataset; we reproduce one of their simulation results in Figure 18.11 (technically, this work makes use of the "soft BART" method of [20]). This simulation takes the error variance σ as a simulation parameter (higher values of σ denoting more difficult problems) and taking the five relevant predictors as a member of the same group, and other predictors randomly assigned to other groups. We see that using a correct grouping structure results in lower prediction error, false positives (FP), and false negatives (FN), and is particularly useful for large σ.

18.7 Discussion

In addition to providing high-quality predictions, we have seen that Bayesian additive regression trees can be leveraged to provide accurate variable selection using a variety of both FB and NFB techniques. We have emphasized that in order to obtain good performance with FB techniques it is necessary to apply a sparsity-inducing prior, or otherwise to estimate the s_j's, and we illustrated a connection between the usual variable importance scores and the EM algorithm to make this point. We then discussed three different priors for implementing sparsity: the DART prior, the spike-and-forest prior, and a hybrid approach which we refer to as the Gibbs prior. We found that the Gibbs prior strikes a good balance between computational feasibility and practical performance across many different values of p and signal dimensions.

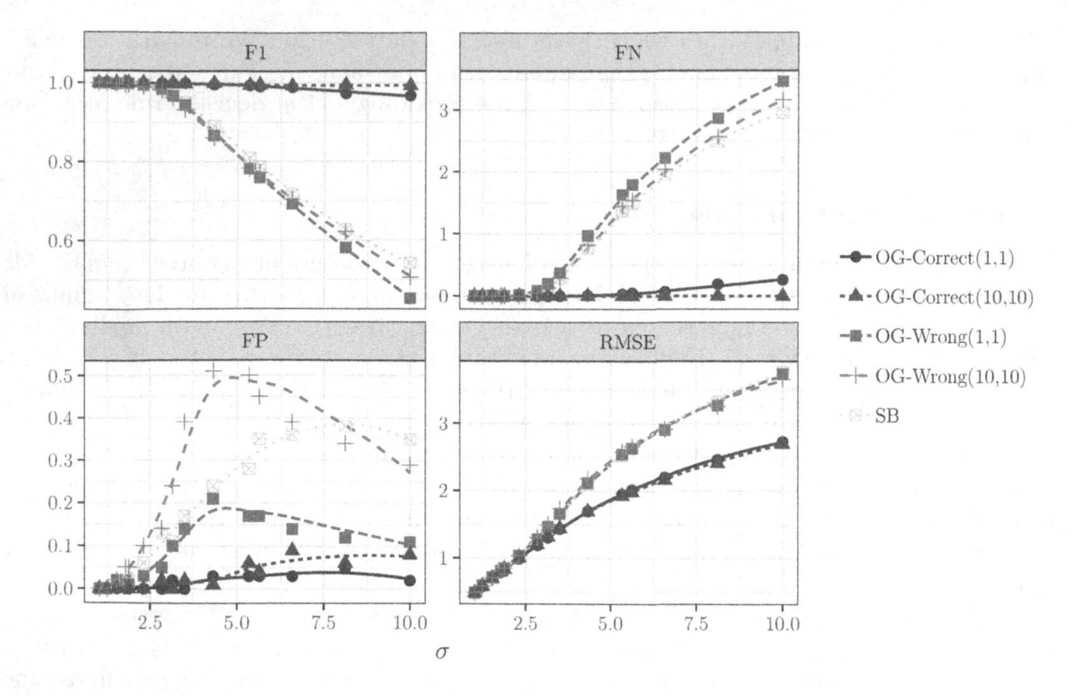

FIGURE 18.11
Results for the simulation study of Section 18.6.2. SB denotes the "soft" BART prior of [20], OG-correct(ξ, ψ) denotes the grouped BART prior with a correct grouping structure with the specified values of ξ and ψ, while OG-Wrong denotes the grouped BART prior with an incorrect grouping structure specified.

We also illustrated that there remains substantial scope for extending these techniques to other settings, and used interaction detection and leveraging grouping structures to illustrate the promise of BART for detecting and leveraging more exotic structures. We showed that DART provides strong prediction benefits on the WIPP dataset, a setting which is thought a-priori to be sparse. We also showed how on the Boston housing dataset that our Dirichlet process forests model is capable of appropriately penalizing the occurrence of interaction structures and identifying them in an FB fashion.

Software

The `dartMachine` and `SoftBart` packages implementing the methods discussed in this chapter are available on GitHub at `www.github.com/theodds/dartMachine` and `www.github.com/theodds/SoftBart`. The `R` scripts using `dartMachine`, which reproduce some of our results and plots, can be found in the online Supplementary Material of this edited book.

Bibliography

[1] P. Basak, A. R. Linero, D. Sinha, and S. Lipsitz. Semiparametric analysis of clustered interval-censored survival datausing soft bayesian additive regression trees (sbart). *arXiv preprint arXiv:02509*, 2020.

[2] D. Blackwell and J. B. MacQueen. Ferguson distributions via Pólya urn schemes. *The Annals of Statistics*, pages 353–355, 1973.

[3] J. Bleich, A. Kapelner, E. I. George, and S. T. Jensen. Variable selection for BART: An application to gene regulation. *The Annals of Applied Statistics*, 8(3):1750–1781, 2014.

[4] V. Bonato, V. Baladandayuthapani, B. M. Broom, E. P. Sulman, K. D. Aldape, and K.-A. Do. Bayesian ensemble methods for survival prediction in gene expression data. *Bioinformatics*, 27(3):359–367, 2010.

[5] H. A. Chipman, E. I. George, and R. E. McCulloch. Bayesian CART model search. *Journal of the American Statistical Association*, 93(443):935–948, 1998.

[6] H. A. Chipman, E. I. George, and R. E. McCulloch. BART: Bayesian additive regression trees. *The Annals of Applied Statistics*, 4(1):266–298, 2010.

[7] P. De Blasi, S. Favaro, A. Lijoi, R. H. Mena, I. Prünster, and M. Ruggiero. Are Gibbs-type priors the most natural generalization of the Dirichlet process? *IEEE Transactions on Pattern Analysis and Machine Intelligence*, 37(2):212–229, 2013.

[8] A. P. Dempster, N. M. Laird, and D. B. Rubin. Maximum likelihood from incomplete data via the EM algorithm. *Journal of the Royal Statistical Society, Series B*, 39(1):1–38, 1977.

[9] D. G. Denison, B. K. Mallick, and A. F. Smith. A Bayesian CART algorithm. *Biometrika*, 85(2):363–377, 1998.

[10] J. Du and A. R. Linero. Incorporating grouping information into Bayesian decision tree ensembles. In *Proceedings of the 35th International Conference on Machine Learning*, pages 1–10, 2019.

[11] J. Du and A. R. Linero. Interaction detection with bayesian decision tree ensembles. In *22nd Proceedings of the International Conference on Artificial Intelligence in Statistics (AISTATS)*, 2019.

[12] Y. Freund, R. Schapire, and N. Abe. A short introduction to boosting. *Journal-Japanese Society For Artificial Intelligence*, 4(5):771–780, 1999.

[13] J. H. Friedman. Multivariate adaptive regression splines. *The Annals of Statistics*, 19(1):1–67, 1991.

[14] D. Harrison and D. L. Rubinfeld. Hedonic housing prices and the demand for clean air. *Journal of Environmental Economics and Management*, 5(1):81–102, 1978.

[15] L. Jacob, G. Obozinski, and J.-P. Vert. Group lasso with overlap and graph lasso. In *Proceedings of the 26th Annual International conference on Machine Learning*, pages 433–440. ACM, 2009.

[16] A. Kapelner and J. Bleich. bartMachine: Machine learning with Bayesian additive regression trees. *Journal of Statistical Software*, 70(4):1–40, 2016.

[17] Y. Li, A. R. Linero, and J. S. Murray. Adaptive conditional distribution estimation with bayesian decision tree ensembles. *arXiv preprint arXiv:2005.02490*, 2020.

[18] A. R. Linero. Bayesian regression trees for high-dimensional prediction and variable selection. *Journal of the American Statistical Association*, 113(522):626–636, 2018.

[19] A. R. Linero, D. Sinha, and S. R. Lipsitz. Semiparametric Mixed-Scale Models Using Shared Bayesian Forests. *arXiv preprint arXiv:1809.08521*, 2018.

[20] A. R. Linero and Y. Yang. Bayesian regression tree ensembles that adapt to smoothness and sparsity. *Journal of the Royal Statistical Society: Series B (Statistical Methodology)*, 80(5):1087–1110, 2018.

[21] Y. Liu, V. Ročková, and Y. Wang. Variable selection with abc bayesian forests. *arXiv preprint arXiv:1806.02304*, 2018.

[22] J. W. Miller and M. T. Harrison. Mixture models with a prior on the number of components. *arXiv preprint arXiv:1502.06241*, 2015.

[23] J. S. Murray. Log-linear Bayesian additive regression trees for categorical and count responses. *arXiv preprint arXiv:1701.01503*, 2017.

[24] R. M. Neal. *Bayesian Learning For Neural Networks.* PhD thesis, University of Toronto, 1995.

[25] J. Pitman. Combinatorial stochastic processes. Technical Report 621, Department of Statistics, University of California, Berkeley, 2002.

[26] P. Radchenko and G. M. James. Variable selection using adaptive nonlinear interaction structures in high dimensions. *Journal of the American Statistical Association*, 105(492):1541–1553, 2010.

[27] C. E. Rasmussen and C. K. I. Williams. *Gaussian Processes for Machine Learning.* MIT Press, Cambridge, 2006.

[28] P. Ravikumar, H. Liu, J. Lafferty, and L. Wasserman. SPAM: Sparse additive models. In *Proceedings of the 20th International Conference on Neural Information Processing Systems*, pages 1201–1208, 2007.

[29] V. Rockova and E. Lesaffre. Incorporating grouping information in Bayesian variable selection with applications in genomics. *Bayesian Analysis*, 9(1):221–258, 03 2014.

[30] V. Rockova and S. van der Pas. Posterior concentration for Bayesian regression trees and their ensembles. *arXiv preprint arXiv:1078.08734*, 2017.

[31] R. A. Sparapani, B. R. Logan, R. E. McCulloch, and P. W. Laud. Nonparametric survival analysis using Bayesian additive regression trees (BART). *Statistics in Medicine*, 2016.

[32] C. B. Storlie, H. D. Bondell, B. J. Reich, and H. H. Zhang. Surface estimation, variable selection, and the nonparametric oracle property. *Statistica Sinica*, 21(2):679, 2011.

[33] Y. W. Teh, M. I. Jordan, M. J. Beal, and D. M. Blei. Hierarchical Dirichlet processes. *Journal of the American Statistical Association*, 101(476):1566–1581, 2006.

[34] M. Yuan and Y. Lin. Model selection and estimation in regression with grouped variables. *Journal of the Royal Statistical Society: Series B (Statistical Methodology)*, 68(1):49–67, 2006.

19

Stochastic Partitioning for Variable Selection in Multivariate Mixture of Regression Models

Stefano Monni
American University of Beirut (Lebanon)

Mahlet G. Tadesse
Georgetown University (USA)

CONTENTS

The relative ease with which high-throughput data can be collected on the same sampling units has raised interest in relating high-dimensional data sets. This problem can be framed as a multivariate regression model with one data set treated as a response matrix and the other as a covariate matrix. The high-dimensionality of both the response and covariate matrices along with the complex correlation structures within each raises statistical challenges. Here, we review a stochastic partitioning method that addresses this problem by fitting a mixture of regression models with variable selection, thereby uncovering in a unified manner group structures and key relationships between the data sets. The problem of variable selection in this context poses added difficulties compared to standard regression settings, as the membership of objects into different components has to be learned simultaneously with the search of component-specific predictors. We emphasize the practical applicability of the method using real data and illustrate features of the associated R package `spavs`.

DOI: 10.1201/9781003089018-19

19.1 Introduction

Technological advances in various fields have made it relatively easy to collect different types of high-dimensional data on the same sampling units. A prominent example is genomic applications where various data types are collected, including epigenomic, genomic, transcriptomic, proteomic and metabolomic data. As a result, research efforts have focused on integrating data from different technologies, with the goal of gaining a better understanding of molecular processes underlying specific phenotypes. Earlier integrative genomic studies were concerned with evaluating the one-to-one association between every possible pair of markers between the two -omic data sets under consideration [22, 35]. This approach has obvious limitations, such as ignoring the correlation between markers and the joint effects of multiple markers, which makes it practically impossible to identify significant associations after suitably correcting for multiple testing. It is also common to fit regression models on each outcome variable separately and implement variable selection. However, the neglected correlation between responses results in suboptimal performance [6]. An improvement over univariate models is offered by multivariate models, especially those that enjoy sparsity. The earlier implementations would select the same subset of regressors for all responses [7, 37] and are thus reasonable mainly in situations where there is a small number of highly correlated outcomes. In the presence of many outcomes with varying degrees of correlation, it is expected that different subsets of covariates would be associated with different subsets of outcomes. In [27], univariate regression models are fit for each outcome and linked by a hierarchical prior specification on the variable selection indicators. Seemingly unrelated regressions (SUR) [39], which allow each response to have a different subset of regressors, are another class of model that has received special attention to address this problem. For variable selection in SUR model, [20] proposed a boosting method with squared error loss, while [3] used a direct Monte Carlo technique in a Bayesian hierarchical framework. In [34], a Bayesian hierarchical model using the variable selection approach for mixture models of [18, 36] is used to identify subsets of brain regions that discriminate between disease types, which in turn are related to a subset of genetic markers. A different class of methods that have been developed to relate two high-dimensional data sets rely on sparse partial least squares (sPLS) [8, 9] and sparse canonical analysis (sCCA) [26, 38]. These methods identify correlated subsets from the two datasets and achieve sparsity using different penalty functions.

This chapter describes a stochastic partitioning method introduced in [21] to relate two high-dimensional data sets by fitting a mixture of regression models with variable selection. This allows the identification of correlated outcomes and their associated subsets of regressors. It is generally reasonable to assume that highly correlated variables are likely to be influenced by the same subset of covariates. In the context of genomic applications, for example, markers with similar expression profiles are believed to share similar regulatory mechanisms, and thus the same regression relationship. In Section 19.2, we provide a brief review of mixture of regression models and variable selection for these models. Section 19.3 presents the stochastic partitioning method for uncovering clusters and identifying cluster-specific relevant covariates. In Section 19.4, we present an application with genomic data and illustrate the functionalities of `spavs`, an R package that implements the stochastic partitioning method. We conclude the chapter with a brief discussion in Section 19.5.

19.2 Mixture of Univariate Regression Models

A mixture of regression models provides an effective tool to understand structures and relationships by identifying clusters of objects that behave similarly and fitting regression models specific to each group. A two-stage approach that clusters the objects in a first step then fits a regression model within each of the identified clusters has several drawbacks. In particular, it ignores the uncertainty in estimating the cluster allocations, thus introducing bias in the estimation process. Both frequentist and Bayesian approaches for fitting univariate mixture of regression models in a unified manner have been proposed. The former rely on maximum likelihood estimation using the expectation-maximization (EM) algorithm, while in the latter model parameters are often estimated using posterior samples obtained by Markov chain Monte Carlo (MCMC) techniques. Variational inference has also been proposed as a computationally efficient alternative to MCMC [25].

19.2.1 Model Fitting

In univariate mixture of regression models, a response variable on n independent observations $\boldsymbol{y} = (y_1, \ldots, y_n)^T$ is related to covariates $\boldsymbol{X} = (\boldsymbol{x}_1, \ldots, \boldsymbol{x}_n)^T$ through a finite mixture model

$$f(y_i|\boldsymbol{x}_i, \boldsymbol{\theta}) = \sum_{k=1}^{K} \pi_k f(y_i|\boldsymbol{x}_i, \boldsymbol{\theta}_k), \tag{19.1}$$

where the number of components K is unknown and needs to be determined, π_k corresponds to the mixture weights, $f(.)$ denotes the probability density function with component parameters $\boldsymbol{\theta}_k$ for the k-th component. When $f(.)$ is from the exponential family, (19.1) is a mixture of generalized linear regression models. In particular, when $f(.)$ is a univariate Gaussian density function, (19.1) becomes a mixture of linear models with component specific parameters $\boldsymbol{\theta}_k = (\boldsymbol{\beta}_k, \sigma_k^2)$, such that an observation with covariates $\boldsymbol{x}_i$ in component k has mean $\boldsymbol{x}_i\boldsymbol{\beta}_k$ and variance σ_k^2.

Model fitting via EM algorithm

In the frequentist framework, a maximum likelihood method using the EM algorithm allows simultaneous estimation of the cluster allocations and the cluster-specific regression parameters [10]. For fixed K, this is accomplished by introducing a latent multinomial cluster allocation vector $\boldsymbol{z}_i = (z_{i1}, \ldots, z_{ik})^T$, such that $z_{ik} = 1$ if observation $(\boldsymbol{x}_i, y_i)$ belongs to class k and $z_{ik} = 0$ otherwise. The complete data log-likelihood is given by

$$\ell(\boldsymbol{\theta}|\boldsymbol{y}, \boldsymbol{Z}, \boldsymbol{X}) = \sum_{k=1}^{K}\sum_{i=1}^{n} z_{ik}\left[\log \pi_k + \log f(y_i|\boldsymbol{x}_i, \boldsymbol{\theta}_k)\right]. \tag{19.2}$$

The E-step consists in taking the expectation of the complete-data log-likelihood, $\mathcal{Q} = E[\ell(\boldsymbol{\theta}|\boldsymbol{y}, \boldsymbol{Z}, \boldsymbol{X})]$. This provides the posterior probability of assigning observation i to component k, which at iteration t is given by

$$\hat{\pi}_{ik}^{(t)} = P\left(z_{ik} = 1|y_i, \boldsymbol{x}_i, \boldsymbol{\theta}_k^{(t-1)}\right) = \frac{\pi_k^{(t-1)} f(y_i|\boldsymbol{x}_i, \boldsymbol{\theta}_k^{(t-1)})}{\sum_{k=1}^{K} \pi_k^{(t-1)} f(y_i|\boldsymbol{x}_i, \boldsymbol{\theta}_k^{(t-1)})}.$$

In the M-step, $\mathcal{Q}$ is maximized for each component separately with respect to the component parameters $(\pi_k, \boldsymbol{\theta}_k)$ using the posterior allocation probabilities as weights. This leads to

$$\begin{aligned}\hat{\pi}_k^{(t)} &= \frac{1}{n}\sum_{i=1}^{n}\hat{\pi}_{ik}^{(t)}.\\ \hat{\boldsymbol{\theta}}_k^{(t)} &= \arg\max_{\boldsymbol{\theta}_k}\sum_{i=1}^{n}\hat{\pi}_{ik}^{(t)}\log f(y_i|\boldsymbol{x}_i, \boldsymbol{\theta}_k).\end{aligned}$$

The EM algorithm is iterated until convergence. Since K is unknown, the mixture of regression models is fit with varying number of components and the value that minimizes an information criterion, such as the Akaike information criterion (AIC) [2], the Bayesian information criterion (BIC) [29] or the integrated completed likelihood (ICL) [4], is chosen.

Model fitting via MCMC algorithms

Some Bayesian strategies for fitting mixtures of regression models are discussed in [16, 19]. The MCMC methods presented in [11] for finite mixture models with fixed K can be extended to the case of finite mixture of regression models. The number of components can subsequently be determined by fitting the finite mixture models with different fixed values of K and evaluating the Bayes factors between these models. For fixed K, the data are augmented by introducing latent multinomial cluster allocation vectors $\boldsymbol{z}_i (i = 1, \ldots, n)$ as above, such that $p(z_{ik} = 1) = \pi_k$. Conjugate prior distributions are specified for the model parameters

$$\begin{aligned}\boldsymbol{\pi} &\sim \mathcal{D}ir(a_1, \ldots, a_K)\\ \boldsymbol{\beta}_k &\sim \mathcal{N}(\boldsymbol{\beta}_{0k}, h\sigma_k^2)\\ \sigma_k^2 &\sim \mathcal{IG}(\nu, \sigma_0^2).\end{aligned} \tag{19.3}$$

Posterior samples are then obtained via Gibbs sampling by successively updating $\boldsymbol{Z}, \boldsymbol{\pi}$ and $\boldsymbol{\theta}$ from their full conditional distributions.

A unified Bayesian approach treats the number of components as an additional model parameter and specifies a prior for K. The mixture model with an unknown number of components can then be fit using reversible jump MCMC [28] or a continuous Markov birth-death process [32]. Alternatively, mixture distributions with a countably infinite number of components may be defined via Dirichlet process mixture (DPM) priors [12, 24]. Various partition priors and posterior sampling strategies are discussed in [19].

Posterior inference for mixture models is complicated due to the non-identifiability that arises from the invariance of the posterior distribution under permutations of the component labels. Namely, conditional on the number of component K, the posterior distribution has $K!$ symmetric modes, each corresponding to one of the possible labels. This is the label switching problem that complicates the inference. Identifiabiliy constraints on the parameters have been proposed to circumvent this problem [11], but were found not always effective. Relabeling algorithms that post-process the MCMC output to minimize the expectation of an appropriate loss function can also be employed [33]. With varying K, the relabeling algorithms are performed by first estimating the number of clusters by the value most frequently visited by the MCMC sampler then focusing on the subset of visited configurations with $\hat{K}$ components. An alternative approach to deal with the label switching phenomenon uses all sampled configurations and evaluates the posterior pairwise probabilities of two samples being assigned to the same component. The entries of the resulting $n \times n$ matrix can be viewed as similarity measures. A single optimal partition can be derived from these posterior pairwise probabilities by defining a loss function and minimizing its posterior expectation [5, 19].

19.2.2 Variable Selection

When the number of covariates p is large or in the presence of multicollinearity between the potential regressors, predictors that are relevant for each cluster need to be selected to fit the model. Thus, a unified method that simultaneously uncovers clusters and selects group-specific relevant covariates is needed. The problem of variable selection in the context of mixture of regression models is rendered more challenging than standard regression settings by the fact that the membership of objects into the different components needs to be learned simultaneously with the search of component-specific predictors.

Penalized mixture of regression models

In the frequentist setting, variable selection and estimation of the mixture components' regression parameters can be performed simultaneously using penalized maximum likelihood methods. The penalized complete log-likelihood is defined as

$$\ell(\boldsymbol{\theta}|\boldsymbol{y}, \boldsymbol{Z}, \boldsymbol{X}) - \eta(\boldsymbol{\theta}), \tag{19.4}$$

where $\ell(\boldsymbol{\theta}|\boldsymbol{y}, \boldsymbol{Z}, \boldsymbol{X})$ is the complete log-likelihood given in (19.2) and $\eta(\boldsymbol{\theta})$ is a penalty function. In [17], an EM algorithm using various penalty was proposed and the properties of the resulting estimators were established assuming the dimension p of the feature space is smaller than the sample size n. The properties of lasso for mixture models in the context of high-dimensional data ($p \gg n$) were further studied in [31]. [23] extended these penalized methods to mixtures of multivariate generalized linear regression models.

Shrinkage priors in mixture of regression models

In the Bayesian framework, [15] incorporates a spike-and-slab prior in the mixture regression model to perform variable selection. A latent binary vector is introduced for each component k ($k = 1, \ldots, K$), resulting in a $K \times p$ indicator matrix, $\boldsymbol{\gamma}$ with elements γ_{kl} taking value 1 if covariate l is relevant for the k-th component. These latent binary indicators are then used to induce a mixture prior on the component specific regression coefficients β_{kl}. The model is fit for fixed K using an MCMC algorithm that iterates between the following steps: (i) the binary variable selection indicator matrix, $\boldsymbol{\gamma}$, is updated using an evolutionary Monte Carlo method; (ii) the cluster allocation matrix $\boldsymbol{Z}$ is updated from its full conditional distribution via Gibbs sampling; (iii) the component parameters, $\sigma_k^2, \boldsymbol{\beta}_k, \pi_k$ ($k = 1, \ldots, K$), are updated from their posterior distributions. This is repeated for varying values of K and the number of components is determined by comparing the different models using Bayes factors evaluated via importance sampling procedures.

The spike-and-slab formulation of [15] provides one possible solution but presents some limitations. First, when there is a large number of objects to cluster, updating the cluster allocation for each object using Gibbs sampling becomes computationally burdensome and the MCMC chain is prone to being stuck at local modes of the posterior distribution. Second, the update of the variable selection indicator vector $\boldsymbol{\gamma}_k$ for each of the K components can be quite expensive when K is large. Third, fitting the model with different values of K to determine the number of components can be computationally expensive when we need to consider a large range of possible values. Finally, drawing inference based on the MCMC output of a single K value fails to capture the uncertainty on the number of clusters. Some of these limitations may be overcome by letting the number of components vary using a reversible jump MCMC or DPM priors.

19.3 Stochastic Partitioning for Multivariate Mixtures

In this section, we review a method for relating two high-dimensional data sets by identifying subsets of correlated responses and their associated subsets of covariates. The data consist of n independent samples with p covariates, $\boldsymbol{X} = (\boldsymbol{x}_1, \ldots, \boldsymbol{x}_p)$, and q outcomes, $\boldsymbol{Y} = (\boldsymbol{y}_1, \ldots, \boldsymbol{y}_q)$. That is, the observations $(\boldsymbol{x}_i, \boldsymbol{y}_i)_{i=1,\ldots,n}$ are realizations of the pair of random variables $(\boldsymbol{X}, \boldsymbol{Y})$ with $\boldsymbol{X} \in \mathbb{R}^{n \times p}$ and $\boldsymbol{Y} \in \mathbb{R}^{n \times q}$. In our context, both p and q are substantially larger than n. Unlike standard mixture models which partition the observations $i = 1, \ldots, n$, our goal is to partition the outcome space. For outcome j $(j = 1, \ldots, q)$, the mixture of regression models with K unknown components is given by

$$f(\boldsymbol{y}_j|\boldsymbol{x}, \boldsymbol{\theta}) = \sum_{k=1}^{K} \pi_k f(\boldsymbol{y}_j|\boldsymbol{x}, \boldsymbol{\theta}_k) = \sum_{k=1}^{K} \pi_k \prod_{i=1}^{n} f(y_{ji}|\boldsymbol{x}_i, \boldsymbol{\theta}_k). \tag{19.5}$$

19.3.1 Model Formulation

The mixture of regression models in (19.5) incorporating variable selection can be formulated as a partition of the data into K components

$$\mathcal{S}_1 \oplus \ldots \oplus \mathcal{S}_K = (\boldsymbol{X}_{I_1}, \boldsymbol{Y}_{J_1}) \oplus \ldots \oplus (\boldsymbol{X}_{I_K}, \boldsymbol{Y}_{J_K}) \tag{19.6}$$

where $I_k \subset \{1, \ldots, p\}$ with $0 \leq |I_k| = p_k \leq p$, $J_k \subset \{1, \ldots, q\}$ with $0 \leq |J_k| = q_k \leq q$ and $\sum_{k=1}^{K} q_k = q$. The $\oplus$ symbol is used to indicate that the union of variables is disjoint for the $\boldsymbol{Y}$ and not necessarily so for the $\boldsymbol{X}$ variables. Indeed, each response $\boldsymbol{y}_j$ is assigned to exactly one component, whereas a predictor $\boldsymbol{x}_l$ may belong to many components or to none, depending on its association with the outcomes in a component, controlling for the other regressors in the component.

The distribution for each element of the q_k outcomes Y_{J_k} of component $\mathcal{S}_k$ is assumed to be

$$Y_{ji}|\mathcal{S}_k \overset{iid}{\sim} \mathcal{N}(\alpha_j + \mu_k, \sigma_k^2), \qquad j \in J_k, \quad i = 1, \ldots, n \tag{19.7}$$

where $\mu_k = \boldsymbol{X}_{I_k}\boldsymbol{\beta}_k$ captures the association between the p_k covariates and q_k outcomes in component $\mathcal{S}_k$. Thus, outcomes in the same component may have different baselines α_j, but have similar profiles with same mean shifts μ_k and variance σ_k^2. Their correlation is captured by their identical dependence on the same subset of regressors. Hence, outcomes associated with the same subset of predictors but with different regression coefficient values would be allocated to distinct components. It is also possible to have components with $p_k = 0$ and $q_k > 0$, which correspond to having no regressor associated with the q_k outcomes in the component. It should be noted that components with $p_k > 0$ and $q_k = 0$ are not allowed as such components would have no contribution to the likelihood function.

19.3.2 Prior Specification

For each configuration or partition of the data, we assign a prior that penalizes large components with stronger penalty for smaller values of ρ $(0 < \rho < 1)$

$$p(\mathcal{S}_1 \oplus \ldots \oplus \mathcal{S}_K) \propto \prod_{k=1}^{K} \rho^{p_k \cdot q_k}. \tag{19.8}$$

We consider conjugate priors for the component parameters and exploit the conjugacy for computational efficiency by integrating them out. For the $(q_k + p_k)$-vector of regression coefficients $\boldsymbol{\theta}_k^T = (\alpha_{t_1}, \ldots, \alpha_{t_{q_k}}, \beta_{s_1}, \ldots, \beta_{s_{p_k}})$ we take

$$\boldsymbol{\theta}_k \sim \mathcal{N}(\boldsymbol{\theta}_{0k}, \boldsymbol{H}_0 \sigma_k^2) \tag{19.9}$$

where $\boldsymbol{\theta}_{0k} = (\alpha_{0t_1}, \ldots, \alpha_{t0_{q_k}}, \beta_{0s_1}, \ldots, \beta_{0s_{p_k}})$ and $\boldsymbol{H}_0 = \text{diag}(h_0 \mathbf{1}_{q_k}^T, h\mathbf{1}_{p_k}^T)$ with $\mathbf{1}_d$ a d-vector of ones. $\boldsymbol{H}_0$ controls the strength of the prior information on the regression coefficients with larger values of h_0 and h corresponding to a wider spread around $\boldsymbol{\theta}_{0k}$. We specify an inverse-gamma prior for the component variances

$$\sigma_k^2 \sim \mathcal{IG}(\sigma_0^2, \nu). \tag{19.10}$$

After integrating out the model parameters, the posterior distribution of a partition $\mathcal{P}$ is given by

$$p(\mathcal{P}) = p(\mathcal{S}_1 \oplus \ldots \oplus \mathcal{S}_K | \boldsymbol{X}, \boldsymbol{Y}) = \prod_{k=1}^{k} f_t(\mathcal{S}_k) \cdot \rho^{p_k \cdot q_k}, \tag{19.11}$$

where $f_t(\mathcal{S}_k)$ is the marginalized likelihood of a component $\mathcal{S}_k$ with q_k outcomes and p_k covariates, which reduces to an nq_k-dimensional multivariate t-distribution with degrees of freedom $2\sigma_0^2$, mean $\boldsymbol{W}\boldsymbol{\theta}_{0k}$, and scale $\left(\boldsymbol{W}\boldsymbol{H}_0\boldsymbol{W}^T + I_{nq_k \times nq_k}\right)\nu/\sigma_0^2$, where $\boldsymbol{W} = (\mathbf{1}_n \otimes \boldsymbol{I}_{q_k \times q_k}, \boldsymbol{X}_{I_k} \otimes \mathbf{1}_{q_k})$ is an $nq_k \times (q_k + p_k)$ matrix of covariates. This can be written in a lower dimensional form involving matrices and vectors of size p_k, which leads to significant computational gains in fitting the model:

$$\begin{aligned} f_t(\mathcal{S}_k) &= \frac{\nu^{\sigma_0^2}}{(\nu + \frac{1}{2}w)^{\sigma_0^2 + nq_k/2}} \frac{\Gamma(\sigma_0^2 + nq_k/2)}{\Gamma(\sigma_0^2)} (2\pi)^{-q_k n/2} h^{-p_k/2} (nh_0 + 1)^{-q_k/2} \\ &\quad \times (\det \boldsymbol{A})^{-1/2}, \end{aligned} \tag{19.12}$$

where the $p_k \times p_k$ matrix $\boldsymbol{A}$ is given by

$$\boldsymbol{A} = \frac{1}{h} I_{p_k} + q_k \cdot \boldsymbol{X}_{I_k}^T \left(I_n - \frac{h_0}{nh_0 + 1} \mathbf{1}_n \mathbf{1}_n^T \right) \boldsymbol{X}_{I_k}$$

and w is the scalar

$$\begin{aligned} w &= \frac{1}{h} \boldsymbol{\beta}_{0,I_k}^T \boldsymbol{\beta}_{0,I_k} + \frac{1}{h_0} \boldsymbol{\alpha}_{0,J_k}^T \boldsymbol{\alpha}_{0,J_k} + \left(\mathbf{1}_n^T \boldsymbol{Y}_{J_k}\right)^T \left(\mathbf{1}_n^T \boldsymbol{Y}_{J_k}\right) \\ &\quad - \frac{h_0}{nh_0 + 1} \left(\frac{1}{h_0} \boldsymbol{\alpha}_{0,J_k} + \boldsymbol{Y}_{J_k}^T \mathbf{1}_n \right)^T \left(\frac{1}{h_0} \boldsymbol{\alpha}_{0,J_k} + \boldsymbol{Y}_{J_k}^T \mathbf{1}_n \right) - \boldsymbol{v}^T \boldsymbol{A}^{-1} \boldsymbol{v} \end{aligned}$$

with the p_k-vector $\boldsymbol{v}$ given by

$$\begin{aligned} \boldsymbol{v} &= \frac{1}{h} \boldsymbol{\beta}_{0,I_k} + (\boldsymbol{X}_{I_k}^T \boldsymbol{Y}_{J_k}) \mathbf{1}_{q_k} - \frac{h_0}{(Nh_0 + 1)} \left(\frac{1}{h_0} \boldsymbol{\alpha}_{0,J_k} + \boldsymbol{Y}_{J_k}^T \mathbf{1}_n \right)^T \\ &\quad \times \left(\frac{1}{h_0} \boldsymbol{\alpha}_{0,J_k} + \boldsymbol{Y}_{J_k}^T \mathbf{1}_n \right) (\boldsymbol{X}_{I_k}^T \mathbf{1}_n). \end{aligned}$$

19.3.3 Model Fitting

The number of components K is unknown. It can range from $K = 1$ corresponding to all outcomes having a similar profile and being in a single component modulated by the same subset of regressors, to $K = q$ with each outcome having a different profile and being associated to its own subset of predictors. The total number of possible partitions with components satisfying $q_k > 0$ is therefore given by $\sum_{k=1}^{q} S_2(q,k)2^{p \cdot k}$, where S_2 are the Stirling numbers of the second kind [1]. It is not possible to fully explore this huge space. A Markov chain is thus constructed to sample from the posterior distribution. This is accomplished by defining split/merge moves and incorporating a parallel tempering algorithm [14]. The marginalization of the model parameters substantially accelerates the MCMC implementation as it avoids updating these parameters from their posterior distributions and reallocating them at every split or merge moves.

Briefly, at each MCMC iteration, the following two types of moves are performed:

1. **Type 1:** Remove/add a regressor from/to a component. This is done by randomly picking a component and proposing to either remove one of its included covariates or add a covariate uniformly selected among those not already in the component. The removal is equivalent to splitting a component with q_k outcomes and p_k covariates into two components, one with q_k outcomes and $p_k - 1$ covariates, and the other with 0 outcome and 1 covariate. The addition is equivalent to merging a component with q_k outcomes and p_k covariates with another component with 0 outcomes and 1 covariate.
2. **Type 2:** Split/merge components in the $\boldsymbol{X}$ and $\boldsymbol{Y}$ spaces by randomly picking between
   - a split move, where a component with $q_k > 2$ is randomly picked and its q_k outcomes are randomly split into two new components. For its covariates, a random subset is assigned to the two new components to account for covariates shared by components, while the remaining covariates are placed in one or the other component. The shared subset of regressors is uniformly sampled from the range $[0, \lfloor p_k/2 \rfloor]$ with probability p_{split} or from the interval $[\lfloor p_k/2 \rfloor, p_k]$ with probability $(1 - p_{split})$.
   - a merge move, in which two randomly selected components are combined.

A parallel tempering algorithm is incorporated to improve the mixing of the MCMC sampler. This consists in running in parallel the MCMC chain defined above for a series of R distributions that interpolate between the posterior distribution of interest and a distribution from which sampling is easier, defined as $p_r(\mathcal{P}) = p(\mathcal{P})^{1/T_r}$, where $1 = 1/T_1 > \ldots > 1/T_R > 0$. After a fixed number of MCMC updates for each chain with stationary distribution $p_r(\mathcal{P})$, a state exchange is proposed to swap the partitions between chains in adjacent temperatures. This improves mixing by making it easier for the MCMC sampler to escape local modes of the posterior probability density where it may get trapped.

19.3.4 Posterior Inference

The MCMC output consists of the partitions of the data visited at each iteration, which are composed of components of pairs of subsets of regressors and outcomes. After discarding samples in the burn-in phase and thinning the MCMC output to reduce the autocorrelations, posterior inference can be drawn by calculating the frequency of co-occurrence of two variables in the same component. In particular, we can evaluate

- the $q \times q$ matrix of posterior pairwise probabilities that the pair $(\boldsymbol{y}_j, \boldsymbol{y}_{j'})$ of outcomes is allocated to the same component. This can be used to identify sets of correlated outcomes.

- the $p \times q$ matrix of posterior probabilities of association between a covariate $\boldsymbol{x}_l$ and an outcome $\boldsymbol{y}_j$. By focusing respectively on the rows or the columns of this matrix, one can locate the outcomes that are affected by a particular covariate or can identify the covariates related to a specific outcome.

The elements of each of these matrices can be viewed as similarity measures and can be visualized with a heatmap or using a network representation. If one is interested in a single configuration, a decision theoretic framework is required, along the lines of [19].

19.4 spavs and Application

The R package **spavs** (stochastic partitioning for variable selection) implementing the method is available at `https://github.com/stefanomonni/spavs.git`. Here, we demonstrate its application with the integration of microRNA (miRNA) and gene expression data on the NCI-60 cell lines. The processed miRNA and Affymetrix HG-U133(A-B) data were downloaded from the NCI-60 database, which contains various -omic data collected on 60 cancer cell lines from nine tissue types [30]. After removing one of the cell lines that did not have gene expression measurements and 30 miRNA markers that had identical values across all samples, there were $n = 59$ samples and $p = 597$ miRNA markers left for analysis. For this illustration, we consider $q = 648$ gene expression probe sets with high coefficient of variation. The R Markdown file for replicating this example along with the data sets are provided in the online supplementary material.

19.4.1 Choice of Hyperparameters and Other Input Values

The inverse-gamma hyperparameters σ_0^2, ν for σ_k^2, can be specified based on the range of variability in the outcomes $\boldsymbol{Y}$. A simple evaluation of the variances for each of the q outcomes gives an upper bound; this would correspond to each outcome being a singleton not associated with any covariates. If outcomes are clustered and related to some covariates, their residual variances will be smaller. By default, α_{0t_j} and β_{0s_l} are set to 0, but can be specified to take other values. h_0 and h control the spread of the normal prior distribution for the regression coefficients, α_{t_j} and β_{s_l}, around their prior means. The results are fairly robust to a wide range of choices for these hyperparameters.

The hyperparameter ρ in the partition prior distribution (19.8) must be chosen such that the moves of Type 1 that add/delete a covariate in a component and the moves of Type 2 that split/merge components are accepted with a similar ratio. If substantially more moves of the first type are accepted compared to moves of the second type, many false positive covariates tend to be included in the components. On the other hand, if the acceptance rate for moves of the second type is much larger, relevant covariates tend to be missed. Table 19.1 gives the acceptance rates for the various moves using different values of ρ, with the other hyperparameters held fixed. In this example, $\rho = 0.001$ leads to fairly similar acceptance rates for the two types of moves.

The algorithm can be run with or without parallel tempering with the option `tempering`. If the MCMC chain does not mix well, it is recommended to include tempering. The number of swaps to be attempted (`tswaps`) between a higher temperature chain and its contiguous one has to be specified along with the number of updates between swaps (`mws`) in each

Moves		$\rho = 0.0005$	$\rho = 0.001$	$\rho = 0.005$	$\rho = 0.01$
Type 1	Delete	6.3%	6.6%	12.8%	19.1%
	Add	1.6%	2.8%	8.8%	14.5%
Type 2	Split	5.8%	5.2%	3.5%	2.8%
	Merge	5.5%	4.9%	3.1%	2.4%

TABLE 19.1
Acceptance rates of the two types of moves in the MCMC algorithm for different values of the hyperparameter ρ.

process; thus, the total number of iterations at each temperature is (`tswaps`+1) $\times$ `mws`. The temperature schedule, that is the number and spacing of the temperature ladders, $T_r \in (0,1], r = 1, \ldots, R$, should be chosen to ensure good acceptance rates for the state exchanges between neighboring processes. This can be monitored at each swap by specifying the option `print.monitoring=TRUE`; the user can thus stop the chain if the choice of temperatures is not adequate. The spacing between the temperature ladders is specified in `inv.temp`. When tempering is used, the computation can be parallelized by specifying the number of cores to use in `parallel.params`.

Various options are provided to initialize the MCMC algorithm. It can be started from a random partition of $\boldsymbol{Y}$ (`init.conf="random"`), from a user-defined initial configuration (`init.conf="predetermined"`), or by letting the algorithm calculate a data-based initialization using k-means clustering (`init.conf="data"`). The software allows multiple chains with possibly different starting values to be run, which are specified in `init.conf.params`. Even when tempering is not implemented, the computation can be parallelized by specifying the number of chains, `tchains`, and the number of cores to use.

19.4.2 Post-Processing of MCMC Output and Posterior Inference

The results presented here are obtained by setting the hyperparameters $\rho = 0.001, h_0 = h = 10, \nu = \sigma_0^2 = 0.1$ and using 10 temperature ladders with a spacing of 0.02 for the parallel tempering, with `tswaps = 1000` swaps and `mws=500` steps between swaps, corresponding to a total of 500,500 MCMC iterations. We set `thin=50` to save the sampled configurations every 50 iterations. Upon running the algorithm, the software outputs the following files:

- `outputfile_ar.txt`, which provides information about the number of proposed and accepted moves in all chains, where `outputfile` is the file name specified for the output when running the function `spavs()`. If a single chain is run, the file will have four rows labeled `m1, m2, s1, s2` corresponding respectively to the merge move of Type 1 (i.e., add a covariate to a component), merge move of Type 2, split move of Type 1 (i.e, remove a covariate from a component) and the split move of Type 2. The first column lists the number of proposed moves and the second column the number of accepted moves. This helps the user monitor whether the choice of the hyperparameter ρ is adequate (see Table 19.1). When there are multiple chains, either from using tempering or running the algorithm from different starting configurations, this information will be repeated for each of the chains, resulting in $2\times$ `tchains` columns. When tempering is used, there will be an additional row providing the proposed and accepted number of swaps between consecutive temperature ladders. Table 19.2 reports the acceptance rates for the state exchanges in this example.

- `outputfile_tc_logpost_size.txt`, a set of files indexed by `tc` corresponding to each of the initial configurations considered. When tempering is used, only the output for the

$\mathcal{P}(T_2) \leftrightarrow \mathcal{P}(T_1)$	$\mathcal{P}(T_3) \leftrightarrow \mathcal{P}(T_2)$	$\mathcal{P}(T_4) \leftrightarrow \mathcal{P}(T_3)$	$\mathcal{P}(T_5) \leftrightarrow \mathcal{P}(T_4)$	$\mathcal{P}(T_6) \leftrightarrow \mathcal{P}(T_5)$
39.1%	31.5%	37.2%	45.8%	40.4%
$\mathcal{P}(T_7) \leftrightarrow \mathcal{P}(T_6)$	$\mathcal{P}(T_8) \leftrightarrow \mathcal{P}(T_7)$	$\mathcal{P}(T_9) \leftrightarrow \mathcal{P}(T_8)$	$\mathcal{P}(T_{10}) \leftrightarrow \mathcal{P}(T_9)$	
32.2%	20.0%	42.7%	21%	

TABLE 19.2
Acceptance rates for state exchanges between adjacent temperatures.

chain with $T_1 = 1$ corresponding to the posterior distribution of interest is saved. Each of these files contain two columns reporting respectively the log posterior probability and the number of components for the saved configurations based on the specified thinning. For example, with 500,500 MCMC iterations using a thinning of 50, each file has 10,010 rows.

- `outputfile_tc_conf.txt`, which is associated to each `outputfile_tc_logpost_size.txt`. Each row corresponds to a sampled configuration partitioning $\boldsymbol{X}$ and $\boldsymbol{Y}$ into components $\mathcal{S}_1 \oplus \ldots \oplus \mathcal{S}_K$ as in (19.6). Components $\mathcal{S}_k$ are separated by a semi-colon. The subsets of covariates $\boldsymbol{X}_{I_k}$ and subsets of outcomes $\boldsymbol{Y}_{J_k}$ making up a component are separated by a comma.

The left panel of Figure 19.1 displays the trace plot of the number of components. The MCMC sampler appears to mix well and stabilize after about 100,000 iterations. We discarded the first 200,000 iterations as burn-in. As specified above, a thinning of 50 is used to minimize the autocorrelation between configurations considered for inference. The right panel of Figure 19.1 shows the posterior distribution of the number of components, which shows support for partitions with 140 to 165 components.

The posterior pairwise probabilities of allocating two outcomes $(\boldsymbol{y}_j, \boldsymbol{y}_{j'})$ to the same component or associating covariate $\boldsymbol{x}_l$ with outcome $\boldsymbol{y}_j$ can be evaluated using the function `pairwise_probs()`. The posterior pairwise probabilities can be visualized using a heatmap or a network representation. For the clustering of the outcome variables, the function `minbinder()` in the R package `mcclust` [13] can be applied to the $q \times q$ matrix of posterior pairwise probabilities to evaluate a single "optimal" clustering using a decision theoretic approach. If one looks for a single configuration of both outcomes and covariates, an appropriate loss function different from Binder's [5] would be required. Figure 19.2 displays a network of the mRNA transcripts allocated to the same component with posterior pairwise probability greater than 0.5. Figure 19.3 shows that genes with similar expression profiles are successfully identified in the same component. The genes in the left panel are characterized by a high transcript abundance in colon cancer and low expression levels in melanoma tissues. The genes in the right panel, on the other hand, have consistently high transcript abundance in the melanoma samples and have low expression levels in other cancer tissues. Figures 19.4 and 19.5 display a network and a heatmap for the association between miRNA markers and mRNA transcripts focusing on those with posterior probability of association greater than 0.5.

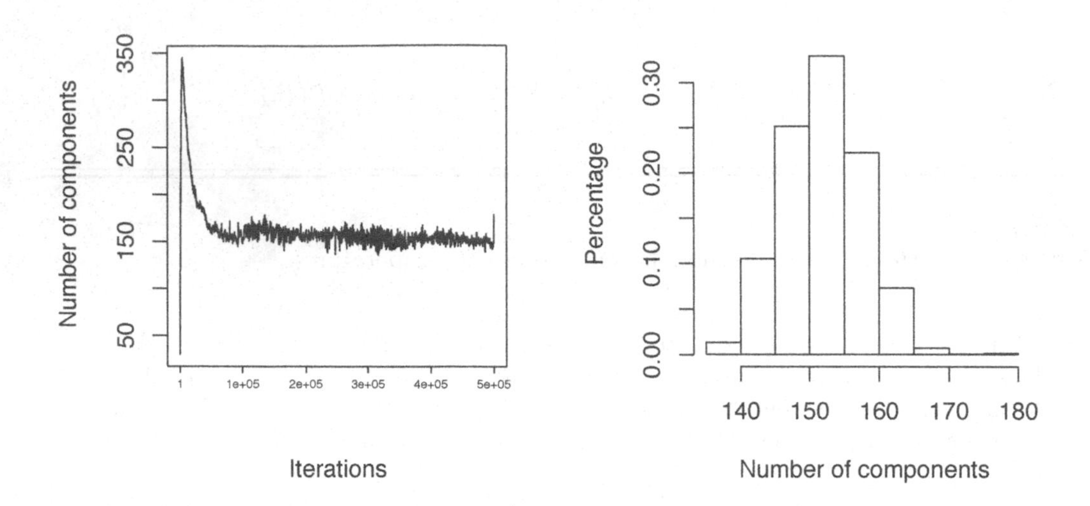

FIGURE 19.1
Trace plot and posterior distribution of number of components.

19.5 Discussion

This chapter reviewed the stochastic partitioning algorithm of [21], which fits mixture of regression models with variable selection to relate two high-dimensional data sets. It also illustrated the associated R package `spavs`. The method identifies sets of correlated outcomes and their associated subsets of covariates using an MCMC sampler that explores the space of pairwise partitions of outcomes and covariates. The underlying premise is that highly correlated outcomes, which get clustered together, are modulated by the same subsets of covariates and share the same regression model. The specification of conjugate priors for the model parameters provides analytical simplification by allowing marginalization over these parameters, thereby providing substantial gain in computational speed and efficiency. An appealing aspect of the procedure is that it is fairly robust to a wide range of hyper-parameter choices and requires minimal user input. This flexibility makes it accessible for application by researchers who may not be familiar with Bayesian inference.

There are a number of possible future directions. The multivariate mixture of regression model considered in [21] assumes a linear relationship between outcomes and regressors in each component. This can be relaxed using nonparametric methods, such as Gaussian processes. Another extension would be to consider non-Gaussian multivariate outcomes. Both of these would require the development of efficient algorithms for identifying structures and relationships between the data sets.

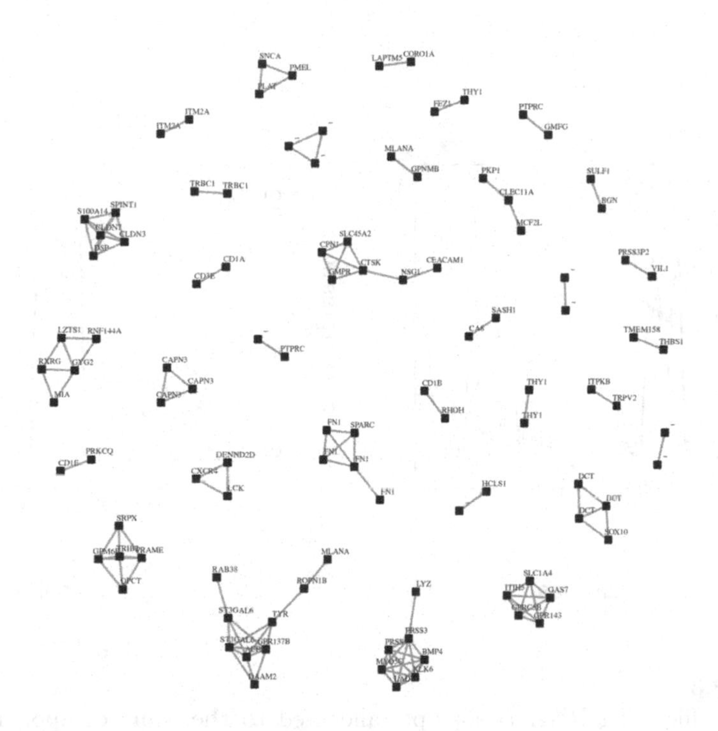

FIGURE 19.2
Network of mRNA transcripts allocated to same components with posterior pairwise probability greater than 0.5.

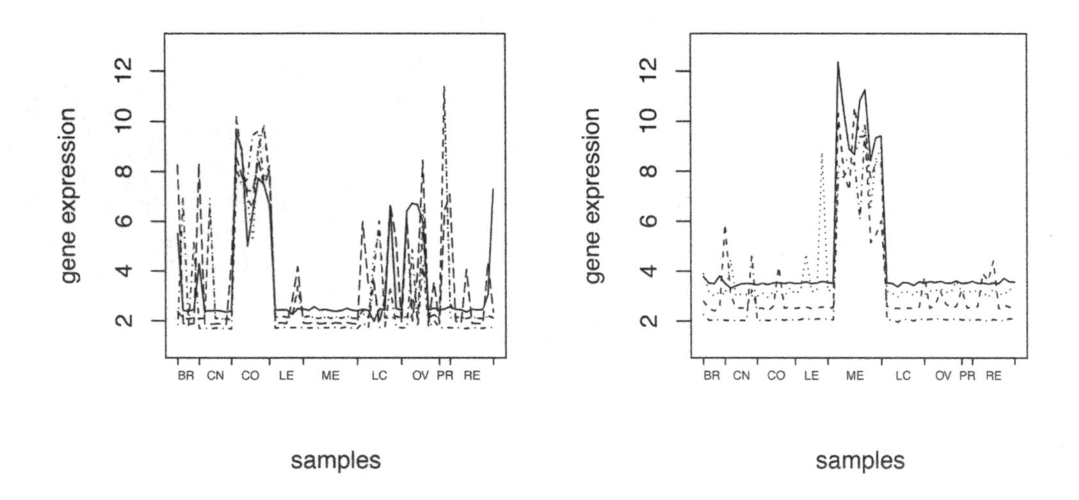

FIGURE 19.3
Expression profiles of mRNA transcripts allocated to the same components with posterior pairwise probability greater than 0.5. The labels on the horizontal axis correspond to the different tissue types: BR – breast; CNS – central nervous system; CO – colon; LE – leukemia; ME – melanoma; LC – lung; OV – ovarian; PR – prostate; RE – renal.

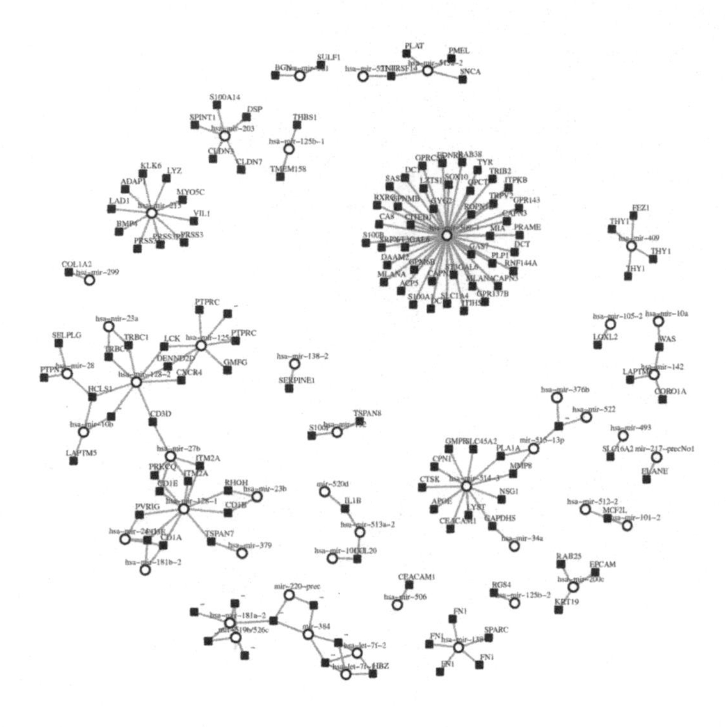

FIGURE 19.4
Network of association between miRNA markers (white circles) and mRNA transcripts (black squares) focusing on posterior probabilities > 0.5.

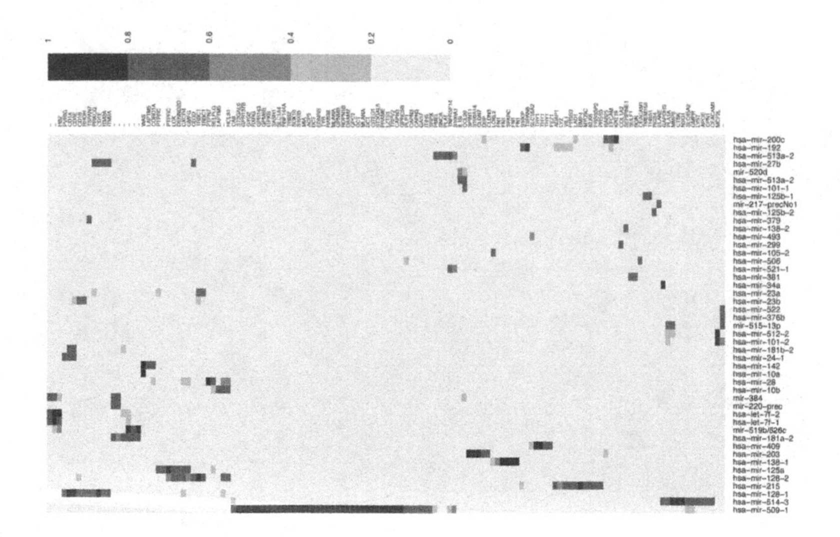

FIGURE 19.5
Heatmap of posterior probability of association between miRNA markers and mRNA transcripts focusing on those with at least one posterior probability > 0.5.

Bibliography

[1] M. Abramowitz and I.A. Stegun. Stirling numbers of the second kind. In M. Abramowitz and I.A. Stegun, editors, *Handbook of Mathematical Functions with Formulas, Graphs, and Mathematical Tables, 9th printing*, pages 824–825, New York, 1972. Dover.

[2] H. Akaike. Information theory and an extension of the maximum likelihood principle. In B. N. Petrov and F. Csaki, editors, *The 2nd International Symposium on Information Theory*, pages 267–281, Budapest, 1973. Akademia Kiado.

[3] T. Ando. Bayesian variable selection for the seemingly unrelated regression models with a large number of predictors. *Journal of the Japan Statistical Society*, 41:187–203, 2011.

[4] C. Biernacki, G. Celeux, and Govaert G. Assessing a mixture model for clustering with the integrated completed likelihood. *IEEE Transactions on Pattern Analysis and Machine Intelligence*, 22:719–725, 2000.

[5] D.A. Binder. Bayesian cluster analysis. *Biometrika*, pages 31–38, 1978.

[6] L. Breiman and J.H. Friedman. Predicting multivariate responses in multiple linear regression. *Journal of the Royal Statistical Society, Series B*, 59:3–54, 1997.

[7] P. Brown, M. Vannucci, and T. Fearn. Bayesian variable selection and prediction. *Journal of the Royal Statistical Society, Series B*, 60:627–641, 1998.

[8] K-A. Lê Cao, D. Rossouw, C. Robert-Granié, and P. Besse. A sparse PLS for variable selection when integrating omics data. *Stat Appl Genet Mol Biol*, 7:37, 2008.

[9] H. Chun and S. Keleş. Sparse partial least squares regression for simultaneous dimension reduction and variable selection. *Journal of the Royal Statistical Society, Series B*, 72:3–25, 2010.

[10] W.S. DeSarbo and W.L. Cron. A maximum likelihood methodology for clusterwise linear regression. *Journal of Classification*, 5:249–282, 1988.

[11] J. Diebolt and C.P. Robert. Estimation of finite mixture distributions through Bayesian sampling. *Journal of the Royal Statistical Society, Series B*, 56:363–375, 1994.

[12] M.D. Escobar and M. West. Bayesian density estimation and inference using mixtures. *Journal of the American Statistical Association*, 90:577–588, 1995.

[13] A. Fritsch and K. Ickstadt. An improved criterion for clustering based on the posterior similarity matrix, *Bayesian Analysis*, 4:367-391, 2009.

[14] C.J. Geyer. Markov chain Monte Carlo maximum likelihood. In E.M. Keramigas, editor, *Computing Science and Statistics*, pages 156–163, Fairfax, 1991. Interface Foundation.

[15] M. Gupta and J.G. Ibrahim. Variable selection in regression mixture modeling for the discovery of gene regulatory networks. *Journal of the American Statistical Association*, 102:867–880, 2007.

[16] M. Hurn, A. Justel, and C. Robert. Estimating mixtures of regression. *Journal of Computational and Graphical Statistics*, 12:1–25, 2003.

[17] A. Khalili and J. Chen. Variable selection in finite mixture of regression models. *Journal of the American Statistical Association*, 102:1025–1037, 2007.

[18] S. Kim, M.G. Tadesse, and M. Vannucci. Variable selection in clustering via Dirichlet process mixture models. *Biometrika*, 93:877–893, 2006.

[19] J.W. Lau and P.J. Green. Bayesian model-based clustering procedures. *Journal of Computational and Graphical Statistics*, 16:526–558, 2007.

[20] R.W. Lutz and P. Bühlmann. Boosting for high-multivariate responses in high-dimensional linear regression. *Statistica Sinica*, 16:471–494, 2006.

[21] S. Monni and M.G. Tadesse. A stochastic partitioning method to associate high-dimensional responses and covariates (with discussion). *Bayesian Analysis*, 4:413–436, 2009.

[22] M. Morley, C.M. Molony, T. Weber, J.L. Devlin, K.G. Ewens, R.S. Spielman, and V.G. Cheung. Genetic analysis of genome-wide variation in human gene expression. *Nature*, 430:743–747, 2004.

[23] F. Mortier, D-Y. Ouédraogo, F. Claeys, M.G. Tadesse, G. Cornu, F. Baya, F. Benedet, V. Freycon, S. Gourlet-Fleury, and N. Picard N. Mixture of inhomogeneous matrix models for species-rich ecosystems. *Environmetrics*, 26:39–51, 2015.

[24] R.M. Neal. Markov chain sampling methods for Dirichlet process mixture models. *Journal of Computational and Graphical Statistics*, 9:249– 265, 2000.

[25] D.J. Nott, S.L. Tan, M. Villani, and R. Kohn. Regression density estimation with variational methods and stochastic approximation. *Journal of Computational and Graphical Statistics*, 21:797–820, 2012.

[26] E. Parkhomenko, D. Trichler, and J. Beyene. Genome-wide sparse canonical correlation of gene expression with genotypes. *BMC Proceedings*, 1(Suppl 1:): S119, 2007.

[27] S. Richardson, L. Bottolo, and J.S. Rosenthal. Bayesian models for sparse regression analysis of high-dimensional data. In J.M. Bernardo, M.J. Bayarri, J.O. Berger, A.P. Dawid, D. Heckerman, A.F.M. Smith, and M. West, editors, *Bayesian Statistics 9*, pages 539–569, New York, 2010. Oxford University Press.

[28] S. Richardson and P. Green. On Bayesian analysis of mixtures with an unknown number of components (with discussion). *Journal of the Royal Statistical Society, Series B*, 59:731–792, 1997.

[29] G. Schwarz. Estimating the dimension of a model. *Annals of Statistics*, 6:461–464, 1978.

[30] U.T. Shankavaram, S. Varma, D. Kane, M. Sunshine, K.K. Chary, W.C. Reinhold, Y. Pommier, and J.N. Weinstein. CellMiner: a relational database and query tool for the nci-60 cancer cell lines. *BMC Genomics*, 10:277, 2009.

[31] N. Städler, P. Bühlmann, and S. van de Geer. $\ell 1$-penalization for mixture regression models. *Test*, 19:209–256, 2010.

[32] M. Stephens. Bayesian analysis of mixture models with an unknown number of components – an alternative to reversible jump methods. *Annals of Statistics*, 28:40–74, 2000.

[33] M. Stephens. Dealing with label switching in mixture models. *Journal of the Royal Statistical Society, Series B*, 62:795–809, 2000.

[34] F.C. Stingo, M. Guindani, M. Vannucci, and V. Calhoun. An integrative Bayesian modeling approach to imaging genetics. *Journal of the American Statistical Association*, 108:876–891, 2013.

[35] B.E. Stranger, M.S. Forrest, M. Dunning, C.E. Ingle, C. Beazley, N. Thorne, R. Redon, C.P. Bird, A. de Grassi, C. Lee, C. Tyler-Smith, N. Carter N, S.W. Scherer, S. Tavaré, P. Deloukas, M.E. Hurles, and E.T. Dermitzakis. Relative impact of nucleotide and copy number variation on gene expression phenotypes. *Science*, 315:848–853, 2007.

[36] M.G. Tadesse, N. Sha, and M. Vannucci. Bayesian variable selection in clustering high-dimensional data. *Journal of the American Statistical Association*, 100:602–617, 2005.

[37] B. Turlach, W. Venables, and S. Wright. Simultaneous variable selection. *Technometrics*, 47:349–363, 2005.

[38] D.M. Witten, R. Tibshirani, and T. Hastie. A penalized matrix decomposition, with applications to sparse principal components and canonical correlation analysis. *Biostatistics*, 10:515–534, 2009.

[39] A. Zellner. An efficient method of estimating seemingly unrelated regression equations and tests for aggregation bias. *Journal of the American Statistical Association*, 57:348–368, 1962.

Index